BEACHES OF THE NORTHERN AUSTRALIAN COAST THE KIMBERLEY, NORTHERN TERRITORY & CAPE YORK

A guide to their nature, characteristics, surf and safety

ANDREW D SHORT

Coastal Studies Unit
School of Geosciences F09
University of Sydney
Sydney NSW 2006

SYDNEY UNIVERSITY PRESS

Australian Beach Safety and Management Program

Coastal Studies Unit
School of Geosciences F09
University of Sydney
Sydney NSW 2006

and

Surf Life Saving Australia Ltd
1 Notts Ave
Locked Bag. 2
Bondi Beach NSW 2026

Short, Andrew D
Beaches of the Northern Australian Coast: The Kimberley, Northern Territory & Cape York 1-920898-16-6
A guide to their nature, characteristics, surf and safety
Published: December 2006

Other books in this series by A D Short:
- *Beaches of the New South Wales Coast,* 1993 0-646-15055-3
- *Beaches of the Victorian Coast and Port Phillip Bay,* 1996 0-9586504-0
- *Beaches of the Queensland Coast: Cooktown to Coolangatta,* 2000 0-9586504-1-1
- *Beaches of the South Australian Coast and Kangaroo Island*, 2001 0-9586504-2-X
- *Beaches of the Western Australian Coast: Eucla to Roebuck Bay,* 2005 0-9586504-3-8
- *Beaches of the Tasmanian Coast and Islands*, March 2006 1-920898-12-3

Forthcoming books:
Beaches of the New South Wales Coast (2nd edition) 1-920898-15-8

Published by:
Sydney University Press
University of Sydney
www.sup.usyd.edu.au

Printed by:
University of Sydney Publishing Service
University of Sydney

Copies of all books in this series may be purchased online from Sydney University Press at:

http://www.sup.usyd.edu.au/marine

Northern Australian beach database:
Inquiries about the Northern Australian beach database should be directed to Surf Life Saving Australia at info@slsa.asn.au

---

Cover photographs: *The University of Sydney research boat 'CSU 3' at the pure white Silica Beach (K 223) on Hidden Island in the western Kimberley; Furze Point (Q 250) on eastern Cape York has a series of crenulate intertidal sand ridges; and Angurubia Inlet (GE 116 & 117) with its extensive flood tide delta separates beaches and dunes on the east coast of Groote Eylandt in the Gulf of Carpentaria (A D Short).*

Table of Contents

Preface

This is the seventh and final book in a series on the beaches of the Australian coast. They have all been produced by the Australian Beach Safety and Management Program (ABSMP), a collaborative project of the Coastal Studies Unit (CSU) University of Sydney and Surf Life Saving Australia (SLSA). The project has compiled a database on everyone of Australia 10 685 mainland beaches, together with 833 beaches on 30 inhabited islands. The books are a byproduct of this database.

The coast of northern Australia is as vast as it is difficult to access. As a consequence time, patience and diligence is required to visit and inspect the 4114 km of coastline. The author's first trip north was in 1976 when he made it as far as Normanton before tropical cyclone Ted prevented him reaching Karumba. In 1980 he co-lead a major CSU field experiment to Cable Beach, Broome, successfully investigating the impact of the 10 m tide range on beach morphodyanmics. The first field study for this program commenced in 1991 when with his family and a caravan he visited all the reasonably accessible beaches in the Broome region from Crab Creek to One Arm Point, the Darwin and Gove Peninsula beaches, and finally made it to Karumba.

In 1994 he flew the entire Queensland coast obtaining low altitude oblique photographs of every beaches, then in 1997 flew from Western Australia right across the top to the Queensland border, photographing all the Kimberley and Northern Territory beaches.

While he made it as far as Cooktown twice while investigating the Queensland coast, he did not get past Cooktown till 1998 when the first of three boat-based trips commenced. The first went from Cooktown to Karumba, the second in 1999 from Borrolooa to Darwin, and third in 2001 from Wyndham to Broome, with a ground-based trip to the greater Darwin region in 2000.

In compiling a book of this magnitude there will be errors and omissions, particularly with regard to the names of beaches, many of which have no official name, and many local factors. If you notice any errors or wish to comments on any aspects of the book please communicate them to the author at the Coastal Studies Unit, University of Sydney, Sydney NSW 2006, phone (02) 9351 3625, fax (02) 9351 3644, email: A.Short@geosci.usyd.edu.au or via Surf Life Saving Australia (02) 9130 7370. In this way we an update the beach database and ensure that future editions are more up to date and correct.

Andrew D Short
Narrabeen Beach, March 2006

Acknowledgements

The northern Australian coast is vast and for the most part difficult to access. In order to undertake this study air, sea and land approaches were made to reach the coast. The entire Queensland coast was flown and photographed in 1994 in an Australian Aerial Patrol (AAP) Cessna piloted Harry Mitchell and Steve Conock. This was followed in 1997 by a flight in an AAP Par Navier, piloted by Harry Mitchell and Dean Franklin, which began in South Australia and flew and photographed the entire Western Australian and Northern Territory coasts. The AAP and Harry Mitchell in particular have assisted the project since the early 1990's and their support is gratefully acknowledged.

On the ground my wife Julia and children accompanied me during the first major field trip in 1991 from Broome across the top to Cairns. The Darwin region was inspected on the ground in 2000, with by wife acting as driver and field assistant.

Most of the coast however is not accessible by vehicle so three boat trips were all undertaken in the university's 8 m boat 'CSU3' driven by Graham Lloyd. Graham is an expert boatman and technician and successfully got the boat from Cairns to Broome in three trips spanning 1998, 1999 and 2001. The biggest problem with small boat travel in the north is obtaining fuel. For assistance with fuel I thank Cape Flattery mines, Point Smith ranger station and Faraway Bay Bush Camp, and the communities of Pormpuraaw, Numbulwar, Milakburra (Bickerton Island), Galiwinku (Elcho Island), Maningrida, Warruwi and One Arm Point. Boat fuel for the Kimberley leg was generously supplied by Paspaley Pearls and Broome Pearls. Malcolm Douglas provided advice and a home for our truck during the Kimberley leg, and his many films and crocodile park are a must for anyone contemplating travel on the north coast.

The project was greatly assisted by the donation of a set of vertical aerial photographs of the entire coast by my former colleague Professor Jack Davies. These provide data on each and every beach and his generosity is much appreciated.

The project has been supported by SLSA since its inception and during the northern Australia phase CEO Greg Nance has provided his full support and encouragement, while Katherine McLeod has expertly maintained the database. All figures were drafted by Peter Johnson. All photographs are by the author, apart from two on page vii.

The project has received the financial support of the Australia Research Council through an ARC Grant (1990-92), ARC Collaborative Research Grant (1996-1998) and ARC SPIRT Grant (1999-2001), and through contract work for the Defence Science and Technology Organisation.

At the University of Sydney thanks to Glen Harris who analyzed most of the beach sediment samples; at University Publishing Service to Josh Fry who expertly oversees production of each of the books, and to Jacqui Owen for the cover design; and at the Sydney University to Press Ross Coleman and Susan Murray-Smith for all their assistance in the production and marketing of this book.

Finally, as the entire beach database was complied and the book was written at my home office, I thank my wife Julia, and children Ben, Pip and Bonnie for putting up with its intrusion into our home life, as well as accompanying me to many parts of the northern Australian coast.

Abstract

This book is about the beach systems of the entire northern Australian coast, from Broome in the west to Cooktown in the east, and includes the entire coast of the Kimberley region, Northern Territory and Cape York, in all 11 879 km of shoreline. This tropical coastline contains 3488 beaches exposed to generally low to at most moderate waves, and tides ranging from 2 to 11 m. It begins with three chapters that provide a background to the physical nature and evolution of the northern Australian coast and its mainland beach systems. Chapter 1 covers the geological evolution of the coast and the role climate, wave, tides and wind in shaping the present coast and beaches. Chapter 2 presents in more detail the sixteen types of beach systems that occur along the northern Australian coast, and then discusses the types of beach hazards along the coast and the role of Surf Lifesaving Western Australia and Northern Territory in mitigating these hazards. The short chapter 3 provides a guide for using the book. Finally, the long chapters 4, 5 and 6 presents a description of every one of the mainland beaches that extend across the top of Australia. The description of each beach covers its name, location, physical characteristics, access and facilities, with specific comments on its surf zone character and physical hazards, as well as its suitability for swimming, surfing and fishing. Based on the physical characteristics each beach is rated in terms of the level of beach hazards from the least hazardous rated 1 (safest) to the most hazardous 10 (least safe). Biological hazards, including crocodiles are major threat in this region, they are not however included in this physical rating system. The book contains 421 original figures which include 365 photographs, which illustrate all beach types, as well as beach maps and photographs of all beaches patrolled by surf lifesavers and many other popular beaches.

Keywords: beaches, surf zone, rip currents, beach hazards, beach safety, Kimberley, Northern Territory, Cape York

Australian Beach Safety and Management Program (ABSMP)

Awards

NSW Department of Sport, Recreation and Racing
Water Safety Award – Research 1989
Water Safety Award – Research 1991

Surf Life Saving Australia
Innovation Award 1993

International Life Saving
Commemorative Medal 1994

New Zealand Coastal Survey
In 1997 Surf Life Saving New Zealand adopted and modified the ABSMP in order to compile a similar database on New Zealand beaches.

Great Britain Beach Hazard Assessment
In 2002 the Royal National Lifeboat Institute adopted and modified the ABSMP in order to compile a similar database on the beaches of Great Britain.

Hawaiian Ocean Safety
In 2003 the Hawaiian Lifeguard Association adopted ABSMP as the basis for their Ocean Safety survey and hazard assessment of all Hawaiian beaches.

Handbook on Drowning 2006
This handbook was product of the World Congress on Drowning held in Amsterdam in 2002. The handbook endorses the ABSMP approach to assessing beach hazards as the international standard.

The author surveys a beach at low tide, Shoal Bay, Northern Territory (J. Short).

The author on beach K 416, near Cape Torrens, Kimberley coast (G. Lloyd)

The Australian Aerial Patrol's Par Navier which was used to fly and photography most of the northern Australian coast, at Wyndham airport.

CSU 3, the truck & tinny launching at Mule Creek, to begin the Northern Territory survey.

CSU 3 moored at Arndeni Inlet, Arnhem Land, NT.

1 THE KIMBERLEY, NORTHERN TERRITORY & CAPE YORK COAST

INTRODUCTION

The northern coast of Australia between Broome and Cooktown covers 11 869 km or 40% of the Australian coast. This coast includes the entire Kimberley region, the Northern Territory coast and most of Cape York (Fig. 1.1). It is located between 10^0 and 18^0S latitude and exposed to a tropical monsoonal climate with hot wet summers and warm dry winters. The warm tropical seas that surround the coast are host to some of the most extensive coral reef systems in the world, including the northern Great Barrier Reef, while in more sheltered locations lie most of Australia's mangrove communities. The summer wet season provides runoff for most of Australia's rivers in terms of both numbers and size. The rivers drain the ancient geology and contribute large volumes of sediment to the coast to build deltas and beaches. The light to moderate southeast trade winds blow across the top end for much of the year, while summer brings the lighter northwest monsoon. Both wind systems generate usually low waves at the coast while the monsoons also bring most of the rain. Tides range from 2-10 m and includes some of the highest tides in the world. The interaction of the wind, waves and tides with the sediments and geology has formed the modern coast, which while dominated by rocky shore has thousands of beaches and hundreds of kilometres of mangrove lined-creeks and estuaries.

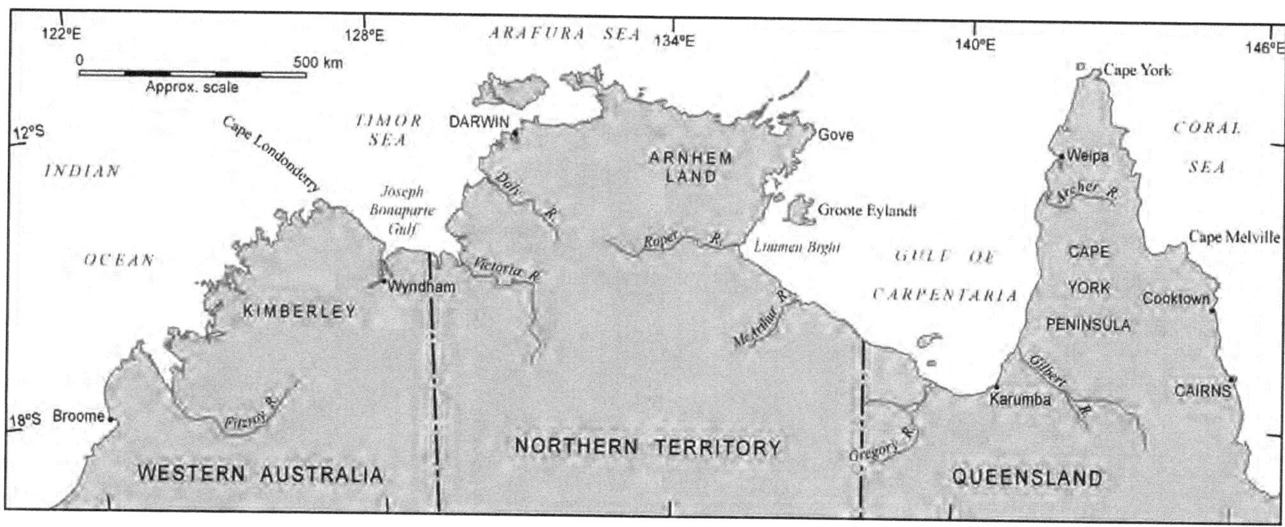

Figure 1.1 Regional map of Northern Australia. This book covers the Kimberley, Northern Territory and Cape York coast and beaches between Broome in the west and Cooktown in the east.

Beaches are spread around the entire northern Australian coast. The Kimberley has 1360 beaches, which occupy only 16%, the smallest proportion of beaches in any Australian coastal region, with much of the coast dominated by rock and mangroves. In the Northern Territory 1488 beaches occupy 38% of the coast, also sharing the coast with extensive mangrove-fringed shore and rocky sections. Cape York has the highest proportion of sand beaches, with 641 beaches spread along 60% of the coast, with mangroves and rocky coast making up the remainder. Overall the 3489 beaches occupy 35% of the coast (Table 1.1), and represent 33% of Australian mainland beach systems by number and 28% by length. This book is about all those beaches, their location, nature and characteristics.

The northern coast also has a significant aboriginal, and on the cape, islander population. They live in generally small towns and communities spread primarily along the Northern Territory and Cape York coast, with aboriginal lands occupying 47% of the coast. The coast has only been lightly penetrated and populated by non-aboriginals since the first unsuccessful attempt at European settlement at Fort Dundas on Melville Island in 1824. Today there is only one major coastal city - Darwin, followed by the coastal towns of Broome, Derby and Wyndham in the west, Nhulunbuy and Karumba in the Gulf, and Cooktown in the east, together with smaller aboriginal communities. What this means is that the coast is lightly populated and lightly impacted with 58% of the coast part of aboriginal land or national parks and reserves. The low level of impact is reflected in the fact that only 44 beaches across the top have a sealed road access, with another 187 with gravel road access and 487 accessible by 4WD (Table 1.2). However 2771 (80%) have no vehicle access and most have no official name. In its present ownership and management the northern Australian coast is likely to increasingly become one of the world's premier pristine tropical coastal regions.

Table 1.1 Regional, northern Australian and Australian proportion of beaches by kilometre length & number

Total coast	Total beach km	% Regional coast	% Northern Australia coast	% Australia coast	Aboriginal Land km	National Parks, etc km
Kimberley	4340	100	37	15	1893	357
Northern Territory	5029	100	42	17	2730	784
Cape York	2501	100	21	8	984	108
Total	11 870		100	40	5607	1249
Total beaches	Number	% beaches	% beaches	% beaches		
Kimberley	1360	100	39	13	596	87
Northern Territory	1488	100	43	14	1117	367
Cape York	640	100	18	6	205	117
Total	3488		100	33	1918	571
Total beaches	Length (km)	% coast	% beaches	% beaches		
Kimberley	702	16	17	5		
Northern Territory	1902	38	46	13		
Cape York	1509	60	37	10		
Total	4113	35	100	28		

Table 1.2 Northern Australia beach access

	Kimberley	Northern Territory	Cape York	Total	%
SLSC	1	3	0	4	
Lifeguards	1	???	0	1	
Sealed road access	5	28	11	44	1.3
Gravel road access	26	141	20	187	5.4
4WD access only	90	253	144	487	14.0
Foot or boat access only	1233	1055	466	2754	78.9
No foot access	6	11	0	17	0.5
Total	1360	1488	641	3489	100

The regions and beaches in this book are presented as they occur clockwise along the coast, beginning at Crab Creek on the northern shore of Roebuck Bay in the west and finishing at Saunders Beach, which terminates on the northern banks of the Endeavour River, opposite Cooktown in the east. In between is a vast array of beaches ranging from small rock- and reef-bound pockets of sand in the Kimberley to the longer sandy beaches of western Cape York. The generally low waves and higher tides have a major impact on the beach systems with 61% being tide-dominated, a further 10% tide-modified, and only 4% similar to the wave-dominated beaches of southern Australia. In addition 18% are fronted by rock flats and 6% fringed by coral reefs.

The book begins with a review of the nature and geological evolution of the northern Australian coast, followed by an overview of the main processes that impact upon the coast and its beaches, namely - climate, waves, tides, sediments and biota. In chapter 2 the types of beaches that occur across northern Australia are presented, followed by an assessment of physical and biological beach hazard; while chapter 3 tells you how to use this book and how to find a beach. Because of the number of beaches and resulting large size of this book these chapters are kept as brief as possible and the reader is referred to earlier books in this series for a more detailed account of these systems, in particular the Western Australian and Queensland books, which have many similar beach systems. These introductory chapters are followed by the longer chapters 4, 5 and 6 which present every beach in the Kimberley, Northern Territory and Cape York respectively.

GEOLOGICAL EVOLUTION OF THE COAST

The northern Australian coast consists of three main geological units that decrease in age from west to east (Fig. 1.2). The oldest is the Northern Australian Craton which extends from the Kimberley across to the western Gulf of Carpentaria (Table 1.3). It is an ancient (>1800 Ma) region of deformed and metamorphosed Palaeoproterozoic rocks that outcrop as the major tectonic units surrounded by younger rocks. Next is the Northeast Orogens, remnants of ancient mountain building, that extend from the southern Gulf to include all of the cape west of Princess Charlotte Bay. Finally, on the east coast, south from Cape Melville is the northern tip of the massive Tasman Fold Belt, a 400 Ma year-long accumulation of accretionary orogens and basins that extends south to Tasmania.

These major units consist of a number of secondary systems (Table 1.3). Starting in the west is the *Kimberley Basin*, which began actively infilling 2000 Ma, with the Fitzroy Trough and King Leopold Orogen bordering it in the west and south respectively, while younger Bonaparte Basin (500 Ma) occupies much of Cambridge Gulf and extends east to the Daly River. The Northern Territory coast is a composite of a series of basins (Bonaparte, Money Shoal, Arafura, McArthur and Carpentaria), with two ancient orogens at Pine Creek and the Arnhem Inlier. The younger *Carpentaria Basin* occupies most of the Gulf of Carpentaria including the entire southern shore and western Cape York. The higher terrain and mountains range of the eastern cape are composite and include the 1500 Ma old Coen Inlier and to the south the Laura and Hodgkinson basins, part of a series of basins which were uplifted during the opening of the Coral Sea by 55 Ma. The geology of each region is briefly described below.

The Kimberley

The Kimberley region is bordered to the west and south by the northern boundary of the larger Canning Basin and the Fitzroy Trough, which occupies the coast between northern Roebuck Bay and Derby. The trough is composed of shallow marine sediments (500-0 Ma), including limestone, which have been deeply weathered forming the bright red pindan of the coastal plain. The *King Leopold Orogen* surrounds the southern Kimberley Basin and consists of Proterozoic (1800 Ma) folded sandstone and volcanics. Today these rocks form the King Leopold Range, which extends to the coast between the highly indented eastern side of King Sound and Walcott Inlet including the islands of the Buccaneer Archipelago.

The *Kimberley Basin* is the oldest of Western Australia's 18 sedimentary basins with rocks dating from 2000-400 Ma. The core of the Kimberley and covering the largest area is a thick central basin sequence of sandstone, shale and basalt that was formed between 1800-1650 Ma and have since been subject to only minor faulting and warping. This now uplifted central region forms a plateau, which has been heavily weathered and dissected. It is surrounded in the east and south by the King Leopold and Halls Creek orogens, both remnants of mountain belts, which ceased tectonic activity by 1800 Ma.

The basaltic lava of the Kimberley was extruded 1800 Ma onto the sea floor covering an area of 250 000 km^2. The area was also glaciated between 700-600 Ma.

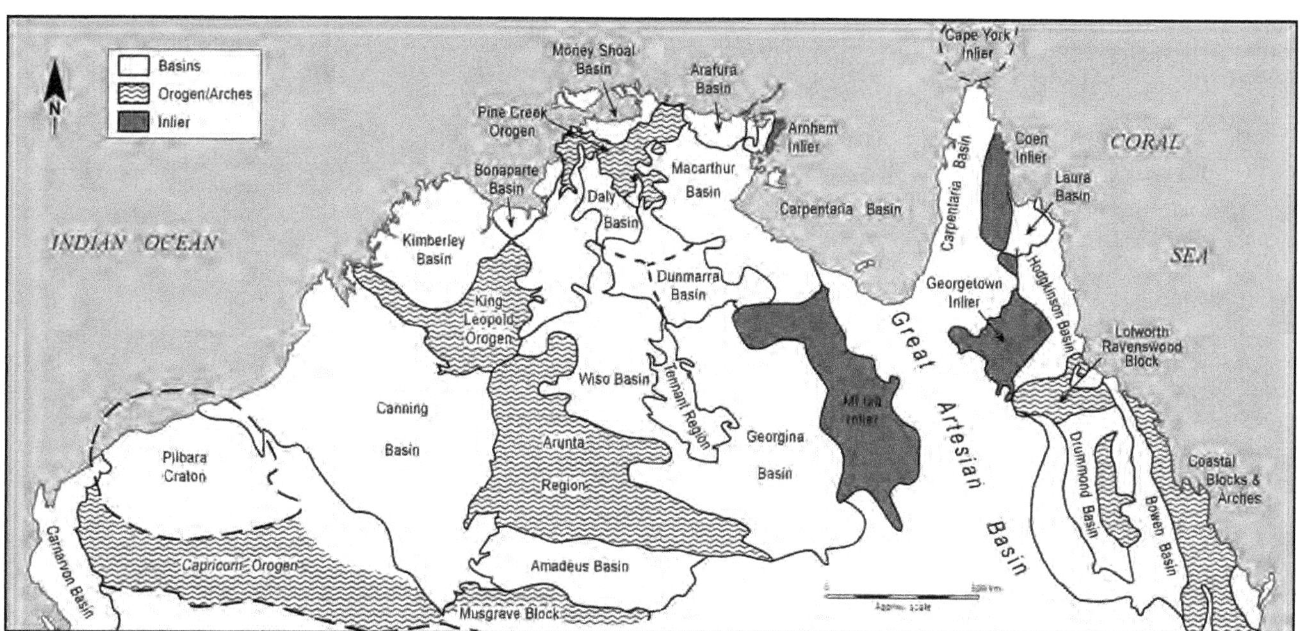

Figure 1.2 Major geological provinces of northern Australia. See Table 1.3 for more detail.

*Table 1.3 Major geological provinces of the Northern Australian coast, their geology and generalised coastal morphology. Orogens in **bold**.*

No.	Region	Age Ma*	Coast location	Geology	Coastal morphology
	Northern Australian Craton				
1	Fitzroy Trough (Canning Basin)	500-0	Broome-King Sound	Limestone	Low-lying coastal pindan plain
2	**King Leopold Orogen**	1800	Eastern King Sound	Folded sandstone & volcanics	Bedrock control, highly indented
3	Kimberley Basin	2000-400	Kimberley coast	Sandstone, shales, basalts	Bedrock control, deeply incised & weathered
4	Bonaparte Basin	500-0	Cambridge Gulf	Sedimentary	Tidal flats & bedrock
5	**Pine Creek Orogen**	2700-140	Darwin-Kakadu	Deeply weathered (laterised) sedimentary rocks	Kakadu escarpment, low bluffs in Darwin region
6	Money Shoal Basin	140-0	Kakadu coast, Van Diemen Gulf, Coburg Pen. Bathurst-Melville Is	Deeply weathered (laterised) sedimentary rocks	Low bedrock bluffs, laterite reefs, tidal flats, beaches
7	Arafura Basin	2500-200	Hall Pt -Buckingham Bay Elcho-Wessel Is.	Deeply weathered (laterised) sedimentary rocks	Low bedrock bluffs, laterite reefs, tidal flats, beaches
8	**Arnhem Inlier**	1860-1800	Melville Bay-Gove to Cape Shield	Deeply weathered (laterised) meta-sedimentary rocks	Low bedrock bluffs, laterite reefs, tidal flats, beaches
9	McArthur Basin	1800-1400	Blue Mud Bay-Robinson R; Groote Eylandt	Deeply weathered (laterised) sedimentary rocks	Low bedrock bluffs, laterite reefs, tidal flats, beaches
10	Carpentaria Basin	180-100	South & east Gulf of Carpentaria	Quaternary continental & coastal sediments, some laterised bluffs	Tidal creeks & flats, cheniers, beach ridges, some bluffs
	Northeast Orogens				
11	**Cape York Inlier**	300	Torres Strait high islands	Granite & metamorphic	High islands, coral reefs
12	**Coen Inlier**	1500	Temple Bay-Cape Sidmouth	Clastic & chemical sediments, volcanics	Headlands & beaches
13	Laura Basin	180-100	Cape Sidmouth-Princess Charlotte & Bathurst bays	Continental & marine sediments	Weathered bluffs, tidal flats, beaches
	Tasman Fold Belt				
14	Hodgkinson Basin	410-350	Cape Melville-Hinchinbrook Is	Volcaniclastic & carbonates	Headlands and beaches

* Ma = million years

By 400 Ma the Kimberley region had moved to the tropics and was covered by shallow seas fringed with coral reefs, which now form narrow limestone regions, including Geikie Gorge. A more southern location and glacial activity prevailed by 250 Ma. The breakup of Gondwanaland began in the northwest 180 Ma (Fig. 1.3), and continued counter-clockwise around the continent, with the south coast separating about 120 ma when the continent began its northward movement (Fig. 1.4). This movement and past warmer global temperatures resulted in the Kimberley having a tropical climate for the past 100 Ma, with more humid conditions prevailing until about 30 Ma when global cooling commenced. Tropical weathering of the basaltic rocks produced the aluminum-rich bauxite deposits in the northern Kimberley and the rich brown soil that fills the valleys. During this time the Prince Regent, Mitchell, Drysdale, King George, Berkeley and other smaller rivers eroded the plateau and cut deep gorges in the sandstone and volcanic rocks with waterfalls a common feature in the resilient rocks. Today the deeply weathered sandstone and basalts dominate much of the coast, while the deeply incised rivers and streams have produced the highly irregular coast and the numerous high bedrock islands, such as the Buccaneer and Bonaparte archipelagos.

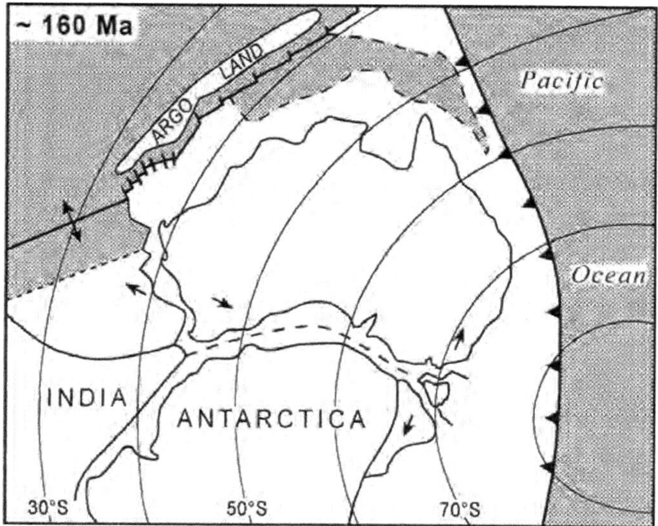

Figure 1.3 The initial separation of Australia from Gondwanland commenced in the northwest and proceeded counter-clockwise around the coast.

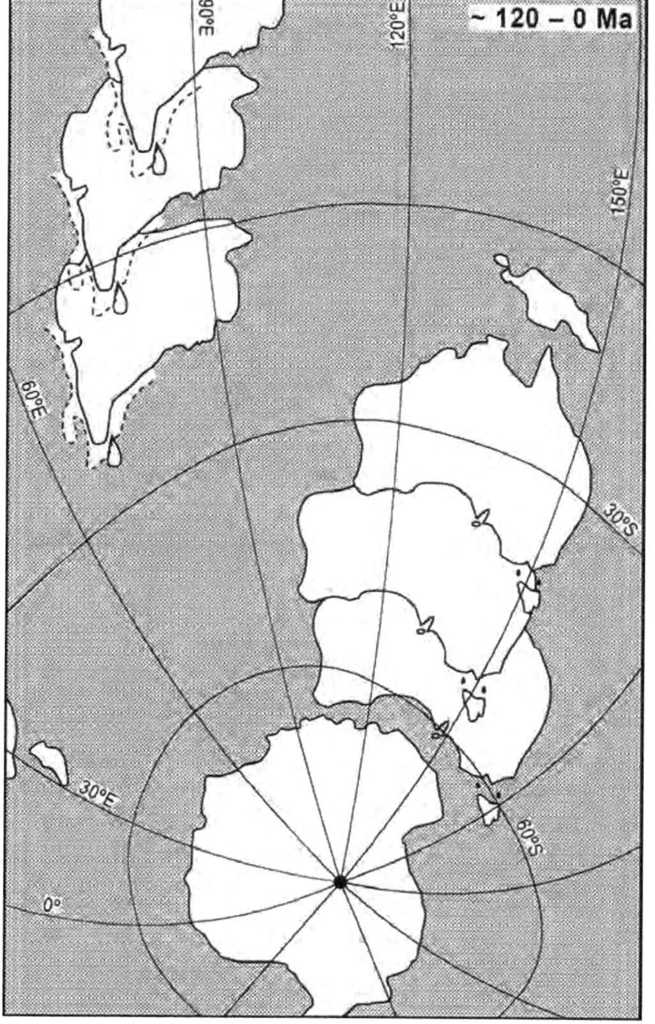

Figure 1.4 Since detaching from Antarctica about 120 Ma the Australian continent has drifted northwards 3000 km at an average rate of 6 cm/yr.

The Northern Territory

The Northern Territory can be divided geologically into the western half, consisting of Proterozoic foldbelts. and the eastern half, a Proterozoic basin (Fig. 1.2). All the rocks have been eroded down to a low peneplain and are deeply weathered. The geology of the coast, which is dominated by the basins, is described from west to east.

The *Bonaparte Basin* is a synclinal basin located east of the Kimberley and extending into the Northern Territory. At the coast it occupies the eastern side of Cambridge Gulf extending east to Fog Bay. It contains a wide range of sediments up to 6000 m thick dating from the Cambrian to Quaternary (500-0 Ma).

The oldest rocks in the Territory are part of the ancient Northern Australia Craton, represented in the Territory by the *Pine Creek Orogen*, the remnants of which occupy the coast in the Darwin region. They also include the rocks of the Kakadu escarpment and along the north coast between Goulburn Island and Hall Point. The basement rocks were laid down as sedimentary rocks, which were intruded with volcanics and folded and metamorphosed between 2500-1800 Ma to form a mountain belt or orogen. The mountains were subsequently eroded exposing the intruded granite. Erosion of the mountains led to deposition of sedimentary layers between 2000-1800 Ma, followed by renewed mountain building and intrusion of granite and metamorphosis of the sedimentary rocks. These rocks were eroded and deeply weathered resulting in laterisation, followed by a renewed period of quartz-rich sedimentation by 1650 Ma. This was followed by 1500 Ma of stability. During the Mesozoic (140 Ma) it was flooded and fossiliferous sandstone and siltstone were deposited over the lowlands. The sandstones of the Arnhem Land escarpment sit on older deformed rocks consisting of the Archaean basement, folded Proterozoic sedimentary and volcanic rocks and Proterozoic granite, which have been eroded down and partially overlain with Cretaceous sandstone. The entire surface has been laterised down to 30 m depth, probably during the Jurassic about 125 Ma, and predates the Cretaceous sediments. The younger sediments and basalts have also been subsequently laterised.

The *Money Shoal Basin* is a pericratonic basin that formed after the continental breakup along the northwestern margin during the Jurassic (Fig. 1.1). It covers an area of 350 000 km^2 much of which lies in the Arafura Sea. In the Northern Territory it includes the Kakadu coast, Van Diemen Gulf, Coburg Peninsula and Bathurst-Melville islands. It consists of a relatively undeformed middle Jurassic to Recent (140-0 Ma) basin containing up to 4500 m of sediment. The sediments consist of marine and clastic sequences overlain by carbonate sequences.

The *Arafura Basin* is a pericratonic basin 500 000 km^2 in area, that extends north of the northern Arnhem Land coast into the Arafura Sea. It occupies the coast between Hall Point and Buckingham Bay and includes the Elcho-

Wessel islands. The basin contains up to 5000 m of Proterozoic to Permian (2500-250 Ma) to possibly Triassic (200 Ma) sediments, consisting of shallow marine sandstone, mudstone and some carbonates. Uplift and folding occurred during the Permo-Triassic, followed by major erosion during the middle Triassic to early Jurassic, which resulted in a planated surface, upon which the Money Shoal basin sediments were deposited.

The small *Arnhem Inlier* extends along the eastern Arnhem Land coast from Melville Bay-Gove to Cape Shield. It consists of Proterozoic sediments, which have been folded and metamorphosed during the Barramundi Orogen (1860-1800 Ma).

The *McArthur Basin* is an intracratonic platform basin covering 180 000 km^2 and consisting of largely unmetamorphosed sedimentary rocks which were deposited 1800-1400 Ma and overlie rocks of the Pine Creek orogen. The rocks include sandstone, shale, carbonate and interbedded volcanic and intrusive igneous deposits, which reach up to 8000 m in thickness. It occupies the coast between Blue Mud Bay and the McArthur-Robinson rivers region, and includes Groote Eylandt.

Cape York/Queensland

The *Carpentaria Basin* is a 560 000m^2 north-south-trending intracratonic basin that covers most of the Gulf of Carpentaria (Figs. 1.2 & 1.5), with 80% located in Queensland waters. The basin was formed as a gentle intracratonic downwarp. At the coast it occupies a section of Northern Territory coast between the northern McArthur Basin centered on Arnhem Bay and the English Companys Islands in the north and Blue Mud Bay in the south. It then extends from just inside the Territory border to include most of the low-lying southern and all the eastern coast of the Gulf of Carpentaria. The basin was formed in the middle Jurassic (180 Ma) and contains Mesozoic (180-100 Ma) clastic sandstone, siltstone and conglomerate sediments up to 1800 m thick.

The *Cape York Inlier* forms the granitic high islands of Torres Strait including Prince of Wales, Thursday and Horn islands, as well as Cape York itself. The granites were intruded during the Carboniferous (300 Ma) and have been exposed by erosion. Flooding of the strait has left the high points as bedrock islands.

On the east coast the *Coen Inlier* is exposed at the coast between Temple Bay and Cape Sidmouth and contains the oldest rocks on the Queensland coast. The rocks date from 1500 Ma and are granitic at the coast grading inland to volcanics, clastic and chemical sediments.

The *Laura Basin* is centred on Princess Charlotte Bay with deposits extending from Cape Sidmouth to Bathurst Bay. It contains continental and marine sediments between 180 and 100 Ma in age.

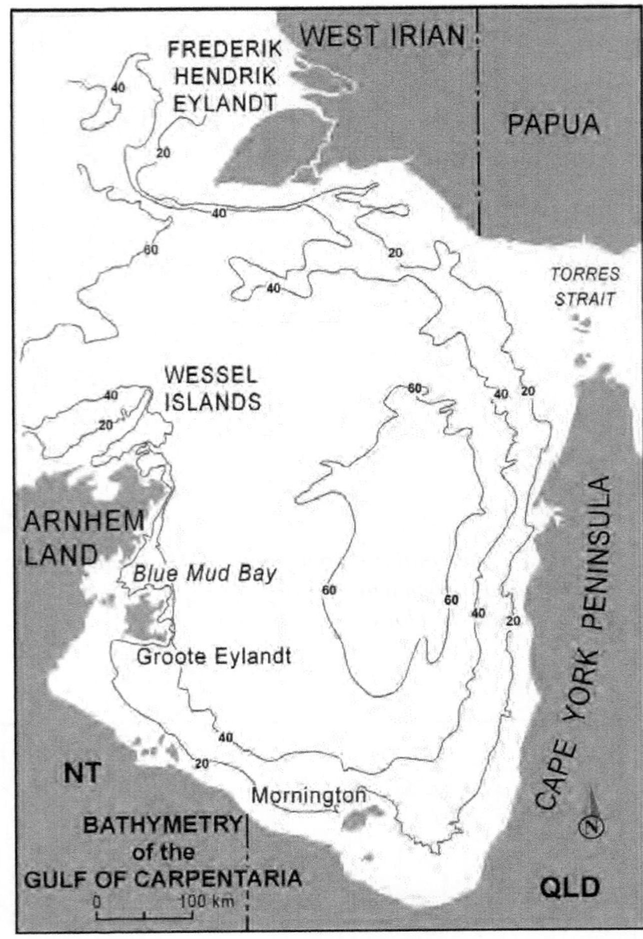

Figure 1.5 The Gulf of Carpentaria is a shallow epicontinental sea, occupying much of the Carpentaria Basin.

The *Hodgkinson Basin*, also known as the Peninsula Ridge, contains north-trending Palaeozoic volcaniclastic and carbonate sediments deposited between 410 and 350 Ma and uplifted during the opening of the Coral Sea about 50 Ma (Fig. 1.6). These now form a ridge of high rocks extending south from Cape Melville to Hinchinbrook Island, and include many prominent headlands including Cape Melville, Cape Flattery, Cape Bedford and Cape Grafton.

The predominantly Cretaceous sedimentary rocks exposed along the coast between Joseph Bonaparte Gulf in the west and Temple Bay on eastern Cape York have been exposed to subtropical Tertiary weathering which has led to the formation of extensive laterised soils reaching 30 m in depth. These soils consist of an upper indurated horizon consisting of red to black ferricrete 0.1-1 m in thickness, underlain by a deeper, less indurated, mottled clay-rich zone which is white to red or white in colour. Whereas the upper ferricrete zone is impervious and resilient to erosion forming a cap-rock along the coastal cliffs and bluffs, the mottled zone is soft and more readily eroded when exposed at the shore. Today the laterite rocks are commonly exposed along the coast and contribute to the predominantly red colour of many of the headlands and bluffs, such as Red Bluff, Red

Cliff, Red Cliffs, Red Island, Red Beach and Red Point. The laterite also outcrops as rocks and reefs, particularly along the Northern Territory coast where 400 beaches are fringed by usually laterite rock reefs. Finally it is mined for bauxite on the Gove Peninsula, on Groote Eylandt and on western Cape York at Weipa, the industry supporting some of the larger towns on the north coast.

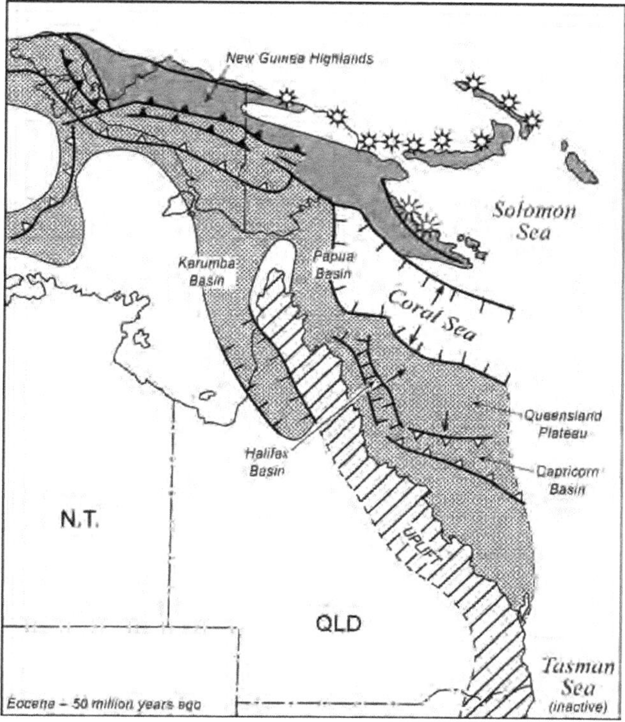

Figure 1.6 Opening of the Coral Sea uplifted the eastern highlands and formed the Queensland coast about 50 Ma.

Formation of the northern Australian coast

The coastline of northern Australia is a product of several factors including the geology, plate tectonics, long term denudation and more recent sea level changes. The geology determines the structure and contributes to the general relief of the coast. The Kimberley and parts of the Territory and Cape York coast are dominated by the geological structure and its denudation. These resilient rocks form headlands and reefs, while the beaches usually occupy the denuded valleys and creek mouths. The eastern Cape York coast resulted from rifting of the Coral Sea between 75 and 55 Ma, which led to the formation of the present coast and shelf. The numerous basins around the coast all tend to favour generally low relief and low gradient sedimentary shorelines. From the west these include the Fitzroy Trough and Bonaparte Basin in the Kimberley; the Money Shoals, Arafura, McArthur and Carpentaria basins dominating the Territory coast; the large Gulf of Carpentaria (Basin) and the smaller Laura Basin in the east. Finally the sea level rise flooded the entire shelf, coming to rest a little above its present level about 6500 years ago (Fig. 1.7). Since then sea level has fallen 1-2 m around much of northern Australia, stranding earlier beaches inland. The fall is a result of slight uplift of the coast due to hydro-isostatic adjustment of the coast and shelf.

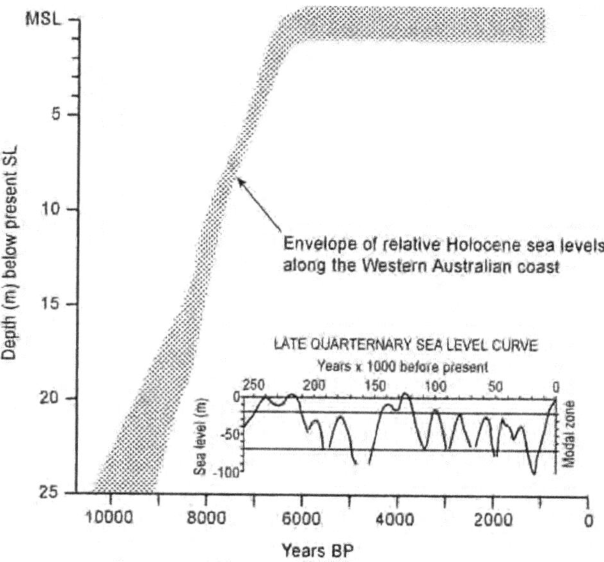

Figure 1.7 Sea level rose rapidly between 18-6.5 ka to reach a little above its present level in northern Australia. There has since been a 1-2 m relative fall in sea level.

CLIMATE

Northern Australia has a tropical monsoonal climate dominated by two pressure systems - the subtropical high and the equatorial low. The high dominates most of the year and its great anticlockwise spiral of winds generates the southeast trades across the northern half of Australia (Fig. 1.8). The high pressure and associated winds persist from April to November, with the wind velocity tending to increase into the winter period. The trades bring some winter rain to the eastern Cape York region, while dry winter conditions dominate the remainder of the top end and coast. During summer the equatorial low moves south and resides across northern Australia centred on the Pilbara and Cloncurry heat lows. Humid, tropical air is drawn south towards these lows bringing afternoon thunderstorms across the top end during the summer wet season. The amount of rain decreases inland and to the south. In addition the intertropical convergence zone, where the northwest monsoons meet the easterly trades (Fig. 1.9), is the breeding ground for tropical cyclones. The cyclones occur between November and April, peaking in February. In the Australian region they form in the Indian Ocean-Timor Sea, the Gulf and Coral Sea.

Once formed they tend to travel west and curve to the south, often resulting in landfalls. The cyclones bring strong winds, heavy rain, high seas and storm surges to affected areas.

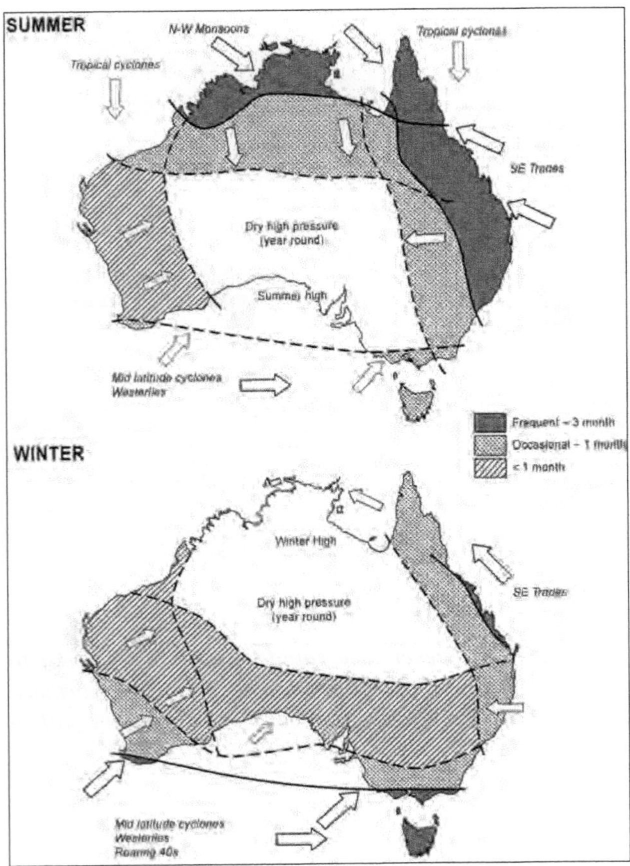

Figure 1.8 Northern Australia is exposed to humid summer northwest monsoonal winds, while during winter the southeast trades bring warm and dry conditions.

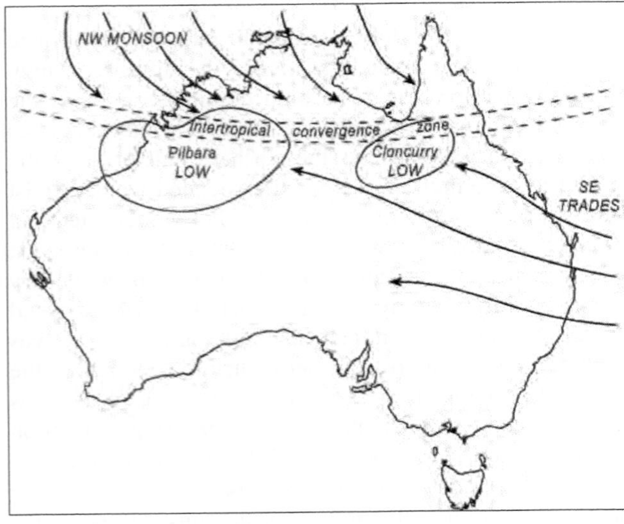

Figure 1.9 The Intertropical Convergence Zone (ITCZ) marks the summer convergence of the northerly monsoons and southerly trades, and the southern boundary of the monsoonal rains.

Wind

There are two major wind systems in northern Australia. The southeast trade winds which blow year round and have the greatest impact between April and November. Figure 1.10 shows the winter wind roses with a dominance of moderate to strong east to southeast winds right around the coast. The trades remain persistent in direction but fluctuate in velocity with the passage of the highs across Australia, tending to be strongest when the highs are centred on Australia, producing a strong ridge of pressure and wind along their eastern-northern side. During the shorter summer period the northwest monsoons bring light to moderate west to northwest winds to the west and north coast, but they do not penetrate across the cape, with light to moderate velocity eastern winds dominating the eastern Cape York coast (Fig. 1.10 summer).

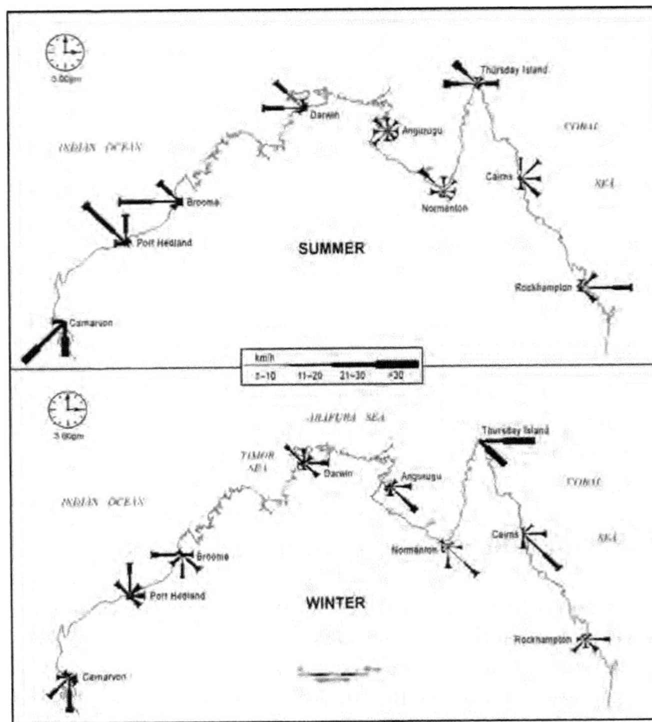

Figure 1.10 Wind roses for 3pm summer with monsoonal westerly winds dominating, and winter with the southeast trades dominating.

Rainfall

Rainfall is highly seasonal across the north coast and has three sources - the trades, the northern monsoons and occasional tropical cyclones, each of which has limited penetration inland as illustrated by the concentration of rain towards the coast in the rainfall pattern (Fig. 1.11). The trades bring rain year round to eastern Cape York, though with a distinct summer maximum. However on crossing the eastern highlands and cape the trades become a dry wind and bring no rain to the rest of northern Australia, as indicated in the winter rainfall map (Fig. 1.11b). The summer northwest monsoon tends to flow onto the north coast and penetrate south between October and April, with rainfall peaking between December and March. The rainfall distribution and degree of penetration are illustrated in the summer rainfall map (Fig. 1.11a). Tropical cyclones can deliver torrential rain when crossing the coast. However they tend to cross south of Broome and Cooktown (see below) and have a relatively low frequency of occurrence across the top end and therefore are not a reliable source of rainfall in this region.

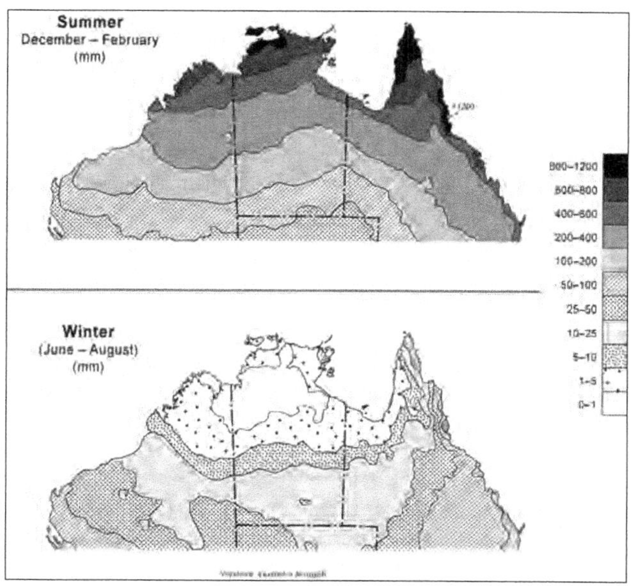

Figure 1.11 Northern Australia receives high summer rainfall (a) and low winter rainfall (b).

Temperature

Temperatures are warm to hot year round across northern Australia. Figure 1.12a shows that during summer the mean maximum along the coast range from 30-40^{0}C with the temperatures increasing southward to peak in the lower Gulf. During winter temperatures range from the mid 20's on the eastern cape to mid 30's increasing both northward and to the west (Fig. 1.12b). Wyndham in Cambridge Gulf lays claim to the hottest town in Australia, with the mean daily maximum exceeding 30^{0}C every month of the year and with a mean daily maximum of 35.6^{0}C, the highest in Australia.

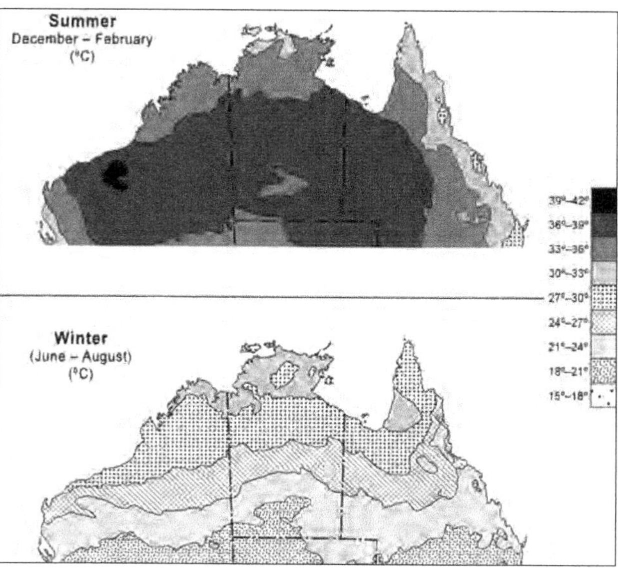

Figure 1.12 Northern Australia has a mean maximum temperature between 30-40^{0}C in summer and 25-30^{0}C in winter.

Tropical cyclones

Tropical cyclones can impact the entire northern Australian coast (Fig. 1.13) and tend to occur between November and April with a pronounced later summer peak (Fig. 1.14). They are most prevalent off the northwest Western Australian coast where they tend to make landfall between Port Hedland and Onslow, centered on 20^{0}S (Fig. 1.15). In the Gulf of Carpentaria they tend to cross the southern Gulf coast, while in the Coral Sea they can land between Cape York and Brisbane, with 40% landing north of Cooktown, and 60% landing between Cooktown and Mackay, with broad maxima either side of Cairns (Fig. 1.15). Therefore most tropical cyclones land south of Broome in the west, in the southern gulf and south of Cooktown in the east, with much of the Kimberley, Northern Territory and Cape York coast receiving a lesser impact. This is because all of the northern coast lies between 9^0 and 18^{0}S, while the cyclones tend to be generated between 15^0 and 20^{0}S, making landfall usually further south, thereby usually missing much of the northern coast. This does not mean these areas are free of tropical cyclones as evidenced by Cyclone Tracy at Darwin in 1975 and more recently Cyclone Ingrid in 2005, which hit the northern Kimberley coast. When tropical cyclones do make landfall they generate very strong winds, heavy rain, high seas and storm surges, all of which can have a devastating impact on the coast through wind damage, river flooding, sea level rise and wave erosion and overwashing. While tropical cyclones have a low frequency of occurrence and impact at any particular location, the results of their impact can persist for long periods, resulting in the formation and preservation of tropical cyclone-generated coastal features, including cheniers, cobble and boulder beach ridges, overwash chutes, and elevated storm deposits (see Nott, 2005).

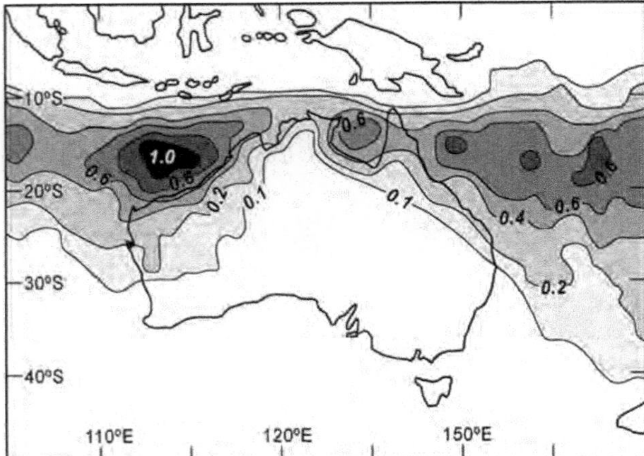

Figure 1.13 Average number of tropical cyclones per year across northern Australia.

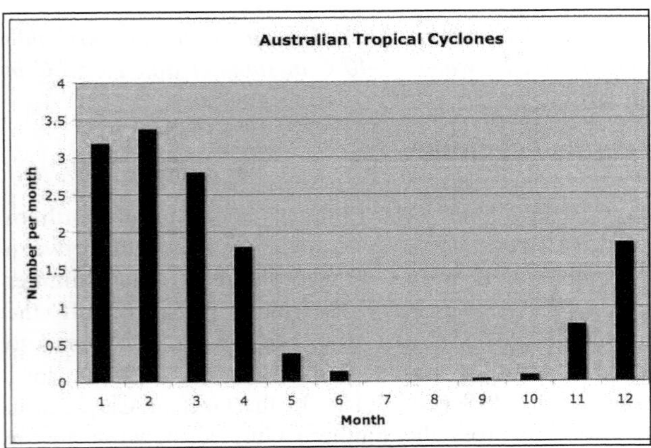

Figure 1.14 Average number of tropical cyclone per month forming off northern Australian. Note the late summer maximum, with none in winter.

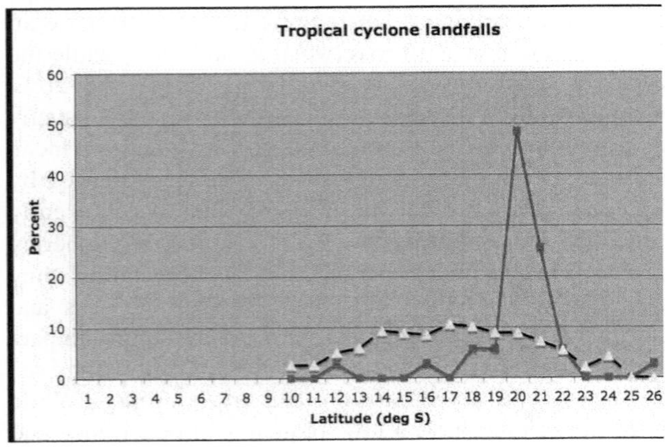

Figure 1.15 Latitudional landfall of tropical cyclones on the Western Australian (solid line) and Queensland coast (dashed). The Western Australia peak centres on the Pilbara coast at 20^{0}S.

OCEAN PROCESSES

North Australia is bordered by the eastern Indian Ocean and southwest Pacific, which off Queensland is known as the Coral Sea. Across northern Australia the two oceans are connected by the Timor and Arafura seas (Fig. 1.1). The Timor Sea extends between the Kimberley-western Northern Territory and Timor and includes the large Joseph Bonaparte Gulf, while the Arafura Sea extends from the Territory's Cape Don to Torres Strait, north to islands of the Indonesian archipelago and New Guinea, and includes the Gulf of Carpentaria. The oceanography of these seas is controlled by six factors.

1. Size and bathymetry, which includes their limited extent, the numerous islands and coral reefs in and bordering the seas, and generally shallow seafloor (<200 m).
2. Seasonal tropical climate, which brings high sea surface temperatures (>26^{0}C), light to moderate northwest and southeast winds, seasonal summer rain and coastal runoff, and occasional tropical cyclones.
3. Wave climate, which consists of locally generated low to moderate seas.
4. Tide regime, which results in Australia's highest tides due to shoaling across the shallow shelf, as well as amplification in King Sound and Cambridge Gulf. The generally high tides also result in strong tidal currents both at the coast and across parts of the shelf.
5. Ocean circulation, which consists of a generally weak east to west flow of water through Torres Strait and down through the Indonesian islands exiting into the Indian Ocean.
6. Rich tropical biota, which is most manifest in the extensive coral reef systems, including the northern Great Barrier Reef, the intertidal mangroves which dominate all low energy shores, and subtidal tropical seagrasses.

Waves

Waves are generally low across the northern coast, a product of the lack of ocean swell, the low to moderate velocity local wind regimes and limited fetch. Ocean swell is precluded from the coast owing to the northerly orientation of most of the coast together with the landlocked nature of the seas, and in the east the blocking of all trade wind waves by the northern Great Barrier Reef. The coast is therefore largely dependent on the prevailing southeast trades and summer northwest monsoonal winds, plus any accompanying sea breezes, for the generation of all waves. The wave climate is further constrained and modified by the limited fetch, the varied orientation of the coast, which results in exposed windward and sheltered lee shores, and the generally shallow nearshore gradients, which lead to substantial wave refraction and attenuation. The overall result is a very low to at most moderate energy coast.

Published wave data for northern Australia is very limited and as yet no comprehensive or even regional wave climate exists, apart from waverider buoys off Weipa and Cairns, maintained by the Queensland Beach Protection Authority.

Table 1.4 summarises the main wave sources and characteristics for the northern coast. The southeast trades generate waves on all exposed coasts including along the eastern Cape York coast, the western Gulf of Carpentaria including eastern Arnhem Land, and a few east-facing sections of the Kimberley coast. Elsewhere the trades blow offshore. Strong trade winds in the gulf can produce waves up to 2-3 m high with a period of 4-5 s, more commonly however waves are less than 1 m, particularly at the shore. The summer northwest monsoon reverses the wave climate. The low to moderate velocity wind result in usually low, short waves with a breaker height less than 1 m and period of 3-4 s. The trades and monsoon waves are supplemented by seabreeze winds and waves, which tend to arrive from the east through north. Tropical cyclone waves are too temporally and spatially infrequent to have a significant impact on the annual wave climate.

Table 1.4 Wave sources and characteristics across the northern Australian coast

	Location	Direction	Season	Characteristics
Sea breeze	entire coast	varies	summer max	short, low seas
Trade winds	entire coast	E-SE	year round, winter max	short, seas to 3 m
Monsoonal winds	Kimberley-NT-Gulf region	W-NW	summer	short, seas to 2 m
Tropical cyclones	entire coast	varies	summer	high seas to several metres

Table 1.5 summarises the estimated breaker wave heights and period compiled from the northern Australian beach database. These represent the wave height at the shore after undergoing wave refraction and attenuation, and as a consequence will be lower than the deepwater seas. The *Kimberley* has the lowest waves with 50% of the beaches receiving waves averaging only 0.1 m, while only 10% receive waves averaging 0.4 m and greater, with the highest waves only reaching 1 m on six beaches. Wave period averages a short 5 seconds.

Table 1.5 Percent occurrence of breaker wave heights. Modal height in bold.

Breaker wave height (m)	Kimberley	Northern Territory	Cape York
0.1	**49.2**	12.5	10.5
0.2	28.3	16.0	7.5
0.3	12.3	**21.2**	17.8
0.4	3.8	15.0	**22.2**
0.5	2.6	10.0	17.9
0.6	1.0	6.0	9.8
0.7	0.4	4.0	4.7
0.8	1.4	3.5	3.6
0.9	0.5	3.5	2.2
1.0	0.4	5.0	3.9
1.1		1.2	
1.2		1.0	
1.3		0.6	
1.4		0.3	
1.5		0.2	
	100	100	100

The *Northern Territory* has a modal wave height of 0.3 m, with waves ranging from a low of 0.1 m to a high of 1.5 m, received by only three beaches, with wave period ranging from 3-7 sec for most of the coast. Beaches receiving waves greater than 1 m are all exposed to the southeast trades and face into the longer fetch of the Gulf. Similarly these beaches have the longest wave period reaching 7 sec.

Cape York beaches receive waves ranging from a low of 0.1 m in the southern gulf and protected east coast bays, to a high of 1 m on the more exposed east coast beaches, which are nonetheless protected from higher deep ocean waves and swell by the Great Barrier Reef, which also limits the fetch within the backing lagoon. Wave period ranges from a low of 2 sec in some of the sheltered fetch-limited bays to 4-5 sec on the open coast exposed to the trades.

To summarise, northern Australia receives low to moderate southeast waves for most of the year averaging 1-1.5 m, with periods of 4-5 sec on the most exposed beaches. The majority of beaches however receive some degree of sheltering from headlands, reefs, islands and shallow nearshore with 26% receiving waves 0.1 m or less, 83% receiving waves 0.5 m or less.

Tides

Tide range across northern Australia varies from 2 m in the southern gulf to Australia's highest tides of 10-11 m in Cambridge Gulf and King Sound (Table 1.6). Figures 1.16 and 1.17 illustrate the tidal regimes around the coast, while Table 1.7 lists the tidal characteristics of particular sites.

Table 1.6 Minimum and maximum regional tides

	Tide min. (m)	Tide max. (m)
Kimberley	2.60	11.20
Northern Territory	1.40	5.80
Cape York	1.60	3.50

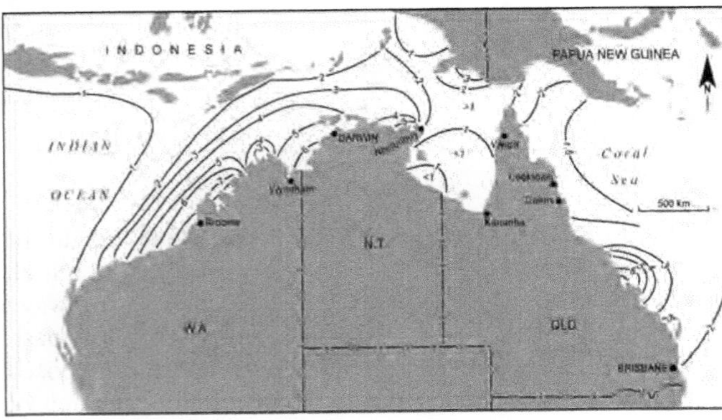

Figure 1.16 Co-range lines for Northern Australia illustrating areas with the same tide range. The tide ranges from less than 1 m in the southern Gulf to more than 8 m in King Sound. See details in Table 1.6.

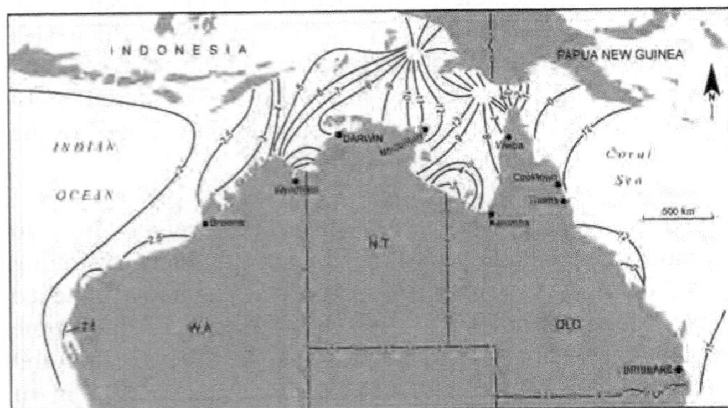

Figure 1.17 Co-tide lines for Northern Australia illustrating the movement of the clockwise movement of the tidal waves around two amphidromic points south of new Guinea, together with others in the Pacific and Indian Ocean (not shown) resulting in a complex series of tidal systems across the northern coast. The lines link areas receiving the tide at the same time.

The highest tides occur along the northwest coast and the *Kimberley* region, with Broome, Derby and Hall Point all exceeding 9 m at spring tide. These high tides are due to amplification of the tide across the broad shallow northwest shelf, as well as additional resonate amplification into the embayments of King Sound and Cambridge Gulf. Tides peak in King Sound, and again in Cambridge Gulf where they reach over 7 m at spring tides at Wyndham. The tidal wave originates in the Indian Ocean with an amphidromic point off Cape Leeuwin. It travels up the Western Australian coast arriving at Broome only 30 minutes after Fremantle, but then slows considerably as it propagates around the Kimberley coast arriving at Wyndham 10 hours later.

In the *Northern Territory*, the tidal wave is associated with a tidal system with an amphidromic point located in the northern Arafura Sea, which propagates in a counter-clockwise direction, moving from west to east across the Territory coast. In comparison to Fremantle the tide arrives at Darwin 8 hours after Fremantle, and Gove Harbour at 11 hours. The tide range is highest on the western Territory coast going into Cambridge Gulf, with the spring range all greater than 6 m, and it varies along the northern coast from a low of 2.4 m at Cape Croker to 4.8 m at Yabooma, dropping to 2.9 m at Gove. In the western Gulf the spring tides range between 2 and 3 m.

The *Gulf of Carpentaria* receives its tide from another separate tidal system with an amphidromic point located just west of Torres Strait, as well as some tidal components arriving from the Coral Sea via Torres Strait. The tide in the Gulf ranges between 2 and 4 m, while the time of arrival is highly variable as indicated in Fig. 1.17. Because of the constriction caused by Torres Strait, very strong easterly tidal currents flow through the strait. In addition the southern gulf occasionally has what are known as 'dodge' tides, when only one low and high will occur each day. Finally the gulf has wind- and pressure-induced seasonal change in sea level, with summer levels being up to 60 m higher than winter levels.

The eastern Cape York-northeast Queensland coast receives its tides from a system located in the Pacific. The tidal wave has to penetrate the Great Barrier Reef and propagates up the coast with a slight south to north lag along the coast. Spring tide range is generally between 2 and 3 m in the south increasing to greater than 3 m close to the Strait.

In summary, tides across northern Australia are all greater than 2 m, with much of the coast receiving spring tides reaching several metres (Table 1.7). The tides and associated tidal currents play a major role in both the coastal and well as nearshore and shelf oceanography. At the coast the high tide ranges cause major oscillations in the shoreline, especially on the low gradient tidal flats, while at the coast and offshore strong tidal currents are required to accommodate the tidal flows in estuaries and inlet, through topographic constrictions and across the generally shallow seafloor

Table 1.7 Tidal characteristics of northern Australian coast

Location	Mean spring high tide (m)	Mean spring low tide (m)	Mean spring tide range (m)	Relative time WST of arrival 0 hr=Perth - = before + = after	Arrival time in CST & EST zones
Kimberley Coast				(WST)	
Broome	9.4	1.1	8.3	+0.5	
Derby	11.2	1.1	10.1	+1.0	
Port Warrender	7.0	0.6	6.4	+3.0	
Cape Voltaire	6.5	0.9	5.4	+3.0	
Hall Pt	9.2	0.4	8.8	+3.5	
Napier Broome Bay	2.6	0.2	2.4	+3.5	
Lesueur Island	2.8	0.1	2.7	+5.5	
Cape Domett	6.9	1.4	5.5	+9.0	
Lacrosse Island	5.9	1.3	4.7	+9.0	
Wyndham	7.7	1.2	6.5	+10.0	
Pelican Island	6.9	1.6	5.3	+10.0	
Northern Territory				(WST)	(CST)
Turtle Pt	6.0	0.8	5.2	+9.5	+0122
Pearce Pt	6.6	0.8	5.8	+9.0	+0104
Daly River	6.0	0.4	5.6	+7.0	- 0054
Tapa Bay	6.5	1.3	5.2	+7.5	- 0019
Darwin	6.9	1.3	5.5	+8.5	0000
Cape Hotham	4.4	1.0	3.4	+8.5	+0058
Cape Don	2.9	0.7	2.2	+8.0	- 0020
Port Essington	2.6	0.5	2.1	+7.0	- 0146
Cape Croker	2.4	0.5	1.9	+7.0	- 0137
North Goulburn Island	2.7	0.5	2.2	+8.5	- 0001
Entrance Island	3.9	1.1	2.8	+8.6	+0012
Yabooma	4.8	1.3	3.5	+10.0	+0135
Mallison Island	4.7	0.6	4.1	+9.5	+0109
Gove Harbour	2.9	0.7	2.2	+11.0	+11.0
Gulf of Carpentaria					(EST)
Cape Grey	1.8	0.4	1.4		- 0033
Rose River	2.4	0.2	2.2		- 0134
Centre Island	2.9	0.6	2.3		- 0031
Queensland (Cape York)					
Karumba	3.8	0.3	3.5		
Staaten River	4.1	1.2	3.1		- 0128
Nassau River	2.4	0.6	1.8		
Archer River	2.3	0.6	1.7		+0015
Weipa	2.9	0.7	2.2		
Cullen Point	3.5	0.8	2.7		- 0102
Vrilya Point	3.6	0.7	2.5		- 0104
Thursday Island	3.0	0.6	2.4		
East Coast					
Albany Island	3.1	0,5	2.6		- 0053
Turtle Head Island	3.5	0.7	2.8		- 0144
Cairncross Islets	3.6	1.0	2.6		+0116
Cape Grenville	3.0	0.8	2.2		+0045
Portland Road	2.5	0.6	1.9		+0022
Cape Melville	2.5	0.6	1.9		+0015
Cape Flattery	2.5	0.5	2.0		+0004
Cooktown	2.3	0.7	1.6		+0003
Cairns	2.7	0.6	2.1		

BIOLOGICAL PROCESSES

Northern Australia has the richest coastal biota on the continent, which includes the subtidal seagrass meadows and coral reef systems and intertidal mangroves. In addition the coast supports a little studied beach fauna and has a wide range of little studied coastal dune vegetation. What is presented in this section is based on what is known about some of these systems, with much still to be investigated.

Coastal dune vegetation

Coastal dunes occur around the entire northern Australian coast. They include the massive transgressive dune systems of Shelburne Bay-Cape Grenville and Cape Flattery, extensive transgressive dunes on Groote Eylandt and eastern Arnhem Land and smaller pockets of dunes along the northern Northern Territory and Kimberley coast. All dunes are vegetated by a predictable succession of plants beginning with grasses, succulents and creepers on the incipient foredunes (Fig. 1.18), grading landward into a combination of sedgelands and shrublands on the foredune and hind dunes, and then into a climax succession of woodlands or forests. While the structure of the dune vegetation is similar around the coast the species vary considerably from west to east as the climate and biogeographical regions change. Little has however been published on the coastal dune vegetation of northern Australia. The following lists some of the major plant communities and species found on parts of the coast.

Figure 1.18 A low vegetated foredune dominated by Triodia pungens (Soft Spinifex) *(K 1146).*

Kimberley

- Incipient foredune/foredune - coastal grasses (*Spinifex longifolius, Ipomoea brasiliensis, Salsola kali, Fimbristylis cymosa, Fimbristylis sericea, Cyperus bulbosus*).
- Foredune - low shrubs (*Acacia bivenosa, Lysiphyllum cunninghamii, Canavalia rosea, Triodia pungens*).
- Hind dune and hollows - dense shrub community of diverse plants, some *Pandanus spiralis*.
- Grades into pindan or rocky vegetation.

Southeastern Cape York

- Incipient foredune - coastal grasses, creepers and succulents (*Ipomoea pes-caprae* (goats foot), *Cyperus pedunculatus* (pineapple sedge), *Thuarea involuta* (tropical beach grass), *Vigna marina* (beach bean), *Wedelia biflora* (beach sunflower)
- Foredune

  Grasses *Cynodon dactylon* (couch grass), *Chloris gayana* (Rhodes grass)

  Shrubs *Casuarina equisetifolia* (she oak), *Cocos nucifera* (coconut tree), *Terminalia catappa* (sea almond), *Hibiscus tiliaceus* (beach hibiscus), *Acacia crassicarpa* (brown salwood), *Calophyllum inophyllum* (beach calophyllum), *Scaevola sericea* (sea lettuce tree)
- Hind dune – *Corymbia tessellaris* (carbeen)

Samphire vegetation

Samphire vegetation in association with algae grows along lower energy sections of coast in the lower swales between beach ridges and in the saline back barrier depressions and dry lagoons. The samphire vegetation is usually low (<1 m) and scrubby and forms a boundary between the shoreline and the landward terrestrial vegetation.

In northern Australia the higher tide ranges produce wide inter- to supratidal zones suitable for samphire vegetation, while the hotter climate including the long winter dry period results in greater climate stress, restricting the development of salt marshes.

In the Kimberley four communities of plants can occur in favourable locations.

1. A more seaward community of succulent samphires *Suaeda arbusculoides* at the seaward fringe, sometimes separated from *Halosarcia halocnemoides* by mud or sand flats covered with dense mats of blue-green algae.
2. A mixed herbaceous and grasses community in the mid to upper marsh level containing *Limonium salicorniaceum*, water couch grass (*Sporobolus virginicus*) and rice grass (*Xerochloa imberbis*).
3. Herbs and low shrubs in higher well drained locations containing *Halosarcia indica*, the halophyte *Frankenia ambita* and *Hemichroa diandra*.
4. The most landward community which can tolerate the high salinity but not waterlogging, including *Neobassia astrocarpa, Trianthema turgidifolia* and some *Triodia* sp.

Mangroves

Mangroves are trees and some palms that grow in the intertidal zone usually between mean sea level and neap high tide. Mangrove woodlands and forest are well developed across northern Australia, which hosts 88% of Australia's 11 500 km^2 of mangroves. Four factors favour the growth and extent of mangroves across the north. First, is the warm tropical climate which permits all 39 mangrove species to be represented across the north. Second, are the generally lower energy shorelines which provide the sheltered habitats mangroves require. Third, are the generally lower gradient sedimentary shorelines and extensive intertidal sand and mud flats that provide a wide intertidal area for mangroves to inhabit; and fourth, the generally higher tide range (2-9 m) which combines with the lower gradient intertidal area to maximise the area suitable for mangrove growth.

The tropical climate not only permits a greater number of mangrove species, but also results in taller mangroves (10-30 m) with a greater biomass (Fig. 1.18). Therefore mangrove communities in the north tend to be diverse, tall and relatively wide. As a consequence of these factors the area of mangroves, number of species and biomass increase substantially from south to north. Australia-wide 96% of mangroves are located in the tropics north of the Tropic of Capricorn. Western Australia has 2430 km^2 of mangroves along the mainland, with another 90 km^2 on islands, which in total comprise 22% of Australia's mangroves, most located on the Kimberley coast (Table 1.8). The Northern Territory has the greatest area of mangroves (29%) followed by the Gulf of Carpentaria (21%) and the tropical northeast Queensland coast (18%). The northern coast between Broome and Cooktown contains 8861 km^2 of mangroves, which represent 88% of Australia's mangrove area. On a global basis Australia has the world's third largest area of mangroves after Brazil and Indonesia.

Table 1.8 Mangrove area in northern Australia

	km^2	%
Kimberley	2256	20
Northern Territory	3360	29
Gulf of Carpentaria	2440	21
Northeast Queensland	805	18
Northern Australia	8861	88
Australia	11 500	100

In terms of mangrove species, the greatest number (35) occurs in north Queensland, with species numbers decreasing to the west and to the south (Table 1.9). The Northern Territory has up to 24 species, the Gulf 19 and the Kimberley 17. In contrast southern Australia has just one species of mangrove (*Avicennia marina*), which extends from southern NSW, across parts of southern Australia and southern Western Australia, up to Kalbarri. On the east coast species number increases to two in southern NSW, eight species by Brisbane and 35 in Cape York.

Figure 1.19 Ten metre high mangroves on the Lockhardt River, Cape York

Table 1.9 Distribution of mangrove species across northern Australia. Some major species in ***bold****.*

Genus	*Species*	*W Kimb*	*E Kimb*	*NT*	*Gulf*	*NE Qld*
Acanthaceae	*Acanthus ebracteatus*		*x*			
	Acanthus ilicifolius			*x*	*x*	*x*
Arecaceae	*Nypa fruticans*					*x*
Avicenniaceae	***Avicennia marina***	*x*	*x*	*x*	*x*	*x*
	Avicennia eucalyptifolia		*x*	*x*		
Bignoniaceae	*Dolichandrone spathacea*				*x*	
Bombaceae	*Camptostemon schultzii*	*x*	*x*	*x*	*x*	*x*
Caesalpiniaceae	*Cynometra iripa*					*x*
	Cynometra ramiflora					*x*
Combretaceae	*Lumnitzera racemosa*		*x*	*x*	*x*	*x*
	Lumnitzera littorea			*x*	*x*	*x*
	Lumnitzera x rosea					*x*
Euphorbiaceae	*Excoecaria agallocha*	*x*	*x*	*x*	*x*	*x*
	Excoecaria ovalis			*x*		
Malvaceae	*Hibiscus tiliaceus*				*x*	
Meliaceae	*Xylocarpus australasicus*		*x*	*x*	*x*	*x*
	Xylocarpus granatum				*x*	*x*
	Xylocarpus mekongensis					*x*
Myrtaceae	***Osbornia octodonta***	*x*	*x*	*x*	*x*	*x*
Myrsinaceae	***Aegiceras corniculatum***	*x*	*x*	*x*	*x*	*x*
Lythraceae	*Pemphis acidula*		*x*			*x*
Plumbaginacae	***Aegialitis annulata***	*x*	*x*	*x*	*x*	*x*
Pteridaceae	*Acrostichum speciosum*					*x*
Rhizophoraceae	*Bruguiera cylindrica*					*x*
	Bruguiera exaristata	*x*	*x*	*x*	*x*	*x*
	Bruguiera gymnorrhiza		*x*	*x*	*x*	*x*
	Bruguiera parviflora			*x*	*x*	*x*
	Bruguiera sexangula			*x*		*x*
	Ceriops australis			*x*	*x*	*x*
	Ceriops decandra			*x*		*x*
	Ceriops tagal	*x*	*x*	*x*		*x*
	Rhizophora stylosa	*x*	*x*	*x*	*x*	*x*
	Rhizophora apiculata			*x*		*x*
	Rhizophora lamarckii					*x*
	Rhizophora mucronata					*x*
Rubiaceae	*Scyphiphora hydrophyllacea*		*x*	*x*	*x*	*x*
Sonneratiaceae	***Sonneratia alba***	*x*	*x*	*x*		*x*
	Sonneratia caseolaris			*x*		*x*
Sterculiaceae	*Heritiera littoralis*					*x*
Number	*39*	*10*	*17*	*24*	*19*	*34*

Seagrass meadows

Seagrasses grow in suitable locations around the Australian coast, and particularly across the tropic north. Their growth is primarily determined by sunlight penetration. They extend from the lower intertidal and to the shallow subtidal zone, reaching a maximum depth of several metres. Seagrasses inhabit all types of seabed, from mud to rock, however the most extensive meadows occur on sand and mud. Seagrasses grow in coastal waters from tropical to temperate regions, with over 30 distinct temperate and tropical species found in Australian waters. The number of species is greater in the tropics with only two species, *Halophila ovalis* and *Syringodium isoetifolium*, occurring in both regions.

Tropical seagrass species begin appearing in Shark Bay where the meadows cover 4000 km^2 and are the largest in the world. They are less prevalent from Carnarvon north along the Pilbara and Canning coasts, due to lack of suitable habitat, while they again flourish in King Sound and around the Kimberley and Northern Territory coast. The most extensive and diverse seagrass communities are in the waters of Torres Strait and north-eastern Queensland where between 10 and 14 species are present (Table 1.10).

Table 1.10 Seagrass species of northern Australia (Kirkman, 1997)

	Kimberley-Northern Territory	Gulf of Carpentaria-Torres Strait	Northeast Queensland
Halodule uninervis		x	x
Halodule pinifolia		x	x
Cymodocea angustata		x	
Cymodocea rotundata	x		x
Cymodocea serrulata	x	x	x
Syringodium isoetifolium	x	x	x
Enhalus acoroides	x	x	x
Thalassodendron ciliatum	x	x	x
Thalassia hemprichii	x	x	x
Halophila ovalis	x	x	x
Halophila ovata	x	x	x
Halophila decipiens	x	x	x
Halophila spinulosa	x	x	x
Halophila tricostata		x	x
Total	10	13	13

Halodule uninervis is the most prominent of the tropical species, particularly where the large tides expose the substrate. *Halodule pinifolia* and *Halophila ovata* prefer the intertidal and shallow subtidal. *Thalassia* sp. is associated with coarser sediments and *Thalassodendron ciliatum* grows directly on coral and calcarenite reefs. Muddy intertidal areas are favoured by *Halophila ovalis*.

Seagrass meadows support a rich epibiota and contribute a high proportion of red algae, foraminifera and bivalve fragments to the beach sediments, as well as seagrass roots and detritus. They also help stabilise nearshore sands, and in the tropics are also grazed by dugongs and green turtles, both of which are most extensive across northern Australia.

Coral reefs

Coral reef systems occur right around the northern Australian coast. Western Australia has the world's most poleward coral reef systems forming the Houtman Abrolhos islands and reefs that extend to 29°S. On the mainland fringing reefs become prominent north from Gnaraloo (24°S), forming the Ningaloo fringing-barrier reef complex between Amherst Point and North West Cape. Between Exmouth and Cape Leveque the coast is dominated by generally low gradient beaches, tidal flats and mangroves, and as a result reefs tend to occur off the coast as atolls and fringing islands. The most extensive Western Australian system surrounds much of the *Kimberley* coast and particularly the adjoining islands which all lie between 15° and 17°S. There are also extensive fringing reefs along the predominantly rocky coast, in addition to 163 usually small beaches fronted by fringing reefs.

In the *Northern Territory* fringing reefs occur along much of the northern coast and islands including Bathurst-Melville, Cape Cockburn, the Goulburn and Wessel islands and around Cape Arnhem, with 35 mainland beaches fronted by fringing reef. They also extend into the western gulf particularly on parts of Groote Eylandt and the Sir Edward Pellew Group.

In the southern and eastern *Gulf* coral reefs occur on Vanderlin, Mornington and Bentinck islands. The reefs begin to dominate the coast and shelf in Torres Strait where they form a number of low islands, and then merge into the northern Great Barrier Reef. On the eastern gulf coast the *northern Great Barrier Reef* dominates the outer shelf and parts of the mid shelf, together with fringing reef forming along parts of the mainland shore, and fronting 20 beaches.

Coral reef systems have two main physical impacts on the backing beaches. First, they cause wave breaking over the reefs, resulting in lower waves at the shore. Barrier reefs usually attenuate all ocean swell, resulting in only fetch-limited wind-generated waves to their lee and resulting low energy shorelines. Fringing reefs also attenuate waves, though sea and swell can reach the beach at mid to high tide. The net result however is a substantially lower energy beach system. On the Queensland coast ocean waves outside the reefs average 1-1.5 m and 9-10 sec, while inside the reef the seas average 0.5 m with 4-5 sec periods, an order of magnitude reduction in wave energy.

Secondly, reefs close to shore and all fringing reefs deliver coral and algal debris to the backing beaches (Fig. 1.20), thereby acting as a major source of beach material. This is apparent on the Kimberley and Northern Territory coasts where the beaches average 50% carbonate material, and some beaches reach 100% (Table 1.11). In contrast western Cape York, free of reefs, averages 19% carbonate. On the eastern cape the mean carbonate content is only 11%. The low percentage to the lee of the Great Barrier Reef is because the reef usually lies some tens of kilometres offshore and cannot physically supply sand to the coast, plus the fact that beach sediments are dominated by quartz-rich sands derived from the numerous coastal rivers and creeks. However on some of the 20 cape beaches fronted by fringing reefs the amount of carbonate material reaches as high as 99%.

In summary northern Australia is bordered in the west by the rugged Kimberley coast, in the east by the moderate relief eastern Cape York, with a predominantly ancient, weathered low lying coast in between - Cambridge Gulf to Cape York. This tropical coast receives an abundance of river-derived sediments, which form extensive tide-dominated beaches and tidal flats. Coral reefs fringe or lie off much of the coast, seagrass meadows cover the shallow seafloor and mangroves cover all low gradient low energy shorelines. The coast is sparsely populated and for the most part difficult to very difficult to access from the land.

Table 1.11 Regional proportion of carbonate beach sands

Region	Mean Carbonate %	Ω	Range (%)
Kimberley	55	32	0.5-98
Northern Territory	47	28	0-100
West Cape York	19	12	0-63
East Cape York	11	23	0-99.7

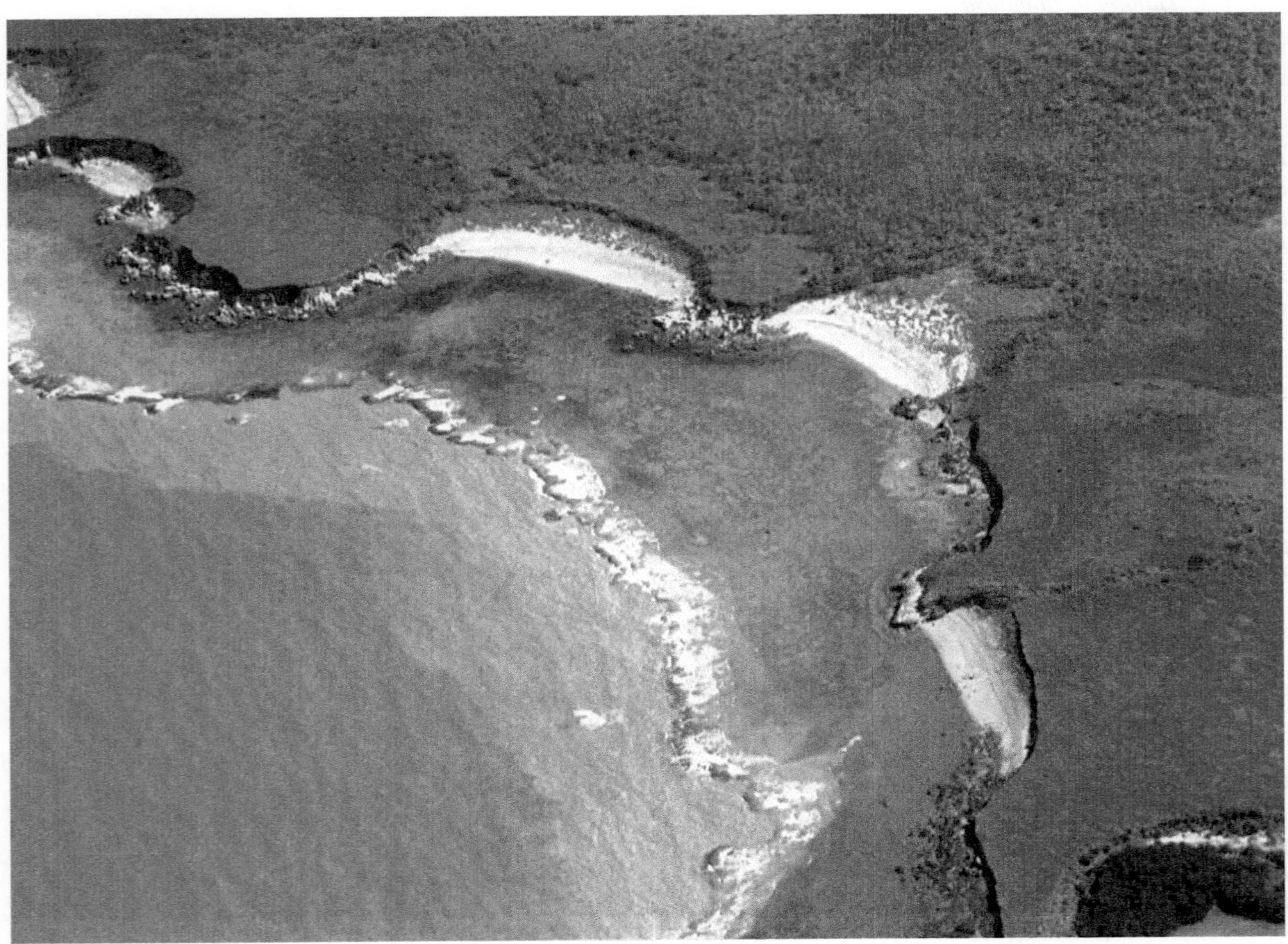

Figure 1.20 Fringing coral reef flats front beaches K 1102-1104 in the eastern Kimberley.

2 BEACH SYSTEMS & HAZARDS

The northern Australian coast extends for 11 869 km between Broome and Cooktown and contains 3489 beach systems, which occupy 4114 km (35%) of the coast. Table 2.1 presents the number of beaches and extent of sandy shoreline for the three northern regions. The beaches occupy only 17% of the rugged Kimberley coast, to 46% of the Territory coast and 37% of the Cape York coast, where mangroves occupy substantial sections, particularly in the southern gulf and some of the larger east coast bays.

Table 2.1 Northern Australia beach-coast characteristics

	No. beaches	%	Sandy coast (km)	%	Total coast (km)	%
Kimberley	1360	39	702	17	4339	37
Northern Territory	1488	43	1902	46	5029	42
Cape York	641	18	1509	37	2501	21
Northern Australia	3489	100	4114	100	11 869	100

BEACH TYPES

Sixteen types of beaches exist around the Australian coast. Theycan be divided into four categories - wave-dominated, tide-modified, tide-dominated and rock/reef fronted. Table 2.2 lists the mean wave height (H_b), spring tide range (TR) and relative tide range (RTR = tide range/wave height)) for each of the twelve beach types that occur across Northern Australia. As expected the highest wave, lowest tides and lowest RTR are associated with the three wave-dominated beach types. The three tide-modified beaches have waves between 0.5-1 m, but higher tide range and RTR. The four tide-dominated beaches, have low waves (<0.4 m), higher tides with an RTR > 7. The two beaches fronted by rocks flats and reefs are not necessarily dependent on H_b and RTR, but tend to have lower waves, and moderate tides and RTR.

The presence and distribution of these sixteen types in northern Australia is presented in Table 2.3, which provides the number in each type followed by the length of sandy shoreline by beach type for the three regions. The three main catagories tend to occur in certain wave-tide combinations, with wave-dominated favouring regions with lower tides and higher waves and where the relative tide range is less than 3. The tide-modified tend to occur where RTR is between 3 and 12, and the tide-dominated between 12 and 50.

Wave-dominated beaches

Only 3.7% of northern Australia beaches are wave-dominated, and then only the lower energy wave-dominated types are present (R, LTT & TBR). Their low representation is a product of the low wave climate, coupled with the high tide ranges, as tides normally need to be less than three times higher than the waves for this beach type to form. The LTT and TBR only occur on the higher energy, lower tide range beaches in the Northern Territory. The three types are illustrated in Fig 2.1.

Table 2.2 Mean wave height, tide range and RTR associated with northern Australian beach types.

Beach type[1]		H_b (m)	Tide Range (m)	RTR
Wave-dom.	1	0.55	2.15	4
	2	0.87	1.40	2
	3	1.33	1.40	1
Tide-mod.	7	0.64	2.99	5
	8	0.90	2.44	3
	9	0.66	5.88	9
Tide-dom.	10	0.40	2.70	7
	11	0.28	4.02	14
	12	0.16	5.03	31
	13	0.16	4.54	28
	14	0.42	3.64	9
	15	0.26	4.32	17

[1] see Table 2.3 for beach type names

The **reflective** (R) beach are characterised by a relatively steep beach face, while usually has a steeper step around low tide, then the relatively deeper water against the shore at high and low tide (Fig. 2.2a). In northern Australia they receive waves averaging 0.55 m (T=4-5 sec) with tides averaging 2.15 m (RTR=4). The low short waves break by surging up the beach face, which is commonly cusped. The surging wave is partly reflected back out to sea, hence the name 'reflective' beach. In northern Australia the 86 reflective beaches average 600 m in length and are usually composed of coarser sands and gravels, including coral debris.

The **low tide terrace** (LTT) usually has a steep reflective high tide beach fronted by an attached bar which is exposed at low tide (Fig. 2.2b). Waves average 0.9 m and tide range 1.4 m (RTR=2), the lowest range in northern Australia. The 28 LTT beaches only occur on the eastern Arnhem Land coast of the Northern Territory, where they are above average length at 1.8 km indicating they occur on longer more exposed beaches.

Table 2.3 Northern Australia beach types by a) number; and b) km coast

a.

	Beach Type[1]	Kimberley	Northern Territory	Cape York	Total	%
1	Reflective	14	61	11	86	2.5
2	LTT	0	28	0	28	0.8
3	TBR	0	14	0	14	0.4
4	RBB	0	0	0	0	0
5	LBT	0	0	0	0	0
6	Dissipative (D)	0	0	0	0	0
7	R+LTT	21	114	83	218	6.2
8	R+LT rips	3	41	56	100	2.9
9	Ultra dissipative (UD)	22	11	14	47	1.3
10	R+sand ridges	25	98	124	247	7.1
11	R+sand flats	417	437	199	1053	30.2
12	R+tidal flats	418	127	52	597	17.1
13	R+tidal flats (mud)	105	120	12	237	6.8
14	R+rock flats/platform	172	402	57	631	18.1
15	R+coral reef	163	35	20	218	6.2
16	R+sand waves	0	0	13	13	0.4
	Total	1360	1488	641	3489	100.0

b.

		km	km	km	km	%
1	Reflective	9.8	33.7	8.9	52.4	1.3
2	LTT	0	50.3	0	50.3	1.2
3	TBR	0	45.4	0	45.4	1.1
4	RBB	0	0	0	0	0
5	LBT	0	0	0	0	0
6	Dissipative (D)	0	0	0	0	0
7	R+LTT	20	182.3	331.7	534.0	13.0
8	R+LT rips	7.1	61.1	315.3	383.5	9.3
9	Ultra dissipative (UD)	78.2	57.6	16.1	151.9	3.7
10	R+sand ridges	12.8	305.8	307.9	626.5	15.2
11	R+sand flats	181.7	503.1	359.3	1044.1	25.4
12	R+tidal flats	208.7	140.7	54.8	404.2	9.8
13	R+tidal flats (mud)	42.8	172.0	57	271.8	6.6
14	R+rock flats/platform	99.3	314.2	23.9	437.4	10.6
15	R+coral reef	41.8	36.4	13.1	91.3	2.2
16	R+sand waves	0	0	21.1	21.1	0.5
	Total	702.2	1902.4	1509.1	4113.7	100.0

[1] see accompanying text for full beach type names

There are 14 **transverse bar and rip** beaches in the Northern Territory all located on the exposed east Arnhem Land coast, where waves can reach 1.5 m. They have an average wave height of 1.3 m and the low tide range of 1.4 m (RTR=1) and tend to occur on the longest most exposed east-facing beaches (mean length=3.2 km), primarily in eastern Arnhem Land south of Cape Arnhem. These are characterised by regularly spaced rips and intervening transverse bar (Fig, 2.2c), both spaced on average 140 m apart (σ=35 m), which result in a total of 500 beach rips in the Territory. Wave energy is insufficient to generate the higher energy wave-dominated beaches, that is the rhythmic bar and beach (RBB), longshore bar and trough (LBT) and dissipative (D).

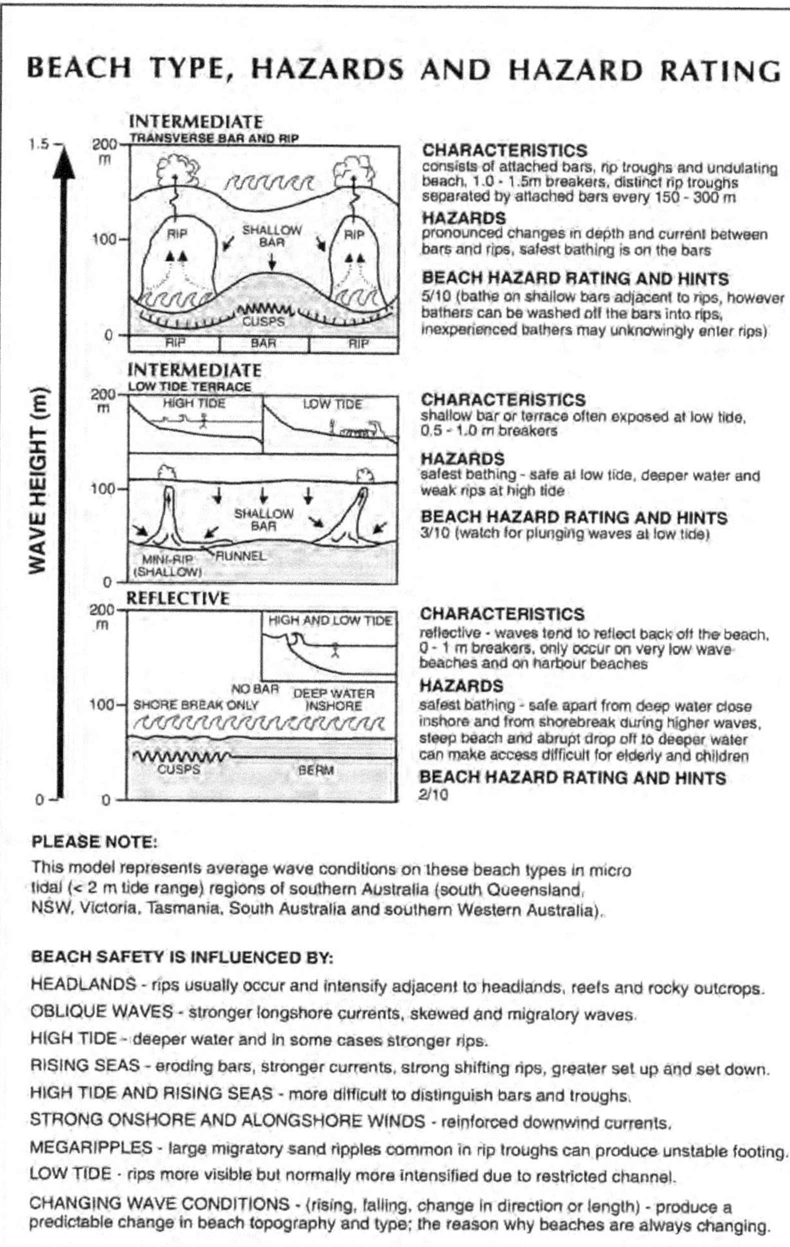

Figure 2.1 Schematic illustration of the three wave dominated beach types that occur in northern Australia. They are the relatively higher energy transverse bar and rip, low tide terrace and reflective.

a.

b.

c.

Figure 2.2 Examples of three northern Australian wave-dominated beaches- a) Reflective beach (K 1216); b) Low tide terrace with wave spilling across the bar (NT 1117); and c) several inactive rip channels along a transverse bar and rip beach on Groote Eylandt (GE 259).

Tide-modified beaches

There are 365 tide-modified beaches spread across the top end, with 46 in the Kimberley, 166 in the Territory and 153 around Cape York, in all 10.4% of the beach systems. Figure 2.3 illustrates the three tide-modified beach types that occur around Australia, principally on northern Australia, as would be expected given the higher tides and lower waves. Figure 2.4 plots the relationship between wave height and period and sand size that determines tide-modified and tide-dominated beach types.

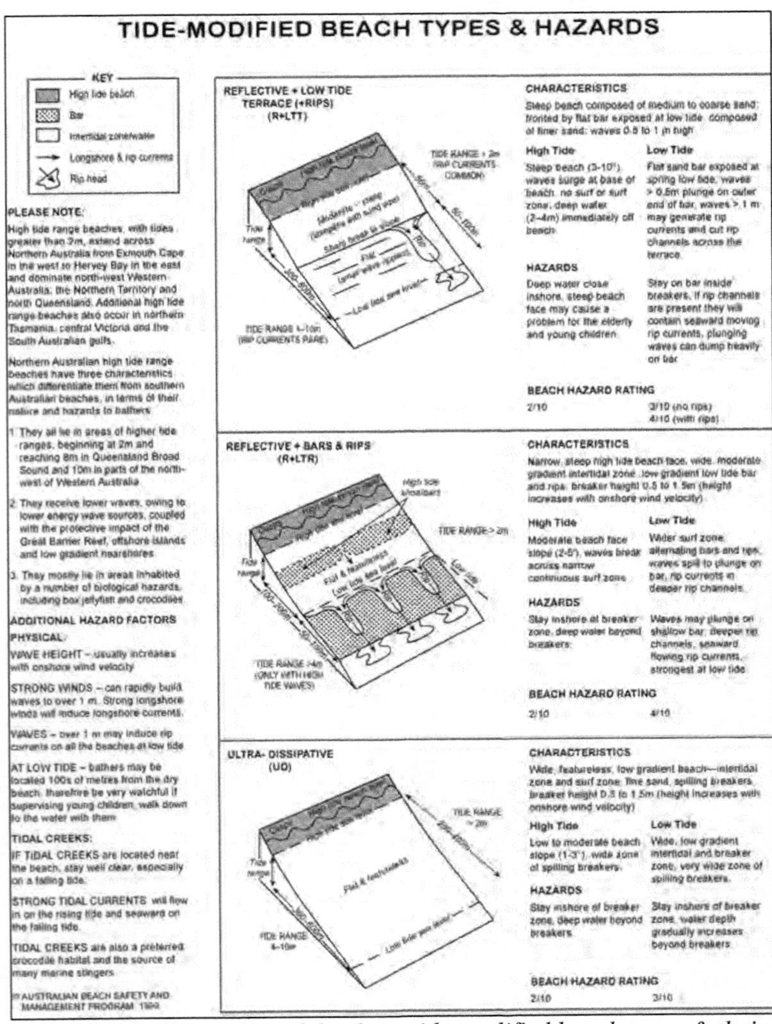

Figure 2.3 Schematic illustration of the three tide modified beach types & their characteristics.

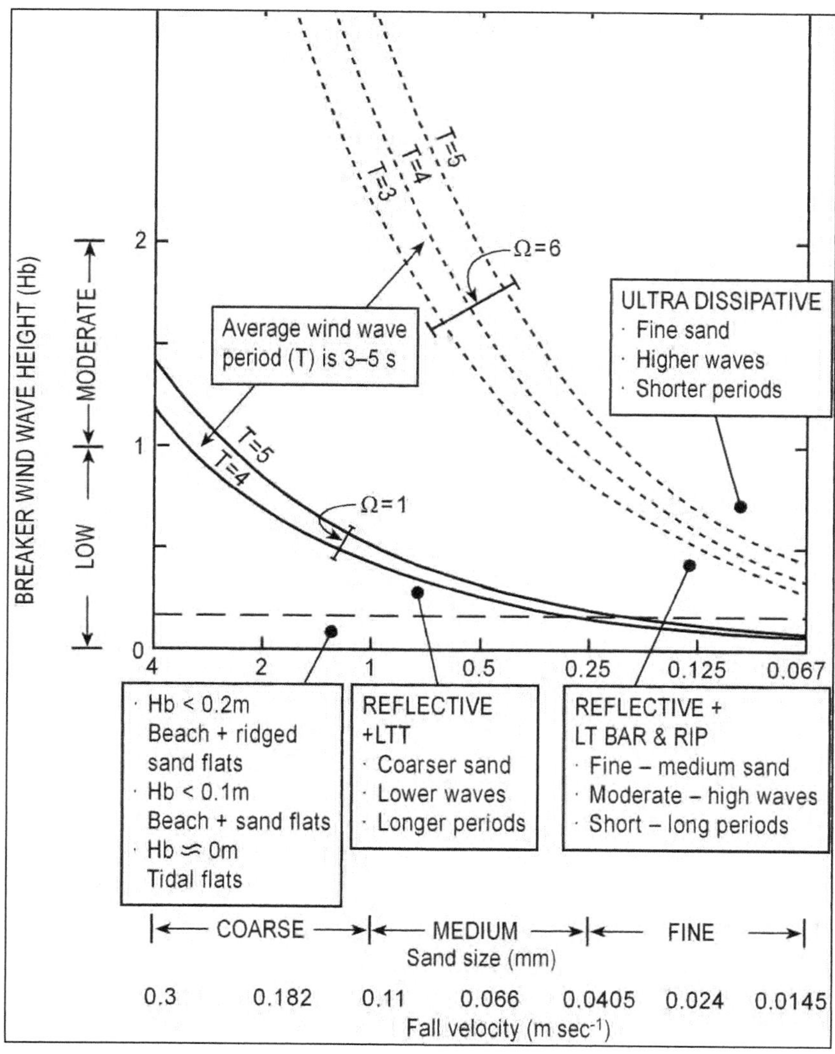

Figure 2.4 A plot of breaker wave height versus sediment size, together with wave period, that can be used to determine approximate Ω (=H_b/TW_s) and beach type on the low to moderate energy northern Australian sea-dominated beaches. To use the chart, determine the breaker wave height, wave period and grain size/fall velocity (W_s) (mm or m/sec). Read off the wave height and grain size, then use the period to determine where the boundary of reflective/intermediate, or intermediate/ultradissipative beaches lies. Ω = 1 along solid T lines and 6 along dashed T lines. Below the solid lines Ω < 1 and the beach is reflective, above the dashed lines Ω > 6 and the beach is ultradissipative, between the solid and dashed lines Ω is between 1 and 6 and the beach is intermediate in the low tide zone.

The **reflective plus low tide terrace** (R+LTT) are the most common tide-modified beach type occurring on 218 beaches in all three regions. These are characterised by a relatively steep cusped reflective high tide beach, usually composed of coarser sand. The beach face slopes to low tide where it abruptly grades into a low gradient, usually finer sand low tide terrace, which can extends 10's metres seaward (Fig. 2.5a). At high tide waves pass unbroken over the terrace and only break on reaching the high tide beach, similar to the reflective tide-dominated beach. As the tide fall waves begin to increasingly break across the terrace and at low tide break on the outer edge producing a wide, shallow surf zone across the terrace. In northern Australia they receive waves averaging 0.6 m, with tide averaging 3 m (RTR=5). They have an average length in the Kimberley of 0.6 km, 1.6 km in the Territory and average 4 km long around Cape York.

The **reflective plus low tide rips** (R+LTR) occur on 100 beaches primarily in the Territory and Cape York, with only three in the Kimberley. These are the highest energy of the tide-affected beaches with waves avenging 0.9 m and tide 2.44 m (RTR=3). They are similar to the R+LTT except that waves are high enough, and tide still low enough, for rip circulation and channels to be cut into the outer edge of the low tide bar (Fig. 2.5b). At high tide the waves pass over the bar without breaking until the beach face, where they usually maintain a relatively steep beach with cusps. As the tide falls wave begin breaking on the bar and at low tide there is sufficient time and wave energy to generate the rips, which have an averaging spacing of 140 m (σ=35 m), the same as the wave dominated TBR. The R+LTR tend to occur on longer beaches with average lengths in the Kimberley, Territory and Cape York of 2.4, 1.5 and 5.6 km respectively. These and the TBR are the best surfing beaches in northern Australia, through the rips do pose a hazard to swimmers, consequently together with TBR they are also the physically most hazardous beaches in northern Australia, particularly at low tide.

The **ultradissipative** (UD) beaches are favoured by beaches with finer sand and higher tides, which result in wide low gradient beaches (Fig. 2.5c). In northern Australia they occur in areas where waves average 0.66 m, tide are higher at 5.88 m and RTR=9, close to the upper threshold of tide-modified beaches. There are a total of 47 ultradissipative beaches spread across all three regions and are the longest beach type in the Kimberley (4 km) and Territory (5.2 km), while the average 1.2 km around Cape York. Ultradissipative beaches are characterised by a moderately steep, usually cusped, high tide beach, with waves surging at the base of the beach at high tide. As the tide falls the surf zone widens as waves break across the low gradient (1°) intertidal zone. The surf zone widens to several tens of metres at low tide, where the beach gradient is very low. The term ultradissipative refers to the fact the waves break across a wide surf zone, particularly at low tide, thereby dissipating much of their energy.

a.

b.

c.

Figure 2.5 Examples of tide-modified beaches: a) steep reflective high tide beach (right) fronted b y low tide terrace (Q 54); b) well developed low tide rips (Q 289); and c) a wide low gradient ultradissipative beach (NT 654).

Tide-dominated beaches

Tide-dominated beaches are the most prevalent across northern Australia with a total of 1887 (61%) spread across all three regions. The combination of higher tides, low waves and in places abundant fine sand and mud all contribute to their dominance. In the Kimberley where they are most prevalent, the highly indented nature of the coast provides additional sheltering and lower waves, with all the Kimberley tide-dominated beaches averaging between 400-500 m in length. Each of the beaches has a reflective high tide sand beach, with the intertidal zone grading from more exposed ridged sand flats through to mud flats (Fig. 2.6). Most of the flats are fringed by tropical seagrass meadows below low water, while mangroves dot many of the lower energy systems.

The highest energy of the tide-dominated beaches in the **reflective plus ridged sand flats** (R+RSF). They tend to occur where waves average 0.4 m and tides are relatively low averaging 2.7 m, resulting in an RTR of 7, below the usual threshold of 10. They are characterised by a relatively coarse steep, occasionally cusped hight tide beach, which abruptly grades into a very low gradient sandy intertidal zone, covered by regularly spaced low amplitude (5-10 cm) sand ridges (Fig. 2.7a). In northern Australia the intertidal flats average 600 m in width and have on average 7 ridges with a mean spacing of 80 m. The flats can range from 50 to 5000 m in width, with up to 22 ridges (Table 2.5). Little is known about the morphodynamics of the ridges, through it is assumed they are formed by waves breaking and reforming over the ridges towards low tide. At spring low they are fully exposed. There are a total of 247 beaches with ridged sand flats occurring in each region, though most prevalent around Cape York.

As wave height drops and tides increase the ridges no longer form and the **reflective plus sand flats** (R+SF) beach dominates. This is the most common beach type in northern Austral particularly in the Kimberley and Territory with a total of 1053 beaches (30%). They are similar to the R+RSF, except waves are lower (mean=0.28 m), tides higher (mean=4 m), resulting in an RTR of 14. They have a relatively small, steep high tide beach, which grades abruptly into intertidal sand flats that average 300 m width (range 10-3000 m) (Fig. 2.7b). The sand flats are flat and featureless apart from wave ripples, wave energy is still sufficiency high to maintain the featureless flats.

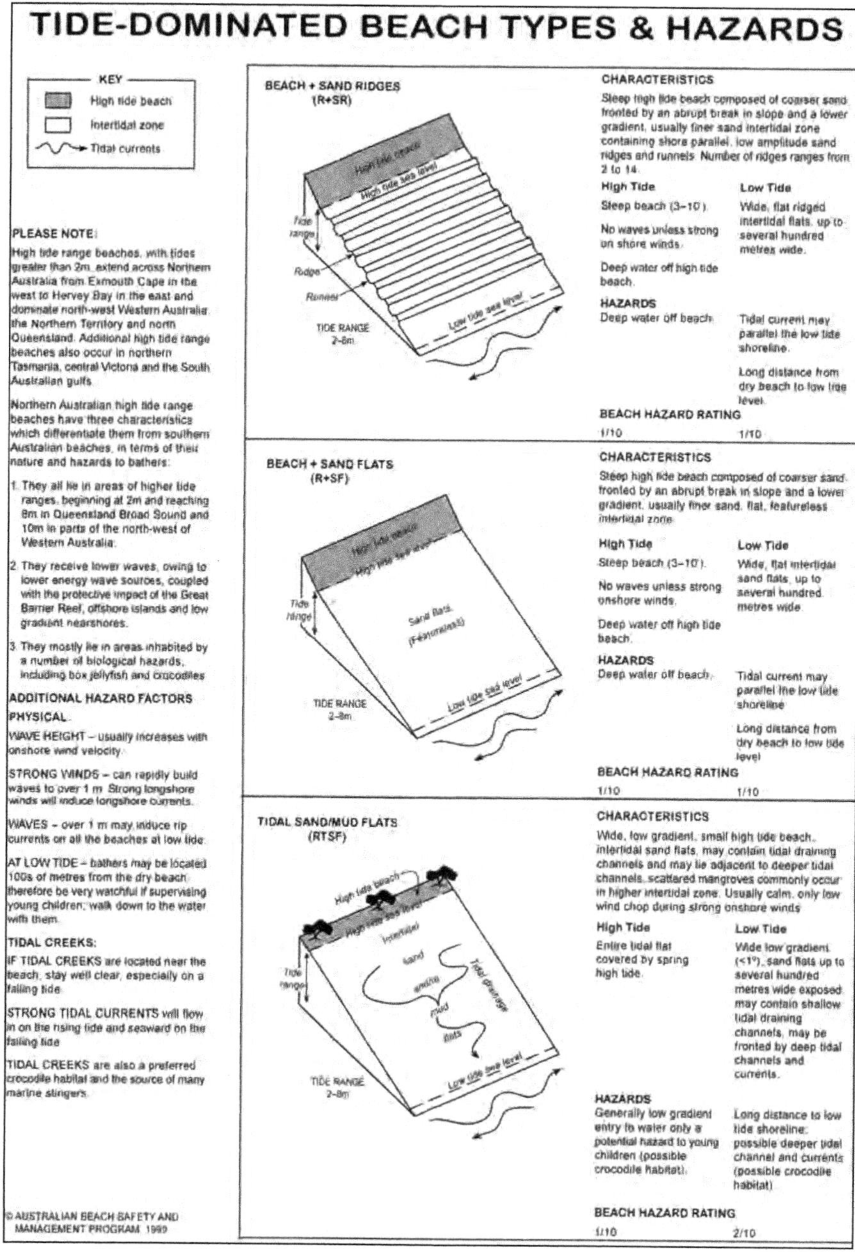

Figure 2.6 Schematic of the three tide-dominated beach types.

Table 2. 5 Some characteristics of tide-dominated beaches with sand/tidal flats and ridges.

	Beach type	Mean intertidal width (m)	σ (m)	No. ridges/ Range (m)	σ
10	R+sand ridges	620	630	7.2 (1-22)	4
				50-5000	
				Range (m)	
11	R+sand flats	300	320	10-3000	
12	R+tidal (sand) flats	345	377	50-2500	
12	R+tidal (sand/mud) flats	300 sand 500 mud		40-5000	
13	R+tidal (mud) flats	500 m	510	50-2000	

a.

b.

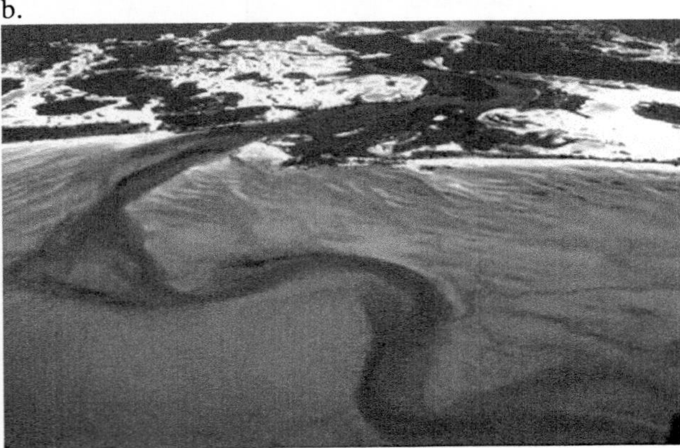
c.

d.

Figure 2.7 Examples of tide-dominated beaches: a) extensive ridge sand flats off beach NT 1488; b) steep reflective beach (right) fronted by a wide low gradient sand flat (K 1333); c) tide dominated sand flats off beach NT 1446 & 1447; and d) high tide sand beach fronted by mangrove fringed mud flats (NT 32).

As wave height continues to fall and tides increases they two lowest energy beach types can form. The **reflective plus tidal sand flats** (R+TSF) have a small, reflective steep usually coarse-grained high tide beach, fronted by intertidal sand flats averaging 350 m width (range 50-2500 m) (Fig. 2.7c). They differ from the R+SF in that they receive lower waves (mean=0.16 m), slightly higher average tides (mean=5 m) producing and RTR of 31. Waves are sufficiently low, and tidal energy sufficiently high for the tidal currents to imprint themselves on the tidal flats, and in some locations for mangroves to be scattered across the upper intertidal zone. These are the second-most common beach type in northern Australia with a total of 597 (17%) occurring in each region, but particularly in the Kimberley. Many of these flats grade from inner sand flats to outer mud flats, with the sand averaging 300 m wide and the mud extending out on average to 500 m.

The final tide-dominated beach type is the **reflective plus tidal mud flats** (R+TMF). These occur in similar wave-tide regimes to the R+TSF, with waves also averaging 0.16 m and tides 4.5 m, providing an RTR of 28. However here in addition to the low waves there is usually a local source of mud such as a river to supply the mud to the shoreline. They have a small high tide beach, usually composed of coarse shelly sand, which grades often very abruptly into wide, very low gradient intertidal mud flats, with mangroves usually scattered in the upper intertidal (Fig. 2.7d). They average 500 m in with ranging from 50-2000 m. There are 237 (7%) of these beach types across northern Australia, with most occurring in the Kimberley and the Territory. Beware of the mud flats as when walking on them at low tide you usually sink to knee depth.

Beaches with rock/reef flats

Beaches fronted by intertidal rock flats (R+RF), and in some cases supratidal rock platforms occur in the Kimberley (172), the Territory where most are fronted by laterite flats (402) and on Cape York (57) in all comprising 18% of the northern beaches. This beach type is dependent more on local geology than wave-tide processes. The intertidal rock flats average 270 m (σ=330 m) in width and range from 50-3000 m wide. They usually consist of a steep high tide beach with the rocks extending seaward from the base of the beach (Fig. 2.8). They tend to the relatively short 400-800 m and are usually bounded by rock headlands or reefs.

The most diagnostic beach type in northern Australia is the beach fronted by a **fringing coral reef** (R+CR). There are a total of 218 (6%) of these beaches primarily in the Kimberley (163). The beaches consist of a usually steep, high tide reflective beach often composed of coral fragments, fronted by reefs, which average 300 m in width (σ=330 m), but can range from 50-2000 m. In the Kimberley these are the shortest beach type averaging only 300 m in length (Fig. 2.9).

Figure 2.8 Beach NT 58 is fringed by continuous intertidal laterite flats.

Figure 2.9 Fringing coral reef along the northern side of Melville island result in low wave conditions at the beach)MI 11) and adjacent mangrove lined shoreline.

The final beach type consists of a reflective high tide beach fronted by inter to subtidal **sand waves** (R+SW), usually oriented perpendicular to the shore. These only occur on 13 beaches which occupy 20 km of shore, all located on the eastern side of Cape York in locations where strong flood tidal currents flow northward close to shore, the currents generate, maintain and slowly migrate the sand waves northward. The sand waves extend from between 100-1500 m seaward (Fig. 2.10).

BEACH HAZARDS

Beach hazards are elements of the beach-surf environment that expose the public to danger or harm. Physical beach hazards are associated with water depth, breaking waves, surf zone and tidal currents and variable seafloor topography, together with local feature such as rocks, reefs and inlets. Biological beach hazards are associated with predatory and poisonous marine animals including sharks, crocodiles, stingers and stingrays.

Figure 2.10 Shore transverse and oblique sand waves along beach GE 265 on Groote Eylandt.

Beach safety is the mitigation of such hazards and requires a combination of common sense, swimming ability and beach-surf knowledge and experience. The following sections highlights the major physical hazards encountered in the surf, followed by the biological hazards.

Physical beach hazards

Each beach type has a particular range of wave, tide, depth and seabed configuration which contribute to an overall level of physical beach hazards. Figure 2.11 provides a matrix for assessing the beach hazard rating for each of the wave-dominated, tide-modified and tide-dominated beach systems across northern Australia. In general terms beaches become more hazardous with increasing wave height, and with changes in tide.

There are six major physical hazards on northern Australian beaches:

1. water depth (deep and shallow)
2. breaking waves
3. surf zone currents (particularly rip currents)
4. tidal currents
5. strong winds
6. rocks, reefs and headlands

In the surf zone, three or four hazards, particularly water depth, breaking waves and currents, usually occur together. In order to swim safely, it is simply a matter of avoiding or being able to handle the above when they constitute a hazard to you, your friends or children.

BEACH HAZARD RATING GUIDE
Impact of changing breaker wave height on hazard rating for each beach type

Wave Dominated Beaches

BEACH TYPE \ WAVE HEIGHT	<0.5 (m)	0.5 (m)	1.0 (m)	1.5 (m)	2.0 (m)	2.5 (m)	3.0 (m)	> 3.0 (m)
Dissipative	4	5	6	7	8	**9**	**10**	10
Long Shore Bar Trough	4	5	6	**7**	**7**	**8**	9	10
Rhythmic Bar Beach	4	5	6	**6**	**7**	8	9	10
Transverse Bar Rip	4	4	**5**	**6**	7	8	9	10
Low Tide Terrace	3	**3**	**4**	5	6	7	8	10
Reflective	**2**	**3**	4	5	6	7	8	10

Tide Modified Beaches
(at high tide - at low tide add 1)

Ultradissipative	1	2	4	6	8	10	10
Reflective + Bar & Rips	1	2	3	5	7	9	10
Reflective + LTT	1	1	2	4	6	8	10

Tide Dominated Beaches
(at high tide - at low tide add 1)

Beach + Sand Ridges	1	1	2	Waves unlikely to exceed 0.5 - 1m
Beach + Sand Flats	1	1		*Note: if adjacent to tidal channel, beware of deep water and strong tidal currents.*
Tidal Sand Flats	1			

BEACH HAZARD RATING	KEY TO HAZARDS
Least hazardous: 1 - 3	Water depth and/or tidal currents
Moderately hazardous: 4 - 6	Shorebreak
Highly hazardous: 7 - 8	Rips and surfzone currents
Extremely hazardous: 9 - 10	Rips, currents and large breakers

NOTE: All hazard level ratings are based on a swimmer being in the surf zone and will increase with increasing wave height or with the presence of features such as inlet, headland or reef induced rips and currents. Rips also become stronger with falling tide.

BOLD gradings indicate the average wave height usually required to produce the beach type and its average hazard rating.

Figure 2.11 Matrix for calculating the prevailing beach hazard rating for wave-dominated, tide-modified and tide-dominated beaches, based on beach type and prevailing wave height and, on tide-modified beaches, state of tide.

Any depth of water is potentially a hazard.

- *Shallow water* is a hazard when people are diving in the surf or catching waves. Both can result in spinal injury if people hit the sand head first.
- *Knee depth* water can be a problem for a toddler or young child.
- *Chest depth* is hazardous to non-swimmers, as well as to panicking swimmers. In the presence of a current, it is only possible to wade against the current when water is below chest depth. Be very careful when water begins to exceed waist depth, particularly if younger or smaller people are present and if in the vicinity of longshore, rip or tidal currents.

Breaking waves

As waves break, they generate turbulence and currents which can knock people over, drag and hold them under water, and dump them on the sand bar or shore. If you do not know how to handle breaking waves (most people do not), stay away from them, stay close to shore and on the inner part of the bar.

Surf zone currents and rip currents

Surf zone currents and particularly *rip currents* are the biggest hazards to most swimmers. They are the hardest for the inexperienced to spot and can generate panic when swimmers are caught by them. The problem with currents, particularly rip currents, is that they can move you unwillingly around the surf zone and ultimately seaward. In moving seaward, they will also take you into deeper water and possibly toward and beyond the breakers. As mentioned earlier, currents are manageable when the water is below waist level, but as water depth reaches chest height they will sweep you off your feet.

Tidal currents

Tidal currents are a major hazards across northern Australia owing to the generally high tide range, numerous inlets and tidal creeks and the associated strong ebb and flood tidal currents, usually flowing in a deep tidal channel. In tidal inlets the adjoining sandy shoreline and beaches may have little or no wave activity, the strong currents flowing in deep water often right off the beach or shore is a major hazard.

When swimming or even boating in or near tidal creeks and inlets, always check the state and direction of the tide and be prepared for strong currents. You should not venture beyond waist deep water.

Strong winds

The dominant wind across the north are the southeast Trades which tend to be light to moderate, only occasionally reaching strong. Depending on the orientation of the coast they may blow onshore, along or offshore. In addition to generating most of the waves that arrive at the shore Table 2.6), moderate to strong trade winds can generate additional hazards (Table 2.7) also listed below.

Whenever strong winds occur, their direction determines the impact they will have on beach and surf conditions, as follows:

- *Longshore winds,* particularly strong trade winds, will cause wind waves to run along the beach, with accompanying longshore and if present rip currents also running along the beach. The waves and currents can very quickly sweep a person alongshore and into mobile rips on lower energy beaches or into deeper rip channels and stronger currents on higher energy beaches.

- *Onshore winds* will help pile more water onto the beach and increase the water level at the shore. They also produce more irregular surf, which makes it more difficult to spot rips and currents.
- *Offshore winds* tend to clean up the surf. However, if you are floating on a surfboard, bodyboard, ski or wind surfer, it also means it will blow the board offshore. In very strong offshore winds, it may be difficult or impossible for some people to paddle against this wind.

Table 2.6 Most waves arriving on the northern Australian coast are generated by local winds, with waves usually below 1 m. Maximum sea height can however reach about 3 m along the western gulf coast. These seas are short and steep, compared to the longer ocean swell of the same height.

On-alongshore wind velocity	Max. sea wave height (m)*
Light	0.3
Moderate	0.5
Strong	1.0
Gale	3.0-4.0

* WAVE PERIOD 2-5 SEC

Table 2.7 Wind hazard rating for northern Australian beaches, based on wind direction and strength. Winds blowing on and alongshore will intensify wave breaking and surf zone currents, with strong longshore winds capable of producing a strong longshore drag. Their impact on surf zone hazards and beach safety is indicated by the relevant hazard rating, which should be added to the prevailing beach hazard rating.

	Light	*Mod*	*Strong*	*Gale*
Longshore	*0*	*1*	*3*	*4*
Onshore	*0*	*1*	*2*	*3*
Offshore	*0*	*1*	*1*	*2*

Rocks, reefs and headlands

Headlands, rock reefs and coral reefs occur right around the northern Australian coast. In association with beaches their major hazards is to produce a hard substrate adjacent to or within a sandy beach system. Where there is wave breaking and surf they cause additional wave breaking, generate stronger topographic rips, and present a hard and often dangerous surfaces. When they occur in shallow water and/or close to shore, they are also a danger to people walking, swimming or diving because of the hard seabed and the fact that they may not be visible from the surface.

Beach Hazard Rating

The *beach hazard rating* refers to the scaling of a beach according to the physical hazards associated with its beach type under normal wave conditions, together with any local physical hazards. It ranges from the lowest least hazardous rating of 1 to a high, most hazardous rating of 10. It does not include biological hazards, which are discussed later in this chapter. The beach characteristics and hazard rating for wave-dominated beaches are shown in Figure 2.1, for tide-modified beaches (Fig. 2.3) and tide-dominated beaches (Fig. 2.7) and all three in Figure 2.11.

The *modal beach hazard rating* indicates the level of hazard under typical or modal wave conditions for each beach type. Figure 2.11 lists the six wave-dominated, and three tide-modified and three tide-dominated beaches, together with their beach hazard rating under waves between less than 0.5 m and greater than 3 m. The modal wave height and modal hazard rating is indicated in BOLD. The rating ranges from a low of 1 on most tide-dominated and some tide-modified beaches, to a high of 10 on high energy dissipative beaches. The figure also indicates how the hazard rating will change under changing wave and beach type conditions, together with the more generalised hazards associated with each.

The *prevailing beach hazard rating* refers to the hazard rating prevailing at a given time as a result of the prevailing wave, tide wind and beach type conditions. Figure 3.3 can be used to determine the prevailing hazard based on beach type and wave height, with Table 3.3 providing the additional wind hazard. If the beach also has local hazards such as rocks, reefs, headlands and inlets an additional 1 is added to the rating. The prevailing beach hazard rating is therefore a function of:

wave height + beach type + wind hazard + other local hazards

What this implies is that beach hazards are a function of some permanent features such as rocks and reefs, as well as more variable factors such as waves, tides and wind, as well as changing beach types, particularly on more energetic wave-dominated beaches. It also means that the hazard rating will change both between beaches as well as over time on any particular beach or part of a beach. These changes can occur quickly as wave, tide and wind conditions change.

Beach Hazard Ratings

1 - least hazardous beach
10 - most hazardous beach

Beach hazard rating is the scaling of a beach according to the physical hazards associated with its beach type and local beach and surf environment.

Modal beach hazard rating is based on the beach type prevailing under average or modal wave conditions, for a particular beach type or beach.

Prevailing beach hazard rating refers to the level of beach hazard associated with the prevailing wave, tide, wind and beach conditions on a particular day or time.

Table 2.8 summarises the rating for all Northern Australian beaches.

NORTHERN AUSTRALIAN BEACH HAZARDS

Beach usage and surf lifesaving

The northern Australian coast has a long history of usage by the aboriginal population who still continue to concentrate in coastal communities and camps. The indigenous population through long association has an affinity with the coast and beaches and is well aware of the physical and biological hazards associated with this environment. Most modern towns and settlements including Broome, Wyndham, Darwin, Nhulunbuy, Weipa and Cooktown are also located on the coast. The growing population of these centres and greater usage of northern Australian beaches by the non-indigenous populations including visitors and tourists, was accompanied by the establishment of surf lifesaving clubs in the Kimberley at Broome in 1988, and three clubs in the Northern Territory at Arafura (Cullen Beach) (the newest established in 2000), Darwin (Casuarina Beach) (1982) and the remote Gove club established at Town Beach in 1974. Interestingly the logos of the three Territory clubs (Fig. 2.12) are a crocodile (Gove), box jelly fish (Darwin) and mud crab (Arafura), the first two the most threatening biological hazards on the coast.

The **Surf Life Saving Northern Territory** (SLSNT) website (www.slsnt.org) provides the following information on SLSNT:

Mission Statement
To provide a safe beach and aquatic environment throughout the Northern Territory.

Surf lifesaving functions have been delivered in the Northern Territory since the mid 1970s, although in 1960 and during WWII voluntary efforts were made to introduce the essence of lifesaving to the Territory. Until around 1989, Northern Territory members were administered from Queensland and it was not until 1989 that SLSNT was established, with equal representation on the national board. All Territory members also belong to Surf Life Saving Australia.

Since 1989 Surf Life Saving Northern Territory has grown into an organisation that has more than 1000 members and consists of three fully affiliated clubs with an extensive amount of resource equipment, including a Lifesaver Offshore Rescue Boat.

Figure 2.12 Logos of the three Northern Territory surf life saving clubs indicate some of the biological hazards associated with northern Australian waters.

In addition to the two Darwin-based clubs, SLSNT operate two jetskis in the Darwin area, and patrols the ebaches between lameroo and Lee Point during the winter beach season (May-October). The jetski also operates during the November to April Wet season when surfers try to make the most of the limited wave season.

SURF LIFE SAVING NORTHERN TERRITORY

PO address:
P O Box 43096
CASUARINA NT 0811
(08) 8941 3501
slsnt@topend.com.au

Location:
DHL Office
301 Bagot Rd
COCONUT GROVE NT 0810

Physical beach hazard ratings

Most northern Australian beaches received relatively low waves, apart from those exposed to periodic strong trade winds and waves, and to very occasional tropical cyclone generated waves. The low wave climate is reflected in the overall low hazard ratings (Tables 2.8 & 2.9). The modal hazard rating for all the regions is 2, with 82% of all beaches rating 1 or 2 and 95% rating 3 or less. This means from a physical perspective most northern Australian beaches have a low to hazard rating under normal wave conditions, which are typically less than 0.5 m. These wave reach the beach at high tide as a low surging waves, only breaking at low tide across the usually shallow low tide bar or flats. The more moderately hazardous beaches rating 4 on Cape York, 4 and 5 in the Kimberley and 4, 5 and 6 in the Territory occur on beaches more exposed to the southeasterly trades where waves may average between 0.5-1 m. While all exposed shore may receive the same wind velocities, the longer fetch across the Gulf of Carpentaria results in the highest waves occurring on the eastern Arnhem Land coast, where rips commonly occur and produce the higher rating of 5 and 6. While the eastern Cape York coast is well exposed to the trades, the restricted fetch between the outer Great Barrier Reef, together with numerous inshore reefs, limits normal waves to less than 1 m and often less then 0.5 m. The individual rating for each beach is provided in chapters 3, 4 and 5.

Table 2.8 Northern Australia regional physical beach hazard ratings

Hazard rating	Number	No. %	Mean length km	Total length km	km %
Kimberley					
1	621	45.7	0.4	246	34.4
2	633	46.5	0.55	348	48.7
3	84	6.2	0.52	44	6.2
4	10	0.7	5.3	54	7.6
5	12	0.9	1.85	22	3.1
6	0				
	1360	100		714	100
Northern Territory					
1	544	36.6	1.38	749	39.4
2	738	49.6	1.23	911	47.9
3	132	8.9	0.94	124	6.5
4	54	3.6	1.14	62	3.2
5	11	0.7	1.88	21	1.1
6	9	0.6	3.97	36	1.9
7	0	0			0
	1488	100		1902	100
Cape York					
1	25	3.9	0.97	24	1.6
2	363	56.6	2.11	767	50.8
3	240	37.4	2.91	698	46.3
4	13	2.0	1.52	20	1.3
5	0	0			0
	641	100		1509	100

Table 2.9 Northern Australian beach hazard ratings, by number of beaches

Hazard rating	Kimberley	Northern Territory	Cape York	Northern Australia	%
1	621	544	25	1190	34.1
2	633	738	363	1734	49.7
3	84	132	240	456	13.1
4	10	54	13	77	2.2
5	12	11	0	23	0.7
6	0	9	0	9	0.3
7	0	0	0	0	0
8	0	0	0	0	0
9	0	0	0	0	0
10	0	0	0	0	0
Total	1360	1488	641	3489	100

Table 2.10 *Beach hazard rating of all Australian beaches, by number of beaches. **Bold** indicates modal hazard rating/s. (Source: Short, 1993, 1996, 2000, 2005, 2006)*

Beach Hazard Rating	Qld	NSW	Vic	Tas	SA	WA	NT	Australia number	Australia %
1	325	0	61	**118**	**320**	**1171**	544	2539	23.8
2	**748**	45	36	101	271	880	**738**	**2819**	**26.4**
3	473	103	90	197	206	415	132	1616	15.1
4	58	134	**92**	242	154	226	54	960	9.0
5	13	85	66	**263**	125	175	11	738	6.9
6	**23**	**232**	109	140	93	119	9	725	6.8
7	9	112	**148**	103	**137**	137		646	6.0
8	1	7	77	78	117	**171**		451	4.2
9			11	23	20	76		130	1.2
10		3	2	4	11	41		61	0.6
Total	1650	721	692	1269	1454	3411	1488	10 685	100.0

Compared to other Australian states (Table 2.10) northern Australia has the lowest overall beach hazard rating, based on each region and for the coast as a whole. The modal hazard rating right across the Top is 2, with no ratings above 4 in Cape York, 5 in the Kimberley and 6 in the Northern Territory. The lower ratings are a product of the low waves, general absence of rip currents and dangerous surf, and also to a degree the high tide range.

Tidal creeks and currents

Most of the northern Australian coast is dominated by high tide ranges (2-10 m) and consequently the associated substantial tidal currents. Tidal currents tend to parallel the coast offshore, flowing east or north with the flooding (rising) tides, and west or south with the ebbing (falling) tides. However their direction is also modified by obstacles such as islands, reefs and headlands, and by all inlets, river and creek mouths. There are 946 creeks, tidal creeks and river mouths adjacent to beach systems across the north, 55% of which are permanently open and another 32% periodically open. Tides must flow into and out of every coastal entrance twice a day, and in doing so generate strong constricted currents, which also maintain deeper tidal channels. As many beaches are located on or adjacent to inlets and river mouths, these strong currents and their deep channels are a very real hazard on all beaches located close to inlets. They are particularly hazardous on a falling tide as the currents flow seaward and can rapidly transport a person or boat a considerable distance offshore. Because of lage tide ranges many of the creeks dry out at low tide leaving exposed often muddy creek beds. These creeks and inlets are also the preferred habitat for saltwater crocodiles which are the greatest threat to swimmers in northern Australia, as is discussed in the following section. For these reasons tidal creeks and inlets must be approached with caution, particularly during an ebbing tide.

When swimming or even boating in tidal creeks, always check the state and direction of the tide and be prepared for strong currents. If swimmers do not venture beyond waist deep water. In addition, crocodiles do inhabit all northern Australian estuaries and are the major hazard in these waters.

COASTAL BIOLOGICAL HAZARDS

This book is not designed to deal with biological hazards. However some mention must be made of sharks and crocodiles, as there have been a number of fatal shark attacks since 2000 in Western Australia at Cottesloe and Cowaramup and in South Australia at Cactus, Elliston and two off Adelaide. There have also been several fatal crocodile attacks in the Kimberley region and on Cape York in recent years including two in Princess Charlotte Bay in 2004 and 2005.

There is no way of avoiding sharks once you enter their territory. If you are concerned about sharks then it is best to stay out of their domain. However all surfers and divers and many swimmers are prepared to spend some time in the ocean with the knowledge that the chances of being attacked are extremely small. On average only one person is fatally attacked in all Australia waters each year, and as unfortunate as they are the above six attacks maintain this average. If you are at all concerned then swim only at patrolled beaches during patrol periods, where lookouts are used to spot and warn swimmers and surfers of the presence of sharks.

Likewise with crocodiles, the best way to avoid them is to stay well clear of their territory, which not only includes all creeks and estuaries in northern Australia but also many freshwater streams and river banks. While crocodile attacks on open beaches are rare, crocodiles commonly visit and come ashore on ocean beaches. I have seen crocodiles and particularly fresh crocodile tracks on many beaches around the Kimberley, Northern Territory and Cape York coast. I strongly recommend anyone working, fishing or recreating in the northern

Australia coastal region to visit the crocodile park in Broome or elsewhere to see crocodiles first hand and learn about their behaviour. Do this before venturing into their domain.

The major reference for biological hazards discussed in this book is:

Venomous and Poisonous Marine Animals - a Medical and Biological Handbook, 504 pp.
by JA Williamson, PJ Fenner, JW Burnett and JF Rifkin published by University of New South Wales Press, Sydney, 1996 ISBN 0 86840 279 6

Available through UNSW Press or the Medical Journal of Australia.

This is an excellent and authoritative text, which provides the most extensive and up-to-date description and illustrations of these marine animals and the treatment for their envenomation.

Marine Stingers

Biological hazards pose the greatest risk to beachgoers in northern Australia owing to the prevalence of a greater number and abundance of venomous, poisonous and dangerous marine animals. Whereas poisonous animals may cause illness in victims, venomous animals are capable of causing fatalities and together with crocodiles and sharks are the great threat to beach users. This section briefly summarises some of the types of poisonous marine organisms together with statistics on the actual number of incidents involving illness and fatalities. For more thorough information see the above book.

Bluebottles and other stinging jellyfish are the most common cause of first aid treatment on all Australian beaches (Table 2.11). In summer, the warm tropical waters of the East Australian Current and onshore winds can deliver them in their thousands to east coast beaches. The tentacles of the bluebottle contain hundreds of minute, poisonous, pressure sensitive harpoons, which upon contact are fired and injected into the skin. They immediately appear as small white beads, which soon swell into a red welt. The pain is felt instantly and usually lasts for about an hour, while the welts may remain for a few days. If stung by a bluebottle swim to shore, remove any tentacles (either pick off with the fingers or wash off with salt (sea) water), seek first aid from the lifesavers and stay calm for the next hour. Recommended first aid is to pack ice or cold packs on the stung area. Do not apply vinegar.

Most local beach users in northern Australia are familiar with the type and occurrence of marine stingers. Table 2.12 indicates their likely period of occurrence of the deadly box jellyfish across northern Australia. They occur year round along the north coast of the Territory (12^{0}S), decreasing to between December to March by Rockhampton (23^{0}S).

In addition to the more well known and feared box jellyfish, there are, however other marine animals that can cause fatalities. Table 2.13 lists the fatalities in Australia and southeast Asia due to jellyfish and box jellyfish, Table 2.14 lists fatalities due to cone fish, while Table 2.15 lists all other marine animals that have caused fatalities in Australia. While box jellyfish are the major risk, so too are puffer fish, and also lethal but less frequent are the blue ringed octopus, stingray, cone shell and ciguatera (fish poisoning).

Table-2.11 Number of jellyfish stings on Australian beaches 1987-1994

Jellyfish	number	percent	average # stings/yr
Physalia Bluebottle	9776	85	1400
Cyanea 'hair jelly'	821	7	120
Catostylus 'blubber'	739	6	110
Carukia 'Irukandji'	119	1	20
Chironex 'box jellyfish'	36	0.3	5
	11 491	99.3	1650

Source: Williamson et al., 1996

Table 2.12 Risk period for Chironex 'box jellyfish' in northern Australia

Lat	PLACE*	July	Aug	Sept	Oct	Nov	Dec	Jan	Feb	Mar	Apr	May	June
12^0	Darwin	x	x	x	x	x	x	x	x	x	x	x	x
16^0	Cairns/Broome					x	x	x	x	x	x	x	x
19^0	Townsville						x	x	x	x	x	x	
21^0	Mackay						x	x	x	x	x	x	
23^0	Rockhampton						x	x	x	x			

Source: Williamson et al., 1996
* crocodiles extend down to Broome on the west coast and Yeppoon on the east coast

While much attention is focused on the occasional fatalities due to marine animals, far more people are affected by illness stemming from contact with various marine organisms. Table 2.16 lists the relative frequency or likelihood of fatalities caused by marine animals, together with those that are likely to cause illness due to infection, wounds and poison. By far the most common are due to cuts and abrasions, together with stings from poisonous, but not venomous, animals.

Table 2.13 Jellyfish fatalities

Location	number
New Guinea	2
Borneo	3
Philippines	9
Malaya	1
Total	*15*
Box jellyfish fatalities in Australia	
EAST QUEENSLAND	30
Gulf & Torres Strait	5
NT - Darwin	8
NT - Islands	9
NT – Arnhem Land	10
Total Australia	*62*

Source: Williamson et al., 1996

Table -2.14 Cone shell fatalities

Location	number
Australia (north Qld)	1
India	1
New Caledonia	1
Fiji	1
Okinawa	4
New Hebrides	1
Total	*9*

Source: Williamson et al., 1996

Table 2.15 lists the cause of all jellyfish stings in Australia. Bluebottles which spread down into the more populous southern states are by far the major causes, averaging 1400 stings (85%) each year, likely a very conservative figure, as most stings are unreported. Fortunately the more deadly box jellyfish, an inhabitant of less populated areas, stings only five people on average each year. This low number is due in part however to the efforts of surf lifesaving clubs across northern Australia who maintain stinger enclosures, encourage the wearing of protective lycra swimming suits and educate the public in the risks associated with coastal swimming.

Table 2.16 provides an overview of the nature of Irukandji stings across northern Australia. Most stings occur on the most populated Queensland coast, where most people stung are male (82%), with most stings occurring in fine weather (when more people go to the beach), and the stings more likely on the upper body (head, neck, arms and trunk - 74%).

Table 2.17 summarises the relative frequency of fatalities and illness due to all marine animals. As can be seen, more fatalities occur worldwide as a result of eating poisonous animals (puffer, shells, ciguatera and turtle) than due to stings. Similarly with illness; most are a result of cuts and abrasions, followed by bluebottle stings. However in Australia as indicated by Table 2.15, marine stingers remain the main cause of death, with the most recent on Umagico Beach near the tip of cape York, in January 2006.

Table 2.15 Summary of poisonous marine fatalities in Australia

Cause	number
Box jellyfish	61
Blue-ringed octopus	2
Stingray	2
Cone shell	1
Ciguatera (fish poisoning)	1
Puffer fish	14
Total	*81*

Source: Williamson et al., 1996

Table 2.16 Irukandji sting records

	Number of stings reported	% stings
Total	301	
morning	45	18%
afternoon	213	82%
Males	185	63%
Females	110	37%
Children under 10 years	42	14%
FINE WEATHER	60	79%
Cloudy weather	16	21%
Site of stings (Total)	234	
Trunk	92	39%
Leg	60	26%
ARMS	58	25%
Head/neck	24	10%
State		
Queensland	238	80%
Northern Territory	35	11%
Western Australia	28	9%

Source: Williamson et al., 1996

Table 2.17 Relative frequency of fatalities and illnesses from marine stinging animals (most to least frequent)

Fatalities	Illness
Pufferfish poisoning	Coral cuts, abrasions. stings
Shellfish poisoning Box jellyfish stings	Jellyfish stings (esp. bluebottle)
Ciguatera poisoning	Ciguatera poisoning
Turtle flesh poisoning	Spiny fish stings
Sea snakes Spine fish injuries Bluebottle stings	Other marine animal poisoning from ingestion

Source: Williamson et al., 1996

Sharks, sting rays and crocodiles

Sharks are the most feared fish in the sea and in most years they attack one or two victims around the Australian coast. Since 1791 there have been 190 recorded fatal shark attacks in Australia. On the east coast, the frequency of attacks has been reduced since the introduction of meshing in 1937. However no beaches in northern Australia are meshed for sharks. Australia wide an average of 1.2 fatal attacks occur each year a very low rate compared to the millions of swimmer and surfer hours in the waters around Australia.

The Department of Environment and Heritage offers the following advice regarding sharks:

Shark attacks occur rarely. Only a few of the 450 or so shark species have been known to attack people. Unfortunately, some attacks are fatal. There are some commonsense precautions to take that can help reduce the risk of a shark attack:

1. Do not swim, dive or surf where dangerous sharks are known to congregate.
2. Always swim, dive or surf with other people.
3. Do not swim in dirty or turbid water.
4. Avoid swimming well offshore, near deep channels, at river mouths or along drop-offs to deeper water.
5. If schooling fish start to behave erratically or congregate in large numbers, leave the water.
6. Do not swim with pets and domestic animals.
7. Look carefully before jumping into the water from a boat or wharf.
8. If possible do not swim a dusk or at night.
9. Do not swim near people fishing or spear fishing.
10. If a shark is sighted in the area leave the water as quickly and calmly as possible.

Stingrays are a common resident of the surf zone, where they usually lie hidden below a veneer of sand, feeding on molluscs and crabs. If disturbed, they speed off in a cloud of sand. However, if you are unfortunate enough to stand directly on one, it might spear your leg with its sharp, serrated barb located below its tail. In extreme circumstances, the barb can lodge in the leg and require surgery to remove it. Fortunately, this occurrence is uncommon. The best way to avoid sting rays is not to run into the water or surf and when in the water slowly shuffle through the sand, this warns them of your approach and they will usually swim rapidly away.

Crocodlies inhabit the estuaries, creeks and coastal waters of northern Australia from Roebuck Bay in the west, right across the top and down the east coast as far as Rockhampton. There are two species, the smaller freshwater crocodile *Crocodylus johnstoni,* which grows to 3 m and lives in freshwater. It is considered timid and harmless. The larger saltwater crocodile *Crocodlyus porsus*, can grow to over 5 m, weight more than one tonne. It lives in salt, brackish and freshwater and is extremely dangerous. Both species are protected and have been slowly increasing in population and expanding their range over the past 30 years. It is estimated there are 20 000 saltwater crocodiles in the Kimberley region, 75 000 in the Northern Territory and 20 000 in Queensland, primarily along the western Cape York coast and rivers, which represents a substantial proportion of the world population of between 200 000 to 300 000.

While crocodiles predominantly live in estuarine and stream environments, they can and do regularly move along the coast, occasionally landing on beaches (Fig. 2.13). Basically you should never swim in rivers and estuaries north of Roebuck Bay in the west and Rockhampton in the east, and use extreme care in known crocodile areas. Most people still use the beaches in the Broome region and north of Yeppoon, particularly in winter, however care must still be taken, as crocodiles will occasionally land on most beaches. Since records began in the 1870's there have been more than 32 fatal and 36 non-fatal crocodile attacks, primarily in the Northern Territory, together with some in the Kimberley and Queensland. Some have occurred on beaches or in shallow water just offshore.

Most attacks occur in the wet season, when saltwater crocodiles do most of their feeding and growing, and many have occurred at evening. They usually do not bother small fishing boats, though small boats and canoes have been attacked, including a fatal attack in August 2005.

Since crocodiles became protected in 1971 there have been 11 deaths in the Northern Territory and 21 throughout Australia. In Queensland there has been 16 attacks since 1985, six of which were fatal. In most cases, the victim was wading, swimming, snorkeling, scuba diving or paddling small canoes in crocodile areas. In virtually all cases, death could have been avoided with correct education and precaution. The low incidence of attacks is due to the fact that most people in Australia are well educated about crocodiles and do not place themselves in or too near the crocodile habitat. In 2005 three fatal attacks occurred at Cape Don and Groote Eylandt in the Northern Territory and in Normanby River on Cape York, with a non-fatal attack in the Kimberley's Doubtful Bay.

Crocodile safety

Follow these guidelines. **Be crocodile smart!**

- Do not disturb crocodile nests as nesting crocodiles may be aggressive.
- Travel quietly in a stable boat when crocodile spotting. Never approach the crocodile too closely, and keep your hands and legs inside the boat.
- Never provoke crocodiles, even small ones.
- Do not encourage wild crocodiles by feeding them. This is illegal and dangerous!
- Camp at least 50m from the water's edge and never prepare food or clean fish at the water's edge.
- Stand back when fishing. Don't stand on overhanging logs.
- Never swim in crocodile territory.

Safety measures also include no swimming, no cleaning of fish on river banks, nor tethering dogs close to the water. Nesting areas should also be avoided.

In north Queensland the sighting of a crocodile or evidence of one landing on a beach will result in closure of the beach for 3 days. The crocodiles usually cruise by beaches enroute to estuaries, but may land on beaches to warm-up. Their tracks are clearly visible when this occurs.

The general rules in avoiding dangerous marine animals in northern Australia are:

- Avoid known crocodile areas
- Swim at beaches meshed for sharks and stingers
- Only swim in stinger enclosures during the stinger season.
- Wear a lycra suit during the stinger season
- Wear footware (booties, etc) if wading on rock shore or coral flats
- Treat all cuts and abrasions in the ocean as potentially infectious
- Look along the swash line for evidence of stranded stingers, bluebottles, jellyfish, etc. If present, stay on the beach.
- If stung, follow recommended treatment guidelines.

Figure 2.13 The two major biological hazards across northern Australian waters are saltwater crocodiles and marine stingers. The above crocodile is on a popular north Queensland beach, while the sign applies across the top end in summer.

3 NORTHERN AUSTRALIAN BEACHES

This brief chapter provides information on how the remainder of this book is organised and how best to use it. The 3488 beaches are presented in three separate chapters. Chapter 4 covers the Kimberley coast, chapter 5 the Northern Territory and chapter 6 Cape York.

Beach number and name

Each chapter provides a description of every beach, beginning in the west of the region/territory and continuing east or clockwise to the end of the region/territory. Every beach in numbered consecutively clockwise. Each beach is also named, however as the majority of northern Australian beaches have no official name, that is, one that appears on a map, local factures are used. In most cases each beach is named after the nearest named feature, which may be a point, headland, creek, reef, islet or island, or some backing features such as a community, hill, mountain or landscape. Wherever possible local aboriginal names were used. This effort was greatly assisted in the Northern Territory by the publication *Northern Territory Aboriginal Communities,* which list every community in the territory. The N, S, E, W suffix following many names indicates the location of the beach relative to the feature, that is, north, south east or west. If followed by a number it indicated how many beaches away from the feature. For example N1 and N2 indicate 1 and 2 beaches north of the feature, while E5 and E6 indicate 5 and 6 beaches east of the feature. None of the given names are official, they are solely provided as a guide and to avoid having numbers only.

- If you know the correct or accepted name of a beach in this book, please provide it to the author so it can be added to future editions.

Coastal regions

The Northern Territory and Cape York are divided into four and three coastal regions respectively and chapters 5 and 6 are subdivided to reflect the regions. The Kimberley is treated as one region. The regions are based on major changes in coastal orientation.

Regional maps

In order to indicate the location of every beach, each chapter contains a series of regional maps locating every beach. These vary in scale depending on the region covered, and some contain inserts for congested areas. Where possible some of the maps follow the regional boundaries, however this was not always possible. The beginning of each chapter listed all the regional maps and the beaches located within each, as well as their page location.

Beach maps

Several beaches patrolled by Surf Life saving Western Australia and Surf Life Saving Northern Territory, as well as a handful of more popular beaches have individual beach maps.

Beach photographs

The three chapters contain 347 photographs of beach systems, The photographs were chosen to provide a representative collection of images of the beaches across the northern coast.

Beach header and descriptions

Most beaches are grouped into a collection of adjacent beaches. Each collection contains the following information:
Header: Beach numbers and name/s
Columns: Beach number | Beach name | Rating (HT LT) | Beach Type | Length
Definitions:

Beach number is the unique consecutive number for the beach (K 1-1360, NT 1-1488, Q 1-460)
Beach name is the unique beach name, usually after some nearby feature
Rating is the *beach hazard rating* (1-10) at high tide (**HT**) and low tide (**LT**), as described on pages 29-30.
Beach type is the type of beach system

wave dominated	(D, LBT, RBB, TBR, LTT, R)	see pages 19-22
tide-modified	(UD, R+LTR, R+LTT)	see pages 22-24
tide-dominated	(R+RSF, R+SF, R+TF. R+MF)	see pages 24-26

Length is the beach length in metres or kilometer

Beach Types

Wave dominated beach types

D	dissipative
LBT	longshore bar & trough (also as outer bar)
RBB	rhythmic bar & beach (also as outer bar)
TBR	transverse bar & rip (also as outer bar)
LTT	low tide terrace
R	reflective

Tide-modified beach types

UD	ultradissipative (none in South Australia)
R+LTR	reflective+bar & rips (none in South Australia)
R+LTT	reflective+low tide bar

Tide-dominated beach types

R+RSF	(HT) beach+sand ridges
R+SF	(HT) beach+sand flats
R+TF	(HT) beach +tidal sand flats
R+MF	(HT) beach +tidal mud flats

Beach plus rock/reef flat

R+RF	reflective+(intertidal) rock flat
R+CR	reflective+(fringing) coral reef

Other comments

sediment size (cobbles, boulders)
+rocks (scattered rocks on beach & in surf)
+rock flats (intertidal rock/laterite flats)
+rock reef (submerged rocks/laterite in surf and/or nearshore)
+coral reef (fringing coral reef)

Finding a beach – there are four ways to find a beach in this book

1. The alphabetical **BEACH INDEX** at the rear of the book lists all beaches.
2. Use the **REGIONAL MAPS** of the particular section of coast to locate a beach, or follow the beaches along the coast until you find the beach.
3. By name of the **SURF LIFESAVING CLUB**. If the beach has a surf club it will be listed in **BOLD** in the BEACH INDEX. If it differs from the beach name, then both are listed in the BEACH INDEX.
4. By name of a surfing break. If a popular surfing break has a name, which may differ from the beach name, then use the **SURF INDEX**.

4 THE KIMBERLEY COAST

The Kimberley region covers 320 000 km^2 of the northwest corner of Australia. It is a distinctive region in terms of its ancient geology, rugged hinterland and rocky indented coast. The uplifted sedimentary rocks of the Kimberley Basin have been heavily dissected and more recently drowned by the sea level rise, producing the rugged interior and irregular coast.

Between Roebuck Bay and the WA-NT border the region has a mainland coastline of 4333 km. The dominant coastal type is rocky shoreline with mangroves lining the protected embayment and usually small beaches tucked into the more exposed open section of coast, often located in joint controlled gaps. In all there are 1375 beaches along the coast which occupy only 713 km (16%) of shore, with an average length of only 529 m.

The coast is also sparsely populated and remote with coastal settlement restricted to the towns of Broome (12000) and Derby (5000) in the west and Wyndham (1000) in the east; and the small coastal aboriginal communities of Beagle Bay, Lombadina, One Arm Point and Kalumburu: the pearl farm base at Kuri Bay; and two tourist camps at Kimberley Coastal Camp and Faraway Bay. The region has a total population of 34 000, half of whom are of aboriginal descent.

Figure 4.1 illustrates the Kimberley coast, and location of the seven maps used to locate the 1375 beaches. These maps are designed solely to indicate beach location and not to divide the coast into subregions. In general terms the Kimberley is divided into the west and east Kimberley, with Cape Londonderry dividing the two on the coast. This division is not used in this chapter, with the entire coast treated as one Kimberley region.

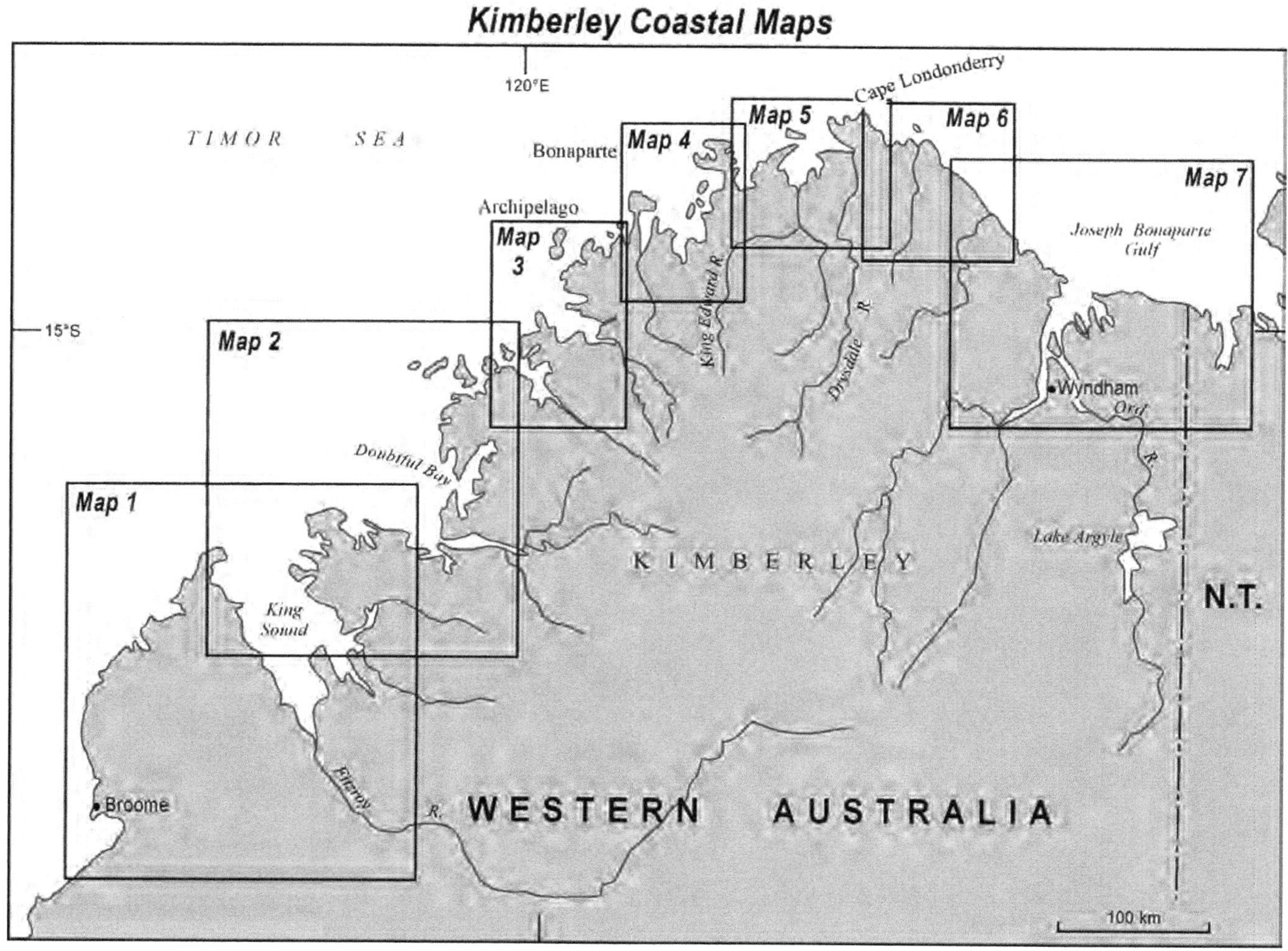

Figure 4.1 The Kimberley region and seven sectional maps used to locate the 1375 Kimberley beaches.

Kimberley sectional maps

Map	Beaches	Figure	Page
1	K 1-214	4.2	40
2	K 215-386	4.29	67
3	K 387-608	4.43	84
4	K 609-932	4.60	103
5	K 933-1149	4.84	131
6	K 1150-1283	4.107	152
7	K 1284-1360	4.120	166

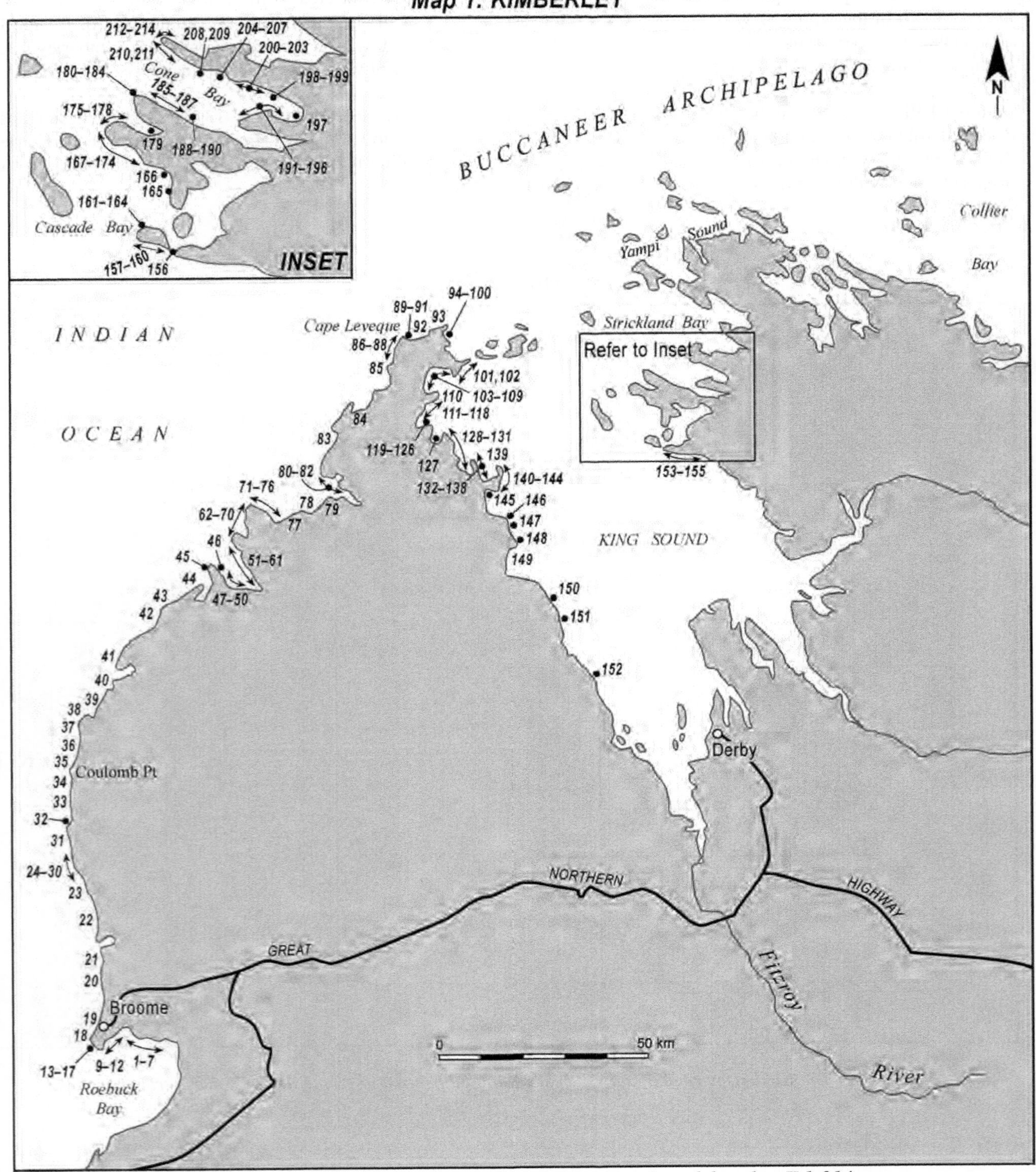

Figure 4.2 *Kimberley map 1: Roebuck Bay to King Sound, beaches K 1-214*

K 1-7 FISHERMENS BEND-CRAB CREEK

No.	Beach	Rating HT	Rating LT	Type	Length
K1	Crab Ck	1	2	R+tidal flats	3.1 km
K2	Fall Pt	1	2	R+tidal+rock flats	700 m
K3	Fall Pt (W)	1	2	R+tidal+rock flats	700 m
K4	Fishermens Bend (E 4)	1	2	R+tidal+rock flats	1 km
K5	Fishermens Bend (E 3)	1	2	R+tidal+rock flats	1.7 km
K6	Fishermens Bend (E 2)	1	2	R+tidal+rock flats	2 km
K7	Fishermens Bend (E 1)	1	2	R+tidal flats	500 m

Spring & neap tide range: 8.3 & 1.9 m

The northern shore of **Roebuck Bay** commences at Crab Creek and extends west for 14 km to Dampier Creek and then southwest for 8 km to Entrance Point, the entrance to Broome harbour, with Broome township located 5 km northeast of the point. The town abuts the western side of Dampier Creek and faces west and south across the bay. All the beaches between Crab Creek and Entrance Point tend to face south and are protected by the limited fetch and shallow tidal flats and shoals within the bay. The first seven beaches (K 1-7) are located along the 11 km of rock-controlled south-facing shore between Crab Creek and Fishermens Bend, the eastern entrance to Dampier Creek. A gravel road runs out from the Broome road across the supratidal flats of Dampier Creek and from Fishermens Bend follows the coast to Crab Creek, providing good access to all the beaches. All the beaches have a reddish hue derived from erosion of the backing red bluffs and reddish Cretaceous Broome sandstones. They are all fronted by a mixture of wide intertidal sand, mud and rock flats.

Crab Creek beach (**K 1**) is a 3.1 km long spit system, which has extended from the red bluffs of Fall Point southeast across the muddy tidal flats of the creek, narrowing from 500 m to 50 m at the tip (Fig. 4.3). The creek mouth is located 1 km west of the mangrove-fringed tip of the spit. The spit consists of a moderately steep 20 m wide high tide beach, fronted by 1 km wide mud flats, which narrow to 500 m off Fall Point.

Figure 4.3 Crab Creek beach (K 1) is the first beach on the northern side of Roebuck Bay, with extensive supratidal salt flats behind and wide mud flats in front.

Beach **K 2** is located along the eastern side of Fall Point and consists of an irregular 700 m long beach dominated by the backing bluffs, protruding sandstone outcrops and rocks as well as muddy tidal flats. The Broome Bird Observatory is located in the woodlands just behind the eastern end of the beach. At Fall Point the shoreline turns and trends west, with beach **K 3** occupying the first 700 m. This is a continuous reddish sand beach, bordered by 5 m high sandstone rocks, bluffs and rock flats, and backed by low eroding red bluffs. Sand-mud flats extend 400 m into the bay.

Beach **K 4** commences on the western side of a 200 m long section of low sandstone bluffs. It is an irregular 1 km long curving strip of high tide sand, wedged in below the bluffs and fronted by predominantly inner rock and outer tidal flats. Beach **K 5** extends to the west for another 1.7 km and is a similar beach with more tidal than rock flats, and backed by scarped 5 m high red bluffs the length of the beach. It becomes more discontinuous toward the west as the bluffs increasingly impinge on the beach.

Beach **K 6** commences at a protruding section of bluff and trends west for 2 km. It is a more continuous high tide beach composed of reddish sand, together with some scattered rocks along the beach. The rocks increase in the inner intertidal area, then grade to 500 m wide mud flats. A solitary small building is located on the bluffs in the centre of the beach. Beach **K 7** commences as a sand beach and after 500 m grades into the mangrove-lined shoreline of Dampier Creek. The Fishermens Bend road meets the shore toward the eastern end of the beach, then turns to run out along the top of the bluffs to Crab Creek.

Broome

Broome is a remote and fascinating town of 12 000 on the shores of Roebuck Bay. The town has a rich history in pearling, cattle and now tourism. Its population reflects a mixture of many cultures that gathered here early last century. The town has boomed again in recent years with the influx of Australian and international travellers and tourists. Its attractions are more than the fact it is the first coastal town north of Port Hedland (510 km by road) and south of Wyndham (1030 km). The town is set in an attractive location, with the mangrove-fringed shores of Roebuck Bay contrasting with the wide sandy surfing beach at Cable Beach, all backed by the bright red pindan sand of the desert. In addition Broome has one of the world's largest tide ranges, reaching 10 m on a spring tide. This has a dramatic impact of the beaches and tidal flats around the bay and on the coast.

The town offers all facilities and accommodation for travellers, including a resort, several hotels and caravan parks, and a crocodile farm out at Cable Beach. In addition the Broome Surf Life Saving Club overlooks Cable Beach and is the northernmost surf club in the west. There are six main beaches around Broome (K 8-13), which range from the tide-dominated beaches of Roebuck Bay to the surfing beaches facing the Indian Ocean (Fig. 4.4).

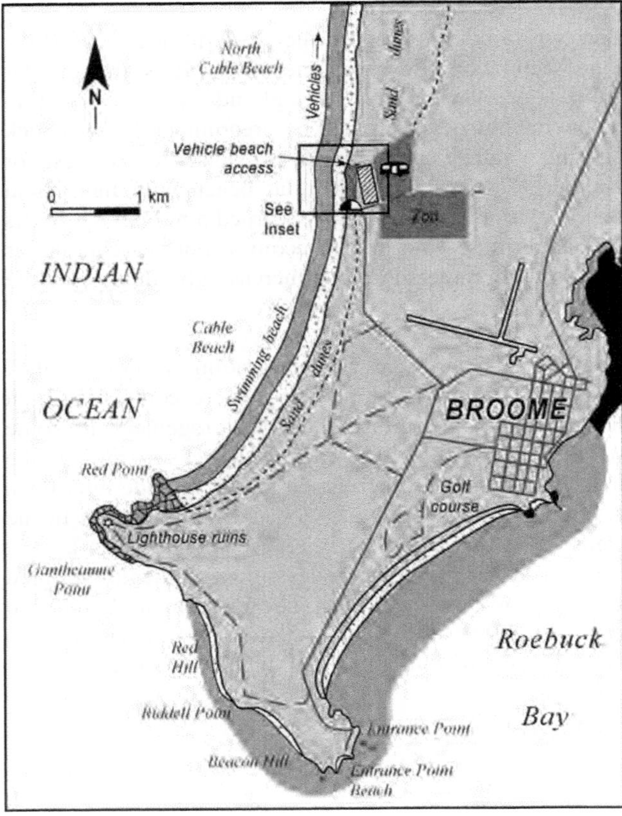

Figure 4.4 Broome is located on the northern shores of Roebuck Bay, with the long beaches of the Indian Ocean located on the western side of the town.

K 8-13 BROOME-ENTRANCE POINT

No.	Beach	Rating HT	Rating LT	Type	Length
K8	Town Beach	1	2	R+tidal flats	200 m
K9	Broome Golf Course	1	2	R+tidal flats	4.7 km
K10	Broome jetty (N)	1	3	R+rock flats	150 m
K11	Broome jetty (S)	1	3	R+tidal flats+rocks	300 m
K12	Entrance Pt	1	3	R+tidal flats+rocks	200 m
K13	Bittern Rock	1	3	R+rock flats	200 m

Spring & neap tide range: 8.3 & 1.9 m

Broome has both bay and ocean beaches. The bay beaches emerge from amongst the mangroves at the western end of town, while the ocean beaches extend north from Entrance Point. The six bay beaches (K 8-13) are all relatively protected from ocean waves, though tidal currents do impinge on the Entrance Point beaches. All are accessible by sealed or gravel road (Fig. 4.5).

Figure 4.5 Entrance Point is the site of the long Broome Jetty (far right), and the three Entrance Point beaches (K 11, 12 & 13) with boat ramps located across the western two beaches.

Town Beach (K 8) is located at the western end of the main town area. It lies adjacent to the remains of Mangrove Point jetty and is backed by a large car park, a park including a kiosk and playground and the Broome caravan park. A long, narrow, steep concrete boat ramp crosses the high tide beach. The beach is 200 m long, faces southeast and is fringed by a small jetty and mangrove-covered rock flats. The beach itself has a steep high tide beach and extensive tidal flats, which on a full moon is the site of the 'stair of the moon' moonlight display, an event which attracts hundred of sightseers every full moon. During the day it is a popular beach for locals and people from the adjacent caravan park. It has a shady picnic area and at high tide is a nice calm swimming beach.

Golf Course Beach (K 9) is located immediately south of Town Beach. The 4.7 km long beach faces southwest across Roebuck Bay and curves round slightly to face east in lee of Entrance Point. The beach is backed by a continuous foredune, with minor vegetated transgressive dunes along the southern half, where the dunes sit atop the red pindan soils. In addition the old meat works is located behind the eastern end, then the rifle range and finally the golf course, with the best access at the southern end of the beach. It consists of a narrow high tide beach fronted by a band of beachrock, then 1 km wide tidal flats, with mangroves fringing each end. It is not a popular beach owing to the more restricted access. The beach terminates on the northern side of **Entrance Point** clustered around which are three small beaches (K 10-12), all fronted by the deeper water and strong tidal currents which sweep around the point.

Broome Jetty beach **K 10** is located amongst the rocks on the north side of the 1 km long jetty, with a commercial slipway forming the northern boundary, while harbour facilities are located on the 10 m high bluff behind the beach. This 150 m long high tide sand beach is fronted by continuous rock flats and a few mangroves. Beach **K 11** is located on the south side of the jetty at the base of 10 m high red bluffs, with stairs leading down to the beach. Sandstone rocks are exposed at the base of the bluffs and scattered along the top of the beach. It consists of a moderately steep, 60 m wide, 300 m long beach fronted by 100 m wide sand flats, then deeper water off the point. A road runs along the bluffs behind the beach to Entrance Point. Immediately to the west is beach **K 12**, a 200 m beach hemmed in by red sandstone ridges, with a boat ramp across the eastern end of the beach and a car park behind. The top of the beach is dominated by sloping beachrock and sandstone cobbles and boulders, with a 100 m wide low gradient sandy intertidal zone. The car park is shared with beach **K 13** which is located on the southern tip of the point. This 200 m long beach has sloping high tide beachrock and sandstone rocks and reefs exposed at high tide with sand in the centre. It also has a boat ramp located across the only section free of low tide rocks, with deeper water and scattered rock reefs, including Bittern Rock, off the end of the beach. These two ramps are the most popular launching location, through they are exposed to strong southeasterlies.

K 14-18 ENTRANCE-GANTHEAUME POINT

No.	Beach	Rating HT	Rating LT	Type	Length
K14	Beacon Hill	1	2	R+LTT	1.4 km
K15	Riddell Pt	1	3	R+rocks	80 m
K16	Riddell Beach	1	2	R+LTT	1.7 km
K17	Riddell Beach (N)	1	3	R+LTT+rocks	400 m
K18	Gantheaume Pt	3	5	R+rock flats	400 m
Spring & neap tide range: 8.3 & 1.9 m					

At Entrance Point the shoreline turns and trends northwest for 5 km to Gantheaume Point. This section of coast contains three more exposed beaches (K 14-16), each backed by bright red 10 m high bluffs and bordered by sandstone rocks and rock flats (Fig. 4.6). A gravel road runs along the top of the bluffs with a climb down the bluffs required to reach the beaches.

Beacon Hill beach (**K 14**) commences at the southwestern tip of Entrance Point, called Bittern Rock, and extends to the northwest for 1.4 km to the 20 m high red bluffs of Riddell Point. The Broome oil storage tanks sit behind the southern half of the beach, with wooded bluffs behind the northern half. The beach consists of a narrow high tide beach and a 200 m wide intertidal sand beach backed by bright red 10 m high pindan bluffs, with some rocks outcropping along the beach and to either end. The beach receives low waves, usually less than 0.5 m, which form a narrow high tide beach below the bluffs, fronted by 200-300 m wide sand flats. A gravel road runs along the top of the northern half of the bluffs with numerous access points to the beach, but no facilities.

Figure 4.6 Riddell Point (left) is bordered by beaches K 14-17, shown here at low tide exposing the wide low tide bar and rocks.

Beach **K 15** is located at the base of 20 m high **Riddell Point**. It consists of an 80 m long strip of sand fronted by a 200 m wide mixture of rocks, rock flats and sand.

Riddell Beach (**K 16**) commences on the northern side of Riddell Point and trends northwest for 1.7 km to a protruding section of red bluffs. The beach is 300 m wide at low tide and has a scattering of sandstone rocks. A gravel road runs along the top of the bluffs, with no formal access to the beach from the road and no facilities.

The northern end of Riddell Beach (**K 17**) is located on the northern side of the bluffs. The bluffs are eroding and both scarped and slumping onto the beach. The narrow high tide beach is fronted by predominant rocky intertidal flats and only patches of sand. The beach extends for 400 m to the southern side of Gantheaume Point.

Gantheaume Point is a 20 m high red sandstone point that protrudes 1 km west at the southern end of Cable Beach. A gravel road runs around the perimeter of the point with views from the bluffs. Beach **K 18** is located on the north side of the point and consists of a 300 m long strip of sand located at the base of the red bluffs, fronted by a 200 m wide mixture of rock and sand flats. The point is famous for the dinosaur footprints preserved in the rock. On top of the point are remnants of desert longitudinal dunes, now partly covered by the Broome Racecourse.

Marine Reserve

The shoreline from Riddell Point to Gantheaume Point and up to Saddle Hill on Cable Beach, and for 1,500 m offshore, is a marine reserve. Shell collecting is prohibited on the beaches.

K 19-21 CABLE BEACH (BROOME SLSC)

No .	Beach	Rating	Type HT LT	Length
Patrols:				
Broome SLSC patrols		??????		
Lifeguard on duty patrols		??????		
K19	Cable Beach	3	4 UD	6 km
K20	Cable Beach (N)	3	4UD+beachrock reef	12 km
K21	Willie Ck (S)	3	4UD+beachrock reef	6.5 km
Spring & neap tide range:		8.3 & 1.9 m		

Between Gantheaume Point and Willie Creek 24 km to the north lie some of the most beautiful and famous beaches in the world. Cable Beach which extends for 6 km north of the point and its 12 km long northern extension are the focus of intense national and international tourist activity and attention, particularly during the winter months. Until the 1970s the beach was at the end of a gravel road with no facilities. A caravan park was developed in the 1970s, but since the 1980s there has been the development of the Cable Beach Resort, the new caravan park, Broome Surf Life Saving Club and a wide range of tourist activities, including Malcolm Douglas's crocodile park and the famous camel rides along the beach.

Cable Beach (K 19) commences on the north side of Gantheaume Point where it initially faces north before swinging round to trend due north for 6 km to terminate at the scattered sandstone rocks that separate it from the northern beach (K 20) (Fig. 4.7). The beach narrows to 20 m on a spring high tide, widening to 300-400 m on the low tide (Fig. 4.8). The beach receives waves averaging about 1 m, higher when summer westerlies blow. The waves are usually higher at high tide, decreasing at low tide owing to the shallow seabed off the beach. It is largely backed by a low foredune and moderately active sand dune that extend up to 300 m inland and have climbed up on top of the backing red bluffs, to reach a maximum height of 30 m. There is access to the beach at the point, with the main access at the northern end where there is a car ramp. Numerous activities including surfing, swimming, camel riding and wind surfing are available along the beach. On the bluffs immediately south of the ramp, is a grassy park and car park, then the surf club, with the main resort and tourist facilities all located on the eastern side of the road. The surf club was founded in 1988 and sits atop the bluffs with an excellent view across the beach.

The northern Cable Beach (**K 20**) extends from the boundary rocks due north for 12 km to so called **Cape Latreille**, a low sand spit that forms the southern side of a shallow creek mouth. This beach is undeveloped and is backed by a 0.5-1 km wide strip of largely stable dunes and foredune ridges. Vehicles are permitted on the beach and numerous vehicles and their occupants spread along the shore, while this is also an optional clothing (i.e. nude) beach. Three kilometres of beachrock reef parallel the southern section of the beach, lowering waves, particularly at low tide.

Figure 4.7 Cable Beach at low tide.

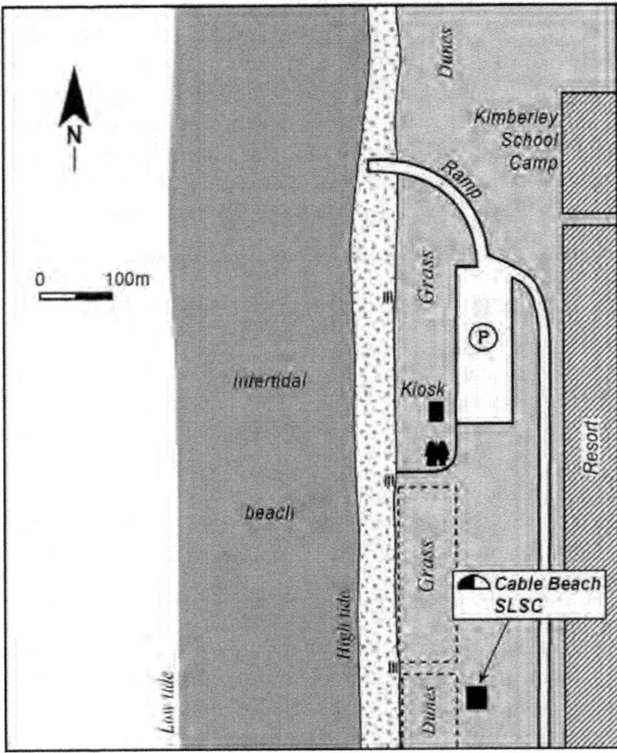

Figure 4.8 Cable Beach map, showing thw wide low tide beach.

Beach **K 21** lies between Cape Latreille and **Willie Creek**. This is a 6.5 km long beach that trends north paralleled by 4 km of attached beachrock, before curving into the mouth of the 500 m wide creek mouth. A 1 km wide foredune ridge plain backs the beach which grades into recurved spits toward the creek mouth. The large tidal range and strong tidal currents have formed an ebb tide delta that protrudes up to 2 km seaward of the creek mouth. The beach is accessible along the sand from Cable Beach and via tracks off McGuilgan Road that reaches the beach either side of Cape Latreille. There are however no facilities at the beach.

Swimming: When using all beaches in the northwest of Australia be mindful of the large tide range, which can reach 11 m at Broome. Always check the tide before using the beach. A rising tide can flood cars and belongings left unattended on the beach, while low tide requires a 200-400 m long walk across the sand to the water (Fig. 4.9), and more chance of encountering rocks and reefs. The best time to swim is either side of high tide.

a.

b.

Figure 4.9 Cable Beach at high (a) and low (b) tide. Spring tide range is up to 10 m, with the beach widening by up to 400 m.

Surfing: There is surf at Cable Beach and along the exposed beach to the north. While it is usually 1 m or less it can reach 2 m with westerly winds, and even higher during tropical cyclones. Again be aware of the tides as the size and location of the breaking waves will change considerably between high and low tide.

Fishing: Broome is famous for its many fishing attractions, with a wide range of fishing opportunities including shore based beach, rock and jetty fishing, as well as boat fishing in the tidal creek, the bay or ocean.

Summary: Cable Beach deserves its reputation as one of the world's top beaches. The wide clean white sand, usually low easy surf, warm water and long warm dry season, together with an abundance of room on the beach (at mid to low tide), excellent tourist facilities on the backing bluffs, and adjacent nude beach, all provide for an idyllic beach environment.

STINGERS: The stinger season is from December to April, so take care if swimming during this period. Lycra swimming suits are advisable and always check with the lifeguards if unsure.

Willie Creek to Coulomb Point

North of Willie Creek is a 45 km section of coast dominated by 14 sandy beaches (K 22-35), most of which are relatively accessible by vehicle along the Manari road, which leaves the Beagle Bay road 17 km north of its junction with the Broome road. Most of the gravel road is accessible by 2WD, with some side tracks to the coast requiring a 4WD. Apart from a pearl farm on the north side of Willie Creek, there are no development or facilities at any of the beaches.

K 22 WILLIE CREEK-CAPE BOILEAU

No.	Beach	Rating HT	Rating LT	Type	Length
K22	Willie Ck (N)	3	4	UD	10.4 km
K23	Cape Boileau	3	4	UD	7 km
Spring & neap tide range:		8.3 & 1.9 m			

North of **Willie Creek** are two longer beaches (K 22 and 23) separated by Cape Boileau. Both beaches are accessible by vehicle tracks at the 3.4 km and 14.5 km mark on the Manari road. Beach **K 22** commences on the north side of Willie Creek and extends due north for 10.4 km, curving round at the northern end in lee of Cape Boileau, and terminating at the mouth of Barred Creek. The beach is ultradissipative with a 300 m wide, low gradient intertidal beach. It is backed by a 1 km wide series of foredune ridges, which widen from 200 m in the south to 500 m in the north and have been transgressed by dunes. There is access to both ends of the beach, with the northern Barred Creek area being a relatively popular picnic and fishing spot.

Cape Boileau is a 2 km long strip of beachrock, which has formed around mangroves during a 2-3 m higher stand of the sea, possibly during the last Pleistocene high stand. The lithified beach and mangrove stump holes form what is locally called the 'petrified forest'. The Barred Creek track runs around the north side of the creek to the 'forest'. The 7 km long beach (**K 23**) begins beneath the steep beachrock bluffs, then continues north as a sandy beach backed initially by boulders and pindan bluffs, then a 4 km long section of active dunes that have extended up over the bluffs and up to 1 km inland. This is a popular area for beach and rock fishing at the mouth of Barred Creek.

K 24-30 QUONDONG POINT

No.	Beach	Rating HT	LT	Type	Length
K24	Quondong Pt (S6)	3	5	R+rock flats	1 km
K25	Quondong Pt (S5)	3	5	UD+rock flats	400 m
K26	Quondong Pt (S4)	3	5	UD+rock flats	200 m
K27	Quondong Pt (S3)	3	5	R+rock flats	200 m
K28	Quondong Pt (S2)	3	5	R+rock flats	400 m
K29	Quondong Pt (S1)	3	5	R+rock flats	300 m
K30	Quondong Beach	3	5	UD+rock flats	1.5 km
Spring & neap tide range:		8.3 & 1.9 m			

Quondong Point is a 20 m high sandstone point located in the centre of a 3 km long section of rocky shore, with seven generally small rock-bound beaches (K 24-30) surrounding the point (Fig. 4.10). The Quondong Point access track is located at the 22 km mark on the Manari road. It leads directly to the coast, with vehicle tracks leading north and south of the shoreline access point at beach K 24. The point area offers basic camping, some small sandy beaches for swimming and rock and beach fishing.

Figure 4.10 Beaches K 24 & 25 receive some low surf which breaks across a low tide bar.

Beach **K 24** is a rock-dominated continuation of beach K 23. The 1 km long beach is backed by continuous red bluffs and large sandstone boulders and fronted by 500 m wide intertidal rock flats. Beach **K 25** is located immediately to the north and is a sandy 400 m long beach, with a central ultradissipative sand section, bordered by rock flats. Its neighbour beach **K 26** is a similar 200 m long beach, with sand extending 100 m down the beach before it is replaced by rock flats. Both beaches are backed by spinifex-covered foredunes and offer the best spots for swimming.

Beaches **K 27, 28** and **29** are three narrow high tide strips of sand fronted by 200-300 m wide rock flats. They are 200 m, 400 m and 300 m long respectively, each separated by small protrusions in the backing sandstone bluffs. A vehicle track leads to the northern end of beach K 29. On the northern side of the track is Quondong Beach (**K 30**), a 1.5 km long, more exposed west-facing beach. It is bordered by 300-500 m wide intertidal rock flats, with a central 500 m long sandy ultradissipative section. Red bluffs back the beach with active dunes blanketing the central section of bluffs and extending 400 m inland. A radio tower is located just beyond the northern end of the beach.

K 31-38 QUONDONG POINT-CAPE BERTHOLET

No.	Beach	Rating HT	LT	Type	Length
K31	Quondong Pt (N)	3	5	UD+rock flats	6.2 km
K32	James Price Pt (S)	3	5	R+rock flats	1.8 km
K33	James Price Pt (N)	3	5	R+rock flats	9.8 km
K34	Coulomb Pt (S)	3	4	UD	4.5 km
K35	Coulomb Pt (N1)	2	3	UD	2.9 km
K36	Coulomb Pt (N2)	2	3	UD	2.7 km
K37	Coulomb Pt (N3)	2	5	UD	5 km
K38	Cape Bertholet (S)	3	5	R+rock flats	3 km
Spring & neap tide range:		8.3 & 1.9 m			

The 28 km of coast between Quondong Point and Cape Bertholet consists of eight west-facing relatively straight sandy beaches, for the most part backed by the low wooded coastal plain, with some sections scarped to reveal the bright red pindan soil. The Coulomb Point track parallels the back of the beaches up to Coulomb Point and 2 km beyond to the first creek crossing, a distance of 52 km from the beginning of the track. Apart from the coastal access tracks and a radio tower, there are no development or facilities along this section of coast.

Beach **K 31** commences 1.5 km north of Quondong Point, immediately north of the radio tower. It is a relatively straight 6.2 km long west-facing beach, backed for the most part by 10-20 m high red bluffs. The bluffs are capped by generally vegetated blufftop dunes extending 200-300 m inland, with only one active section. The beach is wedged in at the base of the bluffs and consists of a mixture of ultradissipative sand beach and rock flats. The vehicle track runs along behind the dunes with no direct vehicle access to the beach.

Beach **K 32** extends for 1.8 km south of James Price Point. It is a crenulate high tide sand beach backed by sand-draped 10-20 m high red bluffs and fronted by rock flats that widen from 300-500 m at the point. A spur off the vehicle track terminates at the point.

Beach **K 33** extends north of **James Price Point** for 9.8 km. It initially consists of a narrow strip of sand below the red eroding bluffs. As the beach turns and trends due north, the bluffs continue, occasionally cut by gullies that provide access to the beach from the track running along the top of the bluffs. The beach for the most part consists of a relatively steep reflective high tide beach fronted by wide intertidal rock flats. It terminates at an inflection in the bluffs, beyond which is 4.5 km long beach **K 34**, which continues due north to Coulomb Point. This beach is backed by an active sand sheet that widens from 200 m in the south to 800 m in lee of the northern point. The dunes increase in height landward forming a transgressive sand ridge up to 36 m high along their inner boundary. The track follows the back of the dunes, with beach access at either end. The beach receives waves averaging less than 1 m and consists of a wide low gradient ultradissipative beach.

Coulomb Point Nature Reserve

Established:	1969	
Area:	28 676 ha	
Beaches:	5	(K 35-39)
Coast length:	20 km	(5955-5975 km)

Coulomb Point, also known as **Manari**, consists of low intertidal sandstone rock flats that extend 1 km seaward of the shore and contain several large tidal pools, with a usually dry sandy creek winding along the northern side of the point (Fig. 4.11). A near continuous sand beach extends 12 km northward to Cape Bertholet with tidal creeks dividing it into four beaches (K 35-38). Beach **K 35** commences at the point and curves toward the northeast then north for 2.9 km to the first tidal creek north of the point. The beach is initially protected by the rocks of the point and is fronted by 1 km wide rocks and sand flats. These narrow northward toward the usually blocked creek mouth, where there is a small tidal delta.

Figure 4.11 Coulomb Point with dune-backed beach K 34 extending to the south (right) separated by the low point and creek from beach K 35 (left).

The vehicle track degenerates to a 4WD track at the creek crossing, with care required if driving north of the creek. Beach **K 36** extends north from the creek for 2.7 km to the northern side of a usually blocked 1 km wide double creek mouth. When flowing the two creeks often converge forming one continuous tidal delta. The intervening sandy beach is up to 400 m wide at low tide.

Beach **K 37** commences on the north side of the second creek and trends due north for 5 km to the first rocks at the southern end of Cape Bertholet. The beach is backed by low active foredunes and overwash flats, which merge into a small creek toward the northern end, just beyond which are the low rocks.

Beach **K 38** commences at the low rocks and trends roughly northeast for 3 km to the end of low sandy **Cape Bertholet**. The shoreline is highly crenulate consisting of a mixture of low intertidal rock flats and high tide beachrock and backing beachrock boulders. These are fronted by wide low gradient beach and sand flats and all backed by moderately active dunes which rise to 24 m in height and extend 500 m inland. The northern end of the beach merges into 1 km wide tidal-sand flats that fill the 2 km wide tidal embayment on its northern boundary.

Beagle Bay Aboriginal Land

Area: 352 151 ha

Coast length (km)		Beaches	
7	(5975-6042 km)	6	(K 39-44)
35	(6105-6135 km)	6	(K 74-77)
66	(6344-6410 km)	7	(K146-152)
Total 168 km		19	

Beagle Bay Aboriginal Land extends from the west coast of the peninsula for 60-130 km to the eastern King Sound shore. It includes the main community of Beagle Bay on the west coast, as well as the eastern shore communities of Maddar, La-Djardarr and Malaburra on King Sound.

K 39-44 CAPE BERTHOLET-CAPE BASKERVILLE -CAMP INLET

No.	Beach	Rating HT	Rating LT	Type	Length
K39	Cape Bertholet (N)	1	1	R+sand flats	7.2 km
K40	Carnot Bay	1	1	R+sand flats	2.2 km
K41	Smirnoff Beach	1	2	R+sand flats/rock	14 km
K42	Red Bluff	1	2	R+sand flats/rock	7.1 km
K43	Low Sandy Pt	1	2	R+rock flats	9.3 km
K44	Camp Inlet	1	2	R+sand flats/rock	5.1 km
Spring & neap tide range:		8.3 & 1.9 m			

North of **Cape Bertholet** the coast turns and trends northeast for 50 km to Sandy Point. In between are six long, low energy beaches, all fronted by wide sand and tidal flats, as well as sections of rock flats. They are separated by five tidal inlets, including Carnot Bay, Baldwin Inlet and Camp Inlet. The only vehicle access to the coast is midway at Red Bluff.

At Cape Bertholet the shoreline turns and trends easterly into a 14 km long, very low energy section of coast that terminates in the extensive tidal flats of Carnot Bay. Between the cape and the bay are two low barrier islands, backed by extensive mangroves, bordered by three tidal creeks and fronted by 1-2 km wide tidal sand flats. The first beach (**K 39**) commences 2 km east of Cape

Bertholet and trends north-northeast for 7.2 km as part of a low crenulate barrier island, of variable width, backed by mangrove forests to either end, together with 3 km wide salt flats in the centre. A 1.5 km wide tidal inlet separates it from beach **K 40**, a low irregular 2.2 km long barrier island, which is backed by 1 km wide mangroves, then 2 km of salt flats, while both beaches are fronted by 1-2 km wide tidal sand flats. Beach K 40 also forms the southern entrance to **Carnot Bay**, a 2,500 ha embayment largely infilled with sand flats, fringed by mangroves, and surrounded by 1-2 km wide salt flats.

Cape Baskerville forms the northern entrance to Carnot Bay. The low bedrock-controlled cape faces west, but curves round into the bay entrance to face south at its southern end. **Smirnoff Beach (K 41)** occupies the cape and trends northeast for a total of 14 km to Red Bluff. The beach is crenulate throughout, owing to outcrops of red bluffs and rock flats, with the high tide beach fronted by a mixture of rock flats and rubble up to 1 km wide. It is backed by a mixture of active and older blufftop dunes, which reach heights of 36 m, together with red bluffs, in the centre and toward the northern end. It terminates at Red Bluff, where there is an easterly inflection in the shoreline. A vehicle track leads to the southern side of the bluff where two small blufftop communities are located.

Beach **K 42** commences on the north side of **Red Bluff** and curves gently toward the northeast for 7.1 km to the mouth of Baldwin Inlet (Fig. 4.12). Rock reefs lie 2-3 km offshore and parallel the low energy reflective sandy beach, which is backed by active sand dunes extending up to 2 km inland and to heights of 35 m. Baldwin Creek is an 800 ha tidal creek, largely dominated by mangroves, fringed by salt flats.

Figure 4.12 Baldwin Creek with beach K 42 to right and Low Sandy Point and beach K 43 to left.

Low Sandy Point forms the northern boundary to the creek. It is a curving bedrock-controlled point that faces southwest at the creek mouth and then curves round to trend northeast for a total of 9.3 km. The entire point and beach (**K 43**) are backed by extensive older and active sand dunes, reaching up to 2 km inland. The beach is bordered by rock flats and reefs at either end, with some patches of rock flats also paralleling the beach. It terminates at a shore-parallel section of beachrock, with a prominent easterly inflection in its lee. Beach **K 44** commences in its lee and trends roughly northeast for 5.1 km to the mouth of Camp Inlet. The beach is dominated by the rock reef and the tidal shoals of the inlet, which extend up to 3 km off the northern end of the beach. The entire beach is backed by active sand dunes, then the mangroves of the inlet. The inlet is 3,000 ha in area and dominated by bare intertidal sand flats and scattered sections of mangroves.

The low **Lacepede Islands** (West, Middle, Sandy and East) lie between 17 and 34 km off Low Sandy Point. They are part of an 8,000 ha coral reef system which blocks the ocean waves contributing to the lower wave energy along this section of coast.

Beagle Bay

Beagle Bay is an open V-shaped northwest-facing bay. The 17 km wide entrance is bordered by South and North Heads, with the bay converging to a 200 m wide tidal creek, 25 km to the southeast. The creek continues on for another 18 km east to the Beagle Bay community. The bay has 34 km of generally low energy shoreline, together with 16 beaches (K 45-60), most fringed by tidal rock, sand and mud flats, with mangroves dominating the apex. There is 4WD vehicle access to much of the bay shore, with the only development being pearl farm buildings on Tooker Point.

K 45-48 SOUTH HEAD-TOOKER POINT

No .	Beach	Rating HT	LT	Type	Length
K45	South Head	1	2	R+rock&tidal flats	5.2 km
K46	Sandy Pt (E)	1	1	Ultradissipative	3.5 km
K47	Ledge Pt	1	2	R+mud flats	2.2 km
K48	Tooker Pt	1	2	R+mud & rock flats	900 m
Spring & neap tide range:				8.3 & 1.9 m	

South Head is a low sandy bedrock-based headland that extends 5 km due north from Camp Inlet and forms the western boundary of Beagle Bay. Beaches are located along both sides of the head, which is largely capped by active sandy dunes (Fig. 4.13).

Beach **K 45** is located along the western side of the head. It is a crenulate, low energy 5.2 km long west-facing high tide beach, fronted by partly sand-covered rock flats, the 500 m wide tidal creek and then the shoals of Camp Inlet, which extend up to 3 km to the west. The central 2-3 km long section is backed by a 1 km wide sheet of active transverse dunes, which reach 20 m in height. There is vehicle access to the head via the beach.

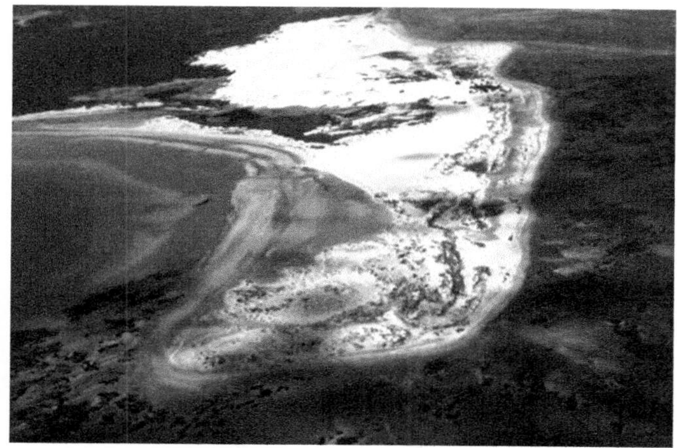

Figure 4.13 The low sandy, dune-capped South Head is bordered by beach K 45 (right) and beach K 46 (left).

The eastern side of South Head is called **Sandy Point** and marks the beginning of 3.5 km long beach **K 46**. The wide, low gradient ultradissipative sandy beach faces initially east, then curves round to face north in lee of Ledge Point, with several hundred metre wide rock flats extending seaward of each boundary point. The northern half of the beach is backed by the dunes spreading over from the South Head beach, while the southern half is crossed by a shallow creek, then vegetated coastal plain, with vehicle tracks paralleling the back of the beach.

Ledge Point forms the western boundary of 2.2 km long beach **K 47,** which curves round into Beagle Bay resulting in a general northeast orientation. The beach is very protected by the rock reefs off Ledge Point and consists of a 50 m wide high tide sand beach, then 500 m wide intertidal mud flats. It is bordered by rock flats at either end with mangroves on the eastern rocks.

Beach **K 48** is a slightly curving 900 m long northeast-facing beach that emerges from the mangrove-fringed rock flats and terminates in the east at **Tooker Point.** The sandy high tide beach is fronted by 300 m wide mud flats, with rock flats converging toward the centre. The entire beach is part of a low vegetated sand spit, which recurves at the point and extends as a narrowing spit for 2 km south into **Alligator Creek**. A gravel road runs out to the tip of Tooker Point where Arrow Pearls has a group of several beachfront buildings and a landing.

K 49-61 BEAGLE BAY

No. Beach	Rating HT	LT	Type	Length
K49 Alligator Ck	1	2	R+mud flats	150 m
K50 Henry Well	1	2	R+mud flats	2.6 km
K51 Mangrove Pt (S2)	1	2	R+sand/rock flats	700 m
K52 Mangrove Pt (S1)	1	2	R+sand/rock flats	1.1 km
K53 Mangrove Pt (N1)	1	2	R+sand/rock flats	1.7 km
K54 Yallet Ck	1	2	R+sand flats	1.4 km
K55 East Sandy Pt (S 1)	1	2	R+sand/rock flats	700 m
K56 East Sandy Pt (N 1)	1	2	R+sand/rock flats	700 m
K57 East Sandy Pt (N 2)	1	2	R+sand/rock flats	400 m
K58 East Sandy Pt (N 3)	1	2	R+sand/rock flats	700 m
K59 Cliff Pt (S 2)	1	2	R+sand/rock flats	1.6 km
K60 Cliff Pt (S 1)	1	2	R+sand/rock flats	400 m
K61 North Head (S)	1	2	R+rock flats	300 m
Spring & neap tide range:			8.3 & 1.9 m	

The 22 km long southern and eastern shores of **Beagle Bay** contain 13 low energy beaches (K 49-61) all dominated by combinations of mud flats in the south and sand and rock flats along the eastern shore. There is 4WD access to most of the beaches, with the only development being a solitary house on Mangrove Point. The Beagle Bay community is located at the infilled apex of the bay, 20 km east of beaches K 50 and 51.

Beach **K 49** is located on the southern side of Alligator Creek, 1.5 km south of Tooker Point. It consists of a 150 m long sandy break in the mangrove-lined shore. It is bordered by mangrove-covered rock flats and fronted by 500 m wide mud flats and tidal channels.

Beach **K 50** is the easternmost beach in the bay, located on the southern side of the creek that continues east toward Beagle Bay community. The 2.6 km long beach is the outermost beach ridge in a 1 km wide series of ten ridges that are associated with the infilling of the inner bay. The low energy ridges are partly fronted by mangroves and 1.5 km wide outer tidal mud flats. Henry Well is located 1.5 km south of the western end of the beach.

Beach **K 51** is the southernmost of the eastern beaches, and represents the terminus for sand moving down the eastern shore to infill the bay. The 700 m long beach is part of a 250 m wide series of recurved spits. Its northern end is fronted by mangrove-covered rock flats, with sand flats extending 500 m into the bay. Beach **K 52** begins on the northern side of the mangroves and trends north for 1.1 km to **Mangrove Point**, where a house is located just behind the beach. The crenulate beach is dominated by partly sand-covered rock flats, and backed by a wooded coastal plain.

Between Mangrove and East Sandy points are three beaches (K 53-55). Beach **K 53** extends for 1.7 km north of Mangrove Point as a crenulate sand beach fronted by 300-400 m wide sand and rock flats. It is separated from beach K 54 by mangrove-covered rock flats. Beach **K 54** is a curving 1.4 km long sand beach fronted by sand flats that widen to 1 km at the northern end. It faces southwest and is part of a 200-400 m wide series of low foredunes that partly block Yallet Creek, which flows out at the northern end of the beach. Beach **K 55** commences on the western side of the small creek mouth, and trends west for 700 m to East Sandy Point. The low energy beach faces south across 1 km wide sand flats. Its exposure to southerly winds has produced some minor dune transgression extending 300 m inland and rising to 21 m.

At **East Sandy Point** the coast turns and trends north toward Cliff Point. Beach **K 56** is a slightly curving west-facing, 700 m long high tide sand beach, fronted by a mixture of sand and rock flats up to 600 m wide. It is backed by a low foredune then wooded coastal plain. Its neighbour, beach **K 57,** is a similar 400 m long beach, with rock flats at both ends, and a clear 300 m wide intertidal sandy section in the centre. Beach **K 58** commences on its northern side and is dominated by irregular 200-300 m wide rock flats for its entire 700 m length. The southern half is backed by coastal plain, with vegetated and active dunes extending 500 m east of the northern half and to heights of 19 m.

Beach **K 59** lies immediately to the north. It is a 1.6 km long west-facing sand beach that curves round at its northern end to face south. The beach consists of a 50 m wide high tide beach fronted by continuous 200 m wide intertidal rock flats. The rock flats continue on to front its neighbour, 400 m long south-facing beach **K 60**. This beach is bordered by the rock flats in the east and the rock flats and 10 m high grassy bluffs of Cliff Point to the west.

North of Cliff Point the coast trends roughly north for 3 km to North Head, the northern end of the bay. The crenulate shoreline is dominated by 10-15 m high bluffs, fronted by patchy 300-500 m wide rock flats. One kilometre south of **North Head** are two adjacent patches of high tide sand (beach **K 61**) totalling 300 m in length, backed by the bluffs and fronted by the rock flats.

K 62-69 TAPPERS INLET-MIDDLE LAGOON

No.	Beach	Rating HT	Rating LT	Type	Length
K62	Tappers Inlet (S)	1	2	R+rock flats	500 m
K63	Tappers Inlet (N1)	1	2	R+rock flats	200 m
K64	Tappers Inlet (N2)	1	2	R+rock flats	100 m
K65	Tappers Inlet (N3)	1	2	R+rock flats	700 m
K66	Middle Lagoon (S2)	1	2	R+rock flats	800 m
K67	Middle Lagoon (S1)	1	2	R+rock flats	500 m
K68	Middle Lagoon	1	2	R+sand flats	1.5 km
K69	Emeriau Pt (S)	1	2	R+sand/rock flats	400 m
Spring & neap tide range:		8.3 & 1.9 m			

Between North Head and Emeriau Point is a relatively straight 8 km long bedrock-controlled section of coast, broken by Tappers Inlet and Middle Lagoon. There are eight high tide beaches, all but one fronted by rock flats. The only vehicle access is to a solitary house on the northern shore of Tappers Inlet and a small community at Emeriau Point (Fig. 4.14).

Beach **K 62** is located on the southern bedrock-controlled entrance to Tappers Inlet. The narrow 500 m long west-facing high tide beach is fronted by 300-400 m wide rock flats, and backed by 200 m of minor dune actively, then the inlet.

Figure 4.14 The curving Middle Lagoon beach (K 68) and dunes, and Emeriau Point (foreground) with its pocket bluff-backed beaches (K 69 & 70).

Tappers Inlet is a 600 ha oval-shaped inlet containing 200 ha of mangroves and extensive sand flats that fill the remainder of the lagoon. It has a 1 km wide bedrock-controlled mouth with the sand flat exposed at low tide in the centre of the bay.

On the north side of the lagoon are two small southwest-facing pocket beaches located between protruding sections of coastal plain and their fronting rock flats. Beach **K 63** is 200 m long and backed by a house linked to a vehicle track, while its neighbour (beach **K 64**) is 100 m long.

Two hundred metres to the west the coast turns and trends north, with beach **K 65** a discontinuous 700 m long high tide sand beach located at the base of a series of crenulate 5-10 m high rocky bluffs, with rock flats extending 200-300 m off the beach. Its neighbour is a more continuous 800 m long narrow high tide beach (**K 66**), also located below rocky bluffs and fronted by 500 m wide rock flats. Finally beach **K 67** is a 500 m long beach lying in lee of the sandy 500 m wide rock flats, with lower backing bluffs permitting some minor dune transgression to the rear. The rock flats also form the southern boundary with Middle Lagoon.

Middle Lagoon is largely occupied by beach **K 68**, a 1.5 km long curving west-facing sandy beach, fronted by rock-protected sand flats at either end, with a central open sandy section. It is backed by a vegetated foredune, then 500 m wide active dunes, which rise inland to a 31 m high transgressive ridge. It is separated at the northern end by a 100 m wide 10 m high grassy headland, from beach **K 69**, a curving 400 m long beach that terminates against **Emeriau Point**. Both the headland and associated rock flats almost encase the 200-300 m wide sandy beach. A small community of about eight houses is located in lee of the headland that separates the two beaches.

Pender Bay

Pender Bay is a relatively open 17 km wide northwest-facing embayment, bordered by Perpendicular Head in the west and Cape Borda to the north. Much of the bay shore consists of scarped red bluffs fronted by a series of sandy beaches, which converge toward two sand spits which partly block the 2 km wide mouth of Kelk Creek, the eastern apex of the bay. The two sand spits extend across the mouth of the creek and represent the terminus for sand moving both east and south into the creek area.

Most of the bay shore is located in the Beagle Bay Aboriginal Reserve. There is vehicle access in the south to Embalgun and Bell Point, where a small community is located, and in the east to Pender.

K 70-76 EMERIAU PT-PERPENDICULAR HEAD

No.	Beach	Rating HT	Rating LT	Type	Length
K70	Emeriau Pt (N)	1	2	UD+rocks	900 m
K71	Chimney Rocks	1	2	UD+rocks	700 m
K72	Chimney Rocks (E)	1	2	R+rocks	250 m
K73	Perpendicular Hd (W)	1	2	R+rocks	800 m
K74	Perpendicular Hd (S 1)	1	2	R+sand/rock flats	700 m
K75	Perpendicular Hd (S 2)	1	2	R+sand flats	800 m
K76	Perpendicular Hd (S 3)	1	2	R+sand flats	1 km
Spring & neap tide range:		8.3 & 1.9 m			

The 5 km of coast between Emeriau Point and Perpendicular Head consist of seven generally north-facing beaches (K 70-76), bordered and backed by red bluffs, with rocks and rock flats dominating the intertidal (Fig. 4.15). The only vehicle access to the area is at Emeriau Point.

Figure 4.15 Beaches K 74-77 extend north to Perpendicular Head.

Beach **K 70** commences on the northern side of Emeriau Point. It is a 900 m long high tide sand beach, bordered and interrupted by 5-10 m high bluffs, with three sections of rock flats, and two more open sandy sections with a few pandanus trees backing the beach. In lee of the sandy section there has been minor dune activity extending up to 200 m inland and to elevations of 20 m. There is vehicle access to the middle of the beach. A 250 m wide, 27 m high bluffed headland separates beach K 70 from **K 71**. This is a crenulate 700 m long sand beach bound in the west and east by craggy bluffs, fronted by intertidal rock flats, as well as rock reefs and small rock outcrops and vegetated islets, which extend from the shoreline up to 500 m seaward. The largest islet is called **Chimney Rocks**. The beach has a low gradient intertidal zone, interrupted by some rocks, and is backed by active dunes extending up to 200 m inland. A 4WD track reaches the centre of the beach.

Beach **K 72** lies 100 m to the east and is a 250 m long high tide sand beach, bordered and backed by continuous 10-20 m high bluffs. They are scarped in the west exposing red soils, while fronted by debris and some vegetation in the east. Bluff debris and boulders extend off each headland and east-west-trending rock flats dominate the low tide area. Beach **K 73** is located another 250 m to the east and consists of a more irregular 800 m long high tide sand beach, backed by continuous fresh bluffs, with some protruding sections, with a scattering of bluff debris and rock flats extending seaward from the mid tide zone. Rocks and rock flats eventually replace the beach extending another 500 m to 20 m high **Perpendicular Head**, which is surrounded by steep red bluffs.

Beaches K 74 and 75 are located immediately south of the point, and consist of adjoining 700 and 800 m long curving east-facing beaches, both backed by 10 m high red bluffs. Beach **K 74** has a high tide sand beach, fronted by a 300 m wide mixture of sand and rock flats, while beach **K 75** has a continuous 300 m wide intertidal sand beach, with rocks exposed in the sub-tidal.

Beach **K 76** commences immediately south of beach K 75 and is a crenulate 1 km long east-facing beach consisting of a series of five high tide pockets of sand, backed and bordered by 10 m high red bluffs. The beaches are linked by a continuous intertidal zone consisting of 200-300 m wide sand and rock flats and debris.

K 77-79 EMBALGUN-KELK CREEK

No.	Beach	Rating HT	Rating LT	Type	Length
K77	Embalgun	1	1	R+sand flats	7 km
K78	Bell Pt	1	2	R+sand/rock flats	2.5 km
K79	Weedong	1	2	R+sand-tidal flats	8 km
Spring & neap tide range:				8.3 & 1.9 m	

Lombardina Aboriginal Land

Area:	153 456 ha	
Coast length (km)		Beaches
30	(6145-6175 km)	4 (K 80-83)
103	(6207-6310 km)	43 (K 89-131)
Total 133		47

Lombardina Aboriginal Land occupies the northern tip of the peninsula, with the Lombardina community located on the western shore, and One Arm Point on the northeastern shore at the mouth of King Sound. It includes the Koojaaman Tourist Resort at Cape Leveque.

The southern shore of Pender Bay consists of three longer north-facing sand beaches, that extend for 17 km from the base of Perpendicular Head at Embalgun to the 2 km wide mouth of Kelk Creek. **Embalgun** beach (**K 77**) commences 3 km south of Perpendicular Head, as the red bluffs give way to the 7 km long sandy beach. The beach initially faces east and is backed by bright red eroding 10 m high bluffs. It then curves round to face north for most of its length where it is backed by low vegetated dunes, finally terminating in the east at the base of 40 m high red bluffs. The beach is protected from westerly waves and is for the most part fronted by 300-400 m wide sand flats, together with scattered rocks and rock flats. A solitary straight vehicle track reaches Embalgun at the western end of the beach.

Beach **K 78** commences on the eastern side of the high bluffs and extends east for 2.5 km to 20 m high **Bell Point**, where a vehicle track terminates at two blufftop houses. The beach curves round and faces generally northwest. It is for the most part backed by scarped red 20-30 m high eroding bluffs, capped by dense woodlands. The beach consists of a relatively narrow 50 m wide strip of high tide sand at the base of the bluffs, fronted by intertidal sandstone rock flats and patches of sand.

Beach **K 79** extends from the eastern side of Bell Point for 8 km to the mouth of Kelk Creek. The beach initially faces east and is backed by sand-draped red bluffs. These give way to a 200-600 m wide, 20 m high sand barrier, backed by Weedong Lagoon, and finally a 4 km long northwest-facing sand spit that has prograded across the mouth of Kelk Creek. It is capped by low active dunes up to 500 m wide and fronted by tidal sand flats and ridges that extend up to 3 km into Pender Bay.

Kelk Creek is a 20 km wide V-shaped tidal embayment. It has bordering sand spits at the entrance and then narrows for 27 km to the east. Mangroves dominate the creek for 15 km, and then high tide salt flats for another 12 km to the head of the tide at Low Balk Bore.

K 80-82 KELK CREEK-CAPE BORDA

No.	Beach	Rating HT	Rating LT	Type	Length
K80	Pender	1	2	R+tidal sand flats	3.7 km
K81	Cape Borda (S)	1	2	R+tidal sand flats	2.6 km
K82	Cape Borda	1	2	R+tidal sand flats	1.2 km
Spring & neap tide range:				8.3 & 1.9 m	

The northern side of Pender Bay extends for 7 km from the mouth of Kelk Creek to Cape Borda. The entire shoreline is dominated by the extensive tidal sand flats and channels of Kelk and an adjacent unnamed creek, with the Kelk Creek tidal channel paralleling the coast to Cape Borda and the sand flats extending another 2-3 km seaward of the cape. The only development in the area is vehicle access to the abandoned Pender settlement on the northern side of Kelk Creek.

Beach **K 80** is a 3.7 km long west-facing barrier island that forms the northern entrance to Kelk Creek, as well as the southern boundary of a neighbouring creek to the north. The beach consists of a high tide beach fronted by the extensive 1 km wide Kelk Creek tidal channel to the south, and a 200 m wide channel of the northern creek along its northern section, with intertidal sand shoals in between extending over 1 km into Pender Bay. Both vegetated and active low dunes extend up to 1.5 km inland and reach 22 m in height. They are backed by an inner vegetated transgressive dune ridge.

Beach **K 81** commences on the northern side of the 1 km wide second creek mouth. The 2.6 km long beach consists of a low 100 m wide overwashed recurved sand spit, that is tied to rock flats at its northern end. It faces southwest across Kelk Creek tidal channel, then several kilometres of tidal sand flats. Beach **K 82** commences on the northern side of the rock flats and extends for 1.2 km due north to the low rocks that continue northeast for 800 m to **Cape Borda**. The beach is fronted by over 1 km wide tidal sand flats. It is backed by 500 m wide unstable overwash and active dunes in the south grading into vegetated dunes and wooded low coastal plain in the north.

K 83 CAPE BORDA (N)-PACKER IS

No.	Beach	Rating HT	Rating LT	Type	Length
K83	Cape Borda	1	1	UD	11.7 km
Spring & neap tide range:				8.3 & 1.9 m	

Beach **K 83** extends 11.7 km north of Cape Borda to the lee of the beachrock reefs of Packer Island (Fig. 4.16). The beach faces generally west-northwest and consists of a continuous sand beach, with scattered sandstone reefs extending over 1 km seaward and 1.5 km north of Cape Borda, with wide sand flats to their lee. Once clear of the rocks is a 7 km long ultradissipative beach, which then grades into 1 km wide tidal sand flats up to the mouth of **Gilbut Creek** in lee of the southern beachrock tip of the island. The beach is backed by an extensive 1,700 ha barrier dune system, which in the centre is 2.5 km wide. The entire dune system has an outer 1-2 km active zone, backed by an inner vegetated transgressive sand ridge, which reaches a maximum height of 38 km toward the south. The dunes are backed by wooded coastal plain, except adjacent to Gilbut Creek, where mangroves back the northern 2 km.

Figure 4.16 The northern end of the long beach K 83 terminates in lee of the beachrock reef (foreground) against Gilbut Creek (left).

Packer Island is a straight, northwest-facing 7.5 km long beachrock island, the beachrock extending as subaqueous reefs for another 4 km to the south, and links with Chile Head 5 km to the north. The island is 5-11 m in height and 100-200 m in width. Toward the northern end it is capped by dunes and a few patches of high tide sand, remnants of the former Holocene beach that may have paralleled the entire island. The beachrock lies above present sea level and has a lower erosion surface, with a higher inner beachrock bluff, suggesting it may be a higher Pleistocene deposit. It is backed by Gilbut Creek in the south and Lombadina Creek to the north. A 3 km long, 0.5-1. km wide section of mangroves and salt flats links the two creeks, with coastal plain to the south. A vehicle track crosses the salt flats to reach the southern end of the island.

K 84-88 CHILE CREEK-CAPE LEVEQUE

No.	Beach	Rating HT	Rating LT	Type	Length
K84	Chile Ck	1	1	R+sand flats	7.5 km
K85	Lombadina	1	1	R+sand flats	9.2 km
K86	Thomas Bay (N1)	1	2	R+sand/rock flats	2.3 km
K87	Thomas Bay (N2)	1	2	R+sand/rock flats	700 m
K88	Cape Leveque (W)	1	2	R+rock reef	5.8 km
Spring & neap tide range:		8.3 & 1.9 m			

Packer Island is the southern part of a straight 30 km long beachrock reef, possibly the longest straight section of reef on the Australian coast. The reef begins south 4 km southwest of Packer Island, emerges the length of the 7.5 km long island, is submerged for 4.5 km between Lombadina Point and Chile Head, where it emerges again for 2 km before extending subaqueously for another 12 km almost to Cape Leveque. The two major breaks in the reef either side of Chile Head are occupied by semicircular bays and two longer beaches (K 84 and 85). North of beach K 85, rock flats and then deeper reef dominate the coast for 9 km to Cape Leveque, and contain three beaches (K 86-88). This section of coast is accessible from Lombadina, located behind Thomas Bay (beach K 85) and in the north at Cape Leveque, where the main road terminates at the lighthouse.

Beach **K 84** is a curving 7.5 km long northwest-facing beach, located in lee of the beachrock arms of Lombadina Point in the south and Chile Head to the north, the two connected by submerged and partly submerged continuous reef. The 1,500 ha embayment is largely filled with intertidal sand flats, with deeper water only adjacent to the entrance. Three mangrove-lined creeks - Lombadina, an unnamed creek and Chile Creek - and their tidal shoals enter the bay in lee of their respective headlands, with the beach strung between the two creek mouths. The beach consists of a high tide beach fronted by the sand flats that are 2 km wide in the centre, grading to tidal shoals towards the creek mouths. It is backed by active dunes extending up to 1 km inland and older vegetated dunes and a transgressive dune ridge extending 1-2 km inland and reaching heights of 35 m.

Thomas Bay is located in lee of the Chile Head beachrock reef and its subaqueous extension. A small creek flows out in lee of the north side of the head, followed by the 9.2 km long curving beach. It faces north behind the head, curving round to face west and finally south at its northern end where it terminates at a reef-controlled inflection in the shoreline. **Lombadina** community is located 1 km behind the beach (**K 85**), 4 km east of Chile Head, with 4WD access to the beach across the intervening dunes. The beach consists of a relatively steep high tide beach, with 1-2 km wide sand flats filling much of the bay out to the beachrock reefs. A 0.5-1 km wide active dune field backs the beach, with vegetated dunes dominating only the inner 20 m high transgressive ridge.

Beach **K 86** commences on the northern side of the rock reef and trends due north for 2.3 km. The low energy beach is bordered by rock reefs and fronted by shallow rock and sand flats that extend 4.5 km offshore to the northern end of the Packer Island-Chile Head beachrock reef. It is backed by moderate active dune transgression extending up to 300 m inland and to heights of 10 m. The Cape Leveque road is located 1.5 km east of the beach. Beach **K 87** is located immediately to the north and consists of a 700 m long high tide beach fronted by continuous intertidal rock reefs, including four lines of beachrock reef. The beach is backed by active dunes extending up to 600 m inland. A small tidal creek flows out between the two northern beachrock reefs. It drains a backing 200 ha mangrove-filled lagoon.

Cape Leveque beach (K 88) commences on the northern side of the creek mouth beachrock and trends north for 5.8 km to the bright red sandstone of the cape (Fig. 4.17). The beach faces north-northwest and is fronted by a 1 km wide band of submerged beachrock reef in the south and rock reefs to the north. The beach itself is relatively steep and reflective at high tide, and backed by a 1 km wide active dune field, which narrows to the north. The beach terminates at the base of the cape's red layered sandstone and provides some of Australia's most spectacular and colourful coastal scenery. The Cape Leveque landing ground backs the northern end of the beach, while the cape area has a lighthouse, small community, and the Aboriginal-run Kooljamin Resort.

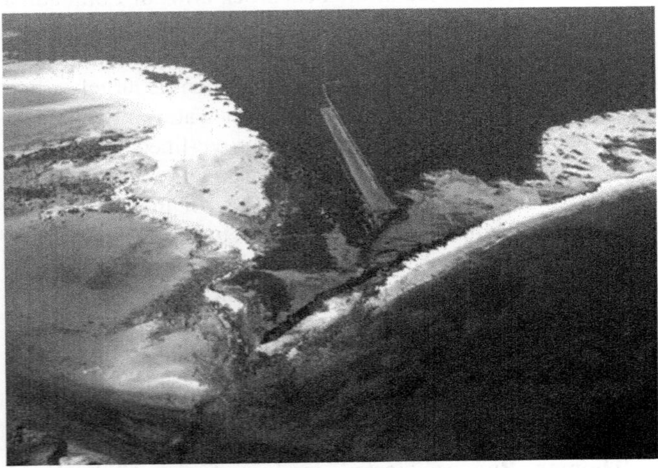

Figure 4.17 Cape Leveque with beach K 88 extending to the right and beaches K 89-92 to the left.

Lombadina Aboriginal Land

Lombadina Aboriginal land encompasses coastal land between Cape Leveque and Cygnet Bay in King Sound. It contains 72 km of coastline including 34 generally low to very low energy beaches. One Arm Point is the main community, with Lombadina community located 3 km outside the southern boundary of the land. In addition there are smaller communities in Catamaran Bay and at Elephant Point.

K 89-93 **CAPE LEVEQUE-SWAN POINT**

No.	Beach	Rating HT	Rating LT	Type	Length
K89	Cape Leveque (E1)	1	2	R+rock flats	100 m
K90	Cape Leveque (E2)	1	2	R+sand/rock flats	700 m
K91	Cape Leveque (E3)	1	2	R+rock flats	600 m
K92	Hunter Ck	1	2	R+sand/rock flats	7 km
K93	Swan Pt (W)	1	2	R+rock flats	4.8 km
Spring & neap tide range:		7.0 & 2.0 m			

The bright red sandstone of the cape is surrounded by a series of white sand beaches with those to the east backed by red desert dunes. The contrasting colours combined with one of the highest tide ranges on the continent in a remote tropical location. with an abundance of usually calm beaches, provide a physically attractive coastal environment which is becoming an increasingly popular tourist destination.

At Cape Leveque the shoreline turns and trends east for 14 km to Swan Point. In between are the three small usually calm cape beaches (K 89-91) and the two longer more exposed beaches either side of Hunter Creek (K 92 and 93).

Beach **K 89** is located on the eastern side of the cape, 200 m south of the tip. It is a 100 m long north-facing strip of high tide sand, wedged in between the 10 m high red bluffs of the cape and rock flats, that at low tide link with Leveque Island, 300 m to the north. Two hundred metres to the south is beach **K 90**, a curving 700 m long sand beach, backed initially by low bluffs that grade into sand dunes. The resort cabins are located on the vegetated slopes behind the northern end of the beach, which also marks the end of the formed road. The beach is fronted by bordering boulders and rock flats, with a central sandy section, which is the most popular swimming area at the cape.

Beach **K 91** is located immediately to the east, and is a curving 600 m high tide sand beach, fronted by continuous rock flats, which extend up to 1.5 km off the beach. The boundary, a rock-controlled sandy point, also marks the boundary of the Lombadina Aboriginal Land.

Beach **K 92** commences on the eastern side of the rock flats and curves round to face predominantly northwest. The beach is relatively steep and backed by active white sand dunes extending up to 600 m inland, where they merge with the red desert sands (Fig. 4.18). It is fronted by a mixture of rock and sand flats, the rocks and reefs extending up to 1.5 km off the beach. It terminates at the 100 m wide mouth of Hunter Creek, with the creek forming a 1 km wide fan of tidal shoals. Beach **K 93** commences on the eastern side of the creek mouth and trends relatively straight to the sandstone rock flats of Swan Point. The beach is backed by active dunes extending up to 80 m inland and reaching heights of 20 m. It is fronted by the tidal shoals and then a continuous sandstone rock flats, bordered by an outer beachrock reef, the lot narrowing from 500 m at the creek mouth to 300 m at the point.

Figure 4.18 Seven kilometre long beach K 92 is a white sand beach backed by the brilliant red Pleistocene sand dunes.

K 94-100 **KARRAKATTA BAY**

No.	Beach	Rating HT	LT	Type	Length
K94	Swan Pt (S)	1	2	R+reef flat	400 m
K95	Nellie Pt (W1)	1	2	R+sand flat	100 m
K96	Nellie Pt (W2)	1	2	R+sand flat	150 m
K97	Karrakatta Bay (1)	1	2	R+sand flat	200 m
K98	Karrakatta Bay (2)	1	2	R+sand flat	100 m
K99	Talboys Pt (N)	1	2	R+sand/reef flat	150 m
K100	Talboys Pt (S)	1	2	R+reef flat	100 m
Spring tide range = 7.0 & 2.0 m					

Karrakatta Bay is a 1 km wide, low energy east-facing bay located in lee of Swan Point. It is bordered by Nellie Point in the north and Talboys Point to the south, and contains 4 km of predominantly mangrove-fringed shoreline. Five small beaches are located in the bay (K 95-99) and one to either side of the points (K 94 and 100). The bay is filled with sand flats and fringed by coral reef flats and faces east toward Sunday and numerous smaller islands. There is no formal vehicle access to and no development in the bay area.

Beach **K 94** is located in lee of Swan Point. It is a 400 m long east-facing beach containing two pockets of high tide sand linked by a strip of intertidal sand, then coral reef flats, which widen to 200 m in the centre. It is backed by low rocky sandstone bluffs.

Nellie Point is located 500 m south of the beach, with beach **K 95** a 100 m long, south-facing pocket beach lying in lee of the low rocky point. It is bordered by low sandstone headland, rock rubble and a clump of mangroves in the east, fronted by 200 m wide sand flats and backed by a 300 m long chute of active dunes. Beach **K 96** is located in the next small embayment 300 m to the west, and is a similar 150 m long pocket beach. It has a 20 m high sandstone headland as its eastern boundary, with a low rocky point, then mangroves to the west and sand flats widening to 400 m off the beach.

Beach **K 97** is located in the westernmost corner of the bay. It is a 200 m long strip of high tide sand fronted by 1.5 km wide sand flats that fill the bay. Mangroves fringe either end of the beach and active dunes that have blown over from beach K 93 back the beach, and probably have supplied sand to the beaches and sand flats of the bay. Beach **K 98** lies 500 m to the southeast in the southernmost section of the bay. It is a 100 m long northeast-facing pocket of sand, bordered by low rocks and mangroves and fronted by the wide bay sand flats.

Beach **K 99** is located just outside the bay on the north side of 10 m high **Talboys Point**. The beach is bordered by low sandstone points and fronted by 100 m wide sand flats, then 200 m wide coral reef flats. Beach **K 100** is located on the south side of the point and consists of a 100 m long pocket of sand bordered by low sandstone bluffs, and fronted by 200 m wide rock and reef flats.

King Sound

King Sound is a large (5,000 km^2) embayment that extends up to 120 km southeast of One Arm Point, the northwestern entrance. The 50 km wide entrance is cluttered with numerous islands, reefs, shoals and deeper tidal channels. The sound averages 50-60 km in width while it is only 3-5 m deep at low tide. It has a 600 km long shoreline, for the most part dominated by wide mangrove-backed tidal mud flats. There is limited access to the coast and a few small communities between One Arm Point and Cunningham Point including land-based pearl farms. Most of the Sound's beaches are located along the 120 km long western shore which is exposed to the prevailing easterly trade winds and assocated waves. Mangroves then dominate for the next 350 km across the southern and along the eastern shore. Derby township and long jetty are located in the southern corner of the Sound, 15 km north of the Fitzroy River mouth.

The sound has the highest tide range in Australia, with a spring range of 10.1 m at Derby, reaching a maximum tide range of 12.1 m. The high tides and congested sound entrance ensure strong daily tidal flows through the entrance channels.

K 101-106 **ONE ARM POINT-MISSION BAY**

No.	Beach	Rating HT	LT	Type	Length
K101	Middle Beach	1	3	R+rock flats	200 m
K102	Jologo Beach	1	1	R+sand flats	3 km
K103	Lugger Cove	1	1	R+sand flats	1.1 km
K104	Riddell Pt (E)	1	1	R+sand flats	300 m
K105	Mission Bay (E)	1	1	R+ridged sand flats	200 m
K106	Mission Bay (W)	1	1	R+ridged sand flats	600 m
Spring tide range = 7.0 & 2.0 m					

One Arm Point is a major Aboriginal community located at the northwestern tip of King Sound. It is the only major settlement on the western side of the sound, and the second largest in the sound after Derby. The community was established in 1972 and is home to the Bardi people, the traditional landowners. The Bardi also live at Lombadina, Broome and Derby. The community runs a super market and service station and also has a small trochus shell industry and craft store.

The community is bordered by two beaches (Fig. 4.19). The northern **Middle** beach (**K 101**) is a steep high tide sand beach, consisting of a 200 m long strip of sand, fronted by continuous 200 m wide jagged rock flats topped with stunted mangroves. The rock flats narrow to 50 m in the centre where they are fronted by 150 m wide sand flats. The beach is backed by the landing ground, while the main boat ramp is located just past the northern end of the beach.

Figure 4.19 One Arm Point community and Middle and Jologo beaches (K 101 & 102).

The main **Jologo** beach (**K 102**) commences at the southern end of the landing ground where there is vehicle access to the 3 km long southeast-facing sandy beach, which is exposed to southerly waves generated across the sound. The moderately steep high tide beach is fronted by 300-500 m wide sand flats, together with patchy rock flats and coral reef beyond the sand flats. The rocks divide the beach into three separate high tide beaches. Minor dune activity backs most of the beach increasing to bare 30 m high dunes in the south, all backed by vegetated precipitation slopes. The beach terminates in the south at a 28 m high vegetated headland called Shenton Bluff.

Lugger Cove (**K 103**) is located in lee of Shenton Bluff, which partly protects it from the dominant southeasterly winds. The 1.1 km long beach faces south toward the bluff. It is bordered by the rocks of the bluff and a small western headland, with mangroves fringing both headlands. Boulders and beachrock are exposed along the intertidal beach, fringed by sand flats extending 500 m into the cove. It is backed with an active 1-3 m foredune with access tracks to both ends of the beach. The cove is still used as an anchorage for the community, with boats anchored beyond the flats in deeper water. The community also operates a pearl farm from the cove, with the farm buildings located at the eastern end of the cove, and floating pearl paddocks spread for a few kilometres south of the bluff.

Beach **K 104** is located 500 m to the west and consists of a 300 m strip of high tide sand fronted by 300 m wide sand flats, then fringing coral reefs extending another 500 m to the south and some beachrock exposed along the beach. An active 2-5 m foredune backs the beach grading into a couple of blowouts at the western end. Mangroves border the eastern end, with 20 m high Riddell Point forming the western boundary.

Mission Bay is located on the western side of 400 m wide Riddell Point. The low energy bay contains two pocket beaches, both bordered and separated by mangroves. Beach **K 105** is a 200 m long strip of high tide sand fronted by 500 m wide sand flats, while beach **K 106** is 600 m long with similar sand flats, then fringing coral reefs off both sets of flats. Both beaches are backed by dune transgression generated by the southeast trades. The dunes extend 100 m inland of beach K 105, and up to 400 m in lee of beach K 106. An access road and a few houses are located at the western end of the bay.

K 107-109 CATAMARAN BAY

No.	Beach	Rating HT	Rating LT	Type	Length
K107	Catamaran Bay (1)	1	1	R+ridged sand flats	400 m
K108	Catamaran Bay (2)	1	1	R+ridged sand flats	200 m
K109	Catamaran Bay (3)	1	1	R+ridged sand flats	300 m
Spring tide range = 7.0 & 2.0 m					

Catamaran Bay is a low energy 3 km wide, east-facing embayment located between Riddell Point-Mission Bay and Bird Rocks. The bay has 5 km of predominantly mangrove-fringed shore, with three small low energy beaches located in amongst the more exposed northern mangroves. Beach **K 107** is a 400 m long strip of high tide sand, bordered by a mangrove-lined rocky point in the east and mangroves to the west. Ridged sand flats extend 300 m off the beach. Beaches **K 108** and **K 109** are adjoining 200 and 300 m long beaches, both fringed by mangroves and fronted by 600 m wide ridged sand flats.

K 110-115 BIRD ROCKS

No.	Beach	Rating HT	Rating LT	Type	Length
K110	Bird Rocks	1	1	R+sand flats	700 m
K111	Bird Rocks (S1)	1	1	R+sand flats	700 m
K112	Bird Rocks (S2)	1	1	R+sand flats	200 m
K 113	Bird Rocks (S3)	1	1	R+ridged sand flats	100 m
K114	Bird Rocks (S4)	1	1	R+ridged sand flats	100 m
K115	Bird Rocks (S5)	1	1	R+ridged sand flats	700 m
Spring tide range = 7.0 & 2.0 m					

Bird Rocks lies 500 m south of the southern entrance to Catamaran Bay and 500 m off the shore. Between the rocks and a 1.5 km wide mangrove-filled bay fed by Chunelarr Creek are 3.5 km of southeast-facing shore, occupied by six low energy beaches (K 110-115) and sand flats. There is no development or vehicle access to the shore.

Beach **K 110** lies in lee of the rocks and consists of a 700 m long east-facing beach, fronted by sand flats that widen to 300 m toward the rocks. The protection afforded by the rocks, fringing reefs and sand flats permit mangroves to fringe the southern end of the beach. Beach **K 111** commences on the southern side of the 300 m long strip of mangroves. It is a straight 700 m long east-southeast-facing beach, fronted by 200 m wide sand flats then fringing reef, all backed by a deflated foredune section, then spinifex-covered slopes rising to 6 m with some minor dune activity.

Beaches **K 112, 113** and **114** are three adjoining pocket beaches, 200 m, 100 m and 100 m long respectively, each separated by small sandstone headlands. They are each fronted by continuous 200 m wide sand flats, widening to 300 m in the south, with fringing coral reefs off the flats.

A low rocky, mangrove-backed point separates the beaches from beach **K 115**. It is a 700 m long southeast-facing strip of high tide sand fronted by 500 m wide ridged sand flats. It has a small rock outcrop in the centre and mangroves to either end.

K 116-119 **SKELETON POINT**

No.	Beach	Rating HT	Rating LT	Type	Length
K116	Skeleton Pt (N3)	1	1	R+sand flats	700 m
K117	Skeleton Pt (N2)	1	1	R+sand/rock flats	900 m
K118	Skeleton Pt (N\1)	1	1	R+sand/rock flats	200 m
K119	Skeleton Pt	1	1	R+sand/rock flats	150 m
Spring tide range = 7.0 & 2.0 m					

Skeleton Point is a 3 km long east-trending 20 m high sandstone headland that separates Chunelarr Creek embayment from the larger Cygnet Bay. Beaches K 116-118 are located along the northern side of the point, with beach K 119 out on the tip of the point. There is 4WD access to the point from Lombadina landing ground and along Chunelarr Creek, with a solitary house at beach K 117.

Beach **K 116** is a curving, bedrock-controlled 700 m long north-facing high tide beach, that faces out across the ridge sand flats of Chunelarr Creek, with low rocks and mangroves fringing each end. Beach **K 117** lies in the next small embayment and is a 900 m long, curving, crenulate generally north-facing high tide beach, fronted by 200 m wide sand flats, together with rock flats and scattered mangroves (Fig. 4.20). The solitary house is located behind the centre of the beach.

Figure 4.20 Beach K 117 is typical of the low energy Skeleton Point beaches and sand flats.

Beach **K 118** is located on the northern tip of the point and is a 200 m long strip of high tide sand, bordered by mangroves, together with a central clump and fronted by a 300 m wide mixture of sand and rock flats. Beach **K 119** lies on the eastern tip of the point and faces east across King Sound. It is a 150 m long strip of high tide sand, backed by the sloping 20 m high wooded point and fronted by 200 m wide rock flats, with a veneer of sand.

K 120-127 **CYGNET BAY (W)**

No.	Beach	Rating HT	Rating LT	Type	Length
K120	Skeleton Pt (S)	1	1	R+ridged sand flats	400 m
K121	Elephant Pt (N)	1	1	R+ridged sand flats	2.3 km
K122	Elephant Pt (1)	1	1	R+sand flats	150 m
K123	Elephant Pt (2)	1	1	R+sand/rock flats	400 m
K124	Milligan Ck	1	1	R+tidal sand flats	1.2 km
K125	Cygnet Bay (S1)	1	1	R+tidal sand flats+rocks	100 m
K126	Cygnet Bay (S2)	1	1	R+tidal sand flats	750 m
K127	Rumbul Bay (E)	1	1	R+tidal sand flats	150 m
Spring tide range = 7.0 & 2.0 m					

Cygnet Bay is a 9 km wide east-facing bay bordered by Skeleton Point in the north and Willie Point to the south. The bay has 28 km of generally low energy shoreline, with eight low energy bay beaches occupying a total of 5.4 km of shoreline, most of the remainder occupied by mangrove-fringed sand and rock flats. There is 4WD access to most of the beaches, and a small community at Elephant Point.

Beach **K 120** is located on the southern side of the bay, 1 km west of Skeleton Point. It is a narrow 400 m long southeast-facing sand beach, backed by vegetated slopes rising to 20 m, and fronted by 300 m wide ridged sand flats, with some rock flats off either end of the beach.

Beach **K 121** lies immediately to the south and is a straight 2.3 km long east-southeast-facing sand beach fronted by 500 m wide ridged sand flats, widening to 700 m in the south at the mouth of a small creek. The beach fronts a low 300-400 m wide dune system, containing some active foredune ridges. There is 4WD access to the northern end of the beach, and the mangroves of the small creek off the southern end.

On the southern side of the creek mouth the shoreline trends southeast for 500 m to Elephant Point, a 20 m high semicircular wooden headland. Beach **K 122** is located on the northern side of the point immediately east of the creek mouth, and consists of a 150 m long strip of high tide sand, bordered by mangroves, and fronted by 400 m wide sand flats. Beach **K 123** curves around the eastern side of the point for 400 m. It is fronted by sand flats widening to 1 km in lee of offshore rocks and mangroves, with scattered rocks and mangroves along the beach. There is a solitary house on the point linked by a 4WD track to the Cape Leveque road.

Beach **K 124** is located 1.5 km west of the point. It is a straight east-southeast-facing 1.2 km long beach fronted by 1 km wide sand flats containing a few scattered rocks. There is a 4WD track to the beach and a house toward the northern end. The southern end of the beach merges into the extensive mangroves of Milligan Creek mouth.

Three kilometres to the south is beach **K 125**, a 100 m long strip of sand in amongst a mangrove-dominated shoreline. The beach is fronted by a 500 m wide mixture of sand and rock flats. One kilometre to the south is a U-shaped 1.5 km wide northeast-facing embayment containing 750 m long beach **K 126**. This is a low energy beach bordered by mangroves, fronted by 500 m wide sand flats, with coral reefs off both headlands, and some minor foredune activity to the lee of the beach.

Rumbul Bay is a 3 km wide mangrove-lined bay, containing one small gap in the mangroves on its northeastern point which contains beach **K 127**. The beach consists of a 150 m long strip of high tide sand, bordered by mangroves and fronted by 200 m wide sand to mud flats.

K 128-131 WILLIE POINT

No.	Beach	Rating HT	Rating LT	Type	Length
K128	Willie Pt (W)	1	1	R+sand flats	2 km
K129	Willie Pt (S1)	1	1	R+sand/rock flats	700 m
K130	Willie Pt (S2)	1	1	R+tidal sand flats	1.7 km
K131	Willie Pt (S3)	1	1	R+tidal sand flats	500 m

Spring tide range = 7.0 & 2.0 m

Willie Point is a 28 m high vegetated sandstone point located midway between Skeleton and Cunningham points. Beach K 128 lies immediately west of the point, with beaches K 129-131 strung out along the coast south of the point. There is no formal vehicle access to the point or beaches.

Beach **K 128** is a curving 2 km long northeast-facing high tide beach, bordered by reef-fringed rock flats extending up to 1 km off either headland, and fronted by 500 m wide sand flats and patches of rock. The beach face is dominated by intertidal sloping beachrock, which has been eroded to form a boulder berm at the top of the beach and a boulder rampart along the base of the beach (Fig. 4.21). It is backed by minor dune activity increasing to the west where there is a deflation basin backed by active dunes rising to 15 m. These in turn are backed by two linear vegetated episodes of dune activity reaching 18 m high.

Figure 4.21 Willie Point beach (K 128) has sand flats backed by a high tide beach composed of in situ beachrock with beachrock boulders along its base.

Beach **K 129** commences 1 km due south of Willie Point. It is a 700 m long east-facing beach, bordered by mangrove-fringed rock flats and fronted by 400 m wide tidal sand and rock flats, then another 200 m wide fringing reef. A sparsely vegetated 2-3 m high active dune and older vegetated foredune back the beach. Beach **K 130** is located 500 m to the south and is a curving 1.7 km long east-facing high tide beach, fronted by 1 km wide tidal sand flats, with 1-1.5 km wide rock flats extending off both mangrove-lined points, both of which are fringed by 200-300 m wide reefs. The mangroves also extend for a few hundred metres along the base of the southern half of the beach. Beach **K 131** lies 2 km to the south and is a very low energy 500 m long strip of high tide sand, fronted by 800 m wide sand to outer mud flats. Mangroves fringe each end of the beach, with a few growing on the beach. It is backed by active dunes extending up to 250 m inland and to heights of 18 m.

K 132-138 DEEP WATER POINT

No.	Beach	Rating HT	Rating LT	Type	Length
K132	Deep Water Pt (S 1)	1	2	R+coral reef flats	150 m
K133	Deep Water Pt (S 2)	1	2	R+sand/coral flats	350 m
K134	Deep Water Pt (S 3)	1	2	R+coral reef flats	200 m
K135	Deep Water Pt (S 4)	1	2	R+coral reef flats	250 m
K136	Deep Water Pt (S 5)	1	2	R+coral reef flats	50 m
K137	Deep Water Pt (S 6)	1	2	R+coral reef flats	80 m
K138	Deep Water Pt (S 7)	1	2	R+coral reef flats	150 m

Spring tide range = 7.0 & 2.0 m

Deep Water Point is a structurally controlled 10-20 m high, horizontally bedded, sandstone point that trends due north for 4 km into Cygnet Bay. It is 500 m wide at its base narrowing to 200 m at its tip. As the name suggests the point provides access to the deep water of the sound. The point is used as a shore base for pearl farms located in the bays to either side of the point. The farm has a cluster of about 10 buildings behind the two northernmost beaches, with boats anchored on the protected western side of the point. A vehicle track winds along the spine of the point providing access to the community and the four northernmost beaches (K 132-135). A landing ground is located just south of the base of the point.

There is a total of seven beaches along the more exposed eastern side of the point (Fig. 4.22). All the beaches face east into the trade winds and a small foredune backs each. The northern four beaches (**K 132, 133, 134** and **135**) are 150 m, 350 m, 200 m and 250 m long respectively and are bordered by cliffed sandstone headlands and surrounding boulder fields, with a continuous 200 m wide fringing reef growing between the headlands. Because of the rocks and reef the beaches tend to narrow towards low tide. The beaches are relatively steep, with beachrock boulders scattered along the lower beach face, and bedrock outcrops on beaches K 133 and 134. They are all backed by spinifex-covered foredunes rising up to 5 m. Beaches **K 136, 137** and **138** are 50 m, 80 m and 150 m in length respectively and are fronted by tidal flats, that widen to 1.5 km east of the southernmost beach. Beachrock boulders are scattered along each of the beaches, while they are each backed by a 1-2 m high spinifex foredune.

Figure 4.22 Four kilometre long Deep Water Point and beaches K132-138.

K 139-144 CUNNINGHAM-AMATANGOORA PTS

No.	Beach	Rating HT	Rating LT	Type	Length
K139	Cunningham Pt	1	2	R+rock/reef flats	100 m
K140	Cunningham Pt (S)	1	2	R+rock/sand flats	600 m
K141	Carlisle Head (N)	1	2	R+sand/reef flats	300 m
K142	Carlisle Head (S)	1	2	R+rock/reef flats	1.4 km
K143	Amatangoora Pt (N2)	1	2	R+rock/reef flats	300 m
K144	Amatangoora Pt (N1)	1	2	R+rock/reef flats	300 m

Spring tide range = 7.0 & 2.0 m

Cunningham Point is a prominent north-trending headland which, combined with Carlisle Head and Amatangoora Point, forms a 5.5 km long east-facing series of 10-50 m high red bluffs, which also separate Cygnet and Goodenough bays. There is a vehicle track to the tip of Cunningham Point and two houses located on the mangrove-fringed western shore, overlooking beach K 142. Six beaches (K 139-144) are located along the east-facing shores of the promontory, most located at the base of the red bluffs and fronted by rock flats then fringing reefs. In 1885 the point was the site of the first Catholic mission in the Kimberley region. It was abandoned two years later when the mission was burnt down.

Beach **K 139** is located at the northern tip of Cunningham Point and is a 100 m long strip of sand fronted by 100 m wide intertidal beachrock and boulder rubble, then reef flats. The beach continues west as a 1 km long recurved beachrock-cemented spit located in lee of mangroves and rock flats. There is 4WD access to the beach and along the crest of the spit.

Beach **K 140** commences on the east side of Cunningham Point and trends south for 600 m. It is a narrow strip of high tide sand, located at the base of eroding red and white 10-15 m high bluffs, and fronted by a narrow band of rock then 200-300 m wide sand flats. Beach **K 141** lies 1 km further south, on the northern side of 48 m high wooded **Carlisle Head**. The beach is a narrow 300 m long strip of sand backed by eroding 10 m high red bluffs, and fronted by 200 m wide sand then reef flats.

Beach **K 142** commences 700 m south of Carlisle Head. It is a straight 1.4 km long high tide sand beach, fronted by a 100 m wide strip of sand, then 500 m wide reef flats (Fig. 4.23). It is backed by red bluffs rising to 20 m at either end, with a solitary house located on the northern bluffs and a minor area of dune activity in a central low point in the bluffs.

Beaches **K 143** and **144** are two adjoining 300 m long beaches located immediately north of **Amatangoora Point**. They are both narrow strips of high tide sand backed by red and white 10 m high bluffs, and fronted by 500 m wide reef flats.

Figure 4.23 Sandy beach K 142 lies between the backing bluffs and low tide reef flats.

K 145-149 GOODENOUGH-DISASTER BAYS

No.	Beach	Rating HT	Rating LT	Type	Length
K145	Goodenough Bay (W)	1	1	R+sand-mud flats	100 m
K146	Murdeh Pt	1	1	R+sand flats	2 km
K147	Foul Pt	1	1	R+sand flats	900 m
K148	Cornambie Pt (N)	1	1	R+sand-mud flats	800 m
K149	Disaster Bay	1	1	R+sand-mud flats	6 km
Spring tide range = 7.0 & 2.0 m					

Goodenough Bay is an open 5 km wide east-facing bay bordered by Amatangoora and Foul points. The intervening 16 km of shoreline are entirely mangroves apart from two low energy beaches (K 145 and 146). Beach **K 145** is located on the western shore of the bay, 2.5 km southwest of Amatangoora Point. It consists of a 100 m long gap in the mangroves which is fronted by 1 km wide sand to mud flats, as well as the tidal channels from an adjacent creek mouth. **Murdeh Point** lies 5 km to the south and is the terminus and outermost of a 4 km long series of seven recurved spits emanating from Foul Point. Beach **K 146** occupies the first 2 km of the spit series, the remaining fringed by mangroves. The beach faces northeast and consists of a low narrow high tide beach, with some scattered mangroves, fronted by 1 km wide sand to mud flats.

Foul Point is a 22 m high red laterised sandstone point. The shore trends south from the point, with beach **K 147** located 1 km to the south. The beach is a 900 m long sand beach, backed by bluffs in the north, then a low foredune, terminating at the mouth of a small tidal creek which is connected to a 10 ha area of salt flats. The beach is fronted by a narrow band of rock flats then a 500 m wide veneer of sand flats and finally reef flats extending 1.5 km offshore. A vehicle track leads to the southern headland where a solitary house is located.

Beach **K 148** is located 2.5 km to the south in lee of 14 m high **Cornambie Point**. The 800 m long beach faces east and is bordered by a wooded 14 m high sandstone headland to the north with mangroves and then Cornambie Point to the south. The beach is fronted by 800 m wide sand to mud flats. It has a small tidal creek at its northern end which drains a 20 ha area of salt flats.

Disaster Bay is located on the southern side of Cornambie Point. The open, east-facing 7 km wide bay contains a slightly curving 6 km long low energy beach (**K 149**), fringed by mangroves at either end. A small creek crosses the northern section of the beach, with a central low red bluff section and a second creek forming the southern boundary. Sand to mud and scattered rock flats extend from 1 km in the north to 3 km off the beach in the south. The Disaster Bay community of about six houses is located in the higher central bluff section, and connected by 4WD track to the Lombadina road.

K 150-152 MALABURRA-HOON CREEK

No.	Beach	Rating HT	Rating LT	Type	Length
K150	Malaburra Spring (N2)	1	2	R+sand-mud flats	2.8 km
K151	Malaburra Spring (N1)	1	2	R+sand-mud flats	3 km
K152	Hoon Ck (S)	1	2	R+sand-mud flats	2.5 km
Spring tide range = 7.0 & 2.0 m					

The three most southern beaches in King Sound are located either side of Hoon Creek, at the beginning of the 30 km wide funnel-shaped entrance to the Fitzroy River mouth. Mangroves dominate the shoreline to either side of the beaches, and south of beach K 152 mangroves dominate the coast for 50 km all the way to the Fitzroy River. The mangroves continue up the eastern shore of the sound for 20 km to Derby and occupy the next 220 km of irregular tidal shoreline up to near Point Usborne, 70 km north of Derby, in all 290 km of mangrove-lined shoreline.

Beach **K 150** is a 2.8 km long relatively straight northeast-facing strip of high tide sand, bordered by two small protruding tidal creeks, with bluffs backing the centre of the beach. It is fronted by 2.5 km wide sand to mud flats. Beach **K 151** commences on the southern side of the southern creek mouth and trends south for 3 km to a 1.5 km long recurved spit opposite Malaburra Spring. The beach is backed for the most part by 10 m high bluffs, with sand spits to either end. Sand to mud flats extend 1-2 km off the beach.

Beach **K 152** is located 7 km southeast of Hoon Creek mouth. It is a 2.5 km long northeast-facing beach backed by a low foredune, with a small recurved spit at the southern end. Three to four hundred metre wide sand to mud flats parallel the beach.

Eastern King Sound

The eastern shore of King Sound is dominated by 260 km of extensive intertidal mud flats, fringed by wide belts of mangroves, and backed at some distance by bedrock. Even at Derby, located on a 15-20 m bedrock high, a 2 km long jetty is required to reach across the tidal flats to the deep waters of King Sound.

Yampi Training Area

	Area:	>500 000 ha		
	Coast length (km)			Beaches
	165	(6660-6825 km)	35	(K 153-187)
	35	(7215-7250 km)	4	(K 284-287)
Total	200		39	

The Yampi Training Area is restricted Defence land that extends from the sheltered eastern shore of King Sound across the Kimbolton, Wyndhm and McLarty ranges to the southern shores of Collier Bay.

K 153-156 **POINT USBORNE**

No.	Beach	Rating HT	Rating LT	Type	Length
K 153	Pt Usborne (W 3)	1	2	R+rock/reef flats	70 m
K 154	Pt Usborne (W 2)	1	2	R+rock/reef flats	50 m
K 155	Pt Usborne (W 1)	1	2	R+rock/reef flats	60 m
K 156	Pt Usborne (N)	1	2	R+rock/reef flats	200 m
Spring & neap tide range: 6.3 & 1.4 m					

The first beaches on the eastern shore are located at the first outcrop of shoreline bedrock, at **Point Usborne**, 50 km north of Derby. The point is composed of Proterozoic sandstone which is arranged in east-southeast-west-northwest-trending ridges, averaging 100-200 m in height, separated by linear drainage systems. In the point area the lightly wooded ridges rise to a maximum height of 54 m. Most of the coast consists of irregular, joint-controlled bedrock shoreline. The four point beaches (K 153-156) are each located in small creek-carved bedrock-controlled embayments, with usually a few mangroves behind the beach, and reef flats exposed at low tide. The beaches are only accessible by boat at high tide.

Beach **K 153** is located 2 km east of the point and consists of two 30 m and 40 m long west-facing pockets of high tide sand. The sand widens at low tide and is backed by a 1 m high grassy foredune. Directly opposite, 300 m to the west is beach **K 154**, a narrow 50 m long strip of sand at the mouth of a small valley. It is backed by a 1 m high grassy foredune and fronted by 200 m wide reef flats. Beach **K 155** lies 1 km east of the point and is a 60 m pocket of sand filling a small valley, which widens to 80m at low tide. Mangrove-fringed rock flats lie to either side and a patch of reef flats is located off the centre of the beach. Finally beach **K 156** is located on the northern side of the point. It is a 200 m long west-facing high tide sand beach, with low tide sand flats grading into 300-400 m wide reef flats. It is backed in the north by a grassy 1 m high foredune rising to 2 m in the south where there is some instability. Five hundred metre long **Point Usborne** forms the southern boundary, and a 100 m long headland, the northern boundary.

K 157-164 **LACHLAN ISLAND**

No.	Beach	Rating HT	Rating LT	Type	Length
K157	Lachlan Is (S1)	1	2	R+reef flats	200 m
K158	Lachlan Is (S2)	1	2	R+reef flats	300 m
K159	Lachlan Is (S3)	1	2	R+sand/reef flats	250 m
K160	Lachlan Is (S4)	1	2	R+reef flats	200 m
K161	Sanderson Pt (E1)	1	2	R+sand/reef flats	50 m
K162	Sanderson Pt (E2)	1	2	R+sand/reef flats	50 m
K163	Sanderson Pt (E3)	1	2	R+sand/reef flats	200 m
K164	Sanderson Pt (E4)	1	2	R+sand/reef flats	100 m
Spring & neap tide range: 6.3 & 1.4 m					

Lachlan Island is a 7 km long bedrock island separated from the mainland at Dimond Head-Point Usborne by a 200 m wide channel. It has a maximum height of 93 m and a shoreline dominated by steep and/or mangrove-fringed bedrock coast. Only along the more exposed southern and western shore are there eight small high tide pockets of sand (beaches K 157-164), all fronted by usually narrow sand flats and wider coral reef flats. All are backed by usually steep vegetated sandstone slopes, cut by small usually dry creeks in the apex of the valleys. There is no habitation on the island and the beaches are only accessible by boat at high tide. The southern side of the bedrock channel between Dimond Head and the island has numerous boab trees dotting the 80 m high north-facing slopes.

Beaches K 157-160 are located along the southern-southwest side of the island, with beach K 157 located 2.5 km due north of Point Usborne. Beach **K 157** is a relatively straight 200 m long strip of sand, backed by steep wooded sandstone slopes, rising to 90 m 1 km inland. It is fronted by 200 m wide reef flats and bordered by bedrock shoreline. Beach **K 158** is located 1 km to the west, and is a 300 m long straight relatively steep high tide beach backed by a low 50 m wide grassy foredune. It has a sloping beachrock and beachrock boulders littering the western half of the beach. At the base it has a fringe of intertidal sand grading into reef flats which widen from 100 m to 300 m in the east. It is bordered by low rocky headlands, the eastern composed of deeply weathered boulder conglomerate, fringed by a narrow, horizontal high tide rock platform.

Beaches K 159 and 160 are located 1.5 and 1 km east of Sanderson Point. Beach **K 159** lies inside a 200 m wide funnel shaped embayment. The irregular 250 m long beach is bordered by mangroves in the east and fronted by 200 m wide sand flats out to the bay mouth, where they grade into reef flats which extend another 100 m outside the bay mouth. The next embayment contains beach **K 160**, a relatively straight 200 m long high tide beach fronted by 100 m wide rock and reef flats.

Beaches K 161-164 are located on the north side of the island between 0.5 and 2 km east of Sanderson Point. They are contained in one 1.8 km wide open north-facing embayment, with a continuous fringing reef linking either end, and each of the beaches occupying one of the four small valleys that feed into the embayment. Beach **K 161** is a 50 m pocket of sand, backed by a small patch of foredune and fronted by 100 m wide reef flats. Beach **K 162** is a similar 50 m pocket of high tide sand fronted by 250 m wide reef flats. Beach **K 163** is the longest at 200 m, with the high tide beach and a veneer of sand spreading out over the reef flats. Beachrock is exposed along the lower beach face and has been eroded to form a boulder berm at the base of the bordering rocks (Fig. 4.24). A 20 m wide low foredune backs the centre of the beach. Beach **K 164** lies at the eastern end of the embayment. It is a 100 m long high tide sand and beachrock beach with the reef flats narrowing to 50 m off its eastern end.

Figure 4.24 Beach K 164 has a veneer of sand over a beachrock base (left), some of which has been eroded to form a storm boulder beach (right).

K 165-168 **CASCADE BAY (NE)**

No.	Beach	Rating HT	Rating LT	Type	Length
K165	Cascade Bay (E1)	1	2	R+sand/reef flats	250 m
K166	Cascade Bay (E2)	1	2	R+sand/reef flats	200 m
K167	Cascade Bay (E3)	1	2	R+sand/reef flats	100 m
K168	Cascade Bay (E4)	1	2	R+reef flats	300 m
Spring & neap tide range:				6.3 & 1.4 m	

Cascade Bay is a an oval-shaped bay up to 17 km (northwest-southeast) in length and 9 km wide (NE-SW). It is bounded by the mainland in the east and north and Pecked, Pasco, Long and Lachlan islands to the north and west, together with a scattering of smaller islands both in the bay and on its perimeter. Beaches in the bay are located on the northern shore of Lachlan Island (K 161-164), on the eastern mainland coast (K 165-168) and along the southern shore of Pecked Island (K 169-174). All the beaches are bedrock-controlled and occupy small indentations in the rocky shore. They are all backed by partly vegetated often steep sandstone slopes, and fronted by a mixture of sand and then reef flats.

Beach **K 165** is the easternmost beach in the bay and is located on the mainland 5 km north of Lachlan Island. It is a 250 m long, west-facing high tide sand beach, bordered by sandstone headlands, and backed by densely vegetated slopes rising to 60 m, while it is fronted by 150 m wide sand to reef flats. A 500 m wide tidal creek is located on the northern side of the northern headland. Beach **K 166** lies 2 km to the north in a south-facing 500 m wide embayment partly occupied by a mangrove-lined tidal creek. The irregular 200 m long beach extends east of the creek mouth and is fronted by 200 m wide, ridged tidal sand flats grading to coral reef flats, which widen and extend up to 500 m off the beach. It is backed by steep slopes and cliffs rising 140 m.

Beach **K 167** lies 2.5 km to the northeast. It is located in a southwest-facing 300 m wide gap in the rocky coast and backed by 20-30 m high cliffs. The 100 m long beach is tucked in the eastern corner of the gap and faces out across a veneer of sand covering 150 m wide reef flats. The next embayment, 300 m to the west, contains beach **K 168**. This is a steep, slightly curving 300 m long beach, which extends at high tide to the backing vegetated sandstone slopes, which rise to 30 m. A few boulders lie on the beach which is fronted by 200 m wide reef flats.

K 169-174 **PECKED ISLAND**

No.	Beach	Rating HT	Rating LT	Type	Length
K169	Pecked Is (S1)	1	2	R+sand/reef flats	100 m
K170	Pecked Is (S2)	1	2	R+reef flats	500 m
K171	Pecked Is (S3)	1	2	R+sand/reef flats	50 m
K172	Pecked Is (S4)	1	2	R+sand/reef flats	50 m
K173	Pecked Is (S5)	1	2	R+sand/reef flats	150 m
K174	Pecked Is (S6)	1	2	R+sand/reef flats	100 m
Spring & neap tide range:				6.3 & 1.4 m	

Pecked Island is a continuation of the sandstone island ridge that backs the bay beaches to the east. It is separated from the mainland by a 500 m wide bedrock channel, while it extends to the west for 6 km. It averages a few hundred metres in width and has a crest of 60 m. All six beaches (K 169-174) are located along the more exposed southern side of the island which faces south across Cascade Bay.

Beach **K 169** is located in a 200 m wide gap between sandstone headlands that extend up to 400 m east of the beach. The beach is a 100 m long strip of east-facing sand, fronted by 200 m wide sand flats, then another 200 m of reef flats. It is bordered by the 40-60 m high ridges, and backed by a grassy foredune, which connects to the side of the neighbouring beach. Beach **K 170** is located immediately to the west in a south-facing gap in the same bedrock ridges. It is a curving 500 m long, south-facing beach fronted by 200 m wide reef flats which link the bordering 40 m high headlands. The foredune from beach K 169 merges with the eastern end of the beach.

Beaches **K 171** and **172** lie 800 m and 1 km further west, and are two 50 m long pockets of sand, wedged in narrow gaps in the 20 m high sandstone. Both are fronted by 50 m wide sand flats edged by reef flats.

Beaches K 173 and 174 are located on an attached islet at the western tip of the island. Beach **K 173** faces west, and is a 150 m long curving strip of sand, bordered by low sandstone ridges, and fronted by 100 m wide sand flats, grading into 300 m wide reef flats. Beach **K 174** is located immediately to the north, and is a 100 m long high tide beach, bordered by rocks to the south and mangroves to the north which partly enclose the beach, with reef flats extending another 300 m beyond the mangroves.

K 175-178 **FAINT POINT**

No.	Beach	Rating HT	Rating LT	Type	Length
K175	Faint Pt (W4)	1	1	R+sand/reef flats	200 m
K176	Faint Pt (W3)	1	1	R+sand/reef flats	100 m
K177	Faint Pt (W2)	1	2	R+sand/reef flats	100 m
K178	Faint Pt (W1)	1	2	R+sand/reef flats	80 m
Spring & neap tide range:				6.3 & 1.4 m	

Faint Point is a 1.5 km ridge of 20 m high sandstone that narrows from 200 m to 50 m in the east. Four small beaches (K 175-178) are located at its western boundary and on its northern shore (Fig. 4.25). Beaches **K 175** and **176** are adjoining 200 m and 100 m long beaches, separated on the high tide beach by a small sandstone outcrop, while bounded by sandstone headlands extending 200 m to the west. The 50 m wide beaches have a sloping beachrock base and are fronted by a veneer of sand over 300 m wide reef flats, while 30-50 m wide grassy foredunes have blown up over the backing rocks.

Beach **K 177** is located 200 m to the east on the northern side of the point. It is a 100 m long north-facing beach, wedged in a narrow gap in the 20 m high sandstone ridge, and fronted by a 200 m wide mixture of sand, rock and reef flats. Beach **K 178** lies 400 m further east, and is a similar 80 m long pocket of sand, backed by a small foredune.

Figure 4.25 Faint Point (right) and the four pocket beaches K 175-178 and fringing reef.

K 179 **CRAWFORD BAY (E)**

No.	Beach	Rating HT	Rating LT	Type	Length
K179	Crawford Bay (E)	1	2	R+reef flats	300 m
Spring & neap tide range:		6.3 & 1.4 m			

Crawford Bay is a 2-3 km wide, 7 km deep U-shaped west-facing bedrock embayment, with beach **K 179** located at its far eastern end at the base of 100 m high cliffs. A 100 m high 1 km long headland forms its northern boundary, with sediment-covered reef flats extending 600 m off the beach, and mangroves fringing either end.

K 180-187 **CONE BAY (S)**

No.	Beach	Rating HT	Rating LT	Type	Length
K180	Cone Bay (S1)	1	1	R+sand flats/reef	200 m
K181	Cone Bay (S2)	1	1	R+sand flats/reef	300 m
K182	Cone Bay (S3)	1	1	R+sand flats/reef	50 m
K183	Cone Bay (S4)	1	1	R+sand flats/reef	100 m
K184	Cone Bay (S5)	1	1	R+sand flats/reef	150 m
K185	Cone Bay (S6)	1	1	R+sand flats/reef	100 m
K186	Cone Bay (S7)	1	1	R+sand flats/reef	400 m
K187	Cone Bay (S8)	1	1	R+tidal flats/reef	150 m
Spring & neap tide range:				6.3 & 1.4 m	

Cone Bay is 3 km wide at its mouth and extends to the southeast for 10 km, splitting into two terminal bays. Small pocket beaches are located along its southern shore (K 180-190), its southeastern extremity (K 191-199) and along its northern shore out to Sir Richard Island (K 200-209). There are two small settlements in the bay, one on an island 1 km north of Porter Hill, and the second, at the far eastern extremity below Cone Hill, which has the only vehicle access to the bay shore.

The five westernmost beaches (K 180-184) occupy a 1.5 km long section of northwest-facing shore at the southwestern entrance to the bay, facing out of the bay toward Barnicoat Island. Beach **K 180** is a 200 m long high tide beach, backed by a small foredune, and fronted by 300 m wide sand flats on top of 400 m wide reef flats. It is bordered by 300 m long sandstone headlands. Beach **K 181** lies to the north of the narrow northern headland, and is a 300 m long continuous strip of high tide sand, backed by irregular protruding sandstone spurs. It is fronted by 50-100 m wide sand flats, then reef flats narrowing from 300 m in the south to 100 m in the north.

Beach **K 182** lies 300 m to the northeast, and is a 50 m pocket of sand wedged into a V-shaped gap in the 20-40 m high sandstone bluffs. It is fronted by 50 m wide sand flats then reef flats to 150 m. Another 300 m to the northeast are beaches **K 183** and **184**. These are adjoining beaches separated by a narrow 150 m long sandstone ridge. They are composed of shelly sand and are relatively steep (Fig. 4.26). They are both backed by small foredunes and fronted by 150 m wide sand flats, then reef flats to 200 m. The dunes behind K 184 are moderately active and have blown up to 70 m inland and to 6 m in height. Clumpy coastal spinifex and *Triodia pungens* (Soft Spinifex) and a few pandanus trees cover the dunes.

Figure 4.26 Beach K 184 consists of a moderately steep beach, backed by a low dune, with the ridge of sandstone.

To the north of beach K 184 the shoreline turns and trends west-southwest into Cone Bay. Beach **K 185** is located 1.5 km to the east in the lee of an attached 200 m long sandstone ridge. The beach faces southeast across 200 m wide sand flats into the bay, and it and the small backing dune have attached the ridge to the mainland. The sand flats continue for 500 m to beach **K 186**, which is a 400 m long discontinuous high tide beach wedged into three gullies in the sandstone. The beach is linked in the lower high tide and has continuous 300 m wide sand flats, then 250-300 m wide reef flats. Beach **K 187** lies 400 m to the southeast in the next small valley. The V-shaped 60 m high valley ridges enclose a small mangrove forest, with the beach-tidal flats fronting the mangroves, then 200 m wide sand flats and the reef edge.

K 188-190 **PORTER HILL ISLAND**

No.	Beach	Rating HT	Rating LT	Type	Length
K188	Porter Hill Is (N1)	1	2	R+sand/reef flats	150 m
K189	Porter Hill Is (N2)	1	1	R+sand/reef flats	100 m
K190	Porter Hill Is (S)	1	1	R+sand/reef flats	200 m
Spring & neap tide range:		6.3 & 1.4 m			

Porter Hill island lies 1.5 km north of 162 m high Porter Hill and 500 m off the steep mainland shore. The island is 700 m long and 100-300 m wide, consisting of lightly vegetated granite rising to a maximum height of 27 m. There are three small beaches clustered toward the eastern end of the island, together with the 10 buildings all located on low overwash-foredunes behind the beaches that link the rocky outcrops. The buildings are part of a pearling operation that has numerous paddocks in the bay.

Beach **K 188** is a 150 m long north-facing high tide cobble-boulder beach, backed by a grassy overwash flat, with one house to the rear. It is bordered by large granite boulders, and fronted by a 200-300 m wide mixture of cobble to sand to reef flats. Beach **K 189** is located 200 m to the south, and is a 100 m long pocket of sand wedged in between 15 m high bare sandstone headlands. It is fronted by rock, sand and reef flats extending 200 m off the beach. The main cluster of houses backs this beach. This beach and K 190 have tied the southern headland to the island. Beach **K 190** faces south toward the mainland shore. It is a 200 m long sand-cobble beach, extending south as a 200 m long sand spit, with reef fringing either side. The reef flats have provided the platforms for beach deposition on an otherwise steep shoreline.

Wotjalum Aboriginal Land

Area: 120 900 ha
Coast length: 390 km (6825-7215 km)
Beaches: 81 (K 191-271)

The Wotjalum Aboriginal land occupies and irregular section of coast between Cone Bay and Yule Entrance, with the islands of the Buccabeer Archipelago ringing the peninsula. There are no official communities on the coast, while Cockatoo and Koolan islands lie just off the coast and outside the Wotjalum Land.

K 191-196 **CONE BAY (SE)**

No.	Beach	Rating HT	Rating LT	Type	Length
K191	Cone Bay (SE1)	1	1	R+tidal sand flats	70 m
K192	Cone Bay (SE2)	1	1	R+tidal sand/mud flats	50 m
K193	Cone Bay (SE3)	1	1	R+tidal mud flats	50 m
K194	Cone Bay (SE4)	1	1	R+tidal sand flats	100 m
K195	Cone Bay (SE5)	1	1	R+tidal sand flats	100 m
K196	Cone Bay (SE6)	1	1	R+tidal sand flats	100 m
Spring & neap tide range:				6.3 & 1.4 m	

The eastern end of Cone Bay is split in two by a 9 km long, 2-4 km wide, 90 m high granite peninsula. Rocky coast and fringing mangroves dominate the very protected southern side of the peninsula, while the northern side contains six small pockets of sand (beaches K 191-196) on an otherwise rock-dominated shoreline. Each beach is located in a joint-controlled gap in the granite, and each is fronted by sand through mud flats and some fringing reefs, with mangroves also bordering some of the beaches.

Beach **K 191** is a 70 m long pocket of sand located 500 m east of the tip of the peninsula. It is surrounded by partly exposed granite on three sides and fronted by 200 m wide sand flats, edged by fringing reef, with two small islands lying 500 m off the beach. Both islands have narrow trailing cobble-sand spits, one of which reaches the coast and is exposed at low tide. Beach **K 192** is located in the third gap to the north. It occupies a 100 m wide break in the rocks, half occupied by mangroves, the other 50 m by the beach. Two hundred metre wide sand-mud flats front the beach, with a fringing reef, and a small rock island 300 m offshore, tied to the coast by a narrow trailing spit.

Beaches K 193, 194 and 195 are three adjoining small beaches (Fig. 4.27) located along a 1 km section of north-facing rocky shore, 3 km east of the tip of the peninsula. Beach **K 193** occupies a 50 m wide gap in the granite, widening to 100 m in lee of the gap. It has fringing mangroves and a 150 m wide mud flat. Beach **K 194** is a 100 m long high tide beach fronted by 200 m long sand flats that extend for 50 m before grading into 100 m wide rock and reef flats. It is separated by a narrow spur of granite from beach **K 195**, a 100 m high tide beach, partly protected by an attached granite islet and mangroves. It is fronted by 300 m wide sand to rock flats.

Beach **K 196** lies 1.5 km further east and is a solitary 100 m long pocket of sand, bordered in the north by a small granite knoll and fronted by 100 m wide tidal flats.

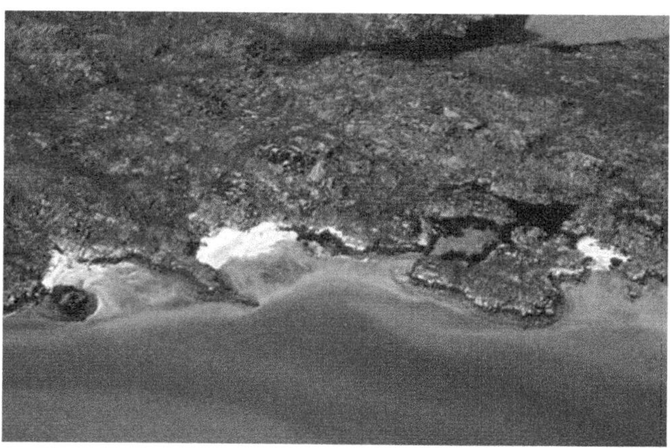

Figure 4.27 The three pocket beaches K 193-195.

K 197-199 **CONE HILL**

No.	Beach	Rating HT	Rating LT	Type	Length
K197	Cone Hill (1)	1	1	R+mud flats	250 m
K198	Cone Hill (2)	1	1	R+mud flats	200 m
K199	Cone Hill (3)	1	1	R+mud/rock flats	300 m
Spring & neap tide range:				6.3 & 1.4 m	

Cone Hill is a 160 m high east-west-trending ridge formed of volcanic rhyodacite. It lies at the northeastern extremity of Cone Bay. Around the base of the hill are three low energy beaches. Beach **K 197** occupies the southern wing of the easternmost deposits in the bay. It is a 250 m long strip of high tide sand and low beach ridge, backed by a small salt flat. Mangroves border each end, as well as a central clump, while mud flats extend 500 m off the beach. It is separated by a 1 km long mangrove-fronted beach ridge/chenier (?) from beach **K 198**. This is a 200 m long strip of high tide sand, with a small rock knoll at its southern end, and mangroves, then a steep rocky incline to Cone Hill at the northern end. It is fronted by 400 m wide mud flats. The beach is the site of a small community of about six houses, which are accessible by a vehicle track that runs along the valley on the south side of Cone Hill to link with the Kimbolton Road.

Beach **K 199** is located 1.5 km to the west at the end of the Cone Hill ridge. It is a 300 m long strip of sand, bordered and backed by lightly vegetated sloping rocks, with rocks increasingly outcropping off the northern end of the beach, amongst the 200-300 m wide mud flats. The beach is moderately steep and extends at high tide right to the backing rocks.

K 199A-209 CONE BAY (N)

No.	Beach	Rating HT	Rating LT	Type	Length
K199A	Cone Bay	1	2	R+sand/mud flats	80 m
K200	Cone Bay (N1)	1	2	R+sand/mud flats	100 m
K201	Cone Bay (N2)	1	2	R+mud flats	200 m
K202	Cone Bay (N3)	1	2	R+mud flats	100 m
K203	Cone Bay (N4)	1	2	R+mud/rock flats	150 m
K204	Cone Bay (N5)	1	2	R+mud flats	150 m
K205	Cone Bay (N6)	1	2	R+mud flats	100 m
K206	Cone Bay (N7)	1	1	R+sand flats	100 m
K207	Cone Bay (N8)	1	1	R+sand flats	50 m
K208	Cone Bay (N9)	1	2	R+sand/mud/reef flats	200 m
K209	Cone Bay (N10)	1	2	R+sand/mud/reef flats	150 m
Spring & neap tide range:		6.3 & 1.4 m			

The northern shore of Cone Bay trends south-southwest to north-northeast from Cone Hill for 20 km to the end of Sir Richard Island, which forms the northern entrance to the bay. The shoreline follows a 120-160 m high ridge of sandstone and dolerite. Steep rocky slopes dominate the shore, with 11 small pockets of sand (K 199A-209) occupying a total of 1.3 km of the shoreline.

Beach **K 199A** is located 3 km west of beach K 199 and lies at the base of the 100 m high escarpment that forms much of the northern shore of the bay. It is an 80 m long pocket of southwest-facing sand backed by a small low grassy foredune then the rocks of the escarpment.

Beach **K 200** is located 6 km west of Cone Hill, and consists of a 100 m long pocket of west-facing sand wedged in between a 200 m long rocky headland and the main ridge. It widens to 120 m at low tide and is bordered by a few mangroves, including one in the centre of the beach, and is fronted by 400 m wide sand to mud flats. Beach **K 201** lies 1 km to the west, and is a straight 200 m long sand beach wedged in a west-facing curve in a 40 m high rocky spur. It is fronted by mud flats widening from 100 m at the point to 200 m in the north.

Beaches K 202 and 203 are adjoining small, low energy beaches located 2 km further west. They connect two small rock outcrops to the shoreline. Beach **K 202** is 100 m long and faces east across 100-200 m wide mud flats, with mangroves both bordering and outcropping in the centre. Beach **K 203** is 150 m long and is located between the two rocks, and faces south over 200 m wide rocks and mud flats with a scattering of low mangroves.

Beaches K 204 and 205 are located on the western side of the only prominent point along this section of coast, a 500 m wide T-shaped ridge of sandstone. Both beaches have beachrock exposed along their base. Beach **K 204** faces west, and is a straight 150 m long beach, backed by a small grassy foredune and fronted by 400 m wide mud flats. Beach **K 205** is located on its immediate north, and 200 m further west. It is 100 m long and shares the same mud flats as its neighbour. The beach is composed of sand over boulders, with a high tide boulder berm.

Beaches K 206 and 207 are located another 1.5 km to the west just outside a linear bedrock-controlled estuary, which is connected through a 100 m wide entrance to the bay (Fig. 4.28). The beaches lie immediately east of the entrance. Beach **K 206** is a 100 m long pocket of south-facing intertidal sand, surrounded on three sides by steeply rising 120 m high sandstone ridges, with a 100 m wide sand flat filling the small pocket. It is submerged at high tide. Beach **K 207** lies 200 m to the west and is a 50 m long southwest-facing intertidal beach, which widens across the 100 m wide sand flats, with bare rocks behind and sandstone ridges to either side.

Figure 4.28 Beaches K 206 & 207 (right) and the narrow entrance to the adjoining estuary.

Beaches K 208 and 209 occupy a 500 m wide south-facing embayment located 1 km from the western end of the mainland peninsula. Beach **K 208** is 200 m long, faces west across the embayment and consists of a low beach fronted by 400 m wide sand-mud-reef flats. Beach **K 209** is 150 m long and located opposite at the western end of the embayment and faces southeast out of the bay across 200 m wide sand-mud-reef flats. It has a relatively steep beach backed by a small grassy foredune and a usually dry creek that crosses the western end of the beach, while beachrock is exposed along the lower beach face.

K 210-214 SIR RICHARD ISLAND

No.	Beach	Rating HT	Rating LT	Type	Length
K210	Sir Richard Is (S1)	1	2	R+reef flats	100 m
K211	Sir Richard Is (S2)	1	2	R+reef flats	100 m
K212	Sir Richard Is (N1)	1	2	R+sand/reef flats	100 m
K213	Sir Richard Is (N2)	1	2	R+sand/reef flats	150 m
K214	Beach K 214	1	2	R+sand/reef flats	100 m
Spring & neap tide range:		6.3 & 1.4 m			

Sir Richard Island is a 3 km long, 500 m wide extension of the mainland ridge, separated from the mainland by a 500 m wide channel. Two beaches are located on its steep bedrock-controlled southern shore. Beach **K 210** is a 100 m long intertidal pocket of southwest-facing sand fronted by 200 m wide reef flats, with fringing reef continuing along the base of the rocky shore. It is located 600 m west of the eastern tip of the island and backed by steep ridges rising to 100 m.

Beach **K 211** lies toward the western tip of the island and is a 100 m long west-facing pocket of sand located in a hollow in the steep 60 m high ridges. It is backed by a low grassy overwash-foredune, and has beachrock exposed along the lower beach face.

Beaches **K 212** and **213** are located on the northern side of the western tip of the island. They are two adjoining slightly more exposed northwest-facing beaches consisting of 100 m and 150 m long high tide beaches, backed by steeply dipping protruding spurs of sandstone, and fronted by 300 m wide sand flats edged by reef flats. Beach K 212 has a small foredune rising to 3-4 m, while K 213 is backed by the steeply sloping rocks.

Beach **K 214** is located 1.5 km to the east on a small island extension of the mainland ridge line. The 100 m long moderately steep pocket beach faces northwest and is backed by a low grassy ridge. It is bordered by 20 m high sandstone ridges with mangroves growing on the boundary rocks and is fronted by 200 m wide sand to reef flats.

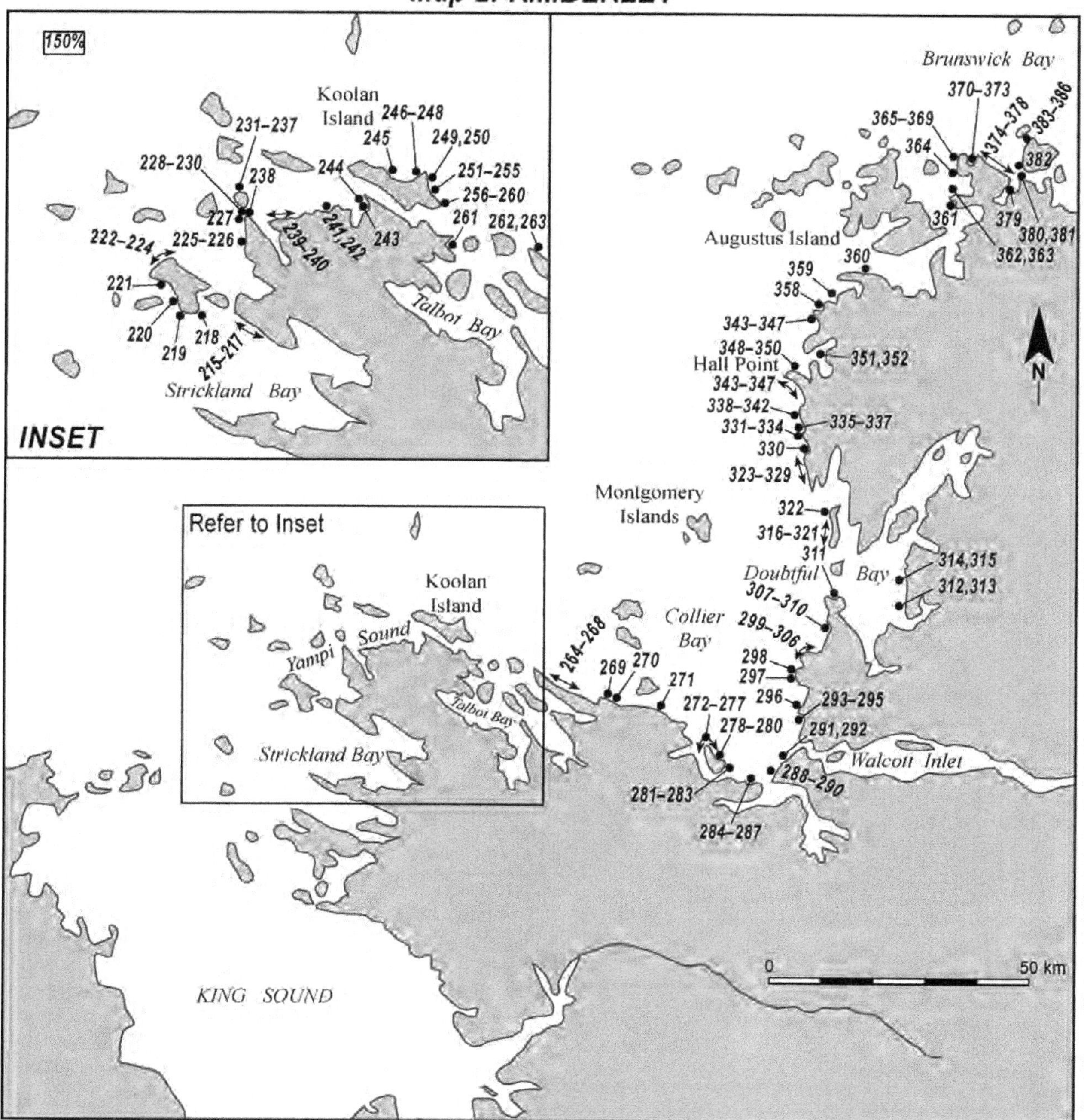

Figure 4.29 Kimberley map 2 locating beaches K 215-386.

Wotjalum Aboriginal Land

Area:	120 900 ha	
Coast length:	390 km	(6815-7215 km)
Beaches:	80	(K 217-271)

K 215-217 **STRICKLAND BAY (E)**

No.	Beach	Rating HT	LT	Type	Length
K215	Strickland Bay (E1)	1	1	R+sand/reef flats	100 m
K216	Strickland Bay (E2)	1	1	R+sand/reef flats	150 m
K217	Strickland Bay (E3)	1	1	R+sand/reef flats	150 m
Spring & neap tide range:		8.5 & 0.9 m			

Strickland Bay is a 22 km deep northwest-southeast trending bedrock-controlled bay. It is 11 km wide toward the island-ringed entrance, narrowing to 3.5 km in the east. The Cecilia, Edeline and Henrietta island groups run for 15 km down the centre of the bay. The entire 45 km of open bay shoreline is bedrock-controlled with fringing mangroves toward the east. There is no habitation in the bay and only three small sand beaches (K 215-217) located within 1.5 km of each other on the northeastern side of the bay.

Beaches K 215 and 216 are located on either side of a low rock spur and reef which protrude 300 m into the bay. Beach **K 215** is a 100 m long south-facing beach fronted by 300 m wide sand then reef flats. It abuts beach **K 216** that faces east across the bay, consisting of two adjoining 80 m and 60 m strips of high tide sand. There is a low grassy foredune behind the longer section, with beachrock exposed on the beach face and rocks and stunted mangroves to either end, while the shorter section is fronted by continuous intertidal rocks and mangroves to the south, then another 100 m of fringing reef.

Beach **K 217** lies 1.5 km to the northwest, immediately south of a small circular-shaped, bedrock-controlled estuary fed by two large upland creeks and largely filled with tidal sand flats. The creeks probably supply the estuarine and beach sand. The beach is a 150 m long east-facing high tide strip of sand wedged in a gap in the rocks. It is backed by a low grassy 40 m wide foredune and has beachrock exposed along the lower beach face (Fig. 4.30). It is fronted by 300 m wide sand flats, together with some rock reefs off the northern end.

K 218-219 **DUNVERT ISLAND**

No.	Beach	Rating HT	LT	Type	Length
K218	Dunvert Is (E)	1	1	R+rock/sand flats	70 m
K219	Dunvert Is (W)	1	1	R+sand/mud flats	250 m
Spring & neap tide range:			8.5 & 0.9 m		

Figure 4.30 Beach K 217 has beachrock outcropping along its base, with a low foredune behind.

Dunvert Island is a 2.5 km long north-south-trending island located toward the northwestern entrance to Strickland Bay. The island is linked to the mainland at low tide via Chambers Island. The island has a 90 m high sandstone ridge and generally steep bedrock shore, fringed by narrow coral reefs. There are two small beaches (K 218 and 219) on the island. Beach **K 218** is located toward the southeastern corner of the island. It is a 70 m long pocket of sand wedged in a gap between two ridges. It is fronted by rock rubble and a 50 m wide sand flat.

Beach **K 219** is located 500 m to the west on the southwestern corner of the island. It is a 250 m long west-facing high tide beach, fronted by 300 m wide sand-mud flats, with fringing boulder beaches, and backed by a partially active foredune rising to 20 m, then steep vegetated 60 m high sandstone slopes.

K 220-224 **HIDDEN ISLAND**

No.	Beach	Rating HT	LT	Type	Length
K220	Hidden Is (W1)	1	2	R+sand/reef flat	300 m
K221	Hidden Is (W2)	1	2	R+sand/reef flat	50 m
K222	Hidden Is (N1)	1	2	R+sand/reef flat	80 m
K223	Silica Beach	1	2	R+sand/reef flat	150 m
K224	Hidden Is (N3)	1	2	R+sand/reef flat	150 m
Spring & neap tide range:		8.5 & 0.9 m			

Hidden Island is a 10 m long sandstone island located at the northern extremity of Strickland Bay. It has a central 100-120 m high northwest-trending ridge, with the entire shoreline characterised by steep sandstone slopes and fringing mangroves on the more protected eastern shores and in numerous small embayments. The only beaches are located in two gaps on the southwestern side of the island (K 220 and 221), and three (K 222-224) along an irregular 2 km long section on the northern shore.

Beach **K 220** lies on the western shore, 1 km north of the southern tip of the island. It occupies part of a semicircular 300 m wide embayment, with the 200 m long intertidal beach occupying the southern part of the bay and a small boulder beach at the northern end. It is fronted by 100 m wide sand flats grading into another 150 m of fringing reef, which links the two sides of the bay.

Beach **K 221** is located 1.5 km to the north in a narrow 100 m wide V-shaped gap in the 60 m long ridge line. The intertidal beach is 50 m long at its apex, widening to 100 m across the fronting tidal flats, which in turn open onto 200 m wide reef flats.

The three northern beaches are each located at the base of small embayments in the rocky coast. Each is bordered by 20 m high headlands extending 200-500 m seaward of the beaches, with sand to reef flats in between. Beach **K 222** is an 80 m long strip of high tide sand widening to 100 m at low tide, with 2-3 m high rocks to either end and steep slopes rising to 44 m behind the beach. It is fronted by 300 m wide sand flats grading into reef flats.

Silica Beach (K 223) lies 1 km to the east and is an unusual 150 m long beach, composed of pure white silica sand, the only such beach on the entire Kimberley coast. As a result it attracts more attention and visitors than most beaches, and is used for day visits and occasional camping. It is backed by a partially active foredune rising to 3-4 m and vegetated by clumpy *Triodia pungens* (Soft Spinifex), then 20 m high bedrock slopes. The white sand slopes offshore between the bordering headlands to deeper rock reefs, permitting easy small boat access, particularly at mid to high tide (Fig. 4.31).

Figure 4.31 The white sand beach face of Silica Beach on Hidden Island (K 223).

Beach **K 224** is located 500 m further east and is bounded by the two headlands that form the northern tips of the island. It is a 150 m long beach that occupies two small valleys linked along the base of a central 30 m high headland. Small grassy foredunes rise to 3-4 m in the valleys, while 200 m wide sand to reef flats extend off the beach.

K 225-230 BOONOOK BAY-YUMURRYUMUH BEACH

No.	Beach	Rating HT	LT	Type	Length
K225	Boonook Bay (N1)	1	2	R+reef flat	300 m
K226	Boonook Bay (N2	1	2	R+sand/reef flat	200 m
K227	Goose Channel (S1)	1	2	R+sand/reef flat	50 m
K228	Goose Channel (S2)	1	2	R+reef flat	100 m
K229	Goose Channel (S3)	1	2	R+sand/reef flat	300 m
K230	Yumurryumuh Beach	1	2	R+sand/reef flat	150 m
Spring & neap tide range:	8.5 & 0.9 m				

Between Hidden Island and the mainland are two lines of narrow, linear islands, with **Boonook Bay** in lee of a third cluster of islands. The headland on the northern side of the bay contains two beaches (K 275-276) on its west-facing shore. Beach **K 225** is located 500 m north of the bay entrance and is a 300 m long strip of sand that widens to the south below 20 m high bluffs of quartz porphyry. It is fronted by 200 m wide reef flats. Beach **K 226** is located immediately to the north in the next small embayment. It is a narrow 200 m long high tide beach containing a few rocks and fronted by 300 m wide sand and rock flats grading to reef flats. It is backed by steep, vegetated 20 m high bluffs, with rock platforms and a few mangroves at their base.

Beach **K 227** is located 2 km due north. It occupies a V-shaped west-facing gap in the 40 m high rocks. The steep, sandy beach is backed by a small, low grassy foredune. The beach is 50 m wide at the apex of the gap, widening to 100 m at low tide and 200 m across the edge of the 200 m wide rock and reef flats.

Beach **K 228** lies 300 m to the north and faces northwest into Goose Channel. The beach is 100 m long and bordered and backed by a 4-5 m high grassy foredune, then 20 m high vegetated rocky slopes. Reef flats extend from the low tide base of the beach 150 m into the channel. Beach **K 229** occupies the next embayment 500 m to the east. It is a relatively straight 300 m long north-facing beach, backed by a narrow 4-5 m high foredune, then slopes rising to 40 m. It is fronted by 500 m wide sand flats grading into fringing reef.

Yumurryumuh Beach (K 230) is located adjacent to Koomi Point, which forms its eastern headland and also the western entrance to the 1 km wide bedrock channel called Coppermine Creek. The beach is 150 m long, faces north and is fronted by 300 m sand then reef flats. Two houses are located on the 10 m high ridge behind the beach and boats are sometimes moored off the beach.

K 231-237 MARGARET ISLAND

No.	Beach	Rating HT	Rating LT	Type	Length
K231	Margaret Is (W1)	1	2	R+sand/reef flats	100 m
K232	Margaret Is (W2)	1	2	R+tidal/reef flats	150 m
K233	Margaret Is (W3)	1	2	R+tidal/reef flats	200 m
K233A	Margaret Is (W3A)	1	2	R+tidal/reef flats	50 m
K234	Margaret Is (N1)	1	2	R+sand/reef flats	250 m
K235	Margaret Is (N2)	1	2	R+sand/reef flats	150 m
K236	Margaret Is (E1)	1	2	R+sand/reef flats	150 m
K237	Margaret Is (E2)	1	2	R+sand/reef flats	200 m
Spring & neap tide range:		8.5 & 0.9 m			

Margaret Island is an irregular 2 km by 2 km island located at the southwestern entrance to Yampi Sound. The island is attached to the mainland by 200-500 m wide reef flats along its southeastern shore. It is composed of quartz-rich porphyry and has a maximum height of 80 m. The entire island is ringed by fringing coral reef with eight small low energy beaches (K 231-237) occupying small embayments on the west, north and east shores of the island.

Beach **K 231** is located on the southwestern shore of the island and faces southwest into the mouth of Coppermine Creek. The high tide beach occupies a 100 m wide curving gap in the wooded 60 m high slopes, widening to 200 m at low tide outside the gap, where it is fronted by 100 m wide sand to reef flats.

Beaches K 232 and 233 are adjoining beaches located in a 1 km deep bay on the western shore of the island. Beach **K 232** is a 150 m long high tide beach wedged deep in the embayment. It is fronted by 500 m wide tidal flats, then another 500 m of reef flats. It is almost linked in the east to beach K 233 with a small wooded islet separating them. Beach **K 233** is a similar 200 m long low energy high tide beach with a low grassy foredune located to either side of a central rock outcrop. It is fronted by 500 m wide tidal to reef flats.

Beach **K 233A** is located on the northwestern tip of the island and is a 50 m long pocket of sand backed by a small low grassy foredune. It narrows between the bordering rocks to 30 m at low tide.

Beach **K 234** lies at the western end of a 1 km wide embayment that occupies much of the northern side of the island. The 250 m long beach narrows to the east where it is bordered by Woodhouse Point. It has a few rocks scattered along the beach and is fronted by 500 m wide reef flats. Beach **K 235** is located in a gully at the eastern end of the beach and links with beach K 236 to join Woodhouse Point to the island. The beach is a 150 m long west-facing pocket of sand, backed by a low grassy foredune and fronted by 500 m wide reef flats, with 40 m high Woodhouse Point forming its northern boundary.

Beach **K 236** lies on the northeastern side of the island, immediately south of Woodhouse Point, and is linked to beach K 235 by a 200 m wide barrier. The 150 m long beach faces into the easterly winds which have built a 2-3 m high grassy foredune, with a small creek crossing the northern end of the beach. It is bordered by steeply dipping rocks truncated by a high tide rock platform and is fronted by 200 m wide reef flats. Beach **K 237** lies 300 m due south, and is a narrow 200 m long strip of primarily intertidal sand extending to the base of the cliffs where driftwood covers the top of the beach. It is fronted by 100 m wide reef flats.

K 238-244 MYRIDI BAY-NARES POINT

No.	Beach	Rating HT	Rating LT	Type	Length
K238	Myridi Bay (W)	1	2	R+sand/reef flats	100 m
K239	Myridi Bay (E1)	1	2	R+sand/reef flats	200 m
K240	Myridi Bay (E2)	1	2	R+tidal/reef flats	100 m
K241	Nares Pt (W4)	1	2	R+reef flats	100 m
K242	Nares Pt (W3)	1	2	R+sand/reef flats	200 m
K243	Nares Pt (W2)	1	2	R+sand/reef flats	200 m
K244	Nares Pt (W1)	1	2	R+sand/reef flats	200 m
Spring & neap tide range:				8.5 & 0.9 m	

The southern shore of **Yampi Sound** extends for 15 km from Margaret Island to Nares Point. It faces north across the Sound to the western end of Koolan Island and Cockatoo Island, located 4 km and 6 km to the north respectively. The shore is composed of quartz sandstone and siltstones, generally forming northwest-trending ridges, which results in a highly indented southern shore. Most of the shoreline is dominated by steep rocky slopes rising up to 100 m, with mangroves filling some of the deeper bays and coral reefs fringing all the open coast and most of the bays. There are seven small low energy, reef-fringed sandy beaches (K 238-244) occupying a total of 1.2 km of the 25 km long shoreline.

Beach **K 238** is located 1 km west of the mouth of Myridi Bay. It occupies an indentation on the western side of an open 500 m wide embayment, and consists of a steep 100 m long beach composed of coral debris fronted by boulders, then sand flats and finally reef flats extending a total of 300 m off the beach. Sandstone boulders border the western side and mangrove-lined rocks the eastern side.

Beach **K 239** lies immediately east of the eastern headland of Myridi Bay. It is a 200 m long north-facing high tide beach, backed by a small 1-2 m high foredune, and fronted by 100 m wide sand flats, then 100 m of reef flats. Beach **K 240** lies immediately behind beach K 239, in a protected west-facing V-shaped embayment. The beach is a100 m long beach ridge located at the apex of the bay, widening across 200 m wide sand flats, then another 500 m of reef flats.

Beach **K 241** is located 5.5 km to the east. It is a 100 m long pocket of sand, wedged in a break in a straight section of 90 m high sandstone ridges. It faces north toward Koolan Island, and is fronted by a 100 m wide fringing reef. Beach **K 242** is located 1 km further east, in an open west-facing embayment. The narrow high tide beach is 200 m long and backed by a 10-20 m wide low foredune, then vegetated slopes which rise to 80 m. A small creek crosses the southern end of the beach. Beachrock is exposed along the lower beach face, with 200 m wide sand flats, then 100 m wide reef flats off the beach. It is bordered by steeply dipping limestone, with a rock outcrop and rock platforms off its northern end (Fig. 4.32).

Figure 4.32 Beach K 242 has a wide moderate gradient intertidal beach, backed by a low dune, with prominent sandstone boundaries, together with a small rock islet right off the beach (right).

Beaches K 243 and 244 are located on the western side of Nares Point, 2.5 km further east. Beach **K 243** lies toward the base of the point. It is a 200 m long west-facing high tide beach backed by slopes which rise steeply to 150 m. It has 500 m wide muddy sand flats, then another 500 m of reef flats. Beach **K 244** is located at the northeastern tip of the point and faces due west down the sound. It is a straight 200 m long high tide sand beach, backed by a small foredune, rock debris and the scarped base of steep 70 m high slopes. It is fronted by 50 m wide sand flats, then 200 m wide reef flats.

K 245-248 **KOOLAN ISLAND (N)**

No.	Beach	Rating HT	Rating LT	Type	Length
K245	Koolan Is (N1)	1	2	R+reef flats	150 m
K246	Koolan Is (N2)	1	2	R+reef flats	50 m
K247	Back Beach (1)	1	2	R+sand/reef flats	200 m
K248	Back Beach (2)	1	2	R+sand/reef flats	100 m
Spring & neap tide range:		8.5 & 0.9 m			

Koolan Island is a 14 km long island that widens from 500 m in the west to 5.5 km in the east. It runs parallel to the mainland coast separated by The Canal. This is a 1 km wide east-west-trending bedrock bay that narrows to 200 m in the east at The Drain and The Gutter. The island was the site of an open cut iron ore mine with its wharf located midway along its southern shore. The now abandoned mining town of Koolan was located on a 100 m plus high ridge toward the eastern end of the island, with the landing ground on its northern perimeter, and access roads to some of the surrounding eastern beaches.

The island has about 40 km of shoreline dominated by sandstone slopes rising steeply to between 80 and 150 m, with some sea cliffs along the more exposed northern shoreline. There are no beaches on the protected southern shore, nor much of the steep north shore. There is a cluster of 16 small low energy beaches (K 245-260) located toward the eastern end of the northern shoreline and in bays along the irregular eastern shore.

Beach **K 245** is located on the northern side of the island, 2.5 km west of Back Beach. It lies on the eastern side of a slight protrusion in the 100 m high cliffs. The beach is 150 m long, faces north, is backed by the steep vegetated slopes and fronted by 300 m wide reef flats. Beach **K 246** is located 2 km to the east, immediately south of Back Beach. It is a 50 m pocket of intertidal sand wedged in a creek-carved V-shaped gap in steep 120 m high sandstone slopes. The beach is submerged at high tide, washed out by the creek in the wet season, and fronted by 200 m wide reef flats.

Back Beach was the main recreational beach on the island. It is accessible by a winding overgrown vehicle track that leaves the northern end of the landing ground. The intertidal beach is located in a west-facing 500 m wide bay surrounded by 60-150 m high slopes. The main beach (**K 247**) faces west, is 200 m long and fronted by 200 m of sand flats, then another 200 m of reef flats. It is separated from its neighbour (K 248) by a 50 m strip of low sandstone rocks. Beach **K 248** is a 100 m long southwest facing strip of sand with 100 m wide sand then reef flats.

K 249-260 **KOOLAN ISLAND (E)**

No.	Beach	Rating HT	Rating LT	Type	Length
K249	Sanders Pt (S1)	1	2	R+mud flats	50 m
K250	Sanders Pt (S2)	1	2	R+mud flats	50 m
K251	Koolan (1)	1	2	R+mud/reef flats	50 m
K252	Koolan (2)	1	2	R+mud/reef flats	50 m
K253	Koolan (3)	1	2	R+mud/reef flats	50 m
K254	Koolan (4)	1	2	R+mud/reef flats	50 m
K255	Koolan (5)	1	2	R+mud/reef flats	50 m
K256	Front Beach (N2)	1	2	R+mud/reef flats	50 m
K257	Front Beach (N1)	1	2	R+mud/reef flats	100 m
K258	Front Beach	1	2	R+mud flats	100 m
K259	South East Pt	1	2	R+sand/reef flats	50 m
K260	Catspaw Pt	1	2	R+sand/reef flats	100 m
Spring & neap tide range:				8.5 & 0.9 m	

The eastern end of Koolan Island is marked by Sanders Point in the north and Catspaw Point 5 km to the south, with four small embayments and headlands in between. The backing coast averages 60-100 m in height, with the original township of Koolan located on the ridge crest, midway along the shore. Twelve small low energy beaches (K 249-260) are located in the embayments, all backed by steep slopes.

Beaches **K 249** and **250** are located in the first bay south of Sanders Point, a 500 m wide east-facing semicircular-shaped bay. The two 50 m long pockets of intertidal sand are wedged into gullies in the steep 100 m high slopes. They are fronted by continuous 300 m wide mud flats, and an outer 100 m wide fringing reef that links the boundary headlands.

Beaches **K 251, 252, 253, 254** and **255** occupy the second bay south of the point, immediately below and east of the old Koolan township. The five beaches are each located in small creek-carved valleys strung around the 600 m wide bay, with the backing sandstone layered slopes rising to 100 m. Each beach is intertidal, about 50 m in length and fronted by 50-100 m wide mud flats grading into 300 m wide reef flats, which link the entrance to the bay. Only beach K 255 shows a little sand at high tide.

Beaches K 256 and 257 are located inside the northern headland of the third embayment. A vehicle track runs down the ridge line to the back of the two adjoining beaches. Beach **K 256** is a moderately steep 50 m long pocket of sand with a few boulders and driftwood occupying the high tide beach. It widens to low tide, where it is fronted by 300 m wide mud to reef flats. It is separated from its neighbour by a 20m wide ridge of sandstone. Beach **K 257** is a 100 m long southeast-facing strip of high tide sand, backed by vegetated slopes. It narrows to 70 m at low tide and is fronted by 300 m wide mud flats, then 200 m wide reef flats.

Front Beach (**K 258**) is located at the southern apex of the bay. It is a very low energy east-facing beach, with South East Point forming the 1 km long southern headland. The high tide beach is backed by a low grassy foredune and fronted by 200 m wide sand flats (Fig. 4.33), then 1 km wide mud flats, which extend out to the mouth of the bay. There is an overgrown vehicle track leading to the slopes above the rear of the beach. Beaches K 259 and 260 are located out on the tip of the southern headland, between South East and Catspaw points. Beach **K 259** is a 50 m pocket of sand backed and bordered by red sandstone rocks, with a narrow finger of sandstone separating it from beach K 260. Beach **K 260** is a similar 100 m long strip of sand with Catspaw Point forming a 20 m high hump on the southern headland. Both beaches are almost covered at high tide with only a patch of sand, rocks and driftwood showing. They are fronted by 50 m wide sand flats, then 100 m wide reef flats.

Figure 4.33 Front Beach (K 258) ia a wide lower gradient intertidal beach backed by a low grassy dune.

K 261-271 **TALBOT-SHOAL BAYS**

No.	Beach	Rating HT	LT	Type	Length
K261	Beach K261	1	2	R+reef flats	300 m
K262	Beach K262	1	2	R+sand/reef flats	150 m
K263	Beach K263	1	2	R+sand/reef flats	100 m
K264	Beach K264	1	2	R+mud flats	50 m
K265	Beach K265	1	2	R+mud flats	50 m
K267	Beach K267	1	2	R+mud flats	80 m
K268	Beach K268	1	2	R+mud flats	80 m
K269	Beach K269	1	2	R+mud flats	50 m
K270	Beach K270	1	2	R+sand/mud flats	70 m
K271	Beach K271	1	2	R+rock flats	100 m
Spring & neap tide range:				9.0 & 1.4 m	

Talbot Bay is a large irregular bedrock-controlled bay located immediately east of Koolan Island. The bay has a 12 km wide mouth which is largely filled with numerous islands and reefs. It extends up to 25 km inland and has a shoreline length of approximately 110 km, plus several tributary bays, including the famous narrow gaps leading to two consecutive bays that form waterfalls and rapids when the tide changes. There are no beaches in the bay, the nearest being located either side of the bay entrance on the more exposed north-facing shore.

Beach **K 261** is located at the western entrance to the bay, and lies 4.5 km southeast of Catspaw Point. The beach is 300 m long and is backed by lightly vegetated sandstone slopes rising to 100 m. It is fronted by a 100 m wide fringing reef.

The eastern entrance to the bay is located at the northern end of an 8 km long northwest-trending sandstone ridge. Beaches K 262 and 263 lie in small embayments 2 km east of the tip of the ridge. Beach **K 262** is a 150 m long sandy beach located inside a 200 m wide embayment. It faces north across 300 m wide sand flats, with an outer fringing reef linking the headlands of the bay. Beach **K 263** is located in the next smaller embayment 500 m to the east. It is a 100 m long northwest-facing, moderately steep high tide sand beach grading into cobbles at low tide, then 200 m wide sand flats and a fringing reef.

The coastline between Talbot Bay entrance and Shoal Bay consists of 40 km of predominantly north- to northeast-facing dissected sandstone ridges, rising rapidly to more than 100 m. Most of the shoreline is rocky, with mangroves filling several small protected embayments and a few cobble-boulder beaches in more exposed indentations. Besides beaches K 262 and 263 there are another eight small pockets of sand (K 264-271), all totalling only 800 m in length.

Beach **K 264** lies 7.5 km southeast of the Talbot Bay entrance on the western side of a 500 m long sandstone peninsula. It is a 50 m long pocket of northwest-facing sand, backed by a small foredune. It widens to 150 m at low tide and is fronted by 200 m wide mud flats, with mangroves growing on the southern side of the flats. Beach **K 265** is located on the eastern side of the peninsula, 200 m to the east. It is a similar steeply sloping, 50 m long pocket of largely intertidal sand fronted by 100 m wide rubble flats. It faces across a 500 m wide bay to beach **K 266**, another 70 m long pocket of high tide sand, fronted by a 100 m wide mixture of boulders, rocks and reef flats. It links a small rock outcrop to the shoreline and is capped and backed by coarse coral debris, then mangroves.

Beach **K 267** is located 1 km further east. The 80 m long beach occupies the base of a narrow north-facing embayment, which is filled with 300 m wide mud flats in front of the beach. Beach **K 268** lies 500 m to the east, in a similar northwest-facing embayment with a small island separating the two. It consists of an 80 m long moderately steep, high tide beach, fronted by 200 m wide mud flats.

Beach **K 269** is a further 8 km to the east and lies on the eastern side of a 1 km wide embayment. The 50 m long beach occupies a gap in the centre of the eastern headland and is fronted by 200 m wide reef flats. Finally, beach **K 270** lies 800 m to the east on the eastern side of a 400 m wide mud-filled embayment. The beach is a steep 70 m long cobble beach wedged between 60 m high sandstone slopes, with cliff debris backing the beach. Beyond beach K 270 sandstone dominates the next 20 km of shoreline to the mangrove-lined mouth of Shoal Bay.

Beach **K 271** is a solitary beach another 9 km to the east of Shoal Bay. It is a 100 m long steep, cobble-boulder beach at the base of a 40 m high slumped cliff face, with rock debris backing the beach and 50 m wide rock flats off the beach.

K 272-283 **SHOAL BAY ISLAND**

No.	Beach	Rating HT	Rating LT	Type	Length
K272	Shoal Bay Is (1)	1	2	R+rock/mud flats	80 m
K273	Shoal Bay Is (2)	1	2	R+mud flats	80 m
K274	Shoal Bay Is (3)	1	2	R+rock flats	70 m
K275	Shoal Bay Is (4)	1	2	R+sand/rock flats	80 m
K276	Shoal Bay Is (5)	1	2	R+sand/rock flats	70 m
K277	Shoal Bay Is (6)	1	2	R+sand/rock flats	60 m
K278	Shoal Bay Is (7)	1	2	R+sand flats	100 m
K279	Shoal Bay Is (8)	1	2	R+sand/rock flats	200 m
K280	Shoal Bay Is (9)	1	2	R+sand/rock flats	150 m
K281	Shoal Bay Is (10)	1	2	R+rock flats	700 m
K282	Shoal Bay Is (11)	1	2	R+rock/sand flats	120 m
K283	Shoal Bay Is (12)	1	2	R+rock flats	150 m
Spring & neap tide range:				7.8 & 3.8 m	

Shoal Bay is located in the southwestern corner of the larger Collier Bay. The open ended bay is 1-2 km wide, 10 km long and lies in lee of an 8 km long island. The bay shore is dominated by mangrove-filled smaller embayments and continuous fringing reef. The only beaches in the area are located along the north and eastern shore of the island which has an irregular shoreline composed of sandstone and siltstone. The beaches (K 272-283) are all located in small embayments and usually backed by vegetated sandstone slopes.

Beaches K 272 and 273 are two 80 m long pockets of sand located in an 800 m wide bay on the northwestern tip of the island. Beach **K 272** consists of a high tide cobble beach, backed by 20 m high slopes, and fronted by a 50 m wide band of rock flats, then mud flats, with reef flats extending up to 500 m off the beach. Mangroves are scattered along the southern half of the beach. Beach **K 273** is located at the southern end of the bay. It is a steep shelly beach that faces north across 300 m wide mud flats, then the reef flats, and is backed by a 50 m wide shelly overwash plain (Fig. 4.34).

Figure 4.34 Beach 273 is a relatively steep beach backed by low overwash flats, typical of beaches in this area.

The northeastern peninsula of the island contains four embayed beaches (K 274-277). Beach **K 274** lies at the northern tip of the island. It is a steep, north-facing 70 m long cobble beach, fronted by 50 m wide rock flats. Beach **K 275** is located in an adjacent embayment 200 m to the southeast. It consists of an 80 m pocket of sand and cobbles, fronted by 200 m wide sand flats, which are bordered by boulder fields fronting each 200 m long arm of the bay, with scattered mangroves on the boulders. A small creek crosses the southern end of the beach. Beach **K 276** occupies the next small, narrow embayment. It is a 70 m long cobble beach, with 200 m wide rock flats, and only a 50 m long central sand flat, all bordered by the boulders and mangroves. Immediately south is beach **K 277** a 60 m long high tide beach, with 100 m wide sand flats and bordering boulder fields. All four beaches and sand flats lie in lee of 300-400 m wide reef flats.

Beach K 278 lies 1 km to the south across a mangrove-filled bay. The beach is one of three (K 278-280) occupying the central part of the eastern side of the island. Beach **K 278** is a 100 m long high tide cobble beach, fronted by 100 m wide sand flats, with rock flats bordering either headland, and mangroves along the central-northern section. Beach **K 279** is located at the base of a small 500 m deep embayment and is part of a small 200 m long cobble barrier. It has a central patch of 200 m wide sand flats, with rock flats to either side and mangroves on the northern rocks. Beach **K 280** occupies the next small embayment. It is a curving 150 m long high tide beach, fronted by a continuous 100 m wide band of rock flats, then low tide mud flats that fill the bay. Mangroves parallel the northern half of the beach.

Beaches K 281 to 283 are located at the southeastern tip of the island. Beach **K 281** is a narrow 700 m long high tide sand and cobble beach, backed by 20 m high sandstone bluffs and fronted by 200 m wide rock flats, then mud flats. Beaches K 282 and 283 are adjoining 120 m and 150 m long cobble-boulder beaches. Beach **K 282** is backed by a small V-shaped lagoon, and fronted by 50 m wide rock flats then sand flats. Beach **K 283** is backed by 20 m high bluffs and fronted by 100 m wide rock flats, then mud flats.

K 284-287 THE FUNNEL (W)

No.	Beach	Rating HT	Rating LT	Type	Length
K284	The Funnel (W 4)	1	2	R+rock/mud flats	400 m
K285	The Funnel (W 3)	1	2	R+rock/mud flats	400 m
K286	The Funnel (W 2)	1	2	Cobble+sand flats	300 m
K287	The Funnel (W 1)	1	2	Cobble+rock flats	400 m
Spring & neap tide range:		7.8 & 3.8 m			

The Funnel connects the large circuitous Secure Bay with the open Collier Bay. The bedrock-controlled entrance is 1 km wide and includes two small islets. West of The Funnel the coast trends west for 9 km to the southern end of Shoal Bay, with four beaches in this section (K 284-287).

Beaches K 284 and 285 are located approximately 5 km west of the entrance. Both are 400 m long and face north into Collier Bay. Beach **K 284** grades from sand into boulders littered with driftwood, and is fronted by 300 m wide mud flats. It is tied to scarped bluffs with a 600 m long elongate lagoon behind. Beach **K 285** is similar with a small barrier and backed by a 1 km wide band of mangroves, which separates the beaches and bluffs from the backing high country.

Beaches K 286 and 287 lie 1 km to the east and are part of a similar pair of small cobble-boulder barriers. Beach **K 286** is a 300 m long, steep, cobble-boulder beach, composed of platey sandstone and littered with driftwood (Fig. 4.35). It is backed by a 50 m wide series of low ridges then a 600 m wide band of mangroves. Red cliffs border the eastern end and it is tied to a rock outcrop in the west. A mangrove-shrouded narrow recurved spit runs south of the outcrop. Beach **K 287** is a 400 m long cobble-boulder ridge backed by scarped slopes that rise to 51 m. Both rocky beaches are fronted by 100 m wide rock flats.

Figure 4.35 Beach K 286 is one of two adjoining beaches composed of platey cobbles, which form a steep rocky ridge.

K 288-292 THE FUNNEL (N)

No.	Beach	Rating HT	Rating LT	Type	Length
K288	The Funnel (N1)	1	2	R+mud flats	50 m
K289	The Funnel (N2)	1	2	R+mud flats	50 m
K290	The Funnel (N3)	1	2	R+mud flats	50 m
K291	The Funnel (N4)	1	2	R+rock/mud flats	250 m
K292	The Funnel (N5)	1	2	R+rock/mud flats	250 m
Spring & neap tide range:		7.8 & 3.8 m			

On the eastern side of The Funnel the coast trends northeast for 10 km to Yule Entrance, a similar bedrock-controlled entrance. Five beaches (K 288-292) are located within the first 3.5 km of The Funnel entrance. Beaches K 288, 289 and 290 are three 50 m pockets of high tide sand and cobbles located within the first 1 km. Beach **K 288** lies in a gap in the sandstone slopes. The sand beach faces west and is fronted by 100 m wide sand and rock flats, with rock outcrops and a small islet off the beach. Beaches **K 289** and **290** lie either side of a 30 m high sandstone headland, that is tied by an intertidal rocky tombolo to a small island. Both are steep high tide cobble beaches fronted by intertidal sand then 150 m wide mud flats.

Beaches **K 291** and **292** are located 2 km further north, just to the south of Shale Island. Both sandy beaches are 250 m in length, and bounded by three 10-20 m high sandstone knolls and fronted by a mixture of rock and sand to mud flats. They are linked by their beach ridges and backed by a continuous 200-500 m wide mangrove forest.

K 293-296 **HIGH BLUFF**

No.	Beach	Rating HT	Rating LT	Type	Length
K293	High Bluff (S3)	1	2	R+mud flats	400 m
K294	High Bluff (S2)	1	2	R+mud/rock flats	100 m
K295	High Bluff (S1)	1	2	R+mud/rock flats	900 m
K296	High Bluff (N)	1	2	R+rock/mud flats	400 m
Spring & neap tide range:				7.8 & 3.8 m	

High Bluff is a 48 m high sandstone bluff located 7 km north of Yule Entrance. The bluff protrudes 2 km west of the run of the shoreline and has a series of three southern beaches linking it to secondary bluffs and the main shoreline.

Beach **K 293** is a 400 m long curving spit that is attached to a mainland ridge in the south and terminates in a mangrove forest to the north. It is backed by a few mangroves and 500 m wide salt flats, while it is fronted by 300 m wide mud flats in the south, grading into narrower rock flats in the north. Beach **K 294** is located on the north side of the mangroves, and is a 100 m long pocket beach, bordered by a pair of 30 m high scarped bluffs. It is fronted by a 100 m wide mixture of sand and rock flats.

Beach **K 295** is a curving 900 m southwest-facing high tide beach that links the northern bluff with High Bluff located 1 km to the northwest. It is backed by 500 m wide salt flats in the south, then three sandstone ridges, the last being High Bluff. It is fronted by 100 m wide rock flats grading into low tide mud flats.

Beach **K 296** is located on the northern side of High Bluff and consists of a 400 m long spit that is attached to the northern tip of the bluff. The spit encases a small mangrove forest as it curves around to the southeast.

K 297-306 **EAGLE POINT**

No.	Beach	Rating HT	Rating LT	Type	Length
K297	Eagle Pt (S8)	1	2	R+mud flats	700 m
K298	Eagle Pt (S7)	1	2	R+rock/mud flats	700 m
K299	Eagle Pt (S6)	1	2	R+rock/mud flats	300 m
K300	Eagle Pt (S5)	1	2	R+rock/mud flats	100 m
K301	Eagle Pt (S4)	1	2	R+rock flats	200 m
K302	Eagle Pt (S3)	1	2	R+sand flats	300 m
K303	Eagle Pt (S2)	1	2	R+mud flats	150 m
K304	Eagle Pt (S1)	1	2	R+mud flats	200 m
K305	Eagle Pt (E1)	1	2	R+rock flats	200 m
K306	Eagle Pt (E2)	1	2	R+rock/mud flats	200 m
Spring & neap tide range:				7.8 & 3.8 m	

Eagle Point is a 50 m high sandstone headland located at the northern end of a protruding 7 km long section of irregular sandstone ridges, which are linked to the mainland by mangroves and salt flats. The irregular shoreline contains ten beaches (K 297-306), most dominated by the bordering rocks and intertidal rock flats.

Beach **K 297** lies at the southern end of the section and is a curving 700 m long west-facing moderate gradient sand beach with beachrock outcropping along the intertidal zone, together with some central rocks and 500 m wide mud flats. The southern half of the beach is backed by an inactive, vegetated, 3 m high, imbricated, cobble-boulder ridge (Fig. 4.36), with the whole system backed and bordered by 10-20 m high vegetated bluffs. The ridge appears to be Holocene in age and may have been formed during a slightly higher sea level. Beach **K 298** is located immediately to the north and is a northwest-facing 700 m long high tide strip of sand that runs along the base of two sections of scarped bluffs, together with a central beach ridge-overwash section, backed by a small salt lake. It is fronted by 200 m wide rock flats, then 500 m wide mud flats.

Figure 4.65 An upper vegetated cobble ridge backs beach K 297. The cobbles have been eroded from the lower beachrock (right).

A 2.5 km wide mangrove-filled embayment lies to the north, with the remaining eight beaches occupying the next 5 km of rocky shore up to and around Eagle Point. Beaches K 299 and 300 are two neighbouring west-facing, steeply sloping cobble-boulder beaches bordered by vegetated slopes rising to 20 m. Beach **K 299** is 300 m long and backed by a small salt flat, while 100 m long beach **K 300** lies below low bluffs. Both are fronted by 200 m wide rock flats, then 100 m wide mud flats.

Beaches K 301 and 302 are located on the northern side of the next headland and face north. Beach **K 301** is a 200 m long steep beach with a shelly base and boulder capping. It is backed by scarped vegetated slopes and fronted by 500 m wide rock flats, then a narrow fringing reef. Its neighbour, beach **K 302**, is a 300 m long moderately sloping sand beach with boulders along its base. It is bordered by rock flats, with central 500 m wide muddy-sand flats.

Beaches K 303 and 304 are located on the opposite side of the same embayment a few hundred metres south of

Eagle Point. Beach **K 303** is a 150 m long high tide sand beach, bordered and backed by sandstone bluffs including a large central rock outcrop. It has rock flats to either side with mangroves on the northern rocks and is fronted by 700 m wide mud flats. Beach **K 304** lies 100 m to the north and is a similar 200 m long sand beach with a low foredune along the northern section. The beach narrows to 50 m at low tide as rock flats and rubble dominate, together with some rocks in the centre of the beach, which are all fronted by 500 m wide mud flats.

Beach **K 305** is located on the northern side of Eagle Point below the steep vegetated 40 m high slopes. It is a narrow, steep strip of high tide sand, backed by a northern scarp which is depositing rock debris on the beach, and a low patchy foredune in the south, with 50-100 m wide rock flats paralleling the beach. Beach **K 306** lies 500 m to the east in a 200 m wide U-shaped embayment. The curving beach is 200 m long and fronted by 50 m wide rock flats containing scattered mangroves, with mud flats filling the rest of the embayment.

K 307-311 **RAFT POINT**

No.	Beach	Rating HT	Rating LT	Type	Length
K307	Raft Pt (S4)	1	2	R+mud flats	50 m
K308	Raft Pt (S3)	1	2	R+mud flats	200 m
K309	Raft Pt (S2)	1	2	R+mud flats	150 m
K310	Raft Pt (S1)	1	2	R+mud flats	60 m
K311	Raft Pt	1	2	R+mud flats	150 m
Spring & neap tide range:				7.8 & 3.8 m	

Raft Point forms the southern entrance to Doubtful Bay. It lies at the northern end of a 4.5 km long straight north-south sandstone ridge, that rises inland to a peak of 140 m. Four beaches are located in a 1 km wide bay immediately south of this section, with one on the tip of the point.

Beach **K 307** is a south-facing 50 m long pocket of sand, wedged in between two small sandstone outcrops. It looks south across 1.5 km wide mud flats, which front a 6 km long section of mangroves. Beach **K 308** is located 100 m to the north and faces northwest into the small bay. It is a 200 m long high tide beach backed by a low grassy foredune and fronted by sloping intertidal rock flats, then 50 m wide mud flats which fill the bay.

Beach **K 309** lies at the apex of the bay with a small upland creek draining into its southern end. The beach is a strip of high tide sand and 50 m wide sand flats, fronted by 800 m wide bay mud flats, with mangroves dominating the southern corner. Two hundred metres to the north is beach **K 310**, a 60 m long moderately steep pocket of sand, with intertidal rock flats, fronted by 700 m wide mud flats.

Beach **K 311** is located between the two 200 m long protruding arms of Raft Point. It is a steep 150 m long sand beach, with some beachrock at its base. It is backed by a 50 m wide small area of overwash, and fronted by a 100 m wide sand flat (Fig. 4.37). It has been used as a base camp and some debris litters the camp site.

Figure 4.37 Beach K 311 is a small steep pocket beach located on the tip of Raft Point. It is backed by a low degraded foredune.

K 312-315 **DOUBTFUL BAY**

No.	Beach	Rating HT	Rating LT	Type	Length
K312	Doubtful Bay (E1)	1	2	R+mud flats	50 m
K313	Doubtful Bay (E2)	1	2	R+mud flats	50 m
K314	Doubtful Bay (E3)	1	2	R+mud+rock flats	400 m
K315	Doubtful Bay (E4)	1	2	R+mud+rock flats	300 m
Spring & neap tide range:			7.8 & 3.8 m		

Doubtful Bay and its tributary George Water extends for 50 km north-south, in lee of a 7 km wide entrance between Raft Point and a northern headland. The bay is up to 15 km wide, and for the most part has bedrock-controlled shoreline dominated by rocky coast, with extensive mangrove forests in embayments and more protected areas. There are only four small low energy beaches (K 312-315) located in the bay, all in a cluster 15 km east of Raft Point.

Beaches **K 312** and **313** are two 50 m long northwest-facing pockets of sand bordered by 30-40 m high sandstone slopes. They are both fronted by 1 km and 200 m wide mud flats respectively. Beach **K 314** is located 1.5 km to the north in a 500 m wide southwest-facing embayment. The curving 400 m long beach occupies the back of the bay and faces southwest across 300 m wide mud flats, partly vegetated with mangroves. Beach **K 315** occupies the next small embayment 500 m to the north. The beach is a 300 m long high tide beach fronted also by partly vegetated 400 m wide mud flats. Both beaches are backed by steep sandstone slopes rising to over 100 m.

Kunmunya Aboriginal Land

Area:	238 000 ha	
Coast length:	398 km (7375-7773 km)	
Beaches:	66	(K 316-381)

Kunmunya Aboriginal land extends from the northern side of Doubtful Bay to the entrance of the Prince Regent River, and includes Augustus Island. The only settlement on the land is the pearling farm at Kuri Bay.

K 316-322 **DOUBTFUL BAY (N)**

No .	Beach	Rating HT	LT	Type	Length
K316	Beach K316	1	1	R+sand flats	100 m
K317	Beach K317	1	1	R+sand flats	50 m
K318	Beach K318	1	1	R+sand flats	50 m
K319	Beach K319	1	1	R+sand flats	80 m
K320	Beach K320	1	1	R+sand flats	50 m
K321	Beach K321	1	1	R+sand flats	50 m
K322	Beach K322	1	2	R+mud flats	100 m
Spring & neap tide range:		8.8 & 2.0 m			

The north side of Doubtful Bay is bordered by a series of north-south trending sandstone ridges, which form a mainland peninsula and a southern series of linear islands. The outermost island is 10 km long and between 0.5 and 1.5 km wide. It contains six small beaches along its exposed western side, five located along the southern 4 km of the island. The sloping sandstone produces sloping intertidal rock flats that dominate between beaches K 316 and 344.

Beaches K 316, 317, 318 and 319 and four pockets of sand are located in joint-controlled gullies along a 500 m long section of rocky shore. Beach **K 316** consists of two 60 m and 40 m long pockets of sand that converge at mid-tide, with a creek and a few mangroves behind the northern pocket. Beach **K 317** lies in a 50 m wide gully with a creek on the northern side. Beach **K 318** is 60 m wide at high tide narrowing to 50 m at low tide amongst sandstone columns. It has a usually dry creek on the north side and a patch of foredune at the southern end of the beach. Beach **K 319** lies at the base of a 200 m deep gully that almost bisects the ridge. The beach is 80 m long with a low grassy foredune, then the dense vegetation of the narrow gully. Each of the beaches is bordered by sloping intertidal rocks, fronted by a narrow sand flat and backed by the 60 m high sandstone ridge line.

Beaches K 320 and 321 occupy a 400 m wide embayment, 4 km north of the southern tip of the island. Beach **K 320** is a 50 m long strip of sand, bordered by sandstone slopes. The beach widens towards low tide where it is fronted by 500 m wide sand flats, which fill the bay. Beach **K 321** lies on the northern side of the bay and is a similar 50 m long sand beach, cut by two narrow spurs of sandstone. The beach widens to 80 m at low tide and is fronted by the same sand flats.

Beach **K 322** lies 2 km south of the northern tip of the island. It is wedged in a 100 m wide gully fed by two usually dry creeks. The beach consists of a small foredune, a steep narrow high tide beach, then some intertidal beachrock, grading into 300 m wide sand flats (Fig. 4.38).

Figure 4.38 Beach K 322 is hemmed in between a gap in sandstone points, widening to the rear.

K 323-331 **FRESHWATER COVE**

No.	Beach	Rating HT	LT	Type	Length
K323	Freshwater Cove (S7)	1	2	R+sand/rock flats	400 m
K324	Freshwater Cove (S6)	1	2	R+sand/rock flats	200 m
K325	Freshwater Cove (S5)	1	2	R+rock/reef flats	400 m
K326	Freshwater Cove (S4)	1	2	R+rock/reef flats	100 m
K327	Freshwater Cove (S3)	1	2	Boulder+sand/reef flats	500 m
K328	Freshwater Cove (S2)	1	2	Boulder+sand/reef flats	150 m
K329	Freshwater Cove (S1)	1	2	R+rock flats	100 m
K330	Freshwater Cove	1	2	R+tidal sand flats	200 m
K331	Freshwater Cove (N)	1	2	R+rock flats	80 m
Spring & neap tide range:		8.8 & 2.0 m			

Freshwater Cove is a 300 m wide north-facing bay, located 7 km north of the northern mainland boundary of Doubtful Bay. The cove and the coast that trends due south to Doubtful Bay contains ten small beaches (K 323-331), fronted by a mixture of sand and rock flats, with a fringing coral reef along most of the shore.

Beach **K 323** lies 5 km south of the cove, and is a 400 m long sand beach, fronted by 100 m wide sand flats, with a few rock patches, then 100 m wide flats of the fringing reef. Beach **K 324** is a similar 200 m long beach, with the same sand and reef flats.

Beaches **K 325** and **326** are two neighbouring beaches 400 m and 100 m long respectively. Both consist of narrow high tide beaches, fronted by 200 m wide rock flats, then the 100 m wide fringing reef.

Beach **K 327** is an irregular 500 m long boulder beach, with 200 m wide sand flats, broken by several rock outcrops and a 100 m wide fringing reef extending to either side of a central gap. Beach **K 328** is located 500 m south of the cove, and is a 150 m long boulder beach, fronted by 100 m wide sand flats, bordered by rock flats, then 200 m wide reef flats.

Beach **K 329** is located on the outer point of the cove. It is a pointed, 100 m long convex patch of high tide sand, fronted in both directions by 100-200 m wide rock flats, then the fringing reef.

Beach **K 330** is located inside Freshwater Cove on the western side of the creek, opposite the mangrove-lined eastern side. It is a 200 m long northwest-facing sand beach, backed by a small patch of foredune then small columns of sandstone. A mangrove-lined upland creek drains across the southern end of the beach and flows over the 600 m wide tidal sand flats, which fill the cove (Fig. 4.39). The beach is used to moor boats from the small community located behind beach **K 331**, located immediately to the north. This is an 80 m long low gradient intertidal beach, with sand to low tide then rocks covering the southern half of the beach. A creek flows across the southern end, with two houses located behind the beach.

Figure 4.39 View seaward across Freshwater Cove beach (K 330) showing the steeper high tide beach and wide low tide sand flats.

K 332-337 **LANGGI**

No.	Beach	Rating HT	LT	Type	Length
K332	Langgi (S5)	1	2	R+rock flats	250 m
K333	Langgi (S4)	1	2	R+sand/rock flats	100 m
K334	Langgi (S3)	1	2	R+rock flats	300 m
K335	Langgi (S2)	1	2	R+rock flats	200 m
K336	Langgi (S1)	1	1	R+sand flats	100 m
K337	Langgi	1	1	R+tidal sand flats	200 m
Spring & neap tide range:				8.8 & 2.0 m	

From Freshwater Cove the rocky coast trends due north for 6 km to the next small embayment, called Langgi. Between the two are five beaches, with one in the embayment. All are backed by vegetated sandstone bluffs and slopes, all are bordered by rocky points and most are fronted by continuous 200-300 m wide intertidal sandstone rock flats.

Beach **K 332** lies 200 m to the north and is a curving 250 m long strip of high tide sand fronted by 200m wide rock flats. Beach **K 333** is located 300 m to the north and is a 100 m long low gradient pocket of sand, that has partly covered the boulder flats with 100 m of sand. Beach **K 334** commences 200 m to the north and is a curving 300 m long beach fronted by irregular 300 m wide rock flats.

Beach **K 335** is a straight 200 m long strip of northwest-facing sand, fronted by 100-200 m wide rock flats. Beach **K 336** is a 100 m long pocket of high tide sand located in lee of a 50 m wide gap in the rocks. It faces northwest and is fronted by 200 m wide sand flats that fill the small gap.

Langgi beach (**K 337**) is located on the south side of the creek that flows into the mangrove-dominated 1 km long Langgi embayment. The 200 m long sand beach is backed by a low grassy foredune. It is divided in two by a small rock outcrop, and fronted by 200 m wide tidal sand flats associated with the creek.

K 338-342 **PRIOR POINT (S)**

No.	Beach	Rating HT	LT	Type	Length
K338	Prior Pt (S5)	1	2	R+sand/rock flats	50 m
K339	Prior Pt (S4)	1	2	R+sand/rock flats	50 m
K340	Prior Pt (S3)	1	2	R+rock flats	50 m
K341	Prior Pt (S2)	1	2	R+sand flats	100 m
K342	Prior Pt (S1)	1	2	R+sand/rock flats	70 m
Spring & neap tide range:				8.8 & 2.0 m	

Prior Point is a 2 km long 20 m high sandstone headland, which protects the bedrock-controlled entrance to a small estuary on its east. Its exposed western side trends south for 4.5 km to Langgi, with five small pocket beaches located in gaps in the sandstone rocks.

Beaches **K 338, 339** and **340** are located between 2 and 2.5 km south of the point and are each 50 m long pockets of high tide sand fronted by 100-200 m wide rock flats and rock rubble, with no sand reaching low tide. Beach **K 341** is a 100 m long sand pocket with sand flats partly covering the rocks. Finally beach **K 342** is a 70 m pocket of high tide sand with 100 m wide rock flats.

K 343-347 **HALL POINT (S)**

No.	Beach	Rating HT	LT	Type	Length
K 343	Hall Pt (S 5)	1	1	Ultradissipative	1.3 km
K 344	Hall Pt (S 4)	1	2	R+rock/sand flats	1 km
K 345	Hall Pt (S 3)	1	2	R+sand flats	300 m
K 346	Hall Pt (S 2)	1	2	R+rock/sand flats	300 m
K 347	Hall Pt (S 1)	1	2	R+rock flats	800 m
Spring & neap tide range:				8.8 & 2.0 m	

Hall Point is a roughly circular 2 km wide, 40 m high sandstone headland that protrudes 4 km west of the line of the coast. It forms the northern boundary of a 4 km wide embayment, with Prior Point the southern headland. In between are five more exposed beaches (K 343-347).

Beach **K 343** is a 1.3 km long west-facing beach that receives sufficiently high westerly waves to maintain a 300 m wide ultradissipative beach, backed by a central 20 m high bluff, with low spinifex-covered beach-foredune ridge plains to either side. The northern beach ridge recurves round into a small usually dry creek mouth. There is a 4WD track down the backing slopes to the southern beach ridge plain, where two houses are also located.

Beach **K 344** is located 1 km to the north. It is a west-facing high tide sand beach fronted by a mixture of 200 m wide sand and rock flats, and backed by vegetated bluffs and slopes rising to 40 m.

Beaches 345-347 are located on the more protected southern side of Hall Point. Beach **K 345** is a 300 m long south-facing beach, backed and bordered by sandstone bluffs and low points. It consists of a high tide boulder beach and patchy intertidal sand and rocks, fronted by 200 m wide sand flats. Its neighbour, beach **K 346**, is a similar 300 m long high tide boulder and intertidal patchy sand beach, largely fronted by scattered rocks, and in the west a continuous 200 m wide rock flat. The rock flat turns and trends northwest, with beach **K 347** a narrow high tide cobble-boulder beach located at the rear of the rock flats, with some sand patches. The rocks are paralleled by 100 m wide fringing reefs.

K 348-350 **HALL POINT (E)**

No.	Beach	Rating HT	LT	Type	Length
K348	Hall Pt (E1)	1	1	R+sand flats	400 m
K349	Hall Pt (E2)	1	1	R+sand flats	150 m
K350	Hall Pt (E3)	1	1	R+sand flats	200 m
Spring & neap tide range:				8.8 & 2.0 m	

The eastern side of Hall Point is linked to the mainland by a peninsula with a series of north-trending 20-40 m high ridges. Three north-facing beaches (WA 348-350) are located between the ridges (Fig. 4.40).

Figure 4.40 The three Hall Point beaches (K 348-350) consist of beaches and wide sand flats wedged in between the sandstone arms of the point.

Beach **K 348** is the north side of a tombolo, which ties Hall Point to the rest of the peninsula. It is a 400 m long north-facing beach bordered by the point to the west and the first 1 km long ridge to the east, with a small usually dry creek against the western rocks. It is a relatively steep beach, with beachrock exposed in the lower intertidal, sand flats extending 400 m into the intervening bay and some deeper coral reef further out. The beach is backed by a low 50-100 m wide low foredune plain, covered with *Triodia pungens* (Soft Spinifex) and a few pandanus, then a mangrove-fronted south-facing beach ridge, which forms the southern side of the tombolo.

Beach **K 349** occupies the next embayment on the eastern side of the 200 m wide ridge. The 150 m long beach and low grassy foredune are wedged in between two sandstone headlands and their mangrove-covered rock flats. It is fronted by 400 m wide sand flats, which extend out to the limit of the western headland, with fringing reef linking the headlands. It is backed by a beach ridge, then vegetated slopes. Beach **K 350** is located in the third embayment, and is a 200 m long high tide sand beach, with 100 m wide sand flats, then a band of fringing reef.

K 351-352 **DECEPTION BAY**

No.	Beach	Rating HT	LT	Type	Length
K351	Deception Bay (1)	1	2	R+reef flats	200 m
K352	Deception Bay (2)	1	1	R+tidal sand flats	100 m
Spring & neap tide range:				8.8 & 2.0 m	

Deception Bay is a 9 km long 2-3 m wide bedrock-controlled bay with a 2 km wide entrance in the centre. Directly opposite the entrance are two small low energy beaches. Beach **K 351** is a 200 m long west-facing strip of sand backed by a high tide cobble beach with a 1-2 m high grassy foredune to the east. It is fronted by a 50 m wide band of rock flats, then a 100 m wide fringing reef. Five hundred metres to the north beach **K 352** lies at the base of a 300 m deep, 100 m wide embayment. The 100 m long beach is fronted by 200 m wide sand flats. Both beaches are backed by wooded slopes.

K 353-358 WILSON POINT (S)

No.	Beach	Rating HT	Rating LT	Type	Length
K353	Wilson Pt (S4)	1	1	R+sand/rock flats	600 m
K354	Wilson Pt (S3)	1	1	R+reef flats	50 m
K355	Wilson Pt (S2)	1	1	R+tidal sand flats	100 m
K356	Wilson Pt (S1)	1	1	R+rock flats	100 m
K357	Wilson Pt	1	1	R+sand/reef flats	150 m
K358	Wilson Pt (E)	1	1	R+sand flats	800 m
Spring & neap tide range:				8.8 & 2.0 m	

Wilson Point is a vegetated sloping sandstone headland that rises to 70 m. It is connected at low tide by rock flats to a 1 km wide 40 m high island. Four beaches (K 353-356) are located within 3 km south of the point, and one beach (K 357) on the south side of the island.

Beaches K 353 and 354 are adjoining sand-cobble beaches 200 m and 400 m long respectively, with a 100 m long low rock outcrop separating the two. Beach **K 353** is a discontinuous 600 m long, steep cobble beach, fronted in the southern half by 100 m wide rock flats. A small tidal creek drains out at the southern end of the beach across the rocks. Beach **K 354** is a 50 m long pocket of sand wedged between its neighbours. It is fronted by a 200 m wide fringing reef and backed by vegetated sandstone slopes.

Beach **K 355** lies 1 km to the north at the base of a small V-shaped northwest-trending rocky embayment. It consists of a steep 100 m long cobble beach backed by a small upland creek which drains into the bay. The bay is 500 m long and filled with tidal sand flats and a central tidal channel, and bordered by rock flats and the sandstone slopes. Beach **K 356** is located 500 m to the north. It is a 100 m long pocket of west-facing sand, located in a gully in the rocky shoreline. It has a small partially active foredune and is bordered by rock flats, with a narrow strip of central sand flats, and a fringing reef 150 m offshore.

Beach **K 357** is located in a small circular-shaped embayment on the southern side of the Wilson Point island. It is a 150 m long south-facing beach bordered and backed by sandstone slopes, and fronted by a 100 m wide patch of sand flats, then 200 m wide reef flats which link two low rocky outcrops that form the entrance to the bay.

Beach **K 358** lies 2 km east of Wilson Point below the 70 m high escarpment that forms the northern side of the point. The 800 m long beach faces northwest into Camden Sound and is backed by vegetated slopes rising to 80 m. The beach consists of a moderately steep, high tide beach that extends almost of the rocks, with room for a few turtles to nest. It is fronted by 200 m wide sand flats, with bordering rock flats, and a rock outcrop in the centre. Reefs fringe either headland, separated by a clear central sandy seabed.

K 359-360A SAMPSON INLET, NEEDLE ROCK

No.	Beach	Rating HT	Rating LT	Type	Length
K359	Sampson Inlet (S)	1	1	R+tidal sand flats	200 m
K360	Needle Rock	1	2	R+rock flats	500 m
K360A	Kuri Bay	1	2	R+rock flats	500 m
Spring & neap tide range:				8.8 & 2.0 m	

Beach **K 359** is located 1 km south of the entrance to Sampson Inlet. It is a 200 m long strip of high tide sand, wedged in behind a 50 m gap between rock flats and a protruding southern point, with sand flats widening outside the constriction, and totally 200 m in width.

Beach **K 360** is located on the northern side of Battery Point, immediately west of a sea stack called Needle Rock. The 500 m long beach faces north and is located at the base of a 100 m high sandstone escarpment. The beach consists of a narrow strip of high tide boulders, fronted by 150 m wide rock flats, then fringing reef. A small creek drains across the middle of the beach.

Beach **K 360A** lies on the northwestern side of Kuri Bay. The beach faces east across the 500 m wide bay mouth. It commences in the north amongst the boulders littering the base of the backing slopes and grades into more sand towards its southern end, with a total boulder-sand length of 500 m. A large boab tree is located at the northern end. The Kuri Bay pearl settlement and landing facilities are located at the base of a small 1 km deep bay.

K 360B-369 PORT GEORGE-HIGH BLUFF

No.	Beach	Rating HT	Rating LT	Type	Length
K360B	Port George IV (1)	1	2	R (cobble)	100 m
K360C	Port George IV (2)	1	2	R+sand/rock flats	200 m
K361	Port George (1)	1	2	R+rock flats	150 m
K362	Port George (2)	1	2	R+mud flats	300 m
K363	Port George (3)	1	1	R+tidal sand flats	400 m
K364	Port George (4)	1	2	R+sand flats	300 m
K365	High Bluff (S 5)	1	2	R+LTT	1.1 km
K366	High Bluff (S 4)	1	1	R+sand flats	250 m
K367	High Bluff (S 3)	1	1	R+sand flats	150 m
K368	High Bluff (S2)	1	2	R+rock flats	200 m
K369	High Bluff (S1)	1	1	R+sand/reef flats	100 m
Spring & neap tide range:				8.8 & 2.0 m	

Port George IV is an irregular-shaped bedrock embayment bordered in the west by Augustus Island and in the east by the 15 km long broad peninsula that terminates at High Bluff. Two low energy beaches (K 360B and 360C) are located within the port, while the more exposed western headlands of the High Bluff peninsula receive sufficient wave energy down the port to maintain nine sandy beaches.

Beaches K 360B and 360C are located on the eastern shores of Port George IV 3 km north of Kunmunya Hill. Beach **K 360B** is a 100 m long cobble beach, with a few patches of sand located at the base of steep well vegetated slopes which rise to 100 m. Beach **K 360C** lies 1 km to the east and is a 200 m long north-facing pocket of high tide sand, backed by a low narrow grassy foredune, then tree-covered slopes. It is fronted by 100 m wide sand and rock flats, with mangroves growing on the eastern rocks.

Beaches K 361 and 362 form a grassy 5-8 m high tombolo that links a 46 m high 500 m long sandstone ridge to the mainland (Fig. 4.41). Beach **K 361** is a 150 m long south-facing beach fronted by narrow rubble flats with two boab trees located behind its western end. It backs onto beach **K 362**, a 300 m long north-facing moderately steep, sandy high tide beach with a few slabs of beachrock littering its base. It is fronted by 200 m wide sand flats and backed by one boab tree on the 200 m wide grassy tombolo which links the two beaches. They are both bordered by the sandstone ridge on the west and a 100 m high sandstone escarpment to the east. Beach **K 363** lies in the next embayment 1 km to the east. It is a wide 400 m long north-facing high tide sand beach, fronted by irregular 200 m wide tidal sand flats. It is backed by a 50-100 m wide low grassy foredune and surrounded on three sides by sandstone slopes rising to 100 m.

Figure 4.41 Beaches K 361 & 362 tie the headland in foreground to the mainland, with beach K 363 to rear.

Beaches K 364 and 365 are another pair of beaches linked by a 50 m wide vegetated tombolo to a 100 m high 800 m long sandstone ridge, located 5 km south of High Bluff. Beach **K 364** is a curving 300 m long beach fronted by 100 m wide sand flats, with a pearl paddock located 500 m off the beach. A 50-100 m wide, 5-6 m high foredune separates it from beach **K 365**, a 1.1 km long curving northwest-facing wide sand beach, fronted by a 50 m wide low tide terrace. It has a few slabs of beachrock and a patch of boulders in the southern corner. Both beaches are bordered in the east by a 140 m high sandstone escarpment. A small rock outcrop separates beach K 365 from **K 366**, a 250 m long continuation of the beach. It is more sheltered from waves with west-facing sand flats widening to 150 m at its northern end. A low foredune runs the length of the beach increasing in height and instability towards the southern end, with the sandstone escarpment behind.

Beach **K 367** is a 150 m long pocket of west-facing sand located at the mouth of a small upland creek which crosses the northern end of the beach and has a few mangroves. The beach grades to boulders to the south and is backed by a small foredune in the north, and fronted by 100-200 m wide sand flats, with rock flats to either side. Beach **K 368** is located on the first point south of High Bluff. It is a 200 m long patch of high tide sand, backed by sandstone slopes, and fronted by 200 m wide rock flats, then another 200 m of fringing reef.

Beach **K 369** lies immediately south of **High Bluff** and faces west across the 10 km wide mouth of Port George to Adieu Point. It is a 100 m long high tide beach, widening to 200 m at low tide, bordered by rock flats and fronted by a 50 m wide low tide terrace, then a 100 m wide fringing reef. It is backed by a 3-4 m high grassy foredune, then a 150 m wide rocky ridge that separates it from beach K 370.

K 370-378 **HIGH BLUFF-HANOVER BAY**

No.	Beach	Rating HT	LT	Type	Length
K370	High Bluff (E1)	1	1	R+sand flats	400 m
K371	High Bluff (E2)	1	1	R+sand flats	200 m
K372	High Bluff (E3)	1	1	R+tidal sand flats	250 m
K373	High Bluff (E4)	1	1	R+sand/reef flats	1 km
K374	Hanover Bay (W1)	1	1	R+tidal sand flats	100 m
K375	Hanover Bay (W2)	1	2	R+rock flats	300 m
K376	Hanover Bay (W3)	1	1	R+tidal sand flats	200 m
K377	Hanover Bay (W4)	1	1	R+tidal sand flats	60 m
K378	Hanover Bay (W5)	1	1	R+sand flats	300 m
Spring & neap tide range:		8.8 & 2.0 m			

Immediately east of High Bluff is a 2 km wide, 2 km deep north-facing bay containing four beaches (K 370-373). Beaches **K 370** and **371** are adjoining 400 m and 200 m long beaches separated by a small high tide headland. They face northwest out of the bay, with 1 km long sandstone headlands to either side. They are backed by a low grassy foredune, then slopes rising to 40 m, while they are fronted by 500 m wide sand flats, with a fringing reef linking the two headlands. Both beaches are littered with slabs of beachrock, which increase in prominence towards the southern end.

Beach **K 372** is located in the next embayment to the east and is a more protected 250 m long high tide strip of sand fronted by 300 m wide tidal sand flats then fringing reef. It has beachrock slabs exposed, darker sand than its neighbours, mangroves on the eastern rocks and is backed by vegetated slopes rising to 140 m. Beach **K 373** occupies the southeast corner of the bay and is a straight 1 km long north-facing beach. It is backed by a low continuous foredune, then slopes rising to 40 m, with a small creek draining across the western end of the beach.

It is fronted by 200 m wide sand flats, then 200 m of reefs.

Beaches K 374 to 378 are all located in small embayments along a 9 km long section of the northeast-facing, deep dissected rocky western shore of Hanover Bay. Beach **K 374** occupies the first small embayment 1 km east of the larger High Bluff bay. It is a 100 m long beach widening to 150 m wide sand flats that fill a V-shaped break in the rocks. A small mangrove-lined creek runs along the eastern side of the beach, which is also backed by 100 m wide grassy overwash flats. It is bordered by 20 m high sandstone headland and rock flats, and fronted by 300 m wide sand flats, with beachrock exposed in the lower intertidal, then 200 m wide fringing reef.

Beach **K 375** is a sinuous 300 m long boulder beach located 3.5 km to the east. The boulders impound a small partly infilled 1 ha lagoon, fed by a small upland creek. The lagoon breaks out through the boulders in the wet season. The beach is fronted by 100 m wide rock flats and boulder rubble, with a few patches of sand toward the eastern end.

Beach **K 376** is located in a west-facing gap on the western tip of the next elongate 200 m wide, 2 km deep embayment. The 200 m long low gradient beach faces northwest and is backed by a low grassy foredune and fronted by 100 m wide sand flats. It grades to boulders in the west.

Beach **K 377** is a 60 m pocket of sand, backed by a patch of foredune and rising 60 m high slopes. It is fronted by 200 m wide sand flats, with a small creek and boulder delta forming the western end of the north-facing beach. One kilometre to the east is beach **K 378**, a 300 m long north-facing intertidal beach located in the centre of a small creek-fed embayment and fronted by 200 m wide sand flats.

Prince Regent Nature Reserve

Area:	633 825 ha	
Coast length:	282 km	(7773-8055 km)
Beaches:	62	(K 387-450)

Prince Regent Nature Reserve includes the linear Prince Regent River, as well as 282 km of coast and 62 beaches between the river mouth and the Roe River mouth in Prince Fredrick Harbour.

K 379-386 HANOVER BAY-UNWINS ISLAND

No.	Beach	Rating HT	LT	Type	Length
K379	Hanover Bay (W1)	1	1	R+tidal sand flats	300 m
K380	Hanover Bay (W2)	1	1	R+sand/rock flats	200 m
K381	Hanover Bay (W3)	1	1	R+sand/rock flats	400 m
K382	Unwins Is (1)	1	1	R+sand flats	300 m
K383	Unwins Is (2)	1	1	R+tidal sand flats	50 m
K384	Unwins Is (3)	1	1	R+sand/rock flats	200 m
K385	Unwins Is (4)	1	1	R+sand/reef flats	50 m
K386	Unwins Is (5)	1	1	R+tidal sand flats	100 m
Spring & neap tide range:				8.8 & 2.0 m	

The southern corner of **Hanover Bay** converges into a 1 km wide, 6 km deep southwest-trending embayment (K 379-381). Three beaches are located on the northeastern entrance to the embayment. Beach **K 379** lies 2 km south of the entrance, but faces northwest out of the mouth to receive sufficient wave energy to maintain a 300 m long moderately steep beach composed of a clean quartz sand. It is backed by a low grassy foredune and 20 m high sandstone slopes and fronted by 500 m wide sand flats.

Beaches K 380 and 381 lie on the eastern side of the entrance on a small sandstone peninsula attached at the northern end to the mainland. They are adjoining high tide beaches, backed and bordered by low sandstone slopes. Beach **K 380** is a 200 m long northwest-facing beach backed by a 1-2 m high grassy foredune. It is fronted by a large rock outcrop and a mixture of sand and rock flats. Beach **K 381** is a similar double crenulate more northerly facing 400 m long beach also fronted by rock and sand flats, causing it to narrow to 60 m at low tide. The beach is backed by a 1-2 m high foredune, then a linear tidal creek that exits 1.5 km to the south at the end of the peninsula.

Beaches K 382 to 386 are located on the exposed western side of Unwins Island, a roughly triangular shaped island 8 km at its widest. The beaches all face northwest into Hanover Bay. Beach **K 382** is a 300 m long high tide sand beach located below 20-30 m high sandstone bluffs, which also protrude onto the centre of the beach. Large sandstone boulders back the beach which is bounded by cliff-backed rock flats and fronted by sand flats.

Beaches K 383 and 384 lie 2 km to the north. Beach **K 383** occupies the mouth of a 50 m wide drowned gully, with the creek flowing through a 50 m wide belt of mangroves, and across the 200 m wide sand flats that front the narrow intertidal beach. Sandstone slopes and rock flats border the beach. Beach **K 384** lies 100 m to the north and is a 200 m long beach, backed by a low grassy foredune and fronted by one 50 m wide patch of sand flat in the north, then rock flats and rock outcrops forming the rest of the foreshore, together with some patchy reef off the rocks.

Beach K 385 and 386 share adjoining gullies 2 km south of the northern tip of the island. Beach **K 385** is a 50 m long intertidal boulder beach, bordered by rock flats, with a small central 50 m wide sand flat, then a 200 m wide reef flat. Beach **K 386** is located on the eastern entrance to a 200 m long deeper winding gully, 1 km south of the island tip. The beach is intertidal and deposited on the side and fronted by the creek channel, with a similar tidal sand flat-filled gully opposite (Fig. 4.42).

Figure 4.42 Beach K 386 (left) is a small pocket of sand wedged in an eroded joint line in the sandstone.

Map 3: KIMBERLEY

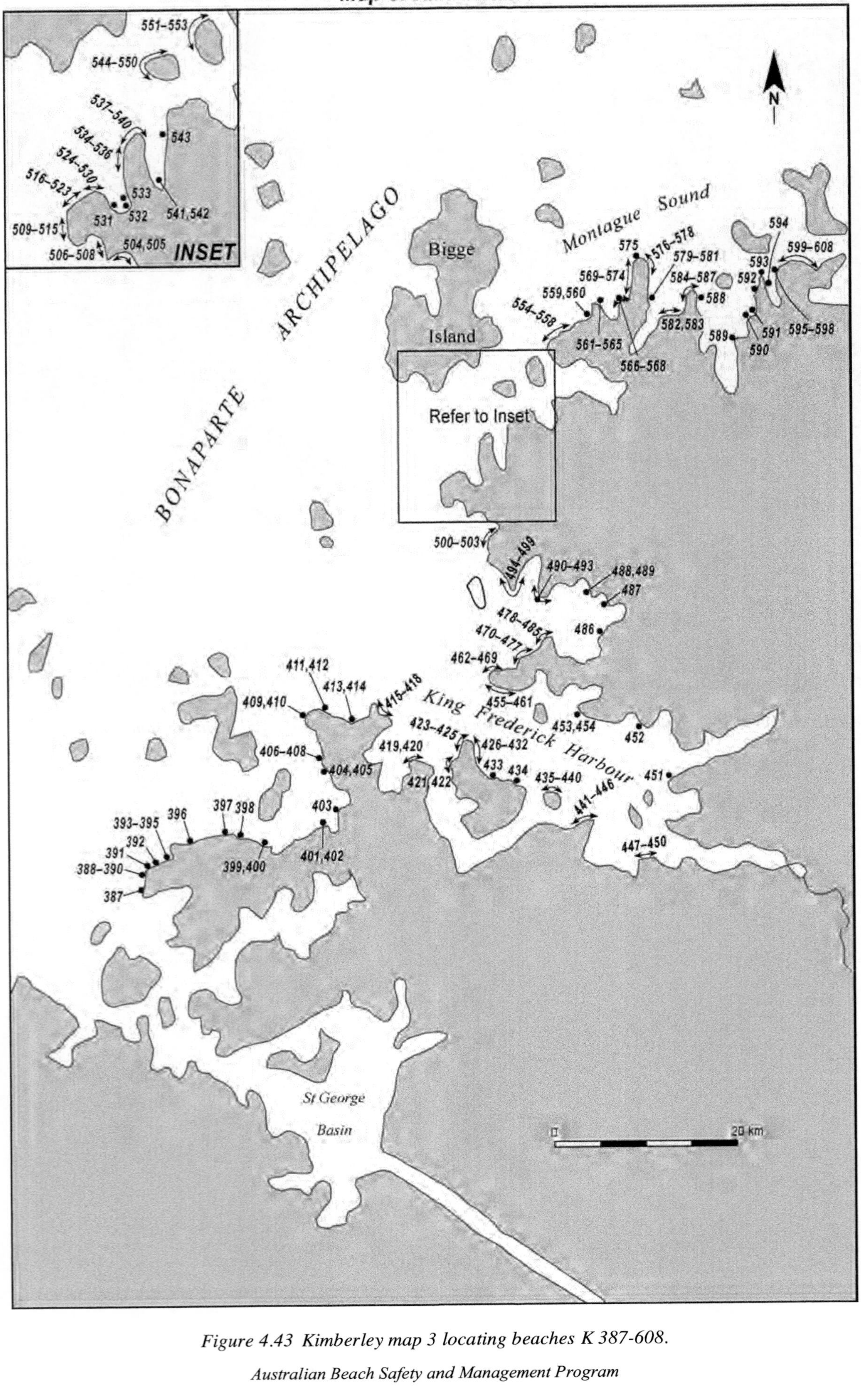

Figure 4.43 Kimberley map 3 locating beaches K 387-608.

Prince Regent Nature Reserve

Area:	635 000 ha	
Coast length:	282 km	(7773-8055 km)
Beaches:	64	(K 387-450)

The Prince Regent Nature Reserve is centred on the straight, joint-controlled Prince Regent River. It also includes the Kings Casades and Enid Falls. The reserve has no road access.

K 387-392 CAPE WELLINGTON

No.	Beach	Rating HT	LT	Type	Length
K387	Cape Wellington (S4)	1	2	R+reef flats	150 m
K388	Cape Wellington (S3)	1	2	R+reef flats	50 m
K389	Cape Wellington (S2)	1	2	R+reef flats	70 m
K390	Cape Wellington (S1)	1	2	R+reef flats	40 m
K391	Cape Wellington (N1)	1	2	R+reef flats	250 m
K392	Cape Wellington (N2)	1	2	R+reef flats	100 m
Spring & neap tide range:		8.8 & 2.0 m			

Cape Wellington is located on the eastern side of 30 km wide Brunswick Bay and looks south across a 6 km wide entrance to Unwins Island. From the cape the coast trends roughly northeast for 12 km to Cape Brewster, with the northwest-trending sandstone ridges, forming a series of headlands and small embayments along the shore. Eleven rock-bound beaches (K 387-397) are located in the small bays between the two capes.

Beach **K 387** is located just inside the west-facing tip of the 10 m high sandstone cape. It is a 150 m long northwest-facing high tide beach, backed by a narrow 2-3 m high foredune, then low sandstone, and fronted by a fringing reef, which narrows from 100 m in the south to 50 m at the northern end of the beach.

Beaches K 388, 389 and 390 are located in the first small bay 2 km north of the cape, on intersecting joint lines in the sandstone (Fig. 4.44). Beach **K 388** is a 50 m long pocket of sand wedged in between the 40 m high sandstone walls of the gully. It faces northwest with a patch of foredune and a small creek running down the eastern side, while it is fronted by 100 m wide sand flats and a fringing reef. Beach **K 389** lies immediately to its east and is a similar slightly wider 70 m long intertidal sand beach, that terminates as the gully pinches. It widens to 100 m at the end of the low tide sand flats, with fringing reef continuing round from its neighbouring beaches. Beach **K 390** lies directly opposite beach K 388. It is a 40 m long pocket of sand, with 50 m wide sand flats and the fringing reef.

Figure 4.44 Beaches K 388-390 (right) and beach K 391 (left), all located in eroded joint lines, and typical of the many small pocket beaches in the area.

Beach **K 391** is located in the same joint-line as beaches K 388 and 390, and backs onto beach K 390. It is a northwest-facing 250 m long beach with patches of beachrock, fronted by a 50 m wide sand flat, then 100 m wide fringing reefs. It is backed by a low 150 m wide foredune plain, which almost reaches the rear of its neighbour. It is bordered by boulders to the west and a rock platform to the east.

Beach **K 392** lies in the next embayment 1.5 km to the east. The beach is located at the base of the 1 km deep embayment. It is 100 m long, backed by a 100 m wide low grassy flat, with a small creek draining along the eastern side.

K 393-397 CAPE BREWSTER (W)

No.	Beach	Rating HT	LT	Type	Length
K393	Cape Wellington (N3)	1	2	R+rock/reef flats	50 m
K394	Cape Wellington (N4)	1	2	R+rock/reef flats	30 m
K395	Cape Wellington (N5)	1	2	R+rock/reef flats	150 m
K396	Cape Brewster (W2)	1	2	R+sand/reef flats	150 m
K397	Cape Brewster (W1)	1	2	R+LTT	1 km
Spring & neap tide range:		8.8 & 2.0 m			

Cape Brewster is a north-pointing sandstone ridge, overlying basal basalt, that is attached at low tide to 400 m long Bat Island. The cape is bordered by 40 m high cliffs and rises inland to 80 m. Beaches K 393-397 are located along the 6 km of shoreline west of the cape.

Beaches K 393-395 occupy the base of a 500 m wide U-shaped north-facing embayment 4 km east of Cape Wellington. Beach **K 393** is a 50 m pocket of sand in a small gully, fronted by 100 m wide sand flats, then irregular rock flats, rocks and fringing reef. Beach **K 394** lies 100 m to the east and is a 30 m wide sliver of sand, and narrow gap of sand flats, with rock flats to either side and rock and reef flats continuing across in front. Beach **K 395** lies 100 m further east and is a 150 m long moderately steep, high tide sand beach, with a 2 m high foredune behind the western half of the beach. A large rock cuts the high tide beach in half, with rock flats to either side and 200 m wide reef flats in between.

Beach **K 396** lies 2 km to the northeast, 4 km from Cape Brewster. It is a 150 m long high tide beach narrowing to 70 m at low tide. It is backed by a small foredune, then sandstone slopes, and fronted by 100 m wide sand flats, then 100 m wide reef flats. It has a deep rock-lined narrow embayment on its immediate east.

Beach **K 397** is immediately west of Cape Brewster. It is a 1 km long slightly curving sandy beach, located at the base of the cape's 40-60 m high cliffs and bluffs. It faces west and is bordered by rock flats and fringing reef. The beach has a central foredune rising to 506 m, an incipient foredune, berm , then 60 m wide intertidal beach fronted by a 50 m wide low tide bar, then a sandy seabed in front.

K 398-401 **CAPE BREWSTER (E)**

No.	Beach	Rating HT	Rating LT	Type	Length
K398	Cape Brewster (E1)	1	1	R+sand flats	400 m
K399	Cape Brewster (E2)	1	1	R+tidal sand flats	400 m
K400	Cape Brewster (E3)	1	1	R+reef flats	100 m
K401	Cape Brewster (E4)	1	1	R+sand/reef flats	100 m
Spring & neap tide range:		8.8 & 2.0 m			

To the east of Cape Brewster the coast trends east for 15 km into Port Nelson. It becomes increasingly protected both by the indented nature of the coast and protection afforded by the large Coronation Island. Three beaches (K 398-400) are located within 4 km of the cape, with a fourth (K 401) 10 km east and in lee of the islands.

Beach **K 398** is located in the first embayment east of the cape, with the cape forming its 1.5 km long western headland. The 400 m long beach faces northwest and lies at the base of a 50 m high escarpment. It is backed by a small grassy foredune, with a creek crossing the western end of the beach. It is fronted by 200 m wide sand-tidal flats, with fringing reef to either side.

Beach **K 399** is located 2 km to the east at the base of the next embayment. It is a 400 m long northwest-facing beach that links a 1 km long headland to the mainland, forming a tombolo with the mangrove-fringed tidal flats on the western side. An 80 m high escarpment lies immediately south of the beach. The beach is backed by a narrow grassy foredune and fronted by 200 m wide sand flats. Beach **K 400** is located at the northwestern tip of the 20 m high headland. It is a 100 m long pocket of cobble and sand that narrows seaward between 100 m wide rock flats, fronted by fringing reef.

Beach **K 401** lies 5 km to the east near the tip of western headland of Careening Bay. The 100 m wide intertidal beach occupies a V-shaped west-facing gap in the rocks and is fringed by mangroves along the northern side, with 200 m wide sand flats filling the bay, and rock flats extending another 50-100 m offshore.

K 402-410 **PORT NELSON-HARDY POINT**

No.	Beach	Rating HT	Rating LT	Type	Length
K402	Careening Bay	1	1	R+tidal sand flats	300 m
K403	Port Nelson (E1)	1	2	R+mud flats	200 m
K404	Port Nelson (E2)	1	1	Tidal sand flats	100 m
K405	Port Nelson (E3)	1	1	R+tidal sand flats	250 m
K406	Port Nelson (E4)	1	1	R+sand/reef flats	400 m
K407	Port Nelson (E5)	1	1	R+sand/reef flats	250 m
K408	Port Nelson (E6)	1	1	R+sand/reef flats	250 m
K409	Hardy Pt (W2)	1	1	R+tidal sand flats	250 m
K410	Hardy Pt (W1)	1	1	R+tidal sand flats	200 m
Spring & neap tide range:		8.8 & 2.0 m			

Port Nelson is a 5 km wide, 15 km deep bay, bounded by Coronation Island on the west and a peninsula terminating at Hardy Point to the east. Nine beaches (K 402-410) are located along the southern and eastern shore, all bordered and backed by basalt headland and slopes.

At the very southern end of the port is **Careening Bay**, an open 1.5 km wide north-facing embayment, with beach **K 402** located at its base. The 300 m long beach is fronted by 200 m wide sand flats with a small creek crossing the eastern end of the beach. There is some dune and beachrock exposed in the creek which has deposited an extensive boulder delta against the eastern rocks. The beachrock is a result of the beach and some of its neighbours containing 80-90% carbonate sands. Just in along the creek is a large historic boab tree into which was carved 'HMC Mermaid 1820' by the crew of the survey ship under the command of Philip Parker King. Another smaller boab at the western end of the beach has been carved more recently by the crew of an Australian Navy ship. The bay continues to be a popular anchorage with cruising yachts.

Beach **K 403** is located 4 km to the northeast, and consists of a solitary 200 m long strip of intertidal sand, fronted by 200 m wide mud flats and backed by basalt slopes rising to 270 m.

Four kilometres to the north is beach **K 404,** a 100 m long pocket of intertidal sand in a V-shaped gully with a clump of mangroves at the mouth of the creek. The beach is inundated at high tide, with 150 m wide sand flats and a line of rock flats exposed at low tide. One kilometre to the north on the same peninsula is beach **K 405**, a west-facing 250 m long beach bordered by rock flats and backed by lightly vegetated slopes. It is fronted by 100 m wide sand flats.

On the next peninsula to the north is beach **K 406**, which faces northwest up the bay and is the most exposed of the port beaches. It is a 400 m long beach backed by 10-20 m high bluffs along the western half, with a small foredune and creek mouth at the eastern end. It has sloping beachrock in the lower intertidal and is fronted by 50-100 m wide sand flats and then 200 m wide fringing reef.

Beaches **K 407** and **408** share the next embayment separated by 400 m of vegetated rocky bluffs. Both beaches are 250 m long and bordered by low basalt points and backed by wooded slopes. They are fronted by continuous 100 m wide sand flats then a 100 m wide fringing reef.

Two kilometres to the north and in lee of **Hardy Point** are beaches K 409 and 410, which occupy a shallow 1 km wide embayment (Fig. 4.45). Beach **K 409** is a 250 m long west-facing sand beach, backed by vegetated steeply rising 100 m high basalt slopes, containing large patches of bare black basalt boulders. Beach **K 410** is 200 m long and faces southwest, with basalt forming its southern point and sandstone of Hardy Point the northern point. A small barrier links it to beach K 411. Both beaches are fronted by 200 m wide sand flats, with fringing reef around the points.

Figure 4.45 Basaltic Hardy Point is backed by beaches K 409-412 to right and beaches K 413 & 414 to left.

K 411-418 HARDY POINT-CAPE TORRENS

No.	Beach	Rating HT	Rating LT	Type	Length
K411	Hardy Pt (E1)	1	1	R+sand/reef flats	450 m
K412	Hardy Pt (E2)	1	1	R+sand/tidal flats	200 m
K413	Hardy Pt (E3)	1	1	R+sand/reef flats	100 m
K414	Hardy Pt (E4)	1	1	R+sand flats	50 m
K415	Cape Torrens (W4)	1	2	R+rock flats	30 m
K416	Cape Torrens (W3)	1	1	R+sand/mud flats	80 m
K417	Cape Torrens (W2)	1	1	R+sand/mud flats	30 m
K418	Cape Torrens (W1)	1	1	R+sand/mud flats	130 m
Spring & neap tide range:		5.4 & 1.2 m			

Hardy Point and Cape Torrens are two sandstone points that are located 8 km apart at the top of a 10 km long peninsula that separates Port Nelson from Prince Frederick Harbour. Between the two points is an indented, unoccupied north-facing rocky shoreline containing eight predominantly small pocket beaches (K 411-418).

Beach **K 411** is located on the eastern side of Hardy Point, and is connected by a 200 m wide barrier to beach K 410. The 450 m long beach faces north into York Sound, and consists of the straight beach backed by a 10 m high grassy foredune with a few trees. It is fronted by 100 m wide sand flats grading into 200 m wide reef flats, with beachrock outcropping along the lower beach with a basaltic eastern headland and surrounding boulder field and the sandstone Hardy Point to the west.

Beach **K 412** occupies the base of the next V-shaped embayment, 1 km to the east. It is a 200 m long north-facing low energy sand beach, backed by a low grassy overwash flat, with a small mangrove-lined tidal creek draining out along the western side. Intertidal sand flats front the creek mouth with low tide sand flats off the eastern half of the beach. The wooded valley sides rise to 90 m on the eastern side of the beach, and on both sides extend northward for 1.5 km as headlands.

Beaches **K 413** and **414** are located 2 km to the east on the western side, toward the entrance of the next V-shaped embayment. They are two small pockets of sand, 100 m and 50 m long respectively, bordered and backed by vegetated slopes that rise to 60 m. South of K 414 the valley progressively narrows and becomes filled with tidal sand flats.

Beaches K 415 and 416 are two 30 m and 80 m long pockets of sand encased in blocky sandstone rising sharply to 100 m. Semi-circular beach **K 415** is also fronted by a ring of sandstone boulders, while beach **K 416** is fronted by sand to mud flats, with rock and boulder fields to either side.

Beaches K 417 and 418 are adjoining slivers of north-facing sand located 1 km west of **Cape Torrens**. Beach **K 417** is a 30 m strip of high tide sand, backed by a higher boulder beach then a 20 m high sandstone cliff, while beach **K 418** is a 130 m long strip of steep sand backed by a low grassy foredune, then steep partly vegetated horizontally bedded slopes rising to 180 m (Fig. 4.46). Both have beachrock exposed on the lower beaches and are fronted by sand to mud flats.

Figure 4.46 Beach K 418 consists of a narrow strip of sand at the base of the steep sandstone slopes.

Prince Frederick Harbour

Coast length: 225 km
Beaches: 43 (K 419-461)

Prince Frederick Harbour is a bedrock-controlled embayment, 10 m wide at its entrance between Cape Torrens and the Anderson Islands, and up to 20 km deep. It has a highly irregular 225 km long shoreline containing 50 beaches which total only 5.4 km or 2% of the bay shoreline. There is no development and no vehicle access to the bay.

K 419-420 CAPE TORRENS (S)

No.	Beach	Rating HT	Rating LT	Type	Length
K419	Cape Torrens (S1)	1	1	R+sand flats	50 m
K420	Cape Torrens (S2)	1	1	R+sand flats	50 m
Spring & neap tide range:		5.4 & 1.2 m			

Beaches **K 419** and **420** are located on a 3 km wide peninsula 5 km south of Cape Torrens. The shoreline between the cape and beaches is dominated by sandstone ridges rising to a maximum of 230 m, with 10 km of mangrove-lined shore in the intervening embayment.

Beach **K 419** is a narrow 50 m long west-facing high tide beach bordered by low protruding sandstone ridges, with the sand flats extending 150 m to the tip of the ridges. Beach **K 420** lies 500 m to the east and faces north into Prince Frederick Harbour. The 50 m long high tide beach is located in a semi-circular gap in the backing 40 m high ridge, with its 100 m wide sand flats expanding outside the run of the rocky shore to 200 m in length.

K 421-425 BOONGAREE ISLAND (W)

No.	Beach	Rating HT	Rating LT	Type	Length
K421	Boongaree Is (W1)	1	1	R+sand flats	50 m
K422	Boongaree Is (W2)	1	1	R+sand flats	150 m
K423	Boongaree Is (W3)	1	1	R+sand flats	20 m
K424	Boongaree Is (W4)	1	1	R+sand flats	300 m
K425	Boongaree Is (W5)	1	1	R+sand flats	40 m
Spring & neap tide range:		5.4 & 1.2 m			

Boongaree Island is located on the southwestern side of Prince Frederick Harbour, with its mangrove-fringed southern shore paralleling the mainland, separated by a 200-500 m wide tidal channel. A wider 2-5 km bedrock channel borders the western side, while the northwestern and northern shore is more open to the waves within the harbour. The entire island consists of massive quartz sandstone and some conglomerate, and is deeply dissected with north-south and southwest-northeast-trending ridges. It has a maximum height of 230 m with most of the shoreline consisting of steeply sloping sandstone. It has approximately 50 km of shoreline and contains 19 beaches. Beaches K 421-425 are located on the northwestern tip of the island, beaches K 426-434 along the northeastern section and beaches K 435-440 on mangrove-attached bedrock islands at the eastern tip of the island.

Beach **K 421** is located in a curving 1 km wide, west-facing bedrock embayment which rises to a crest of 120 m. The 50 m long beach is wedged in at the base of the centre of the bay and is fronted by 100 m wide sand flats.

Beach K 422 and 423 are located in the next embayment 1 km to the west. The 1.5 km deep embayment has mangrove-fringed tidal flats at its apex, with beach **K 422** located immediately to the west. It is a 150 m long north-facing sand beach, backed by a low grassy beach ridge and fronted by 100 m wide sand flats. Beach **K 423** lies on the opposite side of the bay, and is a 20 m pocket of sand and sand flats located in a west-facing gap in the bedrock.

Beaches K 424 and 425 are located 1.5 and 1 km southwest of the northern tip of the island. Beach **K 424** is a curving 300 m long sand beach, bordered and backed by vegetated slopes rising to 100 m, with a central protruding section, which cuts the beach in two. Beachrock outcrops along the lower beach, with 100 m wide sand flats extending off the beach. A small creek drains through a small patch of mangroves in the northern corner. Beach **K 425** lies in the next embayment 500 m to the northeast. It is a 40 m pocket of sand bordered by a western headland and low eastern ridge while it is fronted by 100 m wide sand flats. A small mangrove-filled wetland backs the beach but is drained by a creek in an adjacent gap in the bedrock.

K 426-434 **BOONGAREE ISLAND (NE)**

No.	Beach	Rating HT	LT	Type	Length
K426	Boongaree Is (NE1)	1	1	R+sand/mud flats	50 m
K427	Boongaree Is (NE2)	1	1	R+sand/mud flats	100 m
K428	Boongaree Is (NE3)	1	1	R+sand flats	200 m
K429	Boongaree Is (NE4)	1	1	R+mud flats	200 m
K430	Boongaree Is (NE5)	1	1	R+sand flats	50 m
K431	Boongaree Is (NE6)	1	1	R+sand/rock flats	100 m
K432	Boongaree Is (NE7)	1	1	R+tidal sand flats	150 m
K433	Boongaree Is (NE8)	1	1	R+sand flats	150 m
K434	Boongaree Is (NE9)	1	1	R+sand flats	150 m
Spring & neap tide range:		5.4 & 1.2 m			

At the northern tip of **Boongaree Island** the shoreline trends south for 5 km into a 1 km wide embayment, and then east for another 5 km. Within this section are 14 km of indented shoreline containing nine small bedrock-controlled beaches (K 426-434) which occupy 1.15 km of the shore.

Beaches K 426 and 427 are located immediately east of the northern tip and both lie in lee of a 2 ha rock outcrop which is tied to the coast by beach K 427 (Fig. 4.47). Beach **K 426** is a 50 m long pocket of east-facing sand located at the base of a 20 m high sandstone boulder field, backed by vegetated slopes. It faces across the 300 m wide embayment to beach **K 427**. This is a bare 100 m long sand tombolo that faces west and east and is attached to the 10 m high generally bare rock outcrop.

Figure 4.47 Beach K 426 (right) and K 427 (left) form a small tombolo to link the rocky outcrop to the mainland.

Beaches K 428-43 are located roughly 1 km to the south. Beach **K 428** faces north and consists of a 200 m long strip of high tide sand backed by sandstone bluffs rising to 20 m, with some rocks on the beach, and is fronted by 200 m wide sand flats. Beach **K 429** lies 200 m to the south in a small east-facing embayment. It is a narrow 200 m long strip of high tide sand, also backed by sandstone boulders, and fronted by 200 m wide sand to mud flats. Beach **K 430** lies a further 300 m to the south in an eroded joint line. It is a 50 m long pocket of sand, backed the dissected joint line, and fronted by 100 m wide sand flats.

Beach **K 431** is located 4 km south of the tip of the island. It is a 100 m long north-facing sand beach wedged between the sloping sandstone and a small rocky islet. The beach is fronted by 100 m wide sand flats, which partly cover lines of rocks, and is backed by mangroves to the lee of the beach and islet.

Beach **K 432** lies at the base of the main embayment 5 km south of the northern tip of the island. It is located in the southeast corner of the bay and surrounded by vegetated sandstone ridges rising to 60-80 m. The protected 150 m long beach is fringed by mangroves to the east and fronted by 300 m wide sand flats.

Beach **K 433** is located in the next embayment, 1 km to the northeast. It is a more exposed 150 m long sand beach lying at the base of vegetated slopes that rise to 240 m, 2 km inland. The beach has a small foredune and 100 m wide sand flats grading into deeper mud flats.

Beach **K 434** lies a further 2 km to the east and is located at the base of slopes rising rapidly to 230 m. The straight beach is a 150 m long intertidal beach backed by rocky slopes, with a few mangroves in a dry creek that drains out at the western end of the beach (Fig. 4.48). It is fronted by 100 m wide sand flats.

Figure 4.48 Beach K 434 is a pocket beach crossed by a small creek.

K 435-440 **BOONGAREE ISLAND (E)**

No.	Beach	Rating HT	LT	Type	Length
K435	Boongaree Is (E1)	1	1	R+sand/mud flats	50 m
K436	Boongaree Is (E2)	1	1	R+sand/mud flats	30 m
K437	Boongaree Is (E3)	1	1	R+sand/mud flats	70 m
K438	Boongaree Is (E4)	1	1	R+sand/mud flats	20 m
K439	Boongaree Is (E5)	1	1	R+sand flats	60 m
K440	Boongaree Is (E6)	1	1	R+rock flats	50 m
Spring & neap tide range:		5.4 & 1.2 m			

The eastern tip of **Boongaree Island** consists of a 3 km long island linked to the main island by mangroves and tidal flats. One hundred metres east of the island are two touching 500 m long islets, totalling 1 km in length.

Beaches K 435-438 are located in a single roughly V-shaped, north-facing embayment on the first island. All four beaches are backed and bordered by large sandstone boulders, with vegetated slopes rising to over 100 m in their lee. Beach **K 435** is a 50 m pocket of sand, which is flooded on spring tides. Beaches **K 436** and **437** lie 500 m to the east and are adjoining 30 m and 70 m long pockets of sand, with a small grassy beach ridge in lee of the second beach. Beach **K 438** lies another 100 m to the east in a 20 m wide gap in the rocks, with scattered rocks off the beach. All four beaches are fronted by sand to mud flats.

Beach **K 439** is located at the western end of the first of the islets. It is a 60 m high tide beach, with a sandy intertidal zone and rock flats to either end. Beach **K 440** lies on the eastern end of the second islet, and is a 50 m long high tide beach backed by a 70 m wide grassy beach ridge-overwash flats. The bordering rocks converge toward low tide, with only a narrow sandy gap at the base of the beach.

K 441-446 **PRINCE FREDERICK HARBOUR (S)**

No.	Beach	Rating HT	Rating LT	Type	Length
K441	Prince Frederick Hbr (S1)	1	1	R+mud flats	200 m
K442	Prince Frederick Hbr (S2)	1	1	R+mud flats	40 m
K443	Prince Frederick Hbr (S3)	1	1	R+mud flats	50 m
K444	Prince Frederick Hbr (S4)	1	1	R+mud flats	30 m
K445	Prince Frederick Hbr (S5)	1	2	R+tidal sand flats	20 m
K446	Prince Frederick Hbr (S6)	1	1	R+sand/mud flats	100 m
Spring & neap tide range:		5.4 & 1.2 m			

Two kilometres east of Boongaree Island is a 2 km wide north-trending sandstone peninsula, with six small low energy beaches (K 441-446) located in bedrock gaps along its northern faces. Beach **K 441** lies on the western side of the peninsula. It is a 200 m long north-facing beach, bordered by the run of the peninsula and a 20 m high rock outcrop which it ties to the peninsula. The low energy beach is backed by overwash flats, then the more extensive mangroves of the adjoining embayment.

Beaches K 442-444 are located 1 km to the north in the first small embayment on the north face of the peninsula. Beach **K 442** is a 40 m long pocket of intertidal sand that faces east across the 200 m wide bay. Beach **K 443** is located immediately to the east in the apex of the bay. It is a 50 m long pocket of sand, with both beaches fronted by continuous mud flats. Beach **K 444** lies 200 m to the north in a 30 m wide gap in the rocks. All three beaches are backed by lightly vegetated sandstone slopes rising to 60-80 m.

Beach **K 445** is 1 km further east at the mouth of a small bedrock gully fed by an upland stream, which has a few mangroves at its mouth. The 20 m long intertidal beach faces east across the mouth of the 100 m wide embayment and is fronted by tidal sand flats of the creek.

Beach **K 446** lies at the eastern tip of the peninsula. It is a steep, 100 m long north-facing sand beach, backed by a patch of grass, then exposed sandstone boulders rising to 60 m, with bordering sloping sandstone rocks (Fig. 4.49), while sand to mud flats front the beach.

Figure 4.49 Beach K 446 is a small pocket of sand surrounded by the rugged sandstone slopes.

K 447-450 **PRINCE FREDERICK HARBOUR (S)**

No.	Beach	Rating HT	Rating LT	Type	Length
K447	Prince Frederick Hbr (S7)	1	1	R+mud flats	50 m
K448	Prince Frederick Hbr (S8)	1	1	R+mud flats	100 m
K449	Prince Frederick Hbr (S9)	1	1	R+mud flats	60 m
K450	Prince Frederick Hbr (S10)	1	1	R+mud flats	70 m
Spring & neap tide range:		5.4 & 1.2 m			

Beaches K 447-450 are located along the 1.5 km long northern face of a sandstone peninsula at the very southern extremity of Prince Frederick Harbour. Mangroves completely enclose the back of the peninsula. The four low energy beaches are all bedrock-controlled and face north up the harbour. Beach **K 447** is a 50 m long pocket of sand wedged at the base of a 300 m deep narrow bedrock embayment. Beach **K 448** is a 100 m long strip of sand located in the next 200 m wide embayment, with a rocky headland extending 400 m off the eastern side.

Beach **K 449** is located on the eastern side of the headland and is a 60 m long strip of high tide sand, while beach **K 450** is a similar 70 m long beach a further 100 m to the east, near the eastern end of the peninsula.

Mitchell River National Park

Area: 115 300 ha
Coast length: 140 km (8055-8195 km)
Beaches: 36 (K 451-486)

Mitchell River National Park occupies 140 km coast along the western side of Prince Fredrick Harbour, as well as an elongate northern section along the lower reaches the Mitchell River, which includes the Mitchell Falls.

K 451-454 **PRINCE FREDERICK HARBOUR (E)**

No.	Beach	Rating HT	Rating LT	Type	Length
K451	Prince Frederick Hbr (E1)	1	1	R+sand flats	300 m
K452	Prince Frederick Hbr (E2)	1	1	R+sand flats	600 m
K453	Prince Frederick Hbr (E3)	1	1	R+sand flats	100 m
K454	Prince Frederick Hbr (E4)	1	1	R+sand flats	50 m

Spring & neap tide range: 5.4 & 1.2 m

The eastern side of Prince Frederick Harbour is heavily indented by east-west and north-south-trending sandstone ridges, rising in places to 360 m. The generally low energy shoreline consists of steep bedrock in exposed sections and mangroves in the lower energy embayments, which include the bedrock-controlled mouths of the Roe and Hunter rivers. There are only four small beaches (K 451-454) in the 80 km long eastern shoreline section.

Beach **K 451** is located at the western tip of the northern peninsula that separates the two rivers and looks out across the outer mouth of Hunter River. The 300 m long beach faces northwest and consists of a strip of high tide sand, backed by densely vegetated sandstone and basalt boulder debris and slopes rising steeply to 150 m. It is bordered by rock debris and fronted by sand flats, with some beachrock exposed along the lower beach.

Beach **K 452** lies 6 km to the north past the outer entrance to the river. The 600 m long beach consists of 5-6 patches of high tide sand which are connected at low tide. Each is backed by boulder debris from the backing steep 150 m high sandstone slopes. Boulder debris also separates the beach sections at high tide and is littered on the fronting sand flats.

Beaches K 453 and 454 lie 8 km further west in two bedrock gaps toward the western end of a 3 km wide sandstone peninsula. Beach **K 453** is a 100 m long west-facing beach backed by 50 m high sandstone cliffs, with boulder debris littering the back of the beach and bordering each side. Beach **K 454** lies 200 m to the north and is a similar 50 m long beach, which is all but submerged at low tide.

K 455-461 **PRINCE FREDERICK HARBOUR (N)**

No.	Beach	Rating HT	Rating LT	Type	Length
K455	Prince Frederick Hbr (N1)	1	1	R+sand flats	20 m
K456	Prince Frederick Hbr (N2)	1	1	R+sand flats	50 m
K457	Prince Frederick Hbr (N3)	1	1	R+sand flats	100 m
K458	Prince Frederick Hbr (N4)	1	1	R+sand flats	60 m
K459	Prince Frederick Hbr (N5)	1	1	R+sand flats	30 m
K460	Prince Frederick Hbr (N6)	1	1	R+sand flats	100 m
K461	Prince Frederick Hbr (N7)	1	1	R+sand flats	150 m

Spring & neap tide range: 5.4 & 1.2 m

The northern boundary of Prince Frederick Harbour consists of two northeast-southwest-trending sandstone ridges, terminating in the west as the outliers called the Anderdon Islands. Seven small generally south-facing beaches (K 455-461) are located in gaps in the steep bedrock shore, all within 4 km of the western tip of the harbour boundary.

Beaches K 455, 456 and 457 are three adjoining beaches each occupying a gap in the sandstone slopes that rise to 20-30 m behind. Beach **K 455** occupies a joint line with the 20 m wide beach plugging the southern end, and mangroves running behind to the next embayment. Beach **K 456** lies 200 m to the west and is a 50 m pocket of sand backed and bounded by steeply rising sandstone slopes. Fifty metres to the west is 100 m long beach **K 457** a 100 m long sand beach backed by vegetated slopes, including a gully at the western end.

Beaches K 458 and 459 are located 500 m to the northwest on the outer eastern side of an embayment that bisects the tip of the point. Beach **K 458** is a 60 m long pocket of sand surrounded by vegetated sandstone that rises steeply to 50 m. Beach **K 459** is a 30 m pocket of sand located 100 m to the north.

Beach **K 460** lies on the western side of the embayment and is a 100 m long south-facing strip of high tide sand, backed by an outlier of boulder rubble, with mangroves behind the boulders.

Beach **K 461** lies just inside the western tip of the harbour entrance. It is a 150 m long strip of high tide sand, backed by vegetated slopes rising to 20 m, and bordered by boulder fields of dolerite. It has beachrock exposed in the lower intertidal, together with beachrock and boulder debris at the top and base of the beach.

K 462-473 ANDERDON ISLANDS (W)

No.	Beach	Rating HT	Rating LT	Type	Length
K462	Anderdon Is (1)	1	1	R+sand flats	100 m
K463	Anderdon Is (2)	1	1	R+sand flats	30 m
K464	Anderdon Is (3)	1	2	R+sand flats+rocks	400 m
K465	Anderdon Is (4)	1	2	R+sand flats+rocks	100 m
K466	Anderdon Is (5)	1	1	R+sand flats	70 m
K467	Anderdon Is (6)	1	1	R+sand flats	40 m
K468	Anderdon Is (7)	1	1	R+reef flats	50 m
K469	Anderdon Is (8)	1	1	R+sand flats	30 m
K470	Anderdon Is (9)	1	1	R+sand flats	150 m
K471	Anderdon Is (10)	1	2	R+sand flats+rocks	100 m
K472	Anderdon Is (11)	1	2	R+sand flats+rocks	50 m
K473	Anderdon Is (12)	1	1	R+sand flats+rocks	60 m
Spring & neap tide range:				5.4 & 1.2 m	

The **Anderdon Islands** are a cluster of five small islands spread along 8 km of coast. To the east of the islands the coast consists of highly indented northwest-trending sandstone ridges rising inland to between 40 and 160 m. Spread along the first 5 km of coast are twelve small rock controlled beaches (K 462-473) occupying the outer more exposed indentations, with rocky shore and/or mangroves in the more protected indentations.

Beach K 462 is located on the northern side of the dolerite point that forms the northern entrance to Prince Frederick Harbour, with beach K 461 200 m to the south of the southern side of the point. Beach **K 462** is a 100 m long northwest-facing sand beach, backed by a grassy foredune rising to 3-4 m, then 10-20 m high vegetated slopes, while it is fronted by low tide sand flats. Beach **K 463** lies 200 m further north and is a 30 m long pocket of intertidal sand occupying a gap in the rocks.

Beach **K 464** lies 2 km to the west of the largest of the islands. It is an irregular 400 m long beach consisting of a foreland in lee of rocks and an islet, and a northern segment connected at low tide to the bulk of the beach. The 50 m wide intertidal beach is fronted by sand flats and scattered rocks, with a low grassy foredune in lee of the foreland.

Beaches **K 465** and **466** lie immediately to the north and are two rock-bound 100 m and 70 m long high tide sand beaches, with the run of the rock ridges forming their boundaries (Fig. 4.50), as well as resulting in lines of rocks off beach K 465, and exposed 20 m high rocky ridges behind. Beach K 466 has a moderately steep beach face with exposed beachrock and is backed by a 3-4 m high grassy wedge of foredunes.

Figure 4.50 Beaches K 465 & 466 are wedged into gaps in the linear sandstone ridges.

Beach **K 467** lies toward the entrance of the next embayment and occupies the western end of a joint line that cuts across the headland. The 40 m long beach faces east into the small bay. Beach **K 468** lies at the eastern tip of the same bay and is a 50 m long bare sand tombolo that ties a small islet to the mainland, the islet surrounded by sand and reef flats. Beach **K 469** is a 30 m long pocket of east-facing sand located toward the head of the next 1 km deep embayment, and lies 500 m southeast of the tombolo.

Beaches 470-473 all occupy the same 1 km wide irregular embayment. Beach **K 470** lies at the southern extremity of the embayment and is a 150 m wide double valley filled with 200 m wide sand flats, with a small high tide beach on the western side. Beaches **K 471, 472** and **473** are adjoining 100 m, 50 m and 60 m long pockets of sand located along the eastern side of the embayment. They are each bounded by joint-controlled lines of sandstone which outcrops to the sides and off each of the beaches.

K 474-486 ANDERDON ISLANDS (E)

No.	Beach	Rating HT	Rating LT	Type	Length
K474	Anderdon Is (13)	1	1	R+tidal sand flats	30 m
K475	Anderdon Is (14)	1	1	R+sand flats+rocks	150 m
K476	Anderdon Is (15)	1	1	R+sand flats	100 m
K477	Anderdon Is (16)	1	2	R+rock flats	200 m
K478	Anderdon Is (17)	1	2	R+rock flats	80 m
K479	Anderdon Is (18)	1	2	R+mud flats	50 m
K480	Anderdon Is (19)	1	2	R+sand/rock flats	50 m
K481	Anderdon Is (20)	1	1	R+sand/rock flats	60 m
K482	Anderdon Is (21)	1	1	R+sand flats	100 m
K483	Anderdon Is (22)	1	1	R+sand flats	100 m
K484	Anderdon Is (23)	1	1	R+sand flats	100 m
K485	Anderdon Is (24)	1	1	R+sand flats	80 m
K486	Anderdon Is (25)	1	1	R+sand flats	150 m
Spring & neap tide range:				5.4 & 1.2 m	

The eastern section of the Anderdon islands borders a 5 km long section of highly indented coast dominated by the northwest-trending sandstone ridges. There are thirteen beaches (K 474-486) along this section, all located on the outer, most exposed sections of the bedrock-dominated shoreline, and all located in gaps and small embayments in the sandstone (Fig. 4.51).

Figure 4.51 Beaches K 474-477 are typical of the small pocket beaches along this section of shore.

Beach **K 474** is a 30 m pocket of high tide sand widening to 100 m at low tide. It is located at the base of a narrow 200 m deep V-shaped embayment, with sand flats and scattered mangroves extending 150 m into the bay, as well as rock outcrops toward the bay entrance. Steep slopes rise to 60-80m around the beach.

Beaches K 475 and 476 are located in the next embayment 200 m to the east. Beach **K 475** is a curving 150 m long northeast-facing low energy intertidal beach, backed by rocks and a few scattered mangroves and fronted by sand flats that link its neighbour. Beach **K 476** is a 100 m long north-facing strip of high tide sand, with a small 50 m long headland on its eastern side. A 3-4 m high patch of grassy foredune and sandstone slopes rise to 20-60 m behind the beaches. An intertidal sand ridge extends from the base of beach K 475 across beach K 476 and 100 m offshore, into the adjoining bay.

Beach **K 477** lies 1 km to the east on the tip of the next headland surrounded by lower rocks that protrude out from the steep backing slopes. It is a 200 m long high tide beach, with a central rock outcrop and low grassy foredunes to either side, with some mangroves occupying a depression behind the rocks. Rocks border each end and run across the beach and low tide area.

Beaches K 478-481 occupy four small embayments in the next headland 1 km to the east. Beach **K 478** is an 80 m pocket of intertidal sand bordered by rocks and backed by vegetated slopes rising to 60 m. Beach **K 479** lies 200 m to the northeast and is a 50 m long strip of intertidal sand at the base of a 150 m deep embayment, backed by a steep gully. Beach **K 480** is a 50 m pocket of sand almost encased in rocks, with a narrow gap of sand in the centre of the intertidal slope, and scattered rocks off the beach.

Beach **K 481** is a 60 m long strip of high tide sand, backed by a low sandstone bluff, with a clump of mangroves in lee of the bordering western rocks, with rocks also extending 100 m off the eastern end.

Beaches K 482 to 485 lie 1-2 km to the east on the last of the north-trending sandstone ridges. Beach **K 482** is a 100 m long sand beach and a patch of foredune, bordered and backed by 20 m high sandstone slopes. Beach **K 483** is another 100 m long beach with a large rock outcrop toward the western side, as well as prominent 20 m high cliffed headlands that partly enclose the beach and a small islet off the headlands. Beach **K 484** is a 100 m long strip of sand and small foredune backed and bordered by 20-30 m high sandstone slopes. Beach **K 485** lies 100 m to the east and is an 80 m long pocket of sand backed by vegetated slopes which rise steeply to 50 m.

Beach **K 486** is an isolated beach 5 km due east of beach K 485. It is located on the western side of a 1 km long sandstone peninsula, tied to the mainland by 500 m wide mangroves. The beach is 100 m long at high tide with a low foredune, widening to 150 m at low tide. It is bordered by 40 m high sandstone ridges and fronted by sand flats.

Admiralty Gulf Aboriginal Land

Area:	202 343 ha	
Coast length:	370 km	(8185-8565 km)
Beaches:	195	(K 487-681)

Admiralty Gulf Aboriginal Land occupies 370 km of irregular coast between York Sound and Walmesly Gulf which is part of the larger Admiralty Gulf. It also includes the northern Cape Voltaire.

K 487-499 **YORK SOUND (E)**

No.	Beach	Rating HT	Rating LT	Type	Length
K487	York Sound (1)	1	1	R+sand flats	200 m
K488	York Sound (2)	1	1	R+sand flats	100 m
K489	York Sound (3)	1	1	R+sand flats	100 m
K490	York Sound (4)	1	1	R+sand flats	100 m
K491	York Sound (5)	2	2	Boulder+sand	70 m
K492	York Sound (6)	1	1	R+sand flats	50 m
K493	York Sound (7)	1	2	R+sand flats+rocks	30 m
K494	York Sound (8)	1	2	R+sand flats+rocks	20 m
K495	York Sound (9)	1	2	R+sand flats+rocks	30 m
K496	York Sound (10)	1	2	R+sand flats+rocks	20 m
K497	York Sound (11)	1	2	R+sand flats+rocks	150 m
K498	York Sound (12)	1	2	R+sand flats+rocks	60 m
K499	York Sound (13)	1	2	R+sand/mud flats	20 m
Spring & neap tide range:				5.4 & 1.2 m	

York Sound is a 50 km wide northwest-trending embayment, which in its southern reaches includes Port Nelson and Prince Frederick Harbour. The eastern side of the sound between the Anderdon islands and Augereau Island contains a 16 km wide embayment that narrows for 20 km to the east. Beaches K 462-486 are located along the southern side of this embayment, while beaches K 487 to 499 lie between the eastern extremity and Augereau Island, a direct distance of 25 km, but containing 50 km of heavily indented, joint-controlled sandstone shoreline. The first 14 km into the Sound contain thirteen beaches located on the more exposed outer protrusions of a 20 km long section of indented shore.

Beach **K 487** lies 2 km north of beach K 486, and is also a steep west-facing 200 m long sand beach that extends to the base of the backing sandstone rocks. It is located on the southern side of a sandstone ridge and is bordered by ridges rising to 20 m to the south and up to 120 m in the north, while it is fronted by 100 m wide sand flats.

Beaches K 488 and 489 are located at the base of the next ridge 2.5 km to the west. Beach **K 488** is a bare 100 m long southeast-facing tombolo that links a small bedrock outcrop to the peninsula. Beach **K 489** lies 300 m to the north and is a 100 m long strip of high tide sand at the base of curving 20-30 m high sandstone bluffs.

Beaches K 490-493 are spread around the base of the next peninsula, a 2 km wide 100 m high sandstone ridge located 3 km further west. Beach **K 490** is a south-facing 100 m long pocket of sand, bordered and backed by rugged sandstone terrain rising to 100 m, and fronted by narrow sand flats. Beach **K 491** lies 600 m to the west and is a 70 m long high tide boulder beach, with narrow intertidal sand flats, and backing sandstone slopes. Beach **K 492** lies immediately to the west in a V-shaped joint line. It is a 50 m wide high tide sand beach, widening to 100 m at low tide, with steep 20 m high cliffs to either side. Beach **K 493** is located 500 m to the north on the western tip of the peninsula. It is a 30 m long strip of high tide sand, widening to 80 m at low tide, with sand flats off the beach. It is bordered by sloping sandstone rocks and surfaces, with large rocks also littering the beach and intertidal.

Beaches K 494-499 all occupy the base of the next 2 km wide, 40 m high peninsula (Fig. 4.52), which begins 1 km to the west. Beaches **K 494, 495** and **496** are three 20 m, 30 m and 20 m long pockets of sand toward the southwestern tip of the peninsula. Each is backed and bordered by sloping sandstone surfaces with rocks also outcropping on the beaches, the intertidal and lying offshore as reefs and small outcrops. Beach K 496 lies at the very tip of the peninsula.

Figure 4.52 Beaches K 494-498 are a series of small sand pockets dominated by the irregular sandstone rocks and ridges.

Beaches K497 and 498 are located on the northern side of the tip of the peninsula, immediately east of beach K 496. Beach **K 497** is a irregular 150 m long strip of bare sand that links two small rock outcrops with the peninsula and receives waves from both sides. Rocks litter the beach and inter- to sub-tidal. Beach **K 498** lies 100 m to the east and consists of two adjoining pockets of sand that total 60 m in length and are linked in the intertidal, with rocks also off the beach. Beach **K 499** is located 1 km to the west in a 20 m wide southeast-facing gap in the rocks. It is a very protected low energy beach fronted by sand then mud flats.

K 500-512 AUGEREAU ISLAND (S)

No.	Beach	Rating HT	Rating LT	Type	Length
K500	Augereau Is (S13)	1	1	R+sand flats	150 m
K501	Augereau Is (S12)	1	1	R+sand/mud flats	50 m
K502	Augereau Is (S11)	1	1	R+sand/mud flats	100 m
K503	Augereau Is (S10)	1	1	R+sand/mud flats	40 m
K504	Augereau Is (S9)	1	1	R+mud flats	50 m
K505	Augereau Is (S8)	1	1	R+mud flats	50 m
K506	Augereau Is (S7)	1	1	R+sand flats	40 m
K507	Augereau Is (S6)	1	1	R+sand flats+rock	50 m
K508	Augereau Is (S5)	1	1	R+sand flats+rock	80 m
K509	Augereau Is (S4)	1	1	R+sand/coral reef	50 m
K510	Augereau Is (S3)	1	1	R+sand/coral reef	100 m
K511	Augereau Is (S2)	1	1	R+sand/coral reef	100 m
K512	Augereau Is (S1)	1	1	R+sand/coral reef	100 m
Spring & neap tide range:		5.4 & 1.2 m			

Augereau Island is a 1 km wide sandstone island that forms the most western tip of inflection in the shoreline, with York Sound extending to the west and Scott Strait, between the mainland and Bigge Island to the northeast. The mainland shoreline trends roughly southeast from the island for 9 km and contains 15 km of highly indented sandstone shoreline fronted by numerous small islands, islets and rock reefs, as well a fringing coral reef toward the island. There are thirteen small, pocket, low energy beaches (K 500-512) located along the section of rocky shore that extends southeast of the island.

Beach **K 500** lies 9 km southeast of the island and is a 150 m long tombolo that connects a low 200 m long rock ridge to the mainland. The tombolo is 50 m wide and largely formed by southerly wind waves. It is fronted by sand flats to either side.

Beaches **K 501, 502** and **503** are three adjoining 50 m, 100 m, and 40 m long beaches, located at the mouth of a 2 km deep embayment immediately east of beach K 500. The three beaches are each bordered by 10-20 m high narrow sandstone ridges and fronted by sand to mud flats.

Beach **K 504** lies 1 km to the north and is a narrow 50 m long strip of high tide sand, widening to 100 m at low tide. It is backed by ridges rising to 40 m, with two large rocks lying off the southern end of the beach and fronting mud flats. Beach **K 505** lies on the northern side of the same embayment and is a narrow 50 m long strip of intertidal sand, backed by boulders, cliffs and ridges rising to 100 m and fronted by mud flats.

Beach **K 506** is located on the northern headland of the embayment in a small gap in the rocks in lee of a 300 m long rocky islet. It consists of a 40 m long strip of high tide sand at the base of 20 m high cliffs, with rock flats bordering each end.

Beaches K 507 and 508 lie 1 km to the north at the apex of a small embayment. Beach **K 507** is a 50 m long strip of sand bordered and backed by 10-20 m high sandstone slopes, with a creek entering the bay immediately to its north. Beach **K 508** is located 300 m to the northwest and is a 80 m long sand beach, backed by 5-10 m high sandstone bluffs, with rocks also outcropping across the eastern side of the south-facing beach.

Beach **K 509** lies at the southern tip of a discontinuous 3 km long shore-parallel sandstone ridge. The 50 m long beach faces south, is bordered by 10 m high headlands, and fronted by 50 m wide sand flats grading into deeper coral reef. Beach **K 510** lies 1.5 km to the north and attaches a 300 m long rock ridge to the main ridge. It is a 100 m long south-facing beach, with rock ridges to either side and some rocks off the beach (Fig. 4.53). Two hundred metre wide sand flats front the main beach, with coral reef fringing the outer run of the ridges.

Beach K 511 and 512 lie on the southern side of the northernmost and largest section of the ridge. Beach **K 511** is a narrow 100 m long strip of sand, backed by 10 m high sandstone bluffs and fronted by 100 m wide sand flats, then wider reef flats. Beach **K 512** lies 300 m to the west and is a wider 100 m long sand beach occupying a V-shaped gap in the rocks. It is also fronted by 150 m wide sand flats and some rocks, then the reef flats.

Figure 4.53 Beach K 510 (left-centre) ties a small rock islet to the main rock ridge, which is in turn connected to Augereau Island by sand flats.

K 513-520 **AUGEREAU ISLAND**

No.	Beach	Rating HT	Rating LT	Type	Length
K513	Augereau Is (1)	1	1	R+sand flats+rocks	100 m
K514	Augereau Is (2)	1	1	R+sand flats	150 m
K515	Augereau Is (3)	1	1	R+sand flats/coral	200 m
K516	Augereau Is (4)	1	1	R+sand flats/coral	30 m
K517	Augereau Is (5)	1	2	R+sand/rock/coral	80 m
K518	Augereau Is (6)	1	1	R+sand flats	150 m
K519	Augereau Is (7)	1	1	R+tidal sand flats	60 m
K520	Augereau Is (8)	1	2	R+tidal sand flats	40 m
Spring & neap tide range:				5.4 & 1.2 m	

Augereau Island is a rugged 20-30 m high, relatively bare 1 km long sandstone island, separated from the mainland by a 1 km long tidal channel, that deepens and narrows to 100 m at the northern end. The island has about 3 km of indented shoreline, and contains eight beaches (K 513-520) along its southern, western and northern sides, the tidal channels occupying the eastern side.

Beaches K 513 and 514 lie on the south-facing side of the island. Beach **K 513** is located near the southeast corner of the island and is a narrow curving 100 m long sand beach, with rocks protruding onto the beach, and a line extending out across the western end of the beach. Beach **K 514** occupies the next embayment and is a 150 m long beach, backed by a grassy overwash plain, with rocky headlands extending 200 m off either end of the beach, and a large outcrop off the eastern end of the beach. Both beaches are fronted by 200 m wide sand flats grading into coral.

Beach **K 515** lies at the western end of the island, and is a 200 m long slightly curving southeast-facing wide sand beach, backed by a foredune climbing 6 m up the backing sandstone rocks, and fronted by 150 m wide sand flats then fringing reef.

Beach **K 516** is a 30 m long pocket of sand, backed by a boulder beach on the northwestern tip of the island. This beach is fronted by narrow sand flats then 200 m wide reef flats. Beach **K 517** is located in the adjoining small embayment and is an 80 m long pocket of sand wedged in between blocky sandstone ridges that extend 100 m off the beach (Fig. 4.54), and are linked by sand flats and fringing reef, as well as rock outcrops.

Figure 4.54 Beaches K 517-521, including the tombolo formed by beaches K 519 & 520.

Beach **K 518** is a bare 50 m wide tombolo located at the northern tip of the island. The 150 m long beach faces the northeast, with a smaller beach on the southern side. Both sides are bounded by low rocky outcrops which widen to the north, with sand flats in between and surrounding fringing reef.

Beaches **K 519** and **520** are adjoining 60 m and 40 m long pockets of sand, with their small 100 m deep bays largely filled with sand flats, which extend 100-200 m beyond the headland to the edge of the deep 100 m wide tidal channel which separates the island from the mainland.

K 521-530 **CAPE POND**

No.	Beach	Rating HT	Rating LT	Type	Length
K521	Augereau Is (E1)	1	1	R+sand flats	100 m
K522	Augereau Is (E2)	1	1	R+sand flats	50 m
K523	Cape Pond (W4)	1	1	R+sand flats/coral	100 m
K524	Cape Pond (W3)	1	1	R+sand flats/coral	150 m
K525	Cape Pond (W2)	1	1	R+sand flats/coral	60 m
K526	Cape Pond (W1)	1	1	R+sand flats/coral	100 m
K527	Cape Pond (E1)	1	1	R+sand flats/coral	200 m
K528	Cape Pond (E2)	1	1	R+sand flats/coral	40 m
K529	Cape Pond (E3)	1	1	R+sand flats/coral	80 m
K 530	Cape Pond (E4)	1	1	R+sand flats/coral	300 m
Spring & neap tide range:		5.4 & 1.2 m			

Cape Pond is a 20 high sandstone headland located 3 km east of Augereau Island. The shoreline between the island and the cape is heavy indented with northwest-trending ridges and gullies which contain six pocket beaches (K 521-526). East of the cape the shoreline is similar, with four pocket beaches (K 527-530) contained in the next 3 km.

Beaches K 521 and 522 are located on the north side of the 4 km long ridge that begins at beach K 509, and is linked to the mainland by sand flats immediately east of beach K 522. Beach **K 521** is a 100 m long pocket of sand and beachrock that links the ridge with a smaller rock outcrop, resulting in a 100 m wide beach that is awash at high tide, with rocks and a small channel backing the beach. It is bordered by low sandstone ridges and fronted by 100 m wide sand flats, with the deeper tidal channel between the ridge and Augereau Island, immediately west of the small ridge. Beach **K 522** lies at the northeastern tip of the rocky ridge, and is a 50 m pocket of sand surrounded by 10 m high sandstone slopes, and fronted by a tidal channel largely filled by tidal sand shoals and that extends a few hundred metres along the elongate channel between the ridge and the mainland.

Beach **K 523** is located in the first small north-facing mainland embayment. It is a curving 100 m long beach and patch of foredune, that links two converging joint lines, both plugged with beach sand. The low energy beach is fronted by 500 m wide sand flats that fill the narrow embayment, and then fringing reef.

Beaches K 524 and 525 are located 1 km further east in a 500 m wide embayment. Beach **K 524** is a 150 m long sand beach, backed by a small 1-2 m high foredune then sandstone slopes. It is fronted by 300 m wide sand flats then fringing reef, and a few small rock reefs and islets. Beach **K 525** lies 300 m to the east and is a 60 m long beach occupying a narrow joint line. It is also fronted 150 m wide sand flats then rocks and reef flats, as well as some small exposed rocks.

Beach **K 526** lies immediately east of Cape Pond and occupies the westernmost of two joint lines that converge on the 300 m wide embayment. The beach is 100 m wide at high tide, narrowing as the sandstone ridges pinch out to 20 m wide at low tide. Outside the ridges the bay is filled with a mixture of sand, rock and reef flats. The eastern joint line leads to a 200 ha mangrove woodland, which backs the cape.

Beaches K 527-529 occupy the first embayment east of Cape Pond, with the cape forming the large western headland of beach **K 527**. This is a 200 m long beach backed by two converging joint lines, with a 100 m long overwash plain backing into the eastern line, and linking with the mangrove woodlands that back the cape. Beach **K 528** lies 300 m to the east and is a 40 m pocket of sand backed by a small grassy overwash flat and a patch of mangroves. One hundred metres to the east is beach **K 529**, an 80 m long sand beach occupying the next joint line, with the eastern 300 m long headland of the embayment forming its boundary. All these beaches are fronted by continuous 100-200 m wide sand flats grading into 50-100 m wide reef flats.

Beach **K 530** occupies most of the second embayment east of the cape. It is a curving 300 m long beach, backed by a small grassy overwash flat in the west, with 10-20 m high rocky bluffs behind the eastern half of the beach. The beach is dominated by beachrock, with eroded beachrock boulders also littering the top of the beach. It is fronted by sand flats which narrow from 100 m to join the rocks to the east, then fringing reefs, as well as some rocks on and just off the eastern end of the beach.

K 531-542 **CAPE POND (E)**

No.	Beach	Rating HT	Rating LT	Type	Length
K531	Cape Pond (E5)	1	1	R+tidal sand flats	200 m
K532	Cape Pond (E6)	1	1	R+tidal sand flats	400 m
K533	Cape Pond (E7)	1	1	R+tidal sand flats	100 m
K534	Cape Pond (E8)	1	1	R+sand flats	20 m
K535	Cape Pond (E9)	1	1	R+coral flats	30 m
K536	Cape Pond (E10)	1	1	R+sand flats+coral	200 m
K537	Cape Pond (E11)	1	1	R+sand flats+coral	100 m
K538	Cape Pond (E12)	1	1	R+sand flats+coral	200 m
K539	Cape Pond (E13)	1	1	R+sand flats+coral	300 m
K540	Cape Pond (E14)	1	1	R+sand flats+coral	200 m
K541	Cape Pond (E15)	1	1	R+sand flats+coral	100 m
K542	Cape Pond (E16)	1	1	R+sand flats+coral	150 m
Spring & neap tide range:				5.4 & 1.2 m	

Two kilometres east of Cape Pond is a 1 km wide embayment containing three arms each with a low energy sand beach and sand flats (K 531-533). The coast then trends north for 4 km to beach K 538, and then east again for 3 km to beach K 540. In total the ten beaches (K 531-540) occupy the joint-controlled gaps in an otherwise 8 km of lightly vegetated sandstone-dominated shoreline that rises to 40-60 m inland.

Beach **K 531** is a 200 m wide sand flat occupying a 200 m long east-facing embayment, with a perimeter of 20 m high sandstone bluffs. Immediately to its south is beach **K 532**, a 400 m long and 300 m wide tidal sand flat, backed by a row of mangroves and an upland creek following the joint lines. Rock outcrops border each side of the small sand-filled embayment. Beach **K 533** lies 1 km to the east in a third small 200 m wide embayment. The 100 m long high tide beach occupies the apex of the bay, the sand flats extending 200 m out into the 800 m deep embayment.

Beaches **K 534** and **535** lie 1 km to the north and are two 20 m and 30 m long pockets of sand, located in adjoining east-west-trending joint lines. They are both fronted by a mixture of rock and reefs. Beach **K 536** occupies the next larger joint line and its intersecting north-south joint. As a consequence the semi-circular shaped beach occupies parts of three joints, which combined have a total shoreline length of 200 m and face west. It is fronted by 300 m wide sand flats, then reef.

Beach **K 537** is a 100 m long pocket of sand with a 2-3 m high partly unstable foredune, backed and bordered by low sandstone rocks and slopes, and fronted by 100 m wide sand flats, then reef. Immediately to the north is beach **K 538**, one part of a tombolo that links a 400 m long ridge of sandstone to the coast. The curving 200 m long beach faces south, and is separated from beach K 529 by a 50-100 m wide 5 m high foredune covered with *Triodia pungens* (Soft Spinifex). Beach **K 539** is 300 m long and includes the 200 m long tombolo section as well as another 100 m backed by low sandstone slopes. Both beaches are steep, narrow and fronted by 100 m wide sand flats then coral reefs.

Beach **K 540** lies 1 km to the east and is a 200 m long north-facing sand beach, backed in part by a 150 m wide grassy overwash flat and bordered by low rocky points and backing slopes. Beach **K 541** lies 200 m to the east and is a 100 m long pocket of sand backed by a small grassy overwash flat, and bordered by sandstone headlands that protrude 50-100 m. Beach **K 542** lies immediately to the east, past a 50 m wide headland. It is a 150 m long northwest-facing beach, backed by moderately rising sandstone slopes. All three beaches share the same large embayment, and are fronted by 200-300 m wide sand flats, then fringing reef.

K 543-550 **SCOTT STRAIT**

No.	Beach	Rating HT	Rating LT	Type	Length
K543	Scott Strait (1)	1	1	R+sand flats	100 m
K544	Scott Strait (2)	1	1	R+sand flats+rocks	100 m
K545	Scott Strait (3)	1	1	R+sand flats+rocks	200 m
K546	Scott Strait (4)	1	1	R+tidal sand flats	30 m
K547	Scott Strait (5)	1	1	R+tidal sand flats	150 m
K548	Scott Strait (6)	1	1	R+sand flats/coral	50 m
K549	Scott Strait (7)	1	1	R+sand flats/coral	150 m
K550	Scott Strait (8)	1	1	R+sand flats/coral	20 m
Spring & neap tide range:				5.4 & 1.2 m	

Scott Strait is an approximately 20 km long strait that separates Bigge Island from the mainland. Both sides are dominated by heavily jointed and dissected sandstone. The strait is narrowest between beaches K 545 and 551, where it averages 3 km in width. In addition there are numerous small islands and coral reefs in the main strait, as well as strong tidal currents. Beaches K 543-550 extend for 8 km east of Cape Pond to the centre of the strait at beach K 550. All are dominated by the surrounding sandstone and most occupy small gaps in the joint lines (Fig. 4.55).

Beach **K 543** is an isolated 100 m long strip of high tide sand, backed by sandstone slopes that rise to 120 m. It is bordered by sloping intertidal rocks and fronted by 200 m wide sand flats.

Figure 4.55 Beaches K 544 (right), K 545 (centre) cut by a small creek, and K 546 (left).

Beach **K 544** lies 2 km to the north across a 1 km wide inlet opening. The beach is 100 m long, faces west and consists of a strip of high tide sand, backed by 40 m high slopes, with rocks outcropping on and just off the beach.

Beach **K 545** lies 300 m to the east on the north side of a 40 m high sandstone headland. The beach is 200 m long, with a large rock outcrop off the centre forming a small foreland, with overwash flats and a creek cutting diagonally across the backing 2-3 m high, 100 m wide grassy foredune plain. The beach is composed of 85% carbonate sand which has led to the formation of dune calcarenite. The dunerock is exposed in foredune which is scarped by the creek, while beachrock is also exposed on the steep intertidal beach. Beach **K 546** lies in the next gap in the rocks, 50 m to the east, and is a 30 m long pocket of intertidal sand, backed by a small clump of mangroves at the apex of the gap. Beach **K 547** is another 200 m to the east and is a narrow irregular 150 m long beach, backed by rugged sandstone slopes, with several rocks also outcropping on the sand flats. All three beaches are fronted by 100-200 m wide sand flats, with the deep strait channel linking the intervening bedrock points.

Beaches K 548, 549 and 550 lie on the eastern side of the next embayment, 1 km to the west. Beach **K 548** is a 50 m long pocket of sand, backed and bordered by low sloping sandstone. Beach **K 549** lies 200 m to the east and is a 150 m long west-facing beach backed by gently rising sandstone slopes. Immediately to its north is beach **K 550**, a 20 m pocket of sand, with a 200 m long low sandstone point extending north of the beach. All three beaches are fronted by 200-300 m wide sand flats, with some fringing reef, then the deeper strait channel outside the line of the rocks.

K 551-558 SCOTT STRAIT

No.	Beach	Rating HT	LT	Type	Length
K551	Scott Strait (9)	1	1	R+sand flats	200 m
K552	Scott Strait (10)	1	1	R+sand flats	100 m
K553	Scott Strait (11)	1	1	R+sand flats/coral	250 m
K554	Scott Strait (12)	1	1	R+sand flats/coral	150 m
K555	Scott Strait (13)	1	1	R+sand flats/coral	70 m
K556	Scott Strait (14)	1	1	R+sand flats/coral	60 m
K557	Scott Strait (15)	1	1	R+sand flats/coral	20 m
K558	Scott Strait (16)	1	1	R+sand flats/coral	100 m
Spring & neap tide range:				5.4 & 1.2 m	

Scott Strait is 3 km wide, its narrowest, at beach K 551 and expands northward for 7 km to be 7 km wide at its northern entrance in line with Capstan Island. Between the two are eight bedrock-controlled low energy beaches, all backed by sandstone slopes, with most fronted by sand flats, then fringing reef.

Beach **K 551** is located on a small rock outcrop at the western entrance to a narrow, linear, joint-controlled embayment (Fig. 4.56), which all but cuts off the 4 km long ridge to its west from the mainland. Mangroves back the outcrop and tidal sand flats fill the narrow bay. The 200 m long, north-facing beach is wedged between two small sandstone points, and fronted by 100 m wide sand flats then the deeper waters of the strait. Beach **K 552** is located on the eastern side of the same entrance, a 100 m long pocket of sand in a small curve in the rocks, with sandstone slopes rising to 40 m in its lee, and narrow sand flats off the beach.

Figure 4.56 Beach K 551 is located on the small islet (centre), with the pocket beach K 552 to left.

Beach **K 553** lies 1 km further east and is a 250 m long strip of high tide sand wedged at the base of a 60 m high escarpment. The beach is fronted by 150 m wide sand flats, then a narrow band of fringing reef before the deeper strait waters.

Beaches K 554 to 558 occupy a roughly V-shaped embayment 2 km further east. Beach **K 554** lies on the western arm of the bay and is a 150 m long north-facing strip of sand, backed by low sandstone on the point, with rocks outcropping on and off the beach. Beach **K 555** lies deep in the apex of the bay. It is a 70 m long low beach and sand flats that narrow as the low sandstone boundary rocks converge. It is fronted by the 500 m wide sand flats that fill the bay. Beach **K 556** lies 1 km to the east along the eastern arm of the bay. It is a 60 m long pocket of sand fronted by 100 m wide sand flats, then reef. One hundred metres to the east is beach **K 557**, wedged in behind a small, low rock islet that forms the eastern tip of the bay. The 20 m long beach ties the islet to the mainland. The beach is also littered with rocks, and fronted on both sides by sand flats. Beach **K 558** lies immediately to the east and is a 100 m long beach occupying two adjoining joint lines, with rocks between the joints extending across the beach. Two hundred metre wide sand flats then reef front the beach.

K 559-563 **CAPSTAN ISLAND**

No.	Beach	Rating HT	LT	Type	Length
K559	Capstan Is (N1)	1	1	R+sand flats/coral	200 m
K560	Capstan Is (N2)	1	1	R+sand flats/coral	150 m
K561	Capstan Is (E1)	1	1	R+sand flats+rocks	300 m
K562	Capstan Is (E2)	1	1	R+sand flats+rocks	250 m
K563	Capstan Is (E3)	1	2	R+tidal sand flats	200 m
Spring & neap tide range:				5.4 & 1.2 m	

Capstan Island is an irregular shaped, roughly 2 km by 2 km island that is separated from the mainland by a 2 km long 100-200 m wide joint-controlled channel, which is littered with rocks, and filled with tidal sand flats.

Beaches K 559 and 560 occupy the base of a 1 km wide bay that dominates the north side of the island. Beach **K 559** lies in the southwest corner of the bay and is a 200 m long north-facing beach, backed by an 80 m wide, low grassy overwash plain, then sandstone slopes. It has eroded slabs of beachrock deposited at the top of the moderately steep beach, which is fronted by 200 m wide shallow sand flats, edged by narrow fringing reef. Beach **K 560** lies in the southeast corner of the bay and is a 150 m long beach, cut in the centre by rock outcrops, with rocks also off the centre of the beach. A low foredune backs most of the beach, while a small joint-controlled mangrove-lined creek drains across the centre of the beach. It is also fronted by sand flats then a narrow strip of reef. Midway between the two beaches is a gully containing a 40 m long cobble and boulder beach.

Beaches K 561-563 are located on the eastern side of the island. Beach **K 561** lies 1 km southeast of the northern tip of the island, and is a 300 m long east-facing beach, backed by sandstone slopes rising to 70 m. Sandstone also outcrops off the northern end of the beach forming a small tombolo, and as a clump of rocks toward the southern end of the beach. There is one small 40 m wide patch of low grassy foredune in lee of these rocks, with the beach also fronted by low tide sand flats.

Beach **K 562** lies 200 m to the south and commences at the 50 m wide mouth of a prominent joint line, where it is backed by a low grassy 50 m wide, overwash plain that partly blocks the mouth of the creek. The narrow 250 m long beach then continues south backed by rocky sandstone, with rock debris across the beach, widening slightly at the southern end. A clump of rocks also lies 100 m off the southern end of the beach.

Beach **K 563** is part of the sand shoals that tie the island to the mainland. It is located on the northern side of the 150 m wide tidal channel that flows between the island and mainland, which at its eastern entrance is largely plugged with intertidal sand shoals. The beach is 200 m long and backed by a low 50 m wide grassy overwash plain between the island and rock outcrop, with a small mangrove woodland behind the beach, and the sand flats extending to the south. It is fronted by 500 m wide intertidal sand shoals of the inlet.

K 564-575 **MONTAGUE SOUND**

No.	Beach	Rating HT	LT	Type	Length
K564	Montague Sound (1)	1	1	R+sand flats	150 m
K565	Montague Sound (2)	1	1	R+sand flats	300 m
K566	Montague Sound (3)	1	2	R+sand/rock flats	30 m
K567	Montague Sound (4)	1	2	R+sand/rock flats	70 m
K568	Montague Sound (5)	1	1	R+sand flats+rocks	300 m
K569	Montague Sound (6)	1	1	R+sand flats	80 m
K570	Montague Sound (7)	1	1	R+sand flats	100 m
K571	Montague Sound (8)	1	1	R+sand flats+rocks	70 m
K572	Montague Sound (9)	1	2	R+sand flats	60 m
K573	Montague Sound (10)	1	1	R+sand flats	150 m
K574	Montague Sound (11)	1	1	R+sand flats	150 m
K575	Montague Sound (12)	1	1	R+sand flats/coral	300 m
Spring & neap tide range:				5.4 & 1.2 m	

Montague Sound refers to the relatively large open northwest-facing bay located between Bigge Island and Cape Voltaire, a distance of approximately 45 km, and with a depth between 10 and 30 km. The southern mainland shoreline is heavily indented following joint lines in the deeply dissected Proterozoic sandstone, together with numerous islands, islets and rock reefs scattered along the shore and in the sound. Beaches K 564-575 lie along 20 km of coast between Capstan Island and the prominent point immediately south of Combe Hill Island. All the beaches are located in either joint lines or gaps in the sloping sandstone shoreline, and most are fronted by sand flats and some with rocks as well.

Beach **K 564** is a 150 m long beach located in a narrow, 500 m deep inlet 1 km east of Capstan Island. The intertidal beach lies partly across the mouth of the inlet, with sand flats and a few mangroves filling the remainder of the inlet. Beach **K 565** lies 200 m to the east and is a narrow 300 m long strip of intertidal sand, located at the base of 5-10 m high sandstone bluffs. It is fronted by 100 m wide sand flats then a narrow strip of coral reef.

Beaches K 566-568 are located toward the southwestern side of the next embayment 3 km to the east. **K 566** is a 30 m pocket of east-facing intertidal sand wedged in an east-trending joint line, with rocks to either side and scattered across the fronting sand flats. Beach **K 567** lies 300 m to the south, and is a 70 m long high tide pocket of north-facing sand, fronted by 50-100 m wide intertidal rock flats while steeply rising rocks border and back the beach. Beach **K 568** is located 100 m to the east at the mouth of a major north-trending joint line. The 300 m long beach lies half in the joint and half immediately to the east with a low clump of rocks separating the two sections. The joint section is backed by a 1 m high grassy overwash ridge, then a dry creek with beachrock exposed in the creek and on the beach face. The eastern section has a 50 m wide 1-2 m high grassy foredune and backing sandstone slopes. It is fronted by irregular 200 m wide sand flats.

Beaches K 569 and 570 are two adjoining pockets of sand located on the next headland, where they occupy gaps in the sandstone shoreline, with layered sandstone rising to 60 m behind the beaches. Beach **K 569** is 80 m long with a narrow 1 m high foredune and rock to either end. Beach **K 570** is a discontinuous 100 m long beach, with a small foredune and sandstone rocks on and off the beach. Both beaches lie to the lee of a 1 km long island, with sand flats off the beaches associated with the sides of the 100-200 m wide tidal channel between the island and the mainland.

Beach **K 571** is an isolated 70 m long strip of sand located 1.5 km south of beach K 570 toward the apex of the V-shaped embayment. The beach is bordered by sandstone points backed by sandstone slopes rising to 40 m and fronted by the muddy sand flats of the bay.

Beaches K 572-575 lie along the western side of a 6 km long north-trending 100 m high 0.5-2 km wide sandstone peninsula, that terminates in lee of Combe Hill Island. Beaches K 572 and 573 are adjoining beaches in a small bay 5 km south of the tip. Beach **K 572** occupies an east-trending joint that cuts across the peninsula. The 60 m long beach and patch of a foredune, fills the mouth of the joint, and is fronted by rock and sand flats, both 100 m wide. On its immediate north is beach **K 573**, a 150 m long strip of high tide sand, backed by layered sandstone rising to 100 m, and fronted by 200 m wide sand flats.

Beach **K 574** is located at the base of a small 1 km wide V-shaped west-facing embayment, 2 km south of the tip. The beach is 150 m long and backed by a usually dry, steeply descending, joint-controlled, tree-lined gully. It is bordered and backed by sandstone slopes rising to 130 m and fronted by 200 m wide sand flats and a strip of outer coral.

Beach **K 575** lies on the western tip of the peninsula. It is a more exposed 300 m long beach with beachrock exposed along the beach and dominating the southern end. It is backed by a 100 m wide grassy foredune plain, with sandstone bordering the southern end and backing the plain, while basalt borders the northern end (Fig. 4.57). It is fronted by 50-100 m wide sand flats, then fringing reef, with an islet lying off the northern side of the beach and almost connected to the shore by a 200 m long boulder spit.

Figure 4.57 Beach K 575 (right) and beaches K 576-578 (left).

K 576-587 **MUDGE BAY**

No.	Beach	Rating HT	Rating LT	Type	Length
K576	Mudge Bay (E1)	2	3	Boulder+boulder flats	200 m
K577	Mudge Bay (E2)	1	1	R+sand flats	80 m
K578	Mudge Bay (E3)	1	1	R+sand flats	30 m
K579	Mudge Bay (E4)	1	1	R+sand flats	250 m
K580	Mudge Bay (E5)	1	1	R+sand flats	50 m
K581	Mudge Bay (E6)	1	2	R+tidal sand flats	200 m
K582	Mudge Bay (W1)	1	1	R+tidal sand flats	300 m
K583	Mudge Bay (W2)	1	1	R+tidal sand flats	150 m
K584	Mudge Bay (W3)	1	1	R+sand flats	150 m
K585	Mudge Bay (W4)	1	2	R+sand/rock flats	50 m
K586	Mudge Bay (W5)	1	1	R+sand flats/coral	80 m
K587	Mudge Bay (W6)	1	1	R+sand flats	60 m
Spring & neap tide range:				5.4 & 1.2 m	

Mudge Bay is a 5 km wide V-shaped bay, located on the central-southern shores of Montague Sound. The bay extends for 10 km to the south, narrowing into a joint line that extends southwest for 30 m to York Sound. Six beaches (K 576-581) are located along 6 km of the east-facing western shore of the bay, and six beaches (K 582-587) along 4 km of the more irregular eastern shore of the bay. Both sides of the bay consist of sandstone slope rising to 130 m in the west and 120 m in the east.

Beaches K 576-578 are located on the eastern side of the tip of the peninsula in lee of Combe Hill Island. Beach **K 576** is a 200 m long east-facing boulder beach, with a pocket of high tide sand at its southern end. It is separated by a round 100 m wide headland from beach **K 577**, an 80 m pocket sand beach, wedged in between 20 m high headlands. The beach backs on to the north end of beach K 575, the two linking the northern section of the peninsula. Beach **K 578** lies on the southern side of a 200 m wide headland, at the base of a 100 m deep 50 m wide embayment. The beach is 30 m long at high tide, with the backing small valley connecting with the southern end of beach K 575. It is fronted by 100 m wide sand flats that fill the small bay.

Beach **K 579** lies 2.5 km to the south at the mouth of a northeast-trending joint line. The 250 m long beach occupies a V in the joint, as well as extending 100 m to the south along the base of 100 m high sandstone slopes.

Beaches K 580 and 581 lie another 2 km further south, with beach **K 580** a 50 m pocket of sand at the end of a north-trending joint line, while beach **K 581** is an irregular 200 m long beach, also at the end of a joint line, but extending to either side in lee of low sandstone ridges and reefs. Both beaches are fronted by sand flats.

Beaches K 582 and 583 are located 3 km to the southeast, on the eastern side of the bay. Beach **K 582** is a 300 m long beach bordered and backed by moderate sandstone slopes, with a rock outcrop in the centre of the low tide beach and sand flats offshore. Beach **K 583** lies 300 m to the east, and is a 150 m long high tide beach bordered by 10 m high sandstone points, with some scattered rock reefs off the beach.

Beaches K 584-587 occupy a 1 km long section toward the northeastern entrance to Mudge Bay. Beach **K 584** is a 150 m long west-facing beach occupying a V-shaped joint line, with rocks also extending across the centre of the beach. Beach **K 585** is located to its immediate north and is a 50 m pocket of sand, that narrows towards low tide as the bordering rocks impinge. Rocks also dominate the inter- to sub-tidal. Beach **K 586** occupies a joint line 200 m further north. The V-shaped beach is 80 m wide at high tide, backed by a narrow gully and fronted by 100 m wide sand flats and a strip of fringing coral. Beach **K 587** is a further 200 m to the north and is a 60 m long strip of high tide sand wedged in the mouth of a V-shaped, tree-lined sandstone gully. The beach extends to the backing rocks with the gully sides forming rocky headlands.

K 588-598 MONTAGUE SOUND

No.	Beach	Rating HT	LT	Type	Length
K588	Montague Sound (13)	1	2	R+sand/rock flats	200 m
K589	Montague Sound (14)	1	2	R+tidal sand/rock flats	100 m
K590	Montague Sound (15)	1	2	R+tidal sand flats+rocks	900 m
K591	Montague Sound (16)	1	2	R+rock flats	250 m
K592	Montague Sound (17)	1	1	R+sand flats	100 m
K593	Montague Sound (18)	1	2	R+rock flats	150 m
K594	Montague Sound (19)	1	2	R+rock/sand flats	200 m
K595	Montague Sound (20	1	1	R+tidal sand flat	100 m
K596	Montague Sound (21)	1	2	R+tidal sand flats+rocks	100 m
K597	Montague Sound (22)	1	1	R+sand flats	150 m
K598	Montague Sound (23)	1	1	R+sand flats	30 m
Spring & neap tide range:		5.4 & 1.2 m			

Beaches 588-598 occupy an irregular 9 km wide, 10 km deep bay in the southeastern arm of Montague Sound, immediately east of Mudge Bay. The bay is 14 km deep, narrowing into three joint-controlled inlets, with Wollaston Island lying just off the northeast end of the mainland boundary. The bay has more than 30 km of irregular sandstone shoreline, including several islands within the bay. There are eleven beaches (K 588-598) within the bay, most located along the more exposed northeastern section, the remainder of the bay shore composed of sandstone, with mangroves lining the deeper southern inlets.

Beach **K 588** lies 1 km east of beach K 587 at the northwestern tip of the bay. It is a 200 m long northeast-facing beach that is backed by deeply dissected 50 m high sandstone escarpment, with four rock outcrops along the beach and extending into the intertidal zone. Besides the rocks it is fronted by sand flats.

Beach **K 589** is located 5 km to the south, at the tip of a 60 m high sandstone ridge that forms a promontory within the bay. The 100 m long beach is bordered by red sandstone slopes and backed by a small flat grassy plain, with beachrock exposed on the beach face (Fig. 4.58).

Figure 4.58 Beach K 589 is a small isolated pocket of sand.

Beaches K 590 and 591 are adjoining beaches located on a northwest-facing section of the inner bay that looks out through the bay entrance, 5 km to the northwest. Beach **K 590** is a discontinuous 900 m largely intertidal beach that begins below a 140 m high sandstone ridge, and trends to the east in amongst intertidal rocks and rock flats, with an open lagoon and mangroves behind and an inlet at the eastern end. Beach **K 591** commences on the eastern side of the shallow 200 m wide inlet, and is an irregular 250 m long high tide beach, fronted by rock and sand flats. It is backed by patchy low grassy overwash in amongst the rocks, with a sandstone ridge behind, and a joint-controlled mangrove-lined inlet behind the ridge linking with the lagoon.

Beach **K 592** lies 3 km to the north toward the bay entrance. It is a 100 m long northwest-facing pocket of sand with a small central foredune bordered by 20 m high sandstone slopes. It is fronted by 200 m wide sand and rock flats and some rocks further out.

Beaches K 593 and 594 are located on the northeastern corner of a 2 km wide peninsula that occupies the centre of the bay mouth, and lies 2 km south of Wollaston Island. Beach **K 593** is a discontinuous 150 m long high tide sandy beach, backed by lower sandstone slopes, with sandstone rocks and boulders outcropping along the entire length of the beach and intertidal zone. Its neighbouring beach **K 594** is located at the tip of the peninsula in lee of sandstone rocks and reefs. The 200 m long beach curves round the tip and, while dominated by the rocks, has a 50 m wide grassy overwash flat beyond, then the low sandstone slopes.

Beach **K 595** lies 2 km to the southeast in a 1.5 km wide, 3 km deep V-shaped embayment. The beach is 100 m long and consists of a narrow high tide beach backed by gently rising sandstone slopes and fronted by 100 m wide sand flats littered with rock outcrops. Beach **K 596** lies 200 m to the north and is a 100 m long strip of sand backed by a narrow grassy overwash plain. It is bordered and backed by 5-10 m high sandstone ridges, with some rock outcropping on the beach.

Beach **K 597** is located a further 2 km to the north and is a 150 m long strip of sand extending below 5 m high sandstone bluffs, and fronted by 100-150 m wide sand flats and deeper fringing reef. Beach **K 598** lies 200 m to the north and is a northwest-facing 30 m long pocket of sand, with 20 m high sandstone to the east and rear and a low point on its west. It is fronted by 200 m wide sand flats and some rocks and then fringing reef. It lies 1.5 km due south of Wollaston Island.

K 599-608 **SWIFT BAY (W)**

No.	Beach	Rating HT	Rating LT	Type	Length
K599	Swift Bay (W1)	1	1	R+tidal sand flats	150 m
K600	Swift Bay (W2)	1	1	R+sand flats+rocks	100 m
K601	Swift Bay (W3)	1	1	R+mud flats+rocks	100 m
K602	Swift Bay (W4)	1	1	R+rock flats	200 m
K603	Swift Bay (W5)	1	1	R+rock flats	100 m
K604	Swift Bay (W6)	1	1	R+sand/rock flats	300 m
K605	Swift Bay (W7)	1	1	R+sand flats	50 m
K606	Swift Bay (W8)	1	1	R+sand flats/coral	150 m
K607	Swift Bay (W9)	1	1	R+sand flats/coral	100 m
K608	Swift Bay (W10)	1	1	R+sand flats+rocks	200 m
Spring & neap tide range:				5.4 & 1.2 m	

Swift Bay lies in the southeastern corner of Montague Sound, in lee of Katers Island, with the 38 km long peninsula that leads to Cape Voltarie forming is eastern boundary. Beaches K 599-608 are located along the western arm of the bay in lee of Wollaston and Katers islands. All are protected by the islands and dominated by the sandstone bedrock of the shoreline.

Beach **K 599** is located on the mainland immediately south of the southeastern tip of Wollaston Island. The 150 m long beach faces northeast toward Katers Island and consists of a high tide beach fronted by 200-300 m wide sand flats associate with the 100 m wide tidal channel between the island and mainland.

Beaches K 600-607 all occupy a small 2 km wide, 2 km deep bay located immediately east of Wollaston Island and beach K 599. Beach **K 600** occupies the tip of an easterly-trending sandstone point. The 100 m long beach is bordered and backed by a low sandstone ridge and fronted by sand flats and a scattering of small rocks. Beach K 601 and 602 lie at the southern extremity of the bay. Beach **K 601** is a narrow 100 m long strip of sand backed by low sandstone rocks and fronted by mud flats, with mangroves bordering the southern end. Beach **K 602** lies 200 m to the east at the other side of the mangroves. It is a 200 m long north-facing strip of high tide sand, fronted by irregular 100 m wide rock flats, with mangroves bordering its eastern end.

Beaches K 603 and 604 are located 1 km to the northeast and lie either side of a small mangrove and rock-filled 100 m wide bay (Fig. 4.59). Beach **K 603** lies at the western entrance and is an irregular 100 m long section of sand cut by three linear sand ridges with rock and sand flats to either side. Beach **K 604** is also an irregular 300 m long beach that begins in the small bay, and continues past some entrance rocks for another 200 m partly blocking a wetland at its northern end, that is connected to the main Swift Bay to the east. The beach is backed by rising sandstone slopes, with two patches of grassy foredune and fronted by a mixture of sand and rock flats.

Figure 4.59 Beaches K 603-605 occupy the sides of a small rocky bay.

Beach **K 605** is a 50 m pocket of sand located 100 m to the north. It is backed by 10 m high sandstone bluffs and fronting by sand flats. Beach **K 606** lies another 200 m to the north and is a 150 m pocket beach that has partly filled a small gap in the sandstone. It is backed by a low 100 m wide foredune, then the sandstone ridges, which also form the boundaries. Beachrock is exposed on the beach face with slabs littering the top of the beach The beach is fronted by 100 m wide sand flat then a strip of fringing reef.

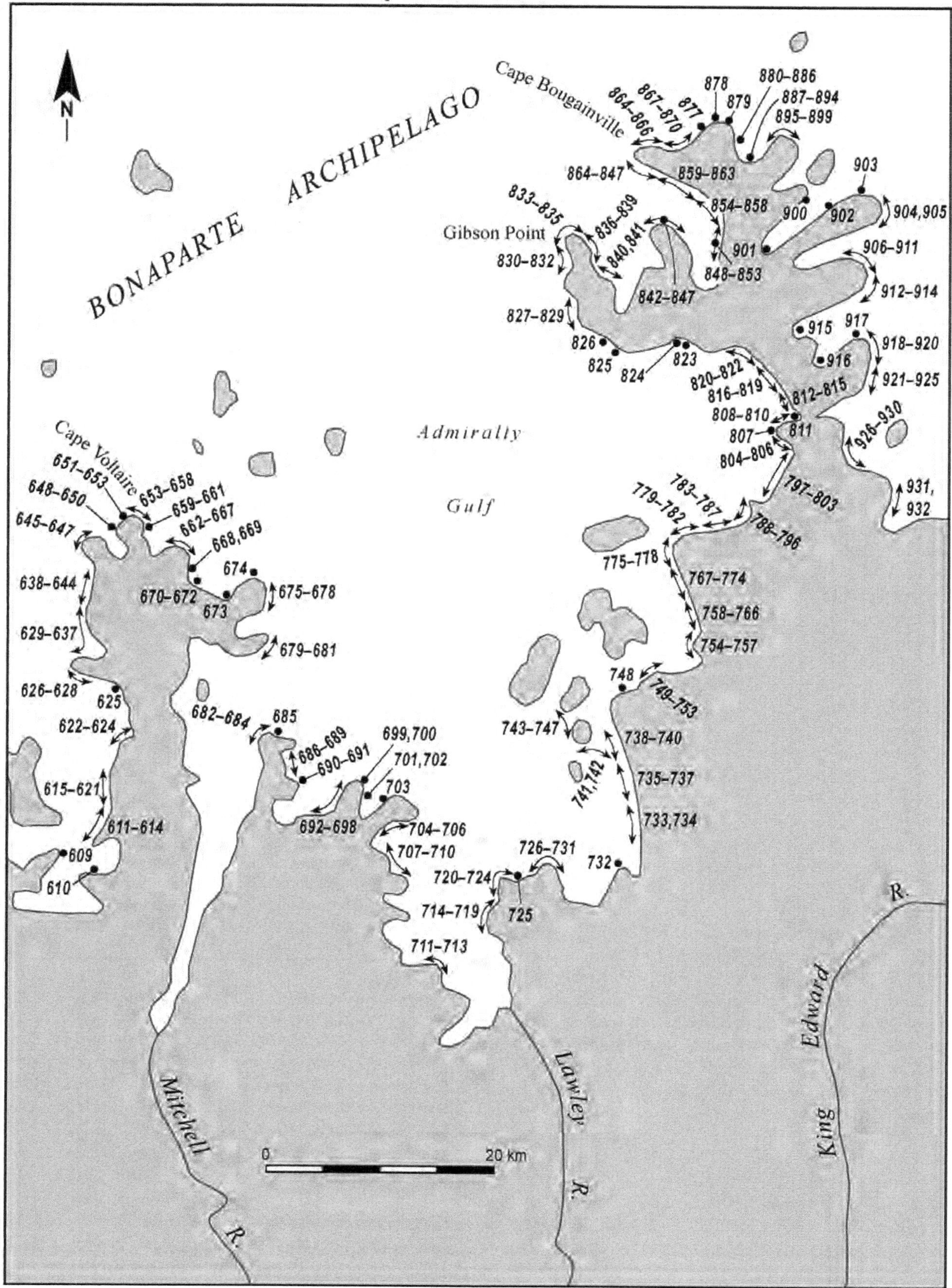

Figure 4.60 *Kimberley map 4, locating beaches K 609-932.*

Beach **K 607** is located at the northeastern entrance to the small bay. It is a 100 m long northwest-facing high tide beach, backed and bordered by low sandstone, and fronted by 50 m wide rock flats a few mangroves then fringing reef. Beach **K 608** lies 1.5 km to the east and occupies the eastern end of a 20-30 m high sandstone ridge. It is a 200 m long sand beach, cut by two sandstone outcrops and fronted by sand flats, with a patch of reef off the northern end.

K 609-614 SWIFT BAY

No.	Beach	Rating HT	Rating LT	Type	Length
K609	Swift Bay 1	1	1	R+sand flats	50 m
K610	Swift Bay 2	1	1	R+sand flats	100 m
K611	Swift Bay (N1)	1	1	R+rock flats	150 m
K612	Swift Bay (N2)	1	1	R+sand flats	100 m
K613	Swift Bay (N3)	1	1	R+sand flats	400 m
K614	Swift Bay (N4)	1	1	R+sand flats	100 m
Spring & neap tide range:	5.4 & 1.2 m				

The main section of **Swift Bay** is an inverted T-shape, which is 2 km wide at its entrance, widening to 8 km along its east-west joint-controlled base. Two beaches (K 609 & 610) lie on either side of the entrance, with the northern beaches (K 611-614) located 4 km northeast of the entrance.

Beach **K 609** lies on the western headland of the bay. It is a 50 m long pocket of sand toward the northern end of the 2 km long headland. It is bordered and backed by rising sandstone slopes and fronted by narrow sand flats then the deeper waters of the bay entrance. Beach **K 610** lies across the entrance 2.5 km to the east. It consists of a double-curved 150 m long bare sand tombolo, tying a low shore-parallel 100 m long sandstone islet to the shore. It is fronted by 50-100 m wide rock flats and is the only sand, on an otherwise 3 km long bedrock headland.

Three kilometres to the north is a J-shaped 2 km long section of mangroves, containing two sub-parallel 20-30 m high sandstone ridges. Beach **K 611** is located on the north side of the southern ridge, and is a 100 m long high tide beach fronted by a rock then sand flats, with a small grassy patch behind. Beach **K 612** links the two ridges with the southern part of a small tombolo, with beach K 614 forming the northern part. The curving, west-facing 100 m long beach is backed by a small mangrove-filled wetland that separates the two beaches. Rocky shore borders the beach, with rocks also outcropping along its centre.

Beach **K 613** runs along the north side of the second ridge. It is a 500 m long high tide sand beach backed by a 20-50 m wide, 1-2 m high wide grassy overwash plain (Fig. 4.61). It is however largely fronted by irregular rock flats, with only one patch of bare sand. Beach **K 614** lies at the junction of the two ridges and the mainland peninsula. The 100 m long beach faces north and is backed by a 100 m wide low grassy plain covered in *Triodia pungens* (Soft Spinifex), then the mangroves and beach K 612. The beach is moderately steep with beachrock exposed along the lower beach face.

Figure 4.61 Beach K 613 (foreground), with beaches K 611 & 612 to right, and beach K 614 to left.

K 615-626 MONTAGUE SOUND (E)

No.	Beach	Rating HT	Rating LT	Type	Length
K615	Montague Sound (E1)	1	1	R+sand flats	100 m
K616	Montague Sound (E2)	1	1	R+sand flats	80 m
K617	Montague Sound (E3)	1	1	R+sand flats	100 m
K618	Montague Sound (E4)	1	1	R+sand flats	120 m
K619	Montague Sound (E5)	1	1	R+sand flats	100 m
K620	Montague Sound (E6)	1	1	R+sand flats	150 m
K621	Montague Sound (E7)	1	1	R+tidal sand flats	250 m
K622	Montague Sound (E8)	1	1	R+tidal sand/rock flats/coral	1 km
K623	Montague Sound (E9)	1	2	R+sand flats+rocks/coral	700 m
K624	Montague Sound (E10)	1	2	R+sand flats+rocks/coral	600 m
K625	Montague Sound (E11)	1	2	R+rock flats+coral	1 km
K626	Montague Sound (E12)	1	1	R+sand flats	100 m
K627	Montague Sound (E13)	1	1	R+sand flats	50 m
K628	Montague Sound (E14)	2	2	R+sand flats+rocks	70 m
Spring & neap tide range:	5.4 & 1.2 m				

The eastern boundary of **Montague Sound** is the 38 km long peninsula that trends north to Cape Voltaire. Swift Bay and beaches K 609-614 lie at its base, with beaches K 615-626 occupying a 12 km long central section. All the beaches face generally west into the sound, while the beaches south of beach K 621 are increasingly protected by Katers Island, which lies 4-8 km off the coast. All the beaches are controlled by the sandstone bedrock, which rises to between 20-60 m along the shore.

Beach **K 615** is located 2 km north of beach K 614. It is a 100 m long pocket of sand wedged in between a southern 100 m long 10 m high point and rocks to the north. It is backed by a patch of grass then sandstone slopes. Beach **K 616** lies 1 km further north at the mouth of a gully following a west-trending joint. The V-shaped beach widens seaward to 80 m and is bordered by rocky sandstone slopes rising to 40 m.

Beaches K 617-621 occupy a 1.5 km long section of shore containing several narrow west-trending sandstone points, with the beaches wedged in between. Beach **K 617** is a 100 m long beach, backed by a narrow 1-2 m high grassy foredune, with a rock outcrop in the centre of the beach face, and 50 m and 200 m long sandstone points to either side. Beach **K 618** is located in lee of the next 150 m long point and faces into a 300 m deep rock controlled inlet that narrows to a small patch of mangroves at its base. The 120 m long beach is located half way along the northern side of the inlet and faces west out the 80 m wide entrance. It is backed by 50 m wide, 1-2 m high grassy foredune.

Beaches **K 619** and **620** occupy the next small bay and are adjoining 100 m and 150 m long pocket beaches, separated by a 100 m wide low sandstone point, with the boundary points extending 100 m into the sound. They are both backed by 50 m wide 1-2 m high grassy foredunes.

Beach **K 621** occupies the mouth of the next joint-controlled gully. The V-shaped beach widen to 250 m at the shoreline, with the creek from the gully winding across the backing sand keeping all but the southern end of the beach free of vegetation. Rocks outcrop on and off the beach.

Beach **K 622** is located 2.5 km to the north in the next northwest-trending joint-controlled gully. The carbonate-rich (70-90%) beach is 1 km long and consists of a southern half backed by a 300 m wide grassy beach-foredune ridge plain, with a creek flowing out along the southern boundary exposing beach- and dune-rock. The northern half abuts sandstone slopes, and is fronted by near continuous rock outcrops. The southern half of the beach is fronted by 200 m wide tidal sand flats and a few rock outcrops, then fringing reef.

Beaches K 623 and 624 are located around the next headland, 1 km to the north. Beach **K 623** is 700 m long backed by a 50 m wide 2-4 m high partly unstable foredune, then low sandstone. There is a small creek and patch of mangroves behind the northern end of beach, while it is fronted by patches of rocks along the beach and on the sand flats. Beach **K 624** is 600 m long with a 1-2 m high grassy foredune then sandstone slopes, and rocks outcropping along and off much of the beach. Beach **K 625** lies 2 km to the north and is a 1 km long strip of high tide sand fronted by continuous irregular rock flats, with fringing coral. It is backed by low sandstone slopes, with two sections of 1 m high grassy foredune.

Beaches K 626, 627 and 628 are three adjoining pocket beaches located 1 km southeast of a 1 km diameter island, 10 km south of Cape Voltaire. The three beaches are backed by sandstone slopes rising to 60 m and are fronted by sand flats and rock outcrops (Fig. 4.62). Beach **K 626** is 100 m long and occupies the mouth of a small gully, with 50 m long rocky points to either side. Beach **K 627** extends for 50 m between two small boulder points, while 70 m long beach **K 628** is backed by continuous boulder rubble, and has a narrow strip of intertidal sand exposed at high tide.

Figure 4.62 Beaches K 626-628.

K 629-637 MONTAGUE SOUND (E)

No.	Beach	Rating HT	LT	Type	Length
K629	Montague Sound (E 15)	1	1	R+sand/rock flats/coral	150 m
K630	Montague Sound (E 16)	1	2	R+coral reef	50 m
K631	Montague Sound (E 17)	1	1	R+tidal sand flats/coral	.300 m
K632	Montague Sound (E 18)	1	1	R+sand flats	100 m
K633	Montague Sound (E 19)	1	1	R+tidal sand flats	300 m
K634	Montague Sound (E 20)	1	1	R+tidal sand flats	200 m
K635	Montague Sound (E 21)	1	1	R+sand flats/coral	60 m
K636	Montague Sound (E 22)	1	1	R+sand flats/coral	80 m
K637	Montague Sound (E 23)	1	1	R+tidal sand flats/coral.	50 m
Spring & neap tide range:	5.4 & 1.2 m				

Seven to ten kilometres south of Cape Voltaire is a 3.5 km wide, 2 km deep V-shaped embayment, with an irregular rocky shoreline containing nine pocket beaches (K 629-637).

Beach **K 629** is a 150 m long pocket of shelly sand located at the mouth of a west-trending joint, with a small creek and patch of mangroves filling the backing gully. The beach is protected by an island 1 km to the west and is fronted by 50 m wide sand flats, then 300 m wide reef flats.

Beach **K 630** lies toward the apex of the embayment. It is a 100 m long pocket of sand surrounded by low rocky sandstone slopes, with an outcrop on the beach. It is fronted by 50 m wide sand flats and 200 m wide fringing coral reef. Beach **K 631** lies immediately to the east in the apex of the bay, with a joint-controlled mangrove-lined creek flowing north onto the sand flats that front the beach. The 300 m long beach begins at the creek mouth, where it is backed by a low 30 m wide grassy 60 m long barrier, with the creek following between the beach and the rocks. The beach then runs north along the base of the sandstone, and is fronted by 200-300 m wide tidal sand shoals, with some fringing reef in the south.

Beaches K 632 to 637 are a cluster of small beaches occupying 1 km of shoreline, toward the northern end of the embayment. Beach **K 632** is a 100 m long pocket of sand wedged in between two small sandstone points. It is backed by a 50 m wide grassy overwash plain, and fronted by 300 m wide tidal sand flats. Beach **K 633** lies 100 m to the north and is an irregular, discontinuous 300 m long beach interrupted by three protruding sandstone bluffs with rocks on the beach and off the southern end, and a mangrove-lined creek bordering the northern end. It is fronted by tidal sand shoals associated with the creek.

Beach **K 634** is located immediately to the north. The 200 m long beach is backed in part by a low sandstone ridge and an opening to a small mangrove-filled wetland behind the northern half of the beach. It is fronted by 300 m wide sand flats that link to a small island, with reef fringing the western side of the island. Beach **K 635** lies in the next small embayment and is a 60 m long pocket of sand bordered by sloping sandstone ridges and fronted by 50 m wide sand flats, then fringing reef which links with the northern end of the island.

Beach **K 636** lies 200 m to the north and is an 80 m long pocket of sand backed by a small patch of mangroves and joint-controlled creek, and fronted by 50 m wide sand flats then 150 m wide fringing reef. Its neighbour 100 m to the north (beach **K 637**) is a similar 50 m long pocket of sand fronted by sand flats then 100 m wide reef flats.

K 638-644 CAPE VOLTAIRE (W)

No.	Beach	Rating HT	Rating LT	Type	Length
K638	Sharp Peak (S3)	1	2	R+coral reef	100 m
K639	Sharp Peak (S2)	1	2	R+coral reef	40 m
K640	Sharp Peak (S1)	1	2	R+coral reef	250 m
K641	Cape Voltaire (S4)	1	2	R+coral reef	200 m
K642	Cape Voltaire (S3)	1	2	R+sand flats/coral	150 m
K643	Cape Voltaire (S2)	1	1	R+sand flats/coral	250 m
K644	Cape Voltaire (S1)	1	2	R+rock flats/coral	500 m
Spring & neap tide range:	5.4 & 1.2 m				

Sharp Peak is a 150 m high dolerite peak located 3 km southeast of Cape Voltaire. The more massive dolerite provides a sharp contrast with the layered sandstone that dominates the surrounding coast, including the cape. Three beaches (K 638-640) are located in the dolerite section all within 1 km south of the peak.

Beaches **K 638** and **639** are adjoining 100 m and 40 m long pockets of west-facing sand overlying boulders. They are backed by 20 m high dolerite bluffs and fronted by low tide rocks, then a 50 m wide band of fringing reef. The reef continues north for 500 m to beach **K 640**. This is a 250 m long sand beach, with some rocks and boulders to either end, and a central sand section, fronted by the same low tide reef. The backing slopes rise steeply to the peak and some rock debris litters the beach.

Beach **K 641** lies in the next embayment, 3 km southeast of the cape. It is a protruding 200 m long high tide sand beach, backed by 10-20 m high sandstone bluffs and slopes, which deliver boulders debris onto the beach together with two waterfalls during the Wet. It is bordered by sandstone over basalt rocks, with some columnar basalt on the southern side of the beach and fronted by a 400 m wide reef flats, with a small islet toward the outer edge.

Beach **K 642** is a south-facing beach 2 km east of the cape. The 150 m long beach is composed of yellow sand, unlike the whiter sand of the neighbouring beaches. It is bordered by 10-20 m high short sandstone points, backed by a 2-3 m high foredune and small sparse wetland, and fronted by 50 m wide sand flats and fringing reef.

Beach **K 643** is the southernmost of a cluster of whiter sand beaches located around the cape. The 250 m long beach has carbonate-rich (80-95%) sand, faces south and is backed by a 300 m wide grassy beach ridge barrier that links with beach K 647, and ties the cape to the mainland. The beach is bordered by 20-30 m high sandstone ridges, and fronted by 300 m wide sand flats, then a fringe of reef.

Beach **K 644** runs along the southern side of the cape for 500 m, with two open 100 m long sandy sections at either end, and a central section in lee of rock flats, with rocks and then reef off the entire beach. It terminates at the 20 m high cape. The northern sandy section forms the southern side a 60 m wide, 3-4 m high grassy tombolo, which, together with beach K 645, ties the cape to the mainland.

K 645-650 CAPE VOLTAIRE-K RAIT BAY

No.	Beach	Rating HT	Rating LT	Type	Length
K645	Cape Voltaire	1	2	R+coral reef	100 m
K646	Davidson Pt 1	1	1	R+sand flats/reef	200 m
K647	Davidson Pt 2	1	2	R+coral reef	100 m
K648	Krait Bay (W)	1	1	R+sand flats	300 m
K649	Krait Bay (E1)	1	2	R+coral reef	200 m
K650	Krait Bay (E2)	1	2	R+coral reef	100 m
Spring & neap tide range:	5.4 & 1.2 m				

Cape Voltaire forms the northwestern tip of the peninsula that separates Montague Sound from Walmesly Bay (Fig. 4.63). The cape is composed of layers of red columnar basalt that rises to 40 m. The top of the peninsula is 17 km wide with Bigge Point forming the northeastern tip.

Figure 4.63 Cape Voltaire (foreground) is surrounded by beaches K 642-644) to right, and beaches K 645-647 to left.

Beach **K 645** is located immediately east of the cape and backs onto beach K 644, with some dune activity on the shared tombolo. The beach is 100 m long at high tide, narrowing to 30 m at low tide, as the bordering sandstone pinches out. Reef flats extend 200 m out from the base of the beach, filling the small bay east of the cape.

Beaches K 646 and 647 are located 1 km east of the cape in a 500 m wide bay that is bordered by **Davidson Point** to the east. Beach **K 646** lies at the base of a 100 m wide, 200 m deep embayment. It curves round for 200 m and faces north across 200 m wide sand flats, with a 100 m wide reef outside the bay boundaries. Beach **K 647** lies immediately to the east and is a 100 m long pocket of sand that backs onto the beach ridge plain of beach K 643. The low energy beach slopes straight down to fringing reef, which extends 200 m out into the bay.

Davidson Point also forms the western entrance to 1 km wide, 2 km deep **Krait Bay,** with a 200 m high point forming the imposing eastern headland. The 3 km long protected bay shore is dominated by sloping sandstone, with mangroves at the southern apex. Beach **K 648** is located on the eastern side 1 km south of the entrance. It is a 300 m long beach backed by a 100 m wide grassy beach ridge plain, with the bordering low sandstone converging to form a 50 m wide gap at the base of the beach, with 100 m wide sand flats then fringing reef.

Beaches K 649 and 650 are located on the eastern side of the bay and face west through the entrance. Beach **K 649** is 200 m long and fronted by 200 m wide reef flats. Its neighbour, beach **K 650** is a 100 m long pocket of sand backed by a 100 m wide overwash flats, bordered by rocks with the same fringing reef.

K 651-661 **VOLTAIRE PASSAGE**

No.	Beach	Rating HT	Rating LT	Type	Length
K651	Voltaire Passage (1)	1	1	R+sand flats/coral	400 m
K652	Voltaire Passage (2)	1	1	R+sand flats/coral	300 m
K653	Voltaire Passage (3)	1	2	R+coral reef	150 m
K654	Voltaire Passage (4)	1	2	R+rock flats/coral	200 m
K655	Voltaire Passage (5)	1	2	R+rock flats/coral	250 m
K656	Voltaire Passage (6)	1	1	R+sand flats/coral	100 m
K657	Voltaire Passage (7)	1	1	R+sand flats	150 m
K658	Voltaire Passage (8)	1	1	R+sand flats/coral	100 m
K659	Voltaire Passage (9)	1	1	R+sand flats/coral	150 m
K660	Voltaire Passage (10)	1	1	Boulder+sand flats/coral	100m
K661	Voltaire Passage (11)	1	1	R+sand flats	80 m
Spring & neap tide range: 5.4 & 1.2 m					

Voltaire Passage crosses above the northern section of the peninsula and passses Lavoisier Island. The mainland coast of the northern peninsula shoreline consists of highly indented sandstone and some dolerite. The 5 km section of coast east of Krait Bay contains 10 km of shoreline and eleven beaches (K 651-661).

Beaches K 651 and 652 are northwest-facing beaches that occupy the base of the first embayment east of K rait Bay. The bay is bounded by a 1.5 km long western headland and a series of rocks and reefs in the east. Within the bay beach **K 651** occupies 400 m of the western side of the bay. In is a moderately steep beach with some slabs of beachrock on the beach face. It is backed by a grassy foredune rising to 6-9 m then the vegetated slopes. A 100 m wide dolerite headland separates it from 300 m long beach **K 652**, which is slightly more exposed and backed by some partly vegetated dunes rising to 10 m in the centre. Both beaches are fronted by 500 m wide sand flats and a band of fringing reef, with a deep channel in the centre of the bay entrance between the reefs.

Beach **K 653** backs onto the eastern end of beach K 652, and forms a rough tombolo that ties a conical 30 m high headland to the mainland, with rocks also outcropping between the two beaches and at the shore. The beach is 150 m long and faces northeast across 300 m wide reef flats that fill the next embayment.

Beach **K 654** lies 500 m to the east at the base of 40 m high sandstone slopes. It is a 200 m long strip of high tide sand wedged in between the sandstone and fronted by rock and boulder flats, then a mixture of sand and reef flats extending 300 m offshore, with tidal pools on the western rock flats. Beach **K 655** is another 300 m to the east and is a similar 250 m long beach wedged between the sandstone slopes and 50 m wide rock flats, with a 20 m gap of sand in the centre. It is fronted by 200 m wide reef flats.

Beach **K 656** lies 200 m around the corner from beach K 655 and faces northeast. It is a 100 m long pocket of sand backed by sandstone slopes, bordered by 50 m wide rock flats and fronted by 50 m wide sand flats, then 100 m wide fringing reef. Beach **K 657** lies a further 500 m to the south, and is a 150 m long pocket of sand backed by sloping sandstone and bordered by rock flats. It is fronted by 200 m wide sand flats then a fringe of reef. It has a small blowout behind the northern end of the beach climbing to 3-4 m over the backing sandstone.

A 300 m wide inlet opening into a bedrock -controlled 50 ha mangrove-lined bay separates beach K 657 and beach **K 658**, which is located on the bays eastern entrance (Fig. 4.64). The beach is a 100 m long pocket of sand and low grassy foredune wedged in between and backed by sandstone slopes rising to 90 m. The rock flats converge toward low tide, with a 50 m wide sandy gap in the centre leading to 100 m wide sand flats then fringing reef.

Figure 4.64 Beaches K 657 & 658 lies either side of a small mangrove-lined bay.

Beach **K 659** lies 300 m to the east and is a 150 m long pocket beach backed by a small grassy foredune and a narrow line of mangroves, then smooth vegetated dolerite slopes rising to 90 m. More tabular sandstone borders the western side, while a dolerite headland with a boulder base forms the eastern boundary, with small mangroves growing on the boulders. It is fronted by 200 m wide sand flats and a fringe of reef.

Beach **K 660** lies on the tip of the headland, 300 m to the east. It is a 100 m long storm beach composed of dolerite boulders, with a veneer of intertidal sand, and a 100 m wide fringing reef. Beach **K 661** lies 200 m further round the headland, and is an 80 m northeast-facing sand beach that occupies a gully formed in the junction of the dolerite and sandstone, with a creek running out across the southern end of the beach. The beach is backed by a 1 m high grassy overwash flat, and fronted by 200 m wide sand flats of the next embayment.

K 662-672 **BIGGE POINT (W)**

No.	Beach	Rating HT	Rating LT	Type	Length
K662	Bigge Pt (W13)	1	1	R+sand flats	100 m
K663	Bigge Pt (W12)	1	2	R+sand/rock flats/coral	80 m
K664	Bigge Pt (W11)	1	2	R+sand/rock flats/coral	400 m
K665	Bigge Pt (W10)	1	1	R+coral reef	150 m
K666	Bigge Pt (W9)	1	2	R+sand flats/coral/rock	300 m
K667	Bigge Pt (W8)	1	2	R+sand flats+rock	400 m
K668	Bigge Pt (W7)	1	1	R+sand flats/coral	60 m
K669	Bigge Pt (W6)	1	1	R+sand flats/coral	100 m
K670	Bigge Pt (W5)	1	1	R+sand flats/coral	150 m
K671	Bigge Pt (W4)	1	1	R+sand flats/coral	60 m
K672	Bigge Pt (W3)	1	1	R+sand flats/coral	100 m
Spring & neap tide range:				5.4 & 1.2 m	

Bigge Point forms the eastern tip of the large peninsula below Voltaire Passage. Extending west of the point is an 8 km wide northeast-facing bay, bordered by a 2 km long point at the eastern end. Beaches K 662-672 are located around the western point and along the western half of the embayment.

Beaches K 662, 663 and 664 lie along the western side of the western headland, a low irregular 1.5 km ridge of sandstone (Fig. 4.65). Beach **K 662** is a 100 m long northwest-facing pocket of sand wedged in between a low dolerite headland to the south and sandstone to the north. A small creek drains into a 1 ha lagoon at the rear of the beach, which has a few mangroves and 50 m wide grassy overwash flats. It is fronted by 200 m wide tidal sand flats. Beach **K 663** is an 80 m long pocket of sand located 500 m to the north, wedged in between a line of sandstone rocks and a low sandstone point, and backed by the 15 m high sandstone ridge. Immediately to the north is beach **K 664**, a double crenulate 400 m long beach bordered to the south by a low sandstone point, with a large low outcrop in the centre and a 20 m high point to the north. It is fronted by some rocks, 200 m wide sand flats, then fringing reef. The northern section backs on to beach K 666 on the eastern side of the point.

Beaches K 665, 666 and 667 occupy the eastern side of the point and face northeast. Beach **K 665** lies out on the eastern tip and is a 150 m long high tide beach, fronted by a narrow strip of sand, then fringing reef. A small rocky headland separates it from beach **K 666**, a curving 300 m long beach with bordering rock points, and some rock outcropping on the beach. Its shares its 50-100 m wide, 15 m high foredune with beach K 664. It is a moderately steep beach, fronted by 150 m wide sand and some rock flats, then reef. Beach **K 667** lies at the eastern base of the point, and is a double crenulate 400 m long high tide sand beach backed by a low foredune, and fronted by 50-100 m wide rock flats, with a 50 m wide sandy gap at the northern end.

Figure 4.65 Beaches K 662-664 are located along the western (right) side of a low sandstone point, with beaches K 665-667 to the left.

Beach **K 668** is located 1.5 km south of the point, near the base of a 1 km deep V-shaped inlet. The beach is 60 m long, facing north out of the bay, with the sandstone bay side to the west and a small sandstone ridge on its east. It is backed by a grassy beach ridge then a patch of mangroves, that are connected via a 50 m wide inlet to the eastern side of the sandstone inlet. It is fronted by the 200 m wide sand flats of the bay and a narrow fringing reef.

Beach **K 669** lies 500 m to the east toward the entrance of the bay. It is a 100 m long, lower gradient north-facing sand beach, that extends to the base of the sandstone slopes. It is bordered by a low sandstone point to the east and a 10 m high dolerite point to the west. It has beachrock outcropping on the beach face and is fronted by 50 m wide sand flats and a strip of reef. A 500 m long dolerite headland separates it from beach **K 670** which is a curving 150 m long northeast-facing sand beach, bordered by the dolerite and sandstone to the east, and fronted by 200 m wide sand flats with a few rock outcrops.

Beach **K 671** is located 1 km to the south. It is a 60 m long pocket of sand located at the junction of three converging joint lines. It is bordered by low sandstone slopes and fronted by 50 m wide sand flats and a strip of reef. Beach **K 672** lies 500 m to the south and is a more protected 100 m long high tide sand beach, wedged in between rock flats to either side, with a central sandy section. It is backed by low sandstone slopes, with mangroves growing on the eastern rocks. It is fronted by 100 m wide sand flats and a fringe of reef.

K 673-681 BIGGE POINT

No.	Beach	Rating HT	Rating LT	Type	Length
K673	Bigge Pt (W 2)	1	1	R+tidal sand flats/coral	300 m
K674	Bigge Pt (W 1)	1	2	R+tidal sand/rock flats	200 m
K675	Bigge Pt	1	2	R+tidal sand/rock flats	100 m
K676	Bigge Pt (S 1)	1	2	R+tidal sand/rock flats	100 m
K677	Bigge Pt (S 2)	1	2	R+tidal sand/rock flats	100 m
K678	Bigge Pt (S 3)	1	2	R+tidal sand/rock flats	250 m
K679	Bigge Pt (S 4)	1	2	R+tidal sand/rock flats	250 m
K680	Bigge Pt (S 5)	1	2	R+tidal sand/rock flats	300 m
K681	Bigge Pt (S 6)	1	1	R+tidal sand flats	200 m
Spring & neap tide range:				5.4 & 1.2 m	

Bigge Point is a low irregular lightly wooded sandstone point surrounded by extensive rock flats and reefs and sections of fringing mangroves. Six beaches are located on the point (K 673-678) (Fig. 4.66), and three on the next point to the south (K 679-681). All are very low energy and largely dominated by intertidal sand and rock flats.

Figure 4.66 Bigge Point is a low irregular sandstone point, with beaches K 674 (right), K 675 (left) and K 676 (top) located in amongst the rocks and mangroves.

Beach **K 673** is located on the western side of the point and is a low curving 300 m long northwest-facing beach, fronted by 500 m wide sand flats, then 200 m wide fringing reef. It is backed by a narrow band of beach ridges and backing 40 ha wetland, which connect the point with the mainland.

Beach **K 674** lies in a 300 m wide embayment at the northwestern tip of the peninsula. The shoreline of the semi-circular shaped embayment is dominated by rock flats and mangroves, with the beach occupying two 150 m and 50 m long sections, separated by a cluster of rocks and mangroves. Both sections are fronted by sand flats, with bordering mangrove-dotted rock flats.

Beach **K 675** is located 500 m to the east, at the base of a 200 m wide embayment, located immediately north of the point. The beach is a 100 m long northeast-facing sand beach, backed by a 100 m wide beach ridge plain and small wetland. It is fronted by a 500 m wide mixture of sand and rock flats, then fringing coral. Beach **K 676** lies immediately to the south, and is a bare 100 m long tombolo that ties a small rock outcrop to the point. The tombolo has a 50 m long sand flat on its northern side, with rock flats to the south.

Beach **K 677** is located 1 km south of the point and faces east. It is a 100 m long opening in an otherwise mangrove-lined shoreline, with a high tide 'beach' ridge continuing to either side behind the mangroves. It is fronted by continuous 200 m wide rock flats. Beach **K 678** lies 1.5 km further south, and is a curving 250 m long beach, bordered by continuous rock flats and fringing mangroves, and fronted by rock , then mud flats. It is backed by a 1 m high, 50 m wide grassy overwash plain.

Beaches K 679-681 are located 3 km to the south on the eastern face of the next 10-20 m high sandstone point. Beach **K 679** is an irregular rock -dominated 250 m long high tide sand beach, with rocks and a few mangroves outcropping along the shoreline and across the 200 m wide sand-mud flats. Beach **K 680** lies 400 m to the south and is a similar 300 m long high tide beach, with substantial rocks on and off the beach, forming two small protrusions in the beach. A small creek and a few mangroves back the southern end of the beach. Beach **K 681** is located 100 m to the south and is a straight 200 m long sand beach, backed by the sandstone slopes and fronted by 100 m wide sand flats.

K 682-691 PICK ERING POINT

No.	Beach	Rating HT	Rating LT	Type	Length
K682	Pickering Pt (W4)	1	1	R+sand flats	200 m
K683	Pickering Pt (W3)	1	1	R+sand flats	150 m
K684	Pickering Pt (W2)	1	1	R+sand flats	250 m
K685	Pickering Pt (W1)	1	1	R+sand flats	250 m
K686	Pickering Pt (S1)	1	1	R+tidal sand flats	150 m
K687	Pickering Pt (S2)	1	2	R+rock flats	100 m
K688	Pickering Pt (S3)	1	1	R+tidal sand flats	300 m
K689	Pickering Pt (S4)	1	1	R+tidal sand flats	300 m
K690	Pickering Pt (S5)	1	1	R+tidal sand flats	200 m
K691	Pickering Pt (S6)	1	1	R+tidal sand flats	250 m
Spring & neap tide range:	6.4 & 0.8 m				

Pickering Point is a 7 km long north-trending sandstone ridge that rises to 110 m at its northern end and varies between 2 and 4 km in width. In the south it is connected to the mainland by a mangrove-filled bay that narrows to 200 m in the west, with a tidal creek linking the bays to either side. Four beaches (K 682-685) are located along the steep northern face of the point, with six beaches (K 686-691) located along the eastern shore.

Beach **K 682** is located at the northwestern tip of the point and consists of a 200 m long north-facing sand beach, with a small tombolo-connected rock outcrop off its centre and patches of mangroves on either side of the tombolo. It is backed by sandstone slopes rising to 60 m. Beach **K 683** lies 300 m to the east at the base of a small 300 m wide bay filled with sand flats. The 150 m long beach faces northwest out of the bay, with bordering and backing 40-60 m high sandstone ridges.

Beach **K 684** lies in the next small embayment another 300 m to the east. It is a 250 m long northwest-facing beach and 100 m wide sand flats, bordered and backed by 20 m high sandstone ridges, with a couple of rocks off the eastern end of the beach and a small sand blowout backing its eastern end. Beach **K 685** lies 1.5 km to the east on the northern side of Pickering Point. It is a straight 250 m long intertidal sand beach located at the base of 60 m high rocky slopes, with rocks protruding across the centre of the beach, and a small creek draining a steep rugged gully at the eastern end.

Beach **K 686** lies 2 km southwest of Pickering Point in the first of two embayments containing six low energy beaches (K 686-691) fronted by sand to mud flats. The beach is a 150 m long northeast-facing strip of high tide sand largely fronted by rocks and rock flats, with a sand section in the centre and mangroves lining the rocky shore to either side. Beach **K 687** lies 300 m to the south and is a 100 m long pocket of sand located between two sections of mangrove-fringed rock flats backed by low sandstone slopes.

Beach **K 688** is located another 100 m to the south. It is a 300 m long beach that blocks a small partly mangrove-filled lagoon behind its centre. It is bordered by low rocky points, and fronted by sand to mud flats, with a rock outcrop on the centre of the beach. Beach **K 689** is another 300 m to the south and is a similar 300 m long beach, with a small, narrow, mangrove-lined creek draining out across its southern end.

Beaches K 690 and 691 occupy the base of a 500 m wide north-facing embayment, located 3 km due south of Pickering Point. The bay is filled with sand to mud flats and contains two beaches (K 690 and 691). Beach **K 690** is a 200 m long northeast-facing beach, with a clump of mangroves in its more protected western corner and bordering sandstone slopes rising to 50 m. Beach **K 691** lies 100 m to the east and is a 250 m long north-facing beach, backed by a low narrow band of rocks, that together with the beach connects the eastern head of the bay with the mainland.

K 692-702 CRYSTAL CREEK (W)

No.	Beach	Rating HT	Rating LT	Type	Length
K692	Crystal Ck (W11)	1	1	R+tidal sand flats	350 m
K693	Crystal Ck (W10)	1	1	R+sand flats/coral	600 m
K694	Crystal Ck (W9)	1	1	R+sand flats/coral	100 m
K695	Crystal Ck (W8)	1	1	R+sand flats/coral	250 m
K696	Crystal Ck (W7)	1	1	R+sand flats/coral	200 m
K697	Crystal Ck (W6)	1	1	R+sand flats/coral	300 m

K698	Crystal Ck (W5)	1	2	R+sand/rocks/coral	100 m
K699	Crystal Ck (W4)	1	2	R+sand/rocks/coral	400 m
K700	Crystal Ck (W3)	1	2	R+sand/rocks/coral	900 m
K701	Crystal Ck (W2)	1	2	R+rock flats	1.4 km
K702	Crystal Ck (W1)	1	2	R+sand/rock flats	200 m
Spring & neap tide range:			6.4 & 0.8 m		

Crystal Creek is a10 km long creek that trends northeast roughly following the boundary between the Proterozoic sandstone to the west and later Proterozoic basalt to the east. The creek reaches the coast at the junction of two peninsulas, and an unnamed 4 km long sandstone ridge to the west, with beaches K 692-702 along its base, and 6 km long Crystal Head to the east (Fig. 4.67).

Figure 4.67 Beaches K 693-700 (foreground), with Crystal Creek (top right) and Head (top left) to rear.

Beach **K 692** lies at the western base of the western headland. It is a low energy west-facing 350 m long strip of sand, bordered by mangroves to the south, scattered mangroves on rock flats to the north and an older recurved spit located in lee of the rock flats, extending another 500 m to the northeast. It lies at the base of the bay and is fronted by 500 m wide sand-mud flats.

Beach **K 693** begins 1.5 km to the east and is a curving 600 m long beach, that emerges from mangrove-covered rock flats and trends northeast to a small dissected sandstone headland. Beach **K 694** is located on the north side of the headland and is another 100 m long section of sand, with beachrock outcropping along the northern end. Both beaches are fronted by 400 m wide sand flats and a fringe of reef that runs north to the top of the peninsula. They are both backed by sandstone slopes rising to 20 m.

Beach **K 695** is located at the head of the next point. It is a 250 m long double-crenulate beach with a small central rocky tombolo fringed by mangroves. Beachrock is exposed along the beach, which is fronted by 100 m wide sand flats then fringing reef that links the two boundary points. Beach **K 696** is located on the next point, and is a similar 200 m long beach, protruding out to a central rock outcrop fringed by mangroves, with the reef attached to the outcrop. The beach curves back to either side with 50-100 m wide sand flats between the beach and reef. The northern section of the beach is backed by a low grassy foredune.

Beach **K 697** lies just below the northwestern tip of the peninsula in a 500 m wide west-facing embayment. The rock -bordered beach occupies the base of the bay and is 300 m long, with the southern head backed by a 50 m wide low grassy overwash flat. Mangroves grow on the southern rocks. A small mangrove-lined creek drains across the centre of the beach, with another patch of grassy foredune behind the northern end of the beach. It is fronted by 300 m wide sand flats, some rocks and then the fringing reef.

Beach **K 698**, located on the northwestern tip, is a 100 m long west-facing beach, tied to sandstone boulder points at each end, with beachrock exposed along the beach, and beachrock boulders to the rear. It is fronted by 100 m wide sand flats then the reef, and backed by a grassy low beach ridge plain which links to the south with beach K 697 and to the east with beach K 699.

Beach **K 699** commences at the tip of the peninsula amongst a 300 m long section of sandstone outcrops, then runs as a slightly curving 400 m long north-facing sand beach, backed by eroded slabs of beachrock and a 50-100 m wide, low, grassy beach ridge plain. It is bordered by a 100 m long low rocky point in the east, and fronted by 200 m wide sand flats, then the fringing reef.

Beach **K 700** commences on the eastern side of the rocky point and continues east for 900 m to a more continuous section of sandstone rocks. It is backed by discontinuous 50-100 m wide beach ridge-overwash plain. It has rock outcrops on the centre of the beach, as well as a few scattered rocks. It is fronted by 500 m wide rock and sand flats narrowing to 200 m wide in the east and fringed by reef in the west.

Beach **K 701** commences on the north side of the peninsula's northeastern point, and trends around the point and then south for a total of 1.4 km, the latter 500 m as an irregular spit enclosing a small wetland. The entire beach is interrupted by several rock outcrops as well as a patch of mangroves on the point. It is fronted by a mixture of rock, sand and mud flats, and backed by low sandstone slopes.

Beach **K 702** lies on the southern side of the small creek that drains the wetland. It is a 200 m long northeast-facing sand beach, backed by a 50-100 m wide beach ridge-overwash plain, and fronted by irregular sand and rock flats, with mangroves fringing both ends.

Lawley River National Park

Area: 17 570 ha
Coast length: 20 km (8650-8670 km)
Beaches: 3 (K 711-713)

Lawley River National Park is a small remote park with no vehicle access or facilities. It is located at the base of Admiralty Gulf at the mouth of the Lawley River, and is designed to protect the estuarine ecosystem.

K 703-713 CRYSTAL HD-PORT WARRENDER

No.	Beach	Rating HT	LT	Type	Length
K703	Crystal Hd (N)	2	3	boulder+rock flats	300 m
K704	Crystal Hd (S1)	1	1	R+sand/mud flats	400 m
K705	Crystal Hd (S2)	1	2	R+sand/mud flats	300 m
K706	Crystal Hd (S3)	1	2	R+sand/mud flats	1.3 km
K707	MacGregor Pt (N4)	1	1	R+sand/mud flats	200 m
K708	MacGregor Pt (N3)	1	2	R+coral reef	150 m
K709	MacGregor Pt (N2)	1	1	R+sand/mud flats	100 m
K710	MacGregor Pt (N1)	1	1	R+sand flats/coral	1.2 km
K711	Port Warrender (S1)	1	1	R+sand/mud flats	250 m
K712	Port Warrender (S2)	1	1	R+sand/mud flats	200 m
K713	Port Warrender (S3)	1	1	R+sand/mud flats	100 m
Spring & neap tide range:	6.4 & 0.8 m				

Crystal Head is a prominent anvil-shaped 190 m high basalt head, with steep slopes on all sides and a 150 m high escarpment along its 2 km long northeastern face (Fig. 4.68). Crystal Creek enters the embayment on its north, while Port Warrender extends for 20 km to the south. There is vehicle access to Crystal Creek at the base of the head, where camping is permitted, and also out to Walsh Point, 10 km due south of the head. This is the only vehicle access to the coast between Cone Bay 320 km to the southwest, and Kalumburu 100 km to the east.

Figure 4.68 One hundred and ninety metre high Crystal Head, with beach K 703 along its northern base (left).

Beach **K 703** lies on the northern side of Crystal Head at the base of a 150 m high escarpment. The beach is composed of basalt boulders and runs for 300 m along the base of the slopes, with steeper rocks to the south and low mangrove-covered rock flats to the north.

Beaches K 704-706 lie on the south side of the head and face southeast into Port Warrender. Beach **K 704** is an irregular, narrow, 400 m long sand beach, which follows the backing bedrock slopes. It also has rocks outcropping on the beach, with inner sand flats and mangroves to either end. It is fronted by sand then mud flats extending 1 km offshore. Beach **K 705** is located 500 m to the southwest, and is a straight 300 m long beach at the base of a small valley. It is bordered by basalt slopes and fronted by sand then 2 km wide mud flats. Beach **K 706** begins 200 m to the east and is a curving 1.3 km long east-facing beach lying at the base of 190 m high slopes. It is also fronted by sand then the wide mud flats of the port.

Beaches K 707 to 710 occupy gaps in the next 4 km of steep northeast-facing basalt shore. Beach **K 707** lies in a 500 m wide embayment at the mouth of a small upland creek. The low energy beach occupies the western side of the bay, with mangroves filling the eastern side, and sand then mud flats extending 1 km into the port. Beach **K 708** lies 1 km to the east in a gap in a protrusion in the basalt slopes. It is a 150 m long strip of sand located below 20 m high basalt bluffs, with higher ground behind, while it is fronted by a narrow fringing reef.

Beach **K 709** lies 1 km to the southeast in the next small creek -fed embayment. It consists of a 100 m long strip of high tide sand bordered by mangroves and fronted by 1 km wide sand to mud flats. Beach **K 710** commences on the eastern side of the embayment headland, and is a 1.2 km long northeast-facing beach, located at the foot of steep 120 m high slopes, with 243 m high Warrender Hill located 2 km west of the beach. It is a relatively steep shelly beach with beachrock exposed on the beach face and deposited as slabs at the top of the beach. It is backed by a narrow vegetated shelly ridge and a boab tree on the basalt slopes. It is bordered in the south by MacGregor Point and fronted by 100 m wide sand flats and a fringing reef.

MacGregor Point forms the western entrance to the southern section of Port Warrender, the port containing 40 km of predominantly mangrove-lined shore. Two kilometres south of MacGregor Point is 120 m high **Walsh Point**, which is free of mangroves and the site of vehicle access to the bay shore. Mangroves continue for another 6 km south to another 150 m high point, on the north side of which are beaches K 711-713. Beaches **K 711** is a 250 m long shelly beach consisting of a 70 m wide band of outer fresh shell ridges and inner grey (weathered) shell ridges. Beach **K 712** is 200 m long with boulders to either end and a shell beach in the centre. Both beaches are located in adjoining gullies, while they face north up the port and are backed by steep slopes rising over 100 m (Fig. 4.69). Beach **K 713** lies a further

1 km to the east and is a 100 m long pocket of sand wedged in a gap at the eastern end of the point. It is bordered by dark basalt boulders with some rocks outcropping along the centre of the beach.

Figure 4.69 Beach K 712 is located at the base of 100 m high basalt slopes.

K 714-723 **PORT WARRENDER (SE)**

No.	Beach	Rating HT	LT	Type	Length
K714	Port Warrender (SE1)	1	2	R+rock flats	300 m
K715	Port Warrender (SE2)	1	2	R+rock flats	600 m
K716	Port Warrender (SE3)	1	2	R+sand flats+rocks	900 m
K717	Port Warrender (SE4)	1	1	R+sand flats/coral	150 m
K718	Port Warrender (SE5)	1	2	R+sand flats	100 m
K719	Kimberley Coastal Camp	1	1	R+sand-rock flats	200 m
K720	Port Warrender (SE7)	1	1	R+sand flats	600 m
K721	Port Warrender (SE8)	1	1	R+sand flats	700 m
K722	Port Warrender (SE9)	1	2	R+sand flats+rocks	100 m
K723	Port Warrender (SE10)	1	2	R+sand flats+rocks	80 m
Spring & neap tide range:	6.4 & 0.8 m				

The southernmost section of Port Warrender is occupied by the mangrove-lined mouth of the Lawley River. Beaches K 711 to 713 are located 5 km to the northwest of the river mouth, while on the eastern side beach K 714 is located 8 km due north of the river, with a mangrove-lined shoreline in between. At Beach K 714 the coast becomes dominated by sandstone slopes rising inland to 200 m. The next 8 km of generally northeast-trending shoreline contains eleven beaches (K 714-723). They are all bordered and backed by the sandstone rocks and slopes.

Beach **K 714** extends out along the base of a low sandstone point. It is a 300 m long west-facing high tide sand beach containing a 1 m high grassy foredune at its eastern end. It is fronted by 100 m wide rock flats, then 200 m wide sand to mud flats. A joint-controlled tidal creek separates it from beach **K 715**, a similar 600 m long beach, fronted by rock flats and some scattered mangroves. The beach is used as a campsite.

Beach **K 716** lies around the corner, and is a northwest-facing 900 m long beach, bordered by low rock points, with some rocks outcropping along the beach. It is fronted by 100-200 m wide sand flats, which widen to the east. Beach **K 717** is located 200 m to the east, past a 20 m high sandstone headland. The beach is 150 m long and bordered by a small low rocky point to the east, beyond which is a small mangrove-filled inlet. The beach faces northwest and is fronted by 200 m wide sand flats.

Beach **K 718** is located at the tip of the next headland and is a 10 m long pocket of sand bordered by low sandstone points, with rocks and a few mangroves along the beach. It is fronted by 100 m wide sand flats, then a fringe of reef. A 1 km wide mangrove-filled bay separates it from the next headland which contains beaches K 719-721.

Beach **K 719** is a curving 200 m long northwest-facing beach, backed by low sandstone slopes, and fronted by 100 m wide sand flats. The Kimberley Coastal Camp is located behind the beach, with the camp's beach facilities and small boats located along the northern half of the beach.

Beach **K 720** is located immediately to the east and is a 600 m long west-facing beach with an unstable 2 m high foredune. It has sandstone rocks and flats fronting the western half, and 100-300 m wide rock flats along the eastern half. Beach **K 721** lies 200 m to the east on the other side of the rock flats. It is a continuous 700 m long north-facing sand beach, backed by slabs of beachrock , a 50-100 m wide, 3-4 m high foredune ridge plain, then the sandstone slopes, with small creek s draining across each end of the beach. It is fronted by 50-100 m wide sand flats, together with some reef off the eastern point

Beaches **K 722** and **723** are neighbouring 100 m and 80 m long pockets of sand, located on the low headland 500 m to the northeast. Both are bordered by low rocky points, and fronted by a mixture of rock and sand flats. Beyond these beaches the coast trends more easterly and becomes heavily indented.

K 724-732 **PORT WARRENDER (SE, E)**

No.	Beach	Rating HT	LT	Type	Length
K724	Port Warrender (SE11)	1	1	R+tidal sand flats	150 m
K725	Port Warrender (SE12)	1	2	R+tidal sand flats	300 m
K726	Port Warrender (E1)	1	2	R+mud flats+rocks	300 m
K727	Port Warrender (E2)	1	2	R+mud flats+rocks	70 m
K728	Port Warrender (E3)	1	2	R+ridged sand flats	60 m
K729	Port Warrender (E4)	1	2	R+rock /sand flats	200 m
K730	Port Warrender (E5)	1	2	R+rock /sand flats	150 m
K731	Port Warrender (E6)	1	2	R+mud flats+rocks	700 m
K732	Port Warrender (E7)	1	2	R+mud flats+rocks	600 m
Spring & neap tide range:	6.4 & 0.8 m				

Beaches K 724 and 725 are located in the first predominantly mangrove-lined embayment 1 km east of beach K 723. Beach **K 724** is a 150 m long strip of sand, bordered and backed by low sandstone slopes and fronted by 1 km wide sand to mud flats. Beach **K 725** lies 300 m further into the bay and is a low 300 m long high tide beach, bordered by mangroves, together with a few scattered along the beach. It is fronted by the same wide sand to mud flats of the bay.

Beach **K 726** lies 2 km to the east, on the eastern side of the next V-shaped embayment. The beach is backed by a narrow irregular 300 m long grassy spit, tied to rock outcrops and enclosing a narrow lagoon backed by sandstone slopes. The beach is fronted by sand to mud flats that fill the 500 m wide bay. Beach **K 727** lies 300 m to the north toward the bay entrance and is a 70 m long pocket of sand at the base of sloping sandstone. It is fronted by 200 m wide tidal flats.

Beach **K 728** is located at the northeast corner of the bay in a north-trending joint line. The 60 m wide beach is wedged in between 20 m high sandstone ridges. It is backed by a mangrove-lined creek , fronted by 100 m wide ridged sand flats, with the eastern sandstone point extending another 200 m northward. Beach **K 729** commences on the western point and is a 200 m long northeast-facing beach that links the low rocky point with the backing bedrock slopes. It has a central rock outcrop, and is fronted by 100 m wide rock and sand flats. Beach **K 730** lies another 150 m to the east and is a similar 150 m long beach, backed entirely by sandstone slopes rising to 40 m with rocks also lying across the beach.

Beach **K 731** lies 500 m to the east on the northwest point of a 4 km wide mangrove-filled bay. The 700 m long beach is interrupted by four rock outcrops and consists of a series of four small sandy sections, backed by a series of sloping sandstone ridges, and fronted by subtidal mud flats of the bay.

Beach **K 732** lies at the eastern entrance to the bay. It is a 600 m long northwest-facing strip of high tide sand, backed by a patch of low sandstone that is tied to the mainland by mangroves to either side. Rocks outcrop along and off the beach (Fig. 4.70), which is fronted by a 500 m wide mixture of rocks, sand and mud flats.

Figure 4.70 Beach K 732 and the adjoining mangroves, shown here at high tide.

K 733-741 **STEEP HEAD ISLAND (E)**

No.	Beach	Rating HT	Rating LT	Type	Length
K733	Steep Head Is (E1)	1	2	R+sand flats+rocks	150 m
K734	Steep Head Is (E2)	1	2	R+sand flats+rocks	200 m
K735	Steep Head Is (E3)	1	2	R+sand flats+rocks	300 m
K736	Steep Head Is (E4)	1	2	R+sand flats+rocks	150 m
K737	Steep Head Is (E5)	1	1	R+sand flats	150 m
K738	Steep Head Is (E6)	1	2	R+sand flats+rocks	150 m
K739	Steep Head Is (E7)	1	1	R+sand flats+rocks	100 m
K740	Steep Head Is (E9)	1	1	R+sand flats	200 m
K741	Steep Head Is (E10)	1	2	R+sand flats+rocks	250 m
Spring & neap tide range:	6.4 & 0.8 m				

Steep Head Island is a 2 km long basalt island, which rises steeply to a 185 m high plateau. It is located 2 km off the sandstone of the mainland coast. The coast immediately east of the island trends south for 6 km to the mouth of a 4 km wide mangrove-lined bay, and 3 km to the north to the junction with a prominent west-trending basalt peninsula. The 9 km of generally west-facing sandstone coast are dominated by rocks on the more exposed section and mangroves in four small embayments, with nine small beaches (K 733-741) wedged in along parts of the more open sections.

Beaches K 733 and 734 are neighbouring low energy beaches located 7 km southeast of the island. Beach **K 733** is a 150 m long pocket of sand, while beach **K 734** is 200 m long. Both are bordered by low sandstone rocky points, with rocks also lying off the beaches.

Beaches **K 735** and **736** are located 4 km south of the island and are two adjoining slightly more exposed beaches. They are 300 m and 150 m long respectively, bordered by low sandstone rocky points and backed by 100 m and 50 m wide grassy overwash plains respectively, which are in turn backed by a 2 km wide bedrock -controlled mangrove-filled wetland, with a bedrock connection to the north. Both beaches are fronted by sand flats, and a few scattered rocks outcropping offshore. Beach **K 737** lies 1 km to the north and is a 150 m long pocket of west-facing sand bordered and backed by sloping sandstone, while fronted by sand to mud flats.

Beach **K 738** lies in lee of the island, 1.5 km to the north of beach K 737 and past a small V-shaped mangrove-lined inlet. The beach is 150 m long and bordered and backed by low sandstone, with an outcrop on the centre of the beach. Beach **K 739** is another 200 m to the north and is a 100 m long pocket beach backed by a 100 m wide salt lake.

Beach **K 740** lies 500 m to the north in a 200 m wide V-shaped north-trending joint line. A mangrove-lined tidal creek fills much of the inlet, with the 200 m long beach located against the eastern sandstone slopes. It is fronted by tidal sand flats and the tidal channel that fill the remainder of the inlet.

Beach **K 741** is another 1.5 km to the north and is a 250 m long strip of high tide sand, backed by sandstone slopes, and fronted by a mixture of rock flats and rock outcrops.

K 742-748 MOON ROCK POINT

No.	Beach	Rating HT	LT	Type	Length
K742	Moon Rock Pt (1)	1	1	R+tidal sand flats	250 m
K743	Moon Rock Pt (2)	2	3	R (boulder)+sand flats	500 m
K744	Moon Rock Pt (3)	1	1	R+sand flats	300 m
K745	Moon Rock Pt (4)	1	1	R+sand flats	150 m
K746	Moon Rock Pt (5)	1	1	R (boulder)+sand flats	100 m
K747	Moon Rock Pt (6)	3	5	R+tidal channel	80 m
K748	Moon Rock Pt (7)	1	1	R+ sand flats	100 m
Spring & neap tide range:				6.4 & 0.8 m	

Moon Rock lies 1.5 km west of an L-shaped basalt peninsula. The peninsula trends west from the north-trending sandstone coast for 4 km, then turns and trends north for 3 km, beyond which lies the basaltic Osborne Island group. The peninsula has a total of 17 km of shoreline, which contains seven generally small pocket beaches (K 742-748), most backed by the steep vegetated slopes rising to as high as 140 m.

Beach **K 742** is located 1 km west of the southern junction with the sandstone (Fig. 4.71). It is a curving 250 m long pocket of sand surrounded by vegetated slopes. Mangroves are scattered along the sandy beach with sand flats extending about 100 m then 400 wide mud flats.

Figure 4.71 The prominent junction between the sandstone (right) and basalt (left), with beaches K 741 located on the sandstone and K 742 on the basalt.

Beaches K 743 to 746 are located along the 3 km long western face of the peninsula. Beach **K 743** is a steep 500 m long 50 m wide cobble-boulder beach fronted by 100 m wide sand flats together with a patch of rocks and mangroves in the northern corner. It is backed by 60 m high vegetated slopes in the south, which decrease to the north, where a small wetland backs the northern end of the beach. A number of pandanus trees also back the southern end of the beach.

Beach **K 744** occupies the next small embayment and is a steep, 300 m long sand beach backed by a 50 m wide low grassy foredune, with a dry creek behind then vegetated slopes. It is bordered by rocks and boulders, together with a rock outcrop on the beach and fronted by 200 m wide sand flats. Beach **K 745** lies 500 m to the north in the next smaller, deeper embayment. It is a 150 m long pocket of sand at the base of the embayment, fronted by 300 m wide sand flats that fill the bay. It has a few mangroves on the rocks at either end and is backed by a low grassy ridge and then mangroves of the backing bay. Two large boab trees are located at the northern end of the ridge and a smaller one at the southern end. The beach ties the northern tip of the peninsula to the base.

Beach **K 746** lies in a small gap at the northwestern tip of the peninsula. It is a steep 100 m long intertidal sand-boulder beach, backed and bordered by low vegetated slopes, and fronted by 100 m wide sand flats. The northern tip of the peninsula faces into a 300 m wide tidal channel, between the peninsula and the southernmost of the Osborn Islands, which feeds into a 3 km wide open bay. Beach **K 747** lies at the northeastern tip of the peninsula and is a protruding convex beach that parallels the eastern side of the tidal channel for 150 m. The beach faces north into the deep channel, with a low grassy beach ridge, then mangroves growing to the rear.

Beach **K 748** lies across the bay 4 km to the east. It is a 100 m long north-facing pocket of sand and low foredune, in an otherwise mangrove-dominated rocky shoreline.

K 749-757 CLIFFY POINT (S)

No.	Beach	Rating HT	LT	Type	Length
K749	Cliffy Pt (S1)	1	2	R+sand flats+rocks	100 m
K750	Cliffy Pt (S2)	1	2	R+sand flats+rocks	150 m
K751	Cliffy Pt (S3)	1	3	R+sand flats+rocks	100 m
K752	Cliffy Pt (S4)	1	2	R+sand flats+rocks	200 m
K753	Cliffy Pt (S5)	1	2	R+sand flats+rocks	250 m
K754	Cliffy Pt (S6)	1	1	R+tidal sand flats	150 m
K755	Cliffy Pt (S7)	1	1	R+tidal sand flats	150 m
K756	Cliffy Pt (S8)	1	1	R+sand flats	600 m
K757	Cliffy Pt (S9)	1	1	Boulder+sand flats	80 m
Spring & neap tide range:				6.4 & 0.8 m	

Cliffy Point is a steep 190 m high basalt point on the southwest corner of Middle Osborn Island. The basalt island is separated from the sandstone mainland by a 3 km wide channel. The mainland coast in lee of the island trends roughly northeast, and along the 9 km of shore opposite the island and point are six generally small bedrock-controlled beaches (K 749-754).

Beaches **K 749, 750** and **751** are three 100 m, 150 m and 100 m long pockets of northwest-facing sand, that occupy parts of an 800 m long section of the shore, with moderately sloping sandstone between and behind the beaches. They are each fronted by sand flats with a few scattered rocks.

Beaches **K 752** and **753** lie 1 km to the east, 2.5 km due south of Cliffy Point. The beaches are 200 m and 250 m long respectively, and both consist of a low sandy high tide beach set in scattered rocks, with fronting sand flats, as well as more scattered rocks. They are bordered and backed by low rough sandstone terrain.

Beaches **K 754** and **755** occupy the base of a west-facing embayment, located 3 km due east of Cliffy Point. The beaches are both 150 m long with a 100 m wide rock outcrop separating the two. They are backed by rugged sandstone slopes rising to 80 m, with a small creek behind beach K 755 and are fronted by common 100 m wide sand flats.

Beach **K 756** is a steep 600 m long, northwest-facing sand beach that occupies an open embayment immediately to the north. The beach is backed by slabs of eroded beachrock , then a 50-100 m wide grassy beach ridge-foredune plain, that extends to the base of the steep sandstone slopes (Fig. 4.72). Beach **K 757** lies at its northern end, just beyond a 50 m wide patch of the rocks. It is a steep 80 m long west-facing beach, and is composed of sand and boulders.

Figure 4.72 The longer beach K 756, with the small K 757 to left.

K 758-766 **CLIFFY POINT (N)**

No.	Beach	Rating HT	Rating LT	Type	Length
K758	Cliffy Pt (N1)	1	2	R+sand flats	200 m
K759	Cliffy Pt (N2)	1	1	R+sand flats	500 m
K760	Cliffy Pt (N3)	1	2	R+sand/rock flats	100 m
K761	Cliffy Pt (N4)	1	1	R+sand flats	150 m
K762	Cliffy Pt (N5)	1	1	R+sand flats+rocks	200 m
K763	Cliffy Pt (N6)	1	1	R+sand flats	250 m
K764	Cliffy Pt (N7)	1	2	R+sand/rock flats	100 m
K765	Cliffy Pt (N8)	1	2	R+sand/rock flats	250 m
K766	Cliffy Pt (N9)	1	2	R+sand flats	100 m
Spring & neap tide range:				6.4 & 0.8 m	

Cliffy Point on Middle Osborn Island is located 3.5 km due west of the mainland coast. From this point the shoreline trends north-northwest to the next major headland, located 2 km due east of Borda Island. Between these two points are 9 km of sloping bedrock coast containing 21 small bedrock -controlled beaches (K 758-778), all backed by vegetated sandstone slopes rising to over 100 m. Beaches K 758 to 766 occupy the first 3 km.

Beach **K 758** is located on the northern headland of a circular 1 km wide bay. The beach is 200 m long, faces west and consists of two patches of sand, bordered by 10 m high sandstone points, with an outcrop in the centre. Tidal sand shoals associated with the bay entrance extend up to 500 m off the beach. Beach **K 759** commences 500 m to the north at the base of the point. It is a 500 m long northwest-facing beach, bordered and backed by moderate sandstone slopes.

Beaches K 760-763 are four adjoining patches of sand located on a 1 km section between two prominent west-trending joint lines. Beach **K 760** lies at the base of the southern joint, and is a 100 m long pocket of sand, backed by the narrow joint-controlled valley, with rocks and some mangroves on the beach and a small sandy opening. Beach **K 761** lies 100 m to the north and is a 150 m long sand beach, bordered by low sandstone points, with sandstone slopes behind. Beach **K 762** is 200 m further north and is a 200 m long beach, with rocks to either end and bedrock slope behind. Beach **K 763** is a 250 m long sand beach that is bordered in the north by the end of the second joint. All four beaches are fronted by narrow sand flats.

Beach **K 764** commences 500 m to the north, on the northern side of another joint-controlled gully, which is partly flooded to form a narrow 100 m deep mangrove-lined inlet. The narrow high tide beach runs along the base of sandstone slopes immediately north of the inlet with some rocks outcropping on the fronting sand flats. Beach **K 765** lies 100 m to the north and is an irregular 250 m long sand beach, with sandstone protruding onto the beach and outcropping on and just off the southern half of the beach, with sand flats along the northern half. Beach **K 766** lies 200 m to the north at the tip of a small headland. The beach is a 100 m long pocket of sand, bordered by low rocky points, and fronted by sand flats.

Kalumburu Aboriginal Land

Area: 166 000 ha
Coast length: 584 km (8766-9350 km)
Beaches 284 beaches (K 775-1058)

Kalumburu Aboriginal Land extends along 584 km of coast and includes the prominent and irregular Cape Bouganville and Vansittart and Napier Broome bays. The main community of Kalumburu is located a few kilometres from the coast on the King Edward River, which drains into the base of Napier Broome Bay.

K 767-778 BORDA ISLAND (S)

No.	Beach	Rating HT	LT	Type	Length
K767	Borda Is (S12)	1	3	R+rock flats	300 m
K768	Borda Is (S11)	1	1	R+sand/rock flats	250 m
K769	Borda Is (S10)	1	1	R+sand/rock flats	100 m
K770	Borda Is (S9)	1	1	R+rock flats	300 m
K771	Borda Is (S8)	1	1	R+sand flats+rocks	80 m
K772	Borda Is (S7)	1	2	R+sand flats/coral	200 m
K773	Borda Is (S6)	1	1	R+sand flats/coral	150 m
K774	Borda Is (S5)	1	1	R+sand flats	150 m
K775	Borda Is (S4)	1	1	R+sand flats/rocks/coral	700 m
K776	Borda Is (S3)	1	1	R+sand flats/rocks/coral	150 m
K777	Borda Is (S2)	1	3	R+rock flats	200 m
K778	Borda Is (S1)	1	2	R+sand flats/coral	300 m
Spring & neap tide range:				6.4 & 0.8 m	

Beach **K 767** lies 7 km south-southeast of Borda Island, immediately north of a prominent west-trending joint, with a narrow 300 m long inlet at its mouth. The beach extends north for 300 m, as a 50-100 m wide grassy overwash flat, fronted by a narrow high tide beach, then irregular 200-300 m wide rock flats with some mangroves amongst the rocks at the base of the beach. A mangrove-studded rocky shoreline and numerous rock outcrops dominate the next 500 m of shoreline to beach K 768.

Beach **K 768** is a 250 m long beach, bordered by small rocky points with a central rock outcrop forming a small tombolo. One hundred metre wide irregular sand flats lie between the tombolo and the northern rocks. Beach **K 769** is 200 m further north and is a 100 m long pocket of sand bordered by low rocky points, with extensive rock reefs off the beach and a prominent mangrove-lined joint to its immediate north.

Beach **K 770** commences 300 m north of the joint lines and is an irregular 300 m long high tide sand beach backed by sandstone slopes, and fronted by irregular rock flats, with only a small sandy opening at the southern end of the beach. Beach **K 771** lies 500 m to the north and is an 80 m long pocket of sand bordered by large sandstone boulders and backed by a small gully and sandstone slopes. Beach **K 772** is another 300 m to the north and is a 200 m long sand beach, bordered by sloping rocks to the south and a west-trending joint to the north, with a small rock outcrop just off the centre of the beach.

Beach **K 773** lies on the northern side of the joint, and is a 150 m long beach, with sandstone boulders extending off the southern rocky point and a small sandstone bluff to the north. A steep gully runs down to the northern end of the beach, with a strip of mangroves in the creek just behind the beach. Beach **K 774** lies 200 m to the north and is a 150 m long strip of sand also plugging a small steep gully, with a patch of mangroves right behind the northern end of the beach. A more prominent 1 ha patch of mangroves fills a joint-controlled inlet immediately north of the beach. Both beaches are fronted by fringing reef.

Beach **K 775** is a 700 m long beach at the base of a 100 m high sandstone ridge. The beach is bordered by rocks at either end, with two outcrops along the beach resulting in three sandy curves, while a low grassy overwash plan is located behind the southern two curves. It is fronted by 200 m wide sand flats, then fringing reef. Beach **K 776** lies immediately to the north at the mouth of a west-trending joint. The beach and backing 100 m wide low grassy overwash flats plug the mouth of the joint, with a 2 ha patch of mangroves behind (Fig. 4.73). Rock outcrops along the southern section of beach, with 300 m wide sand flats then patchy reef off the beach.

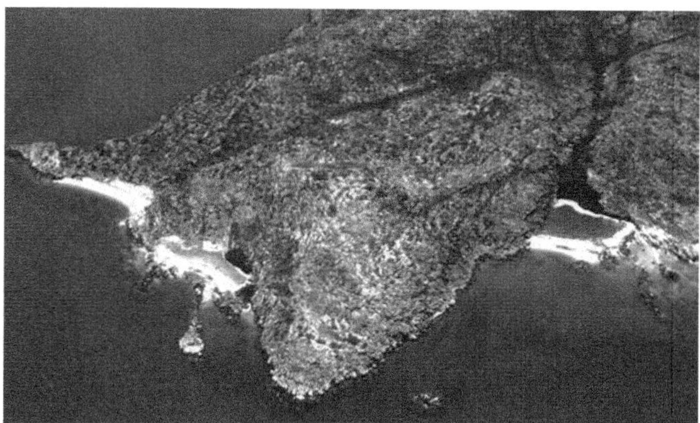

Figure 4.73 Beach K 776 (right) and beaches K 777 & 778 (left).

Beach **K 777** is a curving 200 m long high tide sand beach, backed by a 50 m wide grassy overwash plain, then steeply rising sandstone slopes. It is also fronted by irregular continuous rock flats linking to a small islet off the centre of the beach. A patch of mangroves is wedged in between the rear of the plain and the steep rocky slopes. Its neighbour, beach **K 778**, lies 100 m to the north and is a curving 300 m long sand beach with a small foredune at its southern end blocking a usually dry stream, with rocks and beachrock scattered along the beach. It is backed by irregular rocky slopes and fronted by a mixture of sand and reef flats.

K 779-788 BORDA ISLAND (E)

No.	Beach	Rating HT	LT	Type	Length
K779	Borda Is (E1)	1	1	R+sand flats	400 m
K780	Borda Is (E2)	1	1	R+sand flats	150 m
K781	Borda Is (E3)	1	1	R+sand flats+rocks	500 m
K782	Borda Is (E4)	1	1	R+sand flats+rocks	400 m
K783	Borda Is (E5)	1	1	R+sand flats+rocks	150 m
K784	Borda Is (E6)	1	1	R+sand flats	500 m
K785	Borda Is (E7)	1	1	R+tidal channel	800 m
K786	Borda Is (E8)	1	1	R+sand flats+rocks	600 m
K787	Borda Is (E9)	1	1	R+rock flats	800 m
K788	Borda Is (E10)	1	1	R+tidal sand flats	400 m
Spring & neap tide range:				6.4 & 0.8 m	

East of Borda Island the mainland trends roughly east for 7 km, before turning and trending north to the large peninsula tipped by Cape Bougainville. Beaches K 779-788 occupy the first 7 km of the rocky sandstone shoreline, facing generally north into the upper reaches of Admiralty Gulf.

Beach **K 779** is a curving 400 m long northwest-facing sand beach, located at the mouth of a sloping valley with sandstone ridges to either side. It is fronted by sand flats with several rock reefs off the beach. A low foredune backs the beach with a small mangrove-lined creek running along the back of the foredune and breaking out against the western rocks.

Beaches K 780, 781 and 782 occupy parts of a 1.5 km long section of northwest-facing rocky shore. Beach **K 780** is a 150 m long pocket of sand bordered by low sandstone ridges. Immediately east is 500 m long beach **K 781**, an irregular sand beach, fronted by numerous patchy rock outcrops and reefs which induce crenulations along the shore. It is backed by an unstable dune flat, with a small creek meandering across the centre. Beach **K 782** lies 100 m to the east and is a curving 400 m long beach, located between low sandstone points, with the backing 50-100 m wide beach ridges linking with the mangrove-lined shoreline of the backing bay (Fig. 4.74). It is fronted by sand flats and a few rock outcrops.

Figure 4.74 Beach K 782 (right), with beaches K 783 & 784 extending to the left.

Beaches K 783 and 784 occupy the next small embayment. Beach **K 783** is a curving 150 m long pocket of sand bordered by rock flats, and backed by low slopes. Beach **K 784** commences on the eastern side of the 200 m long rock flats and is a curving 500 m long sand beach with rock flats also bordering the eastern end. It is backed by grassy overwash flats and low sandstone slopes.

Beach **K 785** lies along the eastern entrance to a 1 km deep bedrock -controlled inlet. The irregular, in places very narrow, 800 m long beach is fronted by a mixture of tidal sand shoals and tidal channel, while it is backed by low sandstone slopes to bluffs. Beach **K 786** commences 500 m to the east and is a 600 m long sand beach backed by a 50-100 m wide grassy overwash plain, and fronted by numerous small rock reefs and outcrops, which extend up to 200 m offshore. Beach **K 787** lies 200 m to the east and is a double crenulate 800 m long beach, bordered by low rock flats, with scattered rocks in the centre forming the crenulations. It is backed by irregular grassy overwash flats up to 50 m wide. Beach **K 788** lies immediately to the east in a bedrock -controlled 500 m wide inlet. The 400 m long beach runs along the western side of the inlet and faces east. It is a low, narrow beach fronted by the tidal sand shoals of the inlet.

K 789-795 BORDA ISLAND (E)

No.	Beach	Rating HT	Rating LT	Type	Length
K789	Borda Is (E11)	1	1	R+sand flats	300 m
K790	Borda Is (E12)	1	1	R+tidal sand flats	200 m
K791	Borda Is (E13)	1	1	R+sand flats	350 m
K792	Borda Is (E14)	1	1	R+tidal sand flats	200 m
K793	Borda Is (E15)	1	1	R+sand flats	100 m
K794	Borda Is (E16)	1	1	R+sand flats	250 m
K795	Borda Is (E17)	1	1	R+sand flats	300 m
Spring & neap tide range:				6.4 & 0.8 m	

Beaches K 789 to 795 are located on a northeast-trending section of sandstone coast 8-11 km east of Borda Island. The seven beaches are backed by ridges reaching 100 m in the south decreasing to 20 m in the north.

Beach **K 789** is located on the headland opposite beach K 788, 1 km to the east. It is a narrow, 300 m long high tide beach, located at the base of vegetated slopes, rising to 120 m, 1 km inland. It is bordered by boulder rock flats, and fronted by sand flats.

Beach **K 790** lies 500 m further east in the next small valley. The 200 m long beach and backing low grassy overwash plain plug the mouth of the valley, with a creek running down the northern side, and a few mangroves to the rear. It is fronted by 200 m wide tidal sand shoals.

Beach **K 791** lies 500 m to the east in a creek -carved indentation in the next headland. The beach is 350 m long and backed by a central patch of mangroves, with the creek flowing along the back of the northern half of the beach to exit against the northern rocks.

Beach **K 792** occupies the next small 200 m wide west-facing embayment. The 200 m long beach is backed by a grassy 100-200 m wide beach ridge plain and a small mangrove-wetland, with the creek running along the northern side of the beach. It is fronted by tidal sand flats, with some rocks extending into the bay off the low boundary points. Beach **K 793** is located on the northern point. It is a 100 m long pocket of sand, bordered by low rocks to the south and a 30 m high headland to the north.

Beach **K 794** commences on the northern side of the headland and is a 250 m long beach occupying the base of a sloping valley on the headland, with the vegetated slopes rising to 40 m at the crest of the headland. It is fronted by narrow sand flats. It is bordered in the north by a 300 m long section of rocks and reefs, on the northern side of which is beach K 795.

Beach **K 795** is located between the rocks and the tip of a 2 km long headland. The 300 m long beach faces west and is backed by active dunes, that have blown 200-300 m up and over the 30 m high headland to the backing bay shore. Extensive areas of dunerock are exposed in the dunes, as well as outcrops of the underlying pink sandstone. It is fronted by narrow sand flats.

K 796-807 **BORDA ISLAND (NE)**

No.	Beach	Rating HT LT	Type	Length
K796	Borda Is (NE1)	1 1	R+sand flats	200 m
K797	Borda Is (NE2)	1 1	R+tidal sand flats	100 m
K798	Borda Is (NE3)	1 2	R+sand/rock flats	200 m
K799	Borda Is (NE4)	1 2	R+sand/rock flats	250 m
K800	Borda Is (NE5)	1 2	R+sand/rock flats	900 m
K801	Borda Is (NE6)	1 1	R+sand flats	400 m
K802	Borda Is (NE7)	1 1	R+tidal sand flats	400 m
K803	Borda Is (NE8)	1 1	R+tidal sand flats	100 m
K804	Borda Is (NE9)	1 1	R+tidal sand flats	150 m
K805	Borda Is (NE10)	1 1	R+tidal sand flats	300 m
K806	Borda Is (NE11)	1 1	R+sand flats	500 m
K807	Borda Is (NE12)	1 1	R+sand flats	100 m
Spring & neap tide range:			6.4 & 0.8 m	

A 1 km wide, north-facing V-shaped bay is located 11 km east-northeast of Borda Island. Beaches K 796-799 are located within the bay shore, with beaches K 800-897 extending along the shore for 5 km north of the bay to a prominent sandstone headland. The entire shoreline between beaches K 796 and 807 is backed by moderate sandstone slopes, together with numerous rocks and rock reefs off the shore between beaches K 796 and 805.

Beach **K 796** is located on the eastern side of the headland from beach K 795. The 200 m long beach faces north up the bay and is bordered by low sandstone points, with a few mangroves against the western point that extends for 500 m to the tip of the headland. The beach is backed by a 50 m wide grassy beach ridge plain and a narrow creek wedged between the back of the ridges and the backing slopes.

Beach **K 797** is located 800 m to the south in the apex of the bay. It lies at the mouth of a small creek, with mangroves extending west of the low 100 m long beach. It is fronted by tidal shoals emanating from the creek mouth.

Beaches K 798, 799 and 800 are neighbouring beaches occupying parts of a 2 km long section of rock-dominated shoreline backed by lower sparsely vegetated sandstone slopes. Beach **K 798** is a 200 m long beach bordered by two low sandstone points. Beach **K 799** lies 500 m to the north and is a similar 250 m long beach, while beach **K 800** extends for another 900 m to an easterly inflection in the shoreline. It has several rock outcrops on and off the beach and all three are fronted by sand flats, with rock reefs further offshore.

Beaches K 801 and 802 occupy the next small embayment. Beach **K 801** is a 400 m long beach bordered by low sandstone points, with rising sandstone slopes behind. Beach **K 802** lies 100 m to the north and is a curving 400 m long sand beach, with the northern point inducing the curvature in the beach, and tidal sand flats extending from the point south across the beach.

Beach **K 803** is a 100 m long pocket of sand wedged in a joint line 500 m north of beach K 802. The beach is backed by a small patch of mangroves and fronted by sand flats. A small rock island lies 500 m off the beach.

Beaches K 804 and 805 lie on opposite sides of the next 1 km wide, V-shaped, west-facing embayment. Beach **K 804** is a 150 m long pocket of sand bordered by low sandstone points. Beach **K 805** is a curving 300 m long low energy beach in lee of the northern low rocky headland. Two low rock islets partly block the bay entrance.

Beach **K 806** lies on the north side of the bay. It is a straight 500 m long more exposed west-facing beach, backed by moderately active dunes extending up to 200 m inland and to heights of 30 m (Fig. 4.75), where they are deflated to expose areas of dunerock. The beach face has a low to moderate gradient and is fronted by sand flats containing a shore-parallel sand ridge.

Figure 4.75 Beach K 806 and its backing transgressive dune field.

Beach **K 807** is located 1.5 km to the north just south of the tip of a prominent sandstone headland. The 100 m long sandy beach is bordered by rocks to the south and a low headland to the north. It faces northeast and is backed by a low active foredune that merges with the backing rocks.

K 808-811 **BORDA ISLAND (NE)**

No.	Beach	Rating HT	LT	Type	Length
K808	Borda Is (NE13)	1	2	R+sand/rock flats	150 m
K809	Borda Is (NE14)	1	2	R+sand/rock flats	600 m
K810	Borda Is (NE15)	1	2	R+sand/rock flats	500 m
K811	Borda Is (NE16)	1	1	R+tidal sand flats	300 m
Spring & neap tide range:	6.4 & 0.8 m				

Beaches K 808-811 occupy the 3 km long north-facing northwest section of the sandstone peninsula, which links at beach K 811 to the large highly irregular basaltic peninsula that terminates at Cape Bougainville, 28 km to the north.

Beach **K 808** is located 300 m east of the tip of the point, with beach K 807 just south of the tip. The curving beach is 150 m long and faces north. It is bordered by low sandstone rocky points, with rocks also off the beach. Beach **K 809** is an irregular 600 m long sand beach bordered by numerous rocks on and off the beach, and backed by low sandstone slopes. Beach **K 810** is a curving 500 m long sand beach with low sandstone points to either end, as well as rock outcropping in the centre and off the beach.

Beach **K 811** lies at the apex of the bay that separates the lower sandstone slopes to the south from the steeper, higher basalt slopes to the north. It consists of a curving 300 m long steep low energy beach that faces west along the 2 km deep embayment, with mangrove-covered, laterised, sandstone rocks along the southern side and basalt along the northern side. It is backed by a low flat grassy beach ridge plain with a fence line running across the southern side of the plain. A small mangrove-lined creek drains along the northern side, with beachrock exposed along the side of the creek . The creek drains the 500 m wide backing mangroves and salt flats (Fig. 4.76), that narrow into the junction of the sandstone and basalt.

Figure 4.76 View seaward from beach K 811 across the wide intertidal sand flats.

K 812-816 **GIBSON POINT (S)**

No.	Beach	Rating HT	LT	Type	Length
K812	Gibson Pt (S23)	1	2	R+rock flats	500 m
K813	Gibson Pt (S22)	1	1	R+sand flats	150 m
K814	Gibson Pt (S21)	1	1	R+sand flats	250 m
K815	Gibson Pt (S20)	1	1	R+sand flats	500 m
K816	Gibson Pt (S19)	1	2	R+rock flats	200 m
Spring & neap tide range:	6.4 & 0.8 m				

The large peninsula that terminates at **Cape Bougainville** commences at beach K 811. From here there are 125 km of shoreline along the western half of the peninsula facing into Admiralty Gulf, and another 125 km on the eastern half, facing into Vansittart Bay. The highly irregular shoreline has a total shoreline length of 250 km, with 115 beaches (K 811-925) located in pockets along the generally steep basalt shoreline. They total only 34 km (14%) of the shoreline length. The peninsula is 28 km long north to south, and only 2 km wide at its base. There is no vehicle assess to or development on the entire peninsula. Gibson Point is the westernmost tip of the peninsula and lies 18 km southwest of the cape and 27 km northwest of the narrow base of the peninsula. Between the base and the point are 40 km of bedrock - dominated shoreline containing 23 small beaches (K 812-834), with a total shoreline length of only 6 km (15%).

Beach **K 812** is located 1 km west of beach K 811 and 2 km north of beach K 809. It is a 500 m long double crenulate strip of high tide sand, fronted by near continuous 50-100 m wide rock flats and backed by vegetated slops rising to 180 m.

Beaches K 813 and 814 are located in the first open west-facing embayment to the north. Beach **K 813** is a 150 m long strip of sand, backed by a small central grassy ridge, with basalt slopes rising to 100 m. Beach **K 814** lies 400 m to the north and is a straight 250 m long beach, with a small grassy foredune, then similar slopes.

Beaches K 815 and 816 lie either side of a low 200 m long south-trending basalt point, with the slopes rising to 150 m where the point joins the base as well as behind the beaches. Beach **K 815** is a curving 500 m long southwest-facing beach, backed by a small foredune, then the slopes, with the point affording sufficient protection at the western end to permit shallower sand flats and mangroves on the boundary rocks. Beach **K 816** runs along the west side of the point and is a 200 m long high tide sand beach, fronted by 50 m wide rock flats, with some mangroves on the northern boundary rocks.

K 817-824 **GIBSON POINT (S)**

No.	Beach	Rating HT	LT	Type	Length
K817	Gibson Pt (S18)	1	2	R+sand/rock flats	250 m
K818	Gibson Pt (S17)	2	3	Boulder beach	300 m
K819	Gibson Pt (S16)	2	3	Boulder beach	300 m
K820	Gibson Pt (S15)	1	1	R+sand flats	100 m
K821	Gibson Pt (S14)	1	1	R+sand flats	400 m
K822	Gibson Pt (S13)	1	1	R+sand flats	150 m
K823	Gibson Pt (S12)	1	1	R+sand flats	200 m
K824	Gibson Pt (S11)	1	1	R+sand flats	250 m
Spring & neap tide range:		6.4 & 0.8 m			

Beaches K 817-824 are located south of Gibson Point along the southwestern side of the Cape Bougainville peninsula, between 5 and 13 km northwest of its base with the sandstone. All eight beaches lie at the base of steep, vegetated basalt slopes, and all face west to southwest into Admiralty Gulf.

Beach **K 817** is a 250 m long sand beach located in a moderately steep valley with a creek draining onto the centre of the beach. It has a grassy overwash flat behind the northern half of the beach, with a creek -deposited cobble-boulder delta extending off the beach, while the southern half is backed by the slopes and fronted by sand flats.

Beaches **K 818** and **819** are adjoining narrow boulder beaches, both 300 m long and lying at the base of steep 160 m high slopes, 1 km to the north of beach K 817. Beach K 818 has a patch of high tide sand and a small grassy overwash plain.

Beaches K 820-822 occupy the northern 1.5 km of the next open southwest-facing embayment, with all three backed by steep slopes rising to 140 m. Beach **K 820** is a 100 m long pocket of sand (Fig. 4.77) located another 2 km to the north. Beach **K 821** lies 500 m to the north and is a straight 400 m long sand beach backed by the steep slopes with a small creek and a strip of mangroves backing the northern half of the beach. Beach **K 822** is located 200 m further north at a prominent westerly inflection in the shoreline. The 150 m beach is tucked in the corner with a dry gully behind and steep slopes to either side.

Beach **K 823** is located at the apex of the next open embayment 3 km to the west. It is a solitary 200 m long strip of sand at the base of a steep valley with a stream draining across the beach, and 60 m high slopes to either side. Beach **K 824** lies 1 km further west in the next smaller embayment. It is a steep 250 m long beach at the mouth of two converging gullies, with beachrock exposed along the base of the beach face and rocky tidal flats off the beach. The creek s merge and cross the northern side of the beach, with a small grassy overwash plain, a dry creek and a large boab tree behind. Slopes rising to 100 m border either side.

Figure 4.77 Beach K 820 is a sliver of high tide sand located at the base of 140 m high basalt cliffs.

K 825-834 **GIBSON POINT (S)**

No.	Beach	Rating HT	LT	Type	Length
K825	Gibson Pt (S10)	1	2	R+sand/rock flats	100 m
K826	Gibson Pt (S9)	1	1	R+sand flats	200 m
K827	Gibson Pt (S8)	2	3	R+rocks	250 m
K828	Gibson Pt (S7)	1	2	R+sand flats/coral	300 m
K829	Gibson Pt (S6)	1	1	R+sand flats/coral	300 m
K830	Gibson Pt (S5)	1	1	R+sand flats/coral	400 m
K831	Gibson Pt (S4)	1	1	R+sand flats/coral	300 m
K832	Gibson Pt (S3)	1	1	R+sand flats/coral	400 m
K833	Gibson Pt (S2)	1	1	R+sand flats/coral	150 m
K834	Gibson Pt (S1)	2	3	R+sand flats/coral+rocks	100 m
Spring & neap tide range:		6.4 & 0.8 m			

Gibson Point lies at the northern tip of the 12 km long westernmost arm of the Cape Bougainville peninsula. Beaches K 825-834 are located to the south, along its western side, with beach K 825 located 11 km south-southwest of the point, and beach K 834 at the point. All ten beaches are backed by steep vegetated basalt slopes rising in most places to over 100 m.

Beach **K 825** is an isolated 100 m long pocket of sand located at the base of a steep gully draining 130 m high slopes. It is bordered by steep rocky slopes and fronted by some sand flats and rock debris.

Beach **K 826** lies 2.5 km to the northwest in an open west-facing embayment. It is a 200 m long steep cobble-boulder beach located in the apex of the bay. It is composed of coral debris eroded from a reef located off the beach. The beach face contains a series of mid tide through to high tide storm cusps, with some sand along the base of the beach face. The beach faces southwest, and is backed by a 50 m wide grassy overwash flat composed entirely of coral debris. It is bordered and backed by rocks and slopes including a low 500 m wide saddle of basalt that links with the eastern side of the peninsula.

Beach **K 827** is located on the northern headland. It is an isolated 250 m long beach lying at the base of 100 m high cliffs with no safe foot access. Boulder debris from the cliffs litters the sandy beach and inter- to low tide zone.

Beaches K 828 and 829 occupy adjoining embayments in the next 1 km wide bay, 7 km south of the point. Beach **K 828** is a 300 m long sand beach backed by 100 m high slopes, with some boulder debris on the beach. It is fronted by 50 m wide sand flats, then fringing reef. Beach **K 829** is a similar 300 m long sand beach, backed by a low grassy foredune, with more moderate slopes, and free of rocks. It has a narrow strip of sand flats then reef.

Beach **K 830** is located 2.5 km to the north in the next embayment. It is a slightly more exposed 400 m long steep, cusped, sand beach, backed by a 5 m high, 50 m wide grassy foredune, a narrow tree-lined depression, then basalt slopes of the backing valley rising to 80 m (Fig. 4.78). It is bordered by large boulders and steep slopes, and fronted by 100 m wide sand flats, then reef.

Figure 4.78 Beach K 839 is backed by a small grassy foredune and bordered by the dominant basalt slopes.

Beach **K 831** lies 1.5 km to the north in a small indentation in the steep 60 m high slopes. It is a 300 m long sand beach, with some boulder debris on the southern half, and a small grassy 1-2 m high foredune behind the northern half. It is fronted by 50 m wide sand flats, then fringing reef.

Beach **K 832** is located 2 km south of the point. It is a straight 400 m long beach, backed by slopes decreasing to 20 m in the centre. Rocks protrude onto the northern section of the beach, almost cutting it in two, while the south half is backed by a narrow 1-2 m high grassy foredune. It has a narrow strip of low tide sand, then fringing reef.

Beach **K 833** lies 1 km west of the point. It is a curving 150 m long northwest-facing beach, bordered by low boulder points, and backed by a 50 m wide low grassy foredune then slopes rising to 20 m. Beach **K 834** lies 500 m to the east immediately west of the tip of Gibson Point. It is a 100 m long strip of sand, backed by 5-10 m high basalt boulder bluffs, with boulders and rock debris littering much of the beach.

K 835-841 GIBSON POINT (E)

No.	Beach	Rating HT	Rating LT	Type	Length
K835	Gibson Pt (E1)	1	1	R+sand flats/coral	400 m
K836	Gibson Pt (E2)	1	1	R+sand flats/coral	200 m
K837	Gibson Pt (E3)	1	1	R+sand flats/coral	200 m
K838	Gibson Pt (E4)	1	1	R+tidal sand flats+rocks	300 m
K839	Gibson Pt (E5)	1	1	R+tidal sand flats	250 m
K840	Gibson Pt (E6)	1	2	R+tidal sand flats+rocks	600 m
K841	Gibson Pt (E7)	1	2	R+tidal sand flats+rocks	700 m
Spring & neap tide range:				5.7 & 0.7 m	

Gibson Point marks the northwestern tip of the digitate peninsula that peaks in the north at Cape Bougainville. To the east of Gibson Point is the first in a series of seven radiating elongate, embayments, each separated by several kilometre long linear peninsulas. The entire system is composed of Proterozoic basalt rising to vegetated ridges to 80-150 m in height.

At the point the shoreline turns abruptly and trends south to southeast for 9 km to the base of the first embayment. Beaches K 835-841 are located along the first 6 km of generally northeast-facing shoreline, while the lower reaches of the embayment are fringed in mangroves.

Beach **K 835** commences immediately in lee of the low basalt rocks of the point and curves to the south for 400 m to more low rocks. It is a steep sandy beach, backed by a low grassy 50-100 m wide foredune, a 50 m wide tree-covered depression, and then gently vegetated basalt slopes. Beachrock is exposed along the northern end, with a few boab trees behind the southern end, while it is fronted by 200 m wide sand flats then reef. Beach **K 836** is located in a small east-facing bay 200 m to the south. It is a 200 m long pocket of sand at the base of the bay, backed by 50 m wide grassy overwash flats, and fronted by a mixture of sand to mud flats that fill the 200 m deep bay. Mangroves fringe the rocky bay shore and coral lies outside the bay entrance.

Beach **K 837** lies at the base of the 500 m deep northeast-facing bay 500 m to the south. It is a low energy 200 m long beach that backs onto grassy slopes, part of which protrudes onto the centre of the beach. It has a few mangroves on the rocks to either side, with sand then mud flats filling the bay, and coral outside the entrance.

Beaches K 838 and 839 are located on the south-trending shore outside the bay. Beach **K 838** is a curving 300 m long high tide beach at the base of slopes rising to 40 m. It is fronted by mangrove-covered rock flats in the north, with tidal sand flats off the southern half of the beach. Beach **K 839** lies 500 m to the south at the mouth of a small valley. It is a 250 m long beach, backed by a grassy overwash plain, then the valley sides rising to 40 m. It is fronted by tidal sand flats.

Beaches K 840 and 841 are neighbouring north-facing beaches located 1 km to the southeast. Beach **K 840** is 600 m long, with rocks outcropping along its eastern half. Beach **K 841** is 700 m long with rocks outcropping along much of the beach. Both are backed by a 40 m high vegetated escarpment, and fronted by rocks and sand flats.

K 842-847 **PARRY HARBOUR (S)**

No.	Beach	Rating HT LT	Type	Length
K842	Parry Hbr (S1)	1 1	R+sand flats/coral	250 m
K843	Parry Hbr (S2)	1 1	R+sand flats/coral	300 m
K844	Parry Hbr (S3)	1 1	R+coral reef	150 m
K845	Parry Hbr (S4)	1 1	R+tidal sand flats/coral	200 m
K846	Parry Hbr (S 5)	1 2	R+sand/rock flats/coral	500 m
K847	Parry Hbr (S 6)	1 1	R+sand/rock flats/coral	350 m
Spring & neap tide range:			5.7 & 0.7 m	

Parry Harbour is the name of the irregular, northwest-trending embayment bordered by Gibson Point to the south and Hat Point to the north. It has an 8 km wide entrance, with Hecla Island located in the centre. The bay shore extends for 70 km as a series of two major and several tributary embayments, most lined with mangroves and backed by basalt slopes. The harbour has 15 generally small beaches (K 842-866) which occupy a total of 5.4 km (8%) of the shoreline. Beaches K 842-847 lie along the northern shore of a 5 km wide peninsula, located in the centre of the harbour, and face out of the entrance.

Beach **K 842** is a 250 m long northwest-facing high tide beach, fronted by near continuous rock outcrops with a few mangroves on the eastern rocks. It is backed by a 1 m high grassy ridge, then vegetated slopes. Immediately to the east is a 500 m deep bay containing 300 m long beach **K 843** at its base. This is a steep sandy beach, backed by a narrow grassy 2 m high foredune, then a small mangrove-lined creek that runs the length of the foredune to drain across the western end of the beach. Sand flats extend 200 m off the beach merging into reef flats, while it is surrounded by valley sides rising to 60 m. Mangroves grow along the base of the western boundary rocks.

Beach **K 844** is located 1 km to the north at the tip of the eastern head of the bay. The 150 m long beach and low foredune face west and are bordered by a small rocky point to the north and low slopes to the south. There are a few basalt rocks along the beach, with a few stunted mangroves on the eastern rocks and fringing reef off the beach.

Beach **K 845** is located in the next small open embayment 1 km to the southeast. It is a low energy 200 m long strip of sand, fronted by tidal sand flats, then fringing reef. Beach **K 846** lies 300 m to the south and is a narrow 500 m long northwest-facing beach, backed by slopes rising to 60 m, with rocks outcropping along the centre of the beach, with mangrove-covered rocks to either end. It is fronted by sand then reef flats.

Beach **K 847** is located at the eastern end of the peninsula, 1 km to the east. It is a 350 m long north-facing beach, backed by low vegetated slopes and fronted by a central rock outcrop and sand flats.

K 848-853 **PARRY HARBOUR (E)**

No.	Beach	Rating HT LT	Type	Length
K 848	Parry Hbr (E 1)	1 1	R+tidal sand flats	200 m
K 849	Parry Hbr (E 2)	1 1	R+tidal sand flats	100 m
K 850	Parry Hbr (E 3)	1 1	R+tidal sand flats	150 m
K 851	Parry Hbr (E 4)	1 1	R+tidal sand flats	200 m
K 852	Parry Hbr (E 5)	1 1	R+tidal sand flats	50 m
K 853	Parry Hbr (E 6)	1 1	R+tidal sand flats	250 m
Spring & neap tide range:		5.7 & 0.7 m		

The eastern shore of **Parry Harbour** extends for 18 km to the southeast of Hat Point, with the lower 4 km fringed with mangroves. Beaches begin to form 13 km southeast of the entrance, with beaches K 848-853 located along a 2.5 km long section of northwest-facing basalt shoreline. All are backed by vegetated slopes rising to 80 m.

Beach **K 848** is a 200 m long strip of sand, bordered by a short 20 m high headland, and fronted by tidal sand-mud flats. Beach **K 849** is a 100 m long pocket of sand located in a gap of the northern head, and backed by steep 20 m high slopes.

Beaches **K 850** and **851** are adjoining beaches located in the next 500 m wide embayment. They are 150 m and 200 m long respectively, separated by a central mangrove-covered rock outcrop and backed and bordered by vegetated slopes rising 60-80 m. They are fronted by a continuation of the sand-mud flats that fill the small embayment.

Beach **K 852** is located in a gap on the side of the headland 500 m to the north. It is a 50 m long pocket of sand at the mouth of a small creek.

Beach **K 853** is located on the southern entrance to a 500 m wide inlet. The 250 m long beach faces northwest toward the harbour entrance. It is backed and bordered by low basalt slopes, and fronted by sand flats.

K 854-866 HAT POINT (S)

No.	Beach	Rating HT	LT	Type	Length
K854	Hat Pt (S13)	1	1	R+sand/coral flats	100 m
K855	Hat Pt (S12)	1	1	R+sand/coral flats	150 m
K856	Hat Pt (S11)	1	1	R+sand/coral flats	300 m
K857	Hat Pt (S10)	1	1	R+sand/coral flats	100 m
K858	Hat Pt (S9)	1	1	R+sand/coral flats	100 m
K859	Hat Pt (S8)	1	1	R+sand/coral flats	200 m
K860	Hat Pt (S7)	1	1	R+sand/coral flats	80 m
K861	Hat Pt (S6)	1	1	R+sand/coral flats	350 m
K862	Hat Pt (S5)	1	1	R+sand/coral flats	300 m
K863	Hat Pt (S4)	1	1	R+sand/coral flats	100 m
K864	Hat Pt (S3)	1	2	R+coral flats	200 m
K865	Hat Pt (S2)	1	2	R+coral flats	300 m
K866	Hat Pt (S1)	1	2	R+coral flats	400 m
Spring & neap tide range:				5.7 & 0.7 m	

Hat Point is a 40 m high basalt headland located at the western tip of a narrow 5 km long, 40-60 m high basalt ridge, which connects at its base to the ridge that trends for 6 km north of Cape Bougainville. The point is located 10 km to the southwest of the cape. The point forms the northern entrance to Parry Harbour with the shoreline trending from the point to the southeast for 10 km to beach K 854, and then beyond for another 8 km to the inner reaches of the harbour.

Beach **K 854** is a 100 m long sand beach located on the northern side of the 500 m wide inlet that separates it from beach K 853. The beach is fronted by wide shallow sand-reef flats. Beach **K 855** lies 800 m to the northwest on the south side of a small embayment. The beach is 150 m long and fronted by sand-reef flats with a small point on its northern side then low scarped basalt bluffs separating it from beach K 856. Beach **K 856** is located in a small embayment to the immediate north. The 300 m long beach is fringed with mangroves and fronted by sand flats filling the 300 m deep embayment, then outer coral reefs.

Beaches **K 857** and **858** are neighbouring 100 m long pockets of sand located 500 m northwest of the embayment. Both are bordered by low vegetated basalt points, and fronted by sand-reef flats.

Beaches K 859, 860 and 861 are three adjoining beaches located on the next headland, 2 km to the northwest and 6 km southeast of Hat Point. Beach **K 859** is a 200 m long beach bordered by small 10 m high basalt points, with a narrow foredune, then vegetated slopes rising to 40 m. One hundred metres to the north is an 80 m long pocket of sand (beach **K 860**) wedged in between two small clumps of basalt. Beach **K 861** extends from the northern rocks for 350 m to a vegetated escarpment. The beach is backed by slopes rising steeply to 50 m. All three beaches share sand flats grading into 500 m wide reef flats.

Beach **K 862** is located in a small west-facing embayment 5 km southeast of the point. A mangrove-lined creek drains into its southern corner. The protected beach is fronted by 300 m wide intertidal sand flats, then deeper sand flats and finally fringing reef. Beach **K 863** lies 100 m to the west past a small basalt point. It is a 100 m long strip of sand at the base of 40 m high vegetated slopes. It has a narrow 2 m high foredune behind the eastern end with boulders along the base of the western end, while it is fronted by 500 m wide sand then reef flats.

Beaches K 864, 865 and 866 are located along the 2 km long southern side of Hat Point (Fig. 4.79). These are three slightly more exposed beaches and each is backed by steep vegetated slopes rising to the 50 m high flat crest of the point, while each is fronted by 100-200 m wide fringing reefs. Beach **K 864** is 200 m long, with low backing slopes in the south, which rise to the northern end of the beach. The reef grows right to the base of the beach. Steep slopes and basalt boulders separate it from beach **K 865** 500 m to the north. This is a 300 m long beach, with a strip of low tide sand, then 100 m wide reef flats. Beach **K 866** is a steep 400 m long sandy beach backed at the northern end by the steep 40 m high Hat Point. It forms the western side of a 100 m wide grassy tombolo that links Hat Point with the mainland, with beach K 867 located on the northern side. The beach has a narrow strip of subtidal sand grading into the fringing reef that extends 100 m from the shore in the south, widening to 200 m against the point.

Figure 4.79 Hat Point (foreground) with beaches K 864-866 to right and beach K 867 to left, all fringed by coral reefs.

K 867-875 HAT POINT (E)

No.	Beach	Rating HT	Rating LT	Type	Length
K867	Hat Pt (E1)	1	2	R+coral reef	300 m
K868	Hat Pt (E2)	1	2	R+coral reef	300 m
K869	Hat Pt (E3)	1	2	R+coral reef	50 m
K870	Hat Pt (E4)	1	2	R+coral reef	200 m
K871	Hat Pt (E5)	1	1	R+sand flats/coral	900 m
K872	Hat Pt (E6)	1	2	R+coral reef	200 m
K873	Hat Pt (E7)	1	2	R+coral reef	200 m
K874	Hat Pt (E8)	1	2	R+coral reef	150 m
K875	Hat Pt (E9)	1	2	R+coral reef	1.5 km
Spring & neap tide range:	3.7 & 0.6 m				

At **Hat Point** the shoreline turns and trends east for 3 km before heading to the southeast for 3 km to the base of a 6 km wide open north-facing bay, which is bordered in the east by Cape Bougainville. The entire 15 km of shoreline between the point and cape are fringed by reefs up to 500 m wide, with steep vegetated slopes rising behind to a 50-60 m high plateau. Beaches K 867-875 are located along 7 km of shoreline between the point and the base of the bay.

Beach **K 867** is located immediately east of the point. It is a 300 m long north-facing sand beach, bordered by rising slopes to either end, then a 150 m wide, 20 m high saddle in the centre backing onto beach K 866. It is fronted by 200 m wide reef flats.

Beaches K 868 and 869 lie in the next two small embayments to the east. Beach **K 868** is a low 300 m high tide sand beach, backed by steep slopes, with some rocks outcropping on the beach and to either end. It is fronted by 100 m wide reef flats. Beach **K 869** lies on the eastern side of a 500 m long basalt point. It is a 50 m pocket of high tide sand, widening slightly at low tide, and fronted by a 150 m wide fringing reef.

Beach **K 870** lies 3 km east of Hat Point at the western tip of the adjoining bay. It is a 200 m long high tide sand beach bordered by rock flats and backed by steep vegetated slopes rising to 50 m, while it is fronted by 200 m wide reefs.

Beach **K 871** is located on the southern side of the point, on the western side of the open bay. It is a 900 m long steep, northeast-facing sand beach, backed by a 150 m wide, 2-4 m high grassy foredune, then vegetated slopes rising steeply to 50 m. It is fronted by 100-150 m wide sand flats, then reef flats extending to 500 m.

Beaches **K 872** and **873** are adjoining 200 m long pockets of sand, backed by moderate gradient 40 m high slopes and bordered by 10 m high basalt points. Both are fronted by continuous 200 m wide fringing reef.

Beach **K 874** lies 1 km to the southeast and is a 150 m long high tide beach, with rock outcropping on the western side of the beach. It is backed by 20 m high slopes, and fronted by 50 m wide sand flats, then 500 m of reef. Immediately to the east is the gently curving 1.5 km long beach **K 875**, which occupies the base of the bay. The beach is backed by low to moderate gradient vegetated slopes rising to 40 m, with some rocks outcropping along the beach. It is fronted by 500 m wide fringing reef, with a clump of mangroves growing at the western end in the lee of a 200 m long point.

K 876-878 CAPE BOUGAINVILLE (W)

No.	Beach	Rating HT	Rating LT	Type	Length
K876	Cape Bougainville (W3)	1	2	R+coral reef	400 m
K877	Cape Bougainville (W2)	1	2	R+coral reef	300 m
K878	Cape Bougainville (W1)	1	2	R+coral reef	700 m
Spring & neap tide range:	3.7 & 0.6 m				

Cape Bougainville is the northern tip of a highly irregular, deeply dissected 28 km long basalt peninsula. The flat-topped 50 m high cape consist of a steep-sided remnant of a plateau, whose original surface dominates the peninsula.

Between beach K 875 and the cape is a 6 km long northwest-facing section of steep-sided basalt plateau, which narrows from 2 km in the south to 200 m at the cape. Most of the shoreline consists of basalt slopes fronted by fringing reef, with three beaches (K 876-878) occupying gaps in the slopes.

Beach **K 876** is a 400 m long high tide sand beach, backed by vegetated 40 m high slopes and fronted by 200 m wide reef flats, with sloping points extending 200-300 m out from each end. Beach **K 877** lies 1 km to the north and is a similar 300 m long beach, with rock flats bordering both ends.

Beach **K 878** lies immediately south of the cape and is a straight 700 m long beach backed by steep, vegetated 65 m high slopes, rising to the edge of the grassy flat plateau which narrows to a thin ridge above the northern end of the beach, before expanding again above the cape (Fig. 4.80). Some rock debris litters the high tide sand beach, with a well developed fringing reef extending 200 m off the beach.

K 879-886 CAPE BOUGAINVILLE (E)

No.	Beach	Rating HT	Rating LT	Type	Length
K879	Cape Bougainville (E1)	1	2	R+coral reef	500 m
K880	Cape Bougainville (E2)	1	1	R+sand flats/coral	300 m
K881	Cape Bougainville (E3)	1	1	R+sand flats/coral	500 m
K882	Cape Bougainville (E4)	1	1	R+sand flats/coral	1 km
K883	Cape Bougainville (E5)	1	1	R+sand flats/coral	300 m
K884	Cape Bougainville (E6)	1	1	R+sand flats/coral	400 m
K885	Cape Bougainville (E7)	1	2	R+coral reef	200 m
K886	Cape Bougainville (E8)	1	2	R+coral reef	400 m
Spring & neap tide range:	3.7 & 0.6 m				

Figure 4.80 The flat-topped 50 m high Cape Bougainville, with beach K 878 to right, and beaches K 879 & 880 converging to the left.

To the east of **Cape Bougainville** is a low 500 m wide cuspate foreland formed in lee of **Red Island**, a low basalt outcrop 1.5 km east of the cape. From the tip of the foreland the shoreline trends southwest as the eastern side of the 6 km long peninsula. The first 2 km south of the foreland are occupied by a 2 km wide northeast-facing bay, around the side of which are beaches K 880-886. All the bay beaches are fringed by sand flats and fringing reef.

Beach **K 879** is a curving, crenulate, north-facing, narrow high tide beach, that runs along the north side of the foreland. It is fronted by irregular rock flats, then 500 m wide fringing reef. It is backed by the flat foreland deposits, then slopes rising to the 50 m high plateau. Five hundred metre wide sand, rock flats and reef connect the tip of the foreland with Red Island. Beach **K 880** commences on the southern side of the foreland and curves to the southwest for 300 m, terminating as the steep rocky slopes of the cape replace the sand. It is fronted by 500 m wide sand flats edged by reef.

Beach **K 881** is located 300 m around the rocks on the south-facing side of the 50 m high slopes. It consists of a narrow strip of high tide sand, backed by the slopes and fronted by 500 m wide sand flats. It is separated from beach **K 882** by a 100 m long section of rock outcrop, on the southern side of which is the curving 1 km long beach. The beach occupies a curving valley, which has been filled by up to 300 m of sediment, which forms a small flat grassy coastal plain behind the beach. The beach is fronted by the wide sand flats, then reef.

Beach **K 883** occupies the next small valley on the western side of a low vegetated basalt ridge that separates the two beaches. The 300 m long low energy beach faces south into the bay, and has a creek draining down its western side. It is backed by a mangrove-lined creek and a series of low beach ridges, which have filled the 300 m deep valley.

Beach **K 884** occupies the western apex of the bay. It is a 400 m long, steep, sandy, east-facing beach that has also filled the base of the valley with 500 m wide low grassy *Triodia pungens* (Soft Spinifex) overwash deposits. A small creek with stunted mangroves lies at the back of the plain and drains out along the southern valley side forming a small sand delta against the beach, with sand then reef flats extending another 500 m into the bay.

Beaches K 885 and 886 lie on the southern arm of the bay and face north. Beach **K 885** lies 400 m to the east of beach K 884 and is a 200 m long beach, also filling a small valley with up to 100 m of grass-covered overwash deposits. Beach **K 886** lies 500 m further east and is a similar 400 m long beach, backed by slopes rising to 40 m. Both beaches are fronted by 200 m wide fringing reef.

K 887-897 **CAPE BOUGAINVILLE (E)**

No.	Beach	Rating HT LT	Type	Length
K887	Cape Bougainville (E9)	1 2	R+rock flats	500 m
K888	Cape Bougainville (E10)	1 1	R+sand flats/coral	200 m
K889	Cape Bougainville (E11)	1 1	R+sand flats/coral	300 m
K890	Cape Bougainville (E12)	1 1	R+sand flats/coral	100 m
K891	Cape Bougainville (E13)	1 1	R+sand flats/coral	200 m
K892	Cape Bougainville (E14)	1 1	R+sand flats/coral	400 m
K893	Cape Bougainville (E15)	1 1	R+sand flats/coral	100 m
K894	Cape Bougainville (E16)	1 1	R+sand flats/coral	400 m
K895	Cape Bougainville (E17)	1 2	R+rock flats/coral	1 km
K896	Cape Bougainville (E18)	1 2	R+rock flats/coral	150 m
K897	Cape Bougainville (E19)	1 2	R+rock flats/coral	150 m
Spring & neap tide range: 3.7 & 0.6 m				

Five kilometres southeast of Cape Bougainville is a 2 km wide, 2 km deep, north-facing bay, surrounded by an amphitheatre of steep slopes rising to the plateau crest of between 70 and 80 m. There are eight beaches within the bay (K 888-895), with beach K 887 on the western headland, and beaches K 896 and 897 on the eastern headland.

Beach **K 887** is a 500 m long north-facing high tide beach, located at the base of steep 60 m high vegetated slopes. Basalt rocks protrude onto the beach and outcrop as rock flats, with a narrow fringing reef running the length of the beach and adjoining rock slopes.

Beach **K 888** is located 1 km inside the western entrance to the bay. It is a 200 m long northeast-facing beach, fronted by an inner mixture of sand and reef flats, then reef flats extending out 800 m to the tip of the entrance. Beach **K 889** lies 300 m to the east and is a similar 300 m long beach, fronted by the sand reef flats which narrow to 200 m off the eastern end and continue into the bay.

Beach **K 890** lies 500 m further east and is a 100 m long pocket of sand at the base of 10 m high basalt bluffs, then the higher slopes. It is fronted by 300 m wide reef flats, which continue to beach **K 891,** 300 m to the east. This 200 m long beach is also backed by low bluffs, then the higher slopes, and fronted by 300 m wide reef flats.

Beach **K 892** lies at the southern apex of the bay and faces out the entrance 2.5 km to the north. The beach is 400 m long and bordered and backed by vegetated slopes rising to 80 m, while it is fronted by 200m wide sand flats, then reef flats extending out to 1 km.

Beaches K 893 to 895 lie along the eastern side of the bay and face west across the bay. Beach **K 893** is a 100 m long pocket of sand backed by vegetated slopes and fronted by 1 km wide sand then reef flats. Beach **K 894** is located in the next embayment 300 m to the north. It is a curving 400 m long sand beach, also fronted by the same wide sand to reef flats, with mangroves fringing the northern point.

Beach **K 895** occupies an irregular section of shore that extends for 1 km up to the eastern entrance to the bay. The beach consists of a discontinuous narrow strip of high tide sand dominated by intertidal rock flats and rock outcrops, with some mangroves growing on the rocks. It is fronted by 500 m wide sand to reef flats.

Beach **K 896** lies 500 m east of the eastern bay entrance. It is a 150 m long pocket of sand backed by a steep 60 m high amphitheatre, with prominent basalt rocks bordering each end. It is fronted by a 50 m wide fringing reef, which runs the 2.5 km length of the headland. Beach **K 897** lies 400 m further east and is a similar 150 m long strip of high tide sand, with more rock outcropping on the beach and the continuous fringing reef.

K 898-899 CAPE BOUGAINVILLE (E)

No.	Beach	Rating HT	Rating LT	Type	Length
K898	Cape Bougainville (E20)	1	1	R+sand flats/reef	400 m
K899	Cape Bougainville (E21)	1	1	R+tidal sand flats/reef	200 m
Spring & neap tide range:	3.7 & 0.6 m				

Beaches K 898 and 899 are located in a 500 m wide east-facing bay (Fig. 4.81), 7 km due east of Cape Bougainville. Beach **K 898** occupies the northern side of the bay. It is a 400 m long southeast-facing beach, backed by slopes rising to a 60 m high mesa, with some rocks and slope debris scattered along the beach. It is fronted by reef-fringed sand flats that fill much of the inner bay. Beach **K 899** lies at the southern end of the bay. The moderately steep 200 m long beach faces north up the bay. It is backed by a small grassy overwash flat and one boab tree, with a mangrove-lined creek running along the northern side. It is fronted by 200 m wide tidal sand flats, then some scattered reef.

Figure 4.81 Beaches K 898 & 899 share a small bay bordered by prominent basalt headlands.

K 900-905 CAPE BOUGAINVILLE (E)

No.	Beach	Rating HT	Rating LT	Type	Length
K900	Cape Bougainville (E22)	1	3	R+tidal channel	100 m
K901	Cape Bougainville (E23)	1	1	R+mud flats	300 m
K902	Cape Bougainville (E24)	1	1	R+tidal sand flats	400 m
K903	Cape Bougainville (E25)	2	3	R+rock /reef flats	300 m
K904	Cape Bougainville (E26)	2	3	R+rock /reef flats	400 m
K905	Cape Bougainville (E27)	1	2	R+coral reef	600 m
Spring & neap tide range:	2.9 & 0.8 m				

Beach **K 900** is located 10 km southeast of Cape Bougainville in the lee of an unnamed 5 km long island extension of the second peninsula east of the cape. The 100 m long beach faces east across the 200 m wide tidal channel that separates the island from the mainland. The beach is fronted by a narrow sand flat, then the deep channel. It is backed by vegetated slopes that rise to a 100 m high plateau 2 km to the west.

Beach **K 901** lies 8 km to the southwest, near the base of a 15 km long, 1-2 km wide, northeast-trending bay, which has the island at its northern entrance. The beach is 300 m long and faces east into the lower portions of the bay. It is backed by steep vegetated slopes cresting at a 100 m high ridge.

Beaches K 902-905 lie along the northern side of the southern arm of the bay. Beach **K 902** is located 5 km west of the entrance and 2.5 km southeast of beach K 900. It is a curving 400 m long north-facing beach, backed by vegetated slopes rising steeply to an 80 m high ridgeline. It is fronted by 300 m wide sand flats.

Beach **K 903** is located on the northern side of the eastern tip of the peninsula. It is a north-facing 300 m long veneer of sand over a cobble-boulder beach. It is backed by vegetated slopes rising to 80 m, and fronted by a 50 m wide reef flat.

Beach **K 904** lies on the eastern face of the point and faces east across Vansittart Bay. The beach is 400 m long and consists of a 300 m long section of high tide sand fronted by continuous mid to low tide boulders, with a 100 m long southern section, with only scattered boulders. Slopes rise to 80 m behind the beach.

Beach **K 905** lies along the southern face of the eastern point, and is a 600 m long moderately steep beach, with beachrock exposed along the lower beach face and some beachrock boulders on the upper beach. It is backed by flat grassy overwash flats, then slopes rising to 60 m. A 50-100 m wide fringing reef parallels the northern 500 m of beach (Fig. 4.82).

Figure 4.82 Beach K 905 lies at the northern boundary of Vansittart Bay (left).

Freshwater & Seaflower Bays

Freshwater and Seaflower bays are two of the western arms of the larger **Vansittart Bay**. Vansittart Bay occupies the embayment between the peninsula terminating at beach K 905 to the west and Mary Island and Anjo Peninsula to the east. It is 15 km wide at its mouth, with Long Island located immediately north of the centre of the mouth. It extends for 30 km to the south, and has an irregular shoreline 160 km in length, all dominated by bedrock, with mangroves fringing much of the southern and more deeply embayed shores. There is a total of 75 beaches (K 905A-977) with a total shoreline length of 30 km (19%). There is no habitation in the bay area, though the World War 2 Truscott Air Base was located on its eastern shores.

K 905A-905C **FRESHWATER BAY (N)**

No.	Beach	Rating HT	Rating LT	Type	Length
K905A	Freshwater Bay (N1)	1	2	R+rock flats	300 m
K905B	Freshwater Bay (N2)	1	2	R+sand/rock flats	100 m
K905C	Freshwater Bay (N3)	1	2	R+sand flats	80 m

Spring & neap tidal range: 2.9 & 0.8 m

Freshwater Bay is a roughly V-shaped northeast-facing 9 km deep bay, with a 2 km wide mouth. The bay is bordered by a 1 km wide, 80 m high ridgeline to the north and a 5 km wide peninsula to the south capped by a 120 m high plateau surface. Beaches K 905A-905C occupy parts of the very northern, south-facing shoreline, while beaches K 906-911 face northeast and are located along the northeastern end of the southern peninsula, and beaches K 912-914 lie on the southeastern side of the same peninsula and face into the entrance of **Seaflower Bay**.

Beach **K 905A** is a 300 m long strip of high tide sand located at the base of slopes rising steeply to 80 m. It has exposed beachrock, as well as basalt rocks and boulder debris along the beach, and mangroves to either end. In the west it grades into beach **K 905B**, a 100 m long sandy beach backed by a low foredune, with basalt boulders forming the western boundary.

Beach **K 905C** is located a few hundred metres to the west in a small rocky east-facing embayment, with mangroves lining each side. It is a low energy 80 m long intertidal beach fronted by sand flats.

K 906-914 **FRESHWATER & SEAFLOWER BAYS**

Beach		Rating HT	Rating LT	Type	Length
K906	Freshwater Bay (E1)	1	2	R+sand/reef flats	250 m
K907	Freshwater Bay (E2)	1	2	R+sand/reef flats	100 m
K908	Freshwater Bay (E3)	1	2	R+sand/reef flats	250 m
K909	Freshwater Bay (E4)	1	2	R+sand/reef flats	200 m
K910	Freshwater Bay (E5)	1	2	R+sand flats	150 m
K911	Freshwater Bay (E6)	1	2	R+sand/reef flats	400 m
K912	Seaflower Bay (1)	1	2	R+sand/reef flats	100 m
K913	Seaflower Bay (2)	1	2	R+sand flats	150 m
K914	Seaflower Bay (3)	1	2	R+sand flats	50 m

Spring & neap tide range: 2.9 & 0.8 m

Beach **K 906** is a 250 m long east-facing beach located at the base of a 1 km long circular point. It has a 1-2 m high grassy foredune behind the northern end, with large boulders along the southern end. It is backed by a 20 m high 200 m wide ridge, bordered by rocky shore with mangroves on the northern rocks, and fronted by 300 m wide sand flats, then reef extending up to 500 m off the beach.

Beach **K 907** lies 500 m to the south at the mouth of a narrow densely vegetated valley. The 100 m long beach is backed by a 30 m high escarpment of the valley sides and has scattered boulders on the beach. It is fronted by a 500 m wide mixture of sand and reef flats. Beach **K 908** lies 200 m to the south and is a straight 250 m long east-facing beach, backed by a small central foredune and vegetated slopes rising gradually to the 120 m high plateau. It is fronted by the sand and reef flats, with a small rocky islet located 1 km east of the beach.

Beach **K 909** is located in the next small valley and has a mangrove-lined creek draining across the southern end of the 200 m long beach. The creek drains a small valley that rises to the plateau surface 1.5 km inland. Rocks outcrop along the beach, with coral debris accumulating along the southern end, while sand and reef flats extend 200 m into the bay.

Beach **K 910** is a 150 m long beach consisting of two pockets of sand located toward the eastern tip of the peninsula. It is backed by vegetated slopes rising to 100 m and bordered by basalt rocks. Beach **K 911** lies 400 m to the south in the next small embayment. It is a straight moderately steep, 400 m long east-facing sand beach, backed by 100 m wide flat grassy overwash flats, with a large boab to the rear, then the 100 m high slopes. It is fronted by patchy reef extending 200 m off the beach.

Beaches K 912, 913 and 914 are three adjoining pocket beaches located on the southern side of the end of the peninsula. Beach **K 912** is a curving 100 m long pocket of sand, backed by a narrow, low grassy plain then steeply rising slopes, and bordered by basalt rocks, with a patch of coral off the beach. Beach **K 913** occupies the mouth of a narrow valley. The 150 m long beach is backed by a patch of grass in the southern corner, with the densely vegetated valley sides rising to 60 m to either side. Beach **K 914** lies 100 m to the south and is a 50 m long pocket of sand at the base of steeply rising 60 m high slopes, with some rock debris littering the beach.

K 915-916 SEAFLOWER BAY

No.	Beach	Rating HT	Rating LT	Type	Length
K915	Seaflower Bay (W)	1	2	R+tidal flats	150 m
K916	Seaflower Bay (SW)	1	2	R+tidal flats	150 m
Spring & neap tide range:		2.9 & 0.8 m			

Seaflower Bay is 2.5 km wide at its entrance north of August Point and extends to the west for 8 km as two diverging arms, bordered and separated by ridges rising to between 120 and 140 m. Most of the 25 km long bay shore is dominated by basalt bedrock and mangroves, with only two beaches (K 915 and 916) deep within the bay.

Beach **K 915** lies toward the western end of the northern arm and is a 150 m long strip of east-facing sand, backed by vegetated slopes rising to 100 m. Beach **K 916** lies 3 km to the southeast, but in the southern arm, with the ridge in between. The beach is 150 m long and occupies a small valley, with a grassy patch of sand behind, then the valley sides rising to 140 m to the west.

K 917-925 AUGUST, JULY & JUNE POINTS

No.	Beach	Rating HT	Rating LT	Type	Length
K917	August Pt (W)	1	2	R+rock /reef flats	100 m
K918	August Pt (N)	1	2	R+sand/reef flats	200 m
K919	August Pt (S1)	1	2	R+rock /reef flats	200 m
K920	August Pt (S2)	1	2	R+mud flats	250 m
K921	July Pt (S)	1	2	R+tidal sand flats	200 m
K922	July Pt (S2)	1	2	R+mud flats	250 m
K923	July Pt (S3)	1	2	R+mud flats	150 m
K924	July Pt (S4)	1	2	R+mud flats	450 m
K925	June Pt (N)	1	2	R+mud flats	400 m
Spring & neap tide range:		2.9 & 0.8 m			

The 4 km wide peninsula that forms the southern boundary of Seaflower Bay has along its eastern side four small points named from north to south: August, July, June and September. In amongst the points are nine small pockets of sand (beaches K 917-925), all of which face into the 15 km wide mid-section of Vansittart Bay.

Beach **K 917** is located at the eastern end of Seaflower Bay, 1 km west of August Point. The 100 m long beach occupies a gap in the otherwise basalt shoreline, backed by slopes rising to 40 m. The beach is fronted by scattered rocks, then fringing reef.

Beach **K 918** lies on the eastern tip of August Point and faces northeast up the bay. The 200 m long beach is bordered by 20 m high rounded vegetated basalt points, and fronted by a strip of sand, then fringing reef out to 100 m. Beach **K 919** lies 100 m to the south on the southern side of the point. The point provides sufficient protection for mangroves to grow on the eastern rocks, with rocks also scattered along the curving 200 m long beach.

Beach **K 920** lies 500 m southwest of the point in a small V-shaped valley with sides rising to 40 m on each side. The 250 m long beach and backing small grassy beach ridge barrier has partially filled the valley. A small mangrove-lined creek runs along the southern side of the valley and crosses the southern end of the beach (Fig. 4.83).

Beach **K 921** is located in the next embayment south of July Point. The narrow 200 m long beach is located along the western side of a 50 m wide drowned valley. It is backed by the valley slopes rising to 60 m and fronted by tidal sand shoals and a small tidal channel.

Beach **K 922** lies 600 m to the south and is a 250 m long gap in a steeply sloping ridge that rises to 100 m. The sandy beach is bordered by mangrove-lined rocky points, and fronted by mud flats. Beach **K 923** lies 400 m to the south and is a similar 150 m long beach and mud flats.

Figure 4.83 Beach K 920 occupies a small valley, with a mangrove-lined creek on its southern (left) side.

Beach **K 924** lies at the southern end of the embayment 2 km south of July Point. The 450 m long beach faces northeast into the larger bay. It is bordered by rocky bluffs to either end, with a rocky protrusion near the centre of the beach. The southern half of the beach is backed by a small grassy foredune, with a usually dry creek draining across the southern end of the beach. It is fronted by mud flats and some patchy reef.

Beach **K 925** occupies the small embayment immediately north of July Point. The 400 m long beach and grassy overwash plain, containing two central boab trees, partially fills a small valley. A mangrove-lined creek drains along the northern side and across the beach. The beach is moderately steep, with exposed beachrock , and is fronted by sand-mud flats and patchy reef.

K 926-930 **JAR ISLAND (W)**

No.	Beach	Rating HT	Rating LT	Type	Length
K926	Jar Is (W1)	1	2	R+sand flats+rocks	300 m
K927	Jar Is (W2)	1	2	R+sand flats+rocks	400 m
K928	Jar Is (W3)	1	2	R+sand flats+rocks	600 m
K929	Jar Is (W4)	1	2	R+tidal sand flats	300 m
K930	Jar Is (W5)	1	2	R+sand flats+rocks	400 m
Spring & neap tide range:				2.9 & 0.8 m	

Jar Island is a 200 ha, 60 m high sandstone island located in the southwestern section of Vansittart Bay. It lies across the entrance of a 5 km wide, 4 km deep bay, the rocky shoreline of which contains five beaches (K 926-930).

Beach **K 926** is located 2 km southwest of the northern bay entrance and 4 km due west of Jar Island. The 300 m long beach is located in a gap on the rocky sandstone shoreline and has a small mangrove-lined creek exiting at its northern end, with mangroves also spreading out over the northern rocks. The rest of the beach is backed by low partly vegetated overwash flats, then gently sloping vegetated bedrock. It is fronted by sand flats and scattered rocks.

Beach **K 927** lies 1.5 km to the southwest at the western apex of the bay. The 400 m long beach lies in a small rocky embayment and is broken by rock and mangrove outcrops. It is backed by gently sloping bedrock , and fronted by tidal sand flats with a scattering of rocks and small islets extending out into the main bay.

Beach **K 928** is located 1.5 km to the southeast and lies 4 km southwest of Jar Island. The beach is relatively straight, 600 m long and faces toward the island. It is bordered by a sloping 20 m high northern headland and a low rocky point in the south. It is fronted by sand flats and a few rocks.

Beach **K 929** lies along the southern entrance to a small drowned valley occupied by mangroves, a tidal inlet, the beach and tidal sand shoals. The 300 m long beach runs along the southern side of the entrance with irregularities induced by the attaching sand shoals and rock outcrops. It is fronted by the tidal shoals and the inlet channel which flows out between boundary rocks.

Beach **K 930** commences immediately to the east and is a 400 m long strip of sand, fronted by sand flats and scattered rocks. It is backed by gently sloping vegetated terrain and bordered to the east by mangrove-covered rock flats, backed by a high tide storm beach ridge.

Map 5: KIMBERLEY

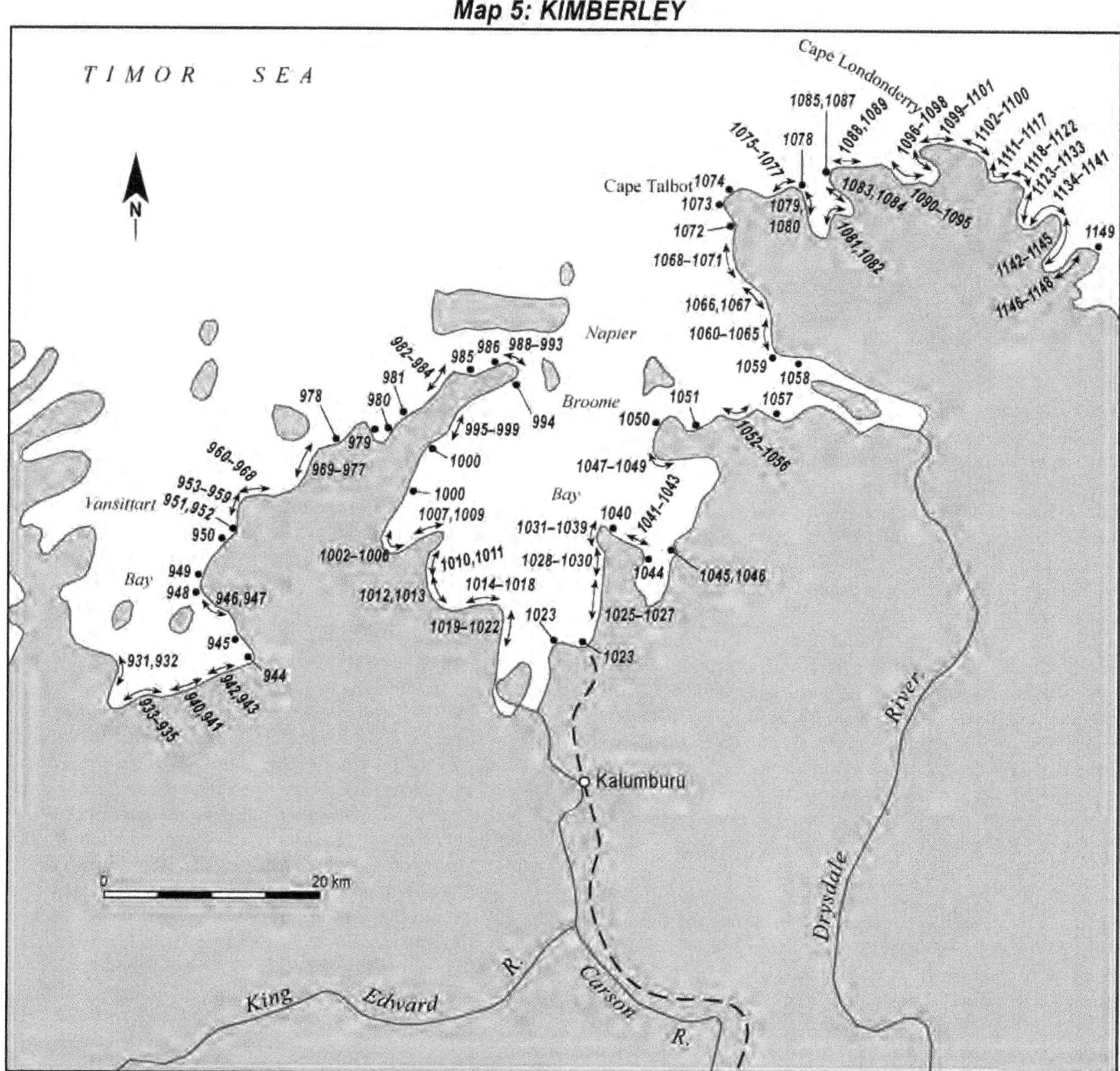

Figure 4.84 Kimberley map 5, locating beaches K 931-1149. They are located either side of Cape Londonderry, the northernmost point of Western Australia.

K 931-939 HILL POINT - MYOLA BLUFF

No.	Beach	Rating HT	LT	Type	Length
K931	Hill Pt (N2)	1	2	R+sand flats+rocks	600 m
K932	Hill Pt (N1)	1	2	R+mud flats	200 m
K933	Myola Bluff (S6)	1	2	R+mud flats	300 m
K934	Myola Bluff (S5)	1	2	R+mud flats	900 m
K 935	Myola Bluff (S4)	1	2	R+mud flats	500 m
K 936	Myola Bluff (S3)	1	2	R+mud flats	150 m
K 937	Myola Bluff (S2)	1	2	R+mud flats	150 m
K 938	Myola Bluff (S1)	1	2	R+mud flats	200 m
K939	Myola Bluff	1	1	R+sand flats	200 m
Spring & neap tide range:	2.9 & 0.8 m				

Hill Point and **Myola Bluff** are two of the southernmost headlands located in the far southwest corner of Vansittart Bay, to either side of Rocky Cove. While much of the southern shoreline is dominated by bedrock and fringing mangroves, there are two beaches just north of Hill Point (K 931-932) (Fig. 4.85) and seven beaches strung out to the south and west of Myola Bluff (K 933-939).

Beach **K 931** is a curving 600 m long east-northeast-facing, low energy beach tied between low mangrove-covered rock outcrops. In the south it is backed by low vegetated overwash deposits extending up to 200 m inland and fronted by tidal mud flats. Dune activity increases towards the northern end, with dunes extending up to 300 m inland and to heights of 10 m. Beachrock is

Figure 4.85 Beaches K 931 & 932 are located immediately north of Hill Point.

also exposed along the northern end of the beach, while considerable driftwood litters the beach crest. Beach **K 932** is located immediately to the east and is a similar 200 m long beach, also fringed by rocks and mangroves and backed by overwash flats, as well as dunes reaching 1-2 m high in the north, which are backed by older higher transgressive dunes.

Beaches **K 933** and **934** are adjoining 300 m and 900 m long beaches located at the western base of Myola Bluff and are the southernmost beaches in the bay. They are backed by gently rising vegetated sandstone slopes and fronted by mud flats. Beach K 934 is moderately steep and backed by a 200 m wide low sparsely vegetated overwash plain, with a small mangrove-lined creek at the western end of the beach.

Beaches K 935-938 lie along the western arm of the narrow 3.5 km long peninsula that terminates in the north at the bluff. All are backed by vegetated sandstone slopes rising to a crest of 40 m and are all are fronted by tidal mud flats and some by scattered rocks. Beach **K 935** is a curving 500 m long beach bordered and to the north fronted by sandstone rocks. It has a low grassy foredune with mangroves growing on the northern rocks. Beach **K 936** lies 1 km to the north and is a 150 m long strip of sand dominated by a central rock outcrop, with a 50 m wide section of dense vegetation behind the beach. Beach **K 937** occupies the next small embayment 500 m to the north and is a 150 m long strip of high tide sand fronted by lines of bedrock, with mangroves on the northern rocks. Beach **K 938** is a slightly curving 200 m long beach and narrow low foredune, with a small rocky point to the south and a northern headland that extends 1 km to the tip of the bluff.

Beach **K 939** occupies the eastern 200 m of the 500 m long north-facing Myola Bluff. It is a straight beach, bordered and underlain by rocky sandstone, and backed by tree-covered slopes rising to the sparsely vegetated 40 m high crest of the bluffs.

K 940-946 **RED BLUFF-PAULINE BAY**

No.	Beach	Rating HT	Rating LT	Type	Length
K940	Red Bluff (S)	1	1	R+tidal sand flats	400 m
K941	Red Bluff	1	1	R+tidal sand flats	400 m
K942	Pauline Bay (S1)	1	1	R+tidal sand flats	800 m
K943	Pauline Bay (S2)	1	1	R+tidal sand flats	500 m
K944	Pauline Bay (S3)	1	1	R+tidal sand flats	1.5 km
K945	Pauline Bay (N1)	1	1	R+tidal sand flats	300 m
K946	Pauline Bay (N)	1	1	R+tidal sand flats	250 m
Spring & neap tide range:		2.9 & 0.8 m			

Pauline Bay is a V-shaped west-facing embayment located on the lower eastern side of Vansittart Bay. It is 7 km wide at its entrance between beach K 947, in lee of Low Island, and Myola Bluff, and extends to the east for 7 km to the mouth of a mangrove-lined tidal creek. Five beaches (K 940-944) are located along its southern shore, with two smaller beaches (K 945 & 946) along the more rock-dominated northern shore.

Red Bluff is a low rocky sandstone outcrop located 3 km due east of Myola Bluff. It has beaches to either side. Beach **K 940** extends south of the bluff for 400 m, as a high tide west-facing sand beach, backed by 200 m of partly vegetated overwash to beach ridges. It is bordered by low rocks in the south then extensive mangroves leading to a creek mouth (Fig. 4.86). At the northern end a small mangrove-lined creek mouth separates it from the rocks of the bluff, with tidal sand shoals extending a few hundred metres off the creek and beach. Beach **K 941** commences on the northern side of the 300 m long bluffs, and is a moderately steep, 400 m long sand beach that faces east into the summer northwest winds and is backed by a climbing foredune, extending 100-200 m inland, where it is covered by a few trees. It is bordered by a low rocky point in the north and fronted by intertidal sand flats.

Figure 4.86 Beach K 940 is located to the north of the low Red Bluff.

Beach **K 942** is located 1 km to the east of beach K 941 on the western side of a small 1 km wide north-facing embayment. The beach faces northwest and consists of a 400 m long beach grading into a 400 m long recurved spit, which is the terminus for sand moving south into the bay. The beach is attached in the north to a low sandstone headland, while much of the beach and spit is backed by salt flats.

Beach **K 943** is located another 2 km to the east. It runs along the north side of a low rounded sandstone projection and consists of a 500 m long northwest-facing beach, with a small creek draining across the western end. There is some minor beach ridge to foredune activity extending up to 100 m to the lee of the beach, then the bedrock. Beach **K 944** lies 1 km further east on the eastern side of the sand projection. It is a 1.5 km long beach grading into a spit system. The beach is attached to the bedrock in the west, and extends 600 m to the west as an attached sand spit, then another 900 m as a detached spit that has slowly been extending to the west, and also represents the terminus for sand moving into the upper reaches of Pauline Bay. The spit is occasionally breached and for the most part is backed by sand flats, with extensive tidal shoals and channels off the beach-spit.

The northern side of the bay is dominated by 6 km of southwest-facing jointed sandstone bedrock. There are two beaches (K 945 & 946) along this section. Beach **K 945** lies midway along the rocky shore and is a narrow 300 m long sand beach located between low sandstone rocks and curving round at its northern end into a small joint-controlled mangrove-lined creek mouth. Tidal sand shoals associated with the creek extend off the beach.

Beach **K 946** is located toward the northern end of the rocky shore. It is a 250 m long sand beach that blocks a small joint-controlled creek. It is bounded by 10 m high sandstone points with some rocks also outcropping on the beach. The small creek periodically drains across the southern end of the beach and has a small lagoon and patch of mangroves, with a low grassy foredune and bedrock backing the central to northern part of the beach. It is fronted by intertidal sand flats.

K 947-952 **TRUSCOTT**

No.	Beach	Rating HT	Rating LT	Type	Length
K947	Truscott (S6)	1	1	R+sand flats	1 km
K948	Truscott (S5)	1	2	R+sand/rock flats	1.8 km
K949	Truscott (S4)	1	2	R+sand/rock flats	2 km
K950	Truscott (S3)	1	2	R+sand/rock flats	1.9 km
K951	Truscott (S2)	1	2	R+rock flats	150 m
K952	Truscott (S1)	1	2	R+rock flats	1.2 km
Spring & neap tide range:				2.9 & 0.8 m	

The abandoned **Truscott Air Base** is located on the 10 km wide peninsula bordered by Vansittart Bay to the west and West Bay to the east. Beach K 947 is located 10 m southwest of the base, while beach K 952 lies 5 km due west. In between are 8 km of near continuous beaches (K 947-952). The shoreline is however also dominated by low sandstone points bordering the beaches and considerable intertidal rock flats and outcrops.

Beach **K 947** represents the terminus for sand moving south from beach K 950. The 1 km long beach is anchored to low bedrock at its eastern end and extends to the west as a series of several recurved spits that have prograded west into the northern part of Pauline Bay (Fig. 4.87). The spits system is up to 400 m wide and covered by grassy vegetation. They are backed by salt flats and mangroves and fronted by relatively narrow sand flats and a tidal channel.

Figure 4.87 Sand moves south along beach K 948 (left) to contribute to the development of a series of recurved spits (right) bordered by beach K 947.

Beach **K 948** begins on the northern side of the rocks that anchor beach K 947 and trends north for 1.8 km to the next rock-controlled low sandy point. It faces west-southwest toward Jar Island. It is backed by a grassy foredune, which is tied to bedrock at either end and in the centre, and which increases in height to 10-15 m towards the southern end. Salt flats back the southern 300 m of foredune, while in the north a 5 ha coastal lagoon, dominated by salt flats, drains across the northern section of the beach. Rocks also outcrop on the shore and across the 300-400 m wide tidal flats.

Beach **K 949** extends for 2 km between the next two low rock-controlled sandy points, both of which have mangroves growing amongst the rocks. The beach faces west-northwest and receives the full brunt of the summer northwest winds, which have built a grassy foredune behind much of the beach, which reaches 20 m high in places, decreasing in height towards the southern point. A shallow 4 ha coastal lagoon drains across the centre of the beach, with mangroves lining the inner mouth and salt flats dominating the lagoon floor. Sand flats lie off the central creek mouth with intertidal rock flats dominating the remainder of the beach.

Beach **K 950** begins on the northern side of beach K 949 and is a similar 1.9 km long west-facing beach, with a central 5 ha salt flat-dominated lagoon drained by a small mangrove-lined creek, with beachrock exposed around the creek mouth, and a small sand delta off the creek. The beach is backed by a vegetated grassy foredune, containing a solitary boab tree (Fig. 4.88), the foredune extending up to 300 m inland and rising in places to 20 m. It is bordered by low rocky points, with rocks extending out across the tidal flats and mangroves to either end.

Figure 4.88 The solitary boab tree located on the low grassy foredune behind beach K 950.

Beach **K 951** is a 150 m north-facing pocket of sand located at a sharp inflection in the shore, where the coast trends away to the west. The beach is bordered by low north-trending ridges of sandstone, which extend across the tidal flats. Beach **K 952** commences immediately to the east and is a crenulate 1.2 km long north-facing high tide beach, dominated by rocks outcropping along the beach and across the tidal flats. In the east it grades into a mangrove-fringed shoreline of the next small embayment.

K 953-958 **TRUSCOTT (W)**

No.	Beach	Rating HT	Rating LT	Type	Length
K953	Truscott (W1)	1	2	R+rock flats/reef	500 m
K954	Truscott (W2)	1	2	R+sand flats/reef	400 m
K955	Truscott (W3)	1	2	R+sand flats/reef	600 m
K956	Truscott (W4)	1	2	R+rock flats/reef	200 m
K957	Truscott (W5)	1	2	R+rock flats/reef	300 m
K958	Truscott (W6)	1	2	R+rock flats/reef	200 m
Spring & neap tide range:		2.9 & 0.8 m			

To the west of Truscott landing ground are 5 km of bedrock-dominated west-facing shoreline. The southern end forms the eastern side of an L-shaped embayment and is dominated by 2 km of mangroves.

Beaches K 953-955 are three adjoining west-facing beaches located midway along the rocky section. Beach **K 953** is a 500 m long beach, bordered by low sandstone points, with sandstone rocks also outcropping the length of the beach and across the 500 m wide tidal flats, which are fringed by coral reef. Beach **K 954** commences 200 m to the north and is a 400 m long sand beach bordered by low sandstone ridges, with a joint-controlled mangrove-lined creek backing the centre of the beach. It is backed by low sandstone terrain and fronted by 500 m wide sand flats, containing a scattering of rocks and edged by fringing reef. Beach **K 955** commences 100 m further north and is a 600 m long sand beach with 500 m wide sand flats fringed by the same reefs. It is backed by low sandstone terrain, with a few rocks outcropping on the beach.

North of beach K 955 is a 1 km wide, rocky, mangrove-lined, small embayment. Beach **K 956** is located on the northern side of the embayment, and is a curving 200 m long beach, locked between low mangrove-covered sandstone rocks to the south and bedrock then a low rocky point to the north. The beach has prograded up to 300 m bayward as a series of low grassy beach ridges (Fig. 4.89). It is fronted by sand and rock flats, with reef 300 m offshore.

Figure 4.89 Beach K 956 is backed by a 300 m wide beach ridge plain and bordered by mangroves.

Beach **K 957** lies immediately to the north and is a 300 m long strip of high tide sand at the base of sloping bedrock, with rocks dominating the shoreline and tidal flats. Beach **K 958** is a similar 200 m long beach, with a prograded convex southern section containing a 100 m wide series of low beach ridges, then the backing bedrock.

K 959-961 **TRUSCOTT (N)**

No.	Beach	Rating HT	Rating LT	Type	Length
K959	Truscott (N1)	1	2	R+rock flats/reef	200 m
K960	Truscott (N2)	1	2	R+rock flats/reef	900 m
K961	Truscott (N3)	1	2	R+sand flats/reef	400 m
Spring & neap tide range:				2.9 & 0.8 m	

At beach K 960 the rocky shoreline turns and trends east for 7 km to the apex of a small mangrove-lined embayment. The roughly north-facing section of shore contains 13 beaches (K 959-971), all located approximately 4 km north of the abandoned Truscott landing ground. The first three beaches (K 959-961) are located along the first 2 km of shore between the point and a small mangrove-filled embayment.

Beach **K 959** is a discontinuous 200 m long north-facing sand beach bordered by low sandstone rocky shore, and backed by a low foredune that has blown up to 100 m inland over the backing gently sloping bedrock. It is fronted by 100 m wide rock flats, then fringing reef.

Beach **K 960** is located 200 m to the east and emerges from a rocky shore to form a 900 m long slightly curving sand beach. Two hundred metre wide rock outcrops and rocky intertidal flats however dominate most of the north-facing beach, with fringing reef paralleling the outer rocks. The reef continues to the east around a 1 km long low rock protrusion.

Beach **K 961** lies immediately to the east of the rocks and is a curving 400 m long beach backed by a low grassy 100 m wide overwash plain. A mangrove-lined creek forms the western end of the beach. It drains a 20 ha lagoon and crosses the beach in lee of a low elongate rocky point. Rocks also border the eastern end, beyond which is a 1 km deep mangrove-filled embayment. The beach faces northeast across 100 m wide sand flats, then another 300 m of fringing reef.

K 962-968 **TRUSCOTT (N)**

No.	Beach	Rating HT	LT	Type	Length
K962	Truscott (N4)	1	1	R+sand flats+reef	70 m
K963	Truscott (N5)	1	1	R+sand flats+reef	250 m
K964	Truscott (N6)	1	1	R+sand flats+reef	60 m
K965	Truscott (N7)	1	1	R+sand flats+reef	50 m
K966	Truscott (N8)	1	1	R+sand flats+reef	50 m
K967	Truscott (N9)	1	1	R+sand flats+reef	150 m
K968	Truscott (N10)	1	1	R+sand flats+reef	150 m
Spring & neap tide range:				2.9 & 0.8 m	

The next 2 km of shoreline is located essentially 3 km due north of the abandoned Truscott landing ground. It is separated by a 1 km wide mangrove-filled embayment from beach K 961. There are seven small north-facing, rock-bordered beaches along this section, all backed by vegetated sandstone slopes rising to 30 m and fronted by sand flats and a continuous fringing reef.

Beach **K 962** is a curving 70 m long pocket of sand bordered by two clumps of sandstone rocks and backed by the sandstone slopes. Immediately to the east is beach **K 963,** a straight 250 m long beach backed by two sandstone gullies and fronted by the continuous sand flats and fringing reef.

Beaches **K 964, 965** and **966** are three adjoining 60 m, 50 m and 50 m long pockets of sand located immediately east of beach K 963. Each is bordered by small sandstone headlands and backed by vegetated slopes. They are all fronted by the continuous sand flats with fringing reefs 200-500 m offshore.

Beach **K 967** lies 200 m east of beach K 966. It is a 150 m long beach, with a small mangrove-lined creek at its western end, and bordering 10 m high sandstone points. Beach **K 968** lies to the east of the 200 m long eastern point. It is a 150 m long strip of sand with a small creek draining across the western end. Both beaches are fronted by the continuous sand flats with the fringing reef 500-600 m offshore.

K 969-971 **TRUSCOTT (N)**

No.	Beach	Rating HT	LT	Type	Length
K969	Truscott (N11)	1	1	R+sand flats+reef	200 m
K970	Truscott (N12)	1	1	R+sand flats+reef	400 m
K971	Truscott (N13)	1	1	R+tidal sand flats+reef	900 m
Spring & neap tide range:				2.9 & 0.8 m	

The three easternmost beaches north of Truscott are located along a 2 km long north-facing section of shore that terminates in the east at a 1 km wide mangrove-lined embayment.

Beach **K 969** is a 200 m long beach with two mangrove-lined creeks converging at the western end of the beach, where they maintain a permanent tidal channel. The remainder of the beach is backed by low vegetated slopes, with a low 200 m long rocky point at the eastern end. Beach **K 970** lies on the eastern side of the point, and is a 400 m long beach, with a small mangrove-lined creek crossing the western end. Rocks increase to the east as the beach grades into mangrove-fringed rock flats. Both beaches are fronted by 500 m wide sand flats, then continuous fringing reef.

Beach **K 971** lies at the eastern end of the 500 m long rock flats. It trends east for 400 m to a series of low rock outcrops, then turns and trends south into the small embayment as a 500 m long spit that terminates amongst mangroves of the bay. Shallow sand flat, rock outcrop, and then fringing reef lie off the beach.

K 972-977 **TRUSCOTT (NE)**

No.	Beach	Rating HT	LT	Type	Length
K972	Truscott (NE1)	1	1	R+tidal flats+reef	250 m
K973	Truscott (NE2)	1	1	R+tidal flats+reef	150 m
K974	Truscott (NE3)	1	1	R+tidal flats+reef	300 m
K975	Truscott (NE4)	1	1	R+tidal flats+reef	250 m
K976	Truscott (NE5)	1	1	R+tidal flats+reef	150 m
K977	Truscott (NE6)	1	1	R+tidal flats+reef	100 m
Spring & neap tide range:				2.9 & 0.8 m	

Four kilometres due north of the Truscott landing ground the shoreline turns and trends north for 3 km to the lee of Mary Island where it then turns to the east. Six small west-facing beaches (K 972-977) are located along the bedrock-dominated 3 km of shore. All are composed of whiter sand and bordered by sandstone rocks and low points and backed by gently sloping vegetated terrain.

Beach **K 972** is a slightly curving 250 m long sand beach bordered at each end by a clump of mangroves and sandstone points extending 100-300 m into the bay. It is backed by 100 m wide low overwash deposits and fronted by 500 m tidal flats, then fringing reef, which runs the length of this section of the bay.

Beach **K 973** is located on the northern headland and is a 150 m long pocket of sand, bordered by small sandstone points and rocks, with some sand blown up onto the backing bedrock slopes. Tidal flats extend 600 m off the beach to the reef. Beach **K 974** lies 300 m to the north and is a double crenulate 300 m long beach, with rocks bordering each end and outcropping in the centre. The tidal flats extend 1 km into the bay.

Beaches K 975, 976 and 977 are three adjoining beaches which commence 500 m north of beach K 974. Beach **K 975** is a straight 250 m long high tide beach dominated by rocks outcropping along the beach and on the tidal flats. Beach **K 976** is 150 m long and separated from beach K 977 by a 150 m long fringe of mangroves surrounding a small creek. It is backed by a 100 m wide grassy overwash flats. Beach **K 977** is a 100 m long pocket of sand backed by a 50 m wide overwash plain, with mangroves to either side and behind. All three are fronted by sand and rock flats grading into fringing reef extending up to 1 km into the bay.

K 978-986 ANJO PENINSULA (W)

No.	Beach	Rating HT	Rating LT	Type	Length
K978	Anjo Peninsula (W1)	1	1	R+tidal sand flats	1 km
K979	Anjo Peninsula (W2)	1	1	R+tidal sand flats	700 m
K980	Anjo Peninsula (W3)	1	1	R+tidal sand flats+reef	2.6 km
K981	Anjo Peninsula (W4)	1	1	R+tidal sand flats+reef	2.4 km
K982	Anjo Peninsula (W5)	1	1	R+tidal sand flats+reef	200 m
K983	Anjo Peninsula (W6)	1	1	R+rock flats+reef	150 m
K984	Anjo Peninsula (W7)	1	1	R+rock flats+reef	300 m
K985	Sharp Point (W2)	1	1	R+rock/sand flats+reef	4.5 km
K986	Sharp Point (W1)	1	1	R+rock flats+reef	500 m
Spring & neap tide range:	2.9 & 0.8 m				

Anjo Peninsula is a 20 km long sandstone peninsula that trends to the northeast and is bordered by Vansittart Bay to the west and Napier Broome Bay to the east. It has approximately 50 km of shoreline and contains 24 beaches (K 978-1001) which occupy 18 km (36% of the shoreline). It terminates at the elongate Anjo Point.

Beach **K 978** is located in a northwest-facing embayment that looks across the 3 km wide channel to Mary Island. The curving 1 km long, moderately steep, white sand beach is protected by the island and as a consequence is fronted by 1 km wide intertidal sand flats with low shore transverse sand ridges, with mangroves to either end (Fig. 4.90). The mid-bay beach is backed by a 100 m wide, partly active series of foredunes that reach 5 m in height, then 1.5 km wide mangrove-salt flats drained by a tidal creek at the western end of the beach. It is bordered by gently sloping vegetated terrain. The beach has prograded about 200 m out into the bay as well as west across the creek mouth.

Figure 4.90 Beach K 978 is fronted by distinctive shore transverse intertidal sand ridges.

Beaches K 979 and 980 are located 2 km to the east in the next 2.5 km wide north-facing embayment and form a near continuous beach-barrier system, separated by a mangrove-fringed inlet and tidal creek which drains a western 30 ha mangrove-salt flat wetland. Beach **K 979** is 700 m long and bordered in the east by the creek, while mangroves line the western end. It is backed by a hummocky 50-100 m wide dune field rising to 2-3 m. Beach **K 980** is a curving 2.6 km long relatively steep beach, with a second creek occasionally breaking out across the centre of the beach to drain a similar size eastern lagoon. It is backed by vegetated foredunes up to 200 m wide and rising to 2-4 m, then the lagoon and an eastern dune-draped bedrock section. The bay is partly protected by a 4 km long reef system located 3 km off the eastern end. Mangroves fringe both ends and tidal sand flats extend up to 1 km into the bay.

Beaches K 981-983 occupy the next 5 km wide northwest-facing embayment. The bay is partly protected by the same reef that shelters beaches K 979 and 980. Beach **K 981** is a curving 2.4 km long beach, backed by a series of low vegetated beach ridges, with a linear 20 ha lagoon draining across the northern end of the beach. The beach is fronted by sand flats then fringing reef. Beach **K 982** lies 200 m to the north past a narrow strip of rocks. It is a 200 m long near continuation of the main beach. It has a 2-3 m high partly active foredune and is also fronted by the sand flats and reef. Beach **K 983** is located on the low northern boundary headland. The 150 m long beach is bordered by mangroves and fronted by sand flats with some rock outcrop, and reefs 600 m off the beach.

Beach **K 984** is located at the tip of the low lightly vegetated 5 m high sandstone headland that separates the two embayments. It is a double crenulate 300 m long strip of high tide sand, that faces north across irregular sandstone rock flats, with patches of sand then fringing reef 500 m off the point. Mangroves and rock flats fringe either side of the 500 m wide point.

Beaches K 985 and 986 occupy the next 5 km wide northwest-facing embayment, that terminates in the north at 10 m high Sharp Point. Beach **K 985** is one of the longer beaches in the region at 4.5 km in length. It is a steep sand beach backed by low vegetated dunes, with older inner transgressive dunes rising to 10-15 m. Behind the centre of the beach are two small 5 ha lagoonal salt flats which drain to either end of the beach. It is fronted by irregular sand flats up to 500 m wide then fringing reef. Beach **K 986** is located immediately north of the low rock outcrops that form the northern end of beach K 985. The straight 500 m long beach faces west and is fringed by scattered rocks, a few mangroves and fringing reef. It is backed by a 50 m wide low grassy foredune, then vegetated terrain.

K 987-993 WOMERAH BAY

No.	Beach	Rating HT	Rating LT	Type	Length
K987	Sharp Pt (E)	1	1	R+sand/mud flats+reef	400 m
K988	Womerah Bay (1)	1	1	R+sand/mud flats+reef	100 m
K989	Womerah Bay (2)	1	1	R+sand/mud flats+reef	400 m
K990	Womerah Bay (3)	1	1	R+sand/mud flats+reef	300 m
K991	Womerah Bay (4)	1	1	R+sand flats+reef	600 m
K992	Womerah Bay (5)	1	2	R+sand flats+reef	500 m
K993	Womerah Bay (6)	2	3	R+sand flats+channel	400 m
Spring & neap tide range:	2.4 & 0.8 m				

Womerah Bay is an open 3.5 km wide north-facing bay, bordered by Sharp Point in the west and Elbow Point to the east. The bay is afforded considerable protection from the 14 km long Sir Graham Moore Islands located across Geranium Harbour 5 km north of the bay. Much of the bay shore as a result is dominated by beaches fronted by mud flats and bordered by near continuous fringing mangroves. The beaches are backed by low gradient vegetated slopes.

Beach **K 987** is a curving 400 m long east-facing beach located in lee of Sharp Point. It is fringed by mangrove-covered rocks and backed by a 50 m section of overwash, while it is fronted by intertidal sand to mud flats, with fringing reef 500 m off the beach (Fig. 4.91).

Figure 4.91 Mangrove-fringed Beach K 987.

Beach **K 988** lies 1.5 km to the east in the centre of the bay. The 100 m long beach is fringed by mangroves and fronted by sand-mud flats, then fringing reef. Beach **K 989** is located 300 m to the east past some mangrove-covered rocks. It is a curving 400 m long beach that terminates in the east at a small mangrove-lined creek mouth. The creek drains a salt flat-filled 2 ha lagoon. Beach **K 990** begins to the east of the creek mouth and is 300 m long, with some rocks outcropping along the shore and mangroves bordering the eastern end. All three beaches share continuous sand-mud flats, with coral reefs 500-600 m offshore.

Beach **K 991** lies 300 m to the east and is bordered by the mangroves to the west and a low sand-backed rocky point in the east. The beach is 600 m long and fronted by 100-200 m wide sand flats then fringing reef. Beach **K 992** continues on the eastern side of the rocks for another 500 m, with some rocks outcropping along the shore and producing a small foreland. The beach is backed by a low active foredune, then a 200 m wide isthmus which backs onto Anjo Bay (beach K 994) on its southern side. It is fronted by 100 m wide sand flats, then a narrow strip of reef. Beach **K 993** extends from the small outcrop of boundary rocks for 400 m to Elbow Point, where the shoreline turns and trends south for 1.5 km to Anjo Point. The beach is backed by a 50 m wide, 2-3 m high foredune-overwash plain, while it slopes steeply to a channel swept by tidal currents with relatively deep water running along the shore. Beachrock is exposed along the beach face.

K 994-1000 ANJO PENINSULA (E)

No.	Beach	Rating HT	Rating LT	Type	Length
K994	Anjo Cove	1	1	R+mud flats	300 m
K995	Anjo Hill (1)	1	1	R+mud flats	100 m
K996	Anjo Hill (2)	1	1	R+tidal sand flats	300 m
K997	Anjo Hill (3)	1	1	R+tidal sand flats	150 m
K998	Mackenzie Anchorage (N3)	1	2	R+rock sand flats	400 m
K999	Mackenzie Anchorage (N2)	1	1	R+tidal flats	1.2 km
K1000	Mackenzie Anchorage (N1)	1	1	R+tidal sand flats	700 m
Spring & neap tide range:	2.4 & 0.8 m				

Anjo Point lies at the eastern tip of the Anjo Peninsula and forms the northwestern entrance to Napier Broome Bay (Fig. 4.92). At the narrow rocky point the peninsula shore turns and trends southwest for 20 km down into West Bay. The first 10 km between the point and Mackenzie Anchorage are occupied by seven low energy beaches (K 994-1000), that face southeast across the 20 km wide bay.

Anjo Cove is a U-shaped 1 km deep southeast-facing bay located immediately west of the point. At the base of the bay is beach **K 994**, a 300 m long strip of sand, bordered by mangroves at each end and backed by a 200 m wide low vegetated sandy isthmus that ties the point to the rest of the peninsula. The arms of the cove consist of 10-20 m high sandstone slopes, fringed by mangroves, with some rocks as well as sand flats off the beach. The beach has a moderate gradient with beachrock exposed along the beach face. The beachrock has been eroded and deposited as a series of distinctive high tide storm boulder cusps spaced about every 10 m.

Figure 4.92 Anjo Peninsula, with beaches K 992 & 993 on the northern side (lower), and mangrove-fringed beach K 994 at the base of Anjo Cove (centre).

One kilometre west of the cove is a second smaller cove, containing a 100 m long south-facing beach (**K 995**) at its apex. The bedrock cove is fringed by mangroves, with a rock reef and islet extending part way across the entrance while the backing vegetated slopes rise to the crest of the 38 m high Anjo Hills.

Immediately west of the cove entrance are two adjoining beaches, **K 996** and **997**, 300 m and 150 m long respectively. They are both bordered by rock flats and scattered mangroves and fronted by a mixture of rock and sand flats. Mangrove-covered rock flats dominate the next 3 km of shore to beach **K 998**, a straight 400 m long high tide sand beach, fronted by continuous 100 m wide rock flats. All three beaches are backed by gently sloping, lightly wooded terrain.

Beach **K 999** is a gently curving 1.2 km long, moderately steep, sand beach that dams a 100 ha wetland consisting of extensive salt flats and a mangrove woodland toward the entrance, with a creek draining across the southern end of the beach and maintaining tidal sand shoals off the mouth. The beach is fronted by 300 m wide sand flats and backed by low partly active outer dunes, and older inner dunes rising from 5-8 m, which have formed a 100-300 m wide barrier system. Low rocks and rock flats border the beach. Beach **K 1000** commences 100 m to the south of the rock flats. It is a curving 700 m long low energy beach, with mangroves outcropping toward the centre and either end, while it is fronted by sand to mud flats. It terminates at the northern entrance to Mackenzie Anchorage, a 4 km wide, east-facing, mangrove-lined embayment.

K 1001-1009 **WEST BAY**

No.	Beach	Rating HT	LT	Type	Length
K1001	West Bay (1)	1	2	R+sand flats+rocks	150 m
K1002	West Bay (2)	1	1	R+sand flats	300 m
K1003	West Bay (3)	1	1	R+sand flats	200 m
K1004	West Bay (4)	1	1	R+sand flats	500 m
K1005	West Bay (5)	1	1	R+sand flats	500 m
K1006	West Bay (6)	1	2	R+sand flats+rocks	300 m
K1007	West Bay (7)	1	1	R+sand flats+rocks	1.8 km
K1008	West Bay (8)	1	1	R+sand flats+rocks	400 m
K1009	Guy Point	1	2	R+sand flats+rocks	250 m
Spring & neap tide range:				2.4 & 0.8 m	

West Bay is a U-shaped, northeast-facing, 4 km wide, 7 km deep bay located on the eastern side of Napier Broome Bay. The low energy bay is only exposed to wind waves generated within the two bays. It has 16 km of generally low shoreline, backed by gently sloping wooded sandstone, with several small beaches and barriers scattered around the bay, usually adjacent to small upland creeks. The bay contains beaches K 1001-1008, with a total shoreline length of 4.3 km (27%), the remainder of the shore a mixture of rock and sand fronted by fringing mangroves.

Beach **K 1001** lies 1 km inside the wooded western entrance to the bay. The beach is 150 m long and faces east out the bay entrance. It is bordered by mangroves and rocky shore to the east and scattered rocks and a small upland-tidal creek to the west. The beach is backed by a 400 m wide mixture of overwash and low active dunes, and fronted by sand flats with scattered rocks.

Beach **K 1002** lies 4 km to the southwest and is located 7 km due east of the old Truscott landing ground. The beach faces east and extends from the southern side of a small tidal creek for 300 m to mangrove-fronted rock flats. The beach is backed by 400 m of low active foredunes and overwash and fronted by the tidal shoals of the creek and tidal sand flats. Beach **K 1003** is a further 1 km to the south and is a straight 200 m long beach with mangroves to either side. Low partly active overwash flats and a small lagoon back the beach with a tidal creek exiting to the east of the beach. It is fronted by 500 m wide tidal flats.

Beaches K 1004 and 1005 are located either side of a tidal creek that enters the southeastern corner of the bay. Beach **K 1004** is a slightly curving 500 m long north-facing beach that is bordered by mangroves to either end, as well as the tidal shoals of the creek to the east. It is backed by low 300 m wide partly active dunes. Beach **K 1005** lies 400 m to the east on the eastern side of the creek mouth. It is a curving northwest-facing 500 m long beach, with tidal shoals off the western end grading into sand flats to the east. It forms a 50-100 m wide sand spit which is backed by a 50 ha mangrove-filled tidal creek.

Beach **K 1006** is located 2 km northeast of the creek mouth, on the northern base of a low wooded 4 km long peninsula that terminates at Guy Point. The 300 m long beach faces northwest and is bordered by low rock flats covered with a 50 m fringe of mangroves, and backed by gently sloping wooded terrain. A mixture of sand flats and scattered rocks extend 500 m off the beach.

Beach **K 1007** lies another 1 km along the peninsula. It is a curving 1.8 km long beach, with low mangrove-covered rocks to the south, and rocks and a small islet forming a sandy foreland on the northern boundary. Beach **K 1008** extends for another 400 m on the northern side of the foreland. Both beaches are fronted by 500 m wide sand flats and scattered rocks. They are backed by overwash flats extending up to 200 m inland, which are in turn backed by a 10 ha salt flat-filled lagoon drained by a small intermittent creek toward the northern end of beach K 1007.

Guy Point forms the tip of the peninsula and the eastern entrance to West Bay. Beach **K 1009** is located on the eastern side of the low sandstone point. It is an irregular 250 m long slightly more exposed northeast-facing sand beach that receives higher wind waves and is fronted by 200 m wide sand flats, with rocks bordering each end, and scattered on the flats and along the western end of the beach.

K 1010-1018 **DEEP BAY (W)**

No.	Beach	Rating HT	Rating LT	Type	Length
K1010	Guy Point (S1)	1	1	R+sand flats	2.6 km
K1011	Guy Point (S2)	1	1	R+sand flats	1.4 km
K1012	Woppinbie Ck (W2)	1	1	R+sand/rock flats	200 m
K1013	Woppinbie Ck (W1)	1	1	R+sand/rock flats	1.2 km
K1014	King Harman Pt (W5)	1	2	R+rock flats	1.2 km
K1015	King Harman Pt (W4)	1	1	R+sand flats	300 m
K1016	King Harman Pt (W3)	1	1	R+sand flats	800 m
K1017	King Harman Pt (W2)	1	2	R+rock flats	400 m
K1018	King Harman Pt (W1)	1	2	R+rock flats	500 m
Spring & neap tide range: 2.4 & 0.8 m					

Deep Bay occupies the southernmost section of Napier Broome Bay and is bordered by Guy Point in the west and Bluff Point 13 km to the east. It extends for up to 20 km to the south into the mouth of the King Edward River. The western part of the bay has a 7 km wide U-shaped embayment between Guy and King Harman points, containing nine beaches (K 1010-1018) which occupy half of the 16 km of bay shoreline, the remainder dominated by low sandstone shore and the mangrove-lined mouth of Woppinbie Creek.

Beach **K 1010** is located 3 km south of Guy Point, and consists of a curving 2.6 km long east-facing, beach-barrier system, with low wooded, mangrove-lined points to either end. The beach is backed by a low 200-400 m wide grassy foredune and backing 30 ha salt flat-filled lagoon, with a small mangrove-lined creek draining across the centre of the beach (Fig. 4.93). A solitary boab tree is located at the southern end of the beach. The beach is fronted by the central sand shoals of the creek and sand-mud flats to either side.

Figure 4.93 Beach K 1010 consists of a barrier backed by a shallow lagoon, which drains across the centre of the beach.

Beach **K 1011** occupies the next small east-facing embayment, 1 km to the south. It contains the straight 1.4 km long beach and low grassy 200 m wide barrier, also backed by a 40 ha salt flat-filled lagoon, with a small mangrove-lined creek crossing the centre of the beach. It is bordered by mangrove-covered rock flats and fronted by sand then mud flats.

Beach **K 1012** is located on a low wooded 1 km wide point that separates beach K 1011 from 1013. The beach lies in lee of low irregular rocky points with rocks scattered between, resulting in a curving 200 m long low energy beach, backed by low grassy overwash flats, and fronted by the sand-mud flats and rocks. Beach **K 1013** commences 200 m to the east and is a 1.2 km long double crenulate beach, backed by a central 50-100 m wide low grassy foredune, with a small creek and tidal shoals protruding in the centre. It has rocks and fringing mangroves to either end as well as some rocks off the beach, with mud flats fronting the beach. The creek drains two interconnected bedrock-controlled lagoons and their upland streams.

The 1 km wide mangrove-lined mouth of Woppinbie Creek lies immediately east of beach K 1013, while beach K 1014 is located on the eastern side of the mouth 1.5 km further east. Beach **K 1014** is an irregular north-facing 1.2 km long shelly beach, backed by low grassy overwash flats, that parallels the backing low wooded bedrock slopes. It has prograded out 100 m in places over a mix of rocks and rock flats, fronted by near continuous rocks and rock flats. It terminates in the east at a 200 m long clump of rocks and mangroves. Beach **K 1015** commences on the eastern side of the mangroves and curves round for 300 m to a low rock outcrop. The small beach is also crossed by a central tidal creek which drains a 20 ha salt flat-filled lagoon that extends east behind the beach (Fig. 4.94). It is backed by a 50-100 m wide low grassy overwash plain. Beach **K 1016** is an 800 m long moderate gradient beach that is backed by a 100-200 m wide grassy foredune with a few scattered trees, then the salt flats. Both beaches are fronted by continuous 500 m wide ridged sand flats.

Figure 4.94 Beach K 1016 (centre) with the smaller beaches K 1014 & 1015 to right, and beach K 1017 (foreground).

Beaches K 1017 and 1018 occupy part of the 1 km long northern side of King Harman Point. Beach **K 1017** is a 400 m long high tide sand beach and low foredune fronted by irregular rocks and rock flats, while 500 m to the east is 500 m long beach **K 1018**, a similar sand beach and foredune with more continuous 500 m wide rock flats. It terminates at the mangrove-fringed point.

K 1019-1022 **KING EDWARD RIVER**

No.	Beach	Rating HT	Rating LT	Type	Length
K1019	King Harman Pt (S1)	1	1	R+tidal flats	700 m
K1020	King Harman Pt (S2)	1	1	R+tidal flats	150 m
K1021	King Harman Pt (S3)	1	1	R+tidal flats	50 m
K1022	King Harman Pt (S4)	1	1	R+tidal flats	50 m
Spring & neap tide range:	2.4 & 0.8 m				

King Harman Point forms the northwestern boundary of the 5 km wide bay into which the King Edward River flows, with the 1 km wide mouth of the river located 10 km south of the point. The Kalumburu community is located a further 9 km upstream (Fig. 4.84). The river mouth and adjoining shoreline are fringed by mangroves, with the nearest beaches located on the western side toward King Harman Point (beaches K 1019-1022) and on the eastern side 6 km north of the river mouth (beaches K 1023-1024).

Beach **K 1019** lies 3 km south of King Harman Point and is a curving 700 m long east-facing beach, backed by a low 300 m wide barrier, with tidal creeks exiting to either end. The northern creek drains a 10 ha area of lagoonal salt flats, while the smaller southern creek is fed by an upland stream. The beach is fronted by sand to mud flats, grading into tidal shoals toward the northern creek. A low rocky outcrop forms the southern boundary, with beach **K 1020** extending 150 m to the south. This beach blocks a small valley, with mangroves, then dense woodlands behind.

Beaches K 1021 and 1022 occupy two gaps in a sandstone protrusion that extends from beach K 1020 for 1 km to the south. Beach **K 1021** is a 50 m pocket of northeast-facing sand bordered by low rocky points, while beach **K 1022** is a similar 50 m pocket, fringed by mangroves then rocks. A low linear islet lies 100 m off the beach.

South of the rocks, mangroves dominate the next 10 km of shore down to the King Edward River mouth, terminating at Kalumburu's Longini river landing. Mangroves then trend north for 8 km to a low bedrock point, immediately east of which begins beach K 1023.

Kalumburu

The Kalumburu community and some of its adjoining beaches are the only locations on the northern Kimberley coast accessible by vehicle. It is however part of Kalumburu Aboriginal Land and a permit must be obtained from the Kalumburu Aboriginal Community before visiting the area. Limited food and fuel are available at the community. There are several beaches spread along 80 km of coast accessible via Kalumburu, with at least two sites providing camping areas.

K 1023-1031 **KALUMBURU BEACHES**

No.	Beach	Rating HT	Rating LT	Type	Length
K1023	Marra Garra	1	1	R+tidal flats	700 m
K1024	Wanganjie	1	1	R+tidal flats	1.7 km
K1025	McGowan Is	1	2	R+sand/rock flats	250 m
K1026	McGowan Is (N1)	1	2	R+sand/rock flats	80 m
K1027	McGowan Is (N2)	1	1	R+sand flats+rocks	700 m
K1028	Jakes Beach (S3)	1	1	R+ridged sand flats	600 m
K1029	Jakes Beach (S2)	1	1	R+ridged sand flats	250 m
K1030	Jakes Beach (S1)	1	2	R+sand/rock flats	800 m
K1031	Jakes Beach	1	1	R+ridged sand flats	800 m
Spring & neap tide range:	2.4 & 0.8 m				

Several beaches are accessible by vehicle from Kalumburu. The nearest are located along the eastern shore of Deep Bay (K 1023-1039), with a few others in Mission Bay (K 1042-1044, K 1046-1049) and the farthest at Lull Bay (K 1053-1055) 35 km to the north.

Marra Garra (beach **K 1023**), is a slightly curving 700 m long north-facing moderate gradient beach, backed by a 100 m wide older foredune area containing a few boab trees, then a 50 ha lagoonal salt flat. A formed road runs out to the eastern end of the beach, which is used as a landing. Rock flats and mangroves fringe either end (Fig. 4.95).

Figure 4.95 Marra Garra Beach (K 1023) is a recurved spit, with the Kalumburu boat landing visible at its eastern end.

Wanganjie (beach **K 1024**) lies 2 km to the east, in a 3 km wide embayment. It is a 1.7 km long northwest-facing sand beach, backed by a 200 m wide low barrier with small creeks draining salt flats to either end, and vehicle access to the eastern end. It is fronted by sand to mud flats.

McGowan Island lies 15 km north of Kalumburu. It is a low small rock outcrop connected to the mainland by a 200 m long shallow reef. The reef connects to a low sandstone headland, with 250 m long beach **K 1025** bordered by this headland in the south and a mangrove-fronted creek mouth to the north. The beach is backed by 100 m wide overwash flats, and fronted by a mixture of sand and rock flats.

Beach **K 1026** is an 80 m long pocket beach bordered by long sandstone points extending 100-200 m into Deep Bay forming a small protected bay. A small creek and a few mangroves are located at the southern end of the beach. There is vehicle access to the beach which is also used to launch boats and a camp site along the rear. Camping is possible with permission. The small McGowan Island community is located immediately north behind beach **K 1027**. This is a 700 m long west-facing sand beach bordered by low sandstone rocks, with some outcropping along its central-northern section, and sand and increasing rock flats off the beach to the north. A vehicle track runs through the trees to the back of the beach where a solitary house is located.

Immediately north of beach K 1027, the shoreline turns and trends north-northeast for 3 km to Jakes Beach, the next accessible beach. In between are three beaches, all separated by low sandstone points. Beach **K 1028** is a 600 m long sand beach, backed in the south by a grassy overwash plain, with a 100 m long section of rocks separating it from a northern 50 m long pocket of sand which in turn is bordered by mangrove-covered rocks. While bordered by the rock points, it is fronted by 500 m wide ridged sand flats. Beach **K 1029** lies 800 m to the north and occupies a small gap in the rocks. The 250 m long beach blocks a usually dry upland creek. It is backed by 50 m wide grassy overwash flats, and fronted by a central clump of mangrove-covered rocks, and then 300 m wide ridged sand flats. There is vehicle access to the rear of the beach. It is bordered at the northern end by a 400 m long section of mangroves and rocks that separates it from beach K 1030.

Beach **K 1030** is an 800 m long relatively straight west-facing beach, bordered by low rocks, with a small central clump of rocks and mangroves. A vehicle track reaches the back of the beach, which is fronted by a mixture of sand and rock flats. **Jakes Beach** (**K 1031**) lies 300 m to the north and is an 800 m long sand beach, with a vehicle track reaching the northern end of the beach. Rocks and some mangroves border the beach, with one central section of intertidal rocks, then 100 m wide ridged sand flats. Both beaches are backed by gently sloping wooded terrain.

K 1032-1039 **FISH ROCK-BLUFF POINT**

No.	Beach	Rating HT	LT	Type	Length
K1032	Fish Rock (S2)	1	1	R+sand flats	600 m
K1033	Fish Rock (S1)	1	1	R+sand flats	900 m
K1034	Fish Rock (N1)	1	2	R+sand/rock flats	200 m
K1035	Fish Rock (N2)	1	1	R+sand flats	200 m
K1036	Fish Rock (N3)	1	2	R+sand/rock flats	100 m
K1037	Fish Rock (N4)	1	2	R+sand/rock flats	200 m
K1038	Bluff Pt (W)	1	2	R+rocks	80 m
K1039	Bluff Pt (E)	2	3	R+rock flats	100 m
Spring & neap tide range:	2.4 & 0.8 m				

To the north of Jakes Beach the shoreline trends north-northwest for 5 km to Bluff Point. The shoreline is dominated by low sandstone points backed by low wooded terrain. There are seven beaches located between Fish Rock and the point (K 1032-1038), with one to the east of the point (K 1039). There is no vehicle access to the beaches.

Beach **K 1032** is located in a 600 m wide open embayment 1.5 km south of Fish Rock. The beach curves round between bordering low mangrove-covered rocky points. A small salt flat-filled lagoon backs the centre of the beach and its 100 m wide foredune ridge plain, with its creek usually blocked by the beach. It is fronted by sand flats, with patchy coral reef offshore. Beach **K 1033** commences 200 m to the north and extends for 900 m to the lee of Fish Rock. It is bordered by low rock flats in the south and a small mangrove-lined creek mouth in the north, with 10-15 m high red bluffs on the headlands backed by gently sloping sparsely vegetated sandstone slopes. Sand then patchy rocks and reef lie off the beach.

To the north of Fish Rock is a 1.5 km series of four small pocket beaches (K 1034-1037), each bordered by small low mangrove-covered sandstone points and backed by lightly wooded gently sloping terrain. Beach **K 1034** is a low 200 m long beach bordered by 10 m high red bluffs in the south, which decrease in slope to the north with the northern boundary rocks covered with mangroves. Beach

K 1035 occupies the next small embayment and is a 200 m long beach extending between low bluffs, rocks and mangroves, and fronted by sand flats and patchy rocks and reefs. Beach **K 1036** is 100 m long, with a band of intertidal rock along much of the beach and mangroves to either end. Beach **K 1037** lies in lee of a small mangrove-covered rock islet, which forms a central foreland on the 200 m long beach. There is a patch of spinifex on the overwashed foreland and a 100 m long section of mangroves in the northern corner of the bay. A usually blocked creek and patch of mangroves lie behind the southern end of the beach.

North of beach K 1037 the shoreline continues as a rocky, mangrove-lined series of crenulations for 1 km to **Bluff Point**. Red sandstone rocks and reefs dominate the rocky shoreline. There are two beaches either side of the point linked by 200 m wide low grassy overwash flats. Beach **K 1038** consists of an 80 m long pocket of sand on the western side of the point, with sandstone rocks littering the fronting sand flats. Beach **K 1039** lies 300 m to the east and is a 100 m long northeast-facing high tide sand beach, backed by a narrow 1-2 m high grassy foredune, then rocks, and fronted by a continuous strip of irregular sandstone rocks and boulders, then deeper water off the point.

K 1040-1044 **HONEYMOON BEACH**

No. Beach	Rating HT	Rating LT	Type	Length
K1040 Bluff Pt (SE1)	1	2	R+sand flats+rocks	50 m
K1041 Bluff Pt (SE2)	1	1	R+sand flats	150 m
K1042 Honeymoon Beach (1)	1	1	R+sand flats	400 m
K1043 Honeymoon Beach (2	1	1	R+sand flats	600 m
K1044 Tamarinda	1	1	R+tidal flats	400 m
Spring & neap tide range:	2.4 & 0.8 m			

Honeymoon Beach is located in a northwest-facing U-shaped bay 4 km to the southeast of Bluff Point and bordered to the east by the elongate 2 km long Tate Point (Fig. 4.96). Two beaches are located along the rocky shore southeast of Bluff Point (K 1040 and 1041) and two more beaches comprise Honeymoon Beach (K 1042 and 1043). A small community is located at the main Honeymoon Beach, which is located 22 km north of Kalumburu.

Beach **K 1040** lies midway between Bluff Point and Honeymoon Beach. It is a 50 m long pocket of northwest-facing sand, wedged into an inflection in the otherwise low sandstone shoreline. It s fronted by a central patch of sand, with rock flats to either side. Beach **K 1041** is located 500 m to the south and is a 150 m long northwest-facing sand beach, backed by a narrow 1-2 m high foredune, and also bordered by low red sandstone points, include a small rock islet off the eastern end, tied to the beach by a strip of sand. It is fronted by continuous sand flats.

Figure 4.96 Honeymoon Beach (K 1042 & 1043) occupies the base of a headland-bound bay and is a popular camping site.

Honeymoon Beach consists of the main 400 m long western half (**K 1042**), backed by a camping area and a house further back amongst the trees, and bordered by two small mangrove-covered clumps of rocks. Small dinghies are launched from the beach and sometimes moored on and off the beach. The eastern half (**K 1043**) is 600 m long and also bordered by mangroves at either end, with Tate Point extending 2 km north of the eastern end of the beach. It is moderately steep and backed by a 100 m wide older weathered foredune. Both beaches are fronted by continuous intertidal sand flats, while a small creek drains salt flats at the western end of the eastern beach.

Tamarinda beach (**K 1044**) is located 1 km south of Honeymoon Beach on the southern side of Tate Point and faces east into Mission Cove. The 400 m long low energy beach is bordered by mangroves and is part of an 800 m long beach ridge that has impounded a narrow lagoon. There is vehicle access to the beach which is also used for camping.

K 1045-1050 **MISSION BAY(E)-GALLEY PT**

No. Beach	Rating HT	Rating LT	Type	Length
K1045 Tidepool Hill (S)	1	1	R+tidal flats	200 m
K1046 Tidepool Hill (N)	1	1	R+sand/rock flats	400 m
K1047 Wargally (S2)	1	1	R+sand/rock flats	300 m
K1048 Wargally (S1)	1	1	R+sand/rock flats	150 m
K1049 Wargally	1	1	R+sand flats+rocks	100 m
K1050 Galley Pt (W)	1	1	R+ridged sand flats	250 m
Spring & neap tide range:	2.4 & 0.8 m			

On the eastern side of Mission Cove the shoreline extends north from the old Pago Mission site for 17 km to Wargally. A vehicle track from Kalumburu reaches Pago, and trends north via Tidepool Hill to Glin Bay and Wargally, 3 km south of Galley Point. Between Pago and Galley Point are 35 km of shoreline dominated by mangroves in Mission Cove and Ian Bay, then deeply jointed sandstone shoreline north to Galley Point. There are only six small beaches along this section totalling 1.4 km in length.

Tidepool Hill is a 20 m high sandstone knoll, with a 10 ha lagoon on its northern side, fronted by beach **K 1045**. The low energy 200 m long beach faces southwest into Mission Cove, with rocks of the hill forming the southern boundary and a mangrove-lined creek mouth draining the lagoon at the northern end. It is fronted by a mixture of sand and rock flats.

Beach **K 1046** is located 500 m north of the hill and is a northwest-facing 400 m long beach, backed by low wooded coastal plain. The beach is bordered by mangrove-fringed rock flats to either end, and fronted by some rocks and sand flats. Mangrove-lined shore dominates the next 20 km of coast which includes the small Glin Bay, where there is vehicle access to the shore, and the larger Ian Bay.

Beach **K 1047** is located 4 km southeast of Red Bluff. It faces south into Ian Bay and is a 300 m long sand beach dominated by large sandstone outcrops. It is tied in the east to the sandstone shoreline and extends west to a low 50 m wide vegetated sand spit, which is also tied to sandstone boulders at its eastern end. A small mangrove-lined creek weaves through the boulders to connect a backing 2 ha shallow lagoon to the bay (Fig. 4.97). A 4WD track reaches the western end of the beach and runs along the back of the beach. Shallow rocks dominate the intertidal zone and offshore.

Figure 4.97 Beach K 1047 is backed and bordered by mangroves.

Beach **K 1048** lies 3 km south of Red Bluff and is a 150 m long pocket of sand encased in a 200 m wide sandstone cove. Scattered mangroves line the rocky shore to either side and rocks dominate much of the floor of the cove, with sand only off the beach. A narrow 1-2 m high grassy foredune then irregular lightly wooded sandstone terrain back the beach.

Wargally beach **K 1049** is located 1 km south of Red Bluff and consists of a 100 m long strip of north-facing sand located on the tip of a north-facing red sandstone point, with large boulders dominating the shoreline and extending up to 500 m off the beach. A 50 m wide patch of grassy overwash flats then large sandstone hummocks back the beach. A vehicle track winds through the sandstone to the beach.

Galley Point forms the boundary between Napier Broome Bay and the mouth of the Drysdale River, 10 km to the east. The western side of the point is dominated by the deeply jointed sandstone, with mangroves fringing the western shore. Beach **K 1050** is tucked inside the western side of the point, 1 km south of the tip. It is a slightly curving, moderately steep 250 m long southwest-facing beach, backed by a 100-200 m wide, older, weathered beach ridge-overwash plain, and fronted by 300 m wide ridged sand flats, with mangrove-covered rock flats bordering each side and extending 200-300 m into the bay.

K 1051-1058 GALLEY POINT-DRYSDALE RIVER

No.	Beach	Rating HT	Rating LT	Type	Length
K1051	Galley Pt (E)	1	1	R+tidal flats	800 m
K1052	Lull Bay (N2)	1	2	R+rocks	300 m
K1053	Lull Bay (N1)	1	1	R+LTT	600 m
K1054	Lull Bay	1	1	R+LTT	500 m
K1055	Lull Bay (S)	1	1	R+sand flats	400 m
K1056	Beauty Pt	1	1	R+sand flats	2.3 km
K1057	Beauty Pt (spit)	1	2	R+tidal sand flats	1.1 km
K 1058	Drysdale River	1	2	R+tidal sand flats	500 m
Spring & neap tide range:		2.4 & 0.8 m			

At Galley Point the shoreline turns and trends east for 12 km to the 5 km wide mouth of the Drysdale River. In between are 25 km of north-facing shoreline containing eight beaches (K 1051-1058). The shore is dominated by mangroves in the west, low rocky coast in the centre, with the extensive sand deposits of the river mouth to the east.

Beach **K 1051** lies 4 km southeast of Galley Point at the base of 3 km wide northwest-facing U-shaped bay. The beach is 800 m long, faces northwest and is bordered by mangrove-covered rock flats. It has prograded 200-300 m into the bay and is backed by a low grassy beach ridge to overwash plain, then low wooded coastal plain.

Beaches K 1052-1055 occupy a 2.5 km long north-facing rocky shoreline. Beach **K 1052** is located in lee of rock reefs at the northern tip of a low sandstone headland. The 300 m long beach is bordered by mangrove-covered rock flats, and runs crenulate longshore in lee of the numerous rocks that extend 500 m off the beach. Beach **K 1053** commences 200 m to the east, past the mangroves, and is a slightly curving 600 m long beach that is bordered in the east by a 100 m wide low sandstone point, with rocks extending 500 m off the point. The beach is backed by an incipient foredune, then a low lightly wooded foredune, which is weathered and probably mid-Holocene in age. It is fronted by a low tide bar and is accessible at the western end by 4WD from adjoining Lull Bay.

Lull Bay beach (**K 1054**) is a curving 500 m long sand beach with a low tide bar running the length of the beach. It is backed by a low vegetated foredune, with an access track to the beach crossing the centre of the dune. A small creek drains out across the western end of the beach and sandstone points border each end, with some rocks scattered off the western half of the beach. Beach **K 1055** extends immediately south of the eastern rocks for 400 m into the mouth of a 100 m wide tidal creek. The beach faces east across sand flats and the creek channel, and has some rocks lying off the beach, which, together with the shifting creek shoals, result in a more crenulate beach. A vehicle track from Lull Bay reaches the back of the beach. Both beaches are used for camping.

Beauty Point beach (K 1056) commences on the eastern side of the mangrove-fringed creek mouth and trends east for 2.3 km to where the beach turns and trends southeast for 1.1 km as a series of recurved spits known as Beauty Point (K 1057). Beach **K 1056** is the outer barrier in a 5 km wide system composed of five inner barriers, each barrier consisting of a few to several beach ridge to recurved spits, separated by mangrove wetlands. The barriers have probably been supplied with river sands from the adjoining Drysdale River. The modern beach faces northwest and is backed by 100-200 m wide wooded beach-foredune ridges, and fronted by extensive river mouth shoals and sand flats, extending a few hundred metres off the eastern end of the beach in particular. Beach **K 1057** contains multiple vegetated recurved spits, fronted by a 500 m wide zone of active intertidal spits. This is an irregular, dynamic beach-spit system, which changes in length and character in response to wave, cyclone and flood events.

The **Drysdale River** mouth is 4 km wide between Beauty Point and the northern sandy Curran Point. In between are extensive tidal and river sand shoals, which become covered with mangroves into the river mouth, as it narrows to 500 m in width 10 km to the southeast. Beach **K 1058** is located on a sandy island midway between the two points. It is a low dynamic 500 m long sand island, bordered by recurved spits and backed by a few beach ridges, then mangroves. Like Beauty Point it is also a very dynamic feature changing in shape and size in response to episodic flood and cyclonic events.

K 1059-1066 **CURRAN POINT**

No.	Beach	Rating HT	LT	Type	Length
K1059	Curran Pt (S)	1	1	R+tidal flats	700 m
K1060	Curran Pt	1	1	R+sand flats	1.3 km
K1061	Curran Pt (N1)	1	1	R+sand flats	150 m
K1062	Curran Pt (N2)	1	1	R+sand flats	500 m
K1063	Curran Pt (N3)	1	1	R+LTT	600 m
K1064	Curran Pt (N4)	1	1	R+LTT	500 m
K1065	Curran Pt (N5)	1	1	R+LTT	400 m
K1066	Curran Pt (N6)	1	1	R+LTT	700 m
Spring & neap tide range:	2.4 & 0.8 m				

Curran Point is a low sand-draped rocky point that forms the northern entrance to the Drysdale River mouth (Fig. 4.98). Mangrove-fringed rocky shore dominates the shoreline that extends for 5 km to the southeast into the river mouth, while to the north predominantly gently sloping sandstone coast extends for 17 km north-northwest to Cape Talbot. Beaches K 1059-1066 are located between the point and the west-facing shore that lies 6 km to the north. Each beach occupies a small sandstone embayment, with joint-controlled creeks draining across each.

Figure 4.98 Curran Point is partly covered by dunes originating from beach K 1060.

Beaches K 1059 and 1060 lie either side of the point. Beach **K 1059** is a narrow curving low energy beach that faces southeast into the river mouth. It is tied to the rock shore in the east, terminating at the point in the west, with the river mouth shoals and deeper channels off the beach.

Beach **K 1060** forms the northern side of the point and extends from the tip of the point, past some low rocks, for 1.3 km to a low rocky point, with a tidal creek crossing the beach against the rocks. The beach faces west and is backed by active dunes that have extended up to 300 m inland and to heights of 15 m. In the east the dunes are moving over wooded sandstone terrain. Beach **K 1061** lies immediately east of the boundary rocks and is a 150 m long pocket of sand bordered by the 100 m wide western rocks and a mangrove-lined creek draining out against a small eastern rocky point. It has a moderately active foredune rising to 3-4 m. Beach **K 1062** commences on the eastern side of these rocks and curves round for 500 m to a low point that protrudes 100 m into the bay, with a small creek against the base of the point. The beach has a moderate gradient and is backed by an active foredune rising to 10-15 m. All three beaches are fronted by sand flats and backed by sparsely vegetated, gently sloping, joint-controlled sandstone terrain.

Beach **K 1063** occupies the next embayment and is a relatively straight 600 m long beach, bordered by low rocky shore, with some rock outcropping on the centre of the beach. It is backed by some low partly active dunes extending up to 300 m inland, with older vegetated dunes continuing up to 400 m inland. A small mangrove-lined creek crosses the northern end of the beach.

Beach **K 1064** lies 500 m to the north in the next embayment. It is a 500 m long beach with a small elongate lagoon and creek behind the centre of the beach and minor dune activity to either side. It is bordered by the low sandstone rocks and rock flats, with a 5 m wide sandy bar along the base of the beach. Beach **K 1065** is another 1 km to the north and is a 400 m long beach, with rocks outcropping off either end. A joint-controlled creek flows across the northern end of the beach, and partly active dunes have spread up to 200 m inland. One hundred metre wide tidal shoals off the creek mouth and a narrow bar front the beach.

Beach **K 1066** occupies the largest of the embayments and is 700 m long, with small rocky points to either end. A creek and 10 ha lagoon back the centre of the beach, the lagoon filled with mangroves, salt flats and scattered sandstone boulders, while the creek mouth meanders across the beach. On the south side dune to overwash flats extend 300 m inland, while several low foredunes back the north side. A narrow bar and some deeper rocks front the beach.

K 1067-1073 **CAPE TALBOT (S)**

No.	Beach	Rating HT	Rating LT	Type	Length
K1067	Curran Pt (N7)	1	2	R+rock flats	300 m
K1068	Curran Pt (N8)	1	2	R+rock flats	150 m
K1069	Curran Pt (N9)	1	2	R+LTT+rocks	400 m
K1070	Curran Pt (N10)	1	2	R+LTT	50 m
K1071	Curran Pt (N11)	1	2	R+LTT	150 m
K1072	Cape Talbot (S2)	1	2	R+LTT	1.3 km
K1073	Cape Talbot (S1)	1	2	R+rock/coral flats	2.5 km
Spring & neap tide range:		2.4 & 0.8 m			

Cape Talbot is a low but prominent headland at the northern entrance to Napier Broome Bay, and just 5 km short of the northernmost point of the Kimberley mainland. To the south of the cape low sandstone shore dominates for 17 km down to Curran Point, with pocket beaches occupying the mouths of the joint-controlled creeks. Beaches K 1067-1071 are five pockets of sand lying along the west-facing shore to the south, located between 5 and 10 km south of the cape.

Beach **K 1067** is a straight 300 m long beach backed by gently sloping jointed sandstone slopes, and a small creek at the northern end. The sandstone also forms irregular rock flats extending up to 100 m off the beach. Beach **K 1068** lies 2 km to the north and is a curving 150 m long pocket of sand at the mouth of a small usually dry creek. It is bordered by 100 m wide rock flats to either end, with sand and rock off the beach.

Beach **K 1069** occupies the next embayment, 1 km to the north. The curving 400 m long beach is bordered by low rocky coast, with one rock outcrop toward the northern end of the beach, and a narrow bar paralleling the beach. A creek and small elongate lagoon flows behind the southern half of the beach, exiting across the centre. Minor 2-3 m high dune activity has spread up to 200 m inland over the backing gently sloping sandstone terrain.

Beach **K 1070** lies 500 m to the north and is a 50 m pocket of sand, backed and bordered by sandstone, including a small usually dry creek. Three hundred metres further north is beach **K 1071**, a 150 m long beach, bordered by low sandstone rocks, with a small creek crossing the northern end of the beach. Both beaches, while bordered by rock, have narrow bars along their bases.

Beach **K 1072** is a curving, west-facing, 1.3 km long, moderate gradient beach which links Cape Talbot to the mainland. It has prograded up to 700 m as a series of up to 15 beach to foredune ridges (Fig. 4.99) covered by clumps of *Triodia pungens* (Soft Spinifex), backed in turn by older dune activity. The beach is bordered by low rocks of the mainland and cape, with a mangrove-lined creek exiting at the southern end. It is backed by the ridges and dunes, with salt flats to either side, and wooden coastal plain behind.

Figure 4.99 Beach K 1072 is backed by a 700 m wide beach-foredune ridge plain.

Beach **K 1073** is a high tide sand beach that extends for 2.5 km along the rocky shore that runs north from beach K 1072 to terminate at the cape. It is fronted by irregular sandstone rock flats and fringing reef.

K 1074-1077 **CAPE TALBOT (E)**

No.	Beach	Rating HT	Rating LT	Type	Length
K1074	Cape Talbot spit	1	2	R+rock/coral flats	2.4 km
K1075	Sandy Island	1	2	R+coral flats	500 m
K1076	Cape Talbot (E1)	1	2	R+sand/coral flats	900 m
K1077	Cape Talbot (E2)	1	2	R+sand/coral flats	250 m
Spring & neap tide range:		2.4 & 0.8 m			

At Cape Talbot the shoreline turns and trends east for 24 km to Cape Londonderry, the northernmost tip of the Kimberley coast. The first 8 km form the northern side of a low sandstone peninsula, with the entire shore fringed by 1-5 km wide coral reef flats. There are four beaches (K 1074-1077) along this section occupying 4 km of the 11 km of shoreline, with mangroves dominating the remainder.

A 2.4 km long spit (beach **K 1074**) is tied to the cape and extends east, where it is backed by older spits and up to 1 km wide mangrove-dominated wetland (Fig. 4.100). The spit is actively building to the east and merges into extensive sand shoals and tidal channel deposits. North and east of the cape extensive fringing coral reef widens from 500 m up to 3 km off the tip of the spit.

Figure 4.100 Beach K 1074 consists of a 2.4 km long spit, backed by an earlier spit, now surrounded by mangroves.

Sandy Island (beach **K 1075**) is a low sandy island located 500 m north of the mainland and 3 km east of the tip of the spit. It consists of a series of beach ridges to recurved spits approximately 500 m in length. It sits atop the extensive coral flats that extend up to 2 km north of the island.

Beach **K 1076** lies 1.5 km southeast of the island and 7 km east of Cape Talbot. It occupies an open 1 km wide embayment. The gently curving beach is 900 m long, faces north and is fronted by sand flats grading into 1 km wide reef flats. A shallow lagoon backs the eastern half of the beach and is drained by a small mangrove-lined creek at its western end. Beach **K 1077** is located 500 m west of the northeastern tip of the peninsula. It consists of a 250 m long pocket of sand bordered by low rock flats with scattered mangroves, fronted by the 1 km wide sand to reef flats.

K 1078-1086 **CAPE TALBOT (E)**

No.	Beach	Rating HT	Rating LT	Type	Length
K1078	Cape Talbot (E3)	1	2	R+sand/reef flats	600 m
K1079	Cape Talbot (E4)	1	2	R+sand/reef flats	1 km
K1080	Cape Talbot (E5)	1	2	R+tidal flats	400 m
K1081	Cape Talbot (E6)	1	2	R+rock/reef flats	1.1 km
K1082	Cape Talbot (E7)	1	2	R+rock/reef flats	800 m
K1083	Cape Talbot (E8)	1	2	R+rock/reef flats	900 m
K1084	Cape Talbot (E9)	1	2	R+rock/reef flats	2.2 km
K1085	Cape Talbot (E10)	1	2	R+reef flats	500 m
K1086	Cape Talbot (E11)	1	2	R+reef flats	400 m
Spring & neap tide range:	2.4 & 0.8 m				

Beaches K 1078-1086 are located in a 3 km wide, 6 km deep bay, 10 km east of Cape Talbot. Beaches K 1078-1080 lie along the western shore, beaches K 1081 and 1082 on the southern shore, and beaches K 1083-1086 along the eastern shore, with the inner reaches of the bay dominated by fringing mangroves. All beaches are low energy systems, with the outer protected by coral reef flats and the inner by tidal flats.

Beach **K 1078** lies around the corner from beach K 1077 and occupies the mouth of a 500 m wide shallow embayment. The curving 600 m long beach blocks the bay forming a 200 m wide beach ridge capped barrier, backed by a 300 m wide salt flats filled lagoon, drained by a southern mangrove-lined creek. Sand then reef flats extend 1.5 km off the beach into the bay, narrowing the bay entrance to 1 km at low tide.

Beach **K 1079** lies 1 km further south and deeper into the bay, and consists of a 1 km long beach fronted by 500 m wide sand flats edged by reef, and bordered by low rocks at each end. It is backed by low foredune to overwash deposits extending up to 200 m inland. Beach **K 1080** is located south of the southern rocks and is a low energy 400 m long beach that faces southeast into the inner part of the bay. It is bordered by mangrove-covered low rocks, and fronted by 500 m wide sand to mud flats.

The southern part of the bay bifurcates into two mangrove-lined embayments. Beaches K 1081 and 1082 lie along the intervening divide. Beach **K 1081** is a 1.1 km long double crenulate beach fronted by sand, rock and patchy reef, with the central rock flats forming a salient in the beach, while a small creek drains out at the western end of the beach. A rocky inflection separates it from 800 m long beach **K 1082**, a long energy crenulate beach fronted by rock flats and a few mangroves and bordered by mangroves extending into the eastern arm of the bay.

Beach **K 1083** lies 1 km to the north and opposite beach K 1082. It is a crenulate 900 m long low energy beach fronted by sand and rock flats. Beach **K 1084** commences at a rocky inflection in the shore and continues north as a rock-controlled crenulate beach for another 2.2 km, with reef flats increasing in width to 500 m off the northern end.

Beach **K 1085** is 500 m long and occupies a small embayment bordered by mangrove-covered rocks. A small creek crosses the northern end of the beach and reef flats extend 500 m offshore. Beach **K 1086** lies immediately to the north and is a similar 400 m long beach, with a northern creek and 500-600 m wide reef flats. It terminates at the eastern entrance to the bay, 3 km east of beach K 1078.

K 1087-1098 CAPE LONDONDERRY (W)

No.	Beach	Rating HT	Rating LT	Type	Length
K1087	Cape Londonderry (W14)	1	2	R+rock/reef flats	1.8 km
K1088	Cape Londonderry (W13)	1	2	R+sand/reef flats	700 m
K1089	Cape Londonderry (W12)	1	2	R+sand/reef flats	2.2 km
K1090	Cape Londonderry (W11)	1	2	R+sand/reef flat	700 m
K1091	Cape Londonderry (W10)	1	2	R+sand/reef flats	500 m
K1092	Cape Londonderry (W9)	1	2	R+sand/reef flats	600 m
K1093	Cape Londonderry (W8)	1	2	R+tidal/reef flats	400 m
K1094	Cape Londonderry (W7)	1	2	R+tidal flats	300 m
K1095	Cape Londonderry (W6)	1	2	R+tidal flats	200 m
Spring & neap tide range: 2.4 & 0.8 m					

Midway between Cape Talbot and Cape Londonderry is a low 5 km wide, wooded peninsula bordered on each side by 5 km deep embayments. Beaches K 1078-1086 occupy the western embayments, while beaches K 1087-1095 line the shore of the northern side of the peninsula and western side of the second embayment.

Beach **K 1087** is a convex, north-facing 1.8 km long beach, backed by a 100 m wide series of beach ridge-overwash flats, then woodlands, while it is fronted by rock then 1.5 km wide reef flats. To the east it curves round into an 800 m wide bay, which contains 700 m long beach **K 1088** at its base (Fig. 4.101). This beach links the two sides of the bay with a 100 m wide series of backing beach ridges in the west, and a mangrove-lined creek mouth to the east, the creek exiting against the western side. The low barrier is in turn backed by a V-shaped salt flat-filled 5 ha lagoon. It is fronted by sand flats emanating from the creek mouth and reef flats filling the bay.

Figure 4.101 Beach K 1088 and its backing barrier partly fill a small embayment.

Beach **K 1089** commences on the eastern entrance to the bay and curves round as a convex north-facing beach, that has prograded out 200-500 m across the fronting 3-5 km wide reef flats. The backing barrier consists of a mixture of beach ridges to foredunes, with some dune activity towards its eastern end. A creek also drains across the eastern side of the beach. Beach **K 1090** commences as a southeast-trending inflection in the sandy shore and continues for 700 m to a sandy foreland, to the east of which is 500 m long beach **K 1091**. Both beaches are fronted by tidal sand shoals up to 500 m wide, then fringing reef flats. A small tidal creek draining a 5 ha area of salt flats borders the southern end of beach K 1091.

Beach **K 1092** lies 1.5 km to the east on the southern shore of the embayment. It is a 600 m long north-facing strip of sand, backed by a small wetland with a creek exiting at the eastern end of the beach, and fronted by 500 m wide sand-mud flats, then fringing reef.

Beach **K 1093** is a further 1 km to the east and emerges from a mangrove-fringed rock flats to continue on for 400 m as a low energy sand beach fronted by rock then mud flats. It terminates at a low rock-controlled inflection in the shore, on the eastern side of which is 300 m long beach **K 1094**, which trends south to a mangrove-lined creek mouth. The creek drains a 20 ha mangrove- and salt flat-filled wetland. It is fronted by sand then mud flats.

Beach **K 1095** is the southern and easternmost beach in the embayment. It is a 200 m long strip of high tide sand, backed by wooded coastal plain, bordered by mangrove-lined shoreline, and fronted by tidal mud flats. Mangroves continue for 2 km to the east to the apex of the bay, where a mangrove-lined 4 km long tidal creek enters the bay.

K 1096-1101 CAPE LONDONDERRY (W)

No.	Beach	Rating HT	Rating LT	Type	Length
K1096	Cape Londonderry (W5)	1	2	R+tidal flats	300 m
K1097	Cape Londonderry (W4)	1	2	R+rock/reef flats	200 m
K1098	Cape Londonderry (W3)	1	2	R+rock/reef flats	400 m
K1099	Cape Londonderry (W2)	1	2	R+rock/reef flats	1 km
K1100	Cape Londonderry (W1)	1	2	R+rock/reef flats	250 m
K1101	Cape Londonderry	1	2	R+rock/reef flats	250 m
Spring & neap tide range: 2.7 & 1.1 m					

Cape Londonderry at 13°45'S latitude is the northernmost point of the Western Australian and Kimberley mainland coast. To the west is 500 km of highly indented coast extending down to King Sound, while to the east the coast forms the western shore of the 250 km wide Joseph Bonaparte Gulf, with Cambridge Gulf and Wyndham at its base, and the Northern Territory forming the eastern shore. The cape itself is 16 m high and backed by a relatively low laterised coastal plain, composed of Proterozoic volcanics and surrounded by extensive fringing reefs. The coast initially extends southeast for 180 km to Cape Dussejour, the western entrance to Cambridge Gulf.

Beaches K 1096-1099 occupy either side of a 5 km long, up to 2 km wide, low wooded promontory that extends west of the cape to form the western boundary of the second of the two embayments between the cape and

Cape Talbot. Beaches K 1096-1098 lie along the protected southern shore of the promontory, with beach K 1099 on the more exposed northern shore. Beach **K 1096** is a convex 300 m long protrusion that faces south across the 1.5 km wide embayment toward beach K 1095. It is bordered by mangrove-fringed rock flats to either end and fronted by tidal flats. Beach **K 1097** is a curving 200 m long high tide sand beach, with rocks outcropping along the shore and rocks and mangroves bordering the northern low point. Beach **K 1098** extends from the northern side of the point for 400 m to the mangrove-fringed lee of the western tip of the peninsula. Rocks outcrop along the shore and rocks and tidal flats front the beach.

Beach **K 1099** lies 2 km to the east on the northern side of the promontory, in lee of coral reef flats that extend up to 6 km to the north to the Stewart Islands. The reef affords considerable protection to the protruding convex-shaped 1 km long beach. It has prograded up to 300 m northward leaving a series of low grassy beach ridges. It is fringed by mangroves to either end and fronted by a 100-200 m wide veneer of sand, then the reef flats.

Beach **K 1100** lies in a 500 m wide mangrove-lined bay immediately west of the cape. The beach is a 250 m long strip of sand bordered and partly shielded by the mangroves. The bay is filled with 300 m wide tidal sand shoals then the reef flats, with a gap in the reefs reaching to the edge of the sand flats off the beach. A 100-300 m wide series of beach ridge-cheniers back the beach and mangroves.

Beach **K 1101** lies on the western tip of the cape. The 250 m long beach faces northwest and has infilled a 300 m deep embayment with a series of low grassy beach ridges. It is fronted by 50 m wide sand flats, then 500 m of reef flats. The low cape and surrounding rock flats and reefs, form the northern boundary of the beach.

K 1102-1110 **CAPE LONDONDERRY (E)**

No.	Beach	Rating HT	Rating LT	Type	Length
K1102	Cape Londonderry (E1)	1	2	R+coral reef flats	150 m
K1103	Cape Londonderry (E2)	1	2	R+coral reef flats	150 m
K1104	Cape Londonderry (E3)	1	2	R+coral reef flats	250 m
K1105	Cape Londonderry (E4)	1	2	R+coral reef flats	150 m
K1106	Cape Londonderry (E5)	1	2	R+coral reef flats	800 m
K1107	Cape Londonderry (E6)	1	2	R+coral reef flats	300 m
K1108	Cape Londonderry (E7)	1	2	R+coral reef flats	400 m
K1109	Cape Londonderry (E8)	1	2	R+coral reef flats	200 m
K1110	Cape Londonderry (E9)	1	2	R+coral reef flats	400 m
Spring & neap tide range:	2.7 & 1.1 m				

Immediately east of **Cape Londonderry** the shoreline begins a 175 km long trend to the southeast, ultimately terminating at Cape Dussejour, the northern entrance to Cambridge Gulf. The first 50 km of coast between the cape and Cape Rulhieres are however moderately irregular, with several north-trending peninsulas composed of Proterozoic volcanics and sandstones, separated by embayments extending 3-10 km to the south. The first 2.5 km of coast east of the cape curve round to the southeast and contain nine pocket beaches (K 1102-1110), all backed by steep bluffs cut into the 10 m high laterised plain. They are also all fronted by continuous 100-300 m wide fringing coral reef.

Beaches K 1102, 1103 and 1104 occupy a curving 700 m wide embayment immediately east of the cape with the beaches lying in three gaps in the bluffs and fronted by continuous reef flats. Beach **K 1102** is a 150 m long strip of high tide sand, at the base of 10 m high bluffs, and fronted by 200 m wide reef flats. Beach **K 1103** lies 200 m to the south and is a 150 m long sand beach, backed by a low 100 m wide foredune, and fronted by 300 m wide reef flats. One hundred metres to the south is beach **K 1104**, a 250 m long beach, backed by a narrow grassy foredune, then the crest of the bluffs. It is fronted by 200 m wide reef flats.

Beach **K 1105** lies 300 m to the south, on the southern side of a 20 m high flat topped, laterised eroding headland. The reefs continue around the headland and are 200 m wide off the beach. The beach is 150 m long, backed by eroding 15 m high bluffs, with bluff debris littering the base of the beach and some of the reef flat. Beach **K 1106** lies on the southern side of a 50 m wide eroding headland and is a curving 800 m long sand beach that has filled two adjoining gaps in the bluffs with low grassy overwash flats, while 200-300 m wide reef flats front the beach (Fig. 4.102).

Figure 4.102 Beaches K 1105 & 1106 are wedged between the laterite bluffs and fringing coral reef.

Beach **K 1107** lies 300 m to the south past a lower section of eroding bluffs. It is a 300 m long high tide beach, backed by both vegetated and eroding 10 m high bluffs, with some sand covering the 300 m wide reef flats. It is separated from beach K 1108 by a 400 m wide 10-20 m high eroding headland. Beach **K 1108** is a 400 m long high tide beach fronting two adjoining sections of densely vegetated 20 m high bluffs, with a protruding section cutting the high tide beach in two. It is fronted by 200 m wide reef flats.

Beaches K 1109 and 1110 lie along the north-facing northern side of a 20 m high V-shaped laterised point. Beach **K 1109** is a 200 m long pocket of sand at the base of eroding and vegetated sections of bluff, while beach **K 1110** is a straight 400 m long high tide beach, backed by bluffs which decrease to grassy slopes to the east. Both beaches are fronted by continuous 100-200 m wide reef flats.

K 1111-1117 **CAPE LONDONDERRY (E2)**

No.	Beach	Rating HT	LT	Type	Length
K1111	Cape Londonderry (E10)	1	2	R+sand/reef flats	700 m
K1112	Cape Londonderry (E11)	1	3	R+coral reef flats	300 m
K1113	Cape Londonderry (E12)	1	2	R+sand/reef flats	200 m
K1114	Cape Londonderry (E13)	1	1	R+sand/reef fats	900 m
K1115	Cape Londonderry (E14)	1	2	R+sand/reef flats	500 m
K1116	Cape Londonderry (E15)	1	2	R+coral reef flats	200 m
K1117	Cape Londonderry (E16)	1	2	R+coral reef flats	200 m
Spring & neap tide range:	2.7 & 1.1 m				

Three kilometres southeast of Cape Londonderry is a 2 km wide bay that extends for 2.5 km to the south narrowing to 1 km. The north-facing embayment is surrounded by 20 m high laterised bluffs, and containing seven beaches (K 1111-1117). Beaches K 1111 and 1112 are located on the east-facing northern side of the bay entrance and exposed to easterly waves. Beach **K 1111** is a curving 700 m long beach located below scarped 20 m high vegetated bluffs, with a partly active dune system extending 200 m in from the southern half of the beach and rising to 10 m. It is fronted by 100-200 m wide sand-mud flats, then fringing reef. A 300 m long scarped bluff and rock debris separate it from beach **K 1112**, a 300 m long high tide sand beach at the base of densely vegetated 20 m high bluffs. It is fronted by 300 m wide reef flats with boulder debris to either end.

Beach **K 1113** is located 2 km to the south in the southwestern corner of the bay. It is a 200 m long low energy high tide beach, bordered by mangroves to the north and a grassy headland to the south. It is fronted by 500 m wide sand flats which extend out over the 1 km wide reef flats. Beach **K 1114** commences on the eastern side of the headland and curves around for 900 m, filling the southern part of the bay. It is backed by a 400 m wide series of beach ridges grading into ridges and salt flats, with a small mangrove-lined creek draining out at the western end of the beach. *Triodia pungens* (Soft Spinifex) covers the ridges. Sand flats extend 500 m north of the beach to the edge of the underlying reef flats (Fig. 4.103).

Beach **K 1115** lies along the eastern shore at the base of steeply sloping 20 m high bluffs. It is a narrow 500 m long crenulate beach with a narrow, 2-3 m high, grassy foredune, that follows the bedrock. A usually dry creek crosses the southern end of the beach with 500 m wide sand then reef flats extending along the shore. Beach **K 1116** is a 200 m long continuation of this system bordered by boulder debris. It has a few mangroves in the centre and is backed by 20 m high densely vegetated bluffs.

Beach **K 1117** lies several hundred metres to the north at the northeast tip of the bay. It is a narrow 200 m long strip of high tide sand at the base of steep 20 m high vegetated bluffs, with reef flats extending 400 m off the beach into the deeper central part of the bay.

Figure 4.103 View of the main beach K 1114, bordered by beach K 1113 (right) and beaches K 1115 & 1116 (left).

K 1118-1126 **CAPE LONDONDERRY (E 3)**

No.	Beach	Rating HT	LT	Type	Length
K1118	Cape Londonderry (E17)	1	2	R+coral reef flats	200 m
K1119	Cape Londonderry (E18)	1	3	R+coral reef flats	300 m
K1120	Cape Londonderry (E19)	1	3	R+coral reef flats	100 m
K1121	Cape Londonderry (E20)	1	3	R+coral reef flats	200 m
K1122	Cape Londonderry (E21)	1	3	R+sand/reef flats	200 m
K1123	Cape Londonderry (E22)	1	3	R+coral reef flats	150 m
K1124	Cape Londonderry (E23)	1	1	R+sand/reef flats	250 m
K1125	Cape Londonderry (E24)	1	1	R+sand/reef flats	350 m
K1126	Cape Londonderry (E25)	1	1	R+sand/reef flats	600 m
Spring & neap tide range:	2.7 & 1.1 m				

A second bay is located 5 km southeast of Cape Londonderry and begins immediately east of beach K 1117. It consists of a 2 km wide northeast-facing entrance, with the northern half of the bay containing beaches K 1118-1123 which are exposed to easterly waves. The southern part of the bay is protected by the eastern headland and contains three more protected beaches (K 1124-1126).

Beach **K 1118** lies on the northwestern tip of the pointed headland that separates the two bays. Like its neighbour it is a 200 m long northeast-facing strip of high tide sand at the base of scarped 20 m high bluffs, with reef flats extending 300 m off the beach and waves averaging about 0.5 m.

Five hundred metres to the south is the beginning of a 3 km long series of small embayments, all containing pocket beaches at the base of 20 m high bluffs and linked by continuous reef flats. Beach **K 1119** is a curving east-facing 300 m long beach with a low narrow foredune, at the base of lightly vegetated 20 m high bluffs with slightly protruding bluffs and boulder debris at either end and 100 m wide reef flats off the beach. Two hundred metres to the south is beach **K 1120**, a 100 m long strip of sand backed by steep densely vegetated bluffs and fronted by 100-200 m wide reef flats.

Beach **K 1121** lies in the next small embayment 200 m to the south. It is a 200 m long beach backed by a 50-100 m wide climbing foredune, then vegetated bluffs. It is bordered by scarped headlands extending eastward 100-200 m, which are linked by 200 m wide reef flats.

Beach **K 1122** lies 300 m to the south on the southern side of a V-shaped 400 m long eroding headland. It consists of a 200 m long pocket of sand backed by a 100 m wide low beach ridge-overwash plain that partly fills the bay, with the 200 m wide reef paralleling the beach. Two hundred metres to the south is a 150 m long pocket of sand (beach **K 1123**), also backed by a 50 m wide grassy overwash plain and fronted by 100 m wide reef flats.

Beach **K 1124** lies in the southwestern corner of the bay and is protected from most waves. It is a low energy 250 m long beach, backed by a low 200 m wide grassy beach ridge plain, with a small creek draining out along the southern side of the beach. It is fronted by a mixture of sand-mud flats blanketing the inner part of the 300 m wide reef flats.

Beaches K 1125 and 1126 occupy the southern and southeastern corners of the bay (Fig. 4.104). Beach **K 1125** is a 350 m long beach that has prograded up to 300 m to partly fill a small valley with a salt flat and beach ridge plain. A vegetated 20 m high, 200 m wide headland separates it from beach **K 1126**, a northwest-facing 600 m long beach, backed by a low grassy barrier up to 200 m wide, then a 5 ha salt flat that is drained by a small creek at the northern end of the beach. Both beaches are fronted by irregular 100-300 wide sand flats, then coral reefs extending up to 500 m into the bay.

K 1127-1129 **CAPE LONDONDERRY (E 4)**

No.	Beach	Rating HT	Rating LT	Type	Length
K1127	Cape Londonderry (E26)	1	2	R+coral reef flats	200 m
K1128	Cape Londonderry (E27)	2	3	R+reef flats+rocks	250 m
K1129	Cape Londonderry (E28)	1	2	R+coral reef flats	300 m

Spring & neap tide range: 2.7 & 1.1 m

Figure 4.104 Beaches K 1125 & 1126 occupy either side of a small bay, with both sand flats and coral reef filling the bay floor.

Seven kilometres southeast of Cape Londonderry is a 3 km long, more exposed, northeast-facing peninsula that separates the second and third embayments east of the cape. The entire front of the peninsula is fringed by reef flats up to 1 km wide, with three beaches located at either end and a mangrove-line bay in the centre.

Beach **K 1127** lies 7.5 km southeast of the cape toward the northern end of the peninsula. It is a 200 m long beach backed and bordered by 20 m high red bluffs, with boulder debris littering the beach and reefs extending 300 m offshore. Beaches K 1128 and 1129 are located 2 km to the southeast at the southern end of the peninsula. Beach **K 1128** is a straight 250 m long northeast-facing beach backed by a grassy overwash plain and fronted by 300 m wide reef flats. It is bordered by irregular rock slopes, with beach **K 1129** commencing 150 m to the east. It is a 300 m long strip of high tide sand, backed by gently vegetated slopes, with rocks outcropping along the beach and on the 200 m wide reef flats.

K 1130-1138 **CAPE LONDONDERRY (E 5)**

No.	Beach	Rating HT	Rating LT	Type	Length
K1130	Cape Londonderry (E29)	1	2	R+sand/reef flats	200 m
K1131	Cape Londonderry (E30)	1	2	R+sand/reef flats	200 m
K1132	Cape Londonderry (E31)	1	2	R+sand/reef flats	200 m
K1133	Cape Londonderry (E32)	1	2	R+sand/reef flats	100 m
K1134	Cape Londonderry (E33)	1	2	R+sand/reef flats	400 m
K1135	Cape Londonderry (E34)	1	2	R+sand/reef flats	350 m
K1136	Cape Londonderry (E35)	1	2	R+sand/reef flats	400 m
K1137	Cape Londonderry (E36)	1	2	R+sand/reef flats+rocks	450 m
K 1138	Cape Londonderry (E37)	1	2	R+sand/reef flats+rocks	100 m

Spring & neap tide range: 2.7 & 1.1 m

The third embayment lies 10 km southeast of Cape Londonderry and has a 1 km wide northeast-facing entrance bordered by beaches K 1129 and 1139. It consists of a low energy double-armed embayment, with the arms extending up to 3 km to the southwest and south respectively. It contains 10 km of irregular rock and mangrove-dominated shoreline, together with nine low energy beaches (K 1130-1138) that total 2.4 km (24%) in length. They are all fronted by tidal sand flats then reef, which surrounds the bay shore.

Beach **K 1130** is located in the lee of the peninsula which houses beaches K 1129 and 1130, lying 700 m south of the former. It is a 200 m long sand beach fronted by tidal sand then reef flats that faces southeast across the bay. It is the only beach on the western shore of the bay, the remainder occupied by rocky shore and mangroves.

Beach **K 1131** lies on the northern side of the middle arm of the bay. It is a 200 m long north-facing strip of sand that fills a small valley. It is bordered and backed by vegetated slopes and fronted by 500 m wide sand-reef flats. Beach K 1132 and 1133 are two pockets of adjoining sand that form a 100 m long isthmus that ties a 200 m wide vegetated hillock to the mainland. Beach **K 1132** is 200 m long and initially runs along the base of vegetated bedrock slope, before forming the western side of the 100 m wide grassy isthmus, with mangroves bordering the northern end. Beach **K 1133** forms the eastern side of the isthmus and is a 100 m long pocket of east-facing sand. Both beaches are fronted by narrow tidal sand flats, then 500 m wide coral reef flats.

Beaches K 1134-1138 occupy the southern arm of the bay. Beach **K 1134** is a 400 m long northeast-facing beach and grassy foredune that has partially filled a small valley behind its western half, with vegetated slopes backing the western half. It is bordered by mangrove-fringed rocky headlands, and fronted by sand-reef flats extending 1 km off the beach.

Beach **K 1135** is the southernmost beach in the southern arm and faces north toward the bay entrance (Fig. 4.105). It is a 350 m long grass-covered spit that has prograded to the west to enclose a mangrove-filled 30 ha lagoon, with a deep tidal creek flowing out against the western end of the beach. It is bordered by mangrove-lined slopes at either end and fronted by 1 km wide sand-reef flats. Beach **K 1136** lies immediately east of the mangroves and is a 400 m long northwest-facing beach, backed by a grassy 200 m wide beach ridge to overwash plain, and fronted by the sand and reef flats that fill the southern arm of the bay.

Beach **K 1137** is an irregular 450 m long strip of high tide sand that runs along the base of gently sloping grass-covered bedrock, with rocks outcropping along the beach and off the western end. It is fronted by sand, rocks and reef flats extending 500 m into the bay. Three hundred metres further round the rocks adjacent to the eastern entrance to the bay is beach **K 1138**, a 100 m long pocket of sand bordered by irregular mangrove-fringed rock flats and fronted by 200 m wide reef flats.

Figure 4.105 Beaches K 1134 (at rear), 1135 & 1136 (left) occupy the southern section of a 3 km deep embayment.

K 1139-1149 **CAPE LONDONDERRY (E 6)**

No.	Beach	Rating HT	LT	Type	Length
K1139	Cape Londonderry (E38)	1	2	R+sand/reef flats	800 m
K1140	Cape Londonderry (E39)	1	2	R+coral reef flats	300 m
K1141	Cape Londonderry (E40)	1	2	R+coral reef flats	200 m
K1142	Cape Londonderry (E41)	1	2	R+coral reef flats	500 m
K1143	Cape Londonderry (E42)	1	1	R+sand/reef flats	700 m
K1144	Cape Londonderry (E43)	1	1	R+sand/reef flats	200 m
K1145	Cape Londonderry (E44)	1	1	R+sand/reef flats	200 m
K1146	Cape Londonderry (E45)	1	1	R+sand/reef flats	2 km
K1147	Cape Londonderry (E46)	1	1	R+sand/reef flats	800 m
K1148	Cape Londonderry (E47)	1	2	R+coral reef flats	100 m
K1149	Cape Londonderry (E48)	1	2	R+coral reef flats	100 m
Spring & neap tide range:	2.7 & 1.1 m				

Beach K 1140 lies at the tip of a 5 km long north-trending peninsula 12 km southeast of the cape, with the protected beaches K 1135-1138 located along parts of its western side, and the more exposed beaches 1139-1141 on its northern tip. The southern half of the eastern side of the peninsula forms the western shore of a 1.5 km wide 2.5 km deep U-shaped embayment, with the more protected beaches K 1142-1145 forming a near continuous 4.4 km long sandy shoreline, around to the eastern side of the bay. Fifty metre high sandstone cliffs then dominate the eastern headland of the bay apart from two more exposed pockets of sand (K 1148 & 1149).

Beach **K 1139** is a curving 800 m long northwest-facing beach, backed by a 70 m wide series of beach ridges and fronted by converging 500 m wide reef flats, together with sand flats off the central part of the beach. The eastern end of the beach continues on for 300 m amongst rocks of the bordering grassy headland. Beach **K 1140** lies 300 m to the east on the northeastern tip of the headland. It is a 300 m long, northeast-facing beach, fronted by 500 m wide reef flats, and backed by unstable dunes extending up to 300 m across the low headland, almost reaching beach K 1138.

Beach **K 1141** lies 500 m to the south in a small east-facing embayment. It is a 200 m long pocket of sand backed by a small foredune then vegetated slopes and bordered by steep grassy slopes rising to 40 m on the southern side. It is fronted by irregular reef flats extending up to 300 m off the beach.

Beach **K 1142** is located in lee of the western entrance to the bay and faces east across the bay mouth. It is a 500 m long beach, bordered by low rocky points, with some rocks also cutting across the northern section of the beach. It is paralleled by 500 m wide reef flats, while behind the beach there is some minor dune activity toward the southern end. Mangroves fringe the lee side of the southern rocks for 300 m to the beginning of beach **K 1143**, a 700 m long east-facing low energy beach that terminates in lee of a rock and sand foreland formed in the lee of a 4 ha island 300 m off the beach. The beach is fronted by tidal sand flats grading into 1 km wide reef flats.

Beach **K 1144** commences on the southern side of the foreland and continues south for 200 m across the mouth of a small infilled valley, the beach bordered by low rocky points. Beach **K 1145** lies immediately to the south and is a 200 m long beach also filling a small valley system. Both beaches are fronted by a continuation of the 1 km wide sand to reef flats.

Beach **K 1146** occupies the southern part of the bay. It is a curving, northeast-facing 2 km long moderate gradient beach backed by a beach-foredune ridge plain, covered in *Triodia pungens* (Soft Spinifex). The plain links the two sides of the bay and is backed by 100-300 m wide beach ridge plain, then up to 500 m of infilled wetlands, with a mangrove-lined creek crossing the centre of the beach. The creek mouth is fronted by tidal sand shoals, then up to 2 km of reef flats that fill the southern half of the bay.

Sloping vegetated bedrock borders the eastern end of the bay and extends northeast for 400 m to the beginning of beach **K 1147**, an 800 m long north-facing beach that is backed by steep vegetated slopes rising to a 50 m high plateau. It is fronted by narrow sand then 500 m wide coral reef flats.

Beach **K 1148** lies 1 km to the north on the outside of the bay. It is a 100 m long pocket of sand, lying at the base of steep, vegetated 50 m high bluffs and fronted by 50-100 m wide reef flats. A steep dry creek is located in the centre of the beach with a low grassy foredune to either side. Beach **K 1149** lies 1.5 km to the east on the northern side of the headland. It is a similar 100 m long pocket of sand occupying a steep-sided indentation in the cliffline, with a usually dry creek entering the centre of the backbeach. It is bordered by the cliffs and boulder debris and fronted by 100 m wide reef flats (Fig. 4.106).

Figure 4.106 Beach K 1149 is a small pocket of sand wedged between the backing bluffs and the fringing coral reef.

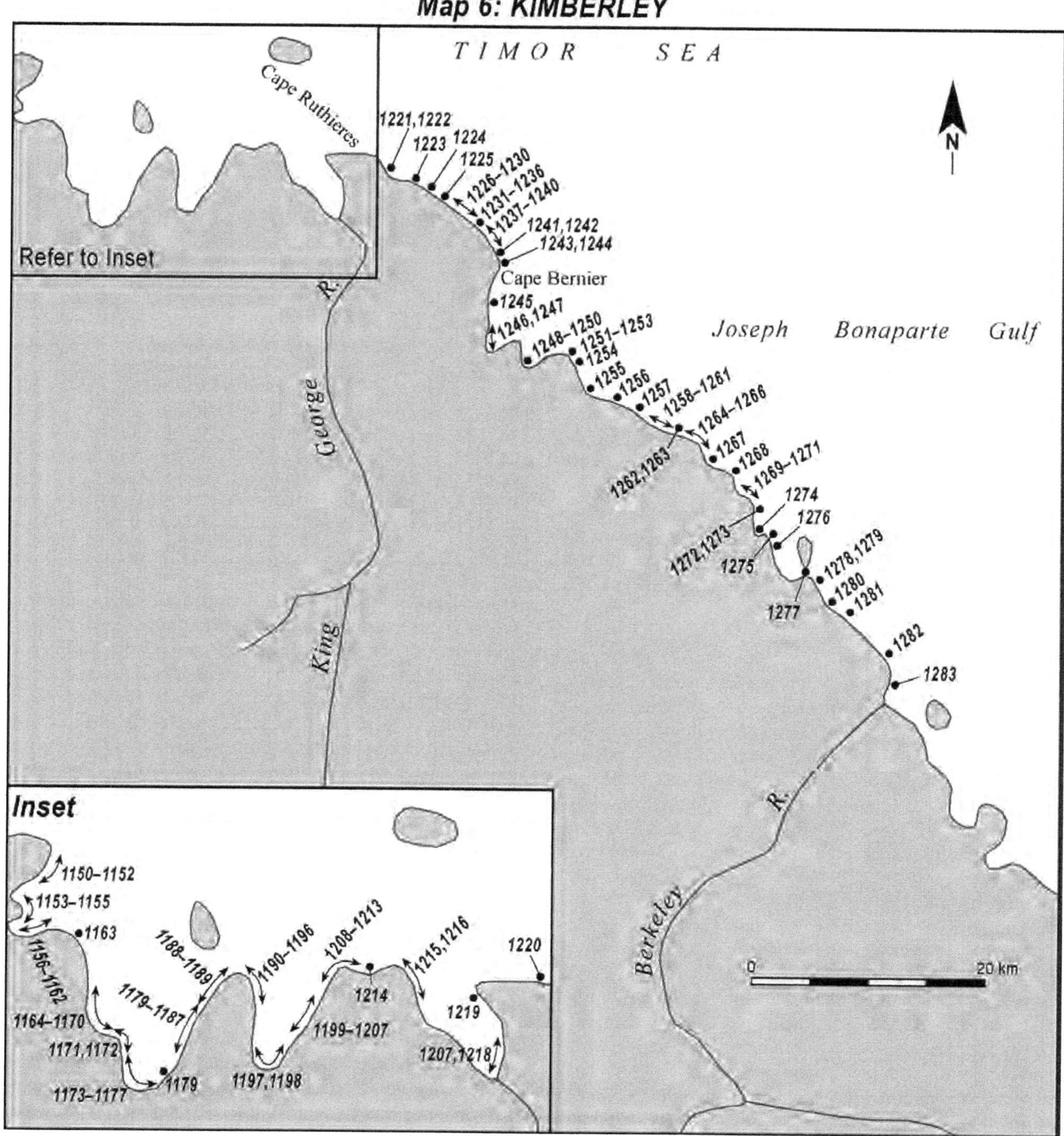

Figure 4.107 Kimberley map 6, locating beaches K 1150-1283.

K 1150-1162 **CAPE LONDONDERRY (E 6)**

No.	Beach	Rating HT	Rating LT	Type	Length
K1150	Cape Londonderry (E49)	1	1	R+sand flats	300 m
K1151	Cape Londonderry (E50)	1	1	R+sand/reef flats	300 m
K1152	Cape Londonderry (E51)	1	1	R+coral reef flats	200 m
K1153	Cape Londonderry (E52)	1	2	R+coral reef flats	200 m
K1154	Cape Londonderry (E53)	1	2	R+coral reef flats	200 m
K1155	Cape Londonderry (E54)	1	1	R+tidal sand flats	1.1 km
K1156	Cape Londonderry (E55)	1	2	R+tidal sand flats	400 m
K1157	Cape Londonderry (E56)	1	1	R+sand/reef flats	400 m
K1158	Cape Londonderry (E57)	1	2	R+coral reef flats	100 m
K1159	Cape Londonderry (E58)	1	2	R+coral reef flats	200 m
K1160	Cape Londonderry (E59)	1	2	R+coral reef flats	50 m
K1161	Cape Londonderry (E60)	1	2	R+coral reef flats	300 m
K1162	Cape Londonderry (E61)	1	2	R+coral reef flats	100 m
Spring & neap tide range:	2.7 & 1.1 m				

Between 17 and 22 km southeast of Cape Londonderry is a 4 km wide northeast-facing embayment, that has northern and southern arms extending to the west and southwest for 2 and 4 km respectively. The bay has

10 km of shoreline and contains 13 beaches totalling 3.9 km (39%) in length, the remainder of the shore dominated by steep sandstone bluffs and cliffs.

The northern arm of the bay has an 800 m wide entrance, extends west for 1.2 km and contains three beaches (K 1150-1152). Beach **K 1150** is a curving 300 m long southeast-facing beach that fills a small valley and is backed by a grassy overwash plain with a few mangroves lining a small creek that drains across the southern end of the beach. It is fronted by sand then reef flats extending 500 m into the bay. Immediately to the south is beach **K 1151**, an east-facing 300 m long unstable sand spit that extends across the 200 m wide mouth of a drowned valley. The valley is bordered by mangrove-lined 20 m high slopes, with a central elongate 300 m long 2 ha lagoon. During the wet season the creek break out across the beach and maintains tidal sand shoals extending a few hundred metres off the beach. The eastern end of the beach is attached to 20 m high sandstone slopes. Five hundred metres to the east is beach **K 1152**, a northeast-facing 200 m long pocket of sand wedged in a small indentation in the 20 m high plateau. It is backed by densely vegetated slopes and fronted by 200 m wide reef flats which extend north to link with the flats off beach K 1150.

Beaches K 1153 and 1154 are located on an exposed east-facing 2 km long section of cliffed sandstone shoreline, which separates the two arms of the bay. Both beaches lie at the base of the 20 m high sandstone plateau. Beach **K 1153** is a 200 m long high tide beach with a few patches of foredune. It is bordered by rock debris and backed by steep vegetated 20 m high slopes. Two hundred metres of debris separate it from 200 m long beach **K 1154**. Both beaches are fronted by continuous 100-200 m wide reef flats.

The southern arm of the bay contains beaches K 1155-1157 at its southern extremity. Beach **K 1155** is a steep, 1.1 km long, east-facing beach backed by a 100 m wide grassy overwash plain which abuts the base of 20 m high vegetated slopes. At its southern end it partly blocks a 1.5 km long drowned valley containing a 30 ha lagoon. A dynamic inlet bordered by beaches K 1155 and 1156 connects the lagoon with the bay. The inlet maintains 500 m wide irregular tidal shoals off beach K 1155, with 100 m wide tidal shoals and the deep tidal channel paralleling north-facing 400 m long beach **K 1156**. Beach **K 1157** lies 300 m to the east in a small partly filled steep-sided valley. It is a curving 400 m long north-facing beach, with a 2 m high well vegetated foredune and a small creek draining across its eastern end.

Beaches K 1158-1162 are located along a 1.5 km long section of the eastern side of the bay. All five are backed by steep vegetated sandstone slopes and fronted by continuous 50-200 m wide reef flats. Beach **K 1158** is a 100 m long pocket of northeast-facing sand, bordered by rock debris. Five hundred metres to the east is beach **K 1159,** a 200 m long strip of high tide sand. Beach **K 1160** lies a further 200 m to the east and is a 50 m pocket of sand. One hundred metres further along is beach **K 1161**, a 300 m long beach backed by a 50-100 m wide, 1-2 m high grassy foredune. Rocks outcrop on the centre of the beach and it is fronted by reef flats which widen to 150 m in the east. Finally beach **K 1162** is a 100 m long pocket of sand with some rock debris on the beach, and fronting reef flats.

K 1163-1172 CAPE LONDONDERRY (E 7)

No.	Beach	Rating HT	Rating LT	Type	Length
K1163	Cape Londonderry (E62)	1	2	R+sand/reef flats	1.7 km
K1164	Cape Londonderry (E63)	1	2	R+sand/reef flats	100 m
K1165	Cape Londonderry (E64)	1	2	R+sand/reef flats	200 m
K1166	Cape Londonderry (E65)	1	2	R+sand/reef flats	300 m
K1167	Cape Londonderry (E66)	1	2	R+sand/reef flats	500 m
K1168	Cape Londonderry (E67)	1	2	R+sand/reef flats	100 m
K1169	Cape Londonderry (E68)	1	2	R+coral reef flats	400 m
K1170	Cape Londonderry (E69)	1	2	R+coral reef flats	100 m
K1171	Cape Londonderry (E70)	1	2	R+sand/reef flats	900 m
K1172	Cape Londonderry (E71)	1	2	R+sand flats	600 m
Spring & neap tide range:	2.7 & 1.1 m				

Twenty-two kilometres southeast of Cape Londonderry is the western entrance to the southernmost and deepest embayment between the cape and Cape Rulhieres, located 27 km due east. The embayment is 7 km wide at its entrance and extends up to 11 km to the south. It is roughly a V-shaped bedrock-controlled bay, with volcanic rocks rising steeply to between 60-100 m around the perimeter to a dissected sandstone plateau surface. The 80 km of bay shoreline are dominated by the boundary bedrock, with 57 beaches (K 1163-1220) usually wedged in an indentation in the rocks or at the mouth of small valleys. The beaches total 24 km in length, occupying 30% of the shoreline. Rocky shoreline dominates the more exposed sections, with mangrove-lined bedrock fringing some of the southernmost protected bays.

Beaches K 1163-1172 all lie in and between three small embayments located along the first 8 km of the western, east-facing side of the bay. Beach **K 1163** occupies the first embayment. It is a curving, moderate gradient, 1.7 km long beach, that in the west has filled a 500 m wide valley with grassy foredune-overwash deposits (Fig. 4.108), and a mangrove-lined creek meandering across the plain and exiting at the western end of the beach. The beach then extends to the east past a former protruding seacliff and then along the northern side of a 1 km long headland. Tidal sand flats front the creek mouth and western end of the beach. These give way to 100-200 m wide fringing reef flats, which in the east link with reef flats extending 2 km to the north to surround an elongate 1 km long island. The island is linked to the beach by the reefs and a narrow intertidal sand spit. Lightly wooded slopes rise to the 80 m high sandstone plateau behind the western-central section of the beach.

Figure 4.108 View along beach K 1163 showing the moderately steep beach and low grassy overwash flats.

Beaches K 1164-1168 occupy indentations in the second embayment, an irregular 1 km wide, 2 km deep bay, bordered in the north by the headland arm of beach K 1163. Beach **K 1164** is a protected 100 m long strip of sand that shares a small east-facing valley with a 100 m line of mangroves. The beach and mangroves are backed by a patch of grassy overwash flats, with a mangrove-lined creek running across the southern end of the beach, and valley slopes rising to 20-30 m around the perimeter. It is fronted by 500 m wide tidal flats, edged by fringing reef that continues into the bay.

Beach **K 1165** occupies the next small valley to the south and is bordered by valley slopes. It is a 200 m long beach and backing 50 m wide grassy barrier, with a small creek exiting at the northern end. Five hundred metre wide tidal sand then the reef flats front the beach. Beach **K 1166** lies 500 m to the south in the next bay. It is a 300 m long east-facing beach, with a narrow grassy barrier with a creek against the southern end. The tidal and reef flats narrow to 300 m off the beach.

Beach **K 1167** occupies the southwestern corner of the embayment. It is a discontinuous 500 m long east-facing sand beach at the base of slopes rising steeply to the 80 m high plateau. It is fronted by 500 m wide tidal flats, edged by the continuous fringing reef. The bays main mangrove-lined creek enters between beaches K 1167 and 1168. Extensive sand shoals extend from the creek mouth several hundred metres into the bay. Beach **K 1168** is located 400 m northeast of the creek mouth and is fronted by the sand shoals. The beach consists of a 100 m long pocket of sand backed by moderately sloping terrain, with rocks and mangroves to either end.

Beaches K 1169 and 1170 occupy part of the 1.5 km long headland that separates the second and third bays. Beach **K 1169** is a 400 m long northeast-facing high tide sand beach, backed by 10-20 m high bluffs and fronted by a continuous 50 m wide fringing reef. Beach **K 1170** lies 300 m to the east and is a narrow 100 m long strip of sand wedged in between 20 m high eroding bluffs and the fringing reef.

Beaches K 1171 and 1172 occupy the third bay, a 1.5 km wide, 1 km deep, more open northeast-facing bay. Beach **K 1171** lies at the base of the bay across the mouth of the main creek. It is a 900 m long northeast-facing beach, backed by a 100-200 m wide grassy barrier then steep vegetated slopes rising to 90 m. The main creek meanders across the centre of the barrier and beach. It is bordered by steep bedrock slopes and fringed by reefs which are up to 600 m wide off the headlands and which narrow to 200 m toward the centre of the bay. Beach **K 1172** commences 500 m to the east, and is a 600 m long strip of sand lying at the base of 20 m high vegetated bluffs, and while fringed by some reef flats is fronted primarily by narrow sand flats.

K 1173-1178 **CAPE LONDONDERRY (E 8)**

No.	Beach	Rating HT	Rating LT	Type	Length
K1173	Cape Londonderry (E72)	1	1	R+sand flats	250 m
K1174	Cape Londonderry (E73)	1	2	R+coral reef flats	200 m
K1175	Cape Londonderry (E74)	1	2	R+coral reef flats	450 m
K1176	Cape Londonderry (E75)	1	1	R+sand/reef flats	400 m
K1177	Cape Londonderry (E76)	1	1	R+sand/reef flats	200 m
K1178	Cape Londonderry (E77)	1	1	R+sand/reef flats	500 m
Spring & neap tide range: 2.7 & 1.1 m					

The southern half of the southernmost embayment between capes Londonderry and Rulhieres consists of an irregular, low energy, 4 km wide, 5 km deep, bedrock-controlled bay containing six beaches (K 1173-1178), together with some elongate rock islets toward the centre of the bay.

Beach **K 1173** is located 700 m south of beach K 1174. It is a 250 m long east-facing beach that occupies a small valley fed by an upland creek. Grassy beach ridge-overwash deposits extend up to 150 m behind the beach, with the creek draining out across the northern end. It is fronted by tidal sand flats with rocky mangrove-lined shore dominating the southwestern arm of the bay.

Beaches K 1174 and 1175 are located in lee of the mid-bay islands and connect to the innermost islet to form a tombolo. Beach **K 1174** forms the western side of the tombolo and is a 200 m long northwest-facing strip of sand. Beach **K 1175** links to the beach at the 50 m wide tip of the tombolo. It is a 450 m long northeast-facing beach backed by a 300 m wide beach ridge plain that has filled a small backing valley, with a small mangrove-lined creek draining out at the eastern end of the beach. Both beaches are fronted by a veneer of sand then 200 m wide rock and reef flats.

Beach **K 1176** is located 300 m to the east in a slight indentation in the 100 m high backing slopes. It is a straighter 400 m long north-northeast-facing beach, backed by the slopes and fronted by narrow sand flats.

Beach **K 1177** lies 10 km south of the entrance at the southern end of a 500 m wide steep side embayment. It is a very low energy 200 m long beach ridge fronted by 500 m wide tidal flats, with a mangrove-lined creek mouth bordering the eastern end of the beach. It is backed by salt flats that extend up to 1.5 km south to the base of the valley.

Beach **K 1178** is located 2 km to the northeast on the eastern side of the southern embayment. It is a 500 m long northwest-facing beach that occupies the base of a small 500 m wide bay. It is backed by a 100 m wide overwash plain, then slopes rising steeply to the 60 m high edge of the plateau. A small upland to tidal creek crosses the northern end of the beach, with sand flats extending 1 km off the beach and filling the bay.

K 1179-1185 **CAPE LONDONDERRY (E 9)**

No.	Beach	Rating HTLT	Type	Length
K1179	Cape Londonderry (E78)	1 2	R+fringing reef	150 m
K1180	Cape Londonderry (E79)	1 2	R+fringing reef	250 m
K1181	Cape Londonderry (E80)	1 2	R+fringing reef	150 m
K1182	Cape Londonderry (E81)	1 2	R+fringing reef	500 m
K1183	Cape Londonderry (E82)	1 2	R+fringing reef	300 m
K1184	Cape Londonderry (E83)	1 2	R+fringing reef	500 m
K1185	Cape Londonderry (E84)	1 2	R+sand/reef flats	300 m
Spring & neap tide range:	2.7 & 1.1 m			

Beaches K 1179-1185 occupy parts of the eastern side of the southernmost bay between the two capes, and lie between 22-20 km west of Cape Londonderry. All face west into the bay which widens from 4 km off beach K 1179 to 7 km at the entrance.

Beaches K 1179-1182 are four neighbouring pockets of sand located along a 2 km long section of coast, each separated by sloping grassy bedrock points and backed by slopes rising 60-80 m to the edge of the plateau (Fig. 4.109). Beach **K 1179** is a 150 m long pocket of sand occupying most of a small sloping valley. It is fronted by 20-50 m wide fringing reef. Beach **K 1180** lies 100 m to the north and is a similar 250 m long beach with the reef widening from 30 m in the south to 80 m off the northern end. Two hundred metres of protruding rocky shoreline separate it from beach **K 1181**, a slightly curving 150 m long beach backed by a patch of grassy overwash flats, with a small mangrove-lined creek draining across the northern end of the beach. A 100 m wide reef parallels the beach. Beach **K 1182** occupies the next small valley 200 m to the north. It is a 500 m long beach, backed initially by scarped 10 m high bluffs, then the valley which is partly filled with a small grassy 50 m wide overwash plain, with a small creek crossing the northern end of the beach and depositing a patch of tidal sand on the fringing reef, which averages 150 m in width.

Figure 4.109 Beaches K 1179-1182 are three small pockets of sand fronted by a continuous fringing coral reef.

Beaches **K 1183** and **1184** are two neighbouring beaches located along a 2 km long northwest-facing section of steep bedrock shoreline. They are 300 m and 500 m long respectively, both backed by vegetated slopes rising steeply to the 60 m high edge of the plateau. They are fronted by continuous 50 m wide fringing reef, with some rocks outcropping along both beaches.

Beach **K 1185** occupies the northeastern corner of a 1 km wide reef-rimmed embayment immediately east of beach K 1184. The 300 m long beach has filled a small valley with a backing 50-100 m wide grassy overwash plain. A small creek drains out across the northern end and sand then reef flats extend 400 m off the beach.

K 1186-1189 **CAPE LONDONDERRY (E 10)**

No.	Beach	Rating HTLT	Type	Length
K1186	Cape Londonderry (E85)	1 2	R+fringing reef	1.2 km
K1187	Cape Londonderry (E86)	1 2	R+fringing reef	400 m
K1188	Cape Londonderry (E87)	1 2	R+fringing reef	200 m
K1189	Cape Londonderry (E88)	1 2	R+fringing reef	600 m
Spring & neap tide range:	2.7 & 1.1 m			

The eastern headland of the southernmost bay between the two capes separates the bay from the adjoining Faraway Bay. The headland is 3 km long, faces north-northwest, and lies in lee of a 2.5 km long island that reaches 64 m in height. Four beaches (K 1186-1189) occupy much of the headland shore.

Beach **K 1186** commences on the western tip of the headland and trends northeast for 1 km to a sandy foreland, then turns and extends east for another 200 m. The beach is backed by up to 300 m wide grassy beach ridge plain and fronted by 50-100 m wide fringing reef which narrows to just a few metres at the tip of the foreland. A few rocks separate the beach from beach **K 1187**, a 400 m long beach that faces the island 500 m to the north. The beach is backed by a central 1-2 m high grassy foredune and fronted by continuous fringing reef that widens from 50 m to 100 m to the east. A second band of rocks separates it from beach **K 1188**, a similar 200 m long beach fronted by up to 200 m wide fringing reef. It has a low foredune along the eastern half, grading into the backing slopes to the west, which rise steeply to a 60 m high plateau surface.

Beach **K 1189** lies 500 m to the east and is a 600 m long north-facing beach that is bent in the middle either side of a small valley (Fig. 4.110). The steep beach is composed of coarse sand and coral debris, deposited as a series of berms-beach ridges. These are backed by an overwash plain extending up to 100 m south and covered in patchy *Triodia pungens* (Soft Spinifex). The reef is 10 m wide towards the end of the beach, pinching out in the centre of the drowned valley, one of the few places to provide relatively deep access to a sandy shore on this section of coast. The eastern arm of the beach is narrow, with some rocks littering the beach.

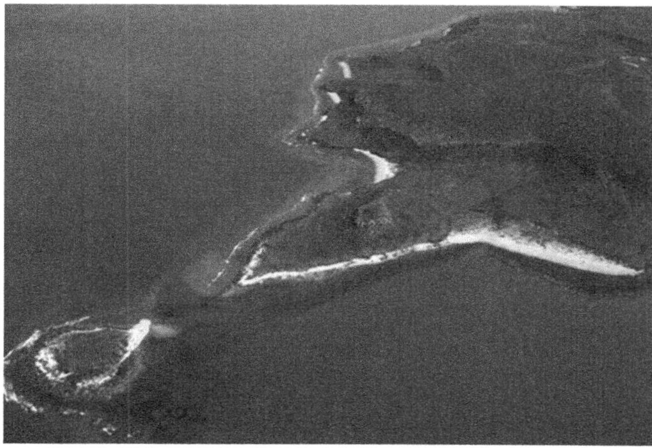

Figure 4.110 Beach K 1189 (right) and the outer beaches of Faraway Bay, K 1190-1193 (left).

K 1190-1198 **FARAWAY BAY (W)**

No.	Beach	Rating HT LT	Type	Length
K1190	Faraway Bay (W1)	1 2	R+fringing reef	250 m
K1191	Faraway Bay (W2)	1 2	R+fringing reef	100 m
K1192	Faraway Bay (W3)	1 2	R+fringing reef	200 m
K1193	Faraway Bay (W4)	1 2	R+fringing reef	300 m
K1194	Faraway Bay (W5)	1 2	R+fringing reef	500 m
K1195	Faraway Bay (W6)	1 2	R+fringing reef	400 m
K1196	Faraway Bay (W7)	1 2	R+fringing reef	1.2 km
K1197	Faraway Bay camp	1 2	R+sand/reef flats	400 m
K1198	Faraway Bay (W9)	1 2	R+sand/reef flats	150 m
Spring & neap tide range:		2.7 & 1.1 m		

Faraway Bay is a 5 km wide, 4-6 km deep, roughly U-shaped north-facing bay, located 12 km west of Cape Rulhieres. The bay has 21 km of generally steeply sloping volcanic shoreline, capped by 60-100 m high sandstone plateau. There are 25 beaches (K 1190-1214) located in pockets in the bedrock and at the mouths of a few small streams, totalling 23 km in length (40%). In addition the Faraway Bay Bush Camp is located on the rocky slopes overlooking beaches K 1197 and 1198. The western shore of the bay runs roughly north-south for 5.5 km, terminating at a 500 m wide, 1.5 km deep bedrock-controlled bay in the south. It has seven beaches along the western side (K 1190-1196) and two at the base of the bay (K 1197-1198).

Beach **K 1190** is located just below the northwest headland of the bay. It is a gently curving 250 m long beach bordered by a bedrock headland and backed by slopes rising steeply to the 60 m high plateau surface. The northeast-facing beach is backed by some minor dune activity extending up to 50 m inland and fronted by 70 m wide reef flat, with waves averaging up to 0.5 m during easterly winds.

Beach **K 1191** lies 500 m to the south and is a 100 m long strip of high tide sand, backed by 60 m high slopes and fronted by 50 m wide reef flats. A 150 m long protrusion in the rocky slopes separates it from beach **K 1192**, a 200 m long northeast-facing sand beach, wedged in between the backing steep, vegetated slopes and a 60 m wide fringing reef. It is bordered in the south by a 200 m wide sloping headland, on the south side of which is 300 m long beach **K 1193**. This beach is bordered by a low rocky headland to the south and a small mangrove-lined creek against the northern headland. It is backed by slopes gradually rising to the plateau surface and fronted by 50-100 m wide reef flats. During easterly winds waves average up to 0.5 m along all three beaches.

Beach **K 1194** commences 500 m to the south and is a discontinuous, curving east-facing beach lying at the base of 60 m high slopes, with some rocks scattered along the beach. One hundred metre wide fringing reefs parallel the beach. Beach **K 1195** commences 100 m to the south and is a similar discontinuous, 400 m long beach, with the rocks, backing slopes and fringing reef.

Beach **K 1196** is located 1 km to the south and occupies the mouth of a small creek, as well as extending south along the base of 80 m high slopes. A low rocky islet lies off the beach. The 1.2 km beach is backed in the northern valley section by a 100 m wide, low, grassy overwash plain, with the creek draining across the centre of the beach, and some mangroves off the northern section. It continues to the south for several hundred metres as a narrow beach wedged between the slopes and fronting reef flats, which widen to 500 m off the valley section.

Beach **K 1197** lies at the base of the bay at the mouth of a mangrove-fringed joint-controlled creek, which flows out the western side of the beach. The moderate gradient beach is 400 m long, faces north up the bay and is backed by a low 300 m wide, grassy, V-shaped beach ridge plain and fronted by 100 m wide tidal sand flats, with reef flats to either side at the base of the valley sides (Fig. 4.111). Steep slopes rising to 80 m border each side of the bay. Beach **K 1198** lies 100 m to the northeast, and is a 150 m long strip of largely intertidal sand lying at the base of the steep slopes of the bay. It is fronted by 100 m wide sand flats then fringing reef.

Figure 4.111 View of Faraway Bay beach (K 1198) from the Faraway Bay Bush Camp.

Faraway Bay Bush Camp is located on a ridge above beaches K 1197 and 1198, with beach K 1197 used to store fuel and small boats, while larger boats sometimes moor off the beach. A steep vehicle track and footpath lead up to the camp which provides a number of cabins and a central dining area and swimming pool all overlooking the beaches and bay.

Faraway Bay received a direct hit from Tropical Cyclone Ingrid, a category 5 cyclone on 16 March 2005. While there was extensive damage to parts of the camp, it was operating again for the 2006 winter season. See the camp at **http://www.farawaybay.com.au/**

K 1199-1207 **GUMBOOT BAY**

No.	Beach	Rating HT	LT	Type	Length
K1199	Gumboot Bay (W1)	1	2	R+fringing reef	50 m
K1200	Gumboot Bay (W2)	1	2	R+fringing reef	150 m
K1200A	Gumboot Bay (S1)	1	2	R+sand flats	50 m
K1200B	Gumboot Bay (S2)	1	2	R+sand/rocks	40 m
K1200C	Gumboot Bay (S3)	1	2	R+sand flats	100 m
K1201	Gumboot Bay (E1)	1	2	R+fringing reef	200 m
K1202	Gumboot Bay (E2)	1	2	R+fringing reef	250 m
K1203	Gumboot Bay (E3)	1	2	R+fringing reef	200 m
K1204	Gumboot Bay (E4)	1	2	R+fringing reef	300 m
K1205	Gumboot Bay (E5)	1	2	R+fringing reef	200 m
K1206	Gumboot Bay (E6)	1	2	R+fringing reef	350 m
K1207	Gumboot Bay (E7)	1	2	R+sand flats	400 m
Spring & neap tide range:	2.7 & 1.1 m				

Immediately east of Faraway Bay is Gumboot Bay, a very scenic narrow bay that extends 3 km to the south. Beaches K 1199-1201 are located either side of the 500 m wide entrance to the bay, while beaches K 1202-1207 occupy the rocky shore extending 2 km east of the bay mouth. The beaches face north up the 4 km deep embayment, with steep bedrock slopes rising 60-80 m to the sandstone plateau on either side.

Beaches K 1199 and 1200 are two pockets of sand located toward the northeastern tip of a 1.5 km long north-trending bedrock peninsula which narrows to 100 m behind beach K 1199. Beach **K 1199** is a 50 m long pocket of west-facing sand backed by low vegetated slopes. Beach **K 1200** lies 400 m to the south and is a slightly curving 150 m long strip of sand and narrow grassy 1-2 m high foredune wedged at the base of slopes rising to 60 m. Both beaches are fronted by 50 m wide fringing reef.

Gumboot Bay continues south of these beaches narrowing before expanding into the lower mangrove-lined bay fed by two waterfalls, with two pockets of sand at the western base of the bay (beaches K 1200A & B). This pristine site has been proposed as a landing for a mining exploration camp. Given the hundreds of kilometres of undeveloped coast along this part of the Kimberley one would hope that a less environmentally significant site could be located for the landing.

Beach **K 1200A** is a 50 m long low gradient sandy beach at the mouth of a small mangrove-lined creek with tidal sand shoals extending 80 m into the bay. The beach looks north up to the bay entrance and is separated from beach K 1200B by a 40 m wide outcrop of metasedimentary rocks. Vegetated 10 m high slopes back both beaches. Beach **K 1200B** is a predominantly boulder beach, with a mixture of sand and boulders exposed at low tide. This is the beach proposed as the landing site for the mining operations.

Beach **K 1200C** lies opposite on the eastern side of the bay. It is a more protected 100 m long strip of sand backed by a narrow grassy overwash flat and bordered by mangroves with vegetated slopes rising to 60 m behind the beach.

Beach **K 1201** lies on the opposite side of the bay mouth and faces toward beach K 1199. It is a curving 200 m long arc of sand occupying a small valley at the base of 60 m high slopes to the south. It is fronted by 100 m wide reef flats. Beach **K 1202** lies 200 m to the north and is a 250 m long north-facing beach, with a narrow, 1-2 m high, grassy foredune grading into rocks at the eastern end, while it is backed by 20 m high slopes, and fronted by 50-100 m wide reef flats. A 300 m wide grassy headland separates it from beaches **K 1203** and **K 1204**, two adjoining 200 and 300 m long beaches that occupy adjoining valleys. Beach K 1203 faces east and with greater exposure to the winds has a grassy foredune rising to 10 m, while K 1204 faces north, with a strip of sand and narrow, low grassy foredune at the base of 10 m high bluffs which link the two beaches. Both are fronted by 50 m wide fringing reef, which narrows off the bluffs.

A 400 m long 10-20 m high eroding headland separates the beaches from beach **K 1205**. This is a 200 m long north-facing beach backed by moderately vegetated slopes and fronted by reef flats, which widen from 50-200 m off the eastern headland. Beach **K 1206** lies 100 m to the east of the eastern side of the small 20 m high headland. The 350 m long beach occupies a small valley curving round in the east to the lee of a boundary 20 m high headland. The valley has been infilled by a low grassy beach ridge plain up 200 m wide in the centre. Fringing reef borders each end of the beach, with a veneer of sand flats in the centre.

Beach **K 1207** commences on the eastern side of the 300 m wide headland. It is a 400 m long north-facing beach capped by high tide cusps. It is backed by 10-20 m high bluffs in the west and a small grassy landward sloping beach ridge in the east, backed by a usually dry tree-lined creek (Fig. 4.112). Steeper slopes behind rise to the 100 m high plateau 2 km to the south.

Figure 4.112 Beach K 1207 has a moderately steep high tide beach backed by grassy overwash flats.

Oombulgurri Aboriginal Land

Area:	1 073 884 ha	
Coast length:	236 km	(9544-9780 km)
Beaches	117 beaches	(K 1204-1321)

Oombulgurri is a community of several hundred people located on the Forrest River 45 km northwest of Wyndham and 30 km west of the Cambridge Gulf. It is surrounded by the Oombulgurri Aboriginal Land, the largest in the Kimberley, which covers an area of 1 073 884 ha and extends along 236 km of coast between Faraway Bay and Thurburn Hill.

K 1208-1214 **GUMBOOT BAY (E)**

No.	Beach	Rating HT	LT	Type	Length
K1208	Gumboot Bay (E8)	1	2	R+fringing reef	400 m
K1209	Gumboot Bay (E9)	1	2	R+fringing reef	150 m
K1210	Gumboot Bay (E10)	1	2	R+fringing reef	200 m
K1211	Gumboot Bay (E11)	1	2	R+fringing reef	400 m
K1212	Gumboot Bay (E12)	1	2	R+fringing reef	700 m
K1213	Gumboot Bay (E13)	1	2	R+fringing reef	300 m
K1214	Gumboot Bay (E14)	1	2	R+fringing reef	700 m
Spring & neap tide range:		2.7 & 1.1 m			

The eastern shore of Faraway-Gumboot bays trends north, then northeast for 4 km to a low basalt islet tied by a tombolo (beaches K 1213 & 1214) to the shore. Continuous reefs fringe the shore with all the beaches backed by slopes decreasing from 80 m in the south to 40 m at the point.

Beaches K 1208-1211 occupy a 1.5 km long section of west-facing rocky shore. Beach **K 1208** is a 400 m long sand beach wedged between slopes rising steeply to 40 m. It has a small creek and a few mangroves at the northern end adjacent to a small grassy foredune, with vegetated debris slope and boulders backing the southern section of beach. A fringing reef parallels the beach widening from 20 to 70 m in the north where a small upland creek drains out across the end of the beach. Beach **K 1209** lies 200 m to the north and is a 150 m long strip of high tide sand at the base of 60 m high slopes with rock and reef flats extending 50 m off the beach. Beach **K 1210** lies 100 m to the north and is a similar 200 m long beach wedged between the steep slopes and the reef.

Beach **K 1211** is the northernmost and largest of the four beaches. It is 400 m long and occupies an indentation in the backing 70 m high slopes. A small creek flows out against the southern rocks, with the remainder of the backshore filled with a low grassy overwash plain, up to 100 m wide in the centre. It is fronted by reefs, which widen from 100 m in the south to 500 m in the north.

Beach **K 1212** lies along the northeast entrance to the bay, 200 m north of beach K 1211. It is a 700 m long northwest-facing beach that occupies two indentations in the backing 40 m high slopes. A few rocks lie along the central-northern section of beach, with the backing foredune rising to 1-2 m and 50 m in width in the north where there is some instability, while it widens to 100 m and becomes more vegetated to the south. A continuous 100-400 m wide coral reef fringes the beach.

Beaches K 1213 and 1214 are located at the tip of the eastern headland and together form a low grass-covered tombolo that ties a low rock islet and rock flat to the shore. Beach **K 1213** is 300 m long, faces north and is backed by a 1-2 m high foredune and sea cliffs in the west, while 700 m long beach **K 1214** faces east. Both are fronted by 400-600 m wide fringing reefs, with easterly winds producing waves up to 1 m and surf on the reef edge. Beach K 1214 is backed by some low dunes along the northern section, which extend up to 300 m inland, rising to 6-8 m. A protruding cliffline and rock debris cut off the southern section of beach at high tide.

K 1215-1219 **KING GEORGE RIVER**

No.	Beach	Rating HT	LT	Type	Length
K1215	King George River (W3)	1	1	R+tidal sand flats	300 m
K1216	King George River (W2)	1	1	R	700 m
K1217	King George River (W1)	1	1	R+tidal sand flats	500 m
K1218	King George River	1	1	R+tidal sand flats	3 km
K1219	King George River (E1)	1	1	R+ sand/reef flats	200 m
Spring & neap tide range:		2.7 & 1.1 m			

The **King George River** is one of the larger rivers draining the rugged Kimberley. It reaches the coast in a 3 km wide, 5 km deep, northwest-facing bedrock embayment, immediately west of Cape Rulhieres. Both the bay and lower reaches of the river are bordered by a 50-100 m high sandstone canyon, which extends to the 300 m wide river mouth and surrounds the bay shore. Five beaches are located within the bay, two either side of the river mouth (K 1217 and 1218) (Fig. 4.113) and three on the rocky bay shore (K 1215, 1216 and 1219).

Figure 4.113 The King George River mouth, bounded by beaches K 1217 & 1218 (left).

Beach **K 1215** occupies the mouths of three smaller bedrock-controlled creeks. The steep, sandy 300 m wide beach is located 1 km into the bay and faces northeast out the bay entrance, with 50-60 m high sandstone headlands bordering each side. The beach itself fronts the central creek, with its 500 m wide tidal sand flats extending north to block the two creeks to either side. A low 5 ha grassy overwash flat backs the beach and partly fills the backing valley, which has a small triangle of mangroves and is bordered by 40 m high sandstone slopes. The eastern creek has a small mangrove-lined lagoon. Waves average less than 0.5 m and break across the sand flats, only reaching the beach at high tide.

On 20 February 1940 the ship *Koolama* was attacked by Japanese bombers off Cape Londonderry. It stopped at Cape Rulhieres and landed survivors at beach K 1215 from where they were rescued. The ship subsequently limped to Wyndham where it sank at the wharf.

Beach **K 1216** is located 2 km to the southeast, halfway into the bay, and is a 700 m long sand beach filling the base of a north-facing box-shaped bedrock bay surrounded by 40-60 m steep sandstone slopes, including a narrow 60 m high, 500 m long eastern headland. The eastern half of the beach is completely overwashed in higher seas and also breached by the backing waterfall-fed creek and small 1 ha lagoon, resulting in bare sand and little vegetation. The western half has unstable 50 m wide dunes rising to 5 m, then a 50 m long pocket of sand at the base of the cliffs, cut off from the main beach at high tide. It fronts the deep water of the bay with a usually steep reflective beach.

Beach **K 1217** occupies the western side of the King George River mouth and extends for 500 m to the west along the base of 80 m high sandstone cliffs. It is fronted by the river mouth shoals that extend up to 1 km into the bay and is bordered to the east by the deep 300 m wide river channel.

Beach **K 1218** is the main beach that extends for 3 km northeast from the river mouth and represents the 'delta' of the King George River. The beach is the outermost ridge in a 1 km wide series of several beach ridge-spits that have prograded into the bay from the backing 80 m high cliffs. Mangroves occupy the first 500 m of the sand ridges. The modern beach is a narrow low bare berm, fronted by 500 m of intertidal sand ridges, then subtidal deposits that are 1.5 km wide off the river mouth. A second creek helps drain the mangroves at the eastern end of the beach. The creek runs along the base of the cliffs and is fronted by a second series of tidal sand shoals in lee of the eastern headland.

Beach **K 1219** is located at the northeastern entrance to the bay in lee of a small 50 m high conical headland that forms the northern tip of the bay entrance. The headland forms the northern side of a 150 m wide embayment that has been partially filled by a grassy 100 m wide 1-2 m high beach ridge plain, with a narrow, usually dry, mangrove-lined creek running along the northern boundary and steep rocky slopes behind. The beach is 200 m long and faces west across the bay mouth. It is fronted by a 50 m wide sand flat edged by a 30 m wide fringe of reef. The moderately protected waters off the beach are a popular anchorage for passing yachts.

K 1220-1225 CAPE RULHIERES (E 1)

No.	Beach	Rating HT	LT	Type	Length
K1220	Cape Rulhieres	2	3	R+rock flats	300 m
K1221	Cape Rulhieres (E1)	2	3	R+rock flats	100 m
K1222	Cape Rulhieres (E2)	2	3	R+rock flats	150 m
K1223	Cape Rulhieres (E3)	2	3	R+rock flats	200 m
K1224	Cape Rulhieres (E4)	2	3	R+rock flats	250 m
K1225	Cape Rulhieres (E5)	2	3	R+rock flats	50 m
Spring & neap tide range:		2.7 & 1.1 m			

Cape Rulhieres is a 60 m high sandstone headland that marks the beginning of a 130 m long relatively straight southeast trend of the coast to Cape Dussejour, the northwest entrance to Cambridge Gulf. Most beaches along this section are located in bedrock-controlled valleys and indentations in the usually high cliffed coast, with coral reef fringing some of the headlands. Beaches K 1220-1225 are located along the first 5 km east of the cape.

Beach **K 1220** lies immediately east of the cape, with the flat-topped 64 m high crest of the cape located 500 m south of the beach. The 300 m long beach faces northeast and curves round at the base of the steep rocky slopes. It is perched behind 50-200 m wide intertidal rock flats, with some tide pools on the inner flats, and narrow reef and surf ringing the outer edge. The beach is littered with rock debris and grades into a boulder beach to the north.

Beaches K 1221 and 1222 occupy a 500 m wide W-shaped bay located 1 km south of the cape. The entire bay is rimmed with rugged 40 m high sandstone cliffs. Beach **K 1221** lies in the northern arm of the bay and is a 100 m long pocket of sand, backed by a narrow strip of grass, then cliffs. It is bordered by sandstone boulder debris and fronted by rock and boulder flats. Three hundred metres to the south is beach **K 1222,** a similar 150 m long north-facing beach, backed by vegetated rocky slopes and also bordered and fronted by the boulder debris and flats. Surf breaks over the boulder flats at the entrance to the bay and around the base of the intervening headland, with usually low waves at the beaches.

Beach **K 1223** occupies the base of the next valley, 2 km to the east. It is a V-shaped 500 m deep north-facing embayment, rimmed by 40 m high sandstone cliffs. The 200 m long beach is backed by a 50-100 m wide low, grassy overwash plain, then a rocky gully splitting the valley sides. It is bordered by 50 m wide rock and boulder flats, with deeper water almost reaching the base of the beach (Fig. 4.114).

Figure 4.114 Beaches K 1223 (right) and 1224 (left) occupy narrow joint-oriented rocky embayments.

Beach **K 1224** lies 1 km to the southeast in the next deeper embayment. This is a 2 km deep, northeast-trending embayment, that narrows to 200 m off the beach and is surrounded by 40 m high cliffs. The beach is 250 m long and lies across the mouth of two bedrock-controlled creeks, with a low 50-100 m wide overwash plain, covered in dense *Triodia pungens* (Soft Spinifex) in the main valley, and mangroves filling the eastern creek mouth. It is a steep sandy beach with small cusp horns formed from coral debris and fronted by 500 m wide sand flats, then another 300 m of fringing reef, which also edges the shore of the bay. Waves break along the edge of the reefs, with low wave conditions at the beach.

Two kilometres to the east is beach **K 1225** is a 50 m long pocket of sand occupying a narrow valley. It is bordered by the steep valley sides and fronted by 50 m wide rock flats and fringing reef.

K 1226-1233 **CAPE RULHIERES (E 2)**

No.	Beach	Rating HT	Rating LT	Type	Length
K1226	Cape Rulhieres (E6)	2	3	R+rock/reef flats	150 m
K1227	Cape Rulhieres (E7)	2	3	R+rock/reef flats	250 m
K1228	Cape Rulhieres (E8)	2	3	R+rock/reef flats	150 m
K1229	Cape Rulhieres (E9)	2	3	R+rock/reef flats	400 m
K1230	Cape Rulhieres (E10)	2	3	R+rock/reef flats	600 m
K1231	Cape Rulhieres (E11)	2	3	R+sand/reef flats	700 m
K1232	Cape Rulhieres (E12)	2	3	R+rock/reef flats	300 m
K1233	Cape Rulhieres (E13)	2	3	R+rock/reef flats	600 m
Spring & neap tide range:	2.7 & 1.1 m				

Between 5 and 8 km southeast of Cape Rulhieres is a 3 km wide, north-facing embayment, which converges on a central 1 km wide bay. Beaches K 1226-1229 are located along the western arm, beaches K 1230 and 1231 in the central bay, and beaches K 1232 and 1233 on the eastern arm.

Beaches **K 1226** and **K 1227** are adjoining 150 and 250 m long beaches occupying indentations at the base of rocky sandstone slopes that rise to 80 m. They are both backed by patches of partly vegetated foredunes and fronted by a 100 m wide mixture of rock outcrops surrounded by fringing reef, with waves averaging 0.5 m at the reef edge.

Beaches K 1228 and 1229 lie 1 km to the southeast and are similar adjoining beaches 150 and 400 m long respectively. Beach **K 1228** has a narrow foredune and is fronted by 80 m wide rock to reef flats, while beach **K 1229** occupies a small valley, with a creek draining along the eastern side. It is fronted by 100 m wide reefs in the west which extend 1 km seaward in the east to link to a low rock outcrop.

Beach **K 1230** lies along the western side of the central bay. It is a crenulate 600 m long east-facing beach that is interrupted by rock outcrops, backed by irregular rocky slopes, and fronted by a mixture of sand and rock flats and fringing reef up to 300 m wide. Beach **K 1231** occupies the base of the bay. It is a 700 m long low energy beach ridge that is spread across the mouth of two mangrove-lined creeks draining two adjoining bedrock-controlled valleys. A few beach ridge-spits extend into both valleys, while tidal sand flats then reefs front the beach and extend 500 m into the bay.

Beach **K 1232** lies along the northeastern side of the embayment. It is a crenulate 300 m long beach, with rock outcrops along and off the beach. It is fronted by a continuous 50-100 m wide fringing reef and backed by sandstone slopes rising to 20 m. Beach **K 1233** lies immediately to the east of the northern end of the beach, where it is separated by a low 100 m wide rocky point. The beach faces northwest and curves round for 600 m between two rocky headlands. It is backed in part by an irregular bedrock valley, which has been partly filled by salt flats, mangroves and outer overwash-beach ridge deposits, totalling up to 400 m in width. The beach is fronted by some sand flats and fringing reef up to 1 km wide.

K 1234-1244 **CAPE BERNIER (W)**

No.	Beach	Rating HT	LT	Type	Length
K1234	Cape Bernier (W10)	2	3	R+rock/reef flats	800 m
K1235	Cape Bernier (W9)	2	3	R+rock/reef flats	100 m
K1236	Cape Bernier (W8)	2	3	R+rock/reef flats	200 m
K1237	Cape Bernier (W7)	1	1	R+tidal sand flats	300 m
K1238	Cape Bernier (W6)	2	3	R+rock/reef flats	500 m
K1239	Cape Bernier (W5)	2	3	R+rock/reef flats	200 m
K1240	Cape Bernier (W4)	2	3	R+rock/reef flats	500 m
K1241	Cape Bernier (W3)	1	2	R+sand/reef flats	50 m
K1242	Cape Bernier (W2)	2	3	R+rock/reef flats	200 m
K1243	Cape Bernier (W1)	1	2	R+sand/reef flats	50 m
K1244	Cape Bernier	2	3	R+rock/reef flats	300 m
Spring & neap tide range:	3.6 & 0.3 m				

Beach K 1234 is located at the tip of a 2 km long north-trending peninsula, roughly midway between Cape Rulhieres and Cape Bernier. Between the peninsula and Cape Bernier 8 km to the east are 10 km of sloping, rocky, indented coastline, containing a total of 11 beaches (K 1234-1244), totalling 3 km in length, the remainder of the shore dominated by heavily jointed sandstone rocks and fringed by rock flats and continuous reefs.

Beach **K 1234** is an 800 m long cuspate beach on the northeastern tip of the 1 km wide peninsula. It has formed in lee of 500 m wide reefs and consists of the convex northeast-facing beach, backed by a 100-300 m wide low grassy overwash plain, then the rocks. At the southern end a small tombolo has formed in lee of a clump of rocks, with beach **K 1235** extending for another 100 m to the south (Fig. 4.115), where it joins the south-trending rocky shoreline. It is also fronted by 300-500 m wide fringing reef.

Beach **K 1236** is located midway down the eastern side of the peninsula and consists of a northeast-facing, 200 m long, curving strip of sand, wedged in a gap in the rocks, and fronted by 200 m wide rock and reef flats, with usually low waves breaking on the reef edge.

Beach **K 1237** lies at the base of the peninsula and extends to the east for 300 m to the mouth of a mangrove-lined creek. The beach is backed by a 200 m wide series of low beach ridges and fronted by 400 m wide tidal sand flats, with reefs to either side. Beach **K 1238** is located 200 m northeast of the creek mouth and is an irregular 500 m long strip of high tide sand, located between low rocky sandstone terrain, and irregular rock flats, then fringing reef extending up to 500 m off the shore.

Figure 4.115 Beaches K 1234 & 1235 form either side of a low sandy cuspate foreland.

Beaches **K 1239** and **K 1240** are located 1 km to the east and consist of 200 and 500 m long, respectively, strips of irregular high tide sand, wedged between rocky terrain rising to 30 m, and 50-100 m wide fringing reef.

Beach **K 1241** lies 1 km to the south. It is a 50 m long pocket of sand at the mouth of a joint-controlled creek that flows through a subterranean tunnel just before reaching the shore on the western side of the beach. The creek has deposited a 100 m wide sand flat off the beach, with fringing reef extending another 100 m to the mouth of the narrow gully. Immediately to the east is 200 m long beach **K 1242**, which occupies a slightly larger gap in the rocks. It is backed by moderately sloping sandstone terrain and fronted by 250 m wide reef flats.

Beach **K 1243** lies 1 km to the south in the next indentation in the shore. It is a narrowing 50 m wide wedge of intertidal sand lying at the mouth of a narrow, joint-controlled mangrove-lined creek, which flows across the beach at high tide. Rocks then reefs fringe either side of the narrow gully, which widens to 200 m at its mouth.

Beach **K 1244** is located at **Cape Bernier**, a right angle turn in the coast backed by slopes gradually rising to 100 m. The 300 m long convex beach lies at the base of the sandstone cape, with a 1-2 m high grassy foredune covering some of the backing rocks. It is fronted by irregular sandstone rocks and flats, encased in fringing reef extending up to 500 m off the shore. A well developed rock platform is located immediately east of the beach.

K 1245-1253 **CAPE BERNIER (E)**

No.	Beach	Rating HT	LT	Type	Length
K1245	Cape Bernier (E1)	3	4	R+LTT	600 m
K1246	Cape Bernier (E2)	2	3	R+fringing reef	250 m
K1247	Cape Bernier (E3)	1	3	R+sand flats	1.5 km
K1248	Cape Bernier (E4)	1	3	R+sand/reef flats	200 m

K1249 Cape Bernier (E5) 1 3	R+sand/reef flats	200 m
K1250 Cape Bernier (E6) 2 3	R+sand/reef flats	300 m
K1251 Cape Bernier (E7) 2 3	R+fringing reef	800 m
K1252 Cape Bernier (E8) 1 2	R+sand/reef flats	200 m
K1253 Cape Bernier (E9) 2 3	R+fringing reef	150 m
Spring & neap tide range: 3.6 & 0.3 m		

Immediately east of Cape Bernier the rocky shore trends south for 9 km to the base of an open 7 km wide, irregular northeast-facing embayment, containing three smaller bays. Beach K 1245 is located in the first bay, beaches K 1246 and 1247 in the second, and beaches K 1248-1250 in the third. All the shoreline is backed by steeply rising sandstone and in the south basalt rocks rising to the 120-160 m high sandstone plateau. Immediately east of the third bay is an irregular 1.5 km wide north-facing embayment containing beaches K 1251-1253.

Beach **K 1245** is located 3 km south of the cape in a northeast-facing 1 km wide bay. The 600 m long beach curves around the base of the bay, which is backed by steep vegetated 80 m high cliffs, with a steep gully in the centre. The beach has some active dunes rising 6-7 m against the backing slopes and is fronted by a low tide bar, which widens to either end, with rock debris and flats and fringing reefs off each headland. Easterly waves which average 0.5-1 m break over the reefs and bar and maintain a permanent rip which drains out the eastern end of the bay.

Beaches K 1246 and 1247 occupy the southeastern side of the second bay. Beach **K 1246** is a curving 250 m long narrow beach that emerges from mangroves and cobbles against the northern headland. It is backed by a narrow, 2-3 m high grassy plain and fronted by a protruding fringing reef up to 500 m wide. Beach **K 1247** commences immediately to the south and is a curving low gradient 1.5 km long beach, backed by a 4-5 m high foredune ridge plain covered in clumpy *Triodia pungens* (Soft Spinifex), then rising slopes. It is fronted by 200 m wide sand flats, across which waves maintain a low, but wide, dissipative surf. A medium sized, mangrove-lined creek borders the southern end of the beach, with tidal sand shoals extending 300 m off the beach. The creek drains a 100 ha mangrove-filled lagoon.

The third bay is 1.5 km wide at the mouth and trends south for 2.5 km to the mouth of a mangrove-lined creek. Beaches K 1248 and 1249 lie on the western side of the inner bay and beach K 1250 on the eastern side. Beach **K 1248** is a 200 m long northeast-facing strip of sand and low narrow foredune at the base of 80 m high slopes, and is fronted by 400 m wide reef flats. Beach **K 1249** lies 100 m to the south and is a similar 200 m long lower energy beach, fringed by mangroves in the north, with some sand then 700 m wide reef flats off the beach, while it is backed by a rising valley. Beach **K 1250** lies 1 km to the east on the eastern shore and faces north toward the bay entrance. It is a 300 m long strip of sand with a 3-4 m high partially active foredune toward the eastern end and mangroves to either end. It is backed by grassy-vegetated slopes rising 30-40 m, with some slope debris littering the centre of the beach. It is fronted by 400 m wide irregular tidal sand flats, with the edge of the fringing reef lining either side of the bay.

Beaches K1251-1253 occupy parts of an irregular north-facing 1.5 km wide embayment, bordered on both sides by rock and coral reefs extending 0.5-1 km off each headland. Beach **K 1251** is located on the western arm of the bay and faces north-northeast, exposing it to refracted easterly waves. The 800 m long beach is backed by sandstone slopes rising steeply to a circular 90 m high 100 ha plateau. Debris from the slopes covers part of the beach, which is bordered by lower sandstone points, and fronted by a narrow fringing reef.

Beach **K 1252** is located 1.5 km to the east in lee of the eastern headland. It is a 200 m long pocket of northwest-facing sand, bordered by mangroves and fronted by 500 m wide tidal sand flats, then fringing reef (Fig. 4.116). It is backed by a grassy 100-200 m wide beach ridge plain, then grassy slopes. Beach **K 1253** lies 300 m to the north on the northern side of the 15 m high sandstone headland. It is a 150 m long pocket of sand fronted by 200 m wide reef flats, with rocky points bordering either end.

Figure 4.116 Beach K 1252 is a sheltered high tide beach fronted by wide sand flats and fringed by dense mangroves.

T 1254-1261 **ROCKY & EVELYN ISLANDS**

No. Beach	Rating HT LT	Type	Length
K1254 Rocky Is (W)	1 2	R+sand flats	500 m
K1255 Seaplane Bay	1 2	R+sand/reef flats	300 m
K1256 Rocky Is (S)	1 2	R+sand/reef flats	300 m
K1257 Evelyn Is (E 1)	1 2	R+fringing reef	100 m
K1258 Evelyn Is (E 2)	1 2	R+sand/reef flats	200 m
K1259 Evelyn Is (E 3)	1 2	R+sand/reef flats	100 m
K1260 Evelyn Is (E 4)	2 3	R+fringing reef	150 m
K1261 Evelyn Is (E 5)	2 3	R+sand/reef flats	200 m
Spring & neap tide range:	3.6 & 0.3 m		

Rocky Island is a 30 m high 15 ha sandstone island located 500 m off the shore, 14 km northwest of Cape Whiskey and 2 km west of Evelyn Island. To the west of the island are two 3 and 4 km deep embayments containing beaches K 1254 and 1255 respectively, while to the east is an irregular coastline extending 8 km to beach K 1261. Most of the coast is dominated by sloping basaltic shoreline with sandstones to the east.

Beach **K 1254** occupies a southern bedrock-backed section of shoreline, between two mangrove-lined creek mouths. The 500 m long northeast-facing beach is fronted by 400 m wide tidal sand flats. It is backed by a 50-100 m wide, low, grassy overwash plain, then vegetated basalt slopes rising to 30 m.

Beach **K 1255** lies 3 km to the southeast deep inside **Seaplane Bay**. The 300 m long beach faces toward the embayment entrance and Rocky Island 3.5 km to the northeast. It occupies the mouth of a small valley with a mangrove-lined creek draining along the western side. It is backed by a low grassy overwash plain and fronted by 500 m wide tidal sand shoals then 1 km wide fringing reefs, which fill the southern half of the 1 km wide bay. Seaplane Bay provides a protected anchorage and was the site of the forced landing of the German seaplane 'Atlantis' in 1932. Following the landing the two pilots wandered the coast for 40 days before being rescued by aborigines while sheltering in the tunnel-cave at beach K 1241.

Beach **K 1256** is located 1.5 km due south of Rocky Island at the base of a 1 km wide 500 m deep reef-fringed embayment. The 300 m long beach is located across the mouth of a small valley, with a creek draining across the centre where it has deposited a gravel delta on the tidal flats. Sand flats, then 500 m wide fringing reef, also front the beach.

Beach **K 1257** is located in lee of a gap in the 1 km wide fringing reefs, 1 km south of the small Evelyn Island, a 10 m high clump of eroding basalt. The 100 m long beach faces east toward the gap in the reefs, and is bordered by mangrove-line shoreline.

Beaches K 1258-1261 occupy a 2 km long section of jointed 20-40 m high sandstone shoreline 5 km east of Evelyn Island and 7 km west of Cape Whiskey. Beach **K 1258** is a 200 m long east-facing beach located in lee of an attached low islet. The protected beach is backed by a 1-2 m high, 50 m wide patch of grass, then the sandstone slopes, and fronted by 100 m wide sand flats, then reef flats. Beach **K 1259** is located at the mouth of a narrow joint-controlled gully 400 m to the south. The 100 m long beach partly blocks the creek mouth and is fronted by 200 m wide sand flats, with reef fringing the bordering 20-40 m high headlands. Beach **K 1260** is located immediately east of the eastern headland, and is a 150 m long strip of sand at the base of 30 m high rocky slopes, while fronted by 150 m wide fringing reef.

Beach **K 1261** lies in the next embayment, 500 m to the southeast. It is a 200 m long pocket beach, bordered by 20-30 m high sandstone cliffs and fringed by 100-200 m wide fringing reefs. The beach is backed by a 50 m wide grassy plain and a small creek that drains a steep-sided valley.

K 1262-1266 **CAPE WHISKEY**

No.	Beach	Rating HT	LT	Type	Length
K1262	Cape Whiskey (W3)	1	2	R+sand/reef flats	250 m
K1263	Cape Whiskey (W2)	1	2	R+sand/reef flats	200 m
K1264	Cape Whiskey (W1)	2	3	R+rock/reef flats	500 m
K1265	Cape Whiskey (E1)	3	4	R+tidal flats/channel	300 m
K1266	Cape Whiskey (E2)	1	2	R+sand/reef flats	500 m
Spring & neap tide range:		3.6 & 0.3 m			

Cape Whiskey is a 12 m high jointed sandstone headland, bordered to either side by similar southeast-trending indented sandstone shoreline. Scattered along the rocky shore either side of the cape are a few pockets of sand containing beaches K 1262-1264 to the west, and K 1265 and 1266 to the immediate south.

Beach **K 1262** lies 4 km north-northwest of the cape and is a low curving 250 m long pocket of sand backed by a small 1-2 m high grassy plain and bounded by mangroves to the south and by low sandstone slopes. It faces east across a 500-600 m wide fringing reef. An incised creek separates it from beach **K 1263**, which is located 1 km to the south. This is a 200 m long north-facing beach, bordered by sloping sandstone points and fringed by mangroves. It is backed by a 1-2 m high, 50 m wide grassy overwash plain, with a usually dry creek draining across the eastern end. It is fronted by 100 m wide sand flats then another 500 m of fringing reef (Fig. 4.117).

Figure 4.117 Beach K 1263 (centre) is a pocket of sand located on a rock- and reef-dominated shoreline.

Beach **K 1264** is located on the northern side of Cape Whiskey. The 500 m long beach extends almost to the cape, and consists of a narrow strip of high tide sand and low grassy overwash plain, backed by sloping sandstone and fronted by some rock flats and 100-200 m wide fringing reef.

Beach **K 1265** occupies the mouth of a 50 m wide mangrove-lined tidal creek 1 km south of the cape. The 300 m long low energy beach faces north out of the creek mouth and across the creek tidal shoals and channel, and is backed by sloping sandstone (Fig. 4.118). Beach **K 1266** is located 500 m to the east and is a curving 500 m long beach, backed by older grassy transgressive dunes rising to 15 m towards the west, with some areas of instability. These are backed by sloping bedrock, with some rock outcrops on and off the beach, all linked by 500 m wide sand flats and outer fringing reef.

Figure 4.118 Beaches K 1265 is located in the creek mouth (right), with beach K 1266 lying just east of the mouth (left).

K 1267-1271 **CAPE WHISKEY (E)**

No.	Beach	Rating HT	LT	Type	Length
K1267	Cape Whiskey (E3)	2	3	R+low tide terrace	1 km
K1268	Cape Whiskey (E4)	2	3	R+tidal sand flats	200 m
K1269	Cape Whiskey (E5)	1	2	R+tidal sand flats	200 m
K1270	Cape Whiskey (E6)	1	2	R+sand/rock flats	300 m
K1271	Cape Whiskey (E7)	1	2	R+sand/rock/reef flats	300 m
Spring & neap tide range:		3.6 & 0.3 m			

Five kilometres southeast of Cape Whiskey is a U-shaped, 1 km wide northeast-facing embayment, bordered on the east by a 500 m long 20 m high sandstone headland, on the eastern side of which is a joint-controlled creek that extends due southwest for 4 km. Beach K 1267 is located in the embayment, while beaches K 1268-1271 line either side of the outer creek mouth.

Beach **K 1267** is a 1 km long northeast-facing beach that fills the base of the embayment. It is backed by active dunes extending up to 500 m inland and increasing in height to 10-15 m in the west. They are the first major transgressive dune field east of Lombadina (beach K 80), 560 km to the southwest. A mangrove-lined creek flows along the western side of the dunes and exits at the western end of the beach. The beach is exposed to refracted easterly waves, which average up to 1 m during easterly conditions. These maintain a narrow low tide bar along the beach whose bounding headlands are fronted by 50-100 m wide reef flats.

Beaches K 1268-1271 lie 500 m to the east, bordering the 1 km long bedrock creek mouth. Beach **K 1268** is located on the western side of the mouth and is a low energy 200 m long pocket of sand bordered and backed by gently sloping irregular sandstone. It faces across the 300 m wide creek mouth and sand shoals to beach **K 1269**, a 200 m long northwest-facing pocket of sand wedged in between protruding sandstone point and rock flats. One hundred metres to the north is 300 m long beach **K 1270**, a similar strip of sand, with rock also outcropping on and along parts of the beach. Beach **K 1271** commences another 100 m to the north and has sand, rock then reef flats off the 300 m long beach. It faces northwest across the 400 m wide creek mouth.

K 1272-1275A **ELSIE ISLAND**

No.	Beach	Rating HT	LT	Type	Length
K1272	Elsie Is (W3)	2	3	R+fringing reef	300 m
K1273	Elsie Is (W2)	1	2	R+sand/reef/rock flats	50 m
K1274	Elsie Is (W1)	1	2	R+sand flats	1.3 km
K1275	Elsie Is (E1)	1	2	R+tidal sand flats	600 m
K1275A	Elsie Is (E2)	1	2	R+tidal sand flats	80 m
Spring & neap tide range:		3.6 & 0.3 m			

Elsie Island is a 1.5 km long, approximately 500 m wide, 20 m high sandstone island. Beaches K 1272 and 1273 lie 3 km northwest of the island. Beach **K 1272** is a 300 m long strip of high tide sand draped over backing low, tabular sandstone rocks and fronted by 100 m wide fringing reef flats. It terminates at a low point on the south side of which is a narrow gully containing a creek, with 50 m long beach **K 1273** located across the mouth. It is fronted by 50 m wide sand flats and then 100 m wide reef flats all wedged between the bordering 5 m high sandstone.

Beach **K 1274** is located immediately southwest of Elsie Island and has filled the mouth of a V-shaped valley with salt flats, inner beach ridges, and the 1.3 km long beach. The beach is backed by active dunes to 5 m high and extending up to 300 m inland, then by three inner barriers, each separated by mangrove-lined depressions, the total system up to 600 m in width. The beach is bordered in the west by a sandstone headland and in the east a cuspate foreland which it shares with beach K 1275. It is fronted by tidal sand shoals to either end. Beach **K 1275** lies on the eastern side of the foreland, formed from wave refraction around Elsie Island, with the tip of the foreland 400 m from the island. The protected beach is fronted by 400 m wide tidal sand flats, with a mangrove-lined tidal creek, flowing out of a bedrock gully at the southern end of the beach. It faces east and is well exposed to the trade winds, which maintain active dunes climbing to 15 m and spilling over onto beach K 1274.

Beach **K 1275A** is located 500 m to the east and due south of Elsie Island. It is an 80 m long pocket of sand wedged in between low rocky points and with backing rocky slopes rising to 60 m. The beach narrows towards low tide and is backed by a low narrow patch of grassy dune.

K 1276-1283 **CAPE ST LAMBERT (W)**

No.	Beach	Rating HT	LT	Type	Length
K1276	Eric Is (W)	2	3	R+low tide terrace	1.4 km
K1277	Eric Is (E1)	2	3	R+ridged sand flats	200 m
K1278	Eric Is (E2)	2	3	R+tidal sand flats	150 m
K1279	Eric Is (E3)	2	3	R+rock flats	200 m
K1280	Cape St Lambert (W2)	3	4	R+low tide bar/rips	1.6 km
K1281	Cape St Lambert (W1)	2	3	R+low tide terraces	1.8 km
K1282	Cape St Lambert (E)	2	3	R+low tide bar/rips	4 km
K1283	Berkeley R (W)	1	2	R+tidal sand shoals	1.8 km
Spring & neap tide range:	4.3 & 0.4 m				

Cape St Lambert is a 30 m high ridge of sandstone, fronted by rock flats and partly covered in active dunes. Between the cape and Eric Island, 6 km to the northwest, is a northwest-trending shoreline containing three larger (K 1276, 1280 and 1281) and three small (K 1277-1279) beaches. To the east of the cape a continuous beach (K 1282-1283) extends for 6 km to the mouth of the Berkeley River.

Beach **K 1276** is located 1 km due west of **Eric Island.** The beach is 1.4 km long, faces northeast and has a slightly convex southern foreland owing to wave refraction around the island. The beach is exposed to easterly waves averaging up to 1 m which maintain a continuous low tide bar, backed by active dunes blowing off the beach and extending up to 600 m inland and climbing up onto the backing 40 m high sandstone slopes (Fig. 4.119). A joint-controlled creek drains along the back of the northern dunes, crossing the northern end of the beach, with sloping sandstone to either end of the beach.

Beach **K 1277** is located 1 km to the south at the mouth of a small joint-aligned creek, and adjacent to two other creeks whose combined sand flats fill a 400 m wide embayment. The 200 m long beach is wedged between sloping 30 m high sandstone headlands and faces north across 400 m wide ridged sand flats, with the creek draining across the eastern end of the beach.

Beach **K 1278** lies 1 km to the east and is a 150 m long pocket of sand located at the mouth of a north-trending joint-aligned creek. The beach forms a small barrier, with the creek draining across the eastern end, and 400m wide sand flats filling the remainder of the small bay, which is bordered by sloping rock flats and sandstone headlands. Beach **K 1279** lies in the next small bay 300 m to the east. It is a curving 200 m long pocket of sand, being fed in part from dunes that have blown over from its eastern neighbour beach K 1280.

Figure 4.119 Section of beach K 1276 showing the sand dune climbing over the red sandstone rocks.

Beach **K 1280** is a more exposed 1.6 km long east to northeast-facing beach. It receives higher easterly waves, which maintain a 100 m wide low tide bar cut by a few rips during winter easterly wave conditions. It is bordered by 20 m high small sandstone headlands to either end, and backed by a dune sheet that has extended up to 1 km inland and climbed the backing sandstone slopes to heights of 60-80 m, as well as crossing the northern headland to spill onto beach K 1279. A dammed creek backs the southern part of the beach, forming a 100 ha salt flat and mangrove wetland, linked by a meandering 500 m long creek that runs along the side of the southern headland to reach the shore.

Beach **K 1281** commences on the southern side of the 100 m wide headland and trends southeast for 1.8 km to Cape St Lambert. Rock and coral reefs extending up to 1 km north of the cape afford some protection from easterly waves, lowering them to less than 1 m. These maintain a 100 m wide low tide bar, usually free of rips. The beach is backed by an active 200 m wide sheet of sand, backed in turn by older vegetated dunes rising to 30 m.

Beach **K 1282** commences on the eastern side of the cape and curves to the southeast for 4 km to a sandy foreland. The low gradient beach is exposed to easterly waves averaging about 1 m, which maintain a 100 m wide bar, cut by low tide rips spaced every 100 m. It is backed by an unstable foredune and active dunes including two large parabolic dunes toward the northern end, which extend up to 500 m inland and climb to just over 100 m in height, with a steep transgressive dune ridge dropping onto the backing lower bedrock terrain.

At the foreland the shoreline turns and trends south for 1.8 km as a crenulate recurved spit (beach **K 1283**) to the 200 m wide mouth of the Berkeley River, one of the larger rivers of the northern Kimberley. The low beach is backed by dunes increasing in height to several metres in the north. It is fronted by migratory sand shoals and channels which narrow to the south.

Map 7: KIMBERLEY

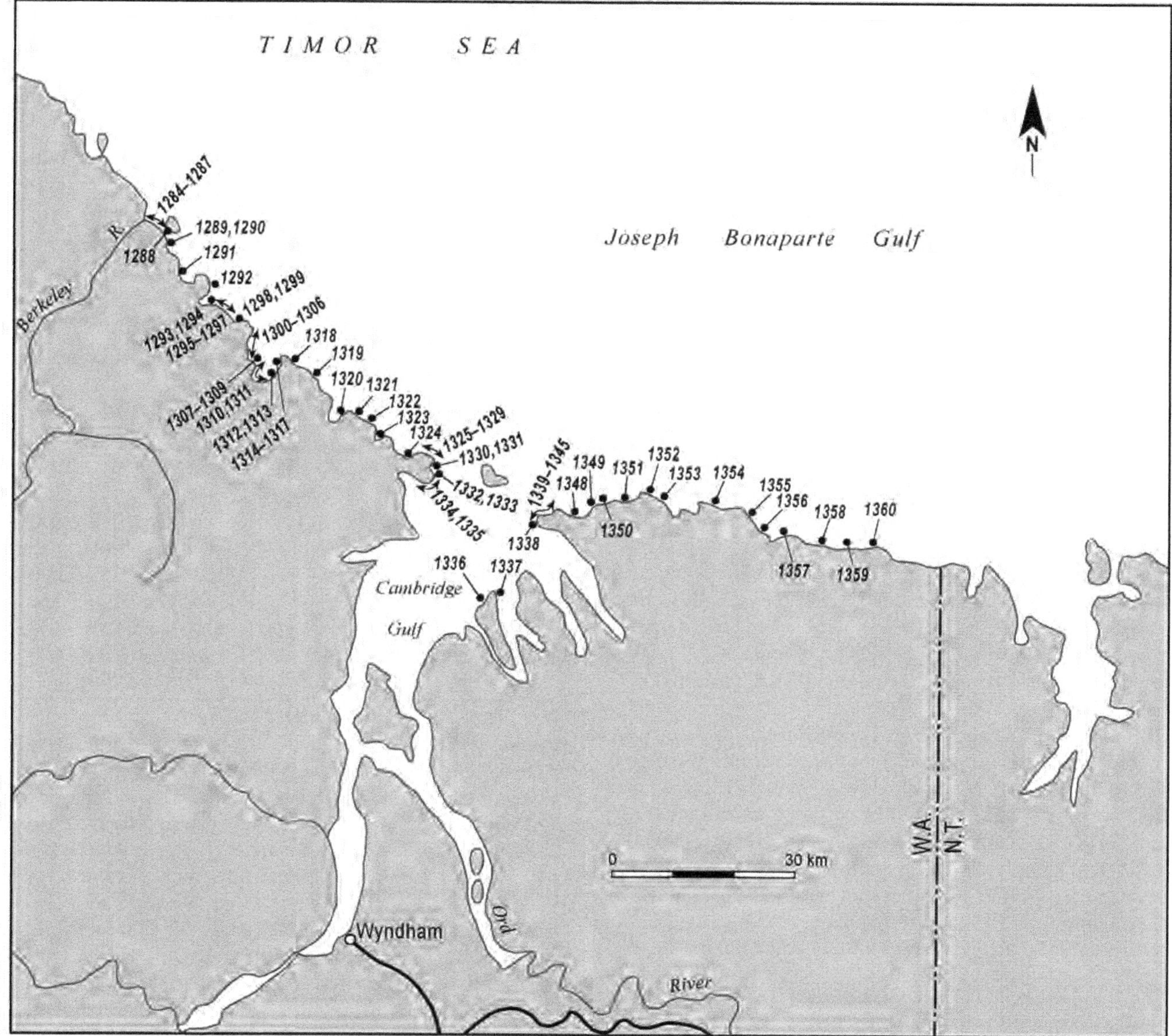

Figure 4.120 Kimberley map 7, locating beaches K 1284-1360.

K 1284-1290 **BERKELEY RIVER (E)**

No.	Beach	Rating HT	LT	Type	Length
K1284	Berkeley R (E1)	1	2	R+tidal sand flats	400 m
K1285	Berkeley R (E2)	1	2	R+sand/rock flats	150 m
K1286	Berkeley R (E3)	1	2	R+sand/rock flats	300 m
K1287	Berkeley R (E4)	1	1	R+tidal sand flats	1.5 km
K1288	Reveley Is (W1)	1	1	R+tidal sand flats	600 m
K1289	Reveley Is (W2)	2	3	R+low tide terrace	1.5 km
K1290	Reveley Is (W3)	2	3	R+sand/rock flats	100 m
Spring & neap tide range:		4.3 & 0.4 m			

The **Berkeley River** flows through a bedrock-bounded funnel-shaped estuary to reach the coast 5 km south of Cape St Lambert, and 3 km west of Reveley Island. The mouth is partly blocked by the sand delivered by the river to the coast during the wet season (Fig. 4.121).

The eastern side of the Berkeley River mouth consists of a 1 km long series of low sandstone points and rock flats, with beaches K 1284 and 1285 located between the rocky points. Beach **K 1284** is a curving 400 m long high tide beach facing north across 300 m wide sand flats and then the deeper river mouth. Rocks outcrop along the beach which is backed in the south by partly active dunes rising to 10-15 m. It is separated from beach **K 1285** by a low 100 m long sandstone point. This beach consists of a 150 m long high tide north-facing beach bordered by low rocky points at either end and fronted by 200 m wide sand flats, then the tidal sand shoals and channels of the river. Dunes from beach K 1286 spill onto the eastern end of the beach. Beach **K 1286** is located immediately to the east and is a 300 m long northeast-facing beach, which receives slightly higher waves which maintain a 200 m wide sand flat. It is exposed to the easterly winds which have blown sand up to 200 m across the backing low sandstone terrain and to heights of 10-15 m.

Figure 4.121 The Berkeley River mouth, with beaches K 1283-1277 located along its western banks (right).

Immediately to the south is beach **K 1287**, a curving 1.5 km long low energy beach composed of white sand. It is fronted by 500 m wide tidal sand flats and scattered mangroves. While the beach faces east, it is protected from waves by 3 km long Reveley Island located 2 km due east, and extensive tidal shoals between the island and the mainland. The beach is backed by a grassy foredune and several older recurved spits with a 50 ha wetland in the north and 40 m high cliffs to the south.

The cliff continues to the next embayment, a 2 km wide system that faces northeast to Reveley Island. The bay is fed by three tidal creeks which have each filled V-shaped bedrock-controlled, steep-sided valleys each containing salt flats and mangroves. A low strip of beach ridge (**K 1288**) links the two western valley mouths to form a very low energy 600 m long beach (Fig. 4.122), fronted by 2 km wide tidal flats and tidal channels that extend out to the island. Mangroves border either end of the beach.

Figure 4.122 Beach K 1288 is located across the mouth of a narrow infilled joint-aligned valley. Shown here at low tide with tidal sand flats exposed.

Beach **K 1289** is located in the next embayment 1.5 km to the east. It occupies the mouth of a 1.5 km wide V-shaped valley, lined by 100 m high sandstone ridges. The curving 1.5 km long east to northeast-facing beach is tied to the base of the ridges in the north. In the centre it has prograded seaward as a series of 12 recurved spits that grade into a low outer 200 m wide grassy foredune ridge plain, terminating at a 5-6 m high active foredune. The ridges impound a 50 ha wetland containing salt flats with mangroves along the creek, which drains out at the southern end of the beach. The beach is low gradient, with waves breaking across a continuous 100 m long low tide bar. At its western end it grades into a boulder beach, backed by slopes which host a few boab trees.

Beach **K 1290** is a 100 m long pocket of sand located 300 m north of the creek mouth, at the base of 100 m high steep slopes. It is fronted by 100 m wide sand and rock flats.

K 1291-1294 **BUCKLE HEAD**

No.	Beach	Rating HT	Rating LT	Type	Length
K1291	Buckle Hd (W)	1	1	R+tidal sand flats	800 m
K1291A	Buckle Hd (1)	1	1	R+sand/rock flats	200 m
K1291B	Buckle Hd (2)	1	1	R+rock flats	200 m
K1292	Buckle Hd (E1)	2	3	R+rock flats	250 m
K1293	Buckle Hd (E2)	1	1	R+tidal sand flats	100 m
K1294	Buckle Hd (E3)	1	2	R+sand/reef flats	600 m
Spring & neap tide range:		4.3 & 0.4 m			

Buckle Head is a 122 m high 400 ha sandstone-capped plateau, surrounded by steep sides and cliffs and linked to the mainland in the south by salt flats. Two embayments lie to either side of the head. The western embayment is a 2 km wide, north-facing, U-shaped bay, bordered by 120-140 m high cliffs along either side, with 1-2 km of tidal flats filling the southern end, which are drained by three tidal creeks. Beach **K 1291** extends for 800 m from the southern cliffline, southwest to the main western creek mouth. It is backed for the most part by steep slopes rising up to 100 m, together with a small upland creek cutting across the centre of the beach. The 100 m wide creek abuts the cliffs in the west with 1 km of mangrove-lined shoreline connecting it to a 10 ha, 70 m high knoll, then another 1 km of mangroves connecting to the western base of Buckle Head. The beach and entire southern section of the bay is filled with 500 m wide tidal sand flats and tidal shoals and channels with refracted easterly waves breaking across the edge of the flats.

Beach **K 1291A** is located on the northeastern tip of Buckle Head at the base of the 100 m high escarpment. It is a 200 m long beach backed by steeply rising vegetated slopes, with some boulders littering the eastern end of the beach. Beach **K 1291B** lies 2 km to the south on the eastern side of the head and faces due east. It is a 200 m long sand beach bounded by rocks, with a small creek at the southern end and a patch of grassy foredune at the northern end. A small dark cobble-boulder beach is located 200 m to the west tucked inside the eastern tip of the head.

Beaches K 1292-1294 are located to the south of the head in the eastern embayment, which is 1 km wide, 3 km deep and faces northeast. It is bordered by two high knolls and Buckle Head on the west and a 100-130 m high cliffline to the east. It is connected by a narrow 3 km long tidal creek to the western bay.

Beach **K 1292** is located on the southeastern corner of Buckle Head with slopes rising steeply to 100 m behind the beach. It is a curving, low gradient, 250 m long, east-facing strip of high tide sand, fronted by 200 m wide rock flats, bordered by mangroves and backed by wooded slopes. Beach **K 1293** is located 2.5 km to the south on the southeastern shore of the bay. It is a 100 m long north-facing pocket of sand that occupies the mouth of a steep upland creek. It is fringed by mangroves, fronted by 200 m wide sand and gravel flats that grade into subtidal mud flats, and bordered by 100 m high cliffs.

Beach **K 1294** lies 1 km due south of beach K 1292 just inside the northeastern entrance to the bay. It is a steep, 600 m long north-facing beach with a small creek backing the centre and beachrock outcrops the length of the beach. It has tabular red sandstone backing the beach and sloping vegetated headland to either end, while it is fronted by sand flats that widen to the east in lee of a low rocky islet 300 m off the beach, that is connected to the shore by sand and reef.

T 1295-1299 **BUCKLE HEAD (E 1)**

No.	Beach	Rating HT	LT	Type	Length
K1295	Buckle Hd (E4)	1	2	R+rock/reef flats	400 m
K1296	Buckle Hd (E5)	1	1	R+tidal sand flats	150 m
K1296A	Buckle Hd (E6)	2	3	R (boulder)	100 m
K1297	Buckle Hd (E7)	1	1	R+tidal sand flats	400 m
K1298	Buckle Hd (E8)	1	1	R+tidal sand flats	400 m
K1299	Buckle Hd (E9)	1	2	R+sand/reef flats	200 m
Spring & neap tide range:		4.3 & 0.4 m			

To the east of Buckle Head is an embayment containing beaches K 1292-1294, followed by 10 km of southeast-trending rocky shore to the next embayment. Beaches K 1295-1299 are located in small gaps towards the centre of the rocky section.

Beach **K 1295** is an irregular, north-facing, 400 m long, high tide beach located at the base of sandstone slopes rising steeply to 60 m. A protruding central rock outcrop divides the high tide beach in two, with rocks also outcropping along the beach, while it is fronted by a mixture of 100-200 m wide rock and outer reef flats. Immediately to the east is a deeply incised north-trending valley with steep red 80 m high valley walls. Beach **K 1296** partially blocks the outer end of the valley and is backed by a narrowing 1 km long mangrove forest. A meandering tidal creek flows through the mangroves and drains across the western side of the beach. The steep, 150 m long beach occupies the rest of the space, together with a low grassy overwash plain on the eastern side. Five hundred metre wide tidal flats and ridges fill the outer end of the valley, with waves averaging over 0.5 m off the edge of the flats (Fig. 4.123).

Figure 4.123 Beach K 1296 occupies the mouth of a narrow elongate valley (see book cover), with a wide low gradient intertidal sand beach.

Beach **K 1296A** lies 500 m to the east in a similar, narrower valley occupied by a steep 100 m long boulder beach. The beach is backed by a narrow mangrove forest, and fronted by intertidal boulders, then 50 m wide rock flats.

Beaches K 1297-1299 are located 2 km to the east at the junction of the next major bedrock creek (Fig. 4.124). Beach **K 1297** is a steep, 400 m long northwest-facing beach, backed by partly active dunes that have blown up to 500 m inland and to heights of 15 m. It has a steep shelly-sandy beach face with beachrock outcropping in places, and is fronted by 500 m wide ridged sand flats. It is bordered by 60 m high sandstone slopes to the west and a low tabular rocky point to the east. Beach **K 1298** is located adjacent to the mangrove-lined creek mouth, 500 m south of beach K 1297 and at the base of the rocky point. It curves to the east of the creek mouth for 400 m to a vegetated 20-40 m high sandstone headland. It is backed by partly active dunes extending about 100 m inland and is fronted by tidal sand flats of the creek mouth that extend 700 m out beyond the headland. Beach **K 1299** lies on the eastern side of the headland and is a 200 m long pocket of west-facing sand, backed by a solitary blowout extending up to 400 m inland and rising to 10-15 m. It is fronted by 200 m wide sand flats, which merge with those of its neighbour, then another 200 m of reef flats.

K 1300-1306 **BUCKLE HEAD (E 2)**

No.	Beach	Rating HT	LT	Type	Length
K1300	Buckle Hd (E10)	1	2	R+rock/reef flats	150 m
K1301	Buckle Hd (E11)	1	2	R+rock/reef flats	200 m
K1302	Buckle Hd (E12)	1	2	R+mud flats	50 m
K1303	Buckle Hd (E13)	1	2	R+mud flats	150 m
K1304	Buckle Hd (E14)	1	1	R+mud flats	200 m
K1305	Buckle Hd (E15)	1	1	R+tidal sand flats	300 m
K1306	Buckle Hd (E16)	1	1	R+tidal sand flats	400 m
Spring & neap tide range:		4.3 & 0.4 m			

Figure 4.124 Beaches K 1297-1299 are three adjoining pockets of sand wedged in between the rocky shore and fringing coral reefs.

Ten kilometres southeast of Buckle Head is a 5 km wide northeast-facing embayment, which contains 20 km of irregular shoreline and is fed by four medium sized upland to tidal creeks. Beaches K 1300-1306 occupy the western side of the bay (Fig. 4.125), beaches K 1307-1311 the southwest base, and beaches K 1312-1317 the eastern side.

Figure 4.125 Beaches K 1300-1306 consist of small pockets of sand along a rock- and reef-dominated shoreline.

The western shore of the bay consists of a series of southeast-trending sandstone ridges, with a few beaches located in the gaps between the ridges. Beach **K 1300** is located just inside the northwestern tip of the bay. It is a 150 m long southeast-facing beach, backed by sloping ridged-sandstone and bordered by protruding low sandstone points. Easterly winds have blown sand from the beach up to 100 m inland over the sandstone. It is fronted by 200-300 m wide rock flats and fringing reef. Immediately to the south is beach **K 1301**, a 200 m long beach backed by 150 m wide dunes, bordered by sandstone ridges and fronted by some rock outcrops and 100 m wide fringing reef.

Beach **K 1302** lies immediately to the south and is a 50 m long pocket of sand backed by 50 m of partly active dunes, and bordered to the south by a small strip of rock, while it is fronted by 200 m wide mud flats. The mud flats fill the small embayment, which includes beach **K 1303** located on the south side of a joint-controlled tidal creek. This beach is 150 m long, backed by 50 m of climbing dunes and fronted by 400 m wide mud flats.

Beach **K 1304** is located in the next embayment. It is a very low energy 200 m long beach wedged in between 30 m high sandstone ridges. It is backed by 70 m of vegetated low dunes and fronted by 700 m wide mud flats.

Beach **K 1305** is located 1 km to the south and is a 300 m long, east-facing beach, backed by two areas of vegetated dunes. It is fronted by sand flats and mangrove-covered rocks, with fringing reef on its eastern side and mud flats to the west. Five hundred metres to the east is beach **K 1306**, a 400 m long east-facing beach, also backed by up to 300 m wide partly active dunes, which is fronted by 600 m wide sand flats grading into mud flats.

K 1307-1312 **BUCKLE HEAD (E 3)**

No.	Beach	Rating HT	Rating LT	Type	Length
K1307	Buckle Hd (E17)	1	1	R+sand/mud flats	1.6 km
K1308	Buckle Hd (E18)	1	1	R+sand/mud flats	200 m
K1309	Buckle Hd (E19)	1	1	R+sand/mud flats	100 m
K1310	Buckle Hd (E20)	1	1	R+sand/rock flats	500 m
K1311	Buckle Hd (E21)	1	1	R+sand/mud flats	1 km
K1312	Buckle Hd (E22)	1	1	R+sand/mud flats	300 m
Spring & neap tide range:	4.3 & 0.4 m				

The south shore of the embayment extends for 5 km in a southeast direction, with a mangrove-fringed tidal creek forming the western boundary and two converging tidal creeks the eastern boundary. Extensive tidal deposits front both creeks with a 20 m high 1 km wide sandstone headland in the centre separating the two deposits. Beach K 1307 is associated with the western creek, beaches K 1308-1310 are located on the headland, while beaches K 1311 and 1312 lie adjacent to the eastern creek.

Beach **K 1307** is a low energy, 1.6 km long, northeast-facing beach set in lee of 1 km wide sand to mud flats. It is bordered by mangrove-lined creek mouths to either end, with mangroves also scattered along the front of the beach. It is backed by a 200-300 m wide series of beach ridges then gently rising sandstone slopes.

Beach **K 1308** is a 200 m long pocket of sand on the northern side of the low central headland. It is bordered by rock flats and a few stunted mangroves and backed by partly vegetated slopes. Continuous sand then mud flats extend over 500 m off the beach. Beach **K 1309** lies 500 m to the east and is a similar 100 m long beach, with more sand then mud flats owing to its proximity near the creek mouth.

Beach **K 1310** commences 500 m to the east and trends south as the eastern side of a 100 m wide tidal channel. It is located amongst protruding rocks, with 100 m wide rock and sand flats fronting the beach, then the deep tidal channel. The mangrove-lined mouth of the creek commences at its southern boundary.

Beach **K 1311** is the outer shoreline of a 2 km wide tidal wetland, with mangrove-lined creek mouths to either side. It consists of a low, 1 km long, low energy beach ridge-chenier, backed by overwash flats. It is fronted by mud flats, then the tidal channel from the eastern creek, then another 2 km of tidal flats. Beach **K 1312** is located on the eastern side of the eastern creek channel and lies at the base of sloping sandstone. The crenulate beach faces west across some mangrove-covered sand and rock flats, then extensive tidal flats that fill this part of the bay.

K 1313-1317 **BUCKLE HEAD (E 4)**

No. Beach	Rating HT	LT	Type	Length
K1313 Buckle Hd (E23)	1	1	R+mud flats	1.5 km
K1314 Buckle Hd (E24)	1	1	R+mud flats	50 m
K1315 Buckle Hd (E25)	1	1	R+mud flats	100 m
K1316 Buckle Hd (E26)	1	1	R+mud/reef flats	50 m
K1317 Buckle Hd (E27)	1	1	R+mud/reef flats	100 m
Spring & neap tide range:	4.3 & 0.4 m			

The eastern side of the embayment commences at a creek mouth bordered by beach K 1312 and extends northeast as 2 km of tidal flats containing beach K 1313, then north as a 2 km long 20 m high sandstone headland with beaches K 1314-1317 located in gaps along the base of the rocky shore.

Beach **K 1313** is located along the shoreline of the tidal flats, which have filled a 4 km deep, V-shaped valley. The beach varies through time. In the 1970s it consisted of an active 1.5 km long recurved spit (chenier) sitting atop the extensive mud flats, while in the late 1990s it was being eroded and fronted by mud flats, extending over 1 km into the bay.

Beaches K 1314-1317 are all located along the outer section of the eastern headland and all occupy the mouths of small gullies flowing down the side of the sandstone headland. Beach **K 1314** is a 50 m pocket of sand, bordered by rock and rock flats and fronted by 200 m wide mud flats. Beach **K 1315** lies 300 m to the north and is a 100 m long pocket of sand with 100 m wide mud flats. Two hundred metres to the north is beach **K 1316**, a 50 m wedge of sand with fronting 300 m wide mud then reef flats. Finally beach **K 1317** is a 100 m long strip of sand backed by a patch of mangroves and upland gully, and bordered by mangrove-covered rock flats, with sand then mud then reef flats extending 400 m off the beach.

K 1318-1324 **THURBURN BLUFF-BARE HILL**

No. Beach	Rating HT	LT	Type	Length
K1318 Thurburn Bluff (W2)	1	1	R+sand/reef flats	400 m
K1319 Thurburn Bluff (W1)	2	3	Boulder beach	150 m
K1320 Thurburn Hill (S1)	1	2	R+mud flats	300 m
K1321 Thurburn Hill (S2)	2	3	Boulder beach	400 m
K1322 Obstruction Hill (E1)	2	3	Boulder beach	400 m
K1323 Obstruction Hill (E2)	1	1	R+tidal sand flats	500 m
K1324 Hope Hill	1	2	Ultradissipative	2.7 km
Spring & neap tide range:	4.7 & 1.9 m			

Thurburn Bluff is an eroding 50 m high section of sandstone cliffs located toward the northern end of a 30 km long southeast-trending section of bedrock shore that turns south at its southern point, Bare Hill, into the large Cambridge Gulf. Sandstone cliffs up to 150 m high dominate most of the shore. The few beaches along this section are either composed of boulders (K 1319, 1321 &1322), or located deep in protected bays (K 1318, 1320 & 1323), with only beach K 1324 an open coast sandy beach. While the coast faces east into the easterly winds and waves, they are lowered, particularly at low tide, by the Medusa Banks extensive linear tidal shoals that emanate from the mouth of Cambridge Gulf, where they are formed by the strong tidal currents.

Beach **K 1318** is located 5 km northwest of Thurburn Bluff. The curving 400 m long northeast-facing beach occupies a gap in the 20 m high wooded sandstone shoreline. The gap is partly filled with rock outcrops and fringing reef, resulting in a 50 m wide channel connecting the beach at low tide to the ocean. The beach has rocks outcropping in the centre and is fronted by 200 m wide sand flats with a few mangroves on the bordering rock flats.

Beach **K 1319** is located 3.5 km to the southeast and close to the bluff. It is a north-facing 150 m long, steep boulder beach, wedged into the mouth of a narrow 50 m deep, northwest-trending valley. The boulders rise to well above high tide and are fronted by 100 m wide sloping intertidal boulders, then rock flats. Well developed rock platforms lie immediately east of the beach.

Beach **K 1320** is located 8 km southeast of Thurburn Bluff and 2 km north of 213 m high Thurburn Hill. The curving 300 m long beach lies on the eastern side of a 2 km wide embayment, the remainder of the bay dominated by cliffs, rocks and inner mangroves. The beach is composed of coarse material, and fronted by rock flats, then 500 m wide mud flats.

Beach **K 1321** is a 400 m long boulder beach that links two sides of a steep V-shaped valley, 1 km east of beach K 1320. The beach consists of high and inter-tidal boulders, and is bordered by eroding 80 m high cliffs.

Beach **K 1322** is located 5 km to the east on the eastern side of 144 m high Obstruction Hill. The 400 m long beach blocks the mouth of a 1.5 km deep V-shaped valley, which is ringed by steep slopes rising to the 140 m high plateau surface (Fig. 4.126). Salt flats occupy the valley floor, with the beach forming a steep high tide beach backed by boulder washover deposits. It is fronted by intertidal boulders and bordered by eroding 20-40 m high cliffs.

Figure 4.126 Boulder beach K 1322 lies across the mouth of a steep V-shaped valley.

Cambridge Gulf

Cambridge Gulf is a north-facing funnel-shaped embayment bordered by Cape Dussejour on the west and Cape Domett 16 km to the east, with Lacrosse Island located in the centre of the entrance. It extends south for 80 km to Wyndham and another 30 km to the mouth of the Pentecost River, as well as linking with the East Arm of the Ord River, the largest river in the Kimberley, and in flood one of the largest in Australia. In addition to the large river inflow, the upper gulf has a tide range up to 8.7 m at Wyndham, resulting in strong tidal currents flowing into and out of the gulf twice a day. These currents maintain long linear gravel tidal shoals on the bed of the gulf. Around the shoreline the rivers have supplied prodigious amounts of sediment that have built the Ord River delta into a series of extensive tidal salt flats, rimmed by one of the largest mangrove systems in northern Australia. In total the gulf has a shoreline of approximately 320 km, mostly lined with mangroves and backed by salt flats up to several kilometres wide. Only at a few places like Wyndham does the usually steep backing bedrock reach the shore.

Beach **K 1323** lies 2 km to the south at the base of a 500 m wide, 1 km deep, northeast-facing embayment. The 500 m long beach is backed by vegetated slopes rising to a 100 m high plateau. It is bordered by mangroves that extend along both sides of the bay, and is fronted by 1 km wide sand-mud flats that fill much of the bay, with some rock flats toward the mouth and a boulder beach along the northeastern headland.

Beach **K 1324** occupies the base of a 4.5 km wide north-facing embayment at the foot of 178 m high Hope Hill. The 2.7 km long low gradient beach commences in the west as a low gradient sandy ultradissipative beach exposed to waves averaging about 1 m. To the east it becomes fronted by intertidal rock flats and the sand gives way to a high tide boulder beach. The beach is backed by a foredune rising to 6 m, then by steep slopes rising to Hope Hill in the centre and 166 km high Faith Hill to the east.

K 1325-1331 **BARE HILL**

No.	Beach	Rating HT	LT	Type	Length
K1325	Bare Hill (W6)	2	3	Boulder+rock flats	250 m
K1326	Bare Hill (W5)	2	3	Boulder+rock flats	300 m
K1327	Bare Hill (W4)	2	3	Boulder+rock flats	400 m
K1328	Bare Hill (W3)	2	3	Boulder+rock flats	300 m
K1329	Bare Hill (W2)	2	3	Boulder+UD	400 m
K1330	Bare Hill (W1)	2	3	Boulder+UD	300 m
K1331	Bare Hill	2	3	Boulder+rock flats	400 m
Spring & neap tide range:				4.7 & 1.9 m	

Bare Hill is a 72 m high grassy hill that together with neighbouring Cape Dussejour forms the northwestern entrance to 16 km wide Cambridge Gulf. Steep sandstone slopes rising to 90 m extend for 4 km west of the hill. The slopes have been dissected by a few creeks with each valley mouth occupied by a north-facing boulder beach (K 1325-1330) (Fig. 4.127), while at the base of the hill beach K 1331 forms a recurved boulder spit.

Figure 4.127 Boulder beaches K 1325-1328 each block the mouth of a series of small valleys cut in the Proterozoic sandstone of Bald Hill.

Beach **K 1325** is a 250 m long boulder beach that blocks the mouth of a small valley, with a 3 ha salt flat-lagoon behind the boulders. It is fronted by 100 m wide rock flats. An eroding 60 m high sandstone cliff separates it from beach **K 1326,** 400 m to the east. This is a similar 300 m long boulder beach backed by a slightly larger valley with a central 10 ha salt flat-lagoon. It is also fronted by continuous 200 m wide rock flats. Beach **K**

1327 is 400 m long and blocks the third valley, which houses a 10 ha lagoon and salt marsh. It is fronted by 200 m wide rock flats, with a veneer of sand toward the east. Beach **K 1328** is located 300 m to the east, is a 300 m long boulder beach at the mouth of a smaller sloping valley, with only a small 1 ha backing infilled lagoon. It is also fronted by the continuous 200 m wide rock flats.

The shoreline turns past beach K 1328, with beach **K 1329** occupying an inflection in the shore as a curving 400 m long east-facing high tide boulder beach, fronted by a 300 m wide ultradissipative intertidal zone, with rock flats bordering each end, and slopes rising to 60 m behind. Beach **K 1330** is a similar boulder and ultradissipative beach 300 m in length, with steeply sloping Bald Hill behind. Waves average about 1 m on both of the east-facing beaches.

Beach **K 1331** extends from the eastern base of Bald Hill as a 400 m long recurved boulder spit, that has prograded both into the adjoining bay, and about 100 m seaward as a series of boulder beach ridges. It terminates at a mangrove-lined creek mouth at the northern end of the bay. In addition sand flats of the bay front the end of the spit and mangroves.

K 1332-1335 **CAPE DUSSEJOUR**

No.	Beach	Rating HT	LT	Type	Length
K1332	Bald Hill (S)	1	1	R+ridged sand flats	400 m
K1333	Cape Dussejour	1	1	R+ridged sand flats	1.4 km
K1334	Cape Dussejour	1	1	Ultradissipative	200 m
K1335	Lazy Rock	1	1	Ultradissipative	100 m
Spring & neap tide range:		4.7 & 1.9 m			

Cape Dussejour is a rounded 45 m high grassy sandstone headland that forms the northwestern entrance to Cambridge Gulf. There are four beaches (K 1332-1335) located on and immediately north of the cape, while to the south mangrove-lined salt flats dominate the gulf shore.

Beach **K 1332** is located immediately southwest of Bald Hill and 2 km north of the cape. An open 1 km wide bay links the hill and cape. The beach is a low 400 m long high tide beach that emerges from the mangroves, which it shares with the end of the boulder spit (beach K 1331) and curves to the south to a 20 m high headland. The beach is backed by moderately sloping vegetated sandstone, with the bedrock-controlled creek mouth at the northern end. It is fronted by ridged sand flats extending 500 m into the bay.

Beach **K 1333** occupies most of the bay shore. It is a curving 1.4 km long northeast-facing beach that extends from the small headland to the northern side of the cape. Two small tidal creeks drain across the beach, one in the north against the headland and one in the centre. The beach is backed by a few beach to foredune ridges rising to 6 m, then bedrock-controlled 50 ha salt flats. It has a moderate gradient high tide beach, with a few mangroves growing along the southern end, then a 700 m wide series of continuous ridged sand flats (Fig. 4.128).

Figure 4.128 Beaches K 1332 (right) and the main beach K 1333 are fronted by continuous intertidal ridged sand flats.

Beaches K 1334 and 1335 are located on the southeastern side of the cape. Beach **K 1334** is a 200 m long ultradissipative beach, located in a curving embayment, with boulder beaches and rock flats to either side, fronted by intertidal sand flats widening to 500 m at low tide. It is backed by tree-covered slopes rising to 40 m. Beach **K 1335** lies 500 m to the south in the next bay, and is a similar 100 m long beach, also bordered by rock flats which widen to 300 m at low tide. The Lazy Rock lie off the southern rock flats.

Wyndham

Wyndham is the port for Cambridge Gulf and was developed in the 1880s both as a landing for the Halls Creek gold rushes and for the export of east Kimberley cattle. While the gold declined long ago, cattle continue to be exported. In recent years the town has suffered the loss of the abattoir, as well as losing out to the newer Kununurra for some services and industry. The geography of the town is defined by the steep sides of the gulf and the adjacent low tidal flats. As a consequence it consists of three parts. The original port area to the north spread along the base of Bastion Point, together with the hotel and a few shops; the Three Mile area where the newer and largest section of the town is located including the main shopping area, school, hospital and caravan park; and the Five Mile, with the country club and a few houses.

Today the town has a population of 1500 and provides a wide range of services for the locals and surrounding population as well an increasing number of tourists who come to visit Australia's hottest town (hotter on average than Marble Bar) and enjoy the spectacular Five Rivers view from Bastion Point, as well as see the giant crocodile and fish the rich gulf waters.

Ord River Nature Reserve

Area: 79 842 ha
Coast length: 120 km (9995-10 115 km)
Beaches: 2 (K 1336-1337)

The Ord River Nature Reserve occupies part of the 100 000 ha Ord River floodplain. This is an area dominated by extensive supratidal salt flats fringed by mangroves and deep tidal channels. The 120 km of primarily mangrove-fringed shore has only two low beaches both cheniers surrounded by mud and mangroves.

K 1336-1337 **BARNETT POINT**

No.	Beach	Rating HT	Rating LT	Type	Length
K1336	Barnett Pt	1	2	R+mud flats	400 m
K1337	Barnett Pt (E)	1	2	R+mud flats	400 m
Spring & neap tide range:				4.7 & 1.9 m	

Barnett Point is located 22 km southeast of Cape Dussejour in the centre of a series of extensive tidal flats dissected by large tidal creeks. The point is bordered by substantial creeks 0.5-2 km wide to either side, and backed by up to 10 km of inter- and supra-tidal flats. Its exposure to northerly waves has winnowed coarser sand and shells from the intertidal mud flats to build a discontinuous series of cheniers (beaches K 1336 & 1337) around the point, backed by a set of older cheniers, then high tide flats.

Beach **K 1336** lies at the western tip of the point and is a northwest-facing 400 m long chenier, consisting of a low high tide beach backed and bordered by mangroves and fronted by a narrow band of sand then mud flats (Fig. 4.129). A 2 km long series of mangrove-fronted cheniers connect it to beach **K 1337**, a 400 m long convex chenier that faces northeast across the 3 km wide eastern creek mouth. It is also bordered and backed by mangroves and fronted by scattered mangroves and mud flats then the deep tidal channel.

K 1338-1344 **CAPE DOMETT**

No.	Beach	Rating HT	Rating LT	Type	Length
K1338	Cape Domett (S)	1	2	R+rock/mud flats	500 m
K1339	Cape Domett (E1)	1	2	R+rock/mud flats	600 m
K1340	Cape Domett (E2)	1	2	R+rock/mud flats	400 m
K1341	Cape Domett (E3)	1	2	R+rock/mud flats	200 m
K1342	Cape Domett (E4)	1	2	R+rock/mud flats	200 m
K1343	Cape Domett (E5)	1	2	R+rock/mud flats	200 m
K1344	Cape Domett (E6)	1	2	R+rock/mud flats	150 m
Spring & neap tide range:				5.5 & 2.1 m	

Figure 4.129 Beach K 1336 is part of a low chenier fronting extensive mangroves and salt flats in lower Cambridge Gulf.

Cape Domett marks the northeastern entrance to Cambridge Gulf. It is 12 m high and composed of deeply weathered Devonian sandstone. One kilometre east of the cape is conical 77 m high Cone Hill. Mangroves and the massive Cambridge Gulf extend south of the cape, while a series of ten small sand beaches (K 1338-1347) extend from the cape to The Needles, 5 km to the east. All the beaches are backed by some bluffs and wooded sandstone terrain and bordered by irregular intertidal sandstone rock flats, with some rocks outcropping along the beaches, while mud flats line the intertidal zone.

Beach **K 1338** lies immediately south of the cape and faces west across the 22 km wide mouth of the gulf. It is bordered and backed by 10 m high sandstone rocks and bluffs, with outcropping rock increasing along the northern end of the beach. It terminates at the cape, where two lines of rocks extend another 300 m to the north. Beach **K 1339** commences at the cape and trends east for 600 m to a 400 m long section of irregular red 15 m high bluffs and fronting rocks. Lower vegetated bluffs back the beach, with some rocks outcropping toward each end and mud flats off the beach.

Beach **K 1340** commences on the east side of the bluffs and is a 400 m long north-facing beach, backed by irregular lower white, then high red bluffs up to 10 m high in the west decreasing in height to the east. Rocks outcrop along the length of the beach.

Beaches K 1341-1344 are four pockets of sand located on and to either side of a 1.5 km long 10-20 m high rocky section. Beach **K 1341** is a 200 m long pocket of north-facing sand, bordered by irregular intertidal rock flats, backed by low vegetated slopes which increase in height to the east, and fronted by mud flats. Beach **K 1342** lies 300 m to the east and is a 200 m long northwest-facing pocket of sand and small grassy foredune, backed by 5-10 m high vegetated bluffs, with rocky points extending out 50 m to either end. Beach **K 1343** is located 500 m to the east and is a 200 m long strip of high tide sand, backed by low bluffs and woodlands, bordered by rock flat and fronted by mud flats. Beach **K 1344** lies at the eastern end of the rocky section and is a 150 m long strip of high tide sand, backed by vegetated bluffs, and bordered and in part fronted by intertidal rock flats.

K 1345-1349 **THE NEEDLES**

No.	Beach	Rating HT LT	Type	Length
K1345	The Needles (W)	1 2	R+mud flats	2 km
K1346	The Needles (E1)	1 2	R+mud flats	100 m
K1347	The Needles (E2)	1 2	R+mud flats	300 m
K1348	The Needles (E3)	1 2	R+mud flats	2.8 km
K1349	The Needles (E 4)	1 2	R+ridged sand flats	600 m
Spring & neap tide range:	5.5 & 2.1 m			

The Needles is a 60 m high, jagged, sandstone sea stack tied to the mainland by the end of beaches K 1345 and 1346. It is backed by a similar sandstone ridge rising to 94 m. Beaches extend both sides of the rocks, as well as two along their base (K 1346 and 1347).

Beach **K 1345** is a straight, 2 km long, north-northwest-facing beach, that is tied to The Needles at its eastern end. It is backed by 20 m high red scarps in the east, grading into densely wooded bluffs, which decline in height to the east and are replaced by a grassy foredune. At the very eastern end 20 m high transgressive dunes from beach K 1346 back the beach. Sand flats with a few scattered rocks at each end front the beach. Beach **K 1346** is located on the southern side of The Needles and is a 100 m long pocket of east-facing sand, wedged between jagged sandstone rocks. Continuous sloping beachrock is exposed along the upper beach face. Because of its aspect the easterly winds have blown sand 200 m in behind the beach, and up to 20 m high over the backing wooded terrain.

Beach **K 1347** is located 1 km to the southeast, at the southern base of the sandstone section. It is bordered by steeply rising sandstone to the west and low linear rock flats to the east. The low gradient 300 m long beach faces north and is backed by low wooded terrain and fronted by sand to mud flats. Mangroves extend to the east for 1 km to the beginning of beach K 1348.

Beach **K 1348** is a 2.8 km long north-northwest-facing beach that emerges from the mangroves in the west and is bordered by tidal sand shoals associated with a 100 m wide creek in the east. Mud flats however front most of the beach. It is backed by a 1 km wide series of up to 30 low beach ridges, with trees lining the ridge crests and the mangrove-lined creek mouth to the east. Beach **K 1349** is located on the eastern side of the creek mouth, and is a crenulate 600 m long beach that is part of a recurved spit that forms the eastern creek mouth. It is tied to low bluffs in the north and merges into mangroves to the south in the creek mouth. It is fronted by several ridged sand flats, then the tidal channel. There is vehicle access to the backing bluffs.

K 1350-1353 **THE NEEDLES (E)**

No.	Beach	Rating HT LT	Type	Length
K1350	The Needles (E5)	1 2	Ultradissipative	1.5 km
K1351	The Needles (E6)	1 2	Ultradissipative	3.2 km
K1352	The Needles (E7)	1 2	R+tidal sand flats	2 km
K1353	The Needles (E8)	1 2	Ultradissipative	3 km
Spring & neap tide range:	5.5 & 2.1 m			

Five kilometres east of The Needles is a second low wooded 1.5 km long bedrock section, the last bedrock on the coast before the Northern Territory border 50 km to the east, with the next bedrock located on the Northern Territory's Pearce Point, 95 km to the northeast. In between are the extensive tidal deposits associated with the mouths of the Keep, Victoria and Fitzmaurice rivers, with a total shoreline length of 560 km.

Beach **K 1350** fronts the bedrock section. It is a double crenulate 1.5 km ultradissipative beach, with a slight central foreland tied to rock flats, with rock flats also bordering each end. In between the beach extends over 200 m at low tide. It is backed by some low vegetated dunes, with vehicle access to the western end.

Beach **K 1351** commences immediately east of the bluffs. The 3.2 km long beach faces northwest, and is fronted by a 300 m wide ultradissipative intertidal beach with a small creek crossing the western end. It is backed by a 150 m wide deflated surface, then extensive beach to foredune ridge plain up to 4 km wide along the eastern creek and containing up to 40 ridges (Fig. 4.130). The innermost ridges commence as several recurved cheniers, which widen to the north. There is a distinct change in form about half way across the ridges, which may be a result of a relative fall in sea level. The outer ridges have been exposed to higher energy conditions and are more foredunes, including a few minor vegetated blowouts. The eastern side of the beach and ridges is bordered by a 1 km wide funnel-shaped creek mouth.

Figure 4.130 Beach K 1351 is a low gradient ultradissipative beach backed by a 4 km wide beach-foredune ridge plain.

Beach **K 1352** is located along the western side of the creek and abuts the end of several of the outer ridges. The beach is up to 2 km long and fronted by sand flats, then tidal channel, with mangroves bordering the southern end. The shape and length of the beach varies over periods of years in response to the changing tidal mouth shoals and the presence and extent of bordering mangroves.

Beach **K 1353** is located on the eastern side of the 1 km wide creek mouth. It trends to the east for 3 km to the western end of the next creek mouth. The beach is ultradissipative, with a 200 m wide intertidal zone. It is bordered by the two wide creek mouths and their tidal deposits, and backed by an irregular series of older beach-foredune ridges. The beach is eroding into the foredune in the west, while backed by mangroves and salt flats in the east.

K 1354-1357 **PELICAN ISLAND (S)**

No.	Beach	Rating HT	LT	Type	Length
K1354	Pelican Is (S1)	1	2	R+mud flats	8.7 km
K1355	Pelican Is (S2)	1	2	R+mud flats	2.8 km
K1356	Pelican Is (S3)	1	2	R+mud flats	1 km
K1357	Pelican Is (S4)	1	2	R+mud flats	1 km
Spring & neap tide range:		5.5 & 2.1 m			

Pelican Island is a 15 m high, few hectares in area, sedimentary island, located 8 km north of beach K 1357. The shoreline to the south and east becomes increasingly low energy, as shoreline gradient decreases, the width of the tidal flats increases causing wave energy to drop toward the three river mouths. The beaches along this section respond with all being fronted by wide mud flats and bordered by large tidal creeks. They are backed by salt flats up to 10 km wide.

Beaches K 1354-1356 are part of a near continuous 14 km long north-facing barrier island composed of chenier through foredune ridges. Beach **K 1354** commences 7 km east of beach K 1353, on the eastern side of a 2 km wide inlet channel. The barrier system has prograded northward during the Holocene and consists of an inner and outer barrier, the inner composed of crenulate, discontinuous recurved-chenier spits, the outer a mixture of inner spits and cheniers, grading into outer foredune and some dune activity. The entire outer system is being eroded in the west with sand moving toward the east. The modern beach is 8.7 km long and consists of the high tide sand and 200-500 m wide intertidal mud flats.

Beach **K 1355** is a discontinuous, 2.8 km long, eastward-migrating section of the barrier. It consists of a low-overwashed high tide beach, fronted by a mixture of sand and mud flats truncated by a few small tidal creeks. It is backed by approximately 1 km of salt flats, then the inner barrier. Beach **K 1356** is located at the distal end of the inner barrier, that has not as yet been overlapped by the outer barrier. It is a 1 km long northeast-facing narrow strip of high tide sand, which also forms the western side of a 500 m wide creek mouth. It is backed by mangroves, then 8 km wide salt flats, and bordered in the east by the tidal channel. In time it will be fronted by the outer barrier, and become an inactive ridge.

Beach **K 1357** is located on the eastern side of the creek mouth, 2 km east of the beach K 1356. It is a 1 km long, low recurved chenier-spit backed by a similar inner system. It borders the creek channel in the west with sand, then mud flats, fronting the beach.

K 1358-1360 **PELICAN ISLAND (E)**

No.	Beach	Rating HT	LT	Type	Length
K1358	Pelican Is (E1)	1	2	R+mud flats	1.3 km
K1359	Pelican Is (E2)	1	2	R+mud flats	1.1 km
K1360	Pelican Is (E3)	1	2	R+mud flats	6 km
Spring & neap tide range:		5.5 & 2.1 m			

The three easternmost beaches on the Kimberley coast begin 12 km southeast of Pelican Island. They extend east for a total of 9 km, with beach K 1360 terminating 10 km from the Western Australia-Northern Territory border.

Beach **K 1358** is a relatively straight, north-facing, 1.3 km long, low beach, fronted by 600 m wide mud flats, and backed by a low vegetated inner chenier, then 12 km wide high tide salt flats. It is bordered to either end by small tidal creeks. Beach **K 1359** is located to the east of the eastern tidal creek. It is a similar 1.1 km long beach ridge, the outermost of a few recurved chenier-spits, with the wide salt flats behind, and 600 m wide mud flats to the north.

A 200 m wide creek mouth separates it from beach **K 1360**, the easternmost in the Kimberley. This is a 6 km long series of recurved spits that are slowly migrating to the east, causing erosion of the western end. In addition there is some minor dune activity toward the central-eastern end, with two blowouts extending 300 m inland and to heights of 16 m. The eastern end of the barrier consists of a series of seven splayed recurved spits that extend up to 2 km inland. They are backed by more than 20 km of salt flats and fronted by 300-400 m wide mud flats with tidal creeks to either end. To the east of the eastern creek are 10 km of shoreline dominated by tidal creeks, mangroves and mud flats, which extend to the Northern Territory border (Fig. 4.131), with the next beach (NT 1) beginning just inside the Northern Territory border.

Figure 4.131 The eastern end of the recurved spits that form beach K 1360, the easternmost beach in the Kimberley, beyond which mangrove-fringed mud flats extend 10 km to the Northern Territory border.

5 NORTHERN TERRITORY COAST

The Northern Territory coast can be divided into four coastal regions: the western Joseph Bonaparte Gulf; the northwest coast including Bathurst and Melville islands; the northern Arnhem Land coast, and the eastern Arnhem Land-Gulf coast including Groote Eylandt and a number of lesser islands (Table 5.1 and Fig. 5.1). This chapter provides a region by region description of every one of the 1488 beaches that occupy the mainland coast of the Northern Territory. No island beaches are included.

Table 5.1 Northern Territory coastal regions

	Region	Boundaries	km	Net km	Beaches NT	No. beaches	Beach km & %
1	West coast	WA/NT border-Charles Pt	0-1052	1052	1-110	110	284 (27%)
		Figure 5.2 page 179					
2	Northwest coast	Charles Pt - Cape Cockburn,	1053-2312	1261	111-451	341	358 (28%)
		Figure 5.16 page 192			109-169		
		Figure 5.19 page 195			135-147		
		Bath & Melville Is Fig. 5.26			*Tiwi Is's*	*89+160*	
		Figure 5.29 page 202			170-414		
		Figure 5.53 page 225			413-542		
3	North Arnhem Land coast	Cape Cockburn-Cape Arnhem	2313-3727	1414	452-1053	602	584 (41%)
		Figure 5.53 page 225			413-542		
		Figure 5.68 page 240			543-687		
		Figure 5.84 page 256			688-958		
4	East Arnhem Land-Gulf coast	Cape Arnhem-Qld border,	3727-5029	1302	1054-1488	435	676 (52%)
		Figure 5.107 page 280			958-1069		
		Figure 5.122 page 291			958-1117		
		Figure 5.128 page 298			1073-1275		
		Figure 5.141 page 313			1275-1414		
		Figure 5.154 page 329			1415-1488		
	Arnhem Land	*Bickerton Island Fig. 5.166*	*0-117*	*117*	*BI 1-55*	*55*	
		Groote Eylandt Fig. 5.168	*0-480*	*480*	*GE 1-285*	*285*	
	Total		5029			1488	1902 (38%)

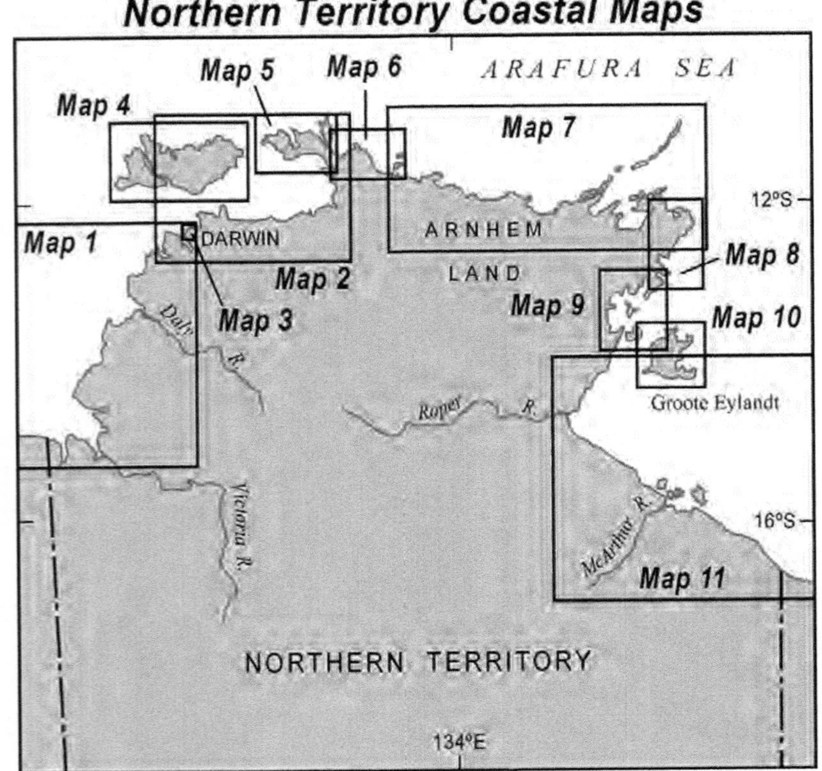

Figure 5.1 Northern Territory coastal regions and coverage of the eleven regional maps.

Region 1 West Coast: Joseph Bonaparte Gulf- Charles Point

Length of coast:	1052 km	(0-1052 km)
Number of beaches:	110	(NT 1-110)
Regional map:	Figure 5.2	
Aboriginal Land:	Daly River-Port Keats; Wagait	
National Parks:	none	
Towns/communities:	Wadeye, Dundee Beach	

The western coast of the Northern Territory commences at the WA/NT border, located roughly in the centre of Joseph Bonaparte Gulf and extends northeast along 1052 km of shoreline to Charles Point. The coastline includes the mangrove-fringed mouths of the Keep, Victoria and Fitzmaurice rivers, which dominate the southeastern shore of the gulf. More exposed shoreline begins at Point Pearce and continues to the north-northeast for 300 km to Charles Point, the northern tip of the Cox Peninsula. Much of the coast is aboriginal land with the main community located 10 km inland at Wadeye, while the only freehold land is at Dundee Beach. This is a relatively low energy coast with the southeast trade winds blowing offshore, resulting in clams at the shore, while the summer northwest monsoonal winds bring low choppy seas. Tides increase from 6.9 m at Darwin to up to 9 m in the lower gulf, amongst the highest in Australia.

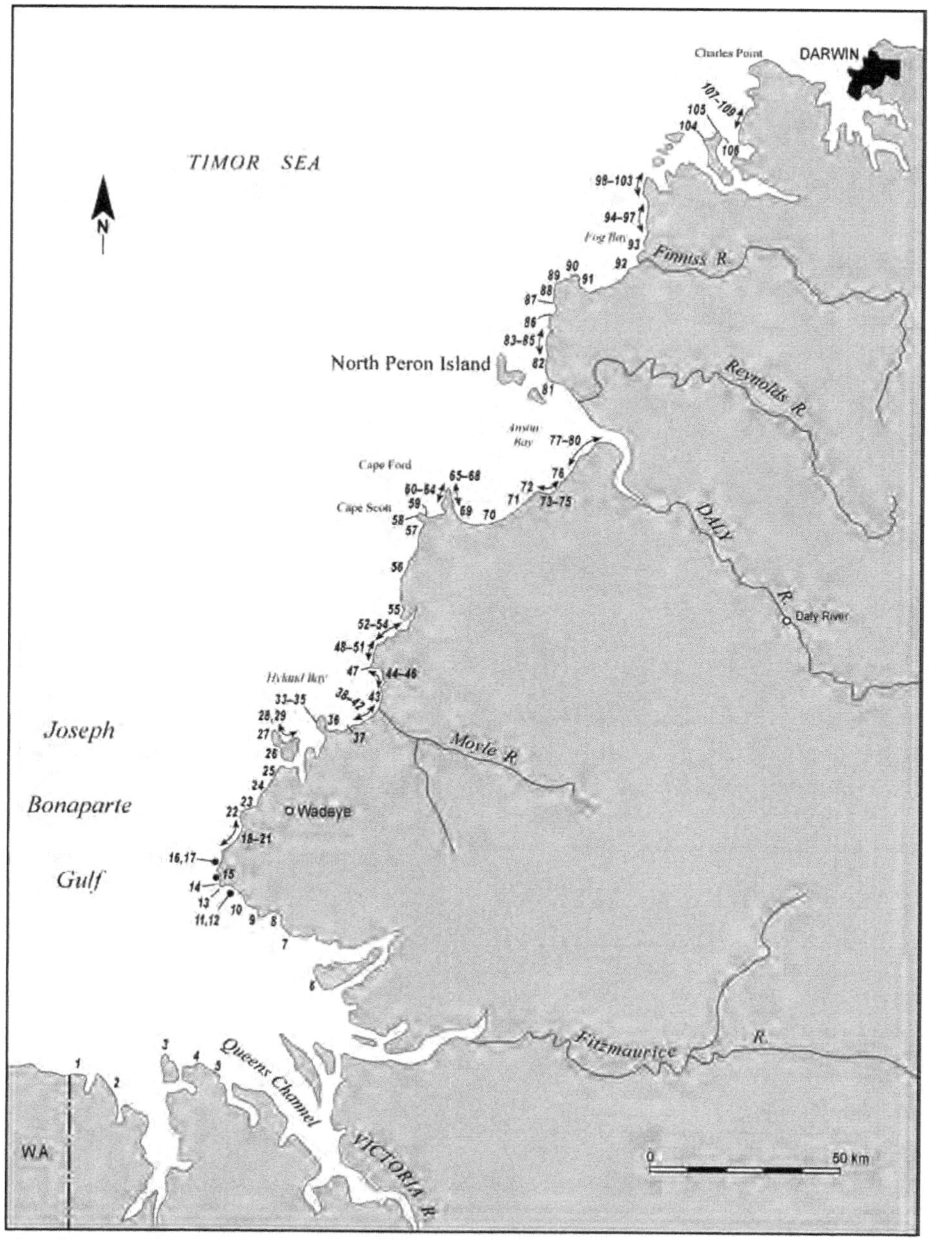

Figure 5.2 Regional map 1: the west coast of the Northern Territory between the WA-NT border and Charles Point, beaches NT 1-110.

NT 1-2 BEACH NT 1 & 2

No.	Beach	Rating HT LT	Type	Length
NT1	Beach NT 1	1 1	R+tidal flats	2.5 km
NT2	Beach NT 2	1 1	R+tidal flats	10.3 km
Spring & neap tide range = 5.2 & 2.0 m				

The most western beach in the Northern Territory is a continuation of a series of low cheniers that extend westward 55 km into Western Australia, and east to the adjoining beach (NT 2). Beach **NT 1** is a low chenier that emerges at the border from the lee of mangroves and extends straight for 2 km before recurving for 500 m into ther mouth of a 200 m wide tidal creek, which forms its eastern boundary. Two smaller tidal creeks which drain backing high tide flats also cross the beach. Beach **NT 2** begins 3.5 km to the east as a 500 m wide series of three vegetated recurved cheniers (Fig. 5.3). The merged chenier extends east for 3 km before narrowing to a beach only, with a 2.5 km long chenier forming the eastern section of the 10.3 km long beach. One and a half kilometre wide tidal creeks border each end of the beach. Both beaches are usually calm, only receiving low waves during onshore summer conditions. The beaches consist of low shelly-sand chenier ridges and are fronted by 1 km wide intertidal sand to mud flats.

Figure 5.3 Beach NT 2 is a low energy beach-chenier fronted by intertidal mud flats, and backed by extensive salt flats.

NT 3-5 TURTLE POINT - FORSYTH CREEK (W)

No.	Beach	Rating HT LT	Type	Length
NT3	Turtle Pt (E 1)	1 1	R+ tidal flats	5.5 km
NT4	Turtle Pt (E 2)	1 1	R+ tidal flats	7.2 km
NT5	Forsyth Ck (W)	1 1	R+ tidal flats	5 km
Spring & neap tide range = 5.2 & 2.0 m				

Beaches NT 1-3 are a series of three cheniers lying between the mouths of the Keep River and Forsyth Creek. The Turtle Point chenier (**NT 3**) starts on the eastern side of the 12 km wide Keep River mouth. It faces north and runs east for 5.5 km. It is cut by two small streams and terminates at the mouth of an 800 m wide tidal creek. On the other side of the creek is a 7.2 km long north-facing gentle convex chenier (**NT 4**) that is crossed by four small creeks. It is separated by 3 km of mangroves from beach **NT 5** that extends southeast for 5 km to the mouth of 1 km wide Forsyth Creek, with two smaller creeks cutting through the chenier. All three cheniers are low, averaging 3 m in height, and alternate between vegetated ridges and low washovers and small creeks. They are backed in places by older cheniers and recurved spits. These in turn are backed by high tide mud flats between 8 and 15 km wide, which are also cut by mangrove-fringed tidal creeks.

Bradshaw Field Training Area

Area: 817 000 ha
Coast length: 140 km (NT 205-345 km)
Beaches: none

The Bradshaw Field Training Area covers an extensive area of hinterland and coastal land between the Victoria and Fitzmaurice rivers. All the coastal land is dominated by tidal flats and mangroves.

Daly River - Port Keats Aboriginal Land

Coast length: 425 km (345-770 km)
Beaches: 75 (NT 6-80)

The Daly River - Port Keats Aboriginal Land occupies 410 km of coast between the mouth of the Fitzmaurice River and the Daly River in Anson Bay. This section contains 75 beaches, with the community of Wadeye (formerly Port Keats) located a few kilometres inland of beaches NT 23-25. Wadeye lies 172 km by road from the Stuart Highway, and much of the coast, while accessible by 4WD tracks, is remote and utilised only by the local aboriginal community. Nine small communities and outcamps are located along the coast, most in lee of beaches.

NT 6 WHALE FLAT

No.	Beach	Rating HT LT	Type	Length
NT6	Whale Flat	1 1	R+ tidal flats	8 km
Spring & neap tide range = 5.8 & 2.2 m				

Whale Flat is the name applied to a 25 km long, up to 15 km wide, high tide mud flat and tidal creek lying between the mouths of Keyling and New Moon inlets. Beach **NT 6** extends for 8 km along the outer western edge of the northern end of the tidal flats, as a low, narrow, discontinuous chenier, cut by one 50 m wide creek and smaller drainage channels. It is fronted by 1 km wide intertidal sand to mud flats, and backed by the 15 km wide high tide mud flats and tidal creeks that extend to the base of the 100 m high escarpment of the Macadam Range.

NT 7 FOSSIL HEAD

No.	Beach	Rating HT LT	Type	Length
NT7	Fossil Head	1 1	R+ tidal sand flats	100 m
Spring & neap tide range = 5.8 & 2.2 m				

Fossil Head is the first occurrence of bedrock on the coast since Cape Domett, in Western Australia, 120 km to the west. The head consists of a 200 m long section of Permian sedimentary rocks (siltstone and sandstone), rising to a maximum height of 20 m, with Fossil Summit, 3 km to the northeast, reaching a height of 76 m. Abutting the western end of the head is a 100 m long beach (**NT 7**). It has a narrow high tide cobble beach, fronted by a 200 m wide section of intertidal sand, then 1 km wide mud flats, which include a few low mangroves (Fig. 5.4). Just behind the beach is a landing strip which services a small aboriginal outstation located in lee of the mangroves on the northern side of the headland. The head is connected to Wadeye via a 35 km 4WD track. The beach is used for launching small boats at high tide, with the ruins of a jetty also crossing the beach.

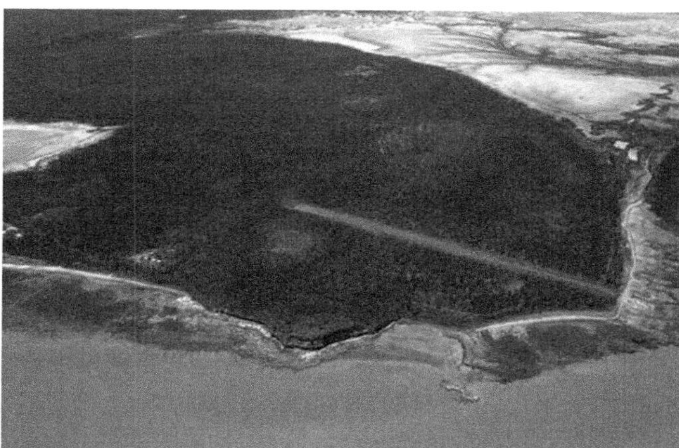

Figure 5.4 Fossil Head with its small pocket beach (NT 7) and backing landing ground.

NT 8-11 TREACHERY BAY

No.	Beach	Rating HT LT	Type	Length
NT8	Treachery Bay (1)	1 1	R+sand flats	800 m
NT9	Treachery Bay (2)	1 1	R+sand flats	500 m
NT10	Treachery Bay (3)	1 1	R+sand flats	800 m
NT11	Treachery Bay (4)	1 2	R+sand/rock flats	100 m
Spring & neap tide range = 5.8 & 2.2 m				

Treachery Bay is an open 15 km wide southwest-facing embayment, with a 20 km long low energy shoreline mainly lined by mangroves, with four small beaches located amongst the mangroves. Beach **NT 8** is an 800 m long west-facing sand beach, bordered by 200 m wide mangrove woodlands, and backed by a 200 m wide beach ridge plain then high tide flats extending 8 km inland. A small creek is located 300 m north of the beach. Beach **NT 9** consists of a curving strip of high tide sand broken by rock flats and mangroves across the intertidal zone. It is backed by a low black soil plain, with a solitary house and vehicle track in lee of the northern end of the beach. Beach **NT 10** is a straight 800 m long sand beach, bordered by mangroves and intertidal rock flats, and backed by six beach ridges. A vehicle track reaches the southern end of the beach. Beach **NT 11** is a small curving 100 m long sand beach located on the north side of a creek mouth. It is part of a recurved spit, which has formed the northern creek boundary. A similar series of beach ridges and recurved spits forms the southern boundary, both systems impounding a 10 km^2 largely infilled tidal creek.

NT 12-15 PEARCE POINT

No.	Beach	Rating HT LT	Type	Length
NT12	Pearce Pt (E2)	1 2	R+rock flats	300 m
NT13	Pearce Pt (E1)	1 2	R+rock flats	700 m
NT14	Pearce Pt (N1)	1 2	R+rock flats	300 m
NT15	Pearce Pt (N2)	1 1	UD	1.7 km
Spring & neap tide range = 5.8 & 2.2 m				

Pearce Point is a low scarped 10 m high bluff composed of weathered Permian siltstone. Woodlands cover the backing low gradient plain. Four low energy beaches (NT 12-15) are located around the point, together with extensive intertidal rock flats and mangroves. Beach **NT 12** is located on the eastern side of the point at the very northern end of Treachery Bay. As the bay mangroves thin out the 300 m long beach continues in lee of the easternmost bluffs. It consists of a narrow high tide beach and a mixture of 200 m wide sand and rock flats together with a scattering of low mangroves. Along the southern side of the point is a curving 700 m long high tide beach (**NT 13**), only open to the bay along the eastern 200 m, where it is backed by the red bluffs and fronted by rock flats, with mangroves fronting the remainder of the beach.

Beach **NT 14** is located on the western tip of Pearce Point. It faces west and consists of a 300 m long high tide beach, backed by a low 50 m wide foredune with several small blowouts. Intertidal rock flats front the entire beach. Beach **NT 15** is located 1.5 km north of the point and is a more exposed west-facing sand beach, consisting of a 200 m wide low gradient intertidal zone, bordered by wider rock flats and backed by low 50-100 m wide active foredune. A vehicle track leads to three houses located behind the northern end of the beach.

NT 16-17 **DITJI BEACH**

No.	Beach	Rating HT LT	Type	Length
NT16	Ditji Beach	1 1	UD	4.1 km
NT17	Ditji (N)	1 1	UD	400 m
Spring & neap tide range = 5.8 & 2.2 m				

Ditji Beach (**NT 16**) is a curving, west-facing 4.1 km long sand beach, bordered by low red bluffs and rock flats, and cut by two small tidal creeks (Fig. 5.5). The entire beach impounds a 250 ha largely infilled tidal creek. A continuous 150 m wide low gradient intertidal bar fronts the beach, while it is backed by low vegetated foredune north of the main creek. South of the creek the dune is active and has blown 500 m inland and to a height of 10 m. A vehicle track reaches the back of the dunes. These dunes represent the first active dune transgression since the Berkeley River 150 km to the west in Western Australia. The northern end of the beach terminates at a low bluff and rock flats, on the northern side of which is 400 m long beach **NT 17**. This is a straight sand beach and 200 m wide low gradient intertidal zone, bordered by rock flats and backed by flat woodland covered plain.

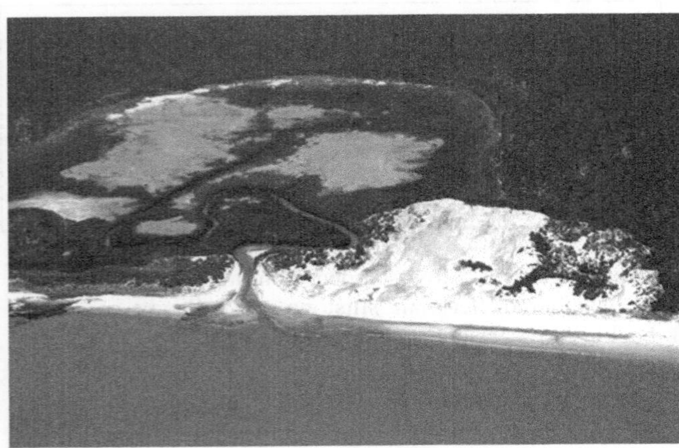

Figure 5.5 Ditji Beach is backed by active dunes, overwash deposits and a largely infilled lagoon.

NT 18-22 **YELCHER BEACH**

No.	Beach	Rating HT LT	Type	Length
NT18	Beach NT 18	1 1	R+LTT	400 m
NT19	Beach NT 19	1 1	R+LTT	1.1 km
NT20	Yelcher Beach	1 1	R+LTB	5.1 km
NT21	Tchindi	1 2	R+tidal shoals	3.7 km
NT22	Tchindi (N)	1 1	UD	1.2 km
Spring & neap tide range = 5.8 & 2.2 m				

Yelcher Beach occupies the centre of an open 12 km wide west-facing embayment, bordered by low bedrock headlands, with a largely infilled 10 km long tidal creek entering the centre of the system. Two rock-bounded beaches occupy much of the southern headland. Beach **NT 18** is a curving 400 m long high tide beach and low tide terrace, backed by a steep 10 m high foredune and bordered by 300 m wide rock flats and mangroves. Five hundred metres to the north is 1.1 km long beach **NT 19**, a straight bedrock-dominated beach, with rock outcropping to form a low bluff along the back of the beach, while intertidal rock flats occupy over half the intertidal area, with sand in between.

Yelcher Beach (**NT 20**) begins inside the northern tip of the low headland, and initially curves to faces west then runs straight for 5.1 km to the dynamic Tchindi Creek mouth. The beach receives low waves (less than 0.5 m) which are sufficient to form a low tide bar system cut by drainage channels. The beach faces into the westerly summer winds and has a moderately unstable foredune system, with dunes extending up to 400 m inland and to a height of 30 m. There is vehicle access to the rear of these dunes and the adjoining beach NT 19. On the north side of the creek mouth is **Tchindi** beach (**NT 21**). This is a 3.7 km long west-facing beach protected by offshore rock reefs and 400-500 m wide tidal shoals. The southern half of the beach is a low series of recurved spits and is overwashed during higher seas. The Tchindi Aboriginal Camping Ground lies behind the centre of the beach, with vehicle access to the camping area and the adjoining beach. Beach **NT 22** lies immediately to the north and is a slightly curving west-facing 1.2 km long low gradient sand beach, bordered by intertidal rock flats and backed by a low foredune.

NT 23-27 **INJIN, MUNDA & DORCHERTY (W)**

No.	Beach	Rating HT LT	Type	Length
NT23	Injin Beach	1 1	UD	6 km
NT24	Untimelli Bch	1 2	R+tidal shoals	800 m
NT25	Munda Beach	1 1	UD	8 km
NT26	Dorcherty (W)	1 1	UD	4.5 km
NT27	Cape Hay (S)	1 2	R+rock flats	2.8 km
Spring & neap tide range = 5.8 & 2.2 m				

Cape Hay forms the northern point of Dorcherty Island. Between the cape and Injin Beach 21 km to the south is an open gently curving embayment containing five generally long beaches (NT 23-27). The main settlement of Wadeye (formerly Port Keats) lies 7 km east of the southern Injin Beach.

Injin Beach (**NT 23**) is a curving 6 km long northwest to west-facing sand beach, bordered by a low siltstone headland and rock flats in the south and a smaller area of mangrove-covered rock flats in the north, with a small tidal creek crossing the centre. The beach receives low waves during westerly conditions and has a 150 m wide intertidal sand bar. Two houses are located at the southern end of the beach, with vehicle tracks from Wadeye reaching both ends of the beach.

On the western side of the rock flats is protected east-facing **Untimelli Beach** (**NT 24**), a narrow, 800 m long, low energy high tide beach, fronted by 800 m wide tidal shoals associated with a small tidal creek, which separates it from Munda Beach. **Munda Beach** (**NT 25**) is a straight 8 km long northwest-facing beach, backed by up to 3 km of 30 low Holocene beach ridges, and fronted by a low gradient 300 m wide intertidal beach. At the northern end a shallow 500 m wide tidal inlet separates it from Dorcherty Island (Fig. 5.6). A vehicle track reaches the northern end of the beach where two houses are located beside the inlet.

Figure 5.6 A wide tidal inlet separates Munda Beach (NT 25, right) from beach NT 26 on Dorcherty Island (left).

Beach **NT 26** extends for 4.5 km along the western side of Dorcherty Island to low bedrock bluffs and rock flats toward the northern end. The southern half is dominated by 400 m wide tidal shoals associated with the main inlet and a second one, while the northern half has a 300 m wide intertidal beach. The entire beach is backed by an irregular 10 km^2 complex of up to 25 beach ridges and recurved spits that extend to Port Keats and form the southern half of the island.

Cape Hay is a low headland composed of weathered Triassic siltstone-shale that forms the northern end of Dorcherty Island. Extending for 2.8 km south of the Cape is a high tide sand beach (**NT 27**), backed by a generally stable foredune that rises to 15 m in the south and is fronted by continuous 200-500 m wide intertidal rock flats. The beach is backed by the foredune, then a 500 m wide mixture of salt flats and three recurved spits.

NT 28-29 **DORCHERTY ISLAND (N)**

No.	Beach	Rating HT	Rating LT	Type	Length
NT28	Dorcherty Is (N)	1	1	R+sand flats	600 m
NT29	Kinmore Pt (W)	1	1	R+sand-mud flats	1.8 km
Spring & neap tide range = 5.8 & 2.2 m					

The north side of **Dorcherty Island** consists of a 4 km wide north-facing bay bordered by Cape Hay and Kinmore Point. Along the southern and eastern shore of the bay are two low energy beaches (NT 28 & 29) fronted by 1 km wide sand to mud flats and separated by a small tidal creek. Beach **NT 28** is a 600 m long high tide sand beach, fringed by mangroves and the tidal creek at its eastern end, backed by a low overwashed foredune, and fronted by the sand then wide mud flats. On the eastern side of the creek is 1.8 km long northwest-facing beach **NT 29**, which consists of a narrow high tide beach fronted by the wide sand to mud flats that fill much of the bay. Mangrove-covered intertidal rock flats form the Kinmore Point boundary. The rock flats consist of fossil-rich Triassic siltstone.

NT 30-35 **PORT KEATS**

No.	Beach	Rating HT	Rating LT	Type	Length
NT30	Kinmore Pt (E1)	1	1	R+sand flats	150 m
NT31	Kinmore Pt (E2)	1	2	R+sand flats	2.3 km
NT32	Tree Pt (W4)	1	1	R+ mud flats	3 km
NT33	Tree Pt (W3)	1	1	R+sand-mud flats	200 m
NT34	Tree Pt (W2)	1	1	R+sand-rock flats	700 m
NT35	Tree Pt (W1)	1	1	R+LTT	600 m
Spring & neap tide range = 5.8 & 2.2 m					

Port Keats is a 20 km long, funnel-shaped, tide-dominated estuary that provides sea access to the Aboriginal community at Wadeye, located at the very tip of the estuary. Towards the mouth of the port it is bordered by the low Kinmore Point on the west and Tree Point 9 km to the east. Mangroves backed by several kilometre wide salt flats and tidal creeks dominate most of the port shoreline. Towards the entrance and in the vicinity of the two low points are six generally low energy beaches (NT 30-35).

Two beaches are located on Dorcherty Island on the eastern side of Kinmore Point. Beach **NT 30** is a 150 m long north-facing sand beach fronted by 500 m wide ridged-sand flats, and bordered by mangroves to the east and intertidal rock flats to the west. Beach **NT 31** faces east into the main Port Keats tidal channel and consists of a curving 2.3 km long beach, initially fronted by 1 km wide sand flats. It narrows to less than 100 m along the southern half of the beach, which is bordered by the deep channel with its strong tidal flows. The beach is backed by a 500 m wide series of beach ridges to recurved spits.

Tree Point forms the eastern entrance to Port Keats. It is a 5 m high, narrow headland composed of Triassic siltstone. Beach **NT 32** begins 8 km south of the point and is a 3 km long west-facing low energy beach, fronted by 600 m wide intertidal mud flats, with mangroves bordering each end and a small tidal creek crossing the centre of the beach. A vehicle track runs from the northern end down to the creek mouth. The remaining three beaches are all located on the west side of Tree Point. Beach **NT 33** is a 200 m long pocket beach bordered by mangrove-covered intertidal siltstone rock flats. A small outstation with six houses backs the northern end of the beach. The high tide beach is fronted by 400 m wide sand to mud flats (Fig. 5.7). Five hundred metres to the north is beach **NT 34**, a 700 m long west-facing, slightly curving sand beach, backed by a low foredune. It is bordered by intertidal rock flats and fronted by a 500 m wide mixture of sand and rock flats. Beach **NT 35** is located at the western tip of the elongated 300 m wide point. The beach is 600 m long and receives slightly higher wind waves (0.5 m), which maintain a moderately steep high tide beach, fronted by a low tide terrace, with rock flats 100 m offshore.

Figure 5.7 Beach NT 33 is located on the western side of Tree Point, and is backed by a small outstation and fringed by extensive mangroves.

NT 36-42 **HYLAND BAY**

No.	Beach	Rating HT	LT	Type	Length
NT36	Hyland Bay (1)	1	1	R+mud flats	100 m
NT37	Hyland Bay (2)	1	1	R+mud flats	2.3 km
NT38	Hyland Bay (3)	1	1	R+LTT	1.6 km
NT39	Hyland Bay (4)	1	1	R+LTT	700 m
NT40	Hyland Bay (5)	1	1	R+LTT	1.4 km
NT41	Hyland Bay (6)	1	1	R+LTT	1.4 km
NT42	White Cliff Pt	1	1	R+sand flats	200 m
Spring & neap tide range = 5.8 & 2.2 m					

Hyland Bay is a 20 km wide, semicircular, northwest-facing bay containing 35 km of shoreline, which is a mixture of low energy mud flats and mangroves, as well as some more exposed beaches backed by red bluffs. The southern 16 km of north-facing shore that extend east of Tree Point contain seven beach systems (NT 36-42).

Mangroves front the shore immediately east of Tree Point for 5 km until beach **NT 36.** The 100 m long beach occupies a gap in the mangroves. It consists of a low energy high tide sand beach, fronted by 800 m wide mud flats bordered on both sides by mangroves as well as some beachrock toward the eastern end. The mangroves here and for 2 km to the east front two previous active recurved spits that impound a small central tidal inlet. Beach **NT 37** begins on the eastern side of the mangroves and also forms part of the eastern beach ridge-recurved spit. The beach is 2.3 km long, faces north and is fronted by mud flats up to 1 km wide, which in the east also grade into mangroves.

Beyond the mangroves is the beginning of a relatively straight, northwest-facing, 5 km long section of more exposed shore backed by 15 km high red bluffs, formed of deeply weathered Permian siltstone and shale. Fronting the bluffs are four near continuous beaches **(NT 38-41)** each bordered by protruding sections of the bluffs and rock flats. Each beach consists of a high tide beach and 200 m wide low tide bar (terrace), forming a thin veneer of intertidal rock flats.

Beach NT 41 terminates at White Cliff Point, a 20 m high white siltstone point capped by red laterite. Tucked inside the northern side of the point is a curving, northeast-facing, 200 m long strip of high tide sand (beach **NT 42**), fronted by 200 m wide ridged sand flats which grade into 1 km wide mud flats. Mangroves form the northern boundary and extend for another 6 km along the bay shore to the mouth of the Moyle River.

NT 43-46 **MOYLE RIVER-HYLAND BAY (N)**

No.	Beach	Rating HT	LT	Type	Length
NT43	Moyle R (N)	1	2	R+mud flats	3.5 km
NT44	Hyland Bay (9)	1	1	R+sand-mud flats	2 km
NT45	Hyland Bay (10)	1	1	R+sand-mud flats	1.2 km
NT46	Hyland Bay (11)	1	1	R+sand-mud flats	2 km
Spring & neap tide range = 5.8 & 2.2 m					

The northern half of Hyland Bay extends from the 200 m wide mouth of the Moyle River along 12 km of relatively open west-facing shore to Cape Dombey, the final 2 km curving to face south. The beaches either side of the Moyle River fill a 6 km wide embayment that has been filled with up to 35 Holocene beach ridges, that have prograded 4 km to the present shoreline. The southern beach ridges are fronted by a continuous 300 m wide band of mangroves. Bordering either side of the beaches are the scarped red bluffs of weathered Permian sand and siltstone.

Beach **NT 43** commences on the northern side of the river mouth and extends north for 3.5 km to the first of the low bluffs. The beach consists of a 50 m wide, high tide sand beach, fronted by intertidal mud flats up to 1.5 km wide, together with the river channel cutting across the shore and mud flats (Fig. 5.8). A vehicle track reaches the northern end of the beach where there is a settlement of several houses.

Figure 5.8 Beach NT 43 is part of a series of cheniers and is fronted by wide intertidal mud flats.

Beach **NT 44** is a 2 km long, west-facing high tide beach, bordered by low beach-fronted bluffs, and backed by a 1 km long largely infilled lagoon, which flows across the northern end of the beach. A section of mangroves grows between the creek mouth and the northern point. The beach is fronted by a narrow strip of sand then mud flats. The two northern beaches (NT 45 & 46) each consist of a bluff-backed strip of high tide sand with fronting mud flats. Beach **NT 45** is 1.2 km long and bordered by protruding rock flats, while beach **NT 46** is 2 km long, with mud flats becoming more dominant, as well as a scattering of mangroves. As beach NT 46 begins to curve round to face south toward Cape Dombey, the 500 m wide mud flats become covered by a 100 m wide stand of mangroves, which extend west to the immediate lee of the cape.

NT 47-53 **CAPE DOMBEY**

No.	Beach	Rating HT	Rating LT	Type	Length
NT47	Cape Dombey	1	1	R+rock flats	1 km
NT48	Cape Dombey (N1)	1	1	R+rock flats	500 m
NT49	Cape Dombey (N2)	1	1	R+sand/rock flats	300 m
NT50	Cape Dombey (N3)	1	1	R+sand/rock flats	1.4 km
NT51	Cape Dombey (N4)	1	1	R+LTT	1.5 km
NT52	Cape Dombey (N5)	1	1	R+sand flats	3 km
NT53	Cape Dombey (N6)	1	1	R+sand flats	1.5 km
Spring & neap tide range = 5.8 & 2.2 m					

Cape Dombey forms the northern boundary of Hyland Bay and consists of 5-10 m high scarped red Permian sedimentary rocks, generally fronted by high tide beach and intertidal rock and sand flats. Beach NT 47 surrounds the cape, while beaches NT 48-53 extend almost continuously for 8 km north of the cape, with each separated by more prominent protrusions in the low bluffs.

Beach **NT 47** is a continuous 1 km long high tide sand beach that curves round the cape, facing initially southeast into Hyland Bay and the adjacent mangroves, then south and finally west, with bedrock outcrops and extensive rocks flats forming each of the inflection points. Beach **NT 48** runs for 500 m from the cape due north between two low bluffs, and is fronted by continuous but irregular intertidal rock flats. Beach **NT 49** is a straight west-facing 300 m long high tide sand beach, fronted by a mixture of sand and rock flats, with rocks also outcropping along the central section of the beach. Beach **NT 50**, while bordered by the low bluffs and rock flats, is generally fronted by 100 m wide sand flats, with a small creek across the northern end of the beach draining a small elongate largely infilled lagoon. Beach **NT 51** is the most exposed of these beaches and consists of the high tide sand beach and a narrow low tide bar, backed by eroded bluffs which grade into rock flats toward the northern end. A straight vehicle track terminates towards the centre of the beach.

At the northern end of beach NT 51 the coast trends northeast for 10 km and contains beaches NT 52 and 53. Beach **NT 52** is a slightly curving, northwest-facing, 3 km long, high tide sand beach, backed by a 1 km long infilled lagoon, with a small creek crossing toward the southern end. There is a sharp break in slope of the base of the beach, which is fronted by 200-300 m wide intertidal flats composed of finer sand. Its northern neighbour, beach **NT 53**, has similar beach morphology, and is backed by bright red 10 m high bluffs. The vehicle track from beach NT 51 runs along the back of the beaches and lagoons and along the top of the red bluffs to the northern end of the beach.

NT 54, 55 **DOOLEY POINT (S)**

No.	Beach	Rating HT	Rating LT	Type	Length
NT54	Beach NT 54	2	1	R+sand-mud flats	4 km
NT55	Dooley Pt (S)	2	1	UD to LTT	7 km
Spring & neap tide range = 5.8 & 2.2 m					

Beaches NT 54 and 55 lie either side of a 1 km wide tidal inlet, which drains a 1000 ha mangrove forest. Both beaches are backed by low 1-2 km wide beach-foredune ridge plains, the entire system filling an 11 km wide shallow embayment, which is bordered by low bluffs, of the backing Quaternary black soil plains. Beach **NT 54** faces northwest and is 4 km long, extending from the southern bluffs to the creek mouth, with the low gradient high tide beach fronted by mud flats, then sand flats and finally the 2 km wide inlet sand shoals. On the north side of the inlet beach **NT 55** is a curving 7 km long beach, initially fronted by the sandy inlet shoals, which give way to a low gradient 200 m wide ultradissipative beach (Fig. 5.9), which extends to the rock flats that front the low densely vegetated Dooley Point at the northern end.

Figure 5.9 The southern part of beach NT 55 consists of a wide series of beach ridges, extending out to the present sandy ultradissipative beach.

NT 56-58 **DOOLEY POINT-CAPE SCOTT**

No.	Beach	Rating HT LT	Type	Length
NT56	Dooley Pt (N)	1 1	UD	19.5 km
NT57	Cape Scott (S)	1 1	UD	6.5 km
NT58	Cape Scott	1 1	R+rock flats	800 m
NT59	Cape Scott (E)	1 1	UD	5.8 km
Spring & neap tide range = 5.8 & 2.2 m				

Dooley Point is a low vegetated bluff north of which the low black soil plains parallel the shore for 6 km, until an inflection in the coast, beyond which the coast trends north-northwest and the plains are scarped, exposing a straight 7 km long 10 m high red bluff. A continuous sand beach fronts the entire 13 km long bluff section, before continuing for the final 6.5 km as a 200-300 m wide, low barrier backed by a 1-2 km wide largely mangrove-filled estuary. Most of the 19.5 km long beach (**NT 56**) consists of a 100-200 m wide ultradissipative to low tide terrace beach, with the northern end terminating at a 100 m wide tidal creek. A solitary straight track reaches the coast at the northern end of the bluffs, where a solitary house and clearing are located on the bluffs.

Beach **NT 57** extends for 6.5 km from the northern side of the inlet to **Cape Scott**, terminating in amongst mangroves in lee of Cape Scott. It is a relatively straight west-facing 200 m wide ultradissipative sand beach. It is backed by a 200-400 m wide barrier composed of up to 10 low beach-foredune ridges, then 2 km of mangroves. The cape is a 5 m high, 100 ha expanse of rock flats capped by marine sands. Surrounding the northern end of the cape is a crenulate 800 m long high tide sand beach (**NT 58**) fronted by 50-200 m wide intertidal rock flats, with mangroves fringing the ends of the flats.

To the east of Cape Scott is an open 6 km wide north-facing bay, bordered on the east by Cape Ford, which projects 7 km to the north. Most of the southern shore of the bay is occupied by beach **NT 59**, a 5.8 km long, slightly curving north-facing beach, connected to the base of Cape Scott in the west and bordered by a small 500 m wide tidal inlet and tidal shoals in the east (Fig. 5.10). The beach is for the most part a low gradient 200 m wide ultradissipative system, backed by a low 1-2 km wide beach-foredune ridge plain and then several hundred metres of mangroves.

Figure 5.10 Cape Scott (left of view) forms the boundary between the low curving beaches NT 57 (foreground) and NT 59 (upper).

NT 60-64 **CAPE FORD**

No.	Beach	Rating HT LT	Type	Length
NT60	Cape Ford (W5)	1 2	R+LTT	600 m
NT61	Cape Ford (W4)	1 2	R+LTT	2.5 km
NT62	Cape Ford (W3)	1 2	R+LTT	900 m
NT63	Cape Ford (W2)	1 2	R+rock flats	900 m
NT64	Cape Ford (W1)	1 2	R+rock flats	500 m
Spring & neap tide range = 5.6 & 2.0 m				

Cape Ford is a 6 km long, northern-trending, low area of Permian sedimentary rocks covered by Quaternary sands. It has a maximum width of 3.5 km tapering to a small sandy foreland at the cape. In the south the cape is separated from the mainland by a 2 km wide section of mangroves, with the only foot and possible vehicle access via the eastern shoreline. Along the more exposed western shore are five beaches (NT60-64) occupying most of the shore. Beach **NT 60** is a curving 600 m long sand beach bordered by rock flats, with a central 400 m long low tide terrace section. It is backed by dunes actively blowing up to 200 m inland to a height of 15 m. Beach **NT 61** extends for 2.5 km north of the rock flats as a crenulate beach alternating between open sandy low tide terrace sections and 50-100 m wide rock flats, all backed by moderately 100-200 m wide active dunes that reach a maximum height of 20 m.

Beach **NT 62** is a curving 900 m long predominantly sand beach and shallow bar, intermixed with rock flats up to 200 m wide and backed by a low narrow foredune. Beach **NT 63** consists of a narrow 900 m long high tide beach fronted by continuous 100-200 m wide intertidal rock flats, with low mangroves fringing the northern end.

Beyond the mangroves is 500 m long beach **NT 64**, which occupies the western side of a small sand foreland that sits on the northern rock flats of the cape.

NT 65-68 **CAPE FORD (E)**

No.	Beach	Rating HT LT	Type	Length
NT65	Cape Ford (E1)	1 2	R+LTT	300 m
NT66	Cape Ford (E2)	1 2	R+sand flats	400 m
NT67	Cape Ford (E3)	1 2	R+mud flats	900 m
NT68	Cape Ford (E4)	1 2	R+mud flats	900 m
Spring & neap tide range = 5.6 & 2.0 m				

The eastern side of Cape Ford faces into Anson Bay, with wave height decreasing into the bay. The shoreline contains 7 km of low to very low energy east-facing crenulate coast, with four beaches (NT 65-68) located amongst the mangroves and tidal flats. Beach **NT 65** forms the eastern side of the sandy foreland that forms the tip of the cape. It is a steep, straight, 300 m long beach, bordered by intertidal rock flats, and with deep water off the central section. Beach **NT 66** consists of a curving 400 m long south-facing section of sand and some mangroves, lying at the northern end of a 2 km section of mangroves. It is fronted by 500 m wide sand and some low tide rock flats. Beach **NT 67** lies on the southern side of the mangroves and is an east-facing slightly curving 900 m long moderately steep sand beach, bordered by mangrove-covered rock flats. Beach **NT 68** is a crenulate 900 m long section of east-facing sand, fronted by mangroves sitting on 300 m wide mud and rock flats, with more than half the length covered in mangroves.

NT 69-72 **ANSON BAY**

No.	Beach	Rating HT LT	Type	Length
NT69	Anson Bay (W)	1 1	R+mud flats	8.9 km
NT70	Anson Bay (E)	1 1	R+mud flats	12.3 km
NT71	Pelican Rock	1 1	R+sand-mud flats	5.1 km
NT72	Red Cliff (W)	1 1	R+sand-mud flats	900 m
Spring & neap tide range = 5.6 & 2.0 m				

Anson Bay is a semicircular northwest-facing bay, bounded by Cape Ford in the southeast and the Peron islands 33 km to the northeast. The bay contains 100 km of generally low energy shoreline including the mangrove-fringed mouth of the Daly River. The first 7 km of shore contain the Cape Ford (east) beaches. Then follow 30 km of low energy shore dominated by mud flat-fronted low energy beaches and mangroves.

Beach **NT 69** is the first of four beaches occupying the southern section of the bay. It is an 8.9 km long, low energy, crenulate high tide sand beach, fronted by mud flats up to 1 km wide, with several sections of mangroves fronting the beach. For the most part it is backed by low wooded black soil plains, with a 1.5 km long beach ridge-recurved spit forming the eastern end, and terminating at a small tidal creek. Beach **NT 70** begins on the other side of the creek and is a very low energy 12.3 km long beach, with 1-1.5 km wide mud flats and several sections of mangroves fronting the beach. The beach is backed by a 1-3 km wide series of low beach ridges to cheniers, then salt flats extending up to 10 km inland. Two small creeks cross the beach with a third larger 500 m wide creek forming the eastern boundary.

Beach **NT 71** lies in lee of Pelican Rock. The 5.1 km long beach begins on the eastern side of the larger creek, initially as a 500 m long recurved spit that grades into a straight sand beach fronted by 200 m wide sand to mud flats. It is backed by a 1 km wide sequence of up to 15 beach-foredune ridges then the low wooded plain. It terminates at a small protrusion of the Permian sedimentary bluffs and accompanying rock flats. Beach **NT 72** faces northwest and runs straight for 900 m to 10 m high Red Cliff. It is backed by continuous red bluffs, with the beach consisting of a high tide sand beach fronted by 100 m wide sand to mud flats (Fig. 5.11).

Figure 5.11 Beaches NT 71 & 72 are both fronted by sand then low tide mud flats, and terminate at Red Cliff in foreground.

NT 73-78 **RED CLIFF-CLIFF HEAD**

No.	Beach	Rating HT LT	Type	Length
NT73	Red Cliff (E1)	1 1	R+sand flats	1.2 km
NT74	Red Cliff (E2)	1 1	R+mud flats	1.5 km
NT75	Red Cliff (E3)	1 1	R+sand flats	1.8 km
NT76	Red Cliff (E4)	1 1	R+LTT	2.2 km
NT77	Cliff Head (S2)	1 1	R+LTT	6.3 km
NT78	Cliff Head (S1)	1 1	R+LTT	2.5 km
Spring & neap tide range = 5.6 & 2.0 m				

Red Cliff and Cliff Head are two low scarped exposures of the deeply weathered Permian sedimentary rocks that underlie much of the coastal plain. Between the two headlands are 15.5 km of initially crenulate north-facing shore that gradually becomes an essentially straight northwest-facing beach up to Cliff Head. The entire shoreline is fronted by six high tide sand beaches (NT 73-

78), separated by bluffs and three small tidal creeks, and fronted by intertidal sand flats. Sand is moving northward along these beaches and around the heads forming recurved sand pulses on the north side of the heads, as well as deflecting the creek mouths to the north.

Beach **NT 73** is a curving northeast-facing beach bordered by 10 m high wooded red bluffs and fronting intertidal rock flats, with 300 m wide ridged sand flats in the centre. Beach **NT 74** curves round from the bluffs for 1.5 km, for the most part backed by a low narrow vegetated sand spit that terminates at the mouth of the first tidal creek. Mud flats up to 800 m wide fill the curve. Beach **NT 75** is a straight 1.8 km long sand beach fronted by low gradient sand flats and backed by a low, narrow, vegetated foredune, with tidal creeks to either end and a small section of bedrock plain in the centre. Beach **NT 76** begins on the north side of the second tidal creek and runs for 2.2 km to the mouth of the third tidal creek. It is backed by a low wooded plain in the centre with small foredune-spits to either end.

Beach **NT 77** is a straight, 6.3 km long, northwest-facing beach, initially backed by a 2.5 km long spit, then 10-20 m high red bluffs. It terminates at a small protrusion in the bluffs, which separates it from beach **NT 78**, which runs for another 2.5 km to **Cliff Head**. Both beaches receive waves up to 0.5 m which maintain a high tide beach and 200-300 m wide low tide bar.

NT 79-80 **DALY RIVER (S)**

No.	Beach	Rating HT LT	Type	Length
NT79	Cliff Head (E)	1 1	R+sand flats	1.8 km
NT80	Daly River (S)	1 1	R+mud flats	5 km
Spring & neap tide range = 5.6 & 2.0 m				

Cliff Head marks the southern boundary of a 24 km wide embayment that is occupied by the 4 km wide mouth of the Daly River and an extensive deltaic plain to either side, while the laterite bluffs of Channel Point form the northern boundary. Between the Head and the river mouth are two low energy beaches (NT 79 & 80). Beach **NT 79** lies in lee of Cliff Head and for the most part is backed by a low wooded plain. The curving north-facing beach receives pulses of sand around the point, which have deposited a series of recurved sand ridges, which have merged with beach NT 80.

Beach **NT 80** is part of a 10 km long regressive recurved beach ridge-chenier plain that has prograded up to 3.5 km into Anson Bay. The present shoreline consists of 5 km of low energy beach, which recurves around into the Daly River mouth where it merges with the mangrove-fringed river shoreline. The entire beach is fronted by 400-800 m wide mud flats (Fig. 5.12).

Figure 5.12 The low curving beach NT 80 has prograded over 3 km into Anson Bay, as a series of beach ridges, where it is fronted by wide mud flats.

NT 81-82 **CHANNEL POINT**

No.	Beach	Rating HT LT	Type	Length
NT81	Channel Pt	1 1	R+sand/mud flats	3.4 km
NT82	Channel Pt (N)	1 1	R+mud flats	5.5 km
Spring & neap tide range = 5.6 & 2.0 m				

Channel Point is a low sandy protrusion in the shoreline located at the northern end of Anson Bay and in lee of the equally low Peron Islands. To the east mangrove-fringed shoreline extends for 20 km to the mouth of the Daly River, while to the north predominantly sandy shoreline runs for 30 km up to Point Blaze. Beaches NT 81 and 82 occupy the first 9 km of shore up to the southern boundary of the Wagait Aboriginal Land.

Beach **NT 81** is an outward curving 3.4 km long narrow spit initially backed by mangroves, then wooded laterite plains. A beachfront settlement of about 30 houses is located along the beach in lee of the point, which is fronted by sand then mud flats up to 400 m wide. Access is via a private track from Wangi. A small elongate tidal creek separates it from beach **NT 82**. This is a 5.5 km long beach and narrow barrier that is backed for most of its length by mangroves, and fronted by 400 m wide mud flats, as well as rock flats off a central inflection point. A landing ground parallels the rear of the northern 1 km of the beach.

Wagait Aboriginal Land

Coast length: 52 km (809-861 km)
Beaches: 10 (NT 83-92)

Wagiat Aboriginal land occupies an area of coast and hinterland between Point Blaze and Fog Bay

NT 83-86 WAGAIT (S)

No.	Beach	Rating HT	LT	Type	Length
NT83	Beach NT 83	1	1	R+sand flats	400 m
NT84	Beach NT 84	1	1	R+sand flats	300 m
NT85	Beach NT 85	1	1	R+sand flats	8.8 km
NT86	Beach NT 86	1	1	R+sand flats	5.3 km
Spring & neap tide range = 5.6 & 2.0 m					

The boundary of the **Wagait Aboriginal Land** intersects the coast at the northern end of beach NT 82 and the beginning of beach NT 83. The straight sandy shore to the south protrudes west at beach NT 83 owing to the outcropping of intertidal rock flats along beaches NT 83, 84 and 85. There are two small communities in lee of beaches NT 83 and 84.

Beach **NT 83** is a crenulate, southwest-facing, 400 m long, sandy high tide beach, with a scattering of rocks and rock flats, and 200 m wide sand flats. About 10 houses are located in a clearing along the southern half of the beach. Beach **NT 84** is an adjoining 300 m long, curving, more west-facing beach, bordered by low rock flats, and fronted by 200 m wide sand flats. About six houses are scattered in a clearing behind the beach.

Beach **NT 85** extends for 8.8 km from the rock flats of beach NT 83. It initially faces southwest and runs in lee of scattered intertidal rock flats for 5 km. It then faces west as the rock flats continue, with only the northern 3 km replaced by 200-300 m wide sand flats. It terminates at the mouth of a small tidal creek. Beach **NT 86** begins on the northern side of the creek, faces west and curves slightly for 5.3 km to the southernmost Tertiary laterite of Point Blaze. A small tidal creek flows across the northern end and 200 m wide sand flats parallel the beach.

NT 87-91 POINT BLAZE-JENNY POINT

No.	Beach	Rating HT	LT	Type	Length
NT87	Pt Blaze (S2)	1	1	R+rock flats	1.2 km
NT88	Pt Blaze (S1)	1	1	R+rock flats	3.5 km
NT89	Pt Blaze (E)	1	1	R+sand/rock flats	4.6 km
NT90	Jenny Pt (W)	1	1	R+mud flats	2.6 km
NT91	Jenny Pt (E)	1	1	R+mud flats	1.4 km
Spring & neap tide range = 5.6 & 2.0 m					

Point Blaze and Jenny Point are two boundary points on a 6 km wide protrusion of Tertiary laterite that separates Anson and Fog bays. Around the low wooded promontory are five beaches (NT 87-91), all substantially influenced by intertidal rock flats. There is no vehicle access to any of the beaches and no habitation.

Beach **NT 87** is a relatively straight 1.2 km long west-facing high tide sand beach, located along the western base of **Point Blaze**. It is fronted by continuous 400-600 m wide rock flats, with mangroves growing on the northern flats that separate it from the next beach. Beach **NT 88** is a slightly curving 3.5 km long sand beach, fronted by continuous rock flats that narrow to 200 m along the central section. It terminates at the mangrove-covered rock flats at Point Blaze, a 5 m high wooded plain.

On the eastern side of the 1 km wide point beach **NT 89** emerges from the mangroves to curve for 4.6 km to a low rounded unnamed headland. The high tide sand beach is cut by a central tidal creek and backed by a narrow foredune system, while it is fronted by rock flats to either end, with a central 2 km of sand flats.

Beach **NT 90** begins on the eastern side of the round point and extends for 2.6 km in a curving embayment to the narrow V-shaped tip of **Jenny Point**, the point consisting of a series of vegetated beach ridges. The beach faces north and is backed by the low wooded plain and fronted by rock flats, apart from a 600 m long eastern section, where 400 m wide mud flats front the beach. Rock flats and some mangroves also extend 300 m off the point. Beach **NT 91** begins off the eastern side of the point and runs due south for 1.4 km before it is overwhelmed by mangroves. It is fronted by 400 m wide mud flats, which narrow to the south.

NT 92-93 FINNISS RIVER

No.	Beach	Rating HT	LT	Type	Length
NT92	Finniss River	1	2	R+tidal shoals	2 km
NT93	Five Mile Beach	1	1	R+mud flats	9.1 km
Spring & neap tide range = 5.6 & 2.0 m					

The **Finniss River** is a larger tidal creek, which drains a 10 000 ha area of mangroves fringed by salt flats. The shoreline to either side has prograded up to 2 km during the Holocene, on the south side as a series of low cheniers, while the north side has a more continuous series of up to 25 beach ridge-foredunes backing the beach. The mouth of the river also forms the northern boundary of the Wagait Aboriginal Land.

Beach **NT 92** is located along the south side of the river mouth. The curving 2 km long low energy strip of sand is fringed by mangroves, and fronted by both the 100 m wide tidal creek (Finniss River) and mud flats up to 1 km wide. It faces northwest and is backed by high tide flats.

On the north side of the river is a south-facing mangrove-fringed point, which on its western side forms the southern end of 9.1 km long, west-facing **Five Mile Beach (NT 93)**. This beach is the outermost of the 25 beach-foredune ridges that have formed between the river and the northern Stingray Head (Fig. 5.13). The beach is relatively straight and fronted by 1.5 km wide mud flats that widen slightly to the south. A couple of fishing shacks are located in lee of the beach at the river mouth, with 4WD access along the beach.

Figure 5.13 The long beach NT 93 is backed by a series of parallel beach ridges and fronted by sand then mud flats.

NT 94-97 **DUNDEE BEACH**

No.	Beach	Rating HT	Rating LT	Type	Length
NT94	Stingray Head	1	2	R+rock flats	2.7 km
NT95	Dundee Beach (1)	1	1	R+rock flats	700 m
NT96	Dundee Beach (2)	1	1	R+sand/rock flats	1.4 km
NT97	Dundee Beach (3)	1	1	R+sand flats	2.3 km

Spring & neap tide range = 5.6 & 2.0 m

Dundee Beach is a 1980s beachfront development occupying a 7 km long section of coast between Stingray Head and Native Point (beaches NT 95-103). The development is located on a low black soil plain, with 5-10 m high scarped red bluffs forming much of the shoreline. It contains beachfront residential blocks and a small resort in lee of beach NT 99.

Beach **NT 94** is located immediately south of the settlement and is a 2.7 km long, relatively straight, west-facing sand beach, bordered by 5-10 m high dark laterite bluffs, with a 1 km wide shallow lagoon backing its centre. A 4WD track runs along the top of the bluffs and back of the beach to the creek mouth. The beach is fronted by a mixture of 500 m wide sand and rock flats off the southern Stingray Head, grading to 200 m wide sand flats to the north. Laterite rocks and beachrock outcrop along the beach, the laterite increasing toward the northern bluffs.

Beach **NT 95** is a straight 700 m long high tide sand beach, fronted by continuous rock flats, and backed by 5 m high scarped bluffs, with rocks and debris from the bluffs littering the beach. A 50 m wide grassy reserve then approximately 20 beachfront properties and a public walkway back the bluffs, though access to the beach is difficult. Beach **NT 96** begins in lee of a 500 m long continuation of the scarped bluffs, reserve and beachfront properties, which gives way to a central lower area containing a beachfront car park and eroded and now disused boat ramp. An eroded causeway built along the back of the beach blocks a small wetland that has consequently dried out. The northern side of the beach and wetland is backed by the reserve and several more beachfront properties. The beach consist of a wider high tide sand beach fronted by a 100 m wide low tide bar, with a scattering of rocks and rock flats, as well as some bedrock reefs exposed 500 m off the beach at low tide. Beach **NT 97** consists of a 1 km long southern section containing low eroding red bluffs and several beachfront properties. The bluffs give way to a low sand beach which impounds a 1 km wide shallow lagoon, drained by a small central creek. The beach continues north of the creek for 500 m to the next low rocky bluffs. The 2.3 km long beach is for the most part fronted by a 100 m wide low tide bar, as well as some deeper rock flats, and a small ebb tide delta off the creek mouth.

NT 98-103 **DUNDEE BEACH**

No.	Beach	Rating HT	Rating LT	Type	Length
NT98	Dundee Beach (4)	1	1	R+LTT	100 m
NT99	Dundee Beach (5)	1	1	R+LTT	400 m
NT100	Native Pt (N1)	1	2	R+rock flats	1.2 km
NT101	Native Pt (N2)	1	1	R+LTT	1.2 km
NT102	Native Pt (N3)	1	1	R+rock flats	1.4 km
NT103	Native Pt (N4)	1	1	UD+rocks	1.2 km

Spring & neap tide range = 5.6 & 2.0 m

The **Dundee Beach** development spreads as far north as Native Point. Between Beach NT 97 and the point is a 2 km long section of low bluffs and rock flats, which toward the northern end have two small beaches at their base. Beach **NT 98** is a 100 m long pocket of sand, backed and bordered by low bluffs and rock flats, with two beachfront houses behind. Beach **NT 99** is the site of the *Lodge of Dundee*, the major tourist and holiday destination for the area. The lodge sits on low bluffs overlooking the 400 m long southwest-facing beach (Fig. 5.14). Rock flats border the beach and extend 500 m off Native Point, the northern boundary. In front of the lodge is a small groyne and breakwater built to shelter the concrete boat ramp, with the sand beaches to either side. Sand moving down the coast is building out on the northern side of the breakwater, widening the beach. The best part of the beach for swimming is on the south side of the lodge or in the small swimming pool which looks over the boat ramp and 'harbour'.

At **Native Point** the shoreline turns and runs due north for 5 km to an entrance to Port Patterson. A formed road runs along the back of the beaches to the northern point, however the track is locked and unauthorised vehicle access is prohibited, as these beaches are all protected turtle nesting sites. The shoreline is slowly eroding and is dominated by low eroding bluffs and rock flats. Beach **NT 100** is a 1.2 km long high tide sand beach fronted by continuous low tide rock flats which terminate at a slight protrusion in the rocks. Beach **NT 101** extends north for another 1.2 km to the mouth of a small tidal creek. The beach has a scattering of rocks and rock flats and a 100 m wide low tide bar. On the northern side of the creek is beach **NT 102**, a 1.4 km long, low gradient beach fronted for the most part by intertidal rock flats. It is separated from NT 103 by a more prominent semicircular rock reef, which impounds a tidal pool. The northernmost beach (**NT 103**) extends for another 1.2 km as a low gradient 200 m wide ultradissipative beach. It terminates at the northern point in lee of 500 m wide rock flats, the beach curving round the point to face north, before it is replaced by the mangroves of Port Patterson. The formed road from Dundee Beach terminates at the point.

Figure 5.14 Dundee Lodge overlooks beaches NT 98 (right) and 99, with a boat ramp in centre.

NT 104-106 **INDIAN ISLAND**

No.	Beach	Rating HT	Rating LT	Type	Length
NT104	Hut Pt (E)	1	2	R+mud flats	3 km
NT105	West Pt (W)	1	2	R+mud/rock flats	600 m
NT106	Herbert Pt (N)	1	2	R+mud flats	150 m
Spring & neap tide range = 5.2 & 1.4 m					

Indian Island is a low 16 km long bedrock island that separates Port Patterson from Bynoe Harbour and forms much of the western shore of the harbour. There is no vehicle access to or on the island and no habitation. While most of the island is ringed by extensive mangroves and tidal flats, at the northern extremity there is sufficient wave energy to maintain three low energy beaches (NT 104-106). Beach **NT 104** extends for 3 km east of Hut Point, and faces northwest into the summer winds. The usually low waves have built a relatively straight sand beach, backed by a densely vegetated foredune and fronted by 800 m wide mud flats, with mangrove-fringed rock flats forming the boundaries (Fig. 5.15).

Beach **NT 105** lies at the northern tip of the island and immediately west of West Point. It is a straight 600 m long northwest-facing high tide sand beach, fronted by a 200-300 m wide mixture of rock, sand and mud flats. One kilometre south of the eastern side of the point is beach **NT 106**, a 150 m long east-facing strip of sand, bordered by mangrove-covered rock and sand flats, with Herbert Point forming the southern boundary. It is fronted by 400 m wide mud flats and faces directly across the 5 km wide harbour to East Point.

Figure 5.15 Hut point (foreground) borders beach NT 104, a sandy high tide beach fronted by 800 m wide mud flats.

NT 107-110 **EAST PT (N) & PT MARGARET**

No.	Beach	Rating HT	Rating LT	Type	Length
NT107	East Pt (N1)	1	2	R+rock flats	600 m
NT108	East Pt (N2)	1	2	R+rock flats	500 m
NT109	East Pt (N3)	1	2	R+rock/sand flats	900 m
NT110	Pt Margaret (E)	1	2	R+rock/sand flats	150 m
Spring & neap tide range = 5.5 & 1.8 m					

East Point is a slight projection in the low wooded coastal plain. It lies along the eastern entrance to Bynoe Harbour, facing across the bay to Indian Island. To the south are approximately 100 km of mangrove-fringed bay shoreline. At the point wave energy begins to increase sufficiently to keep the shore largely clear of mangroves, permitting three beaches (NT 107-109) to occupy 3 km of shoreline immediately north of the point. A 4WD track leads along the bluffs to each of the beaches.

Beach **NT 107** is a straight, 600 m long, west-facing, high tide sand beach fronted by 200 m wide rock flats, fringed by mangroves and backed by the densely wooded plain. Its neighbour, beach **NT 108**, is 500 m long with 50 m wide sand flats then deeper rock flats, while the northern beach **NT 109**, while bordered by mangrove-covered rock flats, has more continuous 300 m wide sand flats.

Point Margaret lies 12 km north of East Point. It is a generally low wooded plain, fringed by near continuous mangroves. One kilometre east of the point is a small bay with a 150 m long sand beach (**NT 110**), largely clear of mangroves and fronted by a mixture of rock and sand flats. A vehicle track reaches the beach where a few fishing shacks are located.

Region 2 Northwest Coast: Charles Point -Cape Cockburn

Length of coast:	1260 km	(1053-2313 km)
Number of beaches:	341	(NT 111-451)
Regional maps:	Figure 5.16	page 193
	Figure 5.19	page 195
	Figure 5.26	page 201
	Figure 5.29	page 202
	Figure 5.53	page 225

Aboriginal Land:	Arnhem Land
National Parks:	Kakadu, Cobourg Peninsula (Gurig)
Towns/communities:	Mandorah, Darwin, Nguiu (Bathurst Is), Milikapiti (Melville Is), Minjilang (Crocker Is).

The northwest coast of the Northern Territory extends for 1260 km from Charles Point and the Darwin region around the mud and mangrove-fringed shores of Arnhem Gulf, and includes Bathurst and Melville islands, the Cobourg Peninsula and Crocker Island, to Cape Cockburn. Much of the coast is protected and sheltered by its northerly orientation, islands and indented shoreline, especially around the Cobourg Peninsula.

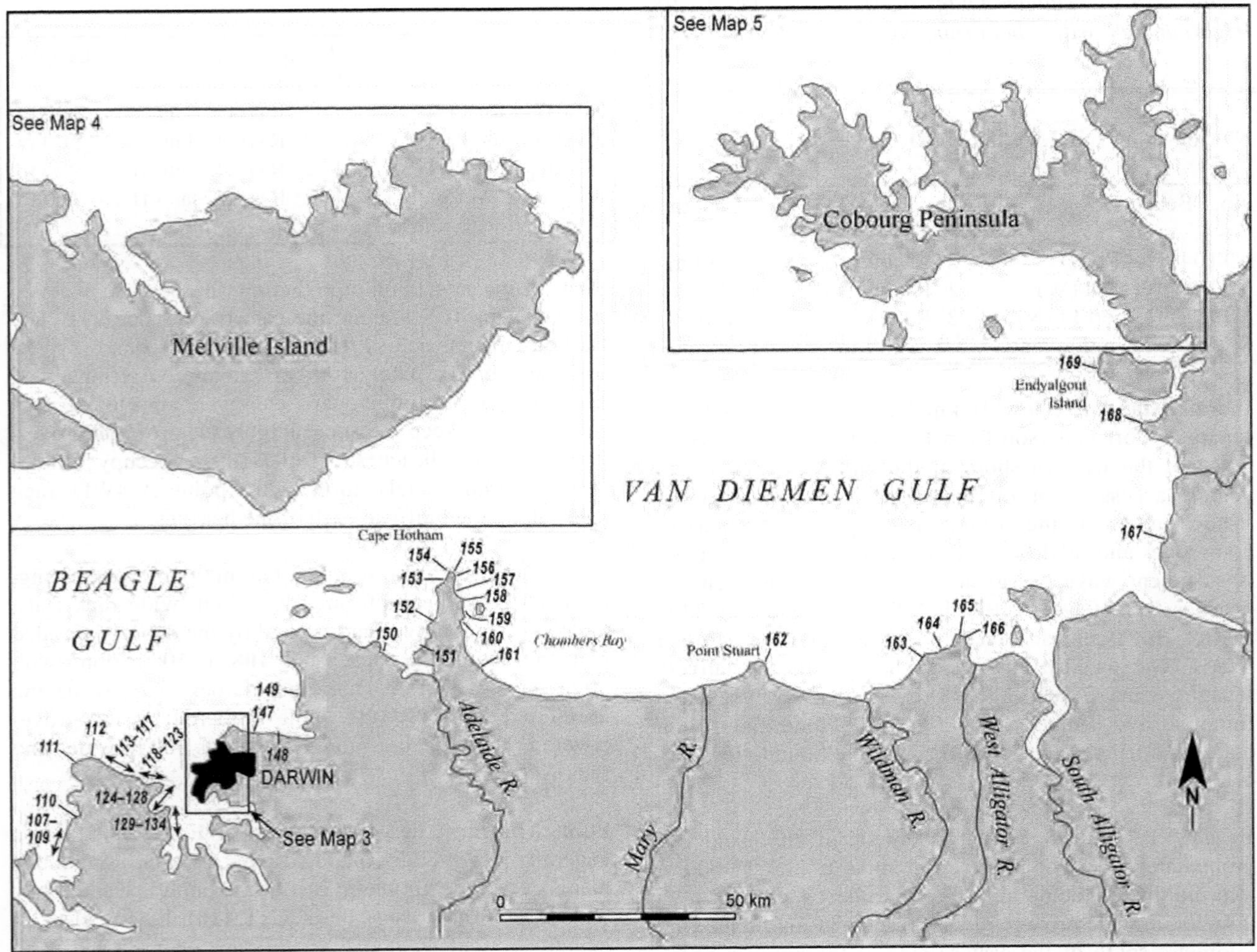

Figure 5.16 Regional map 2 part of the northwest coast between Charles Point and Cape Cockburn. See Figure 5.19 for Darwin region, Figure 5.26 for Bathurst and Melville Island and Figure 5.29 for Cobourg Peninsula.

NT 111-116 **CHARLES POINT**

No.	Beach	Rating HT	LT	Type	Length
NT111	Charles Pt (W)	1	2	R+rock flats	1.5 km
NT112	Charles Pt (E1)	1	2	R+rock flats	1.2 km
NT113	Charles Pt (E2)	1	2	R+rock flats	1.1 km
NT114	Charles Pt (E3)	1	2	R+rock flats	1.2 km
NT115	Harveys Beach (W2)	1	2	R+rock flats	800 m
NT116	Harveys Beach (W1)	1	2	R+rock flats	1.6 km
Spring & neap tide range = 5.5 & 1.8 m					

Charles Point marks the northwest tip of the Cox Peninsula. The Radio Australia transmitting towers are located in lee of the point. Apart from the towers and a lighthouse located 1.5 km east of the point, there is no other development in the area.

Beach **NT 111** is located on the eastern side of the point. It extends to the southwest for 1.5 km before being fronted by a 300 m wide band of mangroves. The beach consists of a moderately steep high tide beach, cut in the centre by a small tidal creek, which drains a backing 5 ha mangrove-filled lagoon. The entire beach is fronted by 500 m wide rock flats, apart from the sandy ebb tide delta of the creek.

Beach **NT 112** begins 400 m east of the point and consists of a curving 1.2 km long, north-facing narrow high tide beach, backed by scarped 10 m high red bluffs and fronted by a 300-500 m wide mixture of sand, mud and rock flats. The Charles Point lighthouse is located on the bluffs toward the eastern end of the beach (Fig. 5.17), with restricted vehicle access to the light and the adjoining beaches.

Figure 5.17 Charles Point lighthouse, with beaches NT 112 to right, and NT 113 to left.

Beach **NT 113** begins 100 m east of the lighthouse and is a 1.1 km long, crenulate, north-facing high tide beach, wedged in below bluffs, which diminish to the east and 400 m wide rock flats, covered in the centre by stunted mangroves. Beach **NT 114** lies immediately to the east and is a curving to crenulate 1.2 km long north-facing high tide beach, backed by a low wooded plain and fronted by rock flats grading into sandy mud flats, together with a few scattered mangroves. A straight vehicle track terminates behind the centre of the beach.

Beaches NT 115 and 116 border a small tidal creek that drains a 100 ha mangrove-filled lagoon. Beach **NT 115** is a curving, northeast-facing, 800 m long, high tide sand beach, which terminates at the creek mouth and is backed by a few low beach ridges and then salt flats. On the eastern side beach **NT 116** extends east for another 1.6 km, interrupted by two expanses of rock flats, with a central 800 m wide sandy-mud flat. A 4WD track runs along the back of the beach to the creek mouth.

NT 117-120 **HARVEYS BEACH-TWO FELLA CK**

No.	Beach	Rating HT	LT	Type	Length
NT117	Harveys Beach	1	1	R+sand flats	2.2 km
NT118	Harveys Beach (E)	1	1	R+LTT	3.2 km
NT119	Two Fella Ck (W)	1	1	R+sand flats	1.5 km
NT120	Two Fella Ck (E)	1	1	R+sand flats	1.5 km
Spring & neap tide range = 5.5 & 1.8 m					

Between Harveys Beach (NT 117) and beach NT 120 are 8 km of north-facing shoreline largely backed by an irregular lagoon containing a 500 m wide strip of mangroves. Vehicle access is restricted to the low bedrock headlands that separate each beach, with the access track to beaches NT 118 and 119 having to cross a section of wetland, which may be flooded during the Wet.

Harveys Beach (**NT 117**) begins at a low protruding sand foreland sitting in lee of 1 km wide intertidal rock flats. The beach sweeps round for 2.2 km to a small section of rock flats that separate it from beach NT 118. In the west it is fronted by 100 m wide sand flats, which narrow to the east. It is backed by a low foredune and a largely infilled 500 m wide lagoon, with a vehicle track running along the foredune. Beach **NT 118** is 3.2 km long, faces north and is bordered by rock flats at either end. Three houses sit on the low foredune adjacent to the western rock flats. The beach consists of a moderately steep high tide beach fronted by a 100 m wide low tide bar and backed by a low foredune, then a mangrove-filled lagoon between the headlands.

Beach **NT 119** occupies the crenulate 1.5 km of shoreline that lies in lee of irregular 100-300 m wide coloured sandstone rock flats, with one 100 m long section of sand flats. It terminates at the eastern end of the rock flats, where the small tidal Two Fella Creek is also located. A vehicle track runs out to the point where there is a solitary house. Beach **NT 120** begins on the eastern side of the creek, extending as a straight beach to the low sandstone rocks that separate it from Imaluk Beach. Tidal sand shoals extend several hundred metres off the creek mouth, narrowing toward the eastern end. In addition a shore-parallel band of intertidal beachrock parallels the base of the sandy high tide beach.

NT 121-124 IMALUK-WAGAIT

No.	Beach	Rating HT LT	Type	Length
NT121	Imaluk Beach	1 1	R+LTT	800 m
NT122	Imaluk (E)	1 1	R+sand flats	1.4 km
NT123	Wagait Beach	1 1	R+LTT	700 m
NT124	Wagait	1 1	R+rock flats	1.3 km
Spring & neap tide range = 5.5 & 1.8 m				

The **Wagait** settlement backs the 3.5 km of coast between Imaluk and Wagait beaches, with a continuous foreshore reserve running between the beaches and the houses, and occasional public walkways to the beach. **Imaluk Beach (NT 121)** is a straight 800 m long moderately steep sand beach fronted by a 100 m wide low tide bar. It is located immediately west of the settlement and is bordered by intertidal rock flats, with sandstone beachrock and boulders increasing to either end (Fig. 5.18). The eastern Imaluk Beach (**NT 122**) has a small tidal creek draining across the western end, with rock flats bordering each end. The sand flats are 200 m wide in front of the creek, narrowing to 100 m in the east. The 50 m wide reserve then several beachfront houses back the beach. Wagait Beach (**NT 123**) lies on the eastern side of the rock flats and is a 700 m long beach bordered by extensive rock flats, with a central 400 m long section fronted by a 100 m wide low tide bar. It is backed by the reserve and houses, with the Wagait water tower located in lee of the eastern end of the beach. The Wagait settlement continues east in lee of beach **NT 124**, a crenulate 1.3 km long northeast-facing beach dominated by irregular sandstone rock flats and boulders extending 200 m seaward. It is backed by the narrow reserve then houses.

Figure 5.18 Imaluk Beach (NT 121) extends east of the low boundary sandstone boulder point.

NT 125-128 MANDORAH

No.	Beach	Rating HT LT	Type	Length
NT125	West Pt	1 1	R+rock/sand flats	1.1 km
NT126	Picnic Pt	1 1	R+ rock flats	600 m
NT127	Picnic Cove	1 1	R+LTT	500 m
NT128	Oak Pt	1 1	R+sand flats	700 m
Spring & neap tide range = 5.5 & 1.8 m				

Mandorah is a small holiday and residential settlement lying 8 km due west of Darwin, to which it is connected by a regular ferry service across Port Darwin. It contains a small marina, jetty and resort and the adjoining settlement of Wagait. The shoreline to either side of the jetty faces west across the port and contains four low energy beaches (NT 125-128). Beach **NT 125** is a 1.1 km long north-facing beach that lies between Wagait and West Point, the easternmost tip of the Cox Peninsula. It consists of a narrow high tide beach with 200 m wide rock flats. A tidal creek at the western end of the beach drains a 30 ha mangrove-filled wetland, which backs the western half of the beach. On the southern side of the point is beach **NT 126**, a 600 m long east-facing narrow strip of high tide sand located below 5 m high coloured sandstone bluffs, with considerable bluff debris scattered along and across the beach which is fronted by continuous 100 m wide rock flats.

The Mandorah jetty crosses the Picnic Point rock flats which form the boundary with Picnic Cove beach (**NT 127**), a 500 m long moderately steep high tide sand beach and 100 m wide low tide bar. The southern end abuts the 100 m long groyne that forms the northern side of the small Mandorah marina and boat ramp. There is sealed road access to the jetty and two large car parks behind the beach as well as the Mandorah hotel-resort behind the southern end of the beach and marina.

Oak Point (beach **NT 128**) is a low energy recurved spit that extends from the scattered rocks and beachrock on the south side of the marina, for approximately 700 m south to the mangroves of Woods Inlet. It is backed by mangroves and is undeveloped, apart from the remains of a World War 2 bunker toward the northern end. The eroded bunker now lies 50 m from the dune line, indicating the degree of erosion since the war.

NT 129-134 TALC HEAD

No.	Beach	Rating HT LT	Type	Length
NT129	Talc Head (W)	1 1	R+tidal flats	500 m
NT130	Mica Beach	1 1	R+tidal flats	400 m
NT131	Kurumba Beach	1 1	R+tidal flats	100 m
NT132	Plater Beach	1 1	R+tidal flats	100 m
NT133	Silversands	1 1	R+tidal flats	200 m
NT134	Hingston Beach	1 1	R+tidal flats	400 m
Spring & neap tide range = 5.5 & 1.8 m				

Talc Head is located at the tip of a 1 km wide peninsula bordered by West Inlet and Port Darwin. It lies 4 km south of Mandorah and 5 km southwest of Darwin. On the tip and along the eastern side of the head are six low energy beaches (NT 129-124), all dominated by bordering rocks and rock flats.

Beach **NT 129** lies immediately west of the 18 m high Talc Head, the head composed of Proterozoic gneiss. It is a 500 m long northwest-facing strip of high tide sand, inter-fingering with a boulder beach and bordering rocks and fronted by the 500 m wide sand to mud flats of Woods Inlet. On the eastern side of the point is **Mica Beach** (**NT 130**) a 400 m long east-facing sand beach, backed by a low foredune, with boulder rubble and 10 m high headlands to either end, as well as cobbles and boulder along the beach. A few houses sit atop the 100 m wide Talc Head ridge that separates the two beaches.

Immediately south of Mica Beach are three small pockets of sand wedged in between four small 10 m wide wooded headlands and linked by low tide sand flats. **Kurumba Beach** (**NT 131**) is a 100 m long pocket of sand, separated by a 50 m wide headland from **Plater Beach** (**NT 132**), a similar 100 m long pocket beach. This is followed by **Silversands** beach (**NT 133**) a 200 m long sand beach, backed by a few houses, with a vehicle track down the backing slopes to the beach.

Hingston Beach (**NT 134**) represents a transition from the beaches to the mangroves of Port Darwin. It is 400 m long, faces east and is bordered by the low southern Silversands beach headland. It has a narrow high tide beach fronted by 500 m wide sand flats, and a scattering of mangroves, which increase in density to the south. A solitary house is located toward the rear of the southern end of the beach.

Darwin

Darwin is the capital and largest city in the Northern Territory. The town was laid out in 1869 as Palmerston on a flat 2 km wide peninsula. The city was severely damaged by bombing raids during World War 2 and by Cyclone Tracy in 1974. It has since expanded well beyond the peninsula to the north and east and today has a population of 70 000. The city is surrounded by the mangrove-lined waters of Frances Bay and Port Darwin in the south, with sandy Fannie Bay to the west and the waters of Beagle Gulf to the north (Fig. 5.19). A number of low energy beaches surround the peninsula, with longest beaches spreading northeast to Lee Point, the northern boundary of the city. The Arafura Surf Life saving Club patrols the popular Mindil Beach, while the older Darwin Surf Life Saving Club is located in the north at Casuarina Beach, one of 12 beaches that extend from the city to Lee Point.

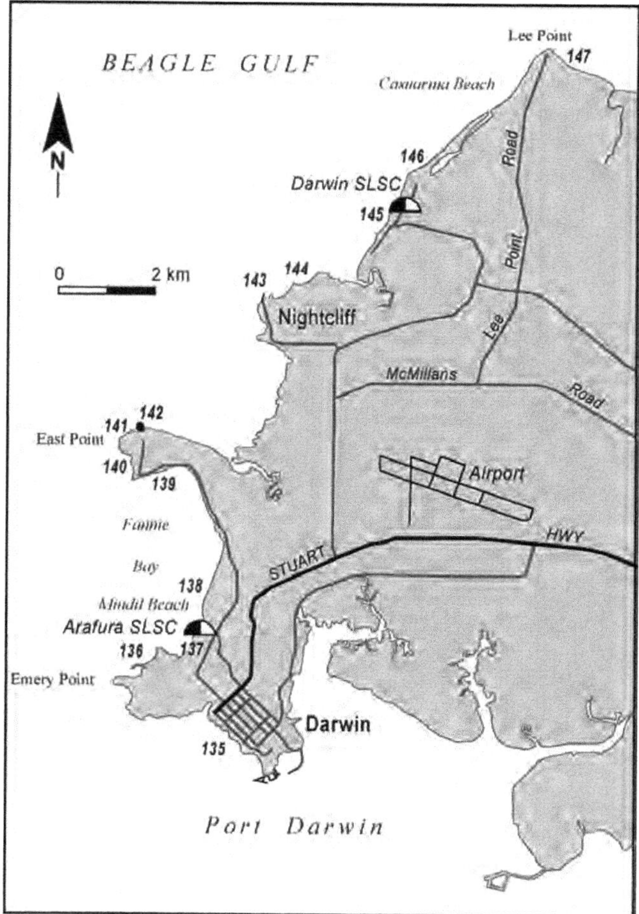

Figure 5.19 Darwin regional map, beaches NT 135-147.

NT 135-142 **DARWIN**

No.	Beach	Rating HT	LT	Type	Length
NT135	Lameroo Beach	1	1	R+sand flats	100 m
NT136	Cullen Beach	1	1	R+LTT	250 m
NT137	**Mindil Beach**	1	1	R+sand flats	500 m
	ARAFURA SLSC				
Patrols: Sundays 1400-1830, May - October					
NT138	Vesteys Beach	1	1	R+sand flats	2.2 km
NT139	Vesteys (N)	1	1	R+sand flats	1.5 km
NT140	Dudley Point	1	1	R+rock flats	400 m
NT141	East Point	1	1	R+rock flats	150 m
NT142	Rocksitters	1	1	R+rock flats	200 m
Spring & neap tide range = 5.5 & 1.8 m					

Darwin is located on a 5 km long, 2 km wide, 20 m high peninsula, which is surrounded by water on three sides. To the east is the protected mangrove-lined shore of Frances Bay, now largely redeveloped as part of the port. To the south is a slightly more exposed 4 km long section facing into Port Darwin and containing the solitary Lameroo Beach backed by the Bicennential Park, while along the 6 km long northern shore are seven more exposed beaches (NT 136-142) between Cullen Bay and East Point (Fig. 5.19).

Lameroo Beach (**NT 135**) is Darwin's solitary downtown beach. It is located at the base of 15 m high densely vegetated bluffs toward the centre of the Bicentennial Park (Fig. 5.20). A graded concrete walkway leads down through tropical rain forest to the beach, where the trees overhang much of the straight, low energy, 100 m long, sand and gravel beach. It faces southwest into the harbour, with rocky tidal flats exposed at low tide.

Figure 5.20 Lameroo Beach (NT 135) is located adjacent to downtown Darwin.

Cullen Beach (**NT 136**) is Darwin's newest beach. It was constructed in the mid-1990s as part of the redevelopment of Cullen Bay, which replaced the tidal flats and mangroves with a lock-controlled marina and waterfront housing estate. The 250 m long beach is located on the northern side of the development, contained between two seawall-groynes (Fig. 5.21) and backed by a continuous rubble wall. It is also backed by a large car park and various commercial facilities, then the Cullen Bay marina facilities. On the eastern side of the lock-controlled Cullen Bay harbour is a 100 m long artificial strip of sand backed by a grassy reserve, called **Kahila Beach**.

Figure 5.21 Cullen Bay marina, with the beach (NT 136) located between the two groynes (left).

Fannie Bay is a 4 km wide, open, west-facing bay located between Myilly and Dudley points. Along the shoreline are four beaches (NT 137-140). **Mindil Beach** (**NT 137**) is Darwin's most famous and popular beach, site of the casino and the Thursday night markets, and backed by the large Mindil beach reserve. The 500 m long beach faces west and is bordered by 15 m high Myilly and Bullocky points. It consists of a moderately steep 100 m wide high tide beach, fronted by 200 m wide low tide sand flats (Fig. 5.22). It is the site of the Arafura Surf Life saving Club which patrols the beach on Sunday afternoons during winter.

Figure 5.22 Mindil Beach (NT 137) with the sun setting over the low tide sand flats.

On the northern side of Bullocky Point is **Vesteys Beach** (**NT 138**), a gently curving 2.2 km long beach. It is home to the Darwin power boat, trailer boat and sailing clubs, and has numerous boats moored off the northern half of the beach, and two large boat ramps crossing the beach. The beach has a 100 m wide moderately steep high tide section, then 400 m wide tidal flats. The main beach terminates at some low scraped bluffs, beyond which is the northern beach **NT 139**, which is part of the East Point Recreation Reserve. This 1.5 km long beach curves round to face southwest and terminates in amongst the mangroves and rocks in lee of Dudley Point.

Beach **NT 140** extends along the south side of **Dudley Point** for about 400 m. The crenulate low energy, low gradient beach is backed by scarped to vegetated 10 m high bluffs, partly covered in rocks and rock outcrops and fronted by 1 km wide tidal flats of Fannie Bay. On the north side of Dudley Point is a 400 m long series of low crenulate bluffs that extend to East Point. At the base of the bluffs are several small, difficult to access pockets of high tide sand, fronted by continuous intertidal rock flats. The park road and recreation facilities are located just behind the bluffs.

On the eastern side of **East Point** is a 150 m long, straight, steep, high tide sand and gravel beach (**NT 141**), with continuous low tide rock flats, and the reserve and road behind. A few stunted mangroves grow on the bordering laterite rock flats. Immediately to the east is **Rocksitters** beach (**NT 142**), a 200 m long 5 m high bluff, with a series of high tide pockets of reddish sand along its base, which are linked at low tide by a continuous rock flat. On the eastern boundary bluff is the site of the Darwin *Rocksitters Club*, with two plaques marking the spot.

NT 143-147 NIGHTCLIFF-CASUARINA-LEE PT

No.	Beach	Rating HT LT	Type	Length
NT143	Nightcliff	1 1	R+sand flats	300 m
Patrols: June-August – Saturdays 0800-1400				
NT144	Rapid Creek	1 1	R+sand flats	300 m
NT145	Dripstone	1 1	R+sand flats	1.2 km
NT146	**Casuarina Beach DARWIN SLSC**	1 1	R+sand flats	4.5 km
Patrols: May-October – Saturday 1500-1830; Sunday 0900-1200, 1500-1830				
NT147	Lee Pt (E)	1 1	R+sand flats	1.5 km
Note: Stinger season October to May				
Spring & neap tide range = 5.5 & 1.8 m				

Darwin's northern coastal suburbs lie on relatively flat terrain bordered at the shore by 5 km of 10 m high claystone bluffs. Below the bluffs are three beaches (NT 143-145), with two more (NT 146 & 147) extending north of the bluffs to the sandy Lee Point.

Nightcliff beach (**NT 143**) is backed by eroding bluffs that are protected by a rubble seawall, with claystone rock platforms at either end of the 300 m long beach. A road runs along the top of the bluffs with two walkways down to the beach, which is patrolled by Darwin SLSC between June and August. The beach is moderately steep and narrow at high tide, while 200 m wide sand and some rock flats are exposed at low tide. It is separated from its neighbouring beach by a 200 m wide 10 m high headland, with a public swimming pool and park located on the headland. On the eastern side of the headland is 300 m long Rapid Creek beach (**NT 144**). This consists of a veneer of high tide sand over protruding rocks, with a mixture of sand and rock flats exposed at low tide. It is backed by steep eroding red bluffs which restrict access to the far eastern end of the beach. The bluffs are backed by a narrow foreshore reserve, road and then apartment buildings.

Rapid Creek separates the suburb of Rapid Creek from the Dripstone Caves, with a 100 m long footbridge across the creek linking the two. Between the creek and the caves is a 1.2 km long northwest-facing beach **(NT 145)**. The beach has a continuous high tide beach, fronted at the western end by the 1.5 km wide creek tidal shoals, which narrow to 1 km in the east. The beach is backed by a 5-10 m high grassy foredune, then a large gravel car park, and further back the Casuarina campus of the Northern Territory University. The caves are located at the base of 10 m high weathered sandstone bluffs which form the eastern boundary of the beach.

Casuarina Beach (**NT 146**) is the site of the **Darwin Surf Life Saving Club**, and is part of the Casuarina Coastal Reserve, a 1180 ha reserve which encompasses the entire 4.5 km long beach as well as Lee Point. The 5 km long reserve offers a wide range of recreation facilities, as well as a section for nude bathing. The continuous sand beach begins on the eastern side of the Dripstone Cliffs and consists of a wide high tide beach, fronted by 200-300 m wide intertidal sand flats containing two low ridges and runnels (Fig. 5.23). The beach terminates at rock flats that lie off sandy Lee Point. The surf club is located toward the southern end of the beach and is surrounded by the grassy reserve.

Figure 5.23 View across Casuarina Beach (NT 146), showing the Darwin Surf Life Saving Club surf patrol and low tide bars.

Beach **NT 147** extends southeast of **Lee Point** for 1.5 km. It is a wide, low gradient, dynamic beach with sand moving eastward along the beach. During the Holocene this movement formed a 6 km long series of sand spits, all but the Lee Point section now fronted by mangroves. The beach is part of the reserve, with vehicle access to the Buffalo Creek boat ramp and car park at the eastern end.

NT 148-149 SHOAL BAY

No.	Beach	Rating HT LT	Type	Length
NT148	Camerons Beach	1 2	R+mud flats	1 km
NT149	Shoal Bay (E)	1 2	R+sand-mud flats	11.1 km
Spring & neap tide range = 5.5 & 1.8 m				

Shoal Bay is a 20 km wide northwest-facing embayment, bordered by Lee Point in the south and Gunn Point to the north, with funnel-shaped Hope Inlet in the apex. The bay has 35 km of shoreline, much in the west and south fronted by mangroves. The southern shore extends for 15 km east of Lee Point, containing the initial sandy Lee Point beach, then a predominantly mangrove-fringed low energy shoreline. Only **Camerons Beach** (**NT 148**) provides a break in the mangroves. The beach is approximately 1 km long, varying in length depending on migrating sand shoals and presence of mangroves. It is accessible via a vehicle track from the backing radio tower facilities. The 'beach' is essentially a low beach ridge, fronted by 500 m wide sand to mud flats and fringed by mangroves.

Tree Point is located on the northern side of 3 km wide Hope Inlet. It forms the southern boundary of a relatively straight, west-facing, 11.1 km long beach (**NT 149**) that occupies much of the eastern shore of Shoal Bay. The beach consists of a high tide sand beach fronted by 300 m wide intertidal sand flats. It is eroding along the central-southern end, exposing mangrove peat. The sand has moved south to build a series of recurved spits that terminate at Tree Point. There is vehicle access to the beach via a 52 km long gravel and bitumen road from Howard Springs. This road services the Gunn Point Prison Farm, located 3 km behind the centre of the beach, and the Tree Point community, which consists of several houses located toward the southern end of the beach, as well as providing access to the Gunn Point Forestry Reserve which begins toward the northern end of the beach. There is a camping reserve at the beach but no facilities.

NT 150-151 **ADAM BAY**

No.	Beach	Rating HT LT	Type	Length
NT150	Pt Stephens	1 2	R+tidal flats	1.1 km
NT151	Ayers Pt (N)	1 2	R+tidal flats	1.2 km
Spring & neap tide range = 3.4 & 1.4 m				

Adam Bay is a 20 km wide, funnel-shaped, northwest-facing bay bordered by Point Stephens in the west and Cape Hotham to the east. It narrows in the centre to the 1.6 km wide entrance of the Adelaide River bordered by Hart and Ayers points. The bay has 36 km of low wave energy, tide-dominated, predominantly mangrove-fringed shoreline, backed in the south by 1-3 km wide high tide salt flats, with low wooded coastal plain in lee of Point Stephens. There are only two strips of sand along the shore, the first along north-facing Point Stephens. The beach (**NT 150**) consists of a relatively straight, narrow, coarse, high tide beach, fringed by rock flats and mangroves, with rock flats exposed at low tide. A few houses are located at the western end of the beach in lee of the point. They are connected via a vehicle track to the Gunn Point road.

Beach **NT 151** is located 3 km north of Point Ayers, and consists of a 1.2 km long strip of west-facing high tide sand, fringed by mangroves and fronted by 500 m wide intertidal mud flats with beach ridges then coastal plain behind.

NT 152-159 **CAPE HOTHAM**

No.	Beach	Rating HT LT	Type	Length
NT152	Escape Cliffs (N)	1 2	R+mud flats	600 m
NT153	Cape Hotham (W)	1 2	R+sand-mud flats	4.5 km
NT154	Cape Hotham (1)	1 2	R+sand-rock flats	900 m
NT155	Cape Hotham (2)	1 2	R+rock flats	300 m
NT156	Cape Hotham (E1)	1 2	R+mud flats	1.1 km
NT157	Cape Hotham (E2)	1 2	R+mud flats	1.4 km
NT158	Cape Hotham (E3)	1 2	R+sand-mud flats	1.2 km
NT159	Cape Hotham (E4)	1 2	R+tidal flats	700 m
Spring & neap tide range = 3.4 & 1.4 m				

Cape Hotham is a 1-7 km wide, 20 km long, northerly-trending cape, composed of a low wooded laterite coastal plain fringed by 40 km of shoreline, consisting of a mixture of mangroves and low energy beaches and tidal flats. The entire cape is part of the Cape Hotham Forestry Reserve. Eight low energy beaches (NT 152-159) are located around the cape, two on the west, two at the cape and four along the eastern side.

Beach **NT 152** is a 600 m long strip of high tide sand, bordered by mangroves, backed by the coastal plain and fronted by 900 m wide mud flats. The southern end continues as a low mangrove-fronted sand spit that helps impound a 400 m mangrove-filled lagoon.

Beach **NT 153** extends for 4.5 km southwest of the cape, as a relatively straight high tide sand beach, cut by a small central creek and backed by a central beach ridge with coastal plain to either side, with mangrove-covered rock flats forming the boundaries. The Cape Hotham lighthouse is located at the western tip of the cape. Beach **NT 154** is located immediately east of the lighthouse. It is a 900 m long slightly crenulate high tide sand beach, with bordering and scattered rock flats, some beachrock and a backing foredune (Fig. 5.24). Its neighbour **NT 155** continues on for another 300 m as two small curving north-facing sections of beach, largely fronted by rock flats and scattered mangroves, with a small strip free of rocks at the eastern end.

Figure 5.24 Cape Hotham lighthouse (right) with beaches NT 154 & 155 extending to the east.

On the eastern side of the cape the shore turns and faces east. Beach **NT 156** is a 1.1 km long high tide sand beach, paralleled by low tide beachrock as well as rock flats and 200 m wide tidal flats, with mangroves bordering each end. Five hundred metres to the south is an open east-facing embayment containing beaches **NT 157** and **NT 158**. These form a northern 1.4 km and southern 1.2 km long beach-spits separated by a small creek. The creek drains a 250 ha largely mangrove-filled lagoon and surrounding salt flats. It has formed a shallow sandy ebb tide delta, which fronts much of beach NT 158.

Six kilometres further south is a 700 m long gap in the mangroves located on the northern side of a small tidal creek which contains beach **NT 159**. It consists of a low sandy beach ridge-chenier, backed by mangroves and fronted by mangrove peat, scattered mangroves and 200 m wide mud flats.

NT 160-161 CHAMBERS BAY

No.	Beach	Rating HT	Rating LT	Type	Length
NT160	Chambers Bay (W)	1	2	R+mud flats	200 m
NT161	Chambers Bay	1	2	R+mud flats	10.9 km
Spring & neap tide range = 3.4 & 1.4 m					

Chambers Bay is a 60 km wide, north-facing open bay, bordered by Cape Hotham in the west and the low cheniers of Point Stuart to the west. The north-facing slowly curving shore of the bay, is for the most part fronted by mangroves and mud flats, and backed by low cheniers and extensive salt flats, with the low coastal plain lying between 3 and 20 m further to the south. Low energy beaches are only located at either extremity of the bay in the west at the base of Cape Hotham (NT 160 & 161) and in the east on Point Stuart (NT 162).

Beach **NT 160** lies at the eastern base of Cape Hotham and consists of a 200 m long east-facing strip of high tide sand, backed by the low wooded coastal plain with 200 m wide mud flats in front. Immediately east begins the 10.9 m long beach **NT 161**, a continuous low energy strip of high tide sand fronted by 200-300 m wide mud flats, with a 200 m wide series of low beach ridges behind. It is backed in turn by extensive mangroves and salt flats up to several kilometres wide. The flats are drained by tidal creeks flowing south and west into the Adelaide River system.

NT 162 POINT STUART

No.	Beach	Rating HT	Rating LT	Type	Length
NT162	Pt Stuart	1	2	R+mud flats	1.5 km
Spring & neap tide range = 3.4 & 1.4 m					

Point Stuart is a low protrusion of the coastal plain that extends 8 km north into Van Diemen Gulf and separates Chambers Bay from Finke Bay, while mangroves fringe the bay shores to each side. The point itself consists of a series of eight cheniers and intervening tidal flats that have prograded up to 2 km into Chambers Bay (Fig. 5.25). On the tip of the point is a 1.5 km long high tide sandy beach (**NT 162**) largely clear of mangroves, fronted by 200 m wide mud and rock flats. There is vehicle access across the backing coastal plain to the eastern side of the point, where the Stuart's Tree Historical Reserve is located. This marks the spot where on 24 July 1862 John McDouall Stuart completed the first successful crossing of the Australian continent.

Figure 5.25 Point Stuart consists of a series of cheniers largely fronted by mangrove-covered mud flats.

Kakadu National Park

Established:	1979, 1984, 1987	
World Heritage Listing:	1981, 1987	
Area:	1 755 200 ha	
Coast length	135 km	(1440-1575 km)
Beaches	4	(NT 163-166)

Kadadu National Park is one of Australia's best known and most frequently visited parks. It contains a wide range of tropical wetland, escarpment and upland environments, including the southern hills and basins bordered by the rugged southern escarpment, together with the lowlands, river floodplains and the coastal tidal flats. The northern coastal boundary extends for 135 km between West Alligator Head and the mouth of the East Alligator River and contains only four very low energy beaches.

NT 163-166 WEST ALLIGATOR HEAD (Kakadu National Park)

No.	Beach	Rating HT	Rating LT	Type	Length
NT163	Pococks (S)	1	2	R+rock flats	1.5 km
NT164	Pococks Beach	1	2	R+mud flats	5.2 km
NT165	West Alligator Hd	1	2	R+mud flats	1 km
NT166	Sandfly Beach	1	2	R+mud flats	400 m
Spring & neap tide range = 3.4 & 1.4 m					

West Alligator Head lies at the western tip of a 10-30 m high ridge of Tertiary laterite, formed around a core of higher Proterozoic sandstone, which rises to 60 m. The head is accessible by 4WD, with vehicle tracks running along the back of the Pococks beaches down to the mouth of the Wildman River. The mangrove-fringed shoreline of Finke Bay extends for 35 km to the west, while 6 km southeast of the head is the mouth of the West Alligator River. Four low energy beaches (NT 163-166) are located along the western side and tip of the head, and lie within Kakadu National Park.

Beach **NT 163** emerges from the Finke Bay mangroves and salt flats as a 1.5 km long strip of northwest-facing high tide sand, backed by low vegetated beach ridges and fronted by rock flats and scattered mangroves. **Pococks Beach (NT 164)** is a 5.2 km long, northwest-facing, sightly curving beach that terminates in lee of West Alligator Head. The beach is fringed by mangroves and backed by a 100-200 m wide series of low foredunes. A vehicle track reaches the northern end of the beach where a small camping area and bore are located.

Beach **NT 165** is a north-facing 1 km long beach bordered by 10 m high West Alligator Head and a second 15 m high headland. Dense woodlands extend down to the back of the beach, with a few mangroves at both ends and mud flats fronting the narrow high tide strip of sand. Its neighbour, **Sandfly Beach (NT 166)**, is a similar 400 m long strip of sand backed by wooded slope and fringed by mangroves and mud flats. Vehicle tracks provide access to both beaches.

Arnhem Land Aboriginal Land

Area:	9 370 700 ha	(inc. Groote Eylandt)
Coastal length:	114 km	(1590-1704 km)
	2324 km	(2261-4585 km)
Total:	2438 km	
Beaches	11	(NT 167-177)
	1021	(NT 419-1439)
Total:	1132	

Arnhem Land occupies the entire northeast section of the Northern Territory, including the adjacent islands. It is a large relatively undeveloped region apart from two larger towns Nhulunby on the Gove Peninsula and Alyangula on Groote Eylandt, both associated with bauxite mining. The population is predominately aboriginal and spread amongst small coastal communities and camps.

NT 167 **ARALAIJ BEACH**

No.	Beach	Rating HT LT	Type	Length
NT167	Aralaij Beach	1 2	R+mud flats	8.1 km

Spring & neap tide range = 3.4 & 1.4 m

Aralaij Beach (NT 167) is an 8.1 km long strip of low energy west-facing sand located 10 km north of the East Alligator River mouth. The beach is bordered by extensive mangroves to the north and south and for the most part backed by salt flats and freshwater marshes up to 5 km wide. Continuous 400-500 m wide mud flats and scattered mangroves front the beach. The area is part of the Murgenella Wildlife Sanctuary, with no vehicle access to the beach.

NT 168 **BEACH NT 168**

No.	Beach	Rating HT LT	Type	Length
NT 168	Beach NT 168	1 2	R+mud flats	150 m

Spring & neap tide range = 2.2 & 0.2 m

Beach **NT 168** is a 150 m long gap in a continuous mangrove-fringed shoreline that extends for 25 km from the mouth of Murgenella Creek in the south to Minimini Creek to the north. In between the coast protrudes 7 km west into Van Diemen Gulf, as a low section of coastal plain surrounded by salt flats, mangroves and mud flats. The narrow sand beach is located on the northern side of the protrusion and has two strips of intertidal bedrock outcropping to either side, linked by intertidal mud flats, which extend 500 m off the beach.

NT 169 **ENDYALGOUT ISLAND (W)**

No.	Beach	Rating HT LT	Type	Length
NT169	Endyalgout Is (W)	1 2	R+mud flats	2.9 km

Spring & neap tide range = 2.2 & 0.2 m

Endyalgout Island is a 15 km wide, 10 km long section of the coastal plain, rising along its northern half to heights of 30 m. The entire northern, eastern and southern shorelines are surrounded by up to 1 km of intertidal salt flats and mangroves. Only along the southern half of the western shore, facing into Van Diemen Gulf, is there a break in the mangroves, where 2.9 km long, west-facing beach **NT 169** is located. It extends from some low scarped red laterite bluffs at the southwestern tip, due north to the mouth of a small tidal creek, beyond which the mangroves resume. The sandy high tide beach is backed in the south by the low coastal plain, then a beach ridge covered with casuarina trees up to the creek. It is fronted by a few scattered mangroves growing on the 500 m wide mud flats.

Gurig National Park & Cobourg Marine Park

Gurig Gunak Barlu National Park Area:		220 700 ha
Cobourg Marine Park	Area:	22 900 ha
Coast length:	557 km	(1704-2261 km)
Beaches:	241	(NT 178-418)

Gurig National Park occupies the entire Cobourg Peninsula, while Cobourg Marine Park surrounds the waters of the peninsula. A permit is required to access the park via Arnhem Land Aboriginal Land. The main Cobourg vehicle track runs up to Black Point where the ranger station is located and offers limited facilities and supplies. There are campsites available at nearby Smith Point.
Permit and information:
Cobourg Sanctuary Board
PO Box 496, Palmerston NT 0831

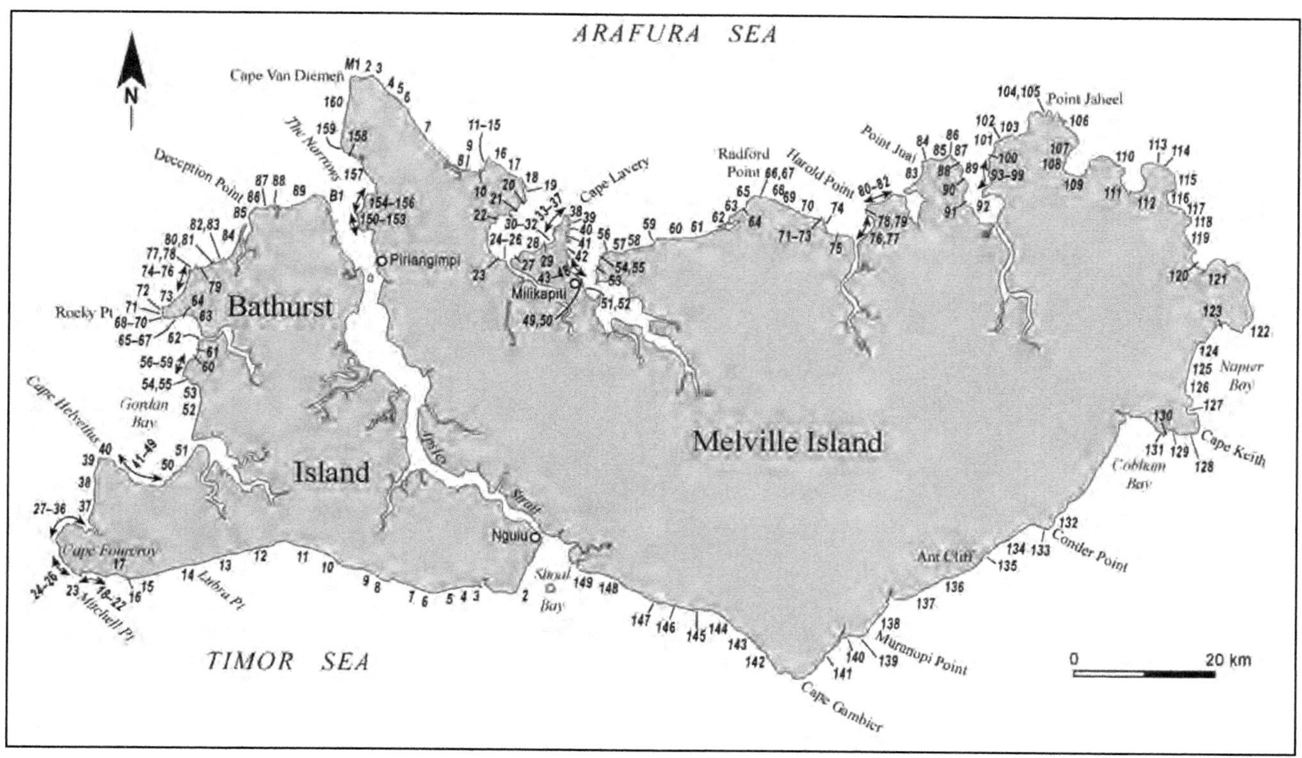

Figure 5.26 Regional map 5: Bathurst and Melville islands, the Tiwi islands.

The Tiwi Islands, as Bathurst and Melville islands are called, are located 70 km due north of Darwin. The two islands are separated by the narrow 60 km long Apsley Strait and have area of 2070 and 6250 km^2 respectively, with a combined area of 8230 km^2. The southern shore of Bathurst and southwest shore of Melville form the northern boundary of Beagle Gulf, while the southeastern shore of Melville together with the Cobourg Peninsula forms the northern shore of Van Diemen Gulf.

Bathurst Island has 251 km of coast containing 89 sandy beaches, while the largest Melville Island has 532 km of shore and 160 beaches (Fig. 5.26). The beaches range from longer sandy beaches, particularly in the south, to more reef-fringed beaches in the northeast (Fig. 5.27 & 5.28)

Figure 5.27 The small beach MI 113 is fringed by coral reefs, bordered by mangroves and backed by salt flats.

Figure 5.28 Beach BI 89 is a long vegetated spit that extends to the northern tip of Bathurst Island at Brace Point (foreground).

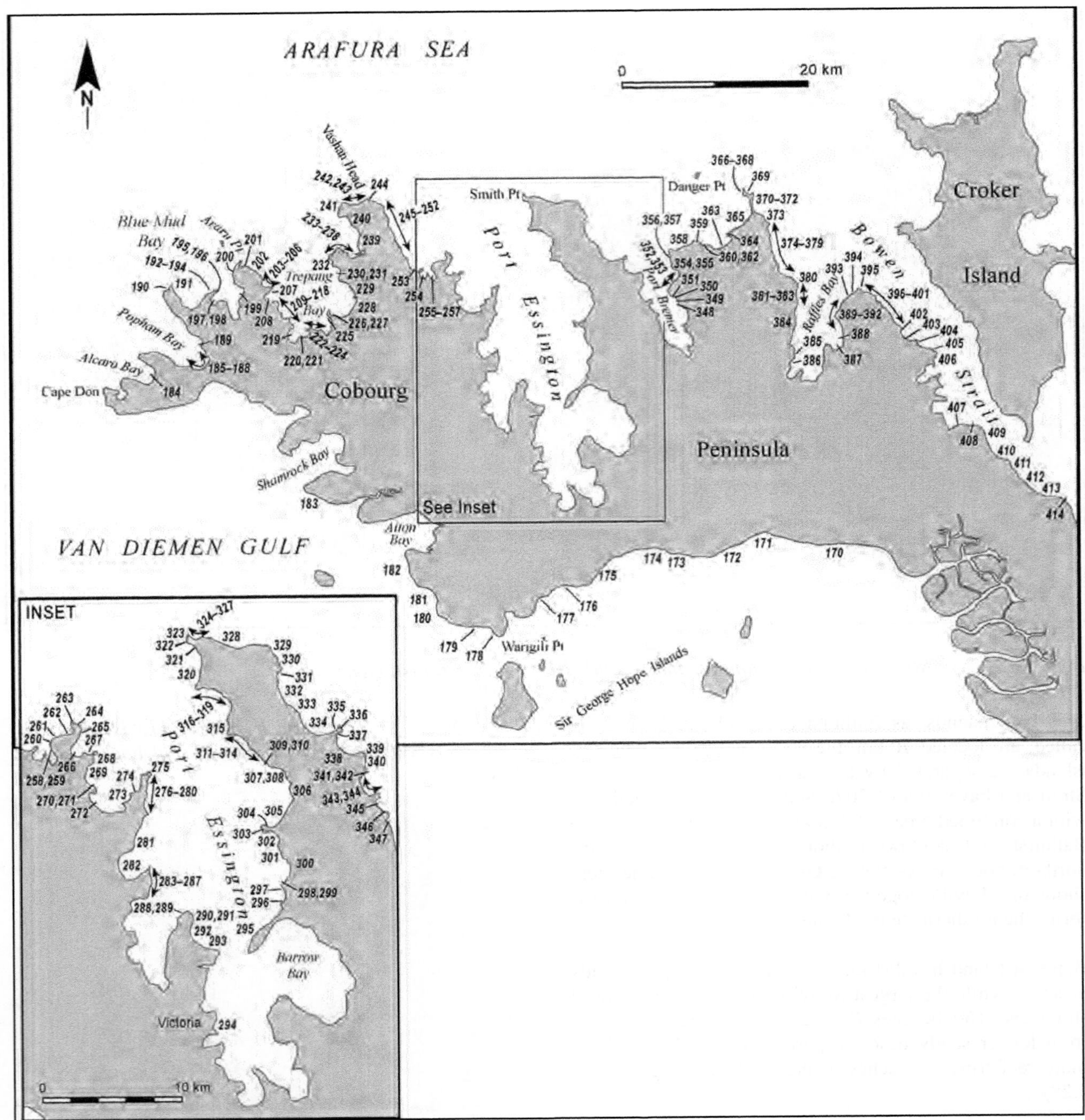

Figure 5.29 Regional map 5: The Cobourg Peninsula, beaches NT 170-414.

NT 170-177 WANGARLU-WURGURLU BAYS

No.	Beach	Rating HT	LT	Type	Length
NT170	Beach NT 170	1	2	R+mud flats	900 m
NT171	Wangarlu Bay	1	2	R+mud flats	2.6 km
NT172	Wangarlu Bay (W)	1	2	R+mud flats	2.5 km
NT173	Wurgurlu Bay (E)	1	2	R+mud flats	1.3 km
NT174	Wurgurlu Bay	1	2	R+mud flats	200 m
NT175	Wurgurlu Bay(W1)	1	1	R+sand ridges	1 km
NT176	Wurgurlu Bay(W2)	1	1	R+sand ridges/coral	600m
NT177	Wurgurlu Bay(W3)	1	1	R+sand ridges/coral	1 km

Spring & neap tide range = 2.2 & 0.2 m

The **Cobourg Peninsula** forms the northern shores of Van Diemen Gulf. This section of coast is part of both the Gurig National Park and Cobourg Marine Park. The southern boundary of the national park begins on the northern side of the 5 km wide Iwalg Creek mouth. A series of large mangrove-fringed creeks occupy the first 20 km of coast until beach **NT 170**, where the coastal plain outcrops at the shore. The 900 m long beach faces due south into the gulf. It is fringed by mangroves and backed by the low wooded plain, with 200 m wide mud flats and a few scattered rock flats and mangroves fronting the beach. Five kilometres to the west is **Wangarlu Bay** which contains two low energy beaches. The first is 2.6 km long beach **NT 171**. This is a relatively straight southeast-facing strip of high tide sand, backed by vegetated 10 m high bluffs of the coastal plain and fronted by 200 m wide mud flats. It is separated by 200 m of mangroves from beach **NT 172**, which extends for another 2.5 km to where the coast swings to face south. This relatively straight beach is also backed by the 10 m high wooded coastal plain and fronted by 200 m wide mud and some rock flats.

To the west of the coastal inflection lies **Wurgurlu Bay**, an open south-facing 20 km long bay containing five beaches. The first 2 km of shore are occupied by mangroves and a tidal creek, with beach **NT 173** a strip of high tide sand largely free of mangroves. It is backed by the coastal plain, with slight protrusions in the plain forming the boundaries. Beach **NT 174** lies on the western side of the next tidal creek. It consists of a 200 m long strip of sand, backed by casuarina trees, with the small tidal creek running out off the eastern end of the beach and tidal sand shoals extending 500 m into the gulf (Fig. 5.30).

Figure 5.30 Beach NT 174 is located adjacent to a small tidal creek that flows into Wurgurlu Bay.

Five kilometres further west the bay shore swings to face into the prevailing southeasterly wind, which permit higher wind waves to reach the shore during strong trade winds. As a consequence the three western beaches of the bay (NT 175-177) are all fronted by ridged sand flats, a product of the higher (short) wind waves. Beach **NT 175** occupies the eastern side of a tidal creek, with much of the sandy beach fronted by a narrow fringe of mangroves, then the 500 m wide sand flats containing 20 low ridges. To the west of the creek mangroves dominate the lee of the ridges. Beaches **NT 176** and **NT 177** lie 5 km further to the southwest. They are neighbouring beaches separated by a small tidal creek, with the coastal plain to either side. They are 0.6 km and 1 km long respectively, and are fronted by 12-14 low sand ridges across the 500 m wide sand flats, which merge on the outer edge with a continuous coral reef.

NT 178-183 **WARIGILI POINT-TWO HILLS BAY**

No.	Beach	Rating HT	Rating LT	Type	Length
NT178	Warigili Pt (E1)	1	1	R+sand flats/coral	150 m
NT179	Warigili Pt (E2)	1	1	R+sand flats/coral	200 m
NT180	Two Hills Bay (S)	1	1	R+sand flats/coral	1.2 km
NT181	Two Hills Bay	1	1	R+sand flats/coral	2.5 km
NT182	Two Hills Bay (N)	1	1	R+sand flats/coral	800 m
NT183	Silvid Bay (N)	1	1	R+sand flats/coral	200 m
Spring & neap tide range = 2.2 & 0.2 m					

Warigili Point is the southernmost tip of the Cobourg Peninsula. It is a 500 m wide headland that projects 1 km to the south and consists of 20 m high weathered Cretaceous sandstone. It is densely wooded but has been scarped around the edges. Immediately west of the point is a 2 km wide mangrove- and salt-flat-filled embayment, partly blocked by beaches NT 178 and 179. Beach **NT 178** is a 150 m long curving sand beach, part of a 50-100 m wide beach ridge-sand spit. It is fringed by mangroves at each end, with a small tidal creek forming the western boundary. Four hundred metre wide sand flats, fringed at the outer edge by coral reefs, front the southwest-facing beach (Fig. 5.31).

Figure 5.31 Beach NT 178 is fronted by sand flats grading into fringing coral reef.

Beach **NT 179** forms a similar beach ridge-spit extending from the western end of the embayment, with a second creek forming its eastern boundary. The beach is partly fronted by a narrow band of mangroves then 200 m wide sand flats, with a prominent outer fringe of coral reef.

Six kilometres west of Warigili Point the coast swings to face west, and three low energy beaches (NT 180-182) occupy the next 4.5 km of coast, up to the mangrove-lined entrance to Aiton Bay. Beach **NT 180** is a straight 1.2 km long southwest-facing narrow strip of high tide sand, fronted by a thin veneer of sand over 400 m wide coral reef flats, with a drop off on the outer reef end. Mangroves are scattered along either end of the beach while the northern end of the beach forms the southern boundary of **Two Hills Bay**. The bay is named after the prominent backing Mount

Roe (160 m) and Mount Bedwell (143 m). The next beach (**NT 181**) is a curving, southwest-facing, 2.5 km long, low energy beach, bordered by low bedrock headlands and scattered rocks. The beach impounds a small (50 ha) tidal lagoon drained by a small shallow creek in the centre of the beach. It is fronted by 1 km wide sand flats, fringed along the edge by coral reef.

Beach **NT 182** extends from the northern headland, as a relatively straight 800 m long southwest-facing beach, bordered by small wooded headlands at each end, with rocks outcropping along the beach and on the sand flats. The sand flats extend for about 50 m before giving way to 300 m wide coral reef, with a deeper gap in the centre of the reef.

Thirteen kilometres to the northwest is an isolated strip of sand (beach **NT 183**) located on a protruding section of the coastal plain that separates Silvid and Shamrock bays. The small 200 m long beach is located adjacent to a small creek that drains a backing 50 ha mangrove-filled lagoon. Mangroves fringe the shore to either side, and the sandy ebb tide creek delta spreads out for 200 m across 400 m wide coral reef flats.

NT 184-189 **ALCARO-POPHAM BAYS**

No.	Beach	Rating HT LT	Type	Length
NT184	Alcaro Bay	1 1	R+sand flats	2.8 km
NT185	Waggali	1 1	R+sand flats	1.3 km
NT186	Ommaridge	1 2	R+sand flats/channel	500 m
NT187	Popham Bay (3)	1 2	R+sand flats/channel	150 m
NT188	Popham Bay (4)	1 1	R+sand flats/coral	250 m
NT189	Popham Bay (5)	1 1	R+sand flats/coral	400 m
Spring & neap tide range = 2.2 & 0.2 m				

Alcaro Bay is the westernmost beach on the Cobourg Peninsula, located just 4 km east of Cape Don. The beach landing for the Cape Don lighthouse is located at the mouth of the creek that forms the western boundary of the 2.8 km long curving, northwest-facing beach (**NT 184**). The beach is protected from all but northwest winds and calms often prevail. It consists of a sandy high tide beach, with an extensive ebb tide delta at the western end grading into 500 m wide sand flats to the east. A casuarina-covered 200 m wide foredune, and then mangrove-filled 100 ha lagoon back the beach (Fig. 5.32).

Four kilometres to the east is **Popham Bay**, a low energy bay protected by headlands that extend 5 km to the northwest on either side of the 2 km wide bay. Deep within the bay are four low energy beaches (NT 185-188), all fronted by wide sand flats. **Waggali** beach (**NT 185**) forms the southern shore of the bay and faces northwest. It is 1.3 km long, backed initially by tree-covered slopes. It terminates at the mouth of a small tidal creek, that cuts off Cape Don from the rest of the Cobourg Peninsula. On the eastern side of the 100 m wide creek is 500 m long **Ommaridge** beach (**NT 186**), a narrow casuarina-covered sand spit that partly impounds the backing 1.5 km wide mangrove forest. It terminates at 10 m high densely vegetated bedrock bluffs. Fifty metre wide shallow sand flats then a deep 200 m wide tidal channel front the beach.

Figure 5.32 Beach NT 184 forms the curving southern shoreline of Alcaro Bay.

Beach **NT 187** begins on the eastern side of the bluffs and is a 150 m long strip of sand running along the base of the tree-covered bluffs. It is fronted by a narrow sand flat then the deep tidal channel. Three hundred metres of rocky shore separate it from beach **NT 188**, a 250 m long west-facing strip of sand backed by a low vegetated beach ridge, that grades into mangroves at the eastern end. Sand-flat-covered coral reef increases in width to 1 km off the eastern end of the beach.

Beach **NT 189** is located 1 km to the north around a 20 m high wooded headland. It is a 400 m long straight west-facing sandy beach, backed by a couple of vegetated beach ridges, with a very small, usually dry, upland creek and small sandy delta crossing the centre, and sand then reef flats extending up to 2 km off the beach.

NT 190-191 **LINGI POINT**

No.	Beach	Rating HT LT	Type	Length
NT190	Lingi Pt	1 2	R+sand/coral flats	1.1 km
NT191	Lingi Pt (E)	1 2	R+sand/coral flats	400 m
Spring & neap tide range = 2.2 & 0.2 m				

Lingi Point is a low, 1 km wide, 3 km long bedrock protrusion, that separates Popham and Blue Mud bays. Mangroves fringe both bay shores, with only the northern tip of the point receiving sufficient wave energy to maintain sandy beaches. The beach and the entire point are also fringed by coral reef flats that widen to 1 km off the point. Beach **NT 190** is a 1.1 km long, northwest-facing sand beach, bounded by protruding laterite rock flats, with a 2 km wide intertidal zone consisting of a mixture of rock, sand and coral reef flats. Low beach ridges, then the wooded coastal plain back the beach. On the eastern side of the point is 400 m long beach **NT 191**, a coarse sand beach bounded by small wooded bluffs and fronted by 1.5 km wide reef flats.

NT 192-196 BLUE MUD BAY

No.	Beach	Rating HT	LT	Type	Length
NT192	Blue Mud Bay (E1)	1	2	R+reef flats	150 m
NT193	Blue Mud Bay (E2)	1	2	R+reef flats	250 m
NT194	Blue Mud Bay (E3)	1	2	R+reef flats	600 m
NT195	Blue Mud Bay (E4)	1	2	R+reef flats	150 m
NT196	Blue Mud Bay (E5)	1	2	R+reef flats	300 m
Spring & neap tide range = 2.2 & 0.2 m					

Blue Mud Bay is a north-facing 4 km wide bay, with 5 km of predominantly mangrove-lined shoreline. Beaches are located only on the western Lingi Point and the unnamed eastern point. The eastern point consists of a 500 m wide section of low wooded coastal plain. Along the 1.5 km long north-facing southern shore are three low energy beaches (NT 192-194) all lying in lee of 1 km wide coral reef flats. Beach **NT 192** is a 150 m long strip of high tide sand, with 100 m wide sand flats, then reef flats, with laterite rocks protruding 50 m off either end. Its neighbour (**NT 193**) is a similar 250 m long beach, with laterite outcropping the length of the beach and with a small grassy beach ridge behind. Beach **NT 194** is a 600 m long continuation, with extensive laterite rock flats exposed along the base of the beach, then some sand and the wider reef flats. It terminates at the northern tip of the point.

On the eastern side of the tip is a curving northeast-facing 150 m long strip of high tide sand (**NT 195**), bordered by prominent laterite rock flats, backed by a few low vegetated beach ridges, with reef flats extending 1 km off the beach. Its neighbour (**NT 196**) begins on the southern side of the laterite and curves for 300 m to another small laterite protrusion, with shallow reefs off the northern half, narrowing toward the southern end of the beach.

NT 197-199 BEACHES NT 197-199

No.	Beach	Rating HT	LT	Type	Length
NT 197	Beach NT 197	1	2	R+mud flats	200 m
NT198	Beach NT 198	1	2	R+mud flats	150 m
NT199	Beach NT 199	1	2	R+mud flats	1.4 km
Spring & neap tide range = 2.2 & 0.2 m					

Beaches NT 197-199 occupy part of a funnel-shaped 4 km deep bay. The 1.5 km wide mouth is fringed by coral reefs, which leave a 500 m wide tidal channel to drain the bay. Beaches NT 197 and 198 are located in a small cove on the western side of the bay and face northeast. Beach **NT 197** occupies a 200 m long gap in the mangroves, as well as being bordered by laterite outcrops and backed by the densely wooded coastal plain. Two hundred metres to the east is **NT 198**, a similar 150 m long beach, with dense mangroves to either side. Both are fronted by 1 km wide mud flats.

On the eastern side of the bay is 1.4 km long west-facing beach **NT 199**. This long low beach is backed by casuarina-covered beach ridges up to 200 m wide, with a small freshwater swamp behind the centre of the beach, then coastal plain. Mangrove-fringed bedrock borders each end, with 1 km wide mud flats off the beach.

NT 200-202 ARARU POINT

No.	Beach	Rating HT	LT	Type	Length
NT200	Araru Pt	1	2	R+reef flats	2.5 km
NT201	Araru Pt (E1)	1	2	R+reef flats	500 m
NT202	Araru Pt (E2)	1	2	R+sand/reef flats	1.9 km
Spring & neap tide range = 2.2 & 0.2 m					

Araru Point is a 2.5 km long sandy foreland that has accumulated along the western side of a protruding section of the coastal plain (Fig. 5.33). The entire point is surrounded by 1-2 km wide coral reef flats that are supplying coarse carbonate sand and gravel to the backing beaches. Beach **NT 200** extends from the point toward the south as an outward curving crenulate sandy beach, terminating with a 200 m long recurved spit that is delivering sand to the quiet waters of the bay off beach NT 199. The beach is backed by shell beach ridges up to 300 m wide near the point, then the wooded coastal plain. A vehicle track reaches the centre of the beach, where three houses and a radio tower are located on the fringe of the coastal plain.

Figure 5.33 Araru Point is a low sandy foreland with beach NT 200 to the right, and NT 201 to the left.

On the eastern side of the point is a dynamic east-facing 500 m long beach ridge (**NT 201**) backed by washover flats and a small lagoon containing some mangroves. The shape of the beach and entrance to the lagoon changes over time, as sand moves into the bay.

The northeastern side of the point consists of a relatively straight 1.9 km long section of northeast-facing shore, consisting of a coarse sand beach (**NT 202**), backed by a few beach ridges, then the coastal plain, with a vehicle track atop the ridges. It is fronted by 1 km wide sand then coral reef flats.

NT 203-206 **TREPANG BAY (1-4)**

No.	Beach	Rating HT	Rating LT	Type	Length
NT203	Trepang Bay (1)	1	2	R+sand/reef flats	400 m
NT204	Trepang Bay (2)	1	2	R+sand/reef flats	1.2 km
NT205	Trepang Bay (3)	1	2	R+rock/reef flats	300 m
NT206	Trepang Bay (4)	1	2	R+rock/reef flats	100 m
Spring & neap tide range = 2.2 & 0.2 m					

Trepang Bay is 6 km wide at its entrance and extends 7 km toward the south, with 35 km of predominantly low energy shoreline, containing 30 largely bedrock-controlled beaches (NT 203-232) and four tidal creeks. Apart from a vehicle track from Araru Point, which reaches the back of the first two beaches, there is no other access or habitation around the bay.

The first small bay east of Araru Point contains beaches NT 203-206. Beaches NT 203, 205 and 206 are located on the opposite headlands, with beach NT 204 filling the bay. Beach **NT 203** is a crenulate, 400 m long east-facing section of sandy shore backed by wooded bedrock slopes, and fronted by sand and mud flats, together with numerous laterite outcrops.

Beach **NT 204** is a curving 1.2 km long northeast-facing beach and barrier, the barrier consisting of about ten coarse sandy beach ridges, that have prograded about 500 m, partly filling the small embayment. A shallow creek at the eastern end drains a small backing lagoon. A vehicle track crosses the beach ridges and terminates at the creek mouth.

Beaches **NT 205** and **NT 206** lie on the southern headland. They are 300 m and 100 m long respectively, and are backed by 5-10 m high wooded slopes and bordered and fronted by numerous outcrops of laterite, then sand and coral flats extending 1 km into the bay (Fig. 5.34).

NT 207-211 **TREPANG BAY (5-9)**

No.	Beach	Rating HT	Rating LT	Type	Length
NT207	Trepang Bay (5)	1	2	R+sand/mud flats	150 m
NT208	Trepang Bay (6)	1	2	R+mud flats	1.3 km
NT209	Trepang Bay (7)	1	2	R+rock/mud flats	50 m
NT210	Trepang Bay (8)	1	2	R+rock/mud flats	50 m
NT211	Trepang Bay (9)	1	2	R+rock/mud flats	50 m
Spring & neap tide range = 2.2 & 0.2 m					

Figure 5.34 Beaches NT 205 & 206 are located on the headland (foreground), with beaches NT 207 & 208 located in the backing bay.

The second bay east of Araru Point has a central beach barrier (NT 208) with four smaller beaches on the bordering headlands. Beach **NT 207** is located on the northern headland and is an east-facing 150 m long strip of sand, at the base of 10 m high wooded slopes, bordered by bedrock shore, and fronted by mud and reef flats.

The main bay beach (**NT 208**) is a slightly curving 1.3 km long beach and backing low barrier, containing about 200 m of prograded beach ridges, then a small elongate marsh. Mangroves dominate the southern shore of the bay.

Mangroves initially fringe the southern headland with three small 50 m long sand beaches (**NT 209, NT 210, NT 211**) located on the most exposed northeast-facing section. They lie in three gaps in the laterite bedrock and are backed by wooded slopes. Each is bordered by protruding laterite reefs, with some fringing coral reef, and all fronted by mud flats.

NT 212-219 **TREPANG BAY (10-17)**

No.	Beach	Rating HT	Rating LT	Type	Length
NT 212	Trepang Bay (10)	1	2	R+mud flats	700 m
NT213	Trepang Bay (11)	1	2	R+mud flats	50 m
NT214	Trepang Bay (12)	1	2	R+mud flats	100 m
NT215	Trepang Bay (13)	1	2	R+mud flats	200 m
NT216	Trepang Bay (14)	1	2	R+mud flats	100 m
NT217	Trepang Bay (15)	1	2	R+mud flats	200 m
NT218	Trepang Bay (16)	1	2	R+mud flats	300 m
NT219	Trepang Bay (17)	1	2	R+mud flats	300 m
Spring & neap tide range = 2.2 & 0.2 m					

The southwestern shores of Trepang Bay become increasing protected with distance from the mouth. The predominantly rocky shoreline is backed by gently wooded slopes. The beaches consist of strips of high tide sand, all fronted by mud flats, in addition to some outcrops of laterite.

Beach **NT 212** is located 5 km south of the mouth and occupies a small bay bordered by low bedrock headlands. It is 700 m long and faces northeast, with mud flats filling most of the bay, apart from bedrock reefs off the headlands. A low beach ridge then wooded slopes back the beach.

The next bay is filled with mangroves, apart from a 50 m gap toward the eastern end where beach **NT 213** is located. Wooded slopes back the small beach. On the adjoining eastern headland is beach **NT 214**, a straight 100 m long northeast-facing strip of high tide sand, backed by 10 m high wooded slopes, and bordered and fronted by laterite reefs and mud flats.

Beaches **NT 215** and **NT 216** lie on the eastern side of the headland and are 200 m and 100 m long respectively. They consist of narrow strips of high tide sand lying at the base of 5 m high scarped bluffs, with mud and rock flats fronting the beach and a prominent low bedrock headland at the eastern end of NT 216.

On the south side of this headland the coast trends south and contains three bedrock-controlled small embayments, which contain beaches **NT 217**, **NT 218** and **NT 219**, each 200 m, 300 m and 300 m long respectively. They all face east and consist of a strip of curving high tide sand, fringed by mud flats and scattered mangroves. They are bordered by bedrock and backed by 10 m high wooded slopes.

NT 220-225 **TREPANG BAY (18-23)**

No.	Beach	Rating HT LT	Type	Length
NT220	Trepang Bay (18)	1 2	R+mud flats	500 m
NT221	Trepang Bay (19)	1 2	R+mud flats	400 m
NT222	Trepang Bay (20)	1 2	R+mud flats	300 m
NT223	Trepang Bay (21)	1 2	R+mud flats	200 m
NT224	Trepang Bay (22)	1 2	R+mud flats	400 m
NT225	Trepang Bay (23)	1 2	R+mud flats	200 m
Spring & neap tide range = 2.2 & 0.2 m				

The southern corner of Trepang Bay contains two adjoining embayments, both with tidal creeks at their apex. The western bay is 1.5 km wide at the entrance and a similar depth. Beaches NT 217-219 line the western shore, followed by 1.5 km of mangroves and the tidal creek, then two small embayments containing beaches NT 220 and 221 on the eastern side of the bay. Beach **NT 220** is a relatively straight 500 m long, north-facing sand beach, backed by a 50 m wide beach ridge plain, with bordering low bedrock headlands and a bedrock island 100 m off the eastern headland. On the eastern side of the island-headland is beach **NT 221**, a 400 m long, north-facing strip of high tide sand. It is bordered by mangroves and 10 m high wooded bedrock headlands, backed by a beach ridge, then a small melaleuca swamp.

The western headland extends for 1 km into the bay, then turns to face north. Along its 1.5 km long, bedrock-controlled northern shore are three low energy beaches (NT 222-224). Beach **NT 222** is a straight 300 m long beach, fronted by mud flats and backed by a scarped 5 m high bluff. Its neighbour (beach **NT 223**) is a curving 200 m long narrow strip of high tide sand intermixed with bedrock rocks and flats, then mud flats, with a low backing scarp. Finally beach **NT 224** is a more continuous 400 m long slightly curving high tide sand beach, backed by the wooded coastal plain, with some bedrock outcrops along the beach, then the mud flats. To the east is the 500 m wide second southern bay, which is dominated by a mangrove-fringed tidal creek.

On the eastern side of the bay a narrow headland extends 1 km northward. On the western side of the headland is a small tidal creek which forms the southern boundary of a very low energy crenulate 200 m long beach, **NT 225**. It is fronted by 500 m wide sand to mud flats, and a scattering of mangroves, and terminates against the 15 m high bluffs of the headlands.

NT 226-229 **ALARU CREEK**

No.	Beach	Rating HT LT	Type	Length
NT226	Trepang Bay (24)	1 2	R+sand/mud flats	100 m
NT227	Trepang Bay (25)	1 2	R+mud flats	150 m
NT228	Alaru Ck	1 2	R+mud flats	2.6 km
NT229	Alaru Ck (E)	1 2	R+mud flats	600 m
Spring & neap tide range = 2.2 & 0.2 m				

The eastern shore of Trepang Bay contains two open west-facing embayments, each containing several beaches and separated by a 3 km long headland, which also has several beaches. The southern of the two bays is 3 km wide and contains four beaches (NT 226-229).

Beaches **NT 226** and **NT 227** are two small pockets of sand located at the base of 20 m high scarps on the eastern side of the narrow headland that separates them from beach NT 225. They are 100 m and 150 m long respectively and consist of a narrow strip of high tide sand fronted by 500 m wide mud flats.

The central portion of the bay has been infilled by up to 1.5 km of low beach ridge-cheniers-salt flat plain. Alaru Creek drains through the centre of the plain, with beach NT 228 to the south and NT 229 to the north. Beach **NT 228** is a 2.6 km long, slightly curving, northwest-facing, low energy strip of high tide sand fronted by 800 m wide mud flats. It is backed by an inner chenier and extensive salt flats (Fig. 5.35) and abuts 20 m high bluffs at the western end, on top of which is a solitary house. Beach **NT 229** begins on the north side of the creek and extends for 600 m as a similar low energy beach and mud flats, with mangroves backing the beach.

Figure 5.35 Beach NT 228 is a strip of sand with salt flats behind and mud flats in front.

NT 230-238 **MIDJARI POINT**

No.	Beach	Rating HT LT	Type	Length
NT230	Trepang Bay (28)	1 2	R+mud flats	900 m
NT231	Trepang Bay (29)	1 2	R+mud flats	100 m
NT232	Midjari Pt (S3)	1 2	R+mud flats	400 m
NT233	Midjari Pt (S2)	1 2	R+rock flats	800 m
NT234	Midjari Pt (S1)	1 2	R+mud flats	1 km
NT235	Midjari Pt (E1)	1 2	R+mud flats	200 m
NT236	Midjari Pt (E2)	1 2	R+mud flats	100 m
NT237	Midjari Pt (E3)	1 2	R+mud flats	500 m
NT238	Midjari Pt (E4)	1 2	R+mud flats	1.2 km
Spring & neap tide range = 2.2 & 0.2 m				

Midjari Point is a 3 km wide protrusion of the 30 m high coastal plain that occupies 6 km of the central shoreline on the eastern side of Trepang Bay. A total of nine beaches (NT 230-238) are located around the shores of the protrusion, with Midjari Point located at the northern tip. In addition sand-mud and coral reef flats surround much of the point.

Beach **NT 230** is a straight, 900 m long, west-facing low energy beach, backed by a 200 m wide beach ridge plain. Mud flats initially front the beach, then coral reefs extend 1.5 km off the shore. It abuts 15 m high 500 m long bluffs at its northern end, which are initially fronted by mangroves. At the northern end of the bluffs is a 100 m long clear section of sandy beach (**NT 231**) backed by the bluffs and fronted by the mud then reef flats.

Beach **NT 232** is a second straight, 400 m long, west-facing beach, backed by a 100 m wide beach ridge plain and then a small shallow lagoon. It is bordered by rock flats in the south and 10 m high wooded bluffs to the north. These bluffs mark the beginning of 800 m long beach **NT 233**, a crenulate, rock-dominated protrusion that is backed by the wooded bluffs, with the beach a narrow high tide strip of sand fronted by irregular rock flats and patches of reef flats.

On the northern side of the rocks is a curving 1 km long northwest-facing sand beach (**NT 234**), backed by a narrow beach ridge plain, and bordered in the north by the rock flats of Midjari Point. It is fronted by mud then reef flats extending 1 km off the point.

On the eastern side of the point the coast trends southeast toward a creek mouth. Between the point and creek are four low energy beaches (NT 235-238) all facing northeast across a 1.5 km wide embayment. Beach **NT 235** is a crenulate 200 m long sand beach and beach ridge backed by wooded slopes, and fronted with a mixture of mud and rock flats, with 100 m of mangroves separating it from the neighbouring beach. Beach **NT 236** begins on the eastern side of the mangroves, and is a 100 m long strip of sand that terminates at a small cuspate foreland formed in lee of rock flats. Beach **NT 237** extends from the eastern side of the foreland and runs for 500 m as a slightly crenulate beach, also influenced by occasional rock flats. Beach **NT 238** begins at the end of the rocks, and curves for 1.2 km as a north-facing beach, backed by a 100 m wide beach ridge plain and a small central lagoon. The northern end of the beach becomes the southern channel of the 50 m wide tidal creek, with the mud flats giving way to the deeper tidal channel.

NT 239-243 **WANARAIJ POINT-VASHON HEAD**

No.	Beach	Rating HT LT	Type	Length
NT239	Wanaraij Pt (S3)	1 1	R+sand flats	3 km
NT240	Wanaraij Pt (S2)	1 1	R+sand flats	1.4 km
NT241	Wanaraij Pt (S 1)	1 1	R+sand flats	2.2 km
NT242	Wanaraij Pt (E)	1 1	R+sand flats	1.7 km
NT243	Vashon Hd (W)	1 1	R+sand flats	700 m
Spring & neap tide range = 2.2 & 0.2 m				

Wanaraij Point and Vashon Head form the northern tip of the peninsula that separates Trepang Bay from Port Essington, and together with Smith and Danger points to the east form the three northernmost fingers of the Cobourg Peninsula.

To the south of Wanaraij Point the coast extends for 6.5 km to the mouth of a tidal creek, with three beaches (NT 239-241) in between. Beach **NT 239** begins as a narrow recurved spit that forms the northern boundary of the creek mouth. It runs due north for 3 km as a low energy sand beach backed by a 200-300 m wide Holocene barrier, consisting of casuarina-covered beach ridges transforming to recurved spits in the south, with a mangrove-lined creek behind. The northern end terminates at a low bedrock-controlled inflection, behind which is a landing ground which is connected by a vehicle track to the Seven Spirit Bay Resort 14 km to the southeast. Two to three hundred metre wide sand flats front the beach.

On the northern side of the inflection begins beach **NT 240**. It is a curving, 1.4 km long, southwest-facing low energy beach, backed by a 100-500 m wide series of beach ridges, upon which the landing ground has been built, roughly parallel to the beach, with a vehicle track reaching the centre of the beach. The beach terminates in lee of an active recurved spit that forms the southern end of the adjoining beach.

Beach **NT 241** is the main **Wanaraij Point** beach, curving round continuously from the southern recurved spit, as a west-facing beach that terminates at the northern sandy point. The beach is backed by a 1.5 km wide plain consisting of up to 30 beach ridges, forming one of the wider beach ridge plains on the Northern Territory coast (Fig. 5.36). The ridges are very distinctive from the air and on the ground form low parallel undulations, which have been largely cleared of casuarina. The steep beach is fronted by 400-1000 m wide sand flats, and then by coral reef flats that extend from 1-2 km off the beach. Vehicle tracks from the landing ground reach the beach.

Figure 5.36 Wanaraij Point is a large sandy foreland composed of multiple beach ridges, and bounded by beach NT 241 (right) and NT 242 (left).

At Wanaraij Point the shore turns and trends east as beach **NT 242**. This is a curving 1.7 km long beach, also backed by the beach ridges that narrow in width to the east where the beach terminates against a low laterite-controlled inflection just south of Vashon Head. Sand flats extend 1 km off the beach.

Vashon Head is a 10-20 m high wooded headland, composed of Tertiary laterite that also extends as shallow reefs off the head. Beach **NT 243** lies immediately west of the head and is a 700 m long northwest-facing sandy beach, bounded by a 200 m long northwest-trending laterite reef at the head and the southern inflection point, with beach NT 242. Scattered reefs extend out up to 500 m, grading into sand flats that continue up to 1 km seaward.

NT 244-252 **VASHON HEAD (E)**

No.	Beach	Rating HT	Rating LT	Type	Length
NT244	Vashon Head (E1)	1	1	R+LTT	2.2 km
NT245	Vashon Head (E2)	1	2	R+sand flat+reefs	100 m
NT246	Vashon Head (E3)	1	2	R+sand flat+reefs	100 m
NT247	Vashon Head (E4)	1	2	R+sand flat+reefs	1.1 km
NT248	Vashon Head (E5)	1	2	R+sand flat+reefs	400 m
NT249	Vashon Head (E6)	1	2	R+sand flat+reefs	600 m
NT250	Vashon Head (E7)	1	2	R+sand flat	1.3 km
NT251	Vashon Head (E8)	1	2	R+sand flat+reefs	2.1 km
NT252	Vashon Head (E9)	1	2	R+sand flat+reefs	700 m

Spring & neap tide range = 2.1 & 0.1 m

At **Vashon Head** the shore turns 90° and trends east, with the adjoining beach (**NT 244**) running as a crenulate north-facing sandy beach for 2.2 km to a low laterite point, where the shore turns to face east. The beach is exposed to easterly waves, which average over 0.5 m and maintain a relatively steep beach and narrow attached low tide bar. Toward the eastern end the bar widens and waves break off the beach (Fig. 5.37).

Figure 5.37 Vashon Head is a low red laterite headland and reef, with surf breaking over the reefs. Beach NT 244 right, and NT 245-247 left.

Beaches **NT 245** and **NT 246** are two neighbouring 100 m long beaches, each bounded by low laterite headlands and protruding reefs, with NT 246 also backed by a scarped 5 m high red laterite bluff. Sand flats and laterite reefs extend up to 400 m off the beaches, with waves breaking over the patchy reefs as well as at the beach.

On the southern side of the headland, beach **NT 247** sweeps away to the south-southeast for 1.1 km, as a relatively straight beach, to a low 50 m long section of laterite rocks, that protrudes 100 m off the shore. The rocks separate it from beach **NT 248**, a straight 400 m long beach, with laterite rocks and reefs bordering each end and some rocks and reefs toward the centre. Both beaches are backed by low foredunes with scattered casuarina, and fronted by 400 m wide sand, rock and reef flats.

Beach **NT 249** is a curving reef and rock-dominated, 600 m long, east-facing beach that separates its two neighbouring beaches. It is backed by 10 m high scarped red bluffs, with a sandy high tide beach, and rocks along the low tide line, and scattered rock reefs extending up to 500 m offshore. Beach **NT 250** is a 1.3 km long, slightly crenulate, northeast-facing sand beach, backed by a low foredune, with some scattered reefs extending up to 500 m off the beach. It terminates in the south at a slight sandy protrusion in lee of some more prominent reefs, with rocks also on the shore. On the southern side of the protrusion is 2.1 km long beach **NT 251**, a northeast-facing sand beach, backed by a low foredune and small freshwater lagoon toward the southern end. It is fronted by sand flats and rock reefs extending up to 500 m off the beach. It terminates at a small sandy foreland formed in lee of a patch of rock.

On the southern side of the rock is beach **NT 252 a** curving 700 m long, northeast-facing beach that extends to a prominent 100 m long, 10 m high, red laterite bluff, which has a vehicle track leading to the bluff edge. The beach is backed by densely wooded slopes and fronted by 500 m wide sand flats merging with coral reef flats that extend 1 km off the beach.

Surf: The higher wind waves and shallow reefs off some of these beaches produce several small reef breaks, especially at the eastern ends of beaches NT 244 and 247, and off NT 248 and 251.

NT 253-259 **CORAL-SEVEN SPIRIT BAYS**

No.	Beach	Rating HT	Rating LT	Type	Length
NT253	Coral Bay (W)	1	2	R+sand-reef flats	2.2 km
NT254	Seven Spirit Bay	1	2	R+sand-reef flats	1 km
NT255	Seven Spirit Bay (head)	1	2	R+rock flats	50 m
NT256	Coral Bay (4)	1	2	R+tidal flats	100 m
NT257	Coral Bay (5)	1	2	R+tidal flats	150 m
NT258	Coral Bay (6)	1	2	R+rock+tidal flats	100 m
NT259	Coral Bay (7)	1	2	R+rock+tidal flats	100 m

Spring & neap tide range = 2.1 & 0.1 m

Coral Bay is a 4 km wide, open, north-facing bay, the southern half of which has a continuous coral reef system extending 3 km across the bay and up to 1.5 km from the shore. Within the reef system are three small embayments containing a total of seven beaches (NT 253-259).

The first 1.5 km wide embayment contains beach **NT 253**, a curving 2.2 km long northeast-facing sand beach, backed by a 300-500 m wide series of low partially cleared beach ridges, with a small melaleuca swamp behind the southern end of the beach. It is bordered by the red bluff to the north, and a wooded 500 m long 200 m wide headland to the south.

On the eastern side of the headland is **Seven Spirit Bay**, containing a deeply embayed, protected, 1 km long beach (**NT 254**) bordered by two headlands (Fig. 5.38), with a jetty and barge landing located on the eastern headland. The Seven Spirit Bay resort is nestled on the wooded bluffs overlooking the eastern corner of the bay. Dense vegetation backs the beach with sand-mud flats and reef flats filling the embayment.

Figure 5.38 Seven Spirit Bay is a deeply embayed low energy bay, with beach NT 254 forming its southern shore.

The eastern headland for the bay is a narrow 500 m long projection that separates Seven Spirit from the easternmost bay. The resort is spread over both sides of the wooded headland, with a jetty-barge landing toward the western tip of the headland. On the tip of the headland is a north-facing 50 m long strip of high tide sand (**NT 255**), bounded by low headlands and rocks, with bedrock rock flats and reef also off the beach.

The eastern bay contains four small low energy beaches (NT 256-259), with intervening bedrock and mangrove-fringed shore, all fronted by shallow sand-mud flats. Beach **NT 256** is a north-facing 100 m long narrow strip of high tide sand, backed by densely wooded 20 m high slopes, with the resort on top. It is bordered by mangroves to the west and bedrock to the east. Deep in the 500 m wide bay is beach **NT 257**, a 150 m long break in the mangroves, with casuarina growing almost to the shore, and a few scattered mangroves along the narrow beach. On the western side of the bay are two 100 m long strips of narrow high tide sand (**NT 258 & NT 259**), each bounded by small bedrock protrusions and backed by wooded slopes, with rock and tidal flats off each beach.

NT 260-263 **WALFORD POINT (W)**

No.	Beach	Rating HT	Rating LT	Type	Length
NT260	Coral Bay (8)	1	2	R+reef flats	800 m
NT261	Walford Pt (W3)	1	2	R+sand flats	100 m
NT262	Walford Pt (W2)	1	2	R+sand flats	800 m
NT263	Walford Pt (W1)	1	2	R+rock flats	100 m

Spring & neap tide range = 2.1 & 0.1 m

Walford Point is a low laterite headland that forms the boundary between Coral and Kennedy bays. The point is surrounded by both laterite and reef flats, as well as several low energy beaches. Four beaches (NT 260-263) are located within 2 km west of the point.

Beach **NT 260** is an 800 m long northwest-facing moderately steep sand beach, backed by gently wooded slopes and fronted by fringing coral reef which extends from 200-400 m off the beach. Beach **NT 261** is a narrow 100 m long strip of high tide sand, bounded and backed by low laterite bluffs, with rock and reef flats fringing each end. Beach **NT 262** is a curving 800 m long medium sand beach, with a moderately steep beach face, and a few partly cleared beach ridges and minor low sand dunes extending 200 m inland. It is fronted by clear 100 m wide sand flats (Fig. 5.39). Beach **NT 263** sits on the western side of the bedrock point, and is a 100 m long narrow high tide beach, backed by a low cleared beach ridge and fronted by a 100 m wide mixture of rock and reefs. The beach ridges extend across the 100 m wide point to beach NT 264.

Figure 5.39 Coral Bay (right) with the curving beach NT 262, and Kennedy Bay (left) with the beach NT 265.

NT 264-267 WALFORD-LOW POINTS

No.	Beach	Rating HT	Rating LT	Type	Length
NT264	Walford Pt (E1)	1	2	R+reef flats	200 m
NT265	Walford Pt (E2)	1	2	R+rock flats	300 m
NT266	Walford Pt (E3)	1	2	R+sand flats	1.5 km
NT267	Low Pt (W)	1	2	R+reef/sand flats	500 m
Spring & neap tide range = 2.1 & 0.1 m					

To the east of **Walford Point** the bedrock shore initially trends south for 2 km into a small 1 km wide embayment, with Low Point, 1.5 km to the southeast, forming the eastern boundary. Between the two points are four beaches (NT 264-267).

Beach **NT 264** lies immediately east of the narrow northern tip of Walford Point. It is backed by the low cleared 100 m wide beach ridge plain, which links it to beach NT 263. However the foreshore is dominated by irregular intertidal rock flats, over which waves break during easterly winds.

Beach **NT 265** is a crenulate 300 m long steep sandy beach, backed by partly cleared beach ridges which link with those of beach NT 262, 300 m to the west. The beach faces east and is fronted by a 200 m wide mixture of rock and reef flats, with sand flats extending up to 500 m off the beach.

Beach **NT 266** lies at the head of the embayment. It is a curving 1.5 km long sandy beach, backed initially by bedrock bluffs in the east, then a 300 m wide, largely cleared beach ridge plain, with a small freshwater marsh behind the centre of the plain. Eight hundred metre wide sand flats fill the embayment. A straight, formed vehicle track reaches the eastern end of the beach and goes on to the top of the bluffs above the western end of beach NT 267.

Beach **NT 267** is located on the north side of the eastern **Low Point** headland. It consists of a 500 m long bedrock- and reef-dominated strip of high tide sand, with the reefs and rocks extending 100 m off either end and patchy along the centre, with sand flats extending up to 300 m off the beach.

NT 268-274 KENNEDY BAY

No.	Beach	Rating HT	Rating LT	Type	Length
NT268	Low Pt (E 1)	1	2	R+sand/reef flats	500 m
NT269	Low Pt (E 2	1	2	R+sand/rock flats	400 m
NT270	Low Pt (E 3)	1	2	R+sand/rock flats	700 m
NT271	Low Pt (E 4)	1	2	R+sand/rock flats	600 m
NT272	Kennedy Bay (1)	1	2	R+sand flats	3.5 km
NT273	Kennedy Bay (2)	1	2	R+sand flats	600 m
NT274	Kennedy Bay (3)	1	2	R+sand/reef flats	500 m
Spring & neap tide range = 2.1 & 0.1 m					

Kennedy Bay is a U-shaped, north-facing, 3 km wide bay bounded by the low, wooded, laterite Low and Turtle points. It is located 6 km southwest of Black Point and provides a good anchorage during easterly winds. It has 9 km of generally low energy shoreline, containing seven beaches.

Beach **NT268** is located immediately south of Low Point. It consists of a curving 500 m long east-facing beach, bounded by rock platforms and reefs, with the central section fronted by sand flats and a few scattered coral reefs. Just to the south is beach **NT 269**, a broken 400 m long series of four pockets of sand separated by patches of rock and mangroves, with a rock spit extending north off the western end. The beach is backed by 10 m high wooded slopes and fronted by a 300 m wide mixture of rock and sand flats.

On the southern side of the spit are beaches **NT 270** and **NT 271**. They are near continuous, narrow, 700 m and 600 m long northeast-facing beaches. They are both backed by 10 m high bright red laterite bluffs and fronted by 500 m wide sand flats containing a few patchy reefs.

Beach **NT 272** is the main central beach for the bay. It occupies the southern shore of the bay and consists of a curving 3.5 km long north-facing beach, bounded by wooded headlands at either end. The beach is backed by a series of beach ridges up to 600 m wide in the centre, where they are backed by two small melaleuca swamps, the larger central one connected to the shore by a meandering tidal creek. Sand flats extend several hundred metres off the beach, together with some reef patches off the centre of the beach.

Beach **NT 273** lies on the western arm of the bay and is a curving, 600 m long, northwest-facing, low energy, gravelly-shelly beach, backed by a 100 m wide beach ridge, then a small melaleuca swamp. A small stand of mangroves fronts the northern end of the beach, with sand flats extending 100 m off the beach.

The northernmost western beach is **NT 274**, a relatively straight 500 m long sand beach, backed by wooded 10-20 m high wooded slopes, which are scarped toward the eastern end. It is fronted by a 400 m wide mixture of sand and reef flats (Fig. 5.40).

Figure 5.40 Beach NT 274 is backed and bordered by laterite bluffs.

NT 275-281 **TURTLE POINTS**

No.	Beach	Rating HT	LT	Type	Length
NT275	Turtle Pt	1	2	R+sand/reef flats	300 m
NT276	Turtle Pt (S1)	1	2	R+sand/reef flats	500 m
NT277	Turtle Pt (S2)	1	2	R+sand/rock flats	300 m
NT278	False Turtle Pt	1	2	R+sand/rock flats	1.1 km
NT279	False Turtle Pt (S)	1	2	R+sand/rock flats	800 m
NT280	White Cliff	1	2	R+sand/rock flats	200 m
NT281	White Cliff (S)	2	2	R+sand flats	1.6 km
Spring & neap tide range = 2.1 & 0.1 m					

Turtle Point lies at the northern end of a 3 km long, 0.5-1.5 km wide peninsula, composed of Tertiary laterite and covered by moderately dense woodlands. Kennedy Bay lies to the west and the more open 10 km wide waters of Port Essington to the east. Two beaches are located on the point. Beach **NT275** lies on the northern tip and is a 300 m long, north-facing, relatively steep beach, backed by a grassy beach ridge, with a small often dry lagoon behind the eastern end. Immediately to the south is beach **NT 276**, a curving, 500 m long east-facing beach, which is well exposed to waves generated within the port by the trade winds. It consists of a steep beach backed by a grassy beach ridge, with 100-200 m wide sand and reef flats off the beach and low laterite rocks to either end.

Beach **NT 277** occupies the next small embayment immediately to the south. It is a curving 300 m long sandy beach, bordered by bright red rocks that extend part way along the back of the beach, with wooded slopes behind. The rocks form reefs off either end with only a small central sand flat section.

Five hundred metres further south is beach **NT 278**, a 1.1 km long slightly curving east-facing beach, which has sand patches at either end, with beachrock dominating much of the central section of beach. It is backed by low wooded slopes and has a mixture of sand and patchy reef flats extending 400-600 m off the beach. It terminates at the low wooded **False Turtle Point**.

On the south side of False Turtle Point the coast trends south for 2.5 km to the north side of Curlew Bay. The shoreline is backed by wooded bedrock slopes, with three bedrock-controlled beaches (NT 279-281) along the shore, each fronted by 200-300 m wide sand flats together with laterite reefs and rocks. Beach **NT 279** begins immediately south of the point and is a slightly curving 800 m long east-facing beach, consisting of a moderately steep narrow beach backed by a casuarina-covered low foredune, then the wooded slopes. It terminates at a small protrusion in the backing slopes, which are scarped, exposing the lower white laterite soil horizons, hence the name 'White Cliff'. At the base of the 15 m high cliffs is beach **NT 280**, a narrow 200 m long curving sand beach, bordered by rock flats and reefs. To its south is beach **NT 281**, a slightly crenulate 1.6 km long sand beach, with bedrock dominating the northern sand flats, together with a few mangroves growing on the rock flats. It terminates at the northern head of Curlew Bay.

NT 282-287 **CURLEW BAY & POINT**

No.	Beach	Rating HT	LT	Type	Length
NT282	Curlew Bay	1	2	R+sand/mud flats	1 km
NT283	Curlew Pt	1	2	R+sand flats+rocks	400 m
NT284	Kangaroo Pt (N2)	1	1	R+sand flats	100 m
NT285	Kangaroo Pt (N1)	1	1	R+sand flats	250 m
NT286	Kangaroo Pt (S1)	1	1	R+sand flats	150 m
NT287	Kangaroo Pt (S2)	1	1	R+sand flats	100 m
Spring & neap tide range = 2.1 & 0.1 m					

Curlew Bay is a 1.5 km wide northeast-facing, low energy bay, with a shoreline dominated by rocks and mangroves, apart from the western head of the bay, which has been infilled by a series of low beach ridges, backed by a small freshwater marsh. The 1 km long, low, narrow beach (**NT 282**) is backed by casuarina trees and fronted by 500 m wide sand to mud flats.

Wooded, 10 m high **Curlew Point** forms the southern headland of the bay and is the northern tip of a 2 km long, 20-30 m high wooded promontory. Along the eastern face of the protrusion are five small, east-facing bedrock-controlled beaches (NT 283-287), centred on 12 m high Kangaroo Point. Beach **NT 283** lies immediately south of Curlew Point and is a 400 m long sand beach, backed by wooded slopes, and bordered by bedrock, with 20 m wide sand flats containing scattered bedrock reefs paralleling the beach. It terminates at a 200 m long east-trending reef. On the south side of the reef the shoreline curves into a small embayment containing 100 m long beach **NT 284**. A small wooded protrusion separates it from beach **NT 285**, a 250 m long beach, terminating at the base of 20 m high wooded slopes, that end at **Kangaroo Point**.

On the south side of the point is a small embayment containing a 150 m long beach-barrier **(NT 286)** backed by a small infilled lagoon. The adjoining embayment contains 100 m long beach **NT 287**, a strip of sand at the base of 20 m high wooded slopes.

NT 288-294 **OYSTER-SPEAR POINTS**

No.	Beach	Rating HT	Rating LT	Type	Length
NT288	Oyster Pt (W2)	1	2	R+rock flats	200 m
NT289	Oyster Pt (W1)	1	2	R+rock flats	100 m
NT290	Oyster Pt (E1)	1	2	R+sand flats+rocks	50 m
NT291	Oyster Pt (E2)	1	2	R+sand+rock flats	50 m
NT292	Spear Pt (W2)	1	1	R+sand flats	1.5 m
NT293	Spear Pt (W1)	1	1	R+sand flats	700 m
NT294	Adam Head	1	1	R+sand flats	700 m

Spring & neap tide range = 2.1 & 0.1 m

Oyster Point is a wooded, 20-30 m high, 500 m wide headland, that separates Knocker and Barrow bays, the two southernmost bays of Port Essington. Both bays are dominated by very low energy mangrove-fringed shoreline, with the only seven beaches (NT 288-294) occurring on the more exposed Oyster and Spear points.

Beach **NT 288** is located 200 m west of Oyster Point. It is a relatively straight 200 m long strip of high tide sand, backed by steep, partly scarped, 10-20 m high bluffs, with patchy rock flats extending up to 200 m off the beach. Immediately west of the point is beach **NT 289**, a 100 m long strip of sand, backed by 15 m high vegetated bluffs, with rock cliff debris spread over the beach and a 100 m wide mixture of sand and rock rubble off the beach. On the immediate east of the 20 m high point is 50 m long beach **NT 290**, a pocket of sand and 100 m wide sand flats contained in a small rock- and reef-bound embayment (Fig. 5.41).

Figure 5.41 Oyster Point with beaches NT 289 (right) and NT 290 (left).

Five hundred metres southeast of the point is beach **NT 291**, a 50 m long north-facing pocket of sand, wedged in at the base of 15 m high red bluffs. A mangrove-fringed point forms the eastern boundary, with bluffs to the west and sandy-mud flats off the centre of the beach.

Spear Point forms the western entrance to Barrow Bay, within which is the historic site of the Victoria Settlement. The western side of the point is exposed to waves generated within the larger Port Essington, and contains two near continuous beaches (NT 292 & 293) and 400 m wide sand flats. Beach **NT 292** is a 1.5 km long sand beach, backed by a low foredune and fronted by stripped sand flats, with some bedrock, rocks and reef toward the centre and eastern end of the beach and a few mangroves at the western end. On the eastern side of the boundary rocks is 700 m long beach **NT 293**, a similar straight beach that terminates at Spear Point. Both beaches are backed by 10-20 m high vegetated bluffs and wooded terrain.

Adam Head is located on the southern side of Victoria Settlement. Beach **NT 294** is a curving 700 m long beach that is located between the head and a southern point. It faces east across Barrow Bay and receives sufficient low easterly waves to maintain the open beach, backed by a low foredune and fronted by 200-300 m wide sand flats.

NT 295-303 **RECORD POINT-TABLE HEAD**

No.	Beach	Rating HT	Rating LT	Type	Length
NT295	Record Pt	1	1	R+sand flats	4.2 km
NT296	Observation Cliff	1	1	R+mud flats	1.4 km
NT297	Observation Cliff (N1)	1	1	R+sand/mud flats	250 m
NT298	Observation Cliff (N2)	1	1	R+sand flats	150 m
NT299	Table Hd (S5)	1	1	R+sand flats	1.8 km
NT300	Table Hd (S4)	1	1	R+mud flats	700 m
NT301	Table Hd (S3)	1	1	R+mud flats	100 m
NT302	Table Hd (S2)	1	1	R+sand/mud flats	700 m
NT303	Table Hd (S1)	1	1	R+sand+rock flats	200 m

Spring & neap tide range = 2.1 & 0.1 m

Record Point is a 3 km long series of recurved spits that has formed over the past 6000 years, as sediment has been eroded from the laterite bluffs of Observation Cliff and moved along the beaches to build the 150 ha point. The point is continuing to grow, gradually narrowing the entrance to Barrow Bay, from its original 5 km to now 1.6 km.

Beach **NT 295** is located along the northern side of the point and faces northwest into the main Port Essington. It consists of a 4.2 km long sand beach, backed initially by scarped red 10 m high bluffs and then a vegetated series of beach ridges to terminal recurved spits, which enclose a small lagoon. The ridge-spits widen from 250 m in the north to a maximum of 800 m in the south. Pulses of sand move along the beach and form a series of intertidal sand ridges along the southern tip (Fig. 5.42).

Figure 5.42 The northern tip of beach NT 295 consists of a migratory series of en echelon intertidal sand ridges.

Between the northern end of Record Point and Table Head is a 5 km section of west-facing shore, containing three south-trending zeta-curved open embayments, all backed by scarped bluffs, with beaches along the base. The first 1.5 km long embayment is backed by the **Observation Cliff** and contains two west-facing beaches (NT 297 & 298). Beach **NT 296** is a 1.4 km long strip of high tide sand wedged in at the base of the 10 m high red bluffs, with sections of the bluffs and debris protruding across the northern few hundred metres of the beach. It is fronted by narrow mud flats.

Beaches NT 297 and 298 occupy small embayments located either side of the northern headland, a 10-15 m high eroding red bluff. Beach **NT 297** lies on the south side and is a curving 250 m long beach, with a narrow strip of sand then deeper mud of the bay. On the west-facing side of the headland is its neighbour, beach **NT 298**, which faces west across the port. The beach is bordered by rock debris from the bluff in the south, with a small tidal creek and rock flats at its northern end.

Beach **NT 299** lies on the northern side of the creek and beach NT 298. It is a 1.8 km long beach backed initially by a low beach ridge-barrier and backing small mangrove-filled lagoon, then by eroding bluffs along its northern half, with 100 m wide sand then mud flats off the beach. Beach **NT 300** is located along the northern part of the embayment and is bordered and backed by low largely vegetated bluffs, with a bedrock reef extending 500 m into the port off the northern end. Two hundred metres to the north is beach **NT 301**, a 100 m long strip of sand located in a slight west-facing curve in the 10 m high bluffs.

Beach **NT 302** occupies the southern half of the northernmost embayment. It is a crenulate 700 m long beach backed by low wooded bluffs and slopes, with laterite reefs outcropping to either end and in the centre. It is followed by 500 m of 10 m high bluffs, then beach **NT 303**, which lies in lee of Table Head. The 200 m long beach faces south and is backed by 10 m high wooded bluffs.

NT 304-313 **BERKELEY BAY**

No.	Beach	Rating HT	Rating LT	Type	Length
NT304	Table Hd (N)	1	2	R+rock flats	100 m
NT305	Caiman Ck	1	1	R+sand flats	2.5 km
NT306	Caiman Ck (N)	1	1	R+sand flats	300 m
NT307	Berkeley Bay (4)	1	1	R+sand/mud flats	1 km
NT308	Berkeley Bay (5)	1	1	R+sand flats	1.1 km
NT309	Berkeley Bay (6)	1	1	R+sand flats	300 m
NT310	Berkeley Bay (7)	1	1	R+sand flats	700 m
NT311	Berkeley Bay (8)	1	1	R+sand flats+rocks	300 m
NT312	Berkeley Bay (9)	1	1	R+sand flats	50 m
NT313	Berkeley Bay (10)	1	1	R+sand flats	200 m
Spring & neap tide range = 2.1 & 0.1 m					

Berkeley Bay is an open 4 km wide west-facing bay, bordered by Table Head in the south and scarped bluffs to the north. A 2.5 km long barrier backed by Caiman Creek occupies the southern half of the bay, with the remainder dominated by 5-15 m high laterite bluffs, with ten beaches (NT 304-313) located along the base of much of the bluffed shore.

Beach **NT 304** is a narrow 100 m long pocket of high tide sand located on the north side of Table Head. It is bordered by laterite reefs with a generally rock-dominated intertidal zone.

Beaches NT 305 and 306 are located either side of the 50 m wide mouth of Caiman Creek. Beach **NT 305** is a curving 2.5 km long west-facing beach, backed by a 300-800 m wide beach ridge barrier, then a 400 ha mangrove-filled lagoon. The creek forms the northern boundary with beach **NT 306**, a 300 m long beach that extends from the creek mouth to the base of 20 m high vegetated bluffs (Fig. 5.43). The creek maintains 400 m wide tidal shoals off the mouth, which narrow to the south. A 2 km long vehicle track from the Black Point road reaches the top of the northern bluffs.

Figure 5.43 Beach NT 305 (foreground) extends to a small creek mouth, with beach NT 306 beyond.

Crenulate bluffs dominate the next 4 km of shore, with seven small beaches (NT 307-313) located along their base. Beaches **NT 307** and **NT 308** are two near continuous crenulate sand beaches 1 km and 1.1 km long respectively, separated by a section of scarped red bluffs. A small freshwater melaleuca swamp is located behind beach NT 307, while densely wooded slopes border each beach and back NT 308. Narrow continuous sand flats lie off the beaches.

Beaches **NT 309, NT 310** and **NT 311** are three curving sand beaches, each located in an indentation in the backing slopes. They are 300 m, 700 m and 300 m long respectively and are bordered by reef-tipped protrusions in the wooded slopes, with the sand continuing around each of the small points, as well as continuous 100 m wide sand flats.

North of beach NT 311 is a 500 m long section of exposed bluffs. Beach **NT 312** is located at the northern end of these bluffs and consists of a 50 m long pocket of sand, backed and bordered by the bluffs. Immediately to the north is the 200 m long beach **NT 313**, a curving strip of sand backed by wooded slopes. A vehicle track off the Black Point road (1 km to the east) reaches the bluffs above these beaches.

NT 314-317 **REEF POINT**

No.	Beach	Rating HT	LT	Type	Length
NT314	Reef Pt (S)	1	1	R+sands flats	1.5 km
NT315	Reef Pt (N)	1	1	R+sands flats	2.8 km
NT316	Black Pt (E4)	1	1	R+sand flats	100 m
NT317	Black Pt (E3)	1	1	R+sand flats	1.3 km
Spring & neap tide range = 2.1 & 0.1 m					

Reef Point is a low projection in the laterite plain, which forms 10 m high bluffs at the shore, while the laterite continues into the bay as a 500 m long reef. To either side of the point are two sand beach-barrier systems. Beach **NT 314** extends east along the southern side of the point, as a curving 1.5 km long sand beach. It has substantial migratory shoreline crenulations, a product of southward migrating sand waves, the crenulations adjacent to the point where the sand waves are attached to the shore. The beach is backed by a 100-200 m wide beach ridge plain, then the wooded coastal plain. A vehicle track from the Reef Point community runs along the largely cleared beach ridges.

On the north side of the point is 2.8 km long beach **NT 315**, a west-facing slightly crenulate beach, backed by a 200-400 m wide beach ridge plain, including a 50 ha melaleuca swamp behind the central part of the barrier. The northern 300 m of beach lie below 10-20 m high vegetated bluffs. A vehicle track from the nearby Black Point road reaches the northern end of the beach, where three houses are located.

Beach **NT 316** is a 100 m long pocket of sand located in the 20 m high bluffs, immediately north of beach NT 315. It is bordered and backed by the bluffs. On its immediate north begins beach **NT 317**, a curving 1.3 km long beach, that begins beneath the bluffs. The bluffs give way to a small barrier-foreland along the western half of the beach, which is partly paralleled by beachrock. A school is located on the bluffs toward the centre of the beach, while the Black Point road runs past, 200-300 m in behind the beach.

NT 318-324 **BLACK-SMITH POINTS**

No.	Beach	Rating HT	LT	Type	Length
NT318	Black Pt (boat ramp)	1	1	R+sand/rock flats	800 m
NT319	Black Pt	1	3	R+rocks	50 m
NT320	Black Pt (N)	1	1	R+sand+reef	2.8 km
NT321	Smith Pt (S2)	2	3	R+rock flats	600 m
NT322	Smith Pt (S1)	2	3	R+rock flats	500 m
NT323	Smith Pt	2	3	R+rock flats	50 m
NT324	Smith Pt (E1)	2	3	R+rock flats	100 m
Spring & neap tide range = 2.1 & 0.1 m					

Black Point is the site of the Black Point ranger station, the main station for the Gurig National Park. Most travellers visiting the park head straight for Black Point to register and to stay at the limited accommodation and camping offered in the area. In addition to the ranger station the point has a small store and kiosk, visitor centre, toilets, fuel, water and a nearby camping area, as well as a jetty and boat ramp across the beach (NT 318).

Beach **NT 318** is the most heavily utilised beach in the area, being the site of the boat ramp and jetty. The curving 800 m long beach faces south into the exposed waters of the port. The 100 m long jetty is located off the rocks at the eastern end, with the boat ramp 100 m to the west (Fig. 5.44). Between the jetty and boat ramp, beachrock and rock flats are exposed at low tide, with a more sandy beach to the east.

Figure 5.44 Black Point with the small settlement and jetty and beaches NT 317-319 to right, and the southern end of beach NT 320 to left.

Figure 5.45 Smith Point, with beach NT 322 in front, and beaches NT 323 & 324 extending east of the low laterite point. Note the rip current plume off beach NT 324.

Beach **NT 319** is located on the south side of Black Point, and consists of a 50 m long strip of high tide sand, completely fronted by intertidal rocks and rubble, making it unsuitable for swimming. Some of the Black Point houses sit atop the bluffs backing the beach.

On the north side of the point is 2.8 km long beach **NT 320**, a west-facing sandy beach, backed by a low foredune, then a 400-800 m wide beach-foredune ridge plain. Beachrock parallels much of the base of the beach, making it unsuitable for safe swimming at low tide. There is deeper water off the beach, then two expanses of coral reef, with a gap in the centre, which is used by boats wishing to anchor in lee of the reef. A dirt road runs along the backing cleared ridges up to Smith Point, with a solitary hunting camp located toward the northern end next to the small creek.

Smith Point is a laterite headland surrounded by 700 m of low bluffs, high tide beaches and rock flats. On the south side of the point beaches **NT 321** and **NT 322** are 600 m and 500 m long respectively. They are both backed by 5 m high scarped laterite bluffs, and fronted by rock flats up to 400 m wide. There is tourist cabin accommodation on the bluffs between the two beaches.

On the tip of Smith Point is an historic beacon initially built at the time of the Port Victoria settlement in 1839. To either side of the point are two bluff- and rock-dominated beaches. Beach **NT 323** is a 50 m long west-facing strip of high tide sand located immediately west of the beacon, with protection afforded by laterite reefs that run 100 m west of the point. On the north side of the point is the more exposed north-facing 100 m long beach **NT 324**. It receives waves averaging 0.5 m, which break over the rocks to either end, and can generate a rip toward the eastern end (Fig. 5.45).

NT 325-328 **SMITH POINT (E)**

No.	Beach	Rating HT LT	Type	Length
NT325	Smith Pt (E2)	1 2	R+LTT	800 m
NT326	Smith Pt (E3)	1 2	R+LTT	350 m
NT327	Smith Pt (E4)	1 2	R+LTT	400 m
NT328	Smith Pt (E5)	1 1	R+sand flats+reef	4 km
Spring & neap tide range = 2.1 & 0.1 m				

Smith Point forms the northern tip (11°08'S) of the 25 km long peninsula that separates Port Essington from Port Bremer. It is also, together with Vashon Head (11°09'S) and Danger Point (11°07'S), one of the most northerly points on the Cobourg Peninsula and on the Northern Territory mainland coast.

At the point the coast trends east for 5.5 km before turning to run southeast for 22 km into Port Bremer. Beaches NT 325-328 all lie immediately east of the point and are more exposed north-facing beaches. All are readily accessible via a 20 m long road and track that run right around the northern few kilometres of the peninsula from Black Point round to Stewart Point and Lizard Bay.

Beach **NT 325** is a curving 800 m long relatively exposed beach, bordered at each end by protruding laterite rocks and reefs, together with a band of beachrock along the western third of the beach, and is backed by a low foredune. It has a moderately steep swash zone, then a low tide terrace, with the waves averaging just over 0.5 m, maintaining a small surf zone. Its neighbour beach **NT 326** is a similar 350 m long curving beach, with more laterite reef off the beach, while the adjoining beach **NT 327** is 400 m long, with prominent reef to either end and a more open central section. There is a low right-hand surf break over the eastern rocks.

Beach **NT 328** occupies most of this section of shore. It is 4 km long and terminates in the east at an outward curving sandy foreland, formed in lee of the low Sandy Island No. 1. This beach is the outermost section of a 2 km wide barrier, consisting of a series of four 100-200 m wide beach ridges, separated by freshwater melaleuca swamps. The Black Point landing ground is located on the outer ridges parallel to the beach, while the Smith Point camping areas are located either side of the small creek that drains the swamp toward the eastern end of the system. A dry weather 4WD track from Black Point also runs around the back of the swamp. The beach is moderately steep, with waves decreasing to the east, owing to a near continuous coral reef, paralleling the eastern half of the beach 100-200 m offshore. At low tide the reef forms a calm shallow lagoon between the reef and beach.

Surf: A right-hand surf break forms over the eastern boundary laterite reef of beach NT 327.

NT 329-334 **SMITH-KUPER POINTS**

No.	Beach	Rating HT	Rating LT	Type	Length
NT 329	Smith Pt (E6)	1	1	R+sand flats+reef	1.1 km
NT330	Smith Pt (E7)	1	1	R+sand flats+reef	1 km
NT331	Kuper Pt (W5)	1	1	R+sand flats+reef	250 m
NT332	Kuper Pt (W4)	1	1	R+sand flats+reef	1 km
NT333	Kuper Pt (W3)	1	1	R+sand flats+reef	2.5 km
NT334	Kuper Pt (W2)	1	1	R+sand flats+reef	3.5 km
Spring tide range = 2.1 & 0.1 m					

At the **Sandy Island** foreland, the coast trends southeast for 7.5 km to Kuper Point. While the six intervening beaches (NT 329-334) all face east into the dominant wind and waves, they are fronted by a mixture of sand flats, laterite and coral reef flats that extend from 1-4 km offshore, thereby reducing wave height at the beaches, particularly at low tide when calm prevails at the shore.

The vehicle track from Smith Point continues on along the back of all the beaches all the way to Kuper Point and beyond. Beach **NT 329** begins on the southern side of the foreland and runs southeast for 1.1 km to a low sandy foreland in lee of some laterite rocks and reef. It is a curving beach, backed by a low foreland and the road, with 1 km wide sand, rock and reef flats off the beach.

On the south side of the foreland is beach **NT 330**, a similar curving, east-facing, 1 km long beach, that extends to a second rock-reef-controlled foreland, backed by wooded slopes, while coral reefs become more prevalent offshore. The track runs around the small point to beach **NT 331**, which is tucked in on the south side. It is a curving 250 m long south-facing beach, which receives pulses of sand moving south around the point, forming at times a highly rhythmic shoreline. The eastern end of the beach is also bounded by red laterite rocks.

On the south side of the rocks is a relatively straight 1 km long, east-facing beach **NT 332**. A low casuarina-covered foredune, then the track and gentle wooded slopes back the beach, with sand flats extending up to 2 km offshore. It terminates at a group of low rocks and accompanying inflection point in the sand, beyond which is the 2.5 km long slightly curving beach **NT 333**. The main track turns inland to go round the back of the 500 m wide largely cleared beach ridge plain that backs this beach. In lee of the plain are two small melaleuca swamps, the southern one connected to the shore via a meandering creek that crosses the southern end of the beach. A few hundred metres past the creek mouth is another bedrock protrusion, with wooded slopes reaching the shore and a low reef off the beach. Beach **NT 334** begins on the southern side of the point and curves round for 3.5 km to Kuper Point, facing initially northeast, then north and finally northwest in lee of the point. The beach is initially backed by a few casuarina-covered beach ridges, then bedrock slopes, with some fronting rocks and rock-reefs, while the vehicle track parallels the back of the entire beach.

NT 335-339 **KUPER-STEWART POINTS**

No.	Beach	Rating HT	Rating LT	Type	Length
NT 335	Kuper Pt (W1)	1	2	R+rock/sand flats	200 m
NT336	Kuper Pt	1	2	R+rock/sand flats	200 m
NT337	Kuper Pt (E1)	1	2	R+rock/sand flats	800 m
NT338	Kuper Pt (E2)	1	1	R+sand flats	2.5 km
NT339	Stewart Pt	1	2	R+sand/rock flats	300 m
Spring & neap tide range = 2.1 & 0.1 m					

Kuper Point is a low wooded bedrock headland surrounded by sandy beaches and rock flats, and connected to the main peninsula by a 200 m wide vegetated tombolo (Fig. 5.46). Between Kuper and Stewart points is a semicircular 2 km wide, northeast-facing embayment, containing two beaches, together with beaches on each point. Beach NT 334 terminates on the low energy western side of the point amongst a small clump of mangroves, with beach **NT 335**, a 200 m long high tide sand beach, wrapping around the western half of the point, with rock flats dominating the wide intertidal zone. Beach **NT 336** is a 200 m long continuation on the eastern half of the point, with the narrow beach wedged in between the backing track and vegetation and the irregular rock flats.

Beach **NT 337** lies immediately south of the point. It is a curving east-facing 800 m long sand beach, fronted by 200-300 m wide scattered rock flats, then 1 km wide sand-mud flats. It is backed by a low foredune, the track, and dense woodlands. It terminates at the final group of rocks, beyond which is 2.5 km long beach **NT 338**, which curves round to face north in lee of Stewart Point. The low energy beach is fronted predominantly by 1 km wide sand-mud flats, together with a few scattered reefs. It is backed by the vehicle track, which continues on right to the point.

Figure 5.46 Kuper Point is surrounded by laterite rocks and reefs and beaches NT 334-337.

Kuper Point beach (**NT 339**) extends for approximately 300 m along the northern side of the point and consists of three pockets of high tide sand, backed by scarped 5 m high red bluffs, with some rocks and reefs extending across the beach, and a mixture of sand and rock flats off the beach.

NT 340-344 **LIZARD BAY**

No.	Beach	Rating HT	LT	Type	Length
NT340	Stewart Pt (S1)	1	1	R+sand flats	300 m
NT341	Stewart Pt (S2)	1	1	R+sand flats	200 m
NT342	Stewart Pt (S3)	1	1	R+sand/rock flats	100 m
NT343	Lizard Bay (1)	1	1	R+sand/rock flats	500 m
NT344	Lizard Bay (2)	1	1	R+sand flats	1.7 km
Spring & neap tide range = 2.1 & 0.1 m					

Stewart Point forms the western headland of Lizard Bay, a 2 km wide northeast-facing embayment. There are five beaches (NT 340-344) between the point and the southern shore of the bay, with bedrock and reefs dominating the eastern shore.

Beach **NT 340** lies immediately south of the eastern tip of Stewart Point. It consists of a curving, east-facing 300 m long strip of high tide sand, backed by low red bluffs and fronted by a mixture of rock and sand flats extending 100 m off the beach.

Five hundred metres to the south are beaches NT 341 and 342, both northeast-facing, 200 m and 100 m long beaches respectively, each bounded and backed by wooded bedrock slopes, rocks and reefs. **NT 341** has trees extending right to the back of the beach, while **NT 342** is backed by a 50 m wide beach ridge then the wooded slopes. They are fronted by the boundary rock reefs and central 100 m wide sand flats. Two hundred metres further on is beach **NT 343**, a crenulate 500 m long east-facing beach consisting of the high tide sand and backing slopes, with rocks and reefs outcropping along the beach, together with a few mangroves and sand-mud flats off the beach.

The main **Lizard Bay** beach (**NT 344**) occupies the southern 1.7 km of the bay shore and faces north-northeast, looking toward the bay mouth. It consists of a low energy high tide beach, with a scattering of mangroves and fronting 200 m wide sand-mud flats.

NT 345-347 **PORT BREMER (S)**

No.	Beach	Rating HT	LT	Type	Length
NT345	Port Bremer (S1)	1	2	R+mud flats	600 m
NT346	Port Bremer (S2)	1	2	R+mud flats	200 m
NT347	Port Bremer (S3)	1	2	R+mud flats	300 m
Spring & neap tide range = 2.1 & 0.1 m					

The eastern side of Port Bremer extends for 11 km south of Edwards Point, terminating in a 2 km wide funnel-shaped bay, with Mawuwu Creek entering at the very southern extremity. Much of the southern shores of the bay is lined with mangroves, with low energy beaches in the western Lizard Bay and along the more exposed east-facing shores opposite East Station Point. Three beaches (NT 345-347) lie almost directly opposite the point.

Beach **NT 345** is a curving 600 m long high tide sand beach, backed by wooded slopes, and fronted by mud flats, with a few mangroves fringing the more protected northern end, and low wooded bedrock headlands to either end. Beach **NT 346** lies immediately south of the southern headland, and is a similar 200 m long beach, also with a few mangroves along the northern shoreline and near continuous mangroves fringing the southern headland. Five hundred metres to the south is beach **NT 347**, the southernmost beach in Port Bremer. It is a 300 m long strip of high tide sand and mud flats, backed by wooded slopes and fringing bedrock headlands.

NT 348-353 **EAST STATION POINT**

No.	Beach	Rating HT	LT	Type	Length
NT348	East Station Pt (S3)	1	2	R+mud flats	300 m
NT349	East Station Pt (S2)	1	2	R+mud flats	200 m
NT350	East Station Pt (S1)	1	2	R+mud flats	300 m
NT351	East Station Pt	1	2	R+rock flats	600 m
NT352	East Station Pt (N1)	1	2	R+mud flats	1.2 km
NT353	East Station Pt (N2)	1	2	R+mud flats	1.8 km
Spring & neap tide range = 2.1 & 0.1 m					

East Station Point is a low wooded bedrock protrusion on the eastern shores of Port Bremer. There is a small settlement of about 15 houses spread around the shore of the point, as well as a jetty, with several boats usually moored off the point. The settlement is accessible via a 6.5 km track off the Point Danger track.

Three low energy beaches (NT 348-350) lie within a kilometre south of the settlement. Beach **NT 348** is a relatively straight, 300 m long, west-facing beach located on the northern side of the next point to the south, beyond which mangroves of the lower bay dominate. It is backed by low wooded slopes and fronted by mud flats. Beaches **NT 349** and **NT 350** are 200 m and 300 m long respectively and lie immediately south of the settlement. They occupy the southern and northern half of a semicircular west-facing embayment which terminates at the rocks that front the point, jetty and settlement.

Beach **NT 351** begins on the western side of the point and extends for 600 m, initially fronted by rock flats, which give way to a small more sandy bay that terminates at the next low rocky point. Beach **NT 352** begins on the northern side of this point and is a narrow 1.2 km long high tide beach, backed by continuous wooded slopes with numerous rock outcrops along the crenulate beach, that initially faces west then curves round to face northwest. It terminates at a larger rock-reef protrusion, on the other side of which beach **NT 353** continues on as a curving, 1.8 km long, more continuous sandy beach, terminating at a small sandy foreland, 2 km south of Edwards Point. Both beaches are fronted by 300 m wide mud flats.

NT 354-357 **EDWARDS POINT (S)**

No.	Beach	Rating HT	Rating LT	Type	Length
NT354	Edwards Pt (S3)	1	2	R+mud flats	300 m
NT355	Edwards Pt (S2)	1	2	R+mud flats	700 m
NT356	Edwards Pt (S1)	1	1	R+sand flats	800 m
NT357	Edwards Pt	1	2	R+rock flats	300 m
Spring & neap tide range = 2.1 & 0.1 m					

Edwards Point forms the eastern boundary of Port Bremer, with beaches NT 354-357 located immediately south of the point (Fig. 5.47). Beach **NT 354** begins on the northern side of the low sandy foreland that separates it from its southern neighbour (NT 353). The beach faces west and curves for 300 m between two similar cleared sandy forelands, both formed in lee of laterite rocks and reefs and fronted by mud flats.

Beach **NT 355** also lies between two low forelands, with additional rocks and crenulations along the 700 m long beach. Its northern neighbour beach **NT 356** forms the northernmost of the foreland-controlled beaches. It is 800 m long and sweeps in a continuous west-facing curve between the grassy forelands.

Beach **NT 357** forms the western side of Edwards Point. It is a relatively straight, northwest-facing strip of high tide sand, fronted by low tide beachrock and rock and sand flats, as well as some patchy coral reefs.

Figure 5.47 Edwards Point, with beaches NT 356 & 357 to right.

NT 358-364 **EDWARDS POINT (E 1-7)**

No.	Beach	Rating HT	Rating LT	Type	Length
NT358	Edwards Pt (E1)	1	2	R+sand flats	2.5 km
NT359	Edwards Pt (E2)	1	2	R+reef flats	600 m
NT360	Edwards Pt (E3)	1	2	R+sand/rock flats	1 km
NT361	Edwards Pt (E4)	1	2	R+rock flats	200 m
NT362	Edwards Pt (E5)	1	2	R+rock flats	400 m
NT363	Edwards Pt (E6)	1	2	R+sand flats	2.5 km
NT364	Edwards Pt (E7)	1	2	R+rock flats	250 m
Spring & neap tide range = 2.1 & 0.1 m					

At **Edwards Point** the coast turns and trends east-northeast toward Danger Point, located 9 km to the northeast. The first 8 km of coast contain seven beaches (NT 358-364) located in two open north-facing embayments. Both bays have central prograding barrier systems, with rocks and some reef flats dominating the areas off the boundary points. There is no vehicle access to or development along this section of coast.

Beach **NT 358** is a 2.5 km long north-northwest-facing beach bounded by Edwards Point in the west and a similar low unnamed point in the east. It consists of a sandy beach protruding in the centre, as a sand foreland in lee of an offshore reef that dominates the eastern half of the foreshore. The beach is backed by a 200-500 m wide series of grassy beach ridges with a freshwater swamp behind (Fig. 5.48).

The eastern point consists of a low 2 km long bedrock-dominated section of shore, containing four rock-dominated beaches. Beach **NT 359** is a straight, north-facing, 600 m long sand beach bounded by laterite rocks and reefs and fronted by sand and reef flats. Beach **NT 360** is located in a small embayment immediately to the east and consists of a curving 1 km long beach, bounded by protruding laterite points and reefs, as well as a central rock outcrop on the beach. It is backed by a grassy beach ridge, with the bay containing a mixture of sand, rock and reef flats. Beach **NT 361** lies on the eastern point and is a curving, 200 m long, north-facing beach, bounded and fronted by laterite rocks and reefs, with a grassy beach ridge area behind. Immediately east is beach **NT 362**, a more crenulate rock-dominated 400 m long high tide sand beach, fronted by rock flats, that faces northeast across the second embayment.

Figure 5.48 Beach NT 358 is a longer sandy beach with a central salient, and is backed by sandy beach ridges and a swamp.

Beach **NT 363** occupies the second embayment and is a 2.5 km long north to northeast-facing beach, backed by a 500-1000 m wide beach ridge plain and two interlinked swamps that drain through a small creek at the eastern end of the beach. Extensive tidal shoals lie off the creek, with sand flats filling the centre of the bay, while rock and reef flats fringe the sides. The beach is accessible via a vehicle track off the main Danger Point track.

Beach **NT 364** lies on the eastern point, a 5 m high laterite bluff, with the beach situated below the bluffs along the west-facing outer edge. The narrow high tide beach is cut into a series of small pockets of sand by rocks and the bluffs and fronted by a mixture of sand and rock flats.

NT 365-368 **DANGER POINT**

No.	Beach	Rating HT LT	Type	Length
NT365	Danger Pt (W3)	1 1	R+sand flats	5 km
NT366	Danger Pt (W2)	1 2	R+rock flats	200 m
NT367	Danger Pt (W1)	1 2	R+rock flats	300 m
NT368	Danger Pt	1 2	R+rock flats	50 m
Spring & neap tide range = 1.9 & 0.2 m				

Danger Point, at 11°07' S, is the most northerly point on the Northern Territory mainland coast. The point itself is a low 1 km wide laterite bluff fronted by rocks and reefs and three small beaches (NT 366-368). It is attached to the mainland by a 2 km long tombolo composed of approximately eight grassy beach to foredune ridges, the western side of which forms the 5 km long main beach **NT 365** (Fig. 5.49). This beach faces northwest in the south, curving round to face west where it attaches to the point. It consists of a lower gradient high tide beach, fronted by sand flats along the southern half and reef flats along the northern half, which widen to 800 m off the point. The Danger Point track terminates in lee of the centre of the beach, adjacent to an old landing ground. Other tracks run from here out to the point and south to the nearer Edwards Point beaches.

Figure 5.49 Beach NT 365 (right) and NT 373 (left) lie either side of the sandy-calcarenite tombolo that links Danger Point (right of picture) to the mainland.

The northern three point beaches are all rock-controlled. Beach **NT 366** is a dynamic 200 m long crenulate strip of sand, which consists of a south-facing recurved spit resulting from sediment transport around the point. The present spit impounds an earlier spit and a small double lagoon. Beach **NT 367** is located on the west-facing section of the point, and consists of a narrow crenulate 300 m long strip of high tide sand and cobbles wedged in between the low red bluffs and fronted by rock and reef flats. Finally, out on the northern tip of the point is beach **NT 368**, a 50 m long strip of sand backed and bordered by laterite bluffs and rocks, with reefs extending offshore.

NT 369-373 **DANGER POINT (E 1-5)**

No.	Beach	Rating HT LT	Type	Length
NT369	Danger Pt (E1)	1 1	R+sand flats	1.3 km
NT370	Danger Pt (E2)	1 1	R+LTT	300 m
NT371	Danger Pt (E3)	1 1	R+LTT	400 m
NT372	Danger Pt (E4)	1 1	R+rock flats	350 m
NT373	Danger Pt (E5)	1 1	R+sand/rock flats	2.9 km
Spring & neap tide range = 1.9 & 0.2 m				

On the eastern side of **Danger Point** the coast trends south for 12 km down to D'Urville Point and then another 11 km into the head of the protected Raffles Bay. The first 5 km form the eastern side of the point and tombolo, and contain five beaches (NT 369-373), which are all accessible by vehicle tracks off the main Danger Point track.

Beach **NT 369** occupies the northern shore of Danger Point. It is a curving 1.3 km long north-facing beach bordered by the northern laterite tip on the west, and more laterite rocks and reefs to the east, the reefs off the point no doubt contributing the name of the point. The beach is steep and narrow, with rock flats exposed at low tide and a small foredune behind sitting atop the bluffs.

Beach **NT 370** lies immediately south of the point and faces toward Croker Island, 15 km to the east. The 300 m long beach begins below low red laterite bluffs and curves round to a low rocky point. Beach **NT 371** extends south of the rocks as a curving 400 m long beach, which terminates at a low laterite point fronted by rock flats. Both beaches are relatively steep and reflective at high tide, with a narrow bar along the base of the low tide beach, and rocks and reefs to either end.

Beach **NT 372** is a relatively straight 350 m long beach fronted by a continuous 200 m wide intertidal rock flat, while it is backed by the northern end of the tombolo-beach ridge plain. The beach narrows and is replaced by a 100 m long section of eroding calcarenite bluffs, on the south side of which is the main beach.

Beach **NT 373** forms the eastern side of the Danger Point tombolo and extends for 2.9 km from the eroding bluffs to southern low laterite rocks and reef. The entire tombolo has been partly lithified into aeolian calcarenite, a result of the combination of carbonate rich (80-90%) beach and dune sands and the seasonally wet and dry climate. The beach is eroding along the eastern shore, cutting into the backing beach ridges and dunes, forming vertical 5-7 m high bluffs. In addition the beach is fronted by an intertidal beachrock flat, as well as some 200-300 m wide sand flats and patchy reef.

NT 374-380 **DANGER-D'URVILLE POINTS**

No.	Beach	Rating HT	Rating LT	Type	Length
NT 374	Danger Pt (E6)	1	1	R+sand flats	1.1 km
NT375	Danger Pt (E7)	1	1	R+sand flats	1.2 km
NT376	Danger Pt (E8)	1	1	R+sand flats	500 m
NT377	Danger Pt (E9)	1	1	R+sand flats	1.4 km
NT378	D'Urville Pt (W3)	1	1	R+sand flats	350 m
NT379	D'Urville Pt (W2)	1	1	R+sand flats	1.6 km
NT380	D'Urville Pt (W1)	1	1	R+sand+rock flats	1.1 km
Spring & neap tide range = 1.9 & 0.2 m					

The eastern base of the Danger Point tombolo is marked by laterite rocks and reefs, and the beginning of a series of seven beaches (NT 374-380) between the rocks and D'Urville Point, 6 km to the south-southeast. All the beaches are slightly zeta-curved to the north and face generally east across Bowen Strait to Croker Island, 12 km to the east. They receive low wind waves and are all fronted by 200-400 m wide sand flats, with each bounded by some form of laterite rocks and reefs. The wave energy is sufficient to gradually move sand south along the beaches and around the small point into Raffles Bay, where it has accumulated as a 3 km long, up to 500 m wide series of recurved spits (beach NT 384). There is no vehicle access to any of the beaches.

Beach **NT 374** is a slightly curving 1.1 km long beach bounded by a 100 m long exposed section of laterite in the north and a small rock-reef point to the south. It is backed by a partly cleared series of beach ridge-foredunes, which widen to 300 m along the southern half of the beach. Its neighbour (beach **NT 375)** begins on the south side of the rocks where a small creek also crosses the beach. A 400 m wide beach ridge plain backs the beach, then a largely infilled freshwater swamp. It is 1.2 km long and terminates at a slight sandy foreland formed in lee of a small reef. Beach **NT 376** extends from the south side of the foreland for 500 m to a second reef and foreland, which is also the site of a small creek mouth. This beach also has a 300 m wide beach ridge plain and small backing infilled swamp. Beach **NT 377** is bounded by the second foreland and a small headland in the south. It is 1.4 km long with 300 m wide beach ridges backing most of the beach and a small central backing melaleuca swamp, which drains across the centre of the beach.

The low wooded bedrock headland marks the western end of a 3 km wide northeast-facing embayment containing the final three beaches (NT 378-380), and terminating at D'Urville Point. Beach **NT 378** is an east-facing 350 m long pocket beach bordered by the headland and a second woody headland. On the south side of the 100 m wide headland is 1.6 km long beach **NT 379**, a curving northeast- to north-facing beach, backed by a 400 m wide beach ridge plain and a continuous 1.5 km long swamp, which drains across the northern end of the beach. It terminates at a rocky outcrop with rock reefs extending 200 m off the beach, and continuing along the length of 1.1 km long beach **NT 380** to D'Urville Point. The low sandy point beach is backed by a narrow series of one to two grassy beach ridges, then densely wooded gentle slopes.

NT 381-386 **D'URVILLE POINT-RAFFLES BAY**

No.	Beach	Rating HT	Rating LT	Type	Length
NT381	D'Urville Pt (S1)	1	1	R+sand flats+rocks	400 m
NT382	D'Urville Pt (S2)	1	1	R+sand flats+rocks	500 m
NT383	D'Urville Pt (S3)	1	1	R+sand flats+rocks	800 m
NT384	D'Urville Pt (S4)	1	1	R+sand flats	3.9 km
NT385	Raffles Bay (W1)	1	2	R+mud flats	200 m
NT386	Raffles Bay (W2)	1	2	R+mud flats	300 m
Spring & neap tide range = 1.9 & 0.2 m					

D'Urville Point forms the western head of Raffles Bay, a 3-4 km wide, 10 km deep, low energy U-shaped bay. The bay contains 28 km of shoreline, much of it fringed by mangroves. Four upland creeks enter the bay, with each bordered by extensive tidal flats and mangroves. Six beaches (NT 381-386) are located south of D'Urville Point on the western bay shore, four immediately south of the point, and two very low energy beaches just south of the Raffles Bay pearling farm.

Beach **NT 381** begins just south of D'Urville Point and is a 400 m long east-facing crenulate beach, located at the base of wooded slopes, with a small scarp at their base. Laterite rocks outcrop on the beach and several reefs extend off the beach, in addition to 400 m wide sand flats. Beach **NT 382** lies immediately to the south and is a relatively straight 500 m long sand beach, facing slightly east-northeast, bordered by small laterite reefs and backed by wooded slopes that extend right down to the back of the beach (Fig. 5.50). It is also fronted by 500 m wide sand flats and a few scattered rock-reefs. Beach **NT 383** faces due east and also runs along the base of 5-10 m high red laterite bluffs, with more rocks and reef off the beach as well as 600 m wide sand flats.

Figure 5.50 Bedrock-dominated beaches NT 381 & 382 extend south of D'Urville Point.

The bluffs terminate at the beginning of beach **NT 384**. A vehicle track runs out to the top of the bluffs overlooking the 3.9 km long beach, with two houses located on the edge of the bluffs and a third down behind the northern end of the beach. The beach represents the net result of 6000 years of sediment movement down into Raffles Bay, past the bluffs and its deposition (perhaps during occasional tropical cyclone events) as a series of multiple recurved spits. The spits have built the shoreline up to 1 km into the bay as well as growing 3.9 km to the south, in total a 200 ha system, backed by a largely blocked 150 ha drowned valley. The beach faces east and is fronted by sand flats that widen from 200 m in the north to 600 m in the south off the mouth of the tidal creek which drains the backing lagoon.

On the south side of the creek mouth is a 20 m high 1.5 km wide wooded peninsula that contains the Raffles Bay pearl farm, which consists of several buildings and a jetty, as well as pearl racks in the bay. On the south side of the farm are two small embayments containing beaches **NT 385** and **NT 386**, 200 m and 300 m long respectively. These two low energy beaches, located deep within the bay, each consist of a narrow high tide beach backed and bordered by wooded slopes, as well as fringing mangroves, with fronting 300 m wide mud flats.

NT 387-388 RAFFLES BAY (E)

No.	Beach	Rating HT	Rating LT	Type	Length
NT387	Raffles Bay (E1)	1	2	R+mud flats	600 m
NT388	Raffles Bay (E2)	1	2	R+mud flats+rocks	3 km
Spring & neap tide range = 1.9 & 0.2 m					

The southern shores of **Raffles Bay** are dominated by mangroves both adjacent to the tidal creeks and fringing the bedrock shore. On the eastern side, opposite the Raffles Bay pearling operation, is a 20 m high wooded headland with the pearl racks located off the headland. The headland forms the western boundary of a 1.5 km wide embayment, the southern shores of which contain a fringe of mangroves and a tidal creek, to the east of which begin the eastern beach systems. Beach **NT 387** extends from the eastern side of the creek mouth for 600 m to the bordering mangrove-fringed bedrock. This is a very low energy, northwest-facing embayed beach, deep within Raffles Bay. It is backed by a low 200 m wide beach ridge plain then wetlands, and fronted by 300 m wide mud flats.

Beach **NT 388** runs for 3 km along the bedrock-controlled eastern shore of the bay, up to a more prominent rock- and reef-controlled inflection in the shore. This beach is highly crenulate with numerous rock and reef outcrops, as well as scattered mangroves particularly on the rock flats, together with 200-300 m wide mud flats.

NT 389-394 HIGH POINT (W)

No.	Beach	Rating HT	Rating LT	Type	Length
NT389	High Pt (W6)	1	2	R+sand/rock flats	100 m
NT390	High Pt (W5)	1	2	R+sand/rock flats	300 m
NT391	High Pt (W4)	1	2	R+sand/rock flats	800 m
NT392	High Pt (W3)	1	2	R+sand flats	350 m
NT393	High Pt (W2)	1	2	R+sand/rock flats	800 m
NT394	High Pt (W1)	1	2	R+sand/rock flats	700 m
Spring & neap tide range = 1.9 & 0.2 m					

High Point is located at the northern tip of the peninsula that separates Raffles Bay from the Bowen Strait, and is the easternmost of the Cobourg Peninsula's six northern peninsulas. The point itself is a 5-10 m high wooded laterite projection, together with a 200 m long island just off the tip. The western side of the point contains 4 km of northwest-facing bedrock-controlled shore, rock-reef and a low island, together with six near continuous beach systems (NT 389-394). There is no vehicle access to the point and no development.

Beach **NT 389** is located on the northern side of the rock-reef-controlled inflection that forms the northern boundary of the Raffles Bay beach. The beach is a 100 m long strip of high tide sand backed by 10 m high wooded bluffs and bordered and fronted by rock and reefs, as well as 300 m wide sand-mud flats. Immediately east is beach **NT 390**, a 300 m long double crenulate beach with backing bluffs, and a strip of beachrock and rock flats along much of its base, then sand-mud flats extending up to 400 m offshore.

Beach **NT 391** is a curving 800 m long sand beach, with bordering rock and reef flats and a largely clear 400 m wide sand-mud flat. It is backed by a narrow beach ridge then wooded 10 m high bluffs, together with a small usually dry creek toward the southern end. A small island lies 500 m off the northern end, with beach **NT 392** running for 350 m as a curving strip of high tide sand, essentially in lee of the island. It is bordered by rock reefs and fronted by sand and rock flats that extend out to the island at low tide.

Rock flats increasingly dominate the northern two beaches. Beach **NT 393** is a relatively straight 800 m long beach, backed by a narrow vegetated beach ridge and a small mangrove- and salt-flat-filled lagoon, drained by a small creek across the centre of the beach. Tidal shoals as well as sand and rock flats extend 800 m off the beach. Low rock reefs separate it from beach **NT 394**, a 700 m long beach that curves gently round to the lee of High Point, the final 100 m fronted by mangroves up to the rocky point.

NT 395-403 **HIGH POINT (E)**

No.	Beach	Rating HT	Rating LT	Type	Length
NT395	High Pt (E1)	1	2	R+rock/sand/reef flats	200 m
NT396	High Pt (E2)	1	2	R+rock/sand/reef flats	700 m
NT397	High Pt (E3)	1	2	R+rock/sand flats	1 km
NT398	High Pt (E4)	1	2	R+rock/sand flats	1.1 km
NT399	High Pt (E5)	1	2	R+rock/sand flats	1.6 km
NT400	High Pt (E6)	1	2	R+rock/sand flats	700 m
NT401	High Pt (E7)	1	2	R+rock/sand flats	1.5 km
NT402	High Pt (E8)	1	1	R+sand flats	1 km
NT403	High Pt (E9)	1	1	R+sand flats	1.8 km
Spring & neap tide range = 1.9 & 0.2 m					

On the eastern side of **High Point** the coast trends away to the southeast with the shoreline facing across 5 km wide Bowen Strait to Croker Island. The point is the tip of a 5 km wide wooded peninsula composed of Tertiary laterite which rises gently to a crest of 30 m. The shoreline is for the most part controlled by laterite protrusions and rock and fringing coral reefs (Fig. 5.51). A near continuous strip of sand does however run from the point for 13 km to Giles Point, with the rocks and reefs separating it into nine exposed, though generally low energy beaches (NT 395-403), with waves averaging less than 0.5 m, owing to the protection of Croker Island.

Beach **NT 395** is located on the eastern tip of High Point and consists of a 200 m long north-facing strip of broken high tide sand, backed by a mixture of rock bluffs and foredune, with rocks and rock reefs fronting the entire beach.

Figure 5.51 High Point, with rock-bound beach NT 395 on the point and NT 396 extending to the south.

Beach **NT 396** lies 500 m to the southeast and marks the beginning of the near continuous series of sand beaches. The beach is 700 m long and faces north. It is backed by a vegetated foredune and fronted by a mixture of sand, rock and coral reef flats, with a low narrow rock reef extending 100 m off the eastern end. On the south side of the reef is beach **NT 397**, a curving 1 km long beach, with a central projection in lee of continuous rock flats, and with rock flats dominating much of the 300 m wide intertidal zone, apart from sand flats in lee of the reef. Dense vegetation extends down to the back of the beach.

Beach **NT 398** begins on the southern side of a bedrock protrusion covered by dense woodlands and fronted by laterite rocks and reefs. The beach meanders for 1.1 km in lee of near continuous rock flats to a second bedrock protrusion. Trees line the back of the beach, and the sand then sand flats extend 200 m off the beach. Another bedrock protrusion forms the southern boundary.

Beach **NT 399** initially curves round in lee of the protrusion as a beach and sand flats. As the 1.6 km long beach straightens out, 100 m wide rock then reef flats front the beach, with the rocks including a crenulate series of scallops along the beach. Toward the southern end, the beach has prograded a few tens of metres with backing beach ridges. Its neighbour (beach **NT 400**) is a crenulate strip of high tide sand, dominated by several rock outcrops and reefs, which induce the crenulations. The beach has prograded out over the reefs, depositing 200-300 m of now densely vegetated beach ridges, with rock reefs still extending 100 m off the beach, followed by 300 m wide sand flats. Beach **NT 401** lies on the south side of a more prominent laterite rock-reef protrusion and extends south for 1.5 km to another bedrock protrusion. The beach is initially backed by the beach ridges, then a 200-300 m wide freshwater swamp and finally bedrock slopes, along the southern half of the beach.

Beach **NT 402** is the northern part of a barrier system that has partly infilled a 1 km wide drowned valley, now occupied by a largely infilled 50 ha estuary. The beach begins amongst the rocks on the southern side of its northern neighbour. As the rocks give way a 200 m wide densely vegetated beach-foredune ridge plain backs the beach, with some minor dune activity on the outer foredune. The beach terminates on the western side of a small tidal creek, which drains the backing lagoon. Beach **NT 403** begins on the eastern side of the creek and runs east for 1.8 km to a low rock point and reef. The beach is backed by the 100-200 m vegetated beach ridges, then the lagoon which contains some mangroves and salt flats (Fig. 5.52). Both beaches are fronted by 300 m wide sand flats.

Figure 5.52 Beach NT 403 is backed by a salt-flat-dominated largely infilled lagoon.

NT 404-406 **GILES POINT**

No.	Beach	Rating HT LT	Type	Length
NT404	Giles Pt (W2)	1 1	R+sand flats	2.2 km
NT405	Giles Pt (W1)	1 3	R+sand flats	800 m
NT406	Giles Pt	1 2	R+sand flats	1.2 km
Spring & neap tide range = 1.9 & 0.2 m				

Giles Point is a low sandy foreland, which represents the terminus of sand moving south along the beaches and rocks from High Point. The point has a wooded bedrock core, blanketed by 100-500 m wide beach-foredune ridges, which extend for 4 km around the curving point as beaches NT 404-406.

Beach **NT 404** is a curving, north-facing, 2.2 km long beach, located between a low wooded rocky headland and some eastern rock flats. The continuous sand beach has a couple of rock patches, with a backing 100-300 m wide beach ridge plain, then a small 100 m wide freshwater swamp. It is fronted by 400 m wide sand flats. On the eastern side of the rock flats is 800 m long beach **NT 405**, a curving northeast-facing beach that terminates at sandy Giles Point. The sand flats narrow off the point as the strong tidal currents of Bowen Strait flow past the tip.

On the south side of the point the sandy shoreline faces due east across the 4 km wide strait to Croker Island. The beach (**NT 406**) is 1.2 km long, running from the point to a 10 m high vegetated bedrock bluff. The sand flats increase in width to 200 m by the southern end of the beach, with deeper tidal channels further out.

Swimming: Beware of strong tidal currents and deep water off the Giles Point beaches.

NT 407-410 **IRGUL POINT**

No.	Beach	Rating HT LT	Type	Length
NT407	Irgul Pt (W2)	1 1	R+sand flats	700 m
NT408	Irgul Pt (W1)	1 1	R+sand flats	1.5 km
NT409	Irgul Pt (E1)	1 1	R+sand flats	1.8 km
NT410	Irgul Pt (E)	1 1	R+sand flats	2.8 km
Spring & neap tide range = 1.9 & 0.2 m				

Irgul Point represents a transition from the more exposed shoreline to the east, to the increasingly protected mangrove-fringed shores of Bowen Strait to the west. The point is part of a gently inclined, 3 km wide, wooded bedrock peninsula that has been largely surrounded by deposition associated with tidal creeks to either side, with beach ridges and sand flats along the more energetic northern and eastern shores. A vehicle track reaches the western end of the point providing access to beaches NT 407 and 408.

Beach **NT 407** is a 700 m long, low energy, north-facing beach backed by low coastal plain that grades into mangroves at its western end. It is fronted by 200 m wide sand flats then the deeper waters of Bowen Strait. A vehicle track from the Cobourg road reaches the eastern end of the beach, with a 4WD track running along the back of the beach to the mouth of the Buffalo River, a medium-sized tidal creek located 1 km west of the beach.

Beach **NT 408** extends for 1.5 km east of the vehicle access point, as the outermost in a series of beach ridge-recurved spits that at the point have prograded the shoreline up to 1 km. The beach is also fronted by 200 m wide sand flats ands then 2 km wide Bowen Strait. The vehicle track runs along the back of the beach to the point.

At Irgul Point the shoreline trends southeast following the bend in the outer recurved spit. Beach **NT 409** begins at the point and runs for 1.8 km as a curving east-facing beach, to a low section of coastal plain. The narrow, low energy beach is backed by beach ridges, then the low bluffs of the plain, with scattered mangroves along the beach. Beach **NT 410** begins on the south side of an inflection in the bluffs and is a curving, 2.8 km long, low energy beach, that faces north towards its eastern end. The beach is backed by a few beach ridges and a 100 ha mangrove-filled drowned valley-lagoon, which drains out across the centre of the beach. A vehicle track reaches the southern section of the beach where the coastal plain lies within 200 m of the beach.

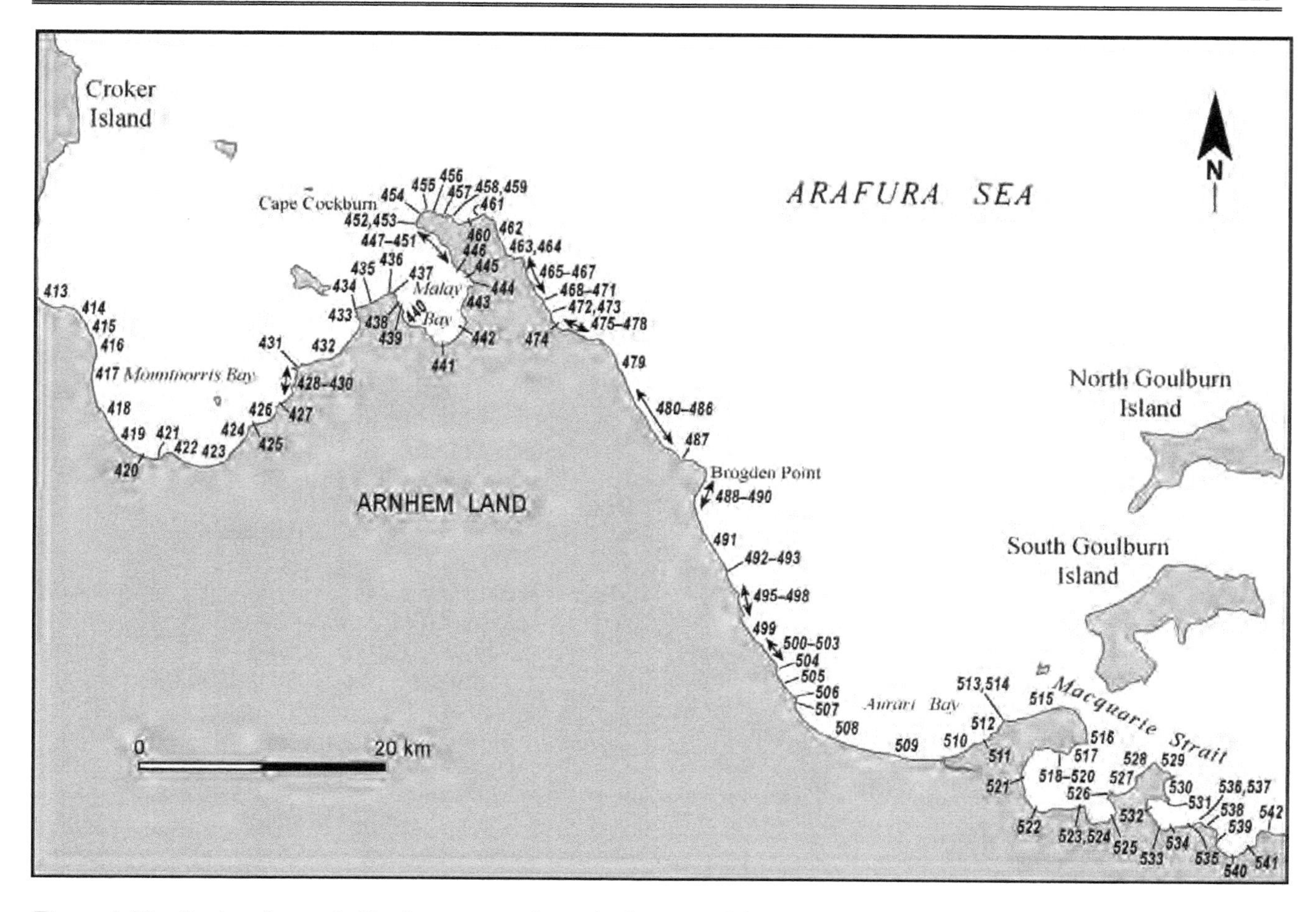

Figure 5.53 Regional map 6: North coast, northern Arnhem Land, beaches NT 413-541.

NT 411-419 **GUIALUNG POINT (E & W)**

No.	Beach	Rating HT	Rating LT	Type	Length
NT411	Guialung Pt (W5)	1	1	R+sand/rock flats	1.3 km
NT412	Guialung Pt (W4)	1	1	R+sand flats	1.5 km
NT413	Guialung Pt (W3)	1	1	R+sand ridges	3.9 km
NT414	Guialung Pt (W2)	1	1	R+sand ridges	1.3 km
NT415	Guialung Pt (W1)	1	1	R+sand ridges	1.2 km
NT416	Guialung Pt (E1)	1	1	R+sand ridges	1.5 km
NT417	Guialung Pt (E2)	1	1	R+sand ridges	500 m
NT418	Guialung Pt (E3)	1	1	R+sand ridges	2 km
NT419	Guialung Pt (E4)	1	1	R+sand ridges	4.5 km

Spring & neap tide range = 1.9 & 0.2 m

Guialung Point is the eastern section of a 6 km wide northeast-facing protrusion in the coastal plain, which forms 10-20 m high bluffs along the shore, and rises to a peak of 56 m within 2 km of the coast. The subaqueous extension of the bedrock has provided a 1-4 km wide platform for coral reef growth. A series of nine beaches (NT 411-419) are spread in a near continuous line around the point, with wave energy gradually increasing from north to south as they emerge from the protection of Croker Island, and south of the point from the lee of coral reefs. The main Cobourg road runs within 2-5 km of the coast and a vehicle track following a fence line reaches the high point of the bluffs above beach NT 415 (Fig. 5.54).

Figure 5.54 Beach NT 415 is backed by low bluffs and a straight access track.

Beach **NT 411** lies 10 m west of the point and is a low energy 1.3 km long northeast-facing beach, lying at the base of 10 m high vegetated bluffs, and fronted by a 200 m wide mixture of sand and rock flats. Two small creeks drain off the coastal plain across the beach. It terminates at an inflection in the bluffs, beyond which is 1.5 km long beach **NT 412**. This northeast-facing beach and beach ridge extend to the mouth of a small creek, which drains a backing 100 ha tidal lagoon containing a mangrove-fringed creek and salt flats. Beach **NT 413** begins on the southern side of the creek and continues on for 3.9 km to the 10 m high bluffs located 2 km north of the point. The beach faces northeast, then north, and receives sufficient wind wave energy to maintain a series of several low intertidal sand ridges across the 300 m wide sand flats, with coral reefs extending up to 500 m off the beach. Several small usually dry creeks drain off the coastal plain along the eastern half of the beach.

Beach **NT 414** begins at a small bedrock outcrop below the bluffs, beyond which the bluffs and beach swing gradually round to face east at the end of the 1.3 km long beach. At the same time the sand flats widen to 600 m and the rock and reef flats extend up to 2 km off the beach, reducing waves sufficiently to permit a few mangroves to grow along the southern end of the beach. At high tide waves cross the flats and maintain some irregular sand ridges. The coastal plain forms one low 100 m wide headland, fringed by 5 m high bluffs at the end of the beach.

Beach **NT 415** lies to the south of the headland, and is a 1.2 km long east-facing beach and beach ridge that blocks a small largely infilled valley. A mangrove-lined creek and salt flats drain the valley, with the creek flowing out across the centre of the beach, with scattered mangroves to either side of the creek mouth. Waves only reach the beach at high tide, as the sand and reef flats extend up to 4 km east of the shore. The beach is bordered at its southern end by Guialung Point, a low 100 m long slight protrusion in the coastal plain, fringed by rock flats and mangroves. On the south side of the point is beach **NT 416**, a relatively straight narrow high tide beach, backed by a low bluff and fronted by the wide sand and reef flats, with a small creek draining a 25 ha patch of salt flats at the southern end.

Beach **NT 417** begins 1.5 km south of the point and represents the beginning of a 12 km long series of beach ridges and tidal creeks separated by low sections of coastal plain. The beach is 500 m long and extends to the first mangrove-fringed creek mouth, which drains a 50 ha coastal salt flat (Fig. 5.55) as well as a small upland catchment. Beach **NT 418** begins on the southern side of the creek and curves slightly to the south for 2 km to the next creek mouth, with a 1 km low section of low wooded coastal plain behind the centre of the beach. Both beaches are fronted by 400 m wide sand flats that contain up to 10 low intertidal sand ridges, produced by wind waves and refracted trade wind swell entering the bay.

Figure 5.55 Beach NT 417 with its wide sand flats and backing beach ridges and drained lagoon.

Beach **NT 419** runs for 4.5 km to the next creek. It gradually swings round to face northeast, with coastal plain backing the central section and the sand ridges continuing across the 400 m wide sand flats. A vehicle track follows a fence line to reach the centre of the beach. The southern end of this beach also marks the easternmost boundary of the Gurig National Park, with Arnhem Land Aboriginal Land commencing to the east.

Arnhem Land Aboriginal Land

Area	9 370 700 ha	
Coast length:	114 km	(1590-1704 km)
	2324 km	(2261-4585 km)
Beaches:	11	(NT 167-177)
	1021	(NT 419-1439)

Major coastal communities:	Population
North coast (West Arnhem Land):	(4000)
Minjilang (Croker Island)	200
Warruwi (South Goulburn Island)	400
Maningrida	2100
Milingimbi	900
Galiwinku (Elcho Island)	1800
Nhulunbuy	4000
Gulf Coast (East Arnhem Land):	(7000)
Yirrkala	1000
Milyakburra (Bickerton Island)	200
Alyangula (Groote Eylandt)	2500
Numbulwar	1200
Total	~20 000

Permit required for entry
For further information contact:
Northern Land Council
Phone: (089) 205 200

Murgenella Wildlife Sanctuary

Coast length	116 km	(2261-2377 km)
Beaches:	89	(NT 420-508)

NT 420-425 MOUNTNORRIS BAY

No.	Beach	Rating HT	Rating LT	Type	Length
NT420	Mountnorris Bay (1)	1	1	R+sand ridges	1.2 km
NT421	Mountnorris Bay (2)	1	1	R+sand ridges	1.4 km
NT422	Mountnorris Bay (3)	1	1	R+sand flats	600 m
NT423	Mountnorris Bay (4)	1	1	R+sand flats	4.3 km
NT424	Mountnorris Bay (5)	1	1	R+sand flats	2.5 km
NT425	Mountnorris Bay (6)	1	1	R+sand flats	300 m

Spring & neap tide range = 1.9 & 0.2 m

Mountnorris Bay is a 16 km wide semicircular north-facing open bay bordered by Guialung Point in the west and Coombe Point to the east. The bay contains 28 km of shoreline, most backed by gently sloping wooded coastal plain, with a mixture of barrier beaches and attached beaches below low bluffs. Numerous small upland streams flow into the bay, several flowing into small estuaries filled with salt flats and mangroves.

The southernmost 10 km of shore face north and contain six beaches (NT420-425) which decrease in exposure to the easterly wind waves as the shoreline swings round to face northeast. The main Cobourg road runs 1-2 km inland of the southern shore with vehicle access to beach NT 423, while in the east the Annesley Point track runs close to the easternmost beach (NT 425).

Beach **NT 420** extends east of the Gurig National Park boundary and into Arnhem Land Aboriginal Land and Murgenella Wildlife Sanctuary. The beach begins on the eastern side of a small tidal creek and extends east for 1.2 km to the next tidal creek. Most of the beach is backed by a 200 m wide series of casuarina-covered beach ridges, with more dynamic spits associated with the creeks at either end, as well as some salt flats in the approximately 100 ha infilled estuary. The beach receives wind waves averaging 0.5 km, which break across a 400 m wide sand flat, containing up to six low sand ridges. Its eastern neighbour, beach **NT 421**, begins on the eastern side of the creek, and extends for 1.4 km to a 10 m high 500 m long bedrock headland. Beach ridges and salt flats initially back the beach, then low wooded bedrock bluffs.

On the eastern side of the headland is beach **NT 422**, a 600 m long north-northeast-facing relatively straight strip of sand, backed by a few casuarina-covered beach ridges and fronted by sand, then rock and reef flats that extend 1.5 km off the headland. The beach terminates at a small tidal creek, which drains a 50 ha salt flat. Beach **NT 423** begins on the eastern side of the creek and runs for 4.3 km as a gently curving north-facing arc. To the west it is backed by beach ridges, which narrow as the low coastal plain reaches the shore along the eastern half of the beach. Three small usually dry upland creeks drain off the plain across the beach. A vehicle track from the main Cobourg road runs out on the eastern side of the creek to reach the beach.

Beach **NT 424** forms part of the eastern, northwest-facing shore of the bay. It is entirely backed by 10-20 m high bluffs of the coastal plain, which dominate the shoreline and are manifest as rock outcrops and rock flats along the crenulate 2.5 km long beach. Wave energy also decreases to the east permitting scattered mangroves to occur, particularly on the rock flats. In addition three small upland creeks drain across the beach. It terminates at a small rectangular headland, along whose north side is 300 m long beach **NT 425**. This beach faces north and is more exposed to the easterly wind waves. It consists of a moderately steep high tide beach, backed by low wooded coastal plain, with 10 m high headlands to either end and 100 m wide sand flats off the beach. There are a couple of houses located on each headland (Fig. 5.56), with vehicle access from these to the Annesley Point track, 4 km to the east.

Figure 5.56 Beach NT 425 and the beginning of NT 426, backed by blufftop houses.

NT 426-433 COOMBE POINT (S & E)

No.	Beach	Rating HT	LT	Type	Length
NT426	Coombe Pt (S5)	1	1	R+sand flats	2.6 km
NT427	Coombe Pt (S4)	1	1	R+sand/reef flats	2.5 km
NT428	Coombe Pt (S3)	1	1	R+sand/reef flats	800 m
NT429	Coombe Pt (S2)	1	1	R+sand/reef flats	500 m
NT430	Coombe Pt (S1)	1	1	R+sand/reef flats	50 m
NT431	Coombe Pt	1	1	R+sand flats	400 m
NT432	Coombe Pt (E1)	1	1	R+sand flats	7.2 km
NT433	Coombe Pt (E2)	1	1	R+sand flats	500 m
Spring & neap tide range = 1.9 & 0.2 m					

Coombe Point is a round, 15 m high, 100 m wide, wooded headland, located on the eastern shore of Mountnorris Bay. Eight beaches lie to either side (NT 426-433). They all face northwest to west across the bay and all are bounded by low bluffs, with some fronted by rock flats and reefs. The Annesley Point road runs 1-2 km in from the coast and provides access tracks to some of the beaches, including a fishing camp at Coombe Point.

Beach **NT 426** is a curving 2.6 km long northwest-facing beach backed by a 100-200 m wide series of beach-foredune ridges and a 10 ha salt flat-filled lagoon behind the centre. Sand flats widen off the beach from 100 m in the west to 400 m in the east, together with some rock flats along the eastern end of the beach. A vehicle track runs down from the western headland along the back of the beach to the small creek mouth. The beach curves round to terminate in lee of a 10 m high 50 m long rocky bluff, on the north side of which is beach **NT 427**. This is a similar 2.5 km long beach, fronted by 500 m wide sand and reef flats, with a narrow beach ridge-foredune behind the beach, then the wooded coastal plain. A vehicle track off the Annesley Point road reaches the centre of the beach, while two small creeks drain off the coastal plain and across the beach.

The next three beaches are increasingly protected by Coombe Point, rock flats and their westerly orientation. Beach **NT 428** is located beyond the next bedrock-induced, mangrove-fringed inflection in the coast. It is a slightly curving, 800 m long, northwest-facing, high tide sand beach, fronted by 200 m wide rock flats, together with scattered mangroves on the inshore flats. Its neighbour (beach **NT 429**) is a similar narrow 500 m long beach that faces due west and is bordered by low white bluffs. Finally the low energy pocket beach **NT 430** is a 50 m long strip of sand lying on the west side of Coombe Point, wedged between the southern white bluffs and mangroves to the north. A vehicle track from Coombe Point runs along the back of the beach, and south for 2 km to beach NT 428.

Coombe Point consists of the main wooded headland and a second southern headland anchored by a rock flat. Between the two is the 400 m long Coombe Point beach (**NT 431**). The main point has actually been tied to the mainland by a small tombolo that links the point beach with its eastern neighbour. The beach faces northwest and receives moderate wind waves, which break across a narrow sand flat and up the moderately steep beach. The beach is accessible by a track off the Annesley Point road and is used as a fishing camp with small boats launched from the beach (Fig. 5.57).

Figure 5.57 Coombe Point with the sheltered beach NT 431 in foreground, and the beginning of the longer more exposed beach NT 432 (left).

Beach **NT 432** begins on the eastern side of the point and curves round for 7.2 km, eventually facing west inside the northern rock-reef-controlled sandy foreland. The beach has prograded along its length from 600 m in the south to 100 m in the north, leaving a series of beach ridges which are cut by five small streams, of which one drains a 5 ha swamp. Two hundred to four hundred metre wide sand flats and some reefs lie off the beach, while the Annesley Point road lies between 0.5 and 2 km east of the shore, with vehicle access to either end.

On the north side of the foreland is 500 m long west-facing beach **NT 433**. This beach curves slightly between two bedrock-reef-controlled forelands. It is backed by a 50 m wide sand beach ridge, and fronted by 200 m wide sand flats. A single house is located behind the northern end of the beach.

NT 434-436 **ANNESLEY POINT (W)**

No.	Beach	Rating HT	Rating LT	Type	Length
NT434	Annesley Pt (W3)	1	1	R+sand/rock flats	400 m
NT435	Annesley Pt (W2)	1	1	R+sand/rock flats	1.5 km
NT436	Annesley Pt (W1)	1	1	R+sand/rock flats	800 m
Spring & neap tide range = 1.9 & 0.2 m					

Annesley Point is an 8 m high wooded bluff, which forms the boundary between Mountnorris and Malay bays. Three northwest-facing beaches (NT 434-436) lie within 3 km west of the point, each controlled by boundary rock reefs, backing wooded bluffs and fronting 100-200 m wide rock flats. Beach **NT 434** is a straight 400 m long sandy beach, that begins on the rock-controlled tip of a sandy foreland and terminates against a beachrock protrusion, with sand then rock and reef flats off the beach.

On the eastern side of the 100 m long section of beachrock is beach **NT 435**, a 1.5 km long beach, backed by a narrow beach ridge then woods and fronted by sand, then reefs. It terminates at a 100 m long protrusion in the coastal plain in lee of a small bedrock reef. On the eastern side beach **NT 436** continues on for 800 m to Annesley Point, with scarped 10 m high bluffs backing the entire beach, and sand and reef flats offshore. The Annesley Point road is located 300 m in from the back of the three beaches, with vehicle tracks leading to each of the beaches.

Malay Bay is a 5 km wide bay located in lee of Cape Cockburn. The bayshore runs south of the cape for 15 km to the Wangaran Swamp, then curves round for 7 km to Annesley Point, which lies 6 km almost due south of the cape. In all the bay has 21 km of generally low energy shoreline, containing 15 beaches (NT 437-451), including five barrier systems, with most of the beaches backed by low bluffs. The beaches occupy 13.5 km (65%) of the bay shoreline. The eastern shores of the bay provide a sheltered anchorage from the southeast trades as well as the low easterly swell that wraps round the cape.

NT 437-440 **ANNESLEY POINT (E)**

No.	Beach	Rating HT	Rating LT	Type	Length
NT437	Annesley Pt (1)	1	2	R+sand/rock flats	100 m
NT438	Annesley Pt (E2)	1	2	R+sand/rock flats	200 m
NT439	Annesley Pt (E3)	1	2	R+sand/rock flats	400 m
NT440	Annesley Pt (E4)	1	1	R+sand flats	1.5 km
Spring & neap tide range = 1.9 & 0.2 m					

At **Annesley Point** the shoreline turns and trends south into Malay Bay. The low laterite bluffs of the point dominate the western 5 km of the bay shore. The first three beaches (NT 437-439) are backed by bluffs and fronted by a mixture of sand and rock flats.

Beach **NT 437** begins immediately east of the point. It is a narrow curving 100 m long strip of high tide sand, wedged in between the backing 10 m high reddish bluffs and the fronting 200 m wide sand and rock flats, with intertidal rock points extending off each end, both hosting a few mangroves. Beach **NT 438** lies in a small curving embayment located 300 m south of the point. It is 200 m long with rock flats extending off both ends and deeper 300 m wide sand flats in the centre. It is backed by curving scarped bluffs, with the Annesley Point road terminating above the northern side of the beach. Beach **NT 439** begins after a 400 m stretch of scattered mangroves. The beach is an outward-curving 400 m long strip of high tide sand, backed by low wooded bluffs and fronted by 100-200 m wide sand and rock flats.

Beach **NT 440** lies across the mouth of a drowned 1.2 km wide valley located between two 10 m highs in the coastal plain. The valley has been infilled by up to 1.6 km of progradation in the form of three series of beach ridge-cheniers, separated by swamps and mangroves totalling 150 ha in area, with a mangrove-lined creek draining the system, and forming the southern boundary of the 1.5 km long beach. Sand flats extend 600 m off the beach.

NT 441-442 **WANGARAN SWAMP**

No.	Beach	Rating HT	Rating LT	Type	Length
NT441	Wangaran Swamp (1)	1	2	R+mud flats	1.2 km
NT442	Wangaran Swamp (2)	1	2	R+mud flats	3 km
Spring & neap tide range = 1.9 & 0.2 m					

The **Wangaran Swamp** is a 700 ha area of swamp, salt flats and chenier-beach ridges that has filled the southern 4 km of Malay Bay. Up to 20 cheniers and beach ridges have prograded 4 km out into the bay, filling the valley with 400 ha of regressive sand and mud deposits. Two narrow, low energy beaches and mud flats front the outermost ridge. Beach **NT 441** occupies the western 1.2 km of the swamp shore, between the mangrove-fringed coastal plain boundary and the creek mouth. It is a narrow high tide beach with scattered mangroves, fronted by 600 m wide mud flats. On the eastern side of the creek is 3 km long beach **NT 442**, which consists of a 2.5 km long outer beach ridge, then continues on as a crenulate sand beach below 10 m high scarped bluffs of the eastern boundary. A vehicle track runs along the top of the bluffs up to the southern end of beach NT 443.

NT 443-451 **MALAY BAY (E)**

No.	Beach	Rating HT	Rating LT	Type	Length
NT443	Malay Bay (E1)	1	2	R+sand/rock flats	1.4 km
NT444	Malay Bay (E2)	1	3	R+sand flats	150 m
NT445	Malay Bay (E3)	1	1	R+sand flats	800 m
NT446	Malay Bay (E4)	1	2	R+sand/rock flats	1.1 km
NT447	Malay Bay (E5)	1	2	R+sand/rock flats	400 m
NT448	Malay Bay (E6)	1	2	R+sand/rock flats	900 m
NT449	Malay Bay (E7)	1	2	R+sand/rock flats	800 m
NT450	Malay Bay (E8)	1	1	R+sand flats	1.1 km
NT451	Malay Bay (E9)	1	2	R+sand/rock flats	500 m
Spring & neap tide range = 1.9 & 0.2 m					

The eastern shores of **Malay Bay** all face west, with the dominant trades blowing offshore resulting in usually calm conditions at the shore. Only low refracted swell and occasional north through westerly wind waves reach the nine low energy beaches (NT 443-451).

Beach **NT 443** is a crenulate 1.4 km long beach backed by a mixture of low wooded bluffs and beach ridges. It consists of a narrow high tide beach fronted by scattered mangrove-lined rock flats and 200-400 m wide sand flats (Fig. 5.58).

Beach **NT 444** lies on the southern side of a tidal creek. It is a 150 m long strip of sand, bordered by mangroves and the creek and fronted by a deep tidal channel then 1 km wide sand flats. Beach **NT 445** forms the northern side of the creek, and is an 800 m long beach ridge-spit, that in the north runs for 200 m along the base of wooded low bluffs. The beach is the outermost in a 1 km wide series of three sets of beach ridges that have filled the 1.5 km wide drowned valley with 200 ha of ridges, mangroves and salt flats.

A protruding section of 15 m high coastal plain forms the northern side of the valley, along the base of which is beach **NT 446.** This is a crenulate 1.1 km long strip of high tide sand, lying at the base of 5-10 m high scarped bluffs, with scattered mangroves and rock flats along the beach, and 300-500 m wide sand flats off the beach. The northern end of the beach terminates at a small protrusion in the bluffs, on the north side of which is 400 m long beach **NT 447**. This beach faces west and extends up to the mouth of a tidal creek, which drains another smaller drowned valley. The beach is initially tied to the coastal plain, then backed by a wooded 100 m wide beach ridge, with the mangrove-lined creek entrance forming the northern boundary. Together with beach NT 448 it encloses a 30 ha area of mangroves and salt flats. Beach **NT 448** emerges from the creek mouth mangroves and runs for 900 m, initially as a beach ridge then along the base of the adjoining coastal plain, until a clump of mangroves. Several hundred metre wide sand flats lie off all three beaches.

Beach **NT 449** is an 800 m long strip of high tide sand, lying at the base of scarped 5 m high bluffs. It extends from the northern side of the mangroves around a small point to the southern side of a third tidal creek. It then continues as a recurved spit blocking a smaller 10 ha drowned V-shaped valley. Beach **NT 450** begins on the north side of the small creek. It extends across the valley mouth as a grassy beach ridge for 500 m, before running along the base of south-facing low bluffs for another 600 m to a stand of mangroves. Both beaches are fronted by 800 m wide sand flats.

Beach **NT 451** is a south-facing 500 m long beach lying just inside Cape Cockburn. It is backed by low wooded bluffs and consists of a narrow high tide beach that faces south across some rock flats and 600 m wide sand flats.

Figure 5.58 The low energy beach NT 443 occupies the sheltered shore of Malay Bay.

Region 3: North Arnhem Land Coast: Cape Cockburn-Cape Arnhem

Length of coast: 1414 km (2313-3727 km)
Number of beaches: 602 (NT 452-1053)

Figure 5.53 page 225 beaches NT 413-542
Figure 5.68 page 240 NT 543-687
Figure 5.84 page 256 NT 688-958
Figure 5.107 page 280 NT 958-1069

Aboriginal Land: Arnhem Land
National Parks: none

Towns/communities: Warruwi (South Goulburn Is), Maningrida, Milingimbi, Galiwinku (Elcho Is), Nhulunbuy, Yirrkala

The northern coast of the Northern Territory west of Cape Cockburn consists of 1414 km of northeast-trending headlands, peninsulas and island chains, which divide the coast into a series of bays, including the larger and sheltered Buckingham and Arnhem bays, while the east-facing sides of the points and islands are more exposed to the trade winds and waves. The entire coast and hinterland are part of Arnhem Land Aboriginal Land, including the mining town and bauxite leases on the Gove Peninsula surrounding Nhulunbuy.

NT 452-458 **CAPE COCKBURN**

No.	Beach	Rating HT	Rating LT	Type	Length
NT452	Cape Cockburn (1)	1	2	R+sand/rock flats	200 m
NT453	Cape Cockburn (2)	2	2	R+rock flats	300 m
NT454	Cape Cockburn (3)	2	2	R+rock flats	100 m
NT455	Cape Cockburn (4)	2	2	R+rock flats	400 m
NT456	Cape Cockburn (5)	2	2	R+rock flats	100 m
NT457	Cape Cockburn (6)	2	2	R+sand/rock flats	600 m
NT458	Cape Cockburn (7)	1	2	R+rock flats	500 m
Spring & neap tide range = 1.9 & 0.2 m					

Cape Cockburn is a 1.5 km wide, 10-15 m high, wooded laterite headland, which forms the northern entrance to Malay Bay. The bedrock-dominated shore extends right around the cape and along the north side for 5 km to De Courcy Head, from where the coast trends southeast. Between the southern tip of the cape and De Courcy Head are seven generally rock- and rock-reef-dominated beaches (NT 452-458). There is no vehicle access to the cape or head and no development.

Beach **NT 452** forms the southern tip of the cape. It is a curving 200 m long west-facing beach, tied to rock reefs at both ends, with a mixture of rock and some sand flats extending 200 m off the beach. A few low beach ridges back the beach. On the northern side of the reef is beach **NT 453**, a 300 m long west-facing beach that receives lowered waves and swell refracted round the cape, which usually maintain a shallow 50 m wide surf zone over the fronting 200-300 m wide rock flats (Fig. 5.59).

Figure 5.59 Beaches NT 453-455 extend south of Cape Cockburn and receive refracted easterly waves, which produce some surf over the laterite reefs.

Beach **NT 454** is located toward the northwestern side of the cape. It is a 100 m long north-facing strip of high tide sand, backed by 10 m high bluffs, bordered by intertidal rock flats, with deeper water off the centre of the beach. Immediately to the north and running up to the tip of the cape is beach **NT 455**, a 400 m long high tide beach fronted by a continuous rock platform and intertidal rock flats. The crenulate beach consists of a strip of sand wedged in between the rocks and the base of the 10 m high red bluffs, with a few mangroves protected by the rocks growing at the base of the beach. Waves averaging over 0.5 m refract round the cape and produce a right-hand break along the edge of the rock flats.

On the eastern side of the tip of the cape the bluffs continue 5 km to De Courcy Head, with beaches NT 456-458 occupying the first three indentations in the bluffs. Beach **NT 456** is a 100 m long pocket of sand bordered by irregular wave-washed rock platforms and flats, with bluffs behind. Two hundred metres further east is beach **NT 457**, a crenulate 600 m long bluff-backed beach, with a 50 m wide band of rock flats then sandy seabed. The surf breaks over the rocks with a permanent rock-controlled rip toward the centre of the beach. A small headland protrudes 100 m north at the eastern end with the waves forming a long right-hand break over the bordering rock flats.

Beach **NT 458** is located in a semicircular north-facing curve in the red bluffs. The 500 m long beach lies along the base of the bluffs, together with some rock debris on the beach. Scattered rock flats fill much of the bay, with waves breaking across the rocks and returning to sea via two permanent rock-controlled rips.

NT 459-461 **DE COURCY HEAD (W)**

No.	Beach	Rating HT	Rating LT	Type	Length
NT 459	De Courcy Hd (W3)	2	3	R+LT bar&rips	1 km
NT460	De Courcy Hd (W2)	1	2	R+rock flats	150 m
NT461	De Courcy Hd (W1)	1	2	R+rock flats	100 m
Spring & neap tide range = 1.9 & 0.2 m					

Beach **NT 459** is a 1 km long northeast-facing beach located 2.5 km west of De Courcy Head. The beach is backed for 500 m by 20 m high red bluffs, then 500 m of transgressive dunes that have climbed up over the bluffs and blown up to 300 m inland. In the south the beach receives waves that are refracted around the head and average just under 1 m to maintain a 50 m wide surf zone. Wave height increases to the north where it maintains a 100 m wide surf zone containing permanent rips to either end and a central beach rip. Reefs, rock platform and bluffs border the northern end.

Between the end of the beach and the head 2 km to the east are two pocket beaches, both occupying small embayments at the base of the bluffs. Beach **NT 460** is a straight 150 m long strip of sand at the base of steep bluffs, with rock platforms and reefs extending 100 m off each end and deeper water in the centre. Waves break along the side of the reefs and flow back out through a permanent central rip.

Beach **NT 461** is located halfway out to the head. It is a 100 m long pocket of sand bordered and fronted by rocks and rock reef, with waves breaking over the reef. The beach can only be accessed by climbing down the steep 30 m high bluffs.

NT 462-471 **DE COURCY HEAD (E)**

No.	Beach	Rating HT	Rating LT	Type	Length
NT462	De Courcy Hd (E1)	3	4	R+LT rips	2.5 km
NT463	De Courcy Hd (E2)	2	3	R+rocks/reefs	100 m
NT464	De Courcy Hd (E3)	2	3	R+rocks/reefs	500 m
NT465	De Courcy Hd (E4)	2	3	R+rocks/reefs	300 m
NT466	De Courcy Hd (E5)	2	3	R+rocks/reefs	100 m
NT467	De Courcy Hd (E6)	3	4	R+LT rips	1 km
NT468	De Courcy Hd (E7)	3	4	R+LT rips	100 m
NT469	De Courcy Hd (E8)	2	3	R+rocks/reefs	300 m
NT470	De Courcy Hd (E9)	2	3	R+rocks/reefs	200 m
NT471	De Courcy Hd (E10)	2	3	R+LTT	250 m
Spring & neap tide range = 2.2 & 0.2 m					

De Courcy Head is a prominent 33 m high rugged headland composed of Tertiary laterite, with red bluffs extending 5 km west to Cape Cockburn. At the head the coast turns and trends southeast for 60 km down into Aurari Bay. While the laterite bluffs dominate much of the way, there are numerous beaches between them and at their base. The first 8 km contain 10 generally small beaches (NT 462-471), including three areas of moderate dune transgression. All the beaches face into the 1 m plus easterly wind waves and all have moderate energy surf zones with rips.

Beach **NT 462** is the longest beach in this section. It begins 1 km southeast of the head and extends for 2.5 km as an exposed straight northeast-facing beach, bordered at each end by 20 m high bluffs. The beach has a 100 m wide surf zone containing several rips spaced about every 150 m, as well as a permanent rip against the northern rocks. It is backed by bluffs along its northern half, and dunes extending 400 m inland and rising to 20 m high in lee of the southern half (Fig. 5.60).

Figure 5.60 The exposed and relatively high energy rip-dominated beach NT 462.

Beach **NT 463** lies just south of its northern neighbour and consists of a 100 m long pocket of sand contained in a near semicircular 20 m high bluff amphitheatre, with reefs extending out either side and a permanent rip out the centre. Two hundred metres to the south is beach **NT 464**, a double curving 500 m long beach, backed by low scarped bluffs, with rock reefs extending 200 m off the beach. Waves wrap around the southern end producing a right-hand point break.

Beach **NT 465** lies on the eastern face of the first headland south of De Courcy Head. It consists of a narrow 300 m long strip of high tide sand, backed by 10 m high bluffs, and fronted by continuous intertidal rock flats and reefs. The beach is contained in a semicircular 100 m wide embayment, with reefs bordering each side of the entrance. The waves produce a left and right break over the reefs, with a permanent rip running out the 100 m wide gap in the middle (Fig. 5.61).

Figure 5.61 The pocket beach NT 465 is bordered by laterite bluffs and reefs, which produce surf breaks to either side of the small bay.

Beach **NT 466** lies 1 km further south and consists of a 100 m long strip of sand almost enclosed in bluffs, rocky points and reefs. Waves break over the reefs forming a protected 'lagoon' in their lee, with low waves to calm conditions at the shoreline. Immediately to the south is the 1 km long exposed beach **NT 467**. This more energetic beach has a 100 m wide surf zone with both beach and boundary permanent rips. It is backed by up to 800 m of active dunes, which rise to a maximum height of 48 m.

The next 1.5 km of coast are dominated by irregular bluffs and reef, and include four reef-controlled beaches (NT 468-471). All four beaches are backed by bluffs and deflated and vegetated clifftop dunes, with the only active dune in lee of the southern beach NT 471. A vehicle track off the Murgenella road runs along the top of the bluffs, as well as providing access to the beach NT 467. Beach **NT 468** is bordered and backed by rock platforms and 20 m high bluffs, with the surf zone a continuation of its northern neighbour. Three hundred metres to the south is beach **NT 469**, a relatively straight 300 m long beach backed by low grassy bluffs and fronted by a reef-dominated 100-150 m wide surf zone, drained by two permanent rips. Beach **NT 470** lies immediately to the south and is a 200 m long strip of sand, backed by low bluffs, with a 100 m wide surf zone dominated by rock reefs. One hundred metres to the south is beach **NT 471**, a 250 m long curving beach that terminates in lee of a small headland where it faces north. The waves break over boundary reefs and a continuous bar, with a permanent rip flowing out past the southern point.

NT 472-478 **GNINGARG POINT (W)**

No.	Beach	Rating HT	Rating LT	Type	Length
NT472	Gningarg Pt (W7)	3	4	R+LT rips	400 m
NT473	Gningarg Pt (W6)	3	4	R+LT rips	400 m
NT474	Gningarg Pt (W5)	3	4	R+LT rips	2.6 km
NT475	Gningarg Pt (W4)	2	3	R+reefs	300 m
NT476	Gningarg Pt (W3)	2	3	R+reefs	200 m
NT477	Gningarg Pt (W2)	2	3	R+LTT	150 m
NT478	Gningarg Pt (W1)	2	3	R+reefs	150 m
Spring & neap tide range = 2.2 & 0.2 m					

Gningarg Point is located 15 km southeast of De Courcy Head. Eighteen beaches (NT 461-478) are located between the two points, three backed by 1-2 km wide transgressive dune systems, the rest for the most part dominated by rock reefs and by 10-30 m high bluffs and cliffs. The first ten beaches (NT 462-471) are described above, the next seven lie within 7 km of the point (NT 472-478). The Murgenella track runs 2-3 km in from the coast, however there is no vehicle access to any of the beaches south of NT 471.

Beaches **NT 472** and **NT 473** are adjoining 400 m long beaches separated by a 50 m wide 10 m high grassy headland, as well as both being bordered by more prominent boundary laterite headlands. Both beaches face northeast and receive waves averaging 1 m which break across a 100 m wide surf zone dominated by permanent rips against the headlands and central point, together with more transitory beach rips. Wave height decreases toward the southern end of beach NT 473, however rock and reefs dominate this end of the beach. Both beaches are backed by 300-400 m long blufftop parabolic dunes.

On the south side of the 100 m wide southern headland is beach **NT 474**, one of the longer beaches on this section of coast. The 2.6 km long northeast-facing beach has an exposed moderate energy surf zone with a permanent rip against the northern rocks and up to 12 beach rips spaced about every 150 m, with waves decreasing slightly in the southern corner. Active dunes extend up to 600 m inland and to heights of 65 m, while older vegetated dunes extend up to 2 km inland. A creek crosses the southern end of the beach, draining 20 ha of salt flats fed by two small upland creeks.

The next section of coast consists of 4 km of north-northeast-facing rocky coast containing four smaller rock- and reef-dominated beaches. Beach **NT 475** faces north and runs along the base of 10 m high grassy bluffs for 300 m. Waves break over reefs lying 50-100 m off the beach, resulting in lower waves at the shore. Immediately east is beach **NT 476**, a 200 m long northeast-facing sandy beach, bordered by rock reefs, and backed by 20 m high bluffs over which a dune has climbed and blown 100 m inland. Two permanent rips drain the 100 m wide surf zone, with rocks and reefs increasing off the southern end of the beach. Its immediate neighbour, beach **NT 477**, is a 150 m long strip of sand located at the base of scarped 15 m high bluffs and bordered by a protruding section of bluffs at the southern end. High tide waves reach the base of the bluffs across a slightly lower energy 50 m wide surf zone.

Beach **NT 478** lies 1 km further east and is located in a 150 m long indentation in the 20-30 m high bluffs. The beach faces due north and consists of a narrow strip of high tide sand backed by the steep vegetated bluffs, with lower waves breaking over scattered rock reefs extending 100 m off the beach.

NT 479-485 **GNINGARG POINT (E)**

No.	Beach	Rating HT LT	Type	Length
NT479	Gningarg Pt (E1)	3 4	R+LT rips	300 m
NT480	Gningarg Pt (E2)	3 4	R+LT rips	200 m
NT481	Gningarg Pt (E3)	3 4	R+LT rips	400 m
NT482	Gningarg Pt (E4)	3 4	R+reefs	300 m
NT483	Gningarg Pt (E5)	3 4	R+reefs	100 m
NT484	Gningarg Pt (E6)	3 4	R+LT rips	200 m
NT485	Gningarg Pt (E7)	3 4	R+LTT+rocks	200 m
Spring & neap tide range = 2.2 & 0.2 m				

Steep 20-30 m high bluffs dominate the 2 km of coast west of 35 m high **Gningarg Point**. At the point the coast turns and trends southeast, again exposing the bluffs and occasional beaches to the prevailing east to southeast winds and waves. The next seven beaches (NT 479-485) lie within 8 km of the point, and all occupy indentations and small valleys in the 30 m high red and white laterite bluffs.

The first beach (**NT 479**) lies 1 km south of the point. It is 300 m long and centred on a V-shaped gully fed by two converging streams, with grassy slopes running down to the beach, and scarped bluffs rising sharply to either side. The beach is well exposed to the easterly waves and has a 100 m wide surf zone with permanent rips to either end and usually two more central beach rips.

Beach **NT 480** lies 2 km further south. It is located in a semicircular 200 m wide embayment, bordered by rock platforms and 20 m high red over white bluffs, and backed by two grassy V-shaped valleys that drain streams onto either end of the beach. The surf extends out to the rocky points with strong rips running out either end and a third central beach rip. Beach **NT 481** lies 200 m around the southern headland. It is a relatively straight 400 m long beach backed by lower 5-10 m high scarped bluffs, with small streams and a valley entering toward the southern end of the beach. Waves break across a 50-100 m wide surf zone, with boundary rips at each end and two to three beach rips in between. Beach **NT 482** is located 400 m to the south around a more prominent, whitish, 30 m high headland that projects 300 m to the east and contains sea caves. The beach is 300 m long and located at the base of 10-25 m high scarped white bluffs, with a hanging valley at the southern end. It consists of three high tide strips of sand separated by rocks, with the 100 m wide surf zone dominated by rock reefs.

Beach **NT 483** is located on the south side of a more prominent bluff that projects 200 m out from the line of the cliffs. The beach consists of a 100 m long narrow strip of sand at the base of the bluffs, fronted by a 100 m wide surf zone, with a prominent slump in the bluffs located 100 m south of the beach (Fig. 5.62).

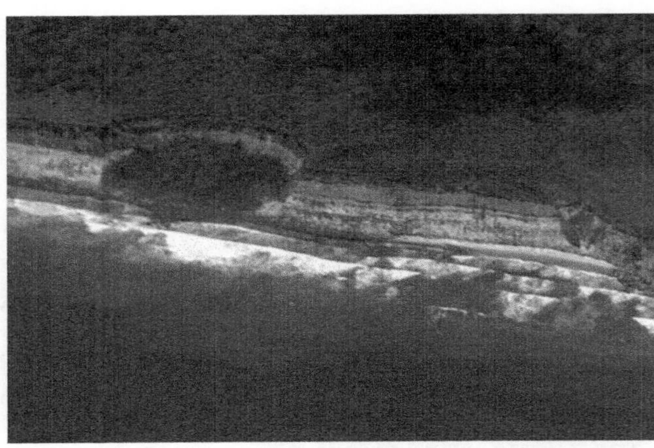

Figure 5.62 Beach NT 483 is a sliver of sand at the base of 29 m high slumping bluffs.

Beach **NT 484** is located another 1 km to the south at the base of two converging streams, which enter at either end of the 200 m long beach. Scarped bluffs lie to either side, with generally low grassy bluffs in between. The surf runs along the bluffs and beach with two to three beach rips occupying the surf off the beach.

One kilometre further south is beach **NT 485**, a 200 m long strip of high tide sand at the base of slumping 30 m high red bluffs, with a mixture of sand and rock reef across the 50-100 m wide surf. There is a right-hand point break off the southern point, which protrudes 100 m to the east.

Surf: There is potential for both beach and a number of point breaks either side of Gningarg Point.

NT 486-491 BROGDEN POINT (W & E)

No.	Beach	Rating HT LT	Type	Length
NT486	Brogden Pt (W2)	3 4	R+LT rips	200 m
NT487	Brogden Pt (W1)	3 4	R+reefs	150 m
NT488	Brogden Pt (E1)	3 4	R+reefs	300 m
NT489	Brogden Pt (E2)	3 4	R+reefs	300 m
NT490	Brogden Pt (E3)	2 3	R+LTT	250 m
NT491	Brogden Pt (E4)	3 4	R+LT rips	5.4 km
Spring & neap tide range = 2.2 & 0.2 m				

Brogden Point is a semicircular-shaped 3 km wide protrusion in the laterite cliffs, that is for the most part surrounded by 5 km of steep vegetated slopes rising to a crest of 30 m. Five small beaches lie either side of the point (NT 486-490), with the first long beach (NT 491) located 3 km south of the point. A vehicle track runs between 0.5-1.5 km in from the coast, however there is no vehicle access to any of the beaches.

To the west of the point are two pocket beaches. Beach **NT 486** lies 5 km west of the tip, and is a 200 m long straight strip of sand backed by 30 m high slumping bluffs, with scarped bluffs to either end. It is fully exposed to the easterly waves and has a 100 m wide rip-dominated surf zone. One and a half kilometres to the east is beach **NT 487**, a 150 m pocket beach located in a small north-facing gap in the bluffs, with creeks entering at either end. Bedrock reefs lie off the beach and maintain a 100-150 m wide surf and a permanent rip exiting toward the western end.

Three kilometres of continuous rocky coast surround the point, with the first beach on the eastern side (**NT 488**) located 1 km south of the tip of the point. This is a discontinuous 300 m long strip of high tide sand wedged in at the base of 10-30 m high bluffs, with rocks and reef dominating the beach and surf. A small steep valley drains to the northern end of the beach, which is fronted by the narrow beach and a high tide rock pool, then surf.

Beach **NT 489** is located around the next small headland, and is a 300 m east-facing strip of sand backed by slumping scarped 20-30 m high red and white bluffs, with no safe foot access to the beach. The beach is fronted by deeper reefs, which lower waves at the shore.

Three kilometres south of the point the bluffs are replaced by one of the longer beaches on this section of coast. It begins as a 250 m long section of beach (**NT 490**) at the base of vegetated slopes, with creeks draining small valleys at each end. This section is protected by deeper offshore reefs, resulting in lower waves at the shore and a continuous low tide bar. Immediately south of a protruding piece of bluffs, 5.4 km long beach **NT 491** runs relatively straight to the south-southeast, terminating at a 35 m high vegetated point. This beach is backed by vegetated bluffs along the northern half, while along the southern half active sand dunes have blown up over the bluffs and 100-300 m inland. Ten small creeks drain out of small valleys along the slopes and across the beach. Waves average about 1 m and maintain a 100-150 m wide surf zone cut by beach rips approximately every 200 m.

NT 492-502 LATERITE POINT (W)

No.	Beach	Rating HT LT	Type	Length
NT 492	Laterite Pt (W11)	3 4	R+rocks/reef	100 m
NT493	Laterite Pt (W10)	3 5	R+rocks/reef	100 m
NT494	Laterite Pt (W9)	3 4	R+LTT	100 m
NT495	Laterite Pt (W8)	3 4	R+LT rips	200 m
NT496	Laterite Pt (W7)	3 4	R+rocks/reef	100 m
NT497	Laterite Pt (W6)	3 4	R+LT rips	400 m
NT498	Laterite Pt (W5)	3 4	R+LTT	300 m
NT499	Laterite Pt (W4)	3 4	R+LT rips	3 km
NT500	Laterite Pt (W3)	3 4	R+LT rips	400 m
NT501	Laterite Pt (W2)	3 4	R+LTT	150 m
NT502	Laterite Pt (W1)	3 4	R+LT rips	150 m
Spring & neap tide range = 2.2 & 0.2 m				

Laterite Point is a densely vegetated 30 m high inflection in the laterite bluffs that occupy much of the western shores of Aurari Bay. The bluffs continue north of the point for 17 km to Brogden Point and south for 4 km to the beginning of the longer Aurari Bay beaches. Gradual erosion of the bluffs is scarping them in many places, as well as eroding the finer sediment from the bluffs and exposed bedrock in the surf zone. The sediment discolours the water and highlights the location of surf zone and rips.

Beaches NT 492-502 lie between the point and 8 km to the northwest. All are backed and largely dominated by the bluffs, with no dunes forming behind any of the beaches, even though they all face into the trades.

Beaches **NT 492** and **NT 493** are adjoining 100 m pockets of east-facing sand, lying at the base of steep vegetated 35 m high bluffs, with rocks bordering each, and rocks and bluff debris littering beach NT 493 in particular. The waves break over the rocks and reefs, with a left-hand point break off the northern end of beach NT 492. Two hundred metres to the south is beach **NT 494**, a similar bluff-backed beach, with rocks and reefs to either side, but with a central 50 m wide sandy surf zone.

Beach **NT 495** lies 1 km further south and consists of a 200 m long narrow straight strip of sand, backed by low vegetated bluffs and a small gently sloping valley, and a creek which crosses the middle of the beach. Waves break across a 100 m wide surf dominated by a permanent southern rip, as well as a more transient beach rip. Three hundred metres to the south is beach **NT 496**, which is located at the base of a scarped point. Bright red and white 20 m high bluffs back the 100 m long beach, which has basal rock rubble to either end, a permanent northern rip and a 100 m wide surf zone.

Beach **NT 497** occupies the next small embayment in the bluffs. It is a relatively straight 400 m long beach at the base of scarped 10 m high bluffs, with rocks to either end as well as outcropping on the centre of the beach. It has a 50-100 m wide surf, containing two permanent rips toward the centre and the northern end of the beach. The next embayment is occupied by beach **NT 498**, a 300 m long narrow strip of sand at the base of eroding 20 m high bluffs, with bordering laterite points, platforms and reefs. The waves break across a continuous bar and rocks, with a permanent northern rip and usually two beach rips.

Beach **NT 499** begins 500 m to the south around the next point, and at 3 km is the longest beach in this section. The entire beach is relatively straight and lies at the base of eroding and slumping 20 m high bluffs, with much debris on the beach and littering the 100 m wide surf zone (Fig. 5.63). A vehicle track reaches the top of the bluffs toward the northern end of the beach. The beach terminates at a slightly protruding 200 m long section of bluffs and rocks, on the southern side of which is 400 m long beach **NT 500**. This relatively straight beach has substantial reef off the northern end and a permanent rip flowing out off the southern end, with the lowered waves breaking across a 50 m wide surf zone. Sloping 20 m high densely vegetated bluffs back the beach.

Figure 5.63 View south along beach NT 499, which is backed by low eroding bluffs.

Beaches **NT 501** and **NT 502** lie 100 m further south and are two adjoining 150 m long beaches, backed by the 30 m high sloping vegetated bluffs on Laterite Point. Rocks and reefs border each beach, together with some rock debris on the beaches and a permanent rip draining each 100 m wide surf zone.

NT 503-507 LATERITE POINT (E)

No.	Beach	Rating HT LT	Type	Length
NT503	Laterite Pt	3 4	R+rocks/reef	100 m
NT504	Laterite Pt (E1)	3 4	R+rocks/reef	50 m
NT505	Laterite Pt (E2)	3 4	R+rocks/reef	1.25 km
NT506	Laterite Pt (E3)	3 4	R+LTT	1.3 km
NT507	Laterite Pt (E4)	3 4	R+rocks/reef	300 m
Spring & neap tide range = 2.2 & 0.2 m				

At Laterite Point the shoreline trends away to the south then slowly swings back toward the southeast by the next point, 3 km away. Between the two points are five bluff-, rock and reef-dominated beaches (NT 503-507).

Beach **NT 503** is located on the western tip of Laterite Point and consists of a 100 m long strip of sand backed by 30 m high vegetated bluffs, and bordered by rocks and reefs, with deeper reefs off the beach. Just around the tip of the point is beach **NT 504**, a 50 m long pocket of high tide sand and cobbles backed by scarped 15 m high red bluffs, with rock debris littering the beach, and reefs off the beach.

One kilometre south of the point the bluffs begin to recede slightly, and beach **NT 505** curves round for 1.25 km as a crenulate sandy beach fronted by a continuous low tide bar, together with some rocks and reef. It terminates at a scattered line of rocks, beyond which is 1.3 km long beach **NT 506**. The Murgenella barge landing is located toward the southern end of the beach, with a clearing on the foredunes behind the landing. Waves begin to decrease in height to less than 1 m along these beaches and into Aurari Bay, owing to protection from the Goulburn Islands.

Beach **NT 507** is located 400 m east of the barge landing. It faces east and winds for 300 m along a rocky section of shore backed by two small sandy forelands and densely vegetated bedrock slopes. Waves break over the rocks and reefs, which extend over 100 m off the beach.

NT 508-514 ANGULARLI CREEK

No.	Beach	Rating HT LT	Type	Length
NT508	Angularli Ck	2 3	R+LTT	6.5 km
NT509	White Rocks	2 3	R+LTT	6.4 km
NT510	Marligur Ck	1 2	R+reef flats	2.1 km
NT511	White Pt (W3)	1 2	R+reef flats	1.5 km
NT512	White Pt (W2)	1 2	R+reef flats	700 m
NT513	White Pt (W1)	1 2	R+rocks/reef flats	200 m
NT514	White Pt	1 2	R+rocks/reef flats	150 m
Spring & neap tide range = 2.2 & 0.2 m				

The southern shores of Aurari Bay consist of a 20 km long series of longer sand beaches (NT 508-510) backed initially by tidal swamp and bordered by tidal creeks. East of Marligur Creek, bedrock again dominates the shore and backs and borders the beaches to White Point (NT 511-514).

The first beach (**NT 508**) begins in lee of the low rocky point east of the barge landing and swings round for 6.5 km to face northeast at **Angularli Creek**. The beach receives waves averaging less than 1 m and has a gentle beach slope and a continuous low tide bar (Fig. 5.64), with tidal shoals extending up to 300 m off the creek mouth. It is backed by a 100-400 m wide series of low foredunes, then salt flats, with mangroves fringing the creek and in lee of the inlet.

Figure 5.64 View west along beach NT 508, with waves breaking across the low tide bar.

Beach **NT 509** begins on the east side of the creek and extends for another 6.4 km to the mouth of **Marligur Creek**. White Rock reef, located 2 km east of Angularli Creek, has lowered waves sufficiently to form a cuspate foreland. The beach is also backed by a 200 m wide series of low foredunes and the salt flats of the two creeks, with mangroves only at the creek mouths.

Marligur Creek flows out against a laterite outcrop on its eastern shore, with a tidal gauge maintained just inside the creek mouth. Beach **NT 510** extends for 2.1 km to the east. It is initially fronted by the creek mouth tidal shoals and sand flats, but after 1 km lies in lee of 400 m wide fringing coral reefs which lower waves at the shore to less than 0.5 m, permitting mangroves to grow at the eastern end of the beach. The beach is backed by a few grassy beach ridges near the creek mouth and low laterite bluffs for most of its length. The bluffs, rocks and 300 m of mangroves form its eastern boundary.

Beach **NT 511** begins on the eastern side of the mangroves and is a relatively straight 1.5 km long northwest-facing beach, fronted by 400 m of fringing coral reef flats. The entire beach is backed by vegetated 10 m high bluffs, which in places protrude onto the beach. It terminates at a mangrove-fringed inflection in the bluffs, beyond which is 700 m long beach **NT 512**. This is a similar reef-fringed beach, which is backed by 50-100 m of grassy beach ridges then the lightly wooded bedrock slopes. The beach is bordered in the east by the bluffs and surrounding mangroves of White Point.

White Point is a 12 m high, 500 m wide, laterite point, which has two small beaches bordered by protruding laterite rocks and reefs on the western side of the point. Beach **NT 513** is a curving 200 m long high tide sand beach, backed by the 10 m high bluffs of the point, and fronted by 200 m wide rocks and reef flats. Its neighbour beach **NT 514** is a similar 150 m long beach, backed by the bluffs with a more prominent rock platform extending 100 m off the eastern point.

NT 515-516 **ROSS POINT**

No.	Beach	Rating HT LT	Type	Length
NT515	Ross Pt	1 2	R+reef flats	7.8 km
NT516	Ross Pt (S)	1 2	R+reef flats	100 m
Spring & neap tide range = 2.2 & 0.2 m				

Ross Point lies 3 km south of South Goulburn Island and is a curving 10 m high bedrock-controlled point, fronted by a continuous beach which forms the boundary between Aurari and Anuru bays.

The main beach (**NT 515**) begins on the eastern side of White Point and runs initially northeast for 4.5 km, before gradually curving round at the point to finally face due east (Fig. 5.65). The first 2 km of beach are free of reef and receive sufficient wave energy to maintain a low tide bar cut by rips every 100 m during higher wave conditions. Three hundred metre wide fringing reef dominates the rest of the beach all the way to the eastern tip. The beach is moderately steep and for the most part backed by a 50-100 m wide strip of grassy beach ridges, then densely vegetated bedrock, with occasional spurs of bedrock crossing the beach and forming small reefs.

Figure 5.65 The eastern end of beach NT 515, with beach NT 516 at the point in the foreground.

At the southeastern tip of the beach is a 100 m long pocket of sand (beach **NT 516**) bordered by two bedrock reefs and fronted by the reef flats. At the end of this beach the shoreline curves round into Anuru Bay and mangroves fringe the next 1 km of shore.

NT 517-521 **ANURU BAY (N)**

No.	Beach	Rating HT LT	Type	Length
NT517	Anuru Bay (N1)	1 1	R+sand flats	1 km
NT518	Anuru Bay (N2)	1 1	R+sand flats	200 m
NT519	Anuru Bay (N3)	1 1	R+sand flats	100 m
NT520	Anuru Bay (N4)	1 1	R+sand flats	400 m
NT521	Anuru Bay (N5)	1 2	R+mud flats	300 m
Spring & neap tide range = 2.2 & 0.2 m				

Anuru Bay is a well protected low energy bay, that faces northeast between Ross and Barclay points. Within the 4.5 km wide entrance are 28 km of predominantly bedrock-controlled, mangrove-fringed shoreline adjacent to three tidal creeks, together with 12 low energy beaches (NT 517-528), which occupy a total of 8.5 km of the shoreline. The only vehicle access to the bay is via a track to Barclay Point, with a small community located 2 km west of the point.

Five beaches (NT 517-521) lie within 5 km of the southeast tip of Ross Point and occupy the northern shores of the bay. The first beach, **NT 517**, is located 1 km west of the tip and faces southeast into the bay, receiving low waves through the bay mouth. It consists of a 1 km long strip of sand, backed by a few grassy beach ridges and fronted by a mixture of sand and 300 m wide reef flats. Around its southern point is beach **NT 518**, a south-facing 200 m long strip of sand, with small mangrove clumps to either end and irregular sand and reef flats off the beach. Immediately to the west is beach **NT 519**, a similar 100 m long strip of high tide sand. Both beaches face south into the bay. A 300 m long strip of mangroves separates it from beach **NT 520**, a crenulate 400 m long low energy recurved spit that represents the terminus of sand movement into the northern part of the bay. The beach is backed by a series of recurved spits separated by dense mangroves, with mangroves extending south along the shores of the bay for 1 km to a tidal creek mouth.

One kilometre south of the creek mouth is beach **NT 521**, a narrow 300 m long strip of high tide sand bordered by mangroves. It is backed by wooded slopes and fronted by 4 km wide sand and mud flats that fill the western half of the bay.

NT 522-528 **ANURU BAY (S)**

No.	Beach	Rating HT LT	Type	Length
NT522	Anuru Bay (S1)	1 2	R+mud flats	300 m
NT523	Anuru Bay (S2)	1 2	R+reef flats	150 m
NT524	Anuru Bay (S3)	1 2	R+reef flats	150 m
NT525	Anuru Bay (S4)	1 2	R+sand flats	300 m
NT526	Anuru Bay (S5)	1 2	R+sand flats	400 m
NT527	Barclay Pt (W2)	1 2	R+sand flats	2.8 km
NT528	Barclay Pt (W1)	1 2	R+sand flats	1.8 km
Spring & neap tide range = 2.2 & 0.2 m				

Two tidal creeks dominate the southern and eastern shores of the bay, with the beaches restricted to the intervening bedrock promontories and the western shores of Barclay Point. Mangroves fringe much of the eastern shore and together with their location within the bay result in low energy beach systems.

Beach **NT 522** is located in the southwest corner of the bay and consists of an isolated 300 m long strip of high tide sand, backed by wooded bluffs and fronted by 4 km wide mud and sand flats. Two and a half kilometres to the east are neighbouring beaches **NT 523** and **NT 524**. They both face north, are 150 m long, bordered by low mangrove-fringed rock points, and backed by gently wooded slopes. A 1.5 km wide mangrove-filled embayment separates these beaches from the next beach.

Beach **NT 525** consists of a 300 m long series of recurved spits, which are actively moving sediment deep into the bay. The dynamic spits are backed by mangroves which become exposed at the shoreline as the spits migrate south. The spits are attached to a low rocky point at their northern end, to the east of which is 400 m long beach **NT 526** (Fig. 5.66). This relatively straight northwest-facing beach is accessible by vehicle and has a camping shelter behind the centre of the beach.

Figure 5.66 Beaches NT 525 with migratory spits, and NT 526 (foreground), lie either side of a low rock-tipped point.

Beach **NT 527** begins at a low rocky point that separates it from its southern neighbour. The shoreline turns and faces north as the 2.8 km long beach slowly curves round to finally face west toward its northern end. The beach is fronted by 300-400 m wide fringing reefs, and mangroves at the low northern rocky point. The vehicle track from the south reaches the centre of the beach, where two houses are located. Beach **NT 528** extends from the north side of the mangrove-rock point for 1.8 km to the western end of Barclay Point. The relatively straight beach is paralleled by a band of beachrock along its southern half, with fringing reefs extending 500 m off the beach in the south, narrowing to 200 m in the north. The vehicle track parallels the back of the beach.

NT 529-531 **BARCLAY POINT**

No.	Beach	Rating HT LT	Type	Length
NT529	Barclay Pt	2 3	R+rock flats	1 km
NT530	Barclay Pt (E1)	2 2	R+sand/rock flats	2.2 km
NT531	Barclay Pt (E2)	2 3	R+rock flats	500 m
Spring & neap tide range = 2.2 & 0.2 m				

Barclay Point is a low rocky point that forms the northeastern tip of the 1.5 km wide peninsula that separates Anuru and Waminari bays. Eight beaches form most of the peninsula shoreline, the western four in Anuru Bay (NT 525-528), beach NT 529 along the northern side of the point, and beaches NT 530-532 on the eastern Waminari Bay side. All the beaches are backed by bedrock and fronted by a mixture of rock and reef flats.

Beach **NT 529** occupies 1 km of the northern side of the point and faces northeast into the prevailing easterly winds, with waves averaging about 1 m. The beach is however fronted by a 100-200 m wide irregular rock reefs, then fringing coral reefs, with waves breaking across the flats and resulting in calmer conditions at the shore. As a consequence mangroves front the eastern 1 km of shore, all the way to the eastern tip of the point. A grassy strip of beach ridges then woodlands back the beach.

Beach **NT 530** begins in the lee of the point and is a curving 2.2 km long east-facing beach, which is fronted by a mixture of sand and rock flats and scattered reefs further out (Fig. 5.67). There is a clump of mangroves in the centre of the beach and a narrow band of grassy beach ridges running behind the beach. The southern end of the beach runs out to a low rock-dominated point, on the southern side of which is 500 m long beach **NT 531**. The entire point and beach are fronted by intertidal rock flats, which are replaced by mangroves toward its western end.

Figure 5.67 Beach NT 530 shown at low tide with a mixture of low tide sand and rock flats.

NT 532-537 **WAMINARI BAY**

No.	Beach	Rating HT LT	Type	Length
NT532	Waminari Bay (1)	1 1	R+sand flats	1.2 km
NT533	Waminari Bay (2)	1 2	R+rock flats	1.3 km
NT534	Waminari Bay (3)	1 2	R+rock flats	250 m
NT535	Waminari Bay (4)	1 2	R+rock flats	400 m
NT536	Waminari Bay (5)	1 2	R+rock flats	1.2 km
NT537	Waminari Bay (6)	1 2	R+rock flats	400 m
Spring & neap tide range = 2.2 & 0.2 m				

Waminari Bay is a funnel-shaped 4 km deep, 2 km wide bay that faces due east. It is bordered on the north side by the southern extension of Barclay Point and to the south by Wamirumu Point. Two barrier beaches (NT 532 & 533) occupy the western end of the bay, with all the beaches along the south shore (NT 534-537) dominated by bluffs and rock flats. There is a vehicle track along the back of beach NT 532, but otherwise no access to the bay beaches.

Beach **NT 532** is located at the western end of the bay. It has acted as a sink for sediment moving into the bay which has prograded it up to 600 m bayward. Today it is a 1.2 km long east-facing beach, backed by the grassy beach ridges, then a tidal swamp that exits west into Anuru Bay. A second tidal creek, originating in lee of Barclay Point forms the northern boundary of the beach while shallow sand flats front the beach.

Five hundred metres to the southeast is beach **NT 533**, a curving 1.3 km long northeast-facing beach, backed by a narrow beach ridge-barrier, then a 50 ha mangrove-filled swamp, which drains across the beach via a small central creek. The beach is bordered by low wooded slopes fringed by mangroves and fronted by sand flats.

Low wooded slopes dominate the next 3 km of shore east to Wamirumu Point. The slopes are fronted by four near continuous beaches (NT 532-537), each separated by rocky protrusions and all fronted by rock flats and patches of sand that widen from 100 m in the west to 300 m at the point. Beach **NT 534** is a curving 250 m long beach hemmed in by low rocky points on either side, as well as rock outcropping on the centre of the beach. Beach **NT 535** is a crenulate 400 m long beach, with several rock outcrops and a few scattered mangroves. Beach **NT 536** is a crenulate 1.2 km long beach containing increasing rock outcrops on the beach and a greater dominance of rock flats off the beach. Finally beach **NT 537** is a relatively straight 400 m long beach fronted by shallow rock flats that extend to the point.

NT 538-544 **WANGULARNI BAY**

No.	Beach	Rating HT LT	Type	Length
NT538	Wamirumu Pt (1)	1 2	R+rock flats	1.5 km
NT539	Wamirumu Pt (2)	1 2	R+rock flats	700 m
NT540	Wangularni Bay	1 1	R+sand flats	2.2 km
NT541	Wangularni Bay (E1)	1 2	R+rock flats	3.5 km
NT542	Wangularni Bay (E2)	1 2	R+sand flats	3.7 km
NT543	Wangularni Bay (E3)	1 2	R+sand/rock flats	50 m
NT544	Wangularni Bay (E4)	1 2	R+rock flats	1.4 km
Spring & neap tide range = 2.2 & 0.2 m				

Wangularni Bay is a 2 km wide northeast-facing bay that is bordered by low wooded rocky points. Wamirumu Point forms the western point and contains two continuous beaches (NT 538 and 539), while the longer Turner Point lies to the east with four beaches (NT 541-544) along its northern shore.

Beach **NT 538** runs for 1.5 km along the north side of Wamirumu Point facing northeast into the prevailing wind waves. Waves break across 300 m wide intertidal rock flats with the crenulate beach backed by wooded slopes. At the point the shoreline turns and runs due south for 700 m as beach **NT 539**. This beach is also highly crenulate, being located to the lee of 200 m wide rock flats, while mangroves border its southern end.

The main Wangularni Bay beach (**NT 540**) begins on the eastern side of the mangroves. It is a curving 2.2 km long sandy beach, backed by a 600 m wide beach-chenier ridge plain containing about seven ridges. These are backed by a mixture of salt flats and mangroves drained by two small creeks at the western end and centre of the beach. Shoals from the tidal creeks and shallow sand flats fill much of the bay in front of the beach.

East of the main beach are 10 km of crenulate bedrock-dominated shoreline containing four beaches (NT 541-544), all backed by low wooded bedrock slopes. The first beach (**NT 541**) is a crenulate 3.5 km long low energy beach, fronted by 100-200 m wide rock flats and scattered mangroves, which curves round to face west at its northern end. Beach **NT 542** continues on for 3.7 km as a gently curving north-facing sandy beach, that is backed toward its western end by several grassy beach ridges and a 50 ha freshwater swamp connected to the beach via a small stream. Rock reefs border each end of the beach, while low ridged sand flats extend up to 1.5 km off the beach.

Beach **NT 543** is a narrow 50 m long high tide beach, backed by wooded slopes and fronted by ridged sand flats up to 500 m wide, together with a scattering of inshore rocks. Bedrock dominates the next 500 m of shore, followed by beach **NT 544**, a crenulate 1.4 km long sand beach, dominated by 100-200 m wide intertidal rock flats.

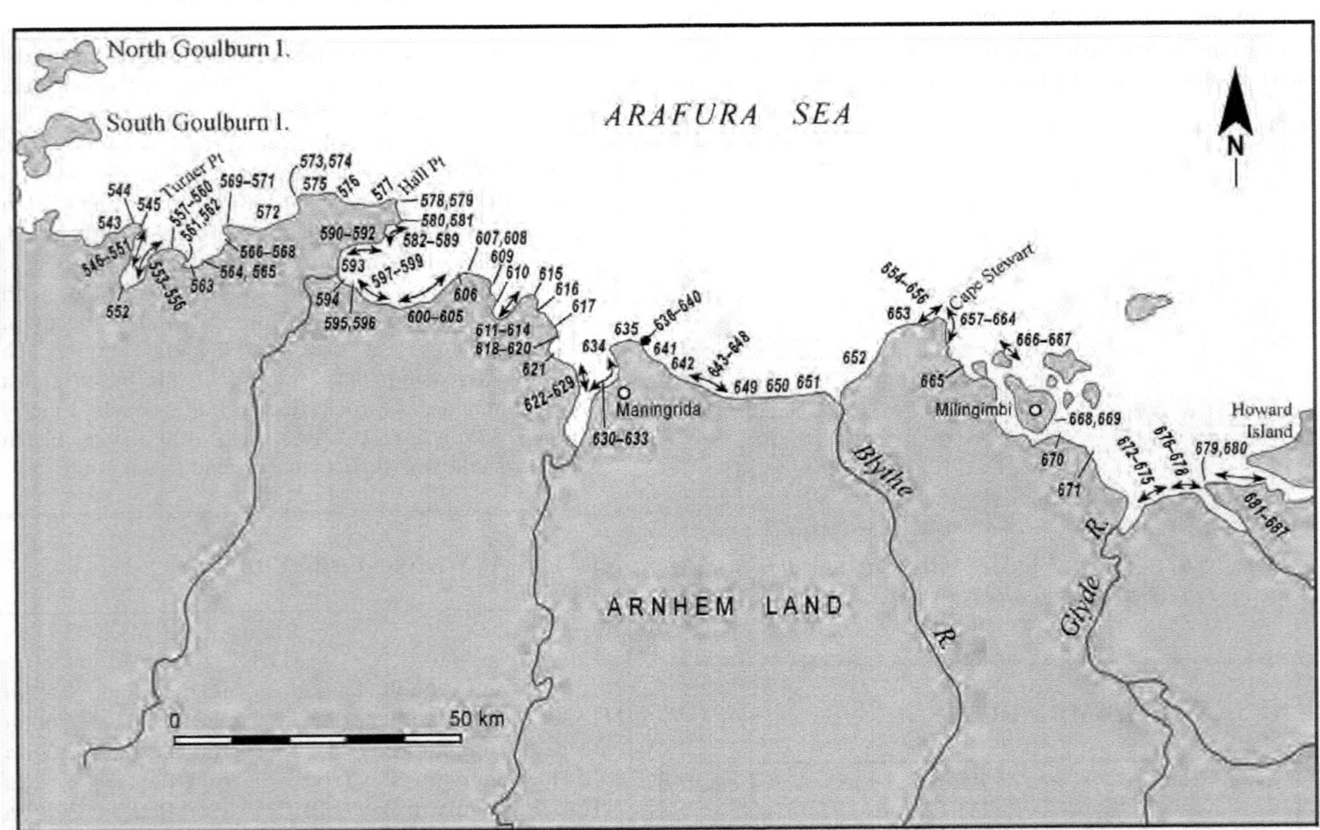

Figure 5.68 Regional map 7a showing a section of the northern Arnhem Land coast between the Goulburn islands and Howard Island, beaches NT 543-687.

NT 545-549 **TURNER POINT (S)**

No.	Beach	Rating HT	LT	Type	Length
NT545	Turner Pt	2	3	R+rock flats	1.1 km
NT546	Turner Pt (S1)	2	3	R+rock flats	150 m
NT547	Turner Pt (S2)	2	3	R+rock flats	700 m
NT548	Turner Pt (S3)	2	3	R+sand/rock flats	300 m
NT549	Turner Pt (S4)	2	3	R+rock flats	1.3 km
Spring & neap tide range = 2.2 & 0.2 m					

Turner Point is a low blunt 2 km wide wooded point, with no vehicle access to the point and no development around its shore. It separates Wangularni Bay from the King River embayment. On the south side of the point is 4 km of rock-dominated shore containing five beaches (NT 545-549).

Beach **NT 545** is located at the northern tip of the point and curves outwards round the point for 1.1 km. The beach has prograded 200-300 m as an irregular series of beach ridges, however rock and 100 m wide rock flats still dominate the shoreline. Beach **NT 546** is located immediately south of the eastern tip of the point. It consists of a double crescentic 150 m long strip of high tide sand fronted by continuous 100 m wide rock flats. Beach **NT 547** lies immediately to the south and is a relatively straight 700 m long high tide sand beach sitting on continuous irregular intertidal rock flats that extend up to 200 m off the beach. Beach **NT 548** is a second 300 m long double crescentic embayment, with the northern section begin partially free of rocks permitting low waves to reach the beach. However rock flats do extend bayward 300 m either side of this section (Fig. 5.69). The final section of beach (**NT 549**) extends south for another 1.3 km amongst crenulate rock- and rock-flat-dominated shore. It terminates at the mouth of a small tidal creek.

Figure 5.69 Beach NT 548 is bordered and dominated by low intertidal rock flats.

NT 550-552 **KING RIVER (W)**

No.	Beach	Rating HT	LT	Type	Length
NT550	King R (W3)	1	1	R+sand flats	1.4 km
NT551	King R (W2)	1	2	R+rock flats	500 m
NT552	King R (W1)	1	1	R+mud flats	300 m
Spring & neap tide range = 2.2 & 0.2 m					

The **King River** is a small upland river feeding into a large tidal creek, hence the name 'river'. The mangrove-lined creek has a bedrock-controlled 600 m wide funnel-shaped entrance. Immediately east of the entrance the shoreline runs roughly due north for 8 km to Turner Point.

Beach **NT 550** is located halfway to Turner Point. It is the only barrier on the western side of the embayment and consists of a 1.4 km long gently curving northeast-facing, relatively steep cusped sand beach. The beach is backed by a 200-300 m wide grassy beach ridge plain, then a 100 ha infilled lagoon containing salt flats and mangroves along the creek. The creek mouth forms the western boundary to the beach, while a low projecting rocky point forms the eastern end.

Beach **NT 551** begins on the eastern side of the rocks and is a double crescentic rock-dominated 500 m long beach that trends south toward the river mouth. Beach **NT 552** is located at the western entrance to the river. It is a 300 m long northeast-facing low energy strip of high tide sand, fronted by 200 m wide tidal flats, then the deep river channel.

NT 553-560 **KING RIVER (E)**

No.	Beach	Rating HT	LT	Type	Length
NT553	King R (E1)	1	1	R+sand flats	1.8 km
NT554	King R (E2)	1	1	R+sand/rock flats	600 m
NT555	King R (E3)	1	2	R+sand/rock flats	900 m
NT556	King R (E4)	1	2	R+sand/rock flats	300 m
NT557	King R (E5)	1	2	R+rock flats	2.4 km
NT558	King R (E6)	1	2	R+sand/rock flats	600 m
NT559	King R (E7)	1	2	R+sand/rock flats	700 m
NT560	King R (E8)	1	2	R+rock flats	400 m
Spring & neap tide range = 2.2 & 0.2 m					

The eastern side of the King River mouth trends to the northeast for 4 km before turning and running due east for 4.5 km to the entrance to Arrla Bay. Along this 9.5 km section of shore are eight beaches (NT 553-560), most dominated by rocks and rock flats.

Beach **NT 553** is a barrier occupying the mouth of a small valley immediately east of the river mouth. The 1.8 km long sand and gravel beach faces north-northeast and is backed by a 200-300 m wide series of grassy beach ridges (Fig. 5.70), then a 100 ha lagoon and salt flats, with mangroves lining the small creek that crosses the western end of the beach. The relatively steep, cusped beach is fronted by shallow sand flats. Along its eastern end it trends north along a bedrock-controlled section of shore and terminates at a low mangrove-lined rocky point.

Figure 5.70 Beach NT 553 has a narrow coarse-grained beach face, backed by low sandy overwash flats.

Beach **NT 554** begins on the north side of the point and is a crenulate 600 m strip of high tide sand fronted by a mixture of sand and rock flats, as well as fringing mangroves. It terminates at a small rock-bound cuspate foreland, on the north side of which is beach **NT 555**, a crenulate 900 m long beach fronted by near continuous rock flats and reef. Woods back the southern half of the beach, while an irregular series of grassy beach ridges backs the northern half and extends east for 4.5 km behind the neighbouring beaches.

Beach **NT 556** is located on the corner of the bedrock promontory where the coast turns and trends due east. It is a curving 300 m long beach bordered by protruding rock reefs that almost encircle the beach, resulting in a protected 'lagoon' in their lee.

Immediately to the east is beach **NT 557**, a 2.4 km long north-facing beach that is fronted by near continuous 100-200 m wide scattered rock reefs, with sand flats further offshore. The rocks and reefs produce crenulations along the shore, with a 100-200 m wide series of grassy beach ridges behind. Beach **NT 558** is located between two more prominent rock reefs, and is a curving 600 m long beach, largely free of rocks between the points. Immediately east is beach **NT 559**, another rock- and reef-dominated 700 m long highly crenulate beach.

Beach **NT 560** is located on the eastern tip of the north-facing shore, and is a curving 400 m long beach located between two rocky points, with mangroves growing on the eastern point.

NT 561-563 **ARRLA BAY (W)**

No.	Beach	Rating HT LT	Type	Length
NT561	Arrla Bay (W2)	1 2	R+rock flats	900 m
NT562	Arrla Bay (W1)	1 2	R+rock flats	500 m
NT563	Arrla Bay	1 2	R+sand flats	1.8 km
Spring & neap tide range = 2.2 & 0.2 m				

Arrla Bay is a 7 km wide northwest-facing embayment, bordered in the east by Guion Point. The western side of the bay contains two rock-dominated beaches (NT 561 and 562) and a barrier beach (NT 563).

Beach **NT 561** begins on the southern side of the western low rocky boundary of the bay. It is a crenulate 900 m long east-facing sandy beach fronted by continuous intertidal rocks and rock flats extending 100 m off the beach, as well as some scattered mangroves. Immediately south is beach **NT 562**, a 500 m long double crescentic beach , with prominent rock flats off each inflection, and some mangroves growing along the shore. South of the beach continuous mangroves fringe the shore to a small tidal creek.

Beach **NT 563** is a 1.8 km long northeast-facing beach and barrier that extend from the tidal creek for 1.8 km to a mangrove-lined bedrock eastern boundary. The beach is fronted by sand flats and backed by a 300 m wide beach ridge plain with scattered casuarina trees, then a 200 ha swamp.

NT 564-569 **ARRLA BAY (E)**

No.	Beach	Rating HT LT	Type	Length
NT564	Arrla Bay (E1)	1 1	R+sand flats	350 m
NT565	Arrla Bay (E2)	1 1	R+sand flats	700 m
NT566	Arrla Bay (E3)	1 1	R+sand ridges	2.5 km
NT567	Arrla Bay (E4)	1 1	R+sand ridges	600 m
NT568	Arrla Bay (E5)	1 1	R+sand ridges	3.6 km
NT569	Arrla Bay (E6)	1 2	R+tidal shoals	300 m
Spring & neap tide range = 2.2 & 0.2 m				

The eastern shore of **Arrla Bay** is a 9 km long northwest-facing section of coast, open to summer westerly winds and waves, but protected by Guion Point from the easterly waves. This side of the bay has six beaches (NT 564-569), each bordered by low bedrock points, together with three tidal creeks.

Beach **NT 564** is a 350 m long north-facing beach located between two low rocky headlands that protrude 100 m out from the beach, with mangroves fringing the eastern head. The beach is backed by wooded bedrock slopes and fronted by 300 m wide sand flats. Its eastern neighbour beach **NT 565** is a similar 700 m long beach backed by a 100 m wide series of beach ridges and then a freshwater swamp. Mangroves fringe both of the low rocky points.

A low slightly convex 1.5 km long headland separates beach NT 565 from the next beach (**NT 566**). This is a 2.5 km long northwest-facing beach that begins amongst rocks on the point, followed by a small tidal creek then a continuous sandy beach. It is fronted by low ridged sand flats that widen from 300 m in the west to 1 km off an eastern boundary tidal creek. The beach is the outermost ridge of a 1 km wide series of beach ridges and recurved spits that have filled the backing embayment with 200 ha of wave deposited sand, together with back barrier salt marshes and mangroves lining the creek. Beach **NT 567** is located on the northern side of the tidal creek and is a 600 m long spit that acts as a conduit for sand moving around the northern low rocky point and into the embayment. It is backed by three recurved spits and terminates at the creek mouth. Tidal shoals and 1 km wide ridged sand flats extend off the beach.

On the north side of the 300 m wide point is 3.6 km long beach **NT 568**. This beach faces northwest and is bordered in the east by a 50 m wide tidal creek. The beach is the outermost of a 200-500 m wide series of beach ridges, that are now being reworked as active sand dunes, reaching 17 m in height at the southern end of the system. Ridged sand flats widen from 200 m off the western end to 1.5 km off the northern tidal creek. On the north side of the creek is 300 m long beach **NT 569**, a low energy beach fronted by the tidal channel then the wide tidal shoals and ridged sand flats. The narrow eroding beach is dominated more by tidal currents than by waves.

NT 570-572 **GUION POINT (E)**

No.	Beach	Rating HT	Rating LT	Type	Length
NT570	Guion Pt (1)	2	3	R+rock flats	100 m
NT571	Guion Pt (2)	2	3	R+rock flats	100 m
NT572	Guion Pt (E)	2	2	R+LTT-sand flats	13.6 km
Spring & neap tide range = 2.2 & 0.2 m					

Guion Point forms the northeastern boundary to Arrla Bay. It consists of a 5 m high, 500 m wide circle of bedrock that is attached to the mainland by the eastern beach-barrier system (NT 572). Rock and rock flats surround the point, with two small beaches forming on the northern side in gaps in the rock flats. Beach **NT 570** is a 100 m long strip of sand bordered by red rocks and reefs, with a 200 m wide sandy-rock flat mixture off the beach. Two hundred metres to the east is beach **NT 571**, a curving 100 m long high tide beach, bordered by protruding low rocky points and reefs, with rock debris filling the small embayment and waves breaking on reefs up to 100 m off the entrance to the small bay (Fig. 5.71).

On the eastern side of the point is one of the longer beaches on this part of the coast. Beach **NT 572** extends for 13.6 km to the east, as a slightly crenulate-curving beach that initially faces north, slowly swinging round to face northwest, and finally west against the low eastern boundary point. The entire beach is backed by a 200-500 m wide low barrier system, containing moderately active low dunes, with a double Holocene barrier behind the first 5 km of beach, the barriers separated by a small meandering tidal creek that drains the 10 000 ha Iliwan Swamp. The swamp is located behind the central section of the beach, with two additional swamps 2500 ha in area behind the western and eastern ends of the beach. The beach is exposed to waves that average about 1 m in the west, decreasing in height to the east in lee of Relief Point. In the west the waves maintain a continuous 100 m wide low tide bar, while, beginning in the centre of the beach, 500-600 m wide ridge sand flats front the beach. In the sheltered west-facing eastern corner of the beach is a temporary fishing camp.

Figure 5.71 Guion Point, with the small rock-bound beaches NT 570 & 571 (foreground), and the beginning of the longer more exposed beach NT 573 (top).

NT 573-579 **RELIEF-CUTHBERT-HALL POINTS**

No.	Beach	Rating HT	Rating LT	Type	Length
NT573	Relief Pt (W2)	1	1	R+sand flats	350 m
NT574	Relief Pt (W1)	1	1	R+sand flats	2.6 km
NT575	Relief Pt	1	1	R+LTT	5.2 km
NT576	Cuthbert Pt	1	1	R+sand/reef flats	4.5 km
NT577	Hall Pt (W)	1	1	R+sand/reef flats	6.3 km
NT578	Hall Pt	1	1	R+sand/reef flats	600 m
NT579	Hall Pt (E)	1	1	R+LTT	3.1 km
Spring & neap tide range = 2.8 & 1.1 m					

Relief Point is the westernmost of a series of four low points that together form the northern shore of a 15 km wide promontory, which on its east and southern side partly encircles the large Junction Bay. Beaches NT 573 and 574 lie immediately west of the point, while between the four points are 24 km of sandy shoreline dominated by three crescentic beaches (NT 575-577) which occupy 16 km of the shore. There is no vehicle access to the peninsula and no development in the region, other than a fishing camp at Relief Point inlet.

Relief Point is a westward-migrating sand spit, to the south and west of which are 3 km of northwest-facing shore containing two beaches. Beach **NT 573** lies at the eastern end of the long Guion Point beach and is a curving west-facing 350 m long beach, bordered by a rock reef in the south and rocks and a low bluff to the north. It is backed by 50-100 m of minor dune activity, with sand flats off the beach. On the north side of the 300 m long bluffs is beach **NT 574**, which trends northeast for 2.6 km to the south side of the Relief Point inlet. Two fishing shacks are located just inside the inlet mouth. Three hundred metre wide sand flats and some patchy reef front the beach, which is backed by 200-400 m of moderately active dunes.

Relief Point is a low spit that continues offshore as a submerged bar and tidal shoal, with a 50 m wide tidal creek exiting at the end of the spit. The beach (**NT 575**) extends for 5.2 km as a curving north-facing arc from the tip of the spit east to Cuthbert Point. The beach receives moderate waves, which maintain a steep cusped beach face and a continuous low tide bar. The beach is backed by moderately active 100-200 m wide sand dunes, then the tidal creek along the western half. The eastern half is backed by older inner barrier ridges and near Cuthbert Point by low wooded bedrock.

Cuthbert Point is a sandy cuspate foreland formed in lee of low rock reefs. The point area is a 70 ha expanse of active sand dunes. Beach **NT 576** begins at the point and curves east as a slightly crenulate 4.5 km long beach, to a small tidal creek. The beach is backed by the sand dunes, which narrow to the east. These are in turn backed by older foredunes and spits extending up to 400 m inland, then an irregular area of salt flats and small lagoons drained by the small creek. The beach (**NT 577**) continues past the creek for another 6.3 km to Hall Point. As rocks outcrop on the beach and as offshore reefs, this section of beach becomes incresingly crenulate culminating at Hall Point, which is fringed by continuous rock flats.

Hall Point is a low point, with sand dunes encroaching from the north and east, to reach a maximum height of 20 m, and are gradually covering a small central area of woodland. Beach **NT 578** lies along the northeast face of the point. It is a 600 m long double crescentic beach, bordered by prominent rock flats, with waves breaking across the rocks. Beach **NT 579** begins immediately to the east and is a curving 3.1 km long east to northeast-facing beach that terminates at the easternmost Braithwaite Point. The northern half of the beach is well exposed to the easterly winds. It consists of a steep, cusped, reflective high tide beach and continuous low tide bar, backed by 200-300 m wide active dunes that reach heights of 15-20 m and are transgressing over the backing salt flats (Fig. 5.72).

Figure 5.72 The curving beach NT 579 is backed by active dunes along its northern half.

NT 580-584 **BRAITHWAITE POINT**

No.	Beach	Rating HT	Rating LT	Type	Length
NT580	Braithwaite Pt (N)	1	2	R+rock flats	100 m
NT581	Braithwaite Pt (E)	1	2	R+rock/sand flats	900 m
NT582	Braithwaite Pt (S1)	1	2	R+tidal flats	200 m
NT583	Braithwaite Pt (S2)	1	2	R+tidal flats	100 m
NT584	Braithwaite Pt (S3)	1	2	R+tidal flats	150 m
Spring & neap tide range = 2.8 & 1.1 m					

Braithwaite Point is a low wooded 50 ha bedrock headland that forms the northern entrance to the 14 km wide mouth of Junction Bay. The point is dominated by rocks and rock reefs along its northern and eastern shores, with the more protected waters of the bay permitting mangroves to grow along the southern shore.

Beach **NT 580** is located on the northern tip of the point and consists of a curving 100 m long high tide beach located in a 50 m wide gap in the rock flats. Waves break over rocks at the entrance to the gap and a 150 m wide rock-floored 'lagoon' lies between the reefs and the beach. Beach **NT 581** lies on the southeast-facing side of the point. It is a crenulate 900 m long sand beach fronted by a mixture of rocks, rock flats, reefs and, toward the south, a section of sand flats.

The point is joined to the mainland by two barriers, the northern in lee of beach NT 579, and the southern consisting of a series of mangrove-lined recurved spits, that contains three small sandy beaches (**NT 582, NT 583** and **NT 584**) and terminates at a tidal creek. The beaches are 200 m, 100 m and 150 m long respectively and all face south into the bay. They are all fringed by mangroves and fronted by low energy tidal flats.

Junction Bay

Area	approx. 25 000 ha	
Coast length	52 km	(2532-2584 km)
Beaches	26	(NT 581-606)

Junction Bay is a large U-shaped east to northeast-facing bay, with a 14 km wide mouth between Braithwaite and Goomadeer points. The Goomadeer, Wurugoij and Majari creeks flow into the southeastern side of the bay, all three associated with extensive salt flats, mangroves and beach ridges totalling 11 500 ha in area. Beaches NT 581-606 occupy 26 km (50%) of the shoreline.

NT 585-592 **JUNCTION BAY (N)**

No.	Beach	Rating HT	Rating LT	Type	Length
NT585	Junction Bay (N1)	1	2	R+sand/rock flats	1.4 km
NT586	Junction Bay (N2)	1	2	R+sand/rock flats	2 km
NT587	Junction Bay (N3)	1	2	R+sand/rock flats	1.3 km
NT588	Junction Bay (N4)	1	2	R+sand/rock flats	100 m
NT589	Junction Bay (N5)	1	2	R+sand flats	1.2 km
NT590	Junction Bay (N6)	1	2	R+sand/rock flats	300 m
NT591	Junction Bay (N7)	1	2	R+sand flats	800 m
NT592	Junction Bay (N8)	1	2	R+sand flats	1.5 km
Spring & neap tide range = 2.8 & 1.1 m					

In lee of Braithwaite Point the shoreline turns and trends west into Junction Bay. Past the creek that borders beach NT 584, the shore turns and faces east for 5 km and contains beaches NT 585-588, before turning and heading west deep into the bay where beaches NT 589-592 are located.

Beach **NT 585** begins on the south side of the 100 m wide mangrove-fringed creek mouth. It trends south-southwest for 1.4 km in lee of a mixture of sand and patchy rock flats, to a slight rock-controlled inflection in the shore. Beach **NT 586** continues on for another 2 km as a gently curving east-facing beach. Finally, at the next rock inflection, beach **NT 587** runs for 1.3 km to the main point. This is a more crenulate beach with several rock-controlled inflections along the continuous sandy beach. All three beaches are fronted by the mixture of sand and rock flats and backed by minor dune activity and older foredunes extending up to 300 m inland, some of which are backed by fresh and salt water wetlands, then the wooded coastal plain.

Beach **NT 588** is located in lee of two prominent rock reefs. The 100 m long curving beach lies between the reefs located 50 m apart, with sand flats extending out between the gap in the reefs. It also marks the inflection where the coast turns and trends to the southwest.

Beach **NT 589** lies on the south side of the reefs and runs to the southwest for 1.2 km as a crenulate strip of high tide. It is initially backed by a foredune, and then by low scarped bedrock bluffs, which swing round to face southeast at the southern end of the beach.

Beach **NT 590** lies 300 m further west at the end of a section of low bluffs. The beach is a 300 m long southeast-facing strip of high tide sand located at the base of 5 m high scarped bluffs, with predominantly sand flats off the beach, together with two close in patches of rock flats, and a bedrock reef off the southern end of the beach. Beach **NT 591** continues immediately south of the beach, and trending west as an 800 m long series of three crescents backed by initially vegetated then scarped blow bluffs. Sand flats front the beach, together with two rock reefs off the inflection points.

Beach **NT 592** commences at the western end of its neighbour as the shoreline turns and trends to the northwest. The 1.5 km long beach begins with backing wooded slopes, then a low foredune, and finally as a crescentic recurved spit that terminates as a low spit overlapping the backing mangroves. The mangroves dominate the next 3 km of shore. The beach and spit represent the active movement of sand deep into the apex of the bay where, as will be described for the next beach, a massive amount of sediment has accumulated over the past 6000 years.

NT 593-596 **GOOMADEER RIVER**

No.	Beach	Rating HT	Rating LT	Type	Length
NT593	Goomadeer R (N)	1	2	R+tidal flats	900 m
NT594	Goomadeer R (S1)	1	2	R+tidal flats	4.3 km
NT595	Goomadeer R (S2)	1	1	R+tidal flats	300 m
NT596	Goomadeer R (S3)	1	1	R+tidal flats	250 m
Spring & neap tide range = 2.8 & 1.1 m					

The **Goomadeer River** is a large tidal creek, fed by several small upland streams and dominated by tidally driven flow. The 'river' occupies the western apex of Junction Bay with two beaches and mangrove-lined shores to either side. Beach **NT 593** occupies the first 900 m north of the river mouth, and is a narrow strip of high tide sand backed by mangroves and fronted by 300-500 m wide tidal shoals, with the deep tidal channel forming the southern boundary.

The beach and 4 km of mangrove-fringed shoreline to the north are the outermost deposits in a 2 km wide plain consisting of a series of several beach ridge-cheniers, each separated by salt flats. These are in turn backed by 500 m wide salt flats and then a series of Pleistocene beach ridges. Together with a similar, though smaller, series on the south side of the river they cover an area of 2000 ha and represent to the order of 50 million cubic metres of sediment that have infilled the western 3-4 km of this low energy bay during the past 6000 m years. To achieve this sediment, to the order of 10 000 m^3 a year would need to be delivered to this end of the bay. There is no development in and no vehicle access to the bay region.

Beach **NT 594** continues on the south side of the river front as a 1.5 km wide series of beach ridges and salt flats. The beach faces east and is 4.3 km long, terminating against a mangrove-fringed 5-10 m high section of the coastal plain. The beach is paralleled by 300 m wide sand flats, which widen to the creek mouth, with a second smaller creek crossing toward the southern end of the beach.

The coastal plain extends 2 km to the east and has two small beaches in amongst the mangroves and low rocky outcrops. Beach **NT 595** is a curving 300 m long beach, backed by wooded coastal plain and fringed by mangrove-covered rock flats including a north-facing clump in the middle. One kilometre further east is beach **NT 596**, a curving 250 m long beach which marks the beginning of a few hundred metre wide series of beach ridges. The beach is however bordered by rock flats and mangroves in the west (Fig. 5.73).

Figure 5.73 Beaches NT 597 (foreground), with rock- and mangrove-bound beaches NT 596 & 594 extending west toward beach NT 594.

NT 597-602 **WURUGOIJ & MAJARI CREEKS**

No.	Beach	Rating HT	LT	Type	Length
NT597	Wurugoij Ck (N)	1	2	R+tidal flats	4.4 km
NT598	Wurugoij Ck (E)	1	1	R+tidal flats	300 m
NT599	Majari Ck (W)	1	1	R+sand ridges	200 m
NT600	Majari Ck (E1)	1	1	R+tidal flats	1.2 km
NT601	Majari Ck (E2)	1	1	R+tidal flats	3 km
NT602	Majari Ck (E3)	1	1	R+sand ridges	3.2 km
Spring & neap tide range = 2.8 & 1.1 m					

Wurugoij and Majari creeks dominate the southern shore of Junction Bay. Both creeks have 200-300 m wide mouths, and tidal shoals and ridged sand flats extending up to 2 km off the creek mouths. They are backed by extensive mangrove forests and surrounding salt flats, totalling 1500 and 2000 ha respectively. In addition sand spits and older beach ridges border either side of each creek.

Beach **597** is a 4.4 km long sand beach, initially backed by older recurved spits, then continuing as a 3 km long solitary sand spit. The northeast-facing beach begins at the low rocky point adjacent to beach NT 596 and terminates as a sandy series of recent recurved spits at Wurugoij Creek mouth. It has irregular 1 km wide sand and rock flats off the northern end, which narrow, then widen to 1 km off the creek mouth. In addition there are two small rock outcrops along the northern half of the beach.

Beach **NT 598** lies on the east side of the creek in lee of the 1 km wide tidal shoals. The beach is 300 m long, faces north and terminates at a mangrove-lined shoreline with a 1 km wide series of three older spit-beach ridges backing the beach. The coastal plain emerges at the shore 1 km east of the beach, with several mangrove-lined low rocky outcrops producing a more crenulate shore. Beach **NT 599** lies in a more open crenulation. It is a 200 m long curving beach bordered by rock flats and mangroves and fronted by 500 m wide ridged sand flats.

The sand flats and mangroves extend east for 1.5 km to the mouth of Majari Creek. Beach **NT 600** lies on the east side of the creek mouth, emerging from the mangroves to form a curving 1.2 km long north-facing low energy beach, fronted by 800 m wide tidal flats. It is backed by a 1 km wide series of recurved spit-beach ridges. The beach terminates at a small tidal creek, beyond which is 3 km long outward-curving beach **NT 601**. This beach begins as a mangrove-backed beach ridge, then ties to the coastal plain along its eastern half, terminating at a tidal creek which enters the bay in lee of a low north-south-trending rock outcrop.

Beach **NT 602** begins at the creek mouth and runs relatively straight for 3.2 km to a small mangrove-fringed rock outcrop, with ridged sand flats up to 700 m wide off the beach. It is backed by a few low sandy beach ridges, then a continuous small tidal creek which drains two separate wetlands and exits at the western end of the beach.

NT 603-608 **GOOMADEER POINT**

No.	Beach	Rating HT	LT	Type	Length
NT603	Goomadeer Pt (W4)	1	2	R+sand flats	350 m
NT604	Goomadeer Pt (W3)	1	2	R+sand flats	200 m
NT605	Goomadeer Pt (W2)	1	2	R+sand flats	3 km
NT606	Goomadeer Pt (W1)	1	2	R+sand flats+rocks	3.5 km
NT607	Goomadeer Pt (E)	2	3	R+sand/rock flats	2.8 km
NT608	Goomadeer Pt (E2)	2	3	R+sand/rock flats	1.8 km
Spring & neap tide range = 2.8 & 1.1 m					

Goomadeer Point is a low wooded bedrock point backed by beach and foredune ridges and fronted by a mixture of sand and rock flats, the latter extending up to 1.5 km off the northern tip of the point. The point is 6 km wide and extends in a gentle crenulate convex curve, separating the large Junction Bay from the smaller Rolling Bay. Along its western and northern shorelines are six beaches (NT 603-608) all bounded by low rock reefs, with rock flats and rock reefs increasing in occurrence toward the point. There is no vehicle access to the point area and no development.

Beach **NT 603** is a curving 350 m long north-northwest-facing beach backed by low wooded coastal plain and bordered by small rock points. Sand flats and occasional rock outcrops extend 400 m off the beach. Its neighbour, beach **NT 604**, is a similar though smaller 200 m long beach, with mangroves fringing each end, including a 50 m wide series of beach ridges behind the northern mangroves.

Beach **NT 605** is a crenulate 3 km long northwest-facing sandy beach that is fronted by both ridged sand flats and patchy rock reefs. In places it is backed by up to 100 m wide beach ridge plains, as well as a 50 ha swamp drained by a small creek toward the centre of the beach. It terminates at a low sandy foreland, beyond which is 3.5 km long beach **NT 606**. This beach continues as a crenulate north-facing sandy beach, backed by 100-200 m wide beach and foredune ridges, some of which are blowing inland. It is fronted by 200-400 m wide patchy rock reefs and sand flats (Fig. 5.74).

Figure 5.74 Beach NT 606 (right) is backed by a sandy, partially active beach ridge plain.

Beach **NT 607** occupies the northern point area and faces generally north. It is a crenulate 2.8 km long beach, which becomes increasingly fronted by rock flats, particularly along the eastern 1 km of beach where they extend from 500-1500 m offshore. The beach terminates at a 500 m long section of mangrove-fringed rock flats. On the south side of the mangroves is beach **NT 608**, a similar 1.8 km long east-facing crenulate sand beach fronted by continuous irregular rock flats with a few mangroves growing on the southern flats.

NT 609-614 **ROLLING BAY**

No.	Beach	Rating HT	Rating LT	Type	Length
NT609	Rolling Bay (W2)	1	1	R+sand ridges	3.5 km
NT610	Rolling Bay (W1)	1	1	R+sand ridges	500 m
NT611	Rolling Bay (E1)	1	1	R+sand flats	1.1 km
NT612	Rolling Bay (E2)	1	2	R+rock flats	1.1 km
NT613	Rolling Bay (E3)	1	1	R+sand flats	700 m
NT614	Rolling Bay (E4)	1	1	R+sand flats	1.5 km
Spring & neap tide range = 2.8 & 1.1 m					

Rolling Bay is a roughly V-shaped north to northeast-facing bay bounded by Goomadeer and Hawkesbury points. The bay is 6 km wide at the mouth, with the main Nungbalgarri Creek filling the southern apex of the bay, 6 km in from the entrance. The bay has 14 km of shoreline containing six beaches (NT 609-614) which occupy 8 km (60%) of the shore. Several low rocky headlands and 5 km of mangroves fringe either side of the creek mouth.

Two beaches lie on the western side of the bay in lee of Goomadeer Point. Beach **NT 609** commences at the southern end of the rock-dominated point. It is a curving 3.5 km long east- to northeast-facing sandy beach, backed by a few low beach ridges including a few sand blowouts, then a narrow elongate swamp, while it is fronted by 500 m wide ridged sand flats. It is bordered in the east by a prominent rock reef extending 800 m north into the bay. This reef also forms the western boundary of 500 m long beach **NT 610**, a similar beach fronted by ridged sand flats, but with prominent rock reefs and scattered mangroves to either side.

To the south of these beaches is the deeper part of the bay occupied by **Nungbalgarri Creek.** The creek occupies a 2 km wide drowned valley and has a 200-300 m wide mouth. It meanders through about 1500 ha of mangrove forests, which extend up to 10 km inland. An older mangrove-fronted series of recurved spits extends for 3 km down the western side of the creek mouth, with a smaller 500 m long series of recurved spits on the eastern side. Both represent the terminus for sediment moving deep into the bay.

Beach **NT 611** lies 1 km east of the creek mouth and consists of a slightly curving 1.1 km long north-facing beach, bounded by low mangrove-fringed rock reefs, and backed by a few vegetated beach ridges, then a 50 ha freshwater swamp with sand flats extending 300-1,000 m off the beach. One kilometre to the northeast is beach **NT 612**, a bedrock-dominated 1.1 km long west-facing high tide beach. The beach is backed by wooded coastal plain and fronted by scattered mangroves, continuous rock flats, then 800 m wide sand flats.

Beach **NT 613** lies around the next point and is a northwest-facing 700 m long sand beach, bordered by low rocky points and backed by densely wooded coastal plain. Its eastern neighbour **NT 614** is a curving 1.5 km long north to northwest-facing beach that terminates in lee of the prominent Hawkesbury Point. The beach is backed by a few vegetated beach ridges, then a small elongate freshwater swamp. Both beaches are fronted by 400-500 m wide sand flats.

NT 615-621 **HAWKESBURY-WEST-GUMERADJI POINTS**

No.	Beach	Rating HT LT	Type	Length
NT615	Hawkesbury Pt	2 3	R+rock flats	150 m
NT616	Hawkesbury Pt	2 3	R+rock flats	2.2 km
NT617	West Pt (W)	1 1	R+LTT	6 km
NT618	West Pt (E1)	1 2	R+rock flats	50 m
NT619	West Pt (E2)	1 3	R+rock flats	200 m
NT620	West Pt (E3)	1 2	R+sand/rock flats	400 m
NT621	Gumeradji Pt(W)	1 1	R+sand flats	1.7 km
Spring & neap tide range = 2.8 & 1.1 m				

Hawkesbury Point is a 7 m high wooded 1 km wide point formed of Tertiary laterite. It is the northernmost of four similar low points that lie along the western entrance to the Liverpool River. West Point lies 9 km to the southeast and Gumeradji Point another 4 km further south, with the river mouth at South West Point another 4 km to the south. Between Hawkesbury and Gumeradji points are 20 km of generally east-facing coast containing six beaches (NT 615-620), the three headlands and 5 km of mangrove-lined shore. There is no vehicle access and no development in the region.

Beach **NT 615** lies at the tip of Hawkesbury Point. It is a curving north-facing 150 m long high tide beach, bordered by prominent rock platforms, flats and reefs, and fronted by a mixture of rock and sand flats, while wooded slopes back the beach. Waves break over the reefs at high tide producing a left-hand surf break. Beach **NT 616** begins 1 km to the southeast and is a continuous strip of highly crenulate high tide sand tied to a series of intertidal rock reefs and backed by a mixture of irregular low foredunes and wooded coastal plain. It is fronted by the rock reefs and patchy sand flats and terminates at a small creek (Fig. 5.75).

The creek has a 500 m wide ebb tide delta and forms the eastern boundary of 6 km long beach **NT 617**. The beach is curving, moderately steep, generally northeast-facing and exposed, with waves averaging about 1 m and maintaining an attached low tide bar. It is backed by a 300-400 m wide series of low foredunes with some minor dune transgression, then a 6 km long 500 m wide mangrove-lined tidal system that drains out the small western creek. It also feeds a larger tidal inlet behind West Point, which forms the eastern boundary of the beach.

Figure 5.75 Crenulate, rock-dominated beach NT 616 extends east of Hawkesbury Point.

West Point (also known as Gumarradadji) is a 6 m high wooded laterite point. It turns right-angle at the tip and runs due south from 2 km. Along this east-facing side are three bedrock-controlled beaches (NT 618-620). Beach **NT 618** is a 50 m pocket of sand located just on the east side of the point. It is bordered by 50-100 m wide rock flats, with some surf breaking along the northern rocks. Five hundred metres to the south are beaches **NT 619** and **NT 620**, two neighbouring 200 m and 400 m long, curving, east to northeast-facing beaches. Both are bordered and partially fronted by rock flats with wooded coastal plain to the rear.

A 2 km wide mangrove-lined bay separates the rocky West Point beaches from beach **NT 621**. This is a 1.7 km long northeast-facing lower energy beach, fronted by 200-300 m wide sand flats. It is bordered by a small rocky point in the west and 1 km long mangrove-lined Gumeradji Point to the east.

NT 622-629 **GUMERADJI-SOUTH WEST POINTS**

No.	Beach	Rating HT LT	Type	Length
NT622	Gumeradji Pt (S1)	1 2	R+rock flats	100 m
NT623	Gumeradji Pt (S2)	1 2	R+rock flats	50 m
NT624	Gumeradji Pt (S3)	1 2	R+rock flats	100 m
NT625	Gumeradji Pt (S4)	1 2	R+rock/sand flats	600 m
NT626	South West Pt (W2)	1 1	R+sand flats	900 m
NT627	South West Pt (W1)	1 1	R+sand flats	1.1 km
NT628	South West Pt (S1)	1 1	R+sand flats	150 m
NT629	South West Pt (S2)	1 1	R+tidal flats	50 m
Spring & neap tide range = 2.8 & 1.1 m				

Gumeradji Point is a 6 m high, 500 m wide, wooded bedrock point, lined with mangroves along its western side. At the point the coast trends south then southeast for 4 km to South West Point, which is located directly opposite Maningrida. Between the two points are six low energy beaches (NT 622-627), with another two (NT 628-629) located just south of South West Point, beyond which are the extensive mangrove-lined shores of the

Liverpool River. There is no vehicle access to and no development on the western side of the river.

The first three beaches **NT 622, NT 623** and **NT 624** lie in three small rocky embayments located immediately south of the point. They are 100 m, 50 m and 100 m long respectively and face northeast across the river mouth. They are each bounded by low rocky points and intertidal rock flats, with a mixture of sand and rock off each beach. Mangroves also fringe the latter two beaches. Most waves break across the rocks and reefs, with only low surging waves at the shore. They are all backed by low wooded slopes.

Beach **NT 625** is a 600 m long northeast-facing beach, bounded by low rocky points and fronted by a mixture of sand and rock flats. Its immediate neighbours, beaches **NT 626** and **NT 627,** are 900 m and 1.1 km in length, and are both bordered by low rocky protrusions terminating at South West Point. They are fronted by 200 m wide sand flats, apart from the eastern tip of NT 627 which has a 200 m long strip of rock flats.

At **South West Point** the shoreline turns and runs south toward the river mouth. Beach **NT 628** is located 500 m south of the point, with mangroves between the point and the beach. The beach faces northeast and is fronted by 200 m wide sand flats. Finally, beach **NT 629** lies beyond another 500 m long stretch of mangroves, and is a 50 m long strip of northeast-facing sand, with mangroves to either side.

Liverpool River

The Liverpool River is a medium sized upland river system, which at the coast flows into a drowned funnel-shaped river mouth. The mouth is 6.5 km wide between Gumeradji and North East Point. The lower reaches of the valley have been flooded and now more than 30 km of mangroves line the river banks south of Maningrida. The lower reaches of the river drain more than 15 000 ha of mangroves-salt flat wetlands. The town is located on the eastern side of the river on the first section of shoreline free of mangroves.

NT 630-633 **MANINGRIDA**

No.	Beach	Rating HT	Rating LT	Type	Length
NT630	Maningrida (S2)	1	1	R+tidal flats	600 m
NT631	Maningrida (S1)	1	1	R+tidal flats	300 m
NT632	Maningrida (boat ramp)	1	1	R+tidal flats	500 m
NT633	Maningrida (N)	1	1	R+tidal flats	1 km
Spring & neap tide range = 2.8 & 1.1 m					

Maningrida is a major aboriginal community located on the eastern banks of the Liverpool River mouth (Fig. 5.76), and 20 km north of the main northern Arnhem Land road. The town, with surrounding community, has a population of 2100, and has all the usual facilities including a barge landing, landing ground, shopping centre, school, sporting facilities, a craft centre, medical centre and administrative centre.

Figure 5.76 Maningrida borders the Liverpool River, with part of the community and the barge landing (centre) shown in this view, together with beaches NT 630-633.

The western side of the community borders the river mouth and consists of 3 km of low wooded shoreline, now partly cleared to accommodate a few houses and the central barge landing. Beach **NT 630** is located 1 km south of the landing and is a northwest-facing, 600 m long, slightly curving beach, backed by low wooded coastal plain, with road access to the northern end where a few houses overlook the beach. It is bordered by low rock flats and fronted by 200-300 m tidal flats.

Beach **NT 631** is a narrow 300 m long sand beach, wedged in at the base of 5 m high bluffs, with a gravel road and houses behind. Mangroves grow on the rocky points and tidal flats front the beach. The main beach (**NT 632**) is a northwest-facing 500 m long beach, that has been cut in two by the barge landing, forming two 150 m and 350 m long sandy high tide beaches. It is backed by the landing facilities and fuel depot in the south, and a few houses to the north, with the town centre 500 m to the east. Low rocky points border each end and a small creek drains across the beach toward the northern end. At low tide 200 m wide tidal flats front the beach, apart from the channel dredged for the barge landing.

Immediately to the north is beach **NT 633**, a narrow crenulate 1 km long beach backed by crenulate bluffs, with some mangrove-covered rocks outcropping along the beach. The beach is backed by a few houses and the cleared area for the landing ground, while a mixture of tidal and rock flats extend 200 m off the beach.

NT 634-640 NORTH EAST-SKIRMISH POINTS

No.	Beach	Rating HT	Rating LT	Type	Length
NT634	North East Pt (W)	1	1	R+tidal flats	200 m
NT635	North East Pt (E)	2	3	R+LTT	4.5 km
NT636	Skirmish Pt	2	4	R+rock flats	250 m
NT637	Skirmish Pt (E1)	2	3	R+LTT	250 m
NT638	Skirmish Pt (E2)	2	3	R+LTT	300 m
NT639	Skirmish Pt (E3)	2	3	R+LTT	400 m
NT640	Skirmish Pt (E4)	2	3	R+LTT	500 m
Spring & neap tide range = 2.8 & 1.1 m					

North East Point (also known as Ndjudda) is located 7 km north of Maningrida, with mangroves occupying much of the shoreline in between. The point is connected to the town by a vehicle track that skirts Gudjerama Creek. The creek maintains a mangrove-lined creek mouth between the town and point.

The point is a 3 m high wooded bedrock protrusion. On its western side is a 200 m long strip of high tide sand (**NT 634**) bordered by rocks and mangroves and fronted by a 100 m wide strip of sand and rock flats. A vehicle track runs along the back of the beach and mangroves.

Curving between North East and Skirmish points is the 4.5 km long north-facing beach **NT 635** (Fig. 5.77). In lee of North East Point a curving reef provides protection to the western 200 m of beach, which for the most part is a more exposed beach receiving waves averaging 1 m in the west. These maintain a relatively steep beach facing a narrow low tide bar. Toward Skirmish Point rocks and rock reef flats increase in prominence, dominating the eastern 2 km of the beach. Since the sea level transgression, a 300-600 m wide series of recurved spits grading to beach-foredune ridges has filled the embayment between the two points, with a V-shaped freshwater swamp extending up to 1 km south of the ridges. The beach is accessible by vehicle and a track runs along its length, with a solitary shack located toward the centre of the beach.

Skirmish Point is a 6 m high wooded bedrock point, with rock reefs extending up to several kilometres off the point. The eastern shoreline consists of a 2 km long series of five bedrock-controlled small embayments bounded by rock reefs, with five sandy beaches (NT 636-640) filling each embayment. A vehicle track runs up to the point and along the back of the beaches.

Figure 5.77 Beach NT 635 curves to the east of North East Point (foreground).

The first beach, **NT 636**, sits right at the point and is a curving, north-facing, 250 m long strip of high tide sand fronted by continuous rock flats and reefs extending offshore. A 100 m wide low rocky point separates it from beach **NT 637**, a curving 250 m long sand beach and bar, bordered by low rocky points and reefs, with some surf over the eastern reef. Low red bluffs back the eastern rocks, with beach **NT 638** lying on the other side. It is a 300 m long beach, with a small cuspate point toward its northern end, and a low rock point to the east. Waves break over reefs off each point and across the narrow low tide bar.

Beach **NT 639** is a 400 m long relatively straight beach, with a solitary house located above the southern point. It has a bar extending up to 100 m off the beach, as well as a few rocks in the surf. On the south side of the house bluff is beach **NT 640**, which is 500 m long and cut by a protruding bluff just south of the house, but with a continuous 50-100 m wide low surf along the low tide bar. Low scarped bluffs border each end.

NT 641-649 BOUCAUT BAY (W)

No.	Beach	Rating HT	Rating LT	Type	Length
NT641	Boucaut Bay(W3)	2	2	R+sand flats	4.4 km
NT642	Boucaut Bay(W2)	2	2	R+sand flats	4 km
NT643	Boucaut Bay(W1)	2	2	R+sand flats	2.3 km
NT644	Nabbarla (W4)	2	2	R+ridged sand flats	700 m
NT645	Nabbarla (W3)	2	2	R+ridged sand flats	200 m
NT646	Nabbarla (W2)	2	2	R+ridged sand flats	700 m
NT647	Nabbarla (W1)	2	2	R+ridged sand flats	200 m
NT648	Nabbarla	2	2	R+ridged sand flats	300 m
NT649	Boucaut Bay (centre)	2	2	R+ridged sand flats	7.5 km
Spring & neap tide range = 2.8 & 1.1 m					

The central section of Boucaut Bay consists of several long sandy beaches (NT 641-643 and 649), terminating in dynamic sand spits and inlets, together with a 5 km long central section consisting of a few low rocky points and shorter intervening beaches (NT 644-648). The small

Nabbarla Kunindabba community is located toward the centre of the bay, immediately east of the central rocky section. A road to this community and a track along the rear of beach NT 641 provide the only vehicle access to these beaches.

Beach **NT 641** commences at the last rocks east of Skirmish Point and trends to the southeast for 4.4 km initially backed by the wooded coastal plain, then by a 50 ha central salt flat dominated lagoon and small tidal creek. Beyond this is a 2 km long vegetated spit that terminates in a series of recurves adjacent to a larger and more dynamic creek mouth. The track from Skirmish point runs along the back of the beach to the first small creek. The beach is fronted by 100-200 m wide sand flats, which widen to several hundred metres off the main creek mouth.

On the eastern side of the creek is 4 km long beach **NT 642**. This beach faces north-northeast and is bordered by two dynamic creeks and their tidal shoals, backed by a near continuous 500 m wide band of mangroves. The beach and low foredunes average about 200 m in width and are fronted by 300-500 m wide ridged sand flats, which widen to 1 km off each of the inlets.

Beach **NT 643** commences on the eastern side of the creek and is a 2.3 km long, north-facing, partly vegetated, low sand spit tied to bedrock in the east. It is backed by a 2 km deep V-shaped mangrove forest in the east, then a low bedrock promontory and the main creek and mangroves in lee of the inlet. Tidal shoals extend 1 km off the inlet and continue at this width along the entire beach, together with a few rock reefs to the east.

The low bedrock section which forms the boundary extends for 5 km to the east. It contains six intervening beaches (NT 644-649) each bordered by low laterite rocky points, together with a scattering of rock reefs off the beaches. In addition irregular ridged sand flats extend 500-700 m offshore. Beach **NT 644** is an arcuate, 700 m long, north-facing, beach, located between low rocky points. Low red bluffs back the narrow eastern end, while the beach widens to the west where it is backed by one to two low beach-foredune ridges.

Two hundred metres to the east is beach **NT 645**, a curving 200 m pocket of sand bordered by rock reefs extending 100 m seaward off each end, and backed by sand blowing up on to the low red laterite bluffs, then the wooded coastal plain. Its immediate neighbour is a curving 700 m long beach, **NT 646**, with prominent boundary bedrock reefs, as well as reefs off the beach. In addition the beach narrows in the centre and in places the red bluffs are exposed at the shore. It widens to the west, while beachrock outcrops along the eastern end.

Beach **NT 647** lies immediately to the east and is a curving, 200 m long, northeast facing pocket of sand, with the western half partly encased in rocks and rock reefs, forming a small 'lagoon', and a rock-bordered but more open eastern half. The vehicle track from the community reaches the eastern end of the beach. Low bluffs extend for 100 m past the beach to the beginning of 300 m long beach **NT 648**. This beach begins where the main track to the Nabbarla community reaches the coast, with the few community houses located in a small clearing toward the eastern end of the beach. Several rock reefs lie off the beach, while the sand flats widen to 1.5 km, becoming increasingly ridged to the east. A 100 m long strip of narrow rock reef forms the eastern boundary.

As the rock and reefs diminish, beach **NT 649** commences and trends essentially due east for 7.5 km to the 2.5 km wide mouth of Berraja Creek. The beach has a few patches of rock at the western end and is initially backed by low wooded coastal plain and a vehicle track. It then continues as a 4 km long series of at least 20 recurved spits that have extended across the mouth of the creek and backing 1200 ha of mangroves and salt flats (Fig. 5.78). The beach is fronted by 1.5 km wide sand flats containing 13 low ridges (average spacing 100 m), which widen to 2 km wide tidal shoals off the inlet mouth.

Figure 5.78 The eastern end of beach NT 649 consists of a series of multiple recurved sandy spits, backed by the dense mangroves of Berraja Creek.

NT 650-654 BOUCAUT BAY-CAPE STEWART

No.	Beach	Rating HT	LT	Type	Length
NT650	Berraja	1	1	R+ridged sand flats	10 km
NT651	Gupanga	1	1	R+tidal flats	800 m
NT652	Yilan	1	1	R+ridged sand flats	14.5 km
NT653	False Point	1	1	R+ridged sand flats	3.2 km
NT654	Cape Stewart(W2)	1	1	R+ridged sand flats	2.8 km
Spring & neap tide range = 2.8 & 1.1 m					

The eastern half of Boucaut Bay is occupied by three beaches (NT 650-652) that extend for a total of 25 km up to False Point, beyond which are two more beaches (NT 653 & 654) before the rocky Cape Stewart coast. Four aboriginal communities are also located along the coast at Berraja, Gupanga, Ji-Marda and Yilan. All four communities are adjacent beaches and are accessible by vehicle, with most tracks having to cross the wide backing salt flats.

Beach **NT 650** begins at Berraja Creek, 1 km west of the small **Berraja** community, which is located just behind the beach. The beach faces north-northwest and extends for 10 km to the dynamic mouth of the Blyth River, where it terminates in a series of large recurved spits. The fine sand beach has a low gradient and is fronted by 1.5 km wide ridged sand flats. It is backed by a series of low beach-foredune ridges, which toward the east splay to an older series of recurved spits extending up to 3 km inland. The beach is eroding in places exposing older soil horizons at the shore. Beyond the spits is the wide floodplain of the Blyth River, which increases from 4 to 20 km in width. The vehicle track to the community follows one of the older recurves.

Just inside the western side of the river mouth is beach **NT 651**. This is a curving, east-facing, 800 m long sand beach, fronted by 50-100 m wide tidal flats, then the deep river channel. The small **Gupanga** community is located just behind the northern end of the beach and connected via tracks following the recurved spits to the main access track.

Beach **NT 652** begins on the east side of the river mouth. The 14.5 km long beach initially faces north-northeast, then gradually swings round to face northeast where it terminates at False Point. The beach has a low gradient and is fronted in the west by the 1 km wide river mouth tidal shoals. These soon narrow to 100-200 m wide sand flats that extend up to the point. The **Ji-Marda** community is located 2 km from the western end of the beach, and the adjacent **Yilan** community 2 km further east and closer to the beach. The beach represents the outermost of a 1-4 km wide beach ridge plain containing up to 30 ridges and recurved spits which is in turn backed by salt flats up to 5 km wide. Access to the communities crosses the salt flats then runs atop older sand spits.

False Point is a low rock-anchored cuspate foreland, at which the shore turns and faces north. The beach (**NT 653**) curves for 3.2 km to a second sandy foreland where it faces west. Rock reefs extend for up to 1 km off each foreland, lowering wave height along the beach. The beach is fronted by 200-400 m wide sand flats which widen toward each end. False Point consists of converging beach ridges from both beach systems, with about six ridges facing north. A solitary house is located at the rear of these ridges, with vehicle access along the beaches.

Beach **NT 654** begins at the second foreland and curves for 2.8 km to the western rocks of Cape Stewart. The beach has the same exposure as its western neighbour, but is clear of reefs, which permit higher waves to reach the shore. These in turn have produced a slightly more unstable beach and foredune system, with sand dunes blowing 100-200 m inland and reaching heights of 12 m. The waves also maintain a few ridges across the 200 m wide sand flats.

NT 655-664 **CAPE STEWART-NGANDADAUDA CK**

No.	Beach	Rating HT	Rating LT	Type	Length
NT655	Cape Stewart (W1)	1	1	R+sand flats	300 m
NT656	Cape Stewart	2	3	R+rock flats/reefs	50 m
NT657	Cape Stewart (E1)	2	3	R+rock flats/reefs	400 m
NT658	Cape Stewart (E2)	2	3	R+rock flats/reefs	200 m
NT659	Cape Stewart (E3)	2	3	R+rock flats/reefs	700 m
NT660	Cape Stewart (E4)	2	3	R+rock flats/reefs	700 m
NT661	Cape Stewart (E5)	2	3	R+rock flats/reefs	100 m
NT662	Ngandadauda Ck (N3)	2	3	R+rock flats/reefs	150 m
NT663	Ngandadauda Ck (N2)	2	3	R+rock flats/reefs	900 m
NT664	Ngandadauda Ck (N1)	2	3	R+rock flats/reefs	200 m
Spring & neap tide range = 3.5 & 1.5 m					

Cape Stewart marks the northwestern tip of a 3 km wide low bedrock protrusion with False Point and the wide Boucaut Bay to the west and the Crocodile Island Group to the west. The 10 km of shoreline between the cape and Ngandadauda Creek contains ten rock and reef-dominated beaches (NT 655-664). There is no vehicle access to the cape or creek area and no settlements.

Beach **NT 655** lies immediately west of the cape. It is a west-facing 300 m long sand beach, backed by low wooded bluffs and fronted by 300 m wide sand and rock flats. It is bordered by protruding bluffs and rocks, the northern rocks running for 200 m up to the tip of the cape. On the eastern side of the tip is a 200 m wide crescent of rock flats, in the centre of which is 50 m long pocket beach **NT 656**. Waves break over the bordering rocks and reefs and only reach the beach at high tide.

One hundred metres to the east around a low rock protrusion is beach **NT 657**, a crenulate 400 m long north-facing strip of sand interrupted by rocks and reefs, with only the two extremities offering a clear beach at high tide. Low red bluffs back the beach, with older sand dunes sitting on top of the bluffs along its eastern half. Three hundred metres of bluffs and rock separate it from beach **NT 658**, which is a curving northwest-facing pocket beach, bordered by prominent rock flats and a central small reef, which results in low waves at the shore. Red bluffs back the western half of the beach, with low dunes, blowing over from its eastern neighbour, which back the eastern half.

Beach **NT 659** lies around a 150 m long, low, sand-capped, rocky point. It is a crenulate 700 m long north-facing sand beach, with rock flats to either end and a few scattered reefs off the beach. These permit more waves to reach the beach which have instigated 50-100 m of minor sand dune transgression.

Beach **NT 660** lies 100 m to the east and is totally dominated by rock flats, which front the entire beach. They produce crenulations along the beach and permit a few mangroves to grow on the rocks, while a low active foredune backs the beach. Its neighbour (**NT 661**) is a 100 m long pocket of sand encased in bordering rock flats, mangroves and rock reefs off the beach over which most waves break (Fig. 5.79). At the eastern end of the beach the rocky shoreline turns and trends south. The next beach, **NT 662**, runs behind continuous mangroves for 600 m before emerging as a 150 m long strip of east-facing sand, bordered at its southern end by some mangrove-covered intertidal rocks. The rocks have formed a sandy foreland, on the south side of which is the curving, 900 m long, east to northeast-facing beach **NT 663**. This beach faces squarely into the easterly waves and has experienced minor progradation with a few low foredunes behind the beach. Mangrove-covered rock flats border the southern end.

Figure 5.79 Beaches NT 660 & 661 are dominated by laterite rocks and reefs.

Beach **NT 664** lies 500 m to the south and is a gradation from beach to tidal flats. The 200 m long strip of high tide sand is bordered by mangrove-covered rock flats, with a 300-400 m wide mixture of rock and sand flats fronting the beach. One kilometre south of the beach is the 500 m wide mouth of Ngandadauda Creek and, beyond, the beginning of the extensive mangrove-lined shores of the Crocodile Island Group.

NT 665-671 **MILINGIMBI**

No.	Beach	Rating HT LT	Type	Length
NT665	Gumugumuk	1 2	R+tidal flats	150 m
NT666	Darbada Is (1)	1 1	R+tidal flats	800 m
NT667	Darbada Is (2)	1 1	R+tidal flats	700 m
NT668	Milingimbi	1 1	R+tidal flats	700 m
NT669	Milingimbi (2)	1 1	R+tidal flats	200 m
NT670	MacKinnon Pt	1 2	R+tidal flats	100 m
NT671	MacKinnon Pt (S)	1 2	R+tidal flats	250 m
Spring & neap tide range = 3.5 & 1.5 m				

Milingimbi township is located on the eastern side of Milingimbi Island which is one of the Crocodile Island Group, an expanse of several low lying, mangrove-fringed islands which lie between Stewart Point and Castlereagh Bay. Most of the mainland and island shores are surrounded by dense mangroves and sand and mud flats bordered by deep tidal channels, and backed by salt flats up to 10 km wide. There are seven 'beaches' in the area, four of which are used as boat ramps, including the main Milingimbi barge landing. They however represent only a small fraction of the mangrove-dominated shoreline.

Gumugumuk beach (**NT 665**) is a 150 m long strip of sand located on the mainland 7 km south west of Cape Stewart. It is bordered by mangroves and has formed as a result of the coastal plain protruding slightly to impinge on a deep 500 m wide tidal creek, whose flow keeps the shore clear of mangroves. There is vehicle access to the small beach, which has narrow tidal flats, then the deep tidal channel, and is used as a boat ramp.

Directly opposite Gumugumuk is **Darbada Island**. The 500 ha uninhabited island has a central core of chenier-recurved spits surrounded by mangroves. The more exposed northern shore of the island has two beaches. Beach **NT 666** is a slightly crenulate 800 m long strip of high tide sand, backed by low casuarina-covered beach ridges and fronted by 1-1.5 km wide sand and some rock flats. Mangroves border either end and a few grow on the tidal flats. Sand moves along the beach to feed its southern neighbour, **NT 667**, which is a 700 m long series of recurved spits, backed by dense mangroves. The beach is a 'flying island' in that it is eroding at its northern end, exposing mangroves, and growing at the southern end with new recurves (Fig. 5.80).

Figure 5.80 Beach NT 667 consists of a series of multiple recurved spits.

Milingimbi Island is about 5000 ha in area. It has a higher central core of about 1200 ha, with the bulk of this island consisting of intertidal salt flats and surrounding mangroves. The town of Milingimbi (population ~900) is located on the more exposed eastern side, where there is sufficient energy to maintain two small beaches (Fig. 5.81), with the township located directly behind. The main town beach (**NT 668**) and site of the barge landing is a 700 m long east-facing strip of high tide sand, backed by a 200 m wide series of low beach ridges and fronted by 600 m wide tidal sand flats. Mangroves dominate the northern end of the beach, with a few to the south, where the beach terminates at a low rocky bluff. The barge landing crosses the centre of the beach with beachfront houses to either side. Immediately south of the small point is beach **NT 669**, a 200 m long beach, bordered by low rocky points and fronted by 600 m wide tidal flats and a few mangroves. A few houses back the beach.

Figure 5.81 Milingimbi township is located adjacent to beach NT 668.

Five hundred metre wide Bennett Creek separates the island from the predominantly mangrove-lined mainland shore, which is in turn backed by 7 km wide salt flats. A vehicle track crosses the salt flats to MacKinnon Point, located 4 km due south of Milingimbi, where there is a small gap in the mangroves, with a 100 m long beach (**NT 670**) and narrow tidal flats. Five hundred metres to the west the mangroves have been cleared for a barge landing. The beach and landing front on to the deep waters of the creek.

Seven kilometres southeast of the point is a second break in the mangroves within which is located 250 m long beach **NT 671**. The beach is fringed by mangroves, with scattered mangroves growing on the 250 m wide tidal flats. A vehicle track across the salt flats provides access to the beach, which can be used for launching boats at high tide.

NT 672-678 **DHIPIRRNJURA**

No.	Beach	Rating HT	Rating LT	Type	Length
NT672	Dhipirrnjura	1	1	R+sand flats	250 m
NT673	Dhipirrnjura (E1)	1	1	R+rock flats	100 m
NT674	Dhipirrnjura (E2)	1	1	R+rock flats	800 m
NT675	Dhipirrnjura (E3)	1	1	R+sand flats	150 m
NT676	Dhipirrnjura (E4)	1	1	R+sand flats	100 m
NT677	Dhipirrnjura (E5)	1	1	R+sand flats	150 m
NT678	Dhipirrnjura (E6)	1	1	R+sand flats	100 m
Spring & neap tide range = 3.5 & 1.5 m					

The small **Dhipirrnjura** community is located on the eastern shores of Castlereagh Bay, 3 km east of the Glyde River mouth. The community consists of four houses, is serviced by a landing ground and has vehicle access. Between the community and the Woolen River mouth, 6 km to the east, is 7 km of north-facing rock and mangrove dominated shoreline containing eight small beaches (NT 672-678), all backed by the woodlands of the coastal plain. Low waves reach the shore breaking against the beaches at high tide and over the sand and rock flats at low tide.

The community lies in lee of beach **NT 672**, a curving 250 m long sandy beach which is used for launching small boats The houses are located amongst the trees of the coastal plain just 50 m for the shoreline. Low rocky points border the beach, with mangroves covering the eastern point (Fig. 5.82). Beach **NT 673** lies immediately to the east and is a 100 m long pocket of high tide sand, bordered and fronted by rock flats.

Figure 5.82 The small Dhipirrnjura community located in lee of beach NT 672.

One kilometre to the east is 800 m long beach **NT 674**. This crenulate beach consists of a sandy high tide beach fronted by near continuous intertidal rock flats, with three small gaps in the rocks and some mangroves on the rocks. At the eastern end of the beach is a larger clump of mangroves, on the other side of which is 150 m long beach **NT 675**. Rocks and mangroves protrude 100 m either side of this beach, with sand flats in between.

Beach **NT 676** lies 2 km to the east and is an isolated 100 m long pocket of high tide sand, bordered by a low 300 m long rocky point to the west and rock flats to the east, all covered with mangroves and with shallow tidal flats off the beach.

Beaches **NT 677** and **NT 678** occupy two small neighbouring embayment that border the western side of the Woolen River mouth. A low 200-300 m long narrow rocky points border each side, with the beaches occupying respectively a 150 m and 100 m long section of the intervening rocky embayments. Sand flats fill the rest of the embayments.

NT 679-687 BANYAN ISLAND

No.	Beach	Rating HT	Rating LT	Type	Length
NT679	Banyan Island (W1)	1	2	R+rock flats	250 m
NT680	Banyan Island (W2)	1	2	R+rock flats	300 m
NT681	Banyan Island (E1)	1	2	R+rock flats	700 m
NT682	Banyan Island (E2)	1	1	R+sand flats	800 m
NT683	Banyan Island (E3)	1	2	R+rock flats	50 m
NT684	Banyan Island (E4)	1	2	R+rock flats	100 m
NT685	Banyan Island (E5)	1	2	R+rock flats	300 m
NT686	Banyan Island (E6)	1	2	R+rock flats	400 m
NT687	Banyan Island (E7)	1	2	R+rock flats	400 m
Spring & neap tide range = 3.5 & 1.5 m					

Banyan Island is a 12 km long and up to 6 km wide island that lies along the eastern side of the Woolen River mouth. It faces into Hutchinson Strait along its northern shore, and as its southern boundary has a 10 m long mangrove and salt flat-lined tidal creek that connects the river and strait. The island is uninhabited with no development. Much of its 40 km of shoreline is very low energy and dominated by mangroves along the west, east and south shores, with rock flats along much of the more exposed northern shore. Between the western rocky tip and extending for 6 km along the northern shore are nine low energy rock-dominated beaches (NT 679-687).

Beaches NT 679 and 680 occupy the western tip of the island and form the eastern head of the Woolen River mouth. Beach **NT 679** is a 250 m long west-facing strip of high tide sand fronted by irregular but continuous intertidal rock flats. The rocky southern end of the beach forms the western entrance to the river. It is separated from its neighbour by 100 m of mangrove-covered rock flats. Beach **NT 680** is a slightly curving, west-facing 300 m long sand beach, bordered by mangrove-covered rock flats, with a small central section free of rocks and fronted by sand flats.

The northern shore of the island is completely dominated by rocky shore backed by gently rising wooded slopes, reaching a maximum height of 22 m, 1 km in from the shore. Beach **NT 681** lies 500 m east of the western tip. It consists of a 700 m strip of high tide sand fronted for the most part by rock then sand flats, with mangroves scattered over the rocks. Beach **NT 682** lies 2 km east of the tip and is the only beach reasonably accessible by boat. The 800 m long north-facing beach is bordered by mangrove-covered rock flats, with 200 m wide sand flats occupying much of the intervening embayment. The beach is moderately steep and backed by a narrow fringe of vegetated overwash deposits.

One kilometre to the east is a more prominent rocky point on the eastern side of which are four beaches. Beach **NT 683** is a 50 m pocket of high tide sand just inside the tip of the point. Beach **NT 684** is a similar 100 m long north-facing strip of sand bordered by wooded rocky points. Beach **NT 685** occupies the apex of the embayment and is a 300 m long sand beach, partly fronted by rock flats. Beach **NT 686** lies 100 m further east past some rocks and is a 400 m long sand beach fronted by near continuous intertidal rock flats, with mangroves scattered over all the intervening rock flats and rocky points.

Beach **NT 687** lies past the next point and represents a transition from beach to rock flats. It is a 400 m long strip of high tide sand fronted by near continuous rock flats, with mangrove-covered rock flats forming the eastern boundary. The shoreline continues for another 5 km into the strait, decreasing in energy, with some patches of high tide sand, but essentially a shore dominated by continuous intertidal rock flats and scattered mangroves.

NT 688-691 POINT GUY (E)

No.	Beach	Rating HT	Rating LT	Type	Length
NT688	Pt Guy (E 1)	1	1	R+sand flats	3.6 km
NT689	Pt Guy (E 2)	1	2	R+rock flats	800 m
NT690	Pt Guy (E 3)	1	2	R+sand/rock flats	1.5 km
NT691	Pt Guy (E 4)	1	2	R+sand flats	250 m
Spring & neap tide range = 3.5 & 1.5 m					

Howard Island is a 40 km long, up to 10 km wide, island separated from the mainland by the sinuous 38 km long mangrove-lined Hutchinson Strait. It averages less than 20 m high with a maximum elevation of 40 m. The only settlement on the island is at Langarra on the north shore. The north side of the island is the only section exposed to waves and the 51 km of shoreline has 21 beaches (NT 668-709) between the western tip at Point Guy and the eastern end at Ganbagawirra Point opposite Elcho Island.

Point Guy is a low rocky point composed of Proterozoic sandstone. Much of the point consists of intertidal rock flats covered by dense mangroves and backed by an older vegetated sand spit. The first beach (**NT 688**) begins as the mangroves thin out 2 km east of the point. The beach is 3.6 km long, faces north and, after 3 km of 600 m wide sand flats, terminates as the shore trends northeast as an irregular series of intertidal rock flats. The beach has a moderate slope and is backed by a 50 m wide series of cleared and wooded beach ridges, which continue as the spit to the west (Fig. 5.83).

Figure 5.83 View east along beach NT 688.

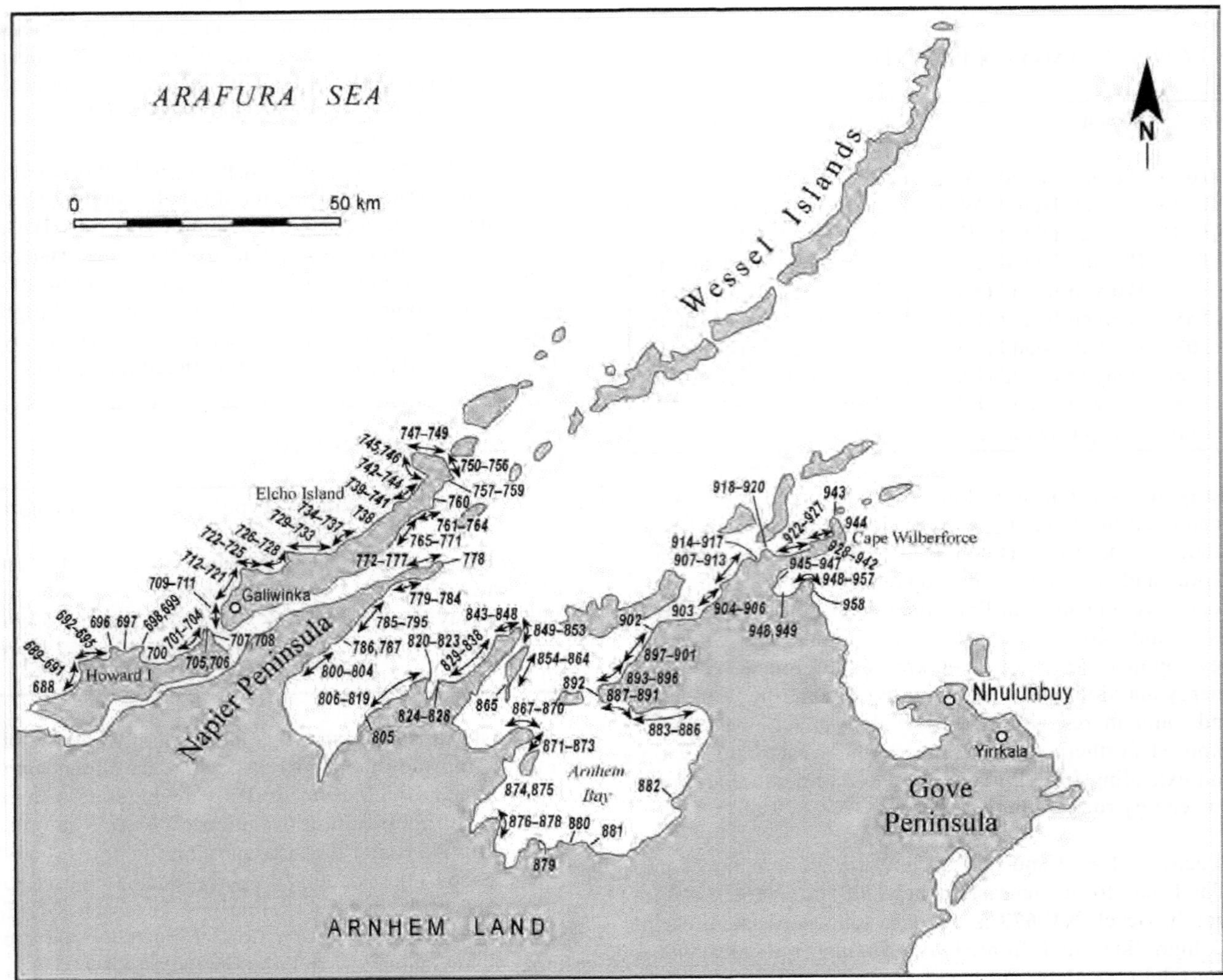

Figure 5.84 Regional map 7b covers the northeastern coast of Arnhem Land and beaches NT 688-958.

Beach **NT 689** is an 800 m long strip of high tide sand fronted by the patchy rock flats and scattered mangroves, with some sections free of rock, then the wide sand flats. It terminates at a low rocky point on the eastern side of which is the curving, northwest-facing 1.5 km long beach **NT 690**. The beach has prominent rock flats extending up to 500 m off the ends of the beach, with a central 300 m long section free of rock. Low foredunes and minor dune activity back the beach.

Five hundred metres to the east, in a small embayment bordered by rock flats and low bluffs, is 250 m long beach **NT 691**. The beach is backed by low scarped bluffs, fronted by sand flats, and faces north across a 3 km wide west-facing mangrove-lined embayment. The northern side of the bay is a low rocky point which forms the southern boundary of beach NT 692. Sand moving around this point has built a 2 km long series of spits along the northern shore of the bay, the most recent fronted by continuous mangroves.

NT 692-699 **LANGARRA**

No.	Beach	Rating HT	Rating LT	Type	Length
NT692	Langarra (W3)	1	2	R+rock flats	900 m
NT693	Langarra (W2)	1	2	R+rock flats	600 m
NT694	Langarra (W1)	1	1	R+rock flats	1.1 km
NT695	Langarra	1	2	R+sand/rock flats	4.8 km
NT696	Langarra (E)	1	2	R+rock flats	1 km
NT697	Jigaimara Pt (E)	2	2	R+sand ridges	2.5 km
NT698	Malga Pt (W2)	2	3	R+rock flats	1.2 km
NT699	Malga Pt (W1)	2	3	R+rock flats	700 m
Spring & neap tide range = 3.5 & 1.5 m					

Beach **NT 692** is a west-facing 900 m long sand beach, backed by a few beach ridges, with intertidal rock flats dominating much of the beach apart from a 200 m long central section. Rock and sand flats extend 400 m off the beach with the rock flats continuing for 300 m north of the beach to a point where the coast turns and trends northeast.

Beach **NT 693** begins just east of the point and is a curving 600 m long sand beach fronted by continuous irregular rock flats, with more extensive rock and rock reefs to either end and with waves breaking across the wide reefs at mid to low tide. A cuspate foreland at the eastern end forms the boundary with its neighbour (beach **NT 694**). This is a 1.1 km long curving beach which, while bordered by rock flats and reefs, is generally open to waves along the central section. Both beaches are backed by a few low beach-foredune ridges, then woodlands, and are accessible by a vehicle track from Langarra.

Langarra community lies at the eastern end of beach **NT 695**. The 4.8 km long beach begins in lee of 200 m wide rock flats, which dominate the western half of the beach. Once the rocks diminish, the moderate waves maintain an attached bar along the eastern half of the beach, which widens to sand flats as well as some rock flats at the community, which is located in the west-facing northeast corner. The community consists of about 12 houses and has a landing ground in lee of the eastern point (Fig. 5.85).

Figure 5.85 The Langarra community, with the landing ground in centre and beaches NT 695 & 696 to either side.

The eastern point consists of low, irregular sandstone rock flats, which continue on for 2 km to Jigaimara Point. Beach **NT 696** occupies the first 1 km of shoreline. It is a crenulate northwest-facing high tide beach, backed by generally cleared ground and fronted by the irregular 100-200 m wide rock flats.

Jigaimara Point forms the western headland of a 2 km wide embayment containing beach **NT 697**. The curving 2.5 km long beach is bordered by the two rock headlands, with Malga Point to the east. The beach faces due north and consists of a moderately steep high tide beach fronted by two to three ridges on the sand flats. It is backed by a 400 m wide series of low grassy beach-foredune ridges. A vehicle track from Langarra runs along the back of the beach to Malga Point.

Malga Point is a 3 km long, 10-15 km high laterite point, which is scarped along much of its northern shore. Along the base of the bluffs are two crenulate sandy high tide beaches. Beach **NT 698** is 1.2 km long and consists of two crescents of sand bordered by extensive rock flats, which narrow toward the centre. Beach **NT 699** is similar, only 700 m long with red bluffs behind. It terminates at the western tip of the point.

NT 700-704 **NGURUBUDANAMIRRI BAY**

No.	Beach	Rating HT	Rating LT	Type	Length
NT700	Ngurubudanamirri Bay(1)	2	3	R+LTT	4.1 km
NT701	Ngurubudanamirri Bay(2)	2	3	R+LT rips	800 m
NT702	Ngurubudanamirri Bay(3)	2	3	R+LT rips	3.3 km
NT703	Worrmi Pt (1)	2	3	R+rock/sand flats	800 m
NT704	Worrmi Pt (2)	2	3	R+rock/sand flat	1.6 km
Spring & neap tide range = 3.5 & 1.5 m					

Ngurubudanamirri Bay is a relatively open 10.5 km wide bay bordered by Malga and Worrmi points There is no vehicle access to and no development in the bay area. The bay contains 19 km of shoreline. The first 4 km in lee of Malga Point are well protected and dominated by mangroves and tidal flats. However once the protection of the point is past moderate waves reach all three north-facing beaches (NT 700-702), maintaining the beaches, bars and surf.

Beach **NT 700** is a relatively straight, north-facing, 4.1 km long, near continuous sand beach with a few low tide rock patches toward the eastern end and a solitary small creek crossing toward the western end. The beach is backed by a few grassy beach-foredune ridges, and fronted by a continuous attached bar, with a permanent rip exiting a small partly protected 'lagoon' against the western small rocky point. It terminates in the east against a 1 km long rocky section, beyond which is beach **NT 701**. This 800 m long sandy beach is backed by destabilised beach and foredune ridges and includes some small deflation basins. It is fronted by an attached bar which, toward the west, receives higher waves and is cut by beach rips, with a permanent rip against the rocks. A circular sandstone headland forms the eastern boundary with a small creek crossing the beach on its western side.

On the eastern side of the 100 m wide headland is 3.3 km long beach **NT 702**. This is the highest energy beach in the bay and during strong onshore wind and wave conditions receives sufficient wave energy to form rips, which cut across the attached bar approximately every 100 m. The beach is backed by a 300-400 m wide series of beach-foredune ridges, which are cut by a small creek toward the eastern end. The beach terminates as rock reefs, and bluffs dominate the shore, which turns and trends north for 2 km to Worrmi Point.

Beaches NT 703 and 704 are moderately protected beaches located along the base of the west-facing bluffs that run out to the point. Beach **NT 703** is 800 m long and contains crenulations induced by the backing red bluffs and fronting intertidal rock flats and reefs. Beach **NT 704** is 1.6 km long, consisting of two shallow crescents backed by bluffs and fronted by irregular rock flats, reefs and islets extending up to 500 m off the beach.

NT 705-708 **WORRMI POINT-NIKAWU**

No.	Beach	Rating HT	Rating LT	Type	Length
NT705	Worrmi Pt (E)	1	1	R+sand flats	600 m
NT706	Ganbagawirra Pt (W)	1	1	R+sand/rock flats	1.6 km
NT707	Ganbagawirra Pt (E)	1	1	R+tidal flats	200 m
NT708	Nikawu	1	1	R+tidal flats	1.5 km
Spring & neap tide range = 3.5 & 1.5 m					

Worrmi Point marks the northern tip of Howard Island and the entrance to Cadell Strait, the 40 km long predominantly mangrove-lined strait that separates Elcho Island from the mainland. At Worrmi Point the coast turns and runs due east for 2 km to Ganbagawirra Point, before the low energy strait shoreline begins.

Beach **NT 705** is bordered by the 5 m high roughly circular Worrmi Point in the west. It curves for 600 m to a small rock outcrop (Fig. 5.86). The beach is moderately steep, fronted by 300 m wide sand flats and backed by vegetated red cliffs, capped by dense woodlands. On the eastern side of the rocks is 1.6 m long beach **NT 706**, a relatively straight continuous sand beach that is fronted by five patches of intertidal rock flats, terminating at Ganbagawirra Point, another low rocky flat. Waves break over the rock flats, and only reach the beach at high tide, with the rock flats inducing a few weak topographic rips. The beach is backed initially by scarped red bluffs, with a foredune covering the central and western bluffs.

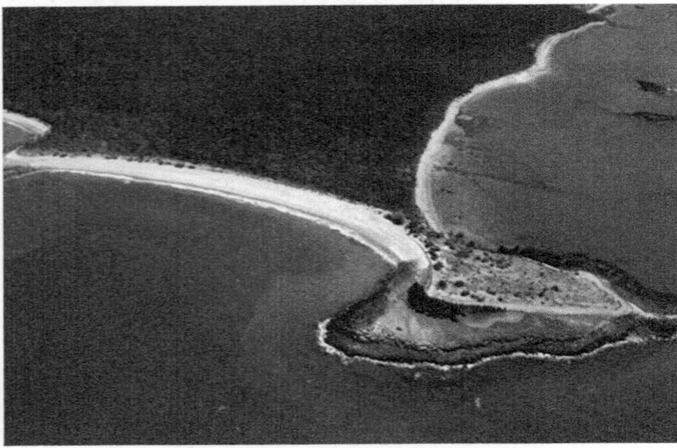

Figure 5.86 Worrmi Point with beaches NT 704 to right and NT 705 to left.

At Ganbagawirra Point the true strait begins as the shoreline turns to trend to the southeast, with Elcho Island and the massive Napier Peninsula shielding the two remaining beaches from ocean waves. Beach **NT 707** lies 1.5 km south of the point and consists of a 200 m long gap in the mangroves. The beach faces north and is fronted by 500 m wide tidal flats, with densely wooded slopes behind.

Beach **NT 708** lies 500 m further east and is the site of the one-house **Nikawu** community, which is located at the northern end of the 1.5 km long beach. The beach faces northeast along the strait and consists of a vegetated barrier, backed by a 10 ha freshwater lagoon, with the creek draining across the southern end of the beach.

Elcho Island

Elcho Island is an elongate 28 000 ha northeast-trending island. It is 52 km long and 3-7 km wide, with a shoreline length of 140 km. The island is long, low and undulating, with a maximum elevation of only 40 m. The southern tip of the island, Matjaganur Point, lies just 2 km off the mainland, with the 40 km long Cadell Strait separating the island from the parallel Napier Peninsula. For much of the strait only 1-3 km separates the island and peninsula, with the narrowest point just 500 m at The Narrows.

Galiwinku, the main settlement, is located near the southern end of the island, with its barge landing in the adjacent strait. Ten smaller communities are scattered at both ends of the island and are all linked by roads and tracks. The island has a population of about 1800.

Shoreline length:	140 km	(2955-3095 km)
No. Beaches	63 beaches	(NT 709-771)

NT 709-717 **GALIWINKU**

No.	Beach	Rating HT	Rating LT	Type	Length
NT709	Matjaganur Pt (N1)	1	2	R+sand/rock flats	500 m
NT710	Matjaganur Pt (N2)	1	2	R+sand/rock flats	1.1 km
NT711	Dhambala	1	2	R+sand flats	600 m
NT712	Pt Bristow	1	2	R+sand/rock flats	1.1 km
NT713	Pt Bristow (N)	1	2	R+sand/rock flats	600 m
NT714	Galiwinku (S)	1	2	R+LT rips	500 m
NT715	Galiwinku	1	2	R+sand/rock flats	400 m
NT716	Gorabi Cliffs (S)	1	2	R+rock flats	100 m
NT717	Gorabi Cliffs	1	2	R+ridged sand flats	150 m
Spring & neap tide range = 3.5 & 1.5 m					

Galiwinku is the major settlement on Elcho Island and is located 5 km north of the southern tip of the island. The township sits atop 20 m high red bluffs looking west out over the Arafura Sea. Beaches begin at the southern Matjaganur Point and dominate much of the more exposed western shore of the 50 km long island. Eleven beaches (NT 709-719) are located either side of Galiwinku and are all accessible by vehicle (Fig. 5.87).

Beach **NT 709** begins at Matjaganur Point and faces west across the 3 km wide southern entrance of Cadell Strait to the mainland. The 500 m long beach is backed by a low foredune, and fronted by a 1 km wide mixture of sand and rock flats, including a few rocks outcropping on the beach and in the strait. A vehicle track reaches the northern end of the beach. Immediately north is beach **NT 710**, a similar 1.1 km long west-facing beach with rock and sand flats, then more rock outcrops. It terminates at a low rocky eastward inflection in the shore, on the north side of which is beach **NT 711**. The small Dhambala community (four houses) is located behind the northern end of this curving 600 m long sandy beach. The beach faces southwest across the strait and is generally free of rocks, but fronted by 1 km wide sand flats, together with some rock reefs out in the strait.

Figure 5.87 Galiwinku is located on the red bluffs, with beaches NT 713-715 extending south of the bluffs.

Point Bristow lies 500 m west of the community and represents the northern entrance to the strait. The low rocky point is surrounded by a outward curving 1.1 km long sandy high tide beach (**NT 712**) and a mixture of irregular intertidal rock flats and intervening sections of sand flats. A road reaches the northern end of the beach and vehicle tracks run along the back of the low foredune. Immediately north is 600 m long west-facing beach **NT 713**, a double arcuate high tide sand beach, fronted by continuous 100 m wide intertidal rock flats, and backed by 15 m high vegetated bluffs.

The main Galiwinku beaches lie to the north of the bluffs. Beach **NT 714** is a 500 m long sand beach bordered by low rock flats and backed by a 5 m high foredune, then an elongate 1.5 km long wetland, that drains across the southern corner of the beach. A road and bridge cross the creek, with houses occupying the bluffs that form the northern boundary. The beach receives sufficient waves to maintain a 50-100 m wide bar, which has permanent rips against the boundary rocks. The rip channels only operate when waves are breaking across the bar, but may remain as a deeper hole during calmer conditions.

On the north side of the rocks is the main beach and boat ramp. This is a 400 m long west-facing pocket of sand (**NT 715**) wedged in below the 20 m high bluffs, and cut by rocks in the south, together with the remains of an old rock jetty with a vehicle access ramp behind. Boats are usually moored on the beach and over the 200 m wide sand flats. The town centre is located on the bluffs at the rear of the beach, while the main barge landing is located 3.5 km to the west in Cadell Strait.

Much of Galiwinku is located above the 20-30 m high red laterite Gorabi Cliffs, with the bluff-top landing ground forming the northern boundary of the town. Two small west-facing beaches are located at the base of the bluffs. Beach **NT 716** is a narrow 100 m long strip of high tide sand wedged in between the base of the bluffs and the fronting 50 m wide rock flats. Beach **NT 717** is located immediately south of the landing ground in a 150 m wide gap in the sloping vegetated bluffs. It consists of a pocket sand beach and 100 m wide ridged sand flats.

NT 718-726 **DHUDUPU-NGAYAWILLI-DHAYIRRI**

No.	Beach	Rating HT LT	Type	Length
NT718	Dhudupu (S)	1 2	R+rock flats	100 m
NT719	Dhudupu	1 2	R+LT rips	500 m
NT720	Ngayawilli	1 2	R+LTT/rocks	600 m
NT721	Ngayawilli (N)	1 2	R+LT rips	3 km
NT722	Warnga Pt (S3)	1 2	R+rock flats	300 m
NT723	Warnga Pt (S2)	1 2	R+LTT	400 m
NT724	Warnga Pt (S1)	1 1	R+LTT+rocks	1.2 km
NT725	Warnga Pt (E)	1 2	R+LT+rips	3.1 km
NT726	Dhayirri	1 2	R+tidal flats	600 m
Spring & neap tide range = 3.5 & 1.5 m				

The 10 km of coast between the Dhudupu and Dhayirri communities contains eight beaches (NT 719-726), plus the Ngayawilli community and the prominent Warnga Point. All the beaches lie just a few kilometres north of Galiwinku and all are accessible by formed tracks.

The **Dhudupu** community (two houses) is located 1.5 km north of the landing ground on top of 20 m high bluffs. Beach **NT 718** is a 100 m long strip of high tide sand located at the base of the bluffs just south of the community. The small community overlooks beach **NT 719,** a 500 m long beach and foredune, backed by a 300 m wide barrier then a 500 m wide wetland. The beach receives occasional higher waves which maintain a 50-100 m wide bar, occasionally cut by a rip. At the northern end a tidal creek drains the backing wetland and maintains 100 m wide tidal shoals.

The small **Ngayawilli** community (two houses) is located 1 km north of the creek mouth at the northern end of beach **NT 720**. This beach begins amongst rock flats that also border the inlet, and extends north for 600 m to a low sandy foreland formed in lee of a 100 m wide rock flat. The beach is fronted by a 50 m wide attached bar together with scattered rock flats.

Beach **NT 721** begins on the north side of the foreland and gently curves for 3 km to the north, terminating as the beach narrows and gives way to 20 m high scarped bluffs. The beach receives sufficient waves in the south to produce an attached bar cut by a few rips, with a continuous bar for most of the length. It is backed by a 5-10 m high foredune which is active in places, with a vehicle track running along the back of the foredune.

Beach **NT 722** is a straight 300 m long sand beach backed by 5-10 m high vegetated bluffs, with rocks bordering each end and extending for 200 m off the beach. The beach terminates in lee of a 2 km long rock-dominated point, that continues on to the northern Warnga Point.

Three beaches (NT 723-725) are located along this northwest-facing section of shoreline (Fig. 5.88). Beach **NT 723** lies immediately to the north and is a semicircular 400 m long beach contained between two rocky points, the northern one 10 m high. The beach is backed by a moderately active 10 m high foredune and fronted by a continuous low tide bar, with a permanent rip against the southern rocks. Beach **NT 724** extends from the northern side of the higher point for 1.2 km to Warnga Point. The sandy beach is interrupted by two rock outcrops which induce crenulations, with a low tide bar in between. It is backed by generally stable foredune which rises to a maximum height of 18 m.

Figure 5.88 The north-facing beaches NT 723-725.

Warnga Point protrudes north for 400 m as a narrow strip of dune-covered rocks surrounded by 100 m wide rock flats, together with a clump of low mangroves on the eastern rock flats. Beach **NT 725** begins as a high tide sand beach just beyond the mangroves. It is initially fronted by rock flats, which give way to a low tide bar and occasional rips. Waves average just over 1 m and break across a 50 m wide surf zone. A continuous foredune backs the beach, rising from 10 m in the west to a height of 23 m in the east.

The one-house **Dhayirri** community is located at the eastern tip of the beach on a sand-covered 100 m wide bedrock point. The point forms the western boundary of a 4 km wide embayment. Beach **NT 426** runs south along the eastern side of the point for 600 m as an irregular beach dominated by rocks, tidal sand shoals and finally mangroves of a tidal creek that flows along the base of the headland. Wave height decreases down the beach into the creek mouth.

NT 727-733 **DHAYIRRI-WATDAGAWUY**

No.	Beach	Rating HT LT	Type	Length
NT727	Dhayirri (E)	1 2	R+LTT	4.9 km
NT728	Ganawa Pt (W)	1 2	R+LTT+rocks	2.7 km
NT729	Ganawa Pt (E1)	1 2	R+LTT	300 m
NT730	Ganawa Pt (E2)	1 2	R+rock flats	400 m
NT731	Watdagawuy	1 2	R+rock flats	4.2 km
NT732	Watdagawuy (E1)	1 2	R+rock flats	800 m
NT733	Watdagawuy (E2)	1 2	R+tidal flats	1.1 km
Spring & neap tide range = 3.5 & 1.5 m				

To the east of Dhayirri are two northwest-facing embayed beaches (NT 727-728) that terminate at Ganawa Point, 5.5 km to the northeast. Beach **NT 727** begins on the eastern side of the tidal creek that exits adjacent to the Dhayirri community. The beach curves round for 4.8 km, the first 3.5 km backed by a 1.3 km wide beach-foredune ridge plain, the latter by 10 m high vegetated bluffs. In addition to the western creek, a second smaller creek drains across the centre of the beach. The beach is initially fronted by 1 km wide tidal shoals adjacent to the inlet, then a continuous attached bar that is occasionally cut by rips and finally by scattered rock flats below the bluffs. A vehicle track parallels the back of the beach down to the main creek mouth.

Beach **NT 728** occupies the second shallow embayment. It is 2.7 km long and for the most part backed by a 10-15 m high grassy foredune, with a small freshwater wetland behind and a creek draining across the northern end of the beach. Low bluffs back the beach north of the creek and scattered rock flats are located along the northern section of beach. The bluffs terminate at Ganawa Point, and a vehicle track parallels the back of the entire beach.

At Ganawa Point the coast turns and trends east toward Gitan Point. The point consists of low laterite bluffs over sandstone rock flats. Five hundred metres to the east is beach **NT 729**, a curving 300 m long beach bordered by rock flats extending 100 m off each end, with the beach continuing along the back of the eastern rock flats. Immediately east is beach **NT 730**, a crenulate 400 m long high tide beach fronted by irregular rock and rock flats, with a low grassy area of beach ridges behind.

The small **Watdagawuy** community (two houses) is located in the centre of beach **NT 731**. The 4.2 km long beach occupies an open north-facing embayment, with a 1 km long central section consisting of a low barrier and backing 2 km long shallow lagoon, drained by a small creek. Most of the beach is however dominated by rocks and rock flats which extend up to 300 m off the beach. Vehicle tracks radiate out from the community and back all the adjoining beaches.

Beach **NT 732** begins at the eastern end of Watdagawuy beach, where the rocks and shore curve round to face northeast. The beach extends for 800 m to a second inflection in the rocks. It consists of a high tide sand beach fronted by continuous rock flats that extend up to 300 m off the beach.

Beach **NT 733** begins in lee of the inflection and runs to the east for 1.1 km, terminating in lee of a sand spit which is part of the next beach. A tidal creek runs along the front of the beach, with the creek and tidal sand shoals extending a few hundred metres offshore.

NT 734-741 GITAN POINT-DHARRWAR

No.	Beach	Rating HT LT	Type	Length
NT734	Gitan Pt (W)	1 2	R+LT rips	3.7 km
NT735	Gitan Pt	1 2	R+rock flats	100 m
NT736	Gitan Pt (E1)	1 2	R+rock flats	200 m
NT737	Gitan Pt (E2)	1 2	R+rock flats	400 m
NT738	Gitan Pt (E3)	1 2	R+LT rips	8.8 km
NT739	Dharrwar (W3)	1 2	R+tidal shoals	900 m
NT740	Dharrwar (W2)	1 2	R+rock flats	500 m
NT741	Dharrwar (W1)	1 2	R+rock flats	900 m
Spring & neap tide range = 3.5 & 1.5 m				

Gitan Point is a 20 m high, 500 m long laterite point, which has been eroded to expose the brilliant red over white colours of the soil. A vehicle track runs along the perimeter of the cliffs linking the beaches to either side with the main north-south island road.

Beach **NT 734** lies immediately south of the point and is a straight, northwest–facing, 3.7 km long beach bordered in the south by an elongate sand point that terminates at the tidal creek. Tidal shoals lie off the creek mouth and spit, with the remainder of the beach having a continuous attached bar that is cut by rips during higher wave conditions. The summer northwest winds have also partly mobilised the foredune, with some dunes reaching a height of 10 m. This is in turn backed by a grassy foredune ridge plain up to 400 m wide.

Three beaches are located around the point. Beach **NT 735** is a 100 m long double pocket of high tide sand, located on the western side of the point at the base of the bluffs, with rock flats dominating the intertidal area (Fig. 5.89). On the north side of the point are beaches **NT 736** and **NT 737**, two neighbouring 200 m and 400 m long high tide strips of sand. They are also backed by red bluffs, with rocks and rock flats lying off both beaches.

The longest beach on the island lies just north of the point. The 8.8 km long beach **NT 738** faces northwest and runs straight between the 300 m wide rock flats off the point, to a tidal creek that forms the northern boundary. The beach is initially backed by 20 m high vegetated bluffs, then for most of its length by active sand dunes extending up to 400 m inland and reaching heights of 20 m, then a mangrove-lined 200 ha tidal system behind the northern 2 km of beach. The beach itself receives moderate waves which maintain a continuous attached bar, which is cut by rips during periods of higher wind waves.

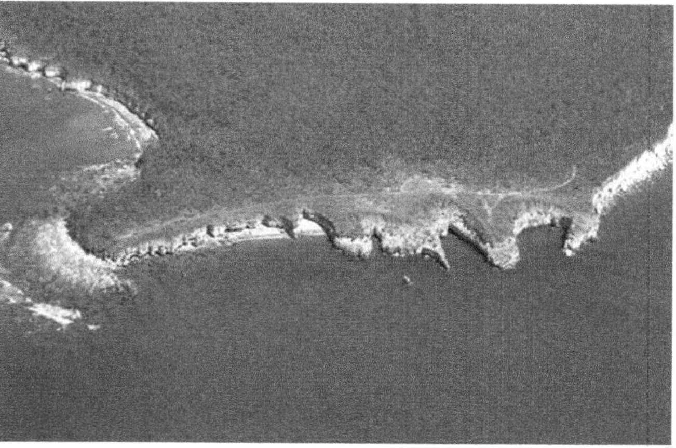

Figure 5.89 Beach NT 735 is located at the base of 20 m high red laterite bluffs.

Three beaches (NT 739-741) are located on the northern side of the inlet. The first, **NT 739**, lies in lee of the tidal shoals and is a dynamic inlet-mouth beach, backed by low foredunes. The 900 m long beach terminates at the beginning of extensive intertidal rock flats, extending 300-400 m off the shore. Beach **NT 740** is located between the first two prominent rock flat protrusions, with no section completely clear of rocks. Its northern neighbour, beach **NT 741**, is a similar 900 m long beach also dominated by the rock flats. A vehicle track runs along the low foredunes behind the beaches.

NT 742-748 REFUGE BAY-NANINGBURA POINT

No.	Beach	Rating HT LT	Type	Length
NT742	Refuge Bay (1)	1 2	R+ rock flats	300 m
NT743	Refuge Bay (2)	1 2	R+rock flats	400 m
NT744	Banthula	1 2	R+sand flats+rocks	7.3 km
NT745	Refuge Bay (4)	1 2	R+tidal flats	1.5 km
NT746	Refuge Bay (5)	1 2	R+rock flats	2 km
NT747	Naningbura Pt (S1)	1 2	R+rock flats	1.8 km
NT748	Naningbura Pt	1 2	R+rock flats	600 m
Spring & neap tide range = 3.5 & 1.5 m				

Refuge Bay is a northwest-facing 6.5 km wide embayment, bordered in the south by a bluffed headland that extends 1.5 km to the northwest, and in the north by the rock-fringed Naningbura Point. The bay contains five beaches (NT 742-746) with the small Dharrwar and Banthula communities located in lee of the main beach.

Beaches **NT 742** and **NT 743** are located in two scallops in the southern headland and face north. They are 300 m and 400 m long respectively, backed by scarped red 5-10 m high bluffs then dense woodlands. They are bordered by rocks and rock debris and fronted by a mixture of sand and rock flats.

The two-house **Dharrwar** community is located at the very southern end of beach **NT 744**, a 7.3 km long northwest-facing beach. In the south the beach is backed by vegetated 5-10 m high bluffs, with sand flats and scattered rocks and reefs off the beach. The **Banthula** community (six houses) is located behind the centre of the beach, with a foredune ridge plain extending from the community to the northern boundary at a 300 m wide tidal creek. The sand flats widen toward the creek where they are over 1 km wide (Fig. 5.90).

Figure 5.90 Beaches NT 744 (right) and NT 745 (left) are located either side of a tidal creek draining into Refuge Bay.

Beach **NT 745** lies on the north side of the creek and is a curving protected west through to south-facing beach, partly fringed by mangroves and fronted by the tidal channels and shoals, with a 250 ha mangrove-filled lagoon behind. It terminates at the beginning of the rocks that dominate the northern coast of the island.

Beach **NT 746** is a high tide sand beach, backed by dense woodlands, and fronted by continuous though irregular rock flats and rock reefs, that in places extend 1.5 km off the beach. The beach initially faces southwest, curving round to face west. It terminates at a 1 km long mangrove-dominated rock section.

On the northern side of the mangroves is beach **NT 747**, a slightly curving 1.8 km long northwest-facing sand beach, backed by a low foredune, then woodlands, and fronted by continuous though irregular rock flats up to 200 m wide. Two houses linked by a vehicle track to the main island track are located in lee of the centre of the beach.

Naningbura Point is the northernmost tip of the island and consists of an outward curving 600 m long sand beach (**NT 748**), backed by a low foredune and fronted by continuous patchy rock flats.

NT 749-755 GAWA-STRETTON STRAIT

No.	Beach	Rating HT	Rating LT	Type	Length
NT749	Gawa (W)	1	2	R+sand flats+rocks	2.8 km
NT750	Gawa	1	2	R+rock-sand flats	2.7 km
NT751	Stretton Strait (S1)	1	3	R+rock flats	1.1 km
NT752	Stretton Strait (S2)	1	2	R+rock flats	150 m
NT753	Stretton Strait (S3)	1	2	R+sand-rock flats	1 km
NT754	Stretton Strait (S4)	1	2	R+sand-rock flats	250 m
NT755	Stretton Strait (S5)	1	2	R+sand flats+rocks	800 m
Spring & neap tide range = 4.1 & 2.8 m					

The northern end of Elcho Island faces across Stretton Strait to Graham Island. The 5 km long strait decreases in width from 3.5-1 km, with the channel narrowing to 200 m at its eastern entrance. The channel is bordered on both sides by extensive intertidal rock and sand flats. Two longer beaches (NT 749 & 750) form the north shore of the island and are accessible by vehicle, with another five east-facing beaches (NT 751-755) lining the eastern side of the strait and not accessible by vehicle. This section of coast and all the beaches are dominated by rocks and rock flats.

Beach **NT 749** lies on the eastern side of a tidal creek that flows out beside Naningbura Point, draining a backing 50 ha mangrove and salt-flat-filled lagoon. The crenulate 2.8 km long north-facing beach is backed by a series of beach ridges to recurved spits that have partly enclosed the backing lagoon. Tidal shoals extend over 1 km off the creek mouth, with a mixture of sand and some rock flats fronting the remainder of the beach. It terminates in the east at a prominent cuspate foreland formed in lee of shallow reefs and flats that extend over 1 km off the shore.

The small **Gawa** community (three houses) is located on the beachrock-dominated eastern side of the foreland. Beach **NT 750** begins adjacent to the community and curves round for 2.7 km to the lee of a low, rocky, mangrove-covered island, adjacent to the narrowest part of the strait. The beach faces north and is fronted by near continuous irregular rock flats, which extend up to 1 km into the strait.

Beach **NT 751** begins in lee of the low island, and trends to the southeast as a crenulate high tide beach dominated by irregular rock flats, with a foredune and some minor dune activity backing the southern half of the beach. In addition tidally driven sand shoals parallel the beach, a product of strong tidal current running through the strait and just off the beach. The beach terminates at a protruding 100 m long low laterite point with a few mangroves growing on the rocks. Beach **NT 752** is located on the tip of the point. It consists of a 150 m long strip of sand bordered by low laterite rocks and reefs, with waves breaking over the rocks at high tide.

Immediately to the south is 1 km long beach **NT 753**, an east-facing crenulate beach consisting of two sandy sections separated and bordered by low rocky points and reefs, with a 5-10 m high vegetated foredune backing the beach. It terminates against a section of vegetated 10-15 m high bluffs. These continue for 1 km to beach **NT 754**, which consists of three narrow strips of sand, totalling 250 m in length, each bordered by protruding low rock flats, and backed by densely vegetated bluffs and the coastal plain.

Beach **NT 755** begins immediately to the south and is a curving, east-facing, 800 m long sand beach. It is backed by dense vegetation and bordered by low rocky points, with a stand of low mangroves on the southern rocks, while it is fronted by a mixture of sand and some scattered rocks and rock reefs.

NT 756-764 **ELCHO ISLAND (E)**

No.	Beach	Rating HT	Rating LT	Type	Length
NT756	Elcho Is (750)	1	2	R+sand flats+rocks	300 m
NT757	Elcho Is (757)	1	2	R+sand flats+rocks	600 m
NT758	Elcho Is (758)	1	2	R+sand flats+rocks	200 m
NT759	Elcho Is (759)	1	2	R+sand flats+rocks	1.7 km
NT760	Elcho Is (760)	1	2	R+sand flats+rocks	2 km
NT761	Elcho Is (761)	1	2	R+sand flats+rocks	400 m
NT762	Elcho Is (762)	1	2	R+sand flats+rocks	1.3 km
NT763	Elcho Is (763)	1	2	R+sand flats+rocks	1.1 km
NT764	Elcho Is (764)	1	2	R+sand flats+rocks	1.6 km
Spring & neap tide range = 4.1 & 2.8 m					

The more exposed northeast coast of Elcho Island is undeveloped with only two access tracks and no communities. While the 20 km of coast between beaches NT 756 and 771 face squarely into the prevailing southeast trades, it is sheltered to some degree by Alger Island, and then the Napier Peninsula, finally grading into the mangrove-lined shores of the Cadell Strait. The 16 beaches (NT 756-771) along this section are all dominated to varying degrees by the sandstone bedrock, which forms all the boundaries and lies off most of the beaches as intertidal rock flats, rocks and reefs. In the 11 km long section between beaches NT 756 and 764 there is only one access track to beaches NT 763 and 764, with none of the others accessible by vehicle.

Beach **NT 756** is located on the southern side of the eastern tip of the island, a 10 m high densely vegetated rocky point (Fig. 5.91). It is a moderately steep, slightly curving, southeast-facing 300 m long beach, bordered by rocky points, and backed by dense wind-trimmed vegetation. It is fronted by sand flats together with a scattering of rocks and rock reefs. Beach **NT 757** lies 100 m further west and is a 600 m long narrow high tide beach, backed by dense vegetation, and fronted by sand flats, rocks and several rock reefs, with narrow rock flats at the southern end. Beach **NT 758** lies in the next small south-facing embayment. It is a 200 m long narrow strip of high tide sand, backed by scarped bluffs, capped by dense wind-blasted vegetation. It is protected by rocks and reefs off the beach, with mangroves growing on rocks at either end and in the centre of the beach.

Figure 5.91 The eastern tip of Elcho Island with beach NT 755 to the north, and beaches NT 756 & 757 in foreground.

Beach **NT 759** begins in lee of narrow rock flats, but for much of its 1.7 km is an open southeast-facing sand beach fronted by 100 m wide sand flats. It is backed by densely vegetated slopes. One kilometre further south is beach **NT 760** a slightly curving 2 km long east-facing sand beach, backed by a 10 m high vegetated foredune. The beach is fronted by narrow sand flats, and occasional rock flats. It terminates at a low, dark, 50 m wide rock point, on the southern side of which is 400 m long beach **NT 761**. This beach is bordered by small rock points, backed by a low vegetated foredune, with rock reefs off the centre and southern end of the beach. The southern reefs enclose a small 'lagoon' and permit mangroves to grow in the lagoon and on the point.

Beach **NT 762** begins on the south side of the small point, and continues relatively straight, but crenulate for 1.3 km. It faces southeast and is backed by a low foredune then dense vegetation. It terminates at an inflection in the shore, beyond which is 1.1 km long beach **NT 763**. This is a rock- and reef-dominated beach that faces into the trades, and is backed by wind-blasted vegetation covering both a low foredune and backing bluffs. A vehicle track from the nearby Djurranalpi community reaches the southern end of the beach.

Beach **NT 764** begins as the rocks diminish and is a relatively straight, southeast-facing 1.6 km long beach. Its exposure to the trades has resulted in minor dune transgression, most likely in the early Holocene. The dunes extend up to 400 m inland and reach a height of 19 m, but are now completely vegetated. A 100 ha area of tropical rain forest lies protected behind the higher dunes.

NT 765-771 ELCHO ISLAND (E)

No.	Beach	Rating HT	LT	Type	Length
NT765	Elcho Is (765)	1	2	R+sand flats+rocks	600 m
NT766	Elcho Is (766)	1	2	R+sand flats	800 m
NT767	Elcho Is (767)	1	2	R+rocks	1.7 km
NT768	Elcho Is (768)	1	2	R+sand flats+rocks	300 m
NT769	Elcho Is (769)	1	2	R+sand flats+rocks	200 m
NT770	Elcho Is (770)	1	2	R+sand flats	150 m
NT771	Elcho Is (771)	1	2	R+ridged sand flats	3.3 km
Spring & neap tide range = 4.1 & 2.8 m					

The 10 km of east coast between beaches NT 765 and 771 represent a transition from the more exposed beaches north of the Napier Peninsula, into the low energy Cadell Strait in lee of the peninsula. This section of coast is backed by bluffs and slopes rising to 40 m. The higher slopes afford greater protection to the backing vegetation, as a consequence of which a 500 ha rain forest parallels the back of the beaches. There is no development on this section of shore and only one access track to the southernmost beach.

Beach **NT 765** lies 100 m south of its northern neighbour. It consists of a 600 m long double arcuate white sand beach, with rock bordering each end as well as outcropping in the centre, together with scattered rocks off the northern half of the beach. It is backed by 30 m high densely vegetated bluffs.

Beach **NT 766** lies 200 m further south and is an 800 m long slightly curving southeast-facing beach, well exposed to the trade winds. It has a foredune which in places has climbed up the 20 m high backing vegetated bluffs. Two hundred metre wide sand flats lie off the beach together with a scattering of rock, with low mangrove-covered rocky points bordering each end. Beach **NT 767** lies immediately to the south and is a 1.7 km long slightly curving sand beach, backed by a low foredune and vegetated 10 m high bluffs and fronted predominantly by sand flats, containing a scattering of rocks and rock reefs. At the southern end of the beach, the rocks increase sufficiently to form a boundary with beach **NT 768,** which consists of a 300 m long strip of sand littered with boulders on and off the beach, while it is backed by a central climbing foredune and grassy 10 m high bluffs.

Beach **NT 769** is located just beyond an inflection in the shore and is a 200 m long pocket of sand bounded and backed by 10 m high grassy bluffs. Beach **NT 770** lies 200 m to the south and is a similar 150 km long pocket beach, with some older vegetated dunes on top of the backing bluffs.

Beach **NT 771** is the longest beach on the east coast and represents the terminus for sand moving down the shore into the low energy waters of the strait. The 3.3 km long beach begins below 10 m high bluffs fronted by narrow sand flats and a few rocks. The sand flats widen to the south, reaching 1 km off the southern end, and become increasingly covered with low sand ridges (Fig. 5.92). The southern kilometre of the beach is backed by a 500 m wide series of beach ridges which grade to the south into terminal spits, separated by mangroves, and finally into mangroves on the strait shore. A vehicle track reaches the beach just above the beach ridge section.

Figure 5.92 The widening sand flats, shown here at high tide, towards the southern end of beach NT 771.

NT 772-778 NAPIER PENINSULA (N)

No.	Beach	Rating HT	LT	Type	Length
NT772	Napier Pens. (W1)	1	2	R+rock flats	1 km
NT773	Napier Pens. (W2)	1	2	R+rock flats	200 m
NT774	Napier Pens. (W3)	1	2	R+rock flats	150 m
NT775	Napier Pens. (W4)	1	2	R+rock flats	2.9 km
NT776	Pt Napier (1)	2	3	R+LTT+rocks	30 m
NT777	Pt Napier (2)	2	3	R+rock flats	50 m
NT778	Pt Napier (3)	2	3	R+rock flats	30 m
Spring & neap tide range = 4.1 & 2.8 m					

The **Napier Peninsula** is a 40 km long northeast-trending sandstone peninsula, which parallels Elcho Island to the north and Flinders Peninsula to the south. The peninsula is 10 km wide at its base, narrowing to 500 m towards its northern tip at Point Napier. Much of the peninsula shore lies protected in Cadell Strait on the north side and in Buckingham Bay on the south side, with most of the shoreline dominated by mangroves. Only the northern tip and parts of the more exposed south shore receive sufficient wave energy to expose the stark sandstone cliffs and to build beaches in crenulations in the bedrock. There is access to the southern section of the peninsula where the small Garriyak and Bularring communities are located, but no vehicle access to the northern 20 km.

Most of the north side of the peninsula is lined with the mangroves of Cadell Strait. Only as the strait widens and finally terminates is sufficient wave energy received to maintain four beaches (NT 772-775) along the northwestern shore. Beach **NT 772** commences 5 km southwest of Point Napier. It is a crenulate 1 km long northwest-facing high tide sand beach fronted by continuous narrow rock flats, with mangroves on the rock flats at each end. Beach **NT 773** lies on the north side of the northern clump of mangroves. It is a straight 200 m long beach bordered by rock flats, with deeper rocks off the beach. Woodlands back both beaches. Three hundred metres to the north is the narrow beach **NT 774**, a strip of high tide sand located below 5 km high scarped bluffs, with rock debris on the beach and dense woodlands on the bluffs.

Beach **NT 775** is a 2.9 km long crenulate northwest-facing beach that continues right up to Point Napier. It consists of a high tide strip of sand with a continuous 50 m wide strip of intertidal rock flats.

Point Napier, the tip of the peninsula, is an 800 m wide jagged sandstone shore containing three small pockets of sand, backed by stunted wind-blasted vegetation. Beach **NT 776** is a 30 m strip of high tide sand fronted by a 40 m wide, 100 m long gap between two rock platforms. Beach **NT 777** is adjacent and is a 50 m long beach with a sandy low tide bar, bordered by rock platforms extending up to 150 m off the beach, which enclose a small 100 m wide bay. It offers the only possible landing on the tip of the peninsula. Beach **NT 778** is located 300 m to the south and is another 30 m pocket of sand almost completely encased in boundary rock platforms and rock reefs, forming a small rocky 'lagoon' off the beach.

NT 779-784 **NAPIER PENINSULA (S 1)**

No.	Beach	Rating HT	Rating LT	Type	Length
NT779	Napier Pens. (S1)	1	2	R+rock flats	100 m
NT780	Napier Pens. (S2)	1	2	R+rock flats	50 m
NT781	Napier Pens. (S3)	1	2	R+rock flats	50 m
NT782	Napier Pens. (S4)	1	1	R+sand flats+rocks	600 m
NT783	Napier Pens. (S5)	1	1	R+sand flats	250 m
NT784	Napier Pens. (S6)	1	1	R+sand flats	1.8 km
Spring & neap tide range = 4.1 & 2.8 m					

The southern shores of the Napier Peninsula are dominated by the deeply jointed sandstone, which extends continuously for 6 km southwest of Point Napier. Only within embayments in the bedrock has sand accumulated to form beaches. The first such embayment is a 4 km wide southeast-facing bay containing six beaches (NT 779-784).

Beaches **NT 779, NT 780** and **NT 781** are all located just inside the northern rocky point of the bay. They face northwest into the bay and are sheltered from the easterly waves by the point and rock reefs. They are 100 m, 50 m and 50 m long respectively, bordered by low sandstone points, with jointed rocks dominating the intertidal zone. Beach **NT 782** lies in the northern corner of the bay and is a slightly curving 600 m long south-facing beach, bordered by mangrove-fringed bedrock (Fig. 5.93). It contains a solitary 10-15 m high foredune, and is fronted by 500 m wide sand flats, that almost fill this northern section of the bay.

Figure 5.93 Beach NT 782 (centre) with the small beaches NT 779-781 to right.

Two kilometres to the west is beach **NT 783**, a 250 m long strip of south-facing sand located below steep, vegetated 10 m high bluffs, with a mixture of sand and rock flats extending 200 m off the beach.

The western corner of the bay is occupied by the curving 1.8 km long beach **NT 784**, with a prominent 15 m high sandstone headland forming the southern boundary border. This beach is exposed to easterly waves and consists of a white sand beach backed by a series of low foredunes and a small freshwater swamp. Sand flats extend 300 m off the beach.

NT 785-796 **NAPIER PENINSULA (S 2)**

No .	Beach	Rating HT	Rating LT	Type	Length
NT785	Napier Pens. (S7)	1	2	R+rock flats	100 m
NT786	Napier Pens. (S8)	1	2	R+rock flats	100 m
NT787	Napier Pens. (S9)	1	2	R+LTT+rocks	250 m
NT788	Napier Pens. (S10)	1	2	R+LTT	250 m
NT789	Napier Pens. (S11)	1	2	R+rock flats	100 m
NT790	Napier Pens. (S12)	1	2	R+LTT	200 m
NT791	Napier Pens. (S13)	1	2	R+LTT	300 m
NT792	Napier Pens. (S14)	1	2	R+LTT+rocks	250 m
NT793	Napier Pens. (S15)	1	2	R+LTT	300 m
NT794	Napier Pens. (S16)	1	2	R+LTT	200 m
NT795	Napier Pens. (S17)	1	2	R+rock flats	50 m
NT796	Napier Pens. (S18)	1	2	R+rock flats	100 m
Spring & neap tide range = 4.1 & 2.8 m					

Ten kilometres southwest of Point Napier and beginning immediately south of the beach NT 784, exposed 30-50 m high heavily jointed sandstone bedrock dominates a relatively straighter section of coast for the next 30 km down to the wide tidal flats of Buckingham Bay. The sandstone has been incised by numerous small streams, which parallel the north-south joint lines and form small valleys. Several of these valleys have been partly infilled with small beaches. This section describes 12 beaches (NT 785-796) along an 8 km long section that begins

5 km south of the bay. All are bordered by sandstone headlands and all face toward the southeast.

Beaches **NT 785** and **NT 786** are two neighbouring sand beaches both 100 m long and occupying small valleys, separated by a 200 m long 20 m high headland. They are bordered by 20 m high sandstone cliffs and rock platforms, with rock debris in the intertidal off both beaches.

One kilometre to the south is beach **NT 787**, a 250 m long sand beach backed by a low vegetated foredune, with the northern headland affording some protection from easterly waves. Thirty metre high cliffs form the south headland. On the south side of the headland is beach **NT 788**, a 250 m long sand beach backed by two small converging valleys, with rock platforms and a 15 m high headland to either side, the platforms on the southern head protruding 200 m seaward. Beach **NT 789** lies on the south side of the headland and is a 100 m long sliver of high tide sand at the base of 20 m high cliffs fronted by rock flats.

Beach **NT 790** lies 300 m further south in a 200 m wide valley. It is a straight beach backed by vegetated slopes, with rock platforms and 15 m high headlands to either side. The next valley south is occupied by the more exposed beach **NT 791**, a 300 m long sandy beach backed by an active foredune rising to 10 m in height, with three small streams converging on the beach. It has rock platforms and 10 m high cliffs to either end. Three hundred metres to the south is a small rock islet connected to the mainland by a narrow 200 m long rock reef. On the south side of the reef is beach **NT 792**, a relatively straight 250 m long sand beach, initially fronted by rock flats, then more open toward the southern end. A small headland at the southern end separates it from beach **NT 793**. This is a straight 300 m long beach, with a small central foredune, all backed by 15 m high cliffs.

Five hundred metres to the south is a prominent straight creek that runs east to the coast forming a small valley occupied by beach **NT 794**. This is a more exposed 200 m long beach with a 50 m wide surf zone, backing low foredune and the creek crossing the northern end. It is bordered by sloping rocks and rock platforms. One hundred metres to the south is the small 50 m long pocket of sand (beach **NT 795**) located in a gap in the bordering rock platforms.

Beach **NT 796** is located another 2.5 km further south on the western side of a more prominent 10 m high headland. The beach faces east and consists of a sand and cobble high tide beach, fronted by a cobble to boulder surf zone, with a small foredune separating the beach from the sandstone slopes behind.

NT 797-804 **NAPIER PENINSULA (S 3)**

No.	Beach	Rating HT	Rating LT	Type	Length
NT797	Napier Pens. (S19)	1	2	R+rock flats	100 m
NT798	Napier Pens. (S20)	1	1	R+LTT	200 m
NT799	Napier Pens. (S21)	1	1	R+LTT+rocks	250 m
NT800	Napier Pens. (S22)	1	1	R+sand flats	800 m
NT801	Napier Pens. (S23)	1	1	R+sand flats	400 m
NT802	Napier Pens. (S24)	1	1	R+sand flats	80 m
NT803	Napier Pens. (S25)	1	1	R+sand flats	60 m
NT804	Napier Pens. (S26)	1	1	R+sand flats	400 m
Spring & neap tide range = 4.1 & 2.8 m					

Beach NT 797 lies on the southern side of the V-shaped rocky headland that separates it from Beach NT 796. Between this point and the tidal flats of Buckingham Bay are 14 km of southeast-facing sandstone shoreline, rising to a 50 m high sandstone plateau behind. The shoreline is dominated by bedrock, apart from a 6 km long section south of the point that contains eight small headland-bound beaches (NT 797-804).

Beach **NT 797** is located immediately in lee of the point and is a protected south-facing strip of high tide sand and cobbles, fronted by a continuous 50 m wide rock flat, and backed by vegetated sandstone slopes. Five hundred metres to the west is beach **NT 798**, a straight 200 m long sand beach occupying a small valley, with a creek draining across the northern end. It is backed by a small foredune and fronted by a shallow attached low tide bar. One hundred metres to the south is beach **NT 799**, a similar straight 250 m long beach, with rocks dominating each end of the beach, and only a smaller central sand bar section. It is backed by sandstone slopes, with rock platforms and 10 m high cliffs to either end.

Beach **NT 800** is bounded by two prominent 20 m high sandstone headlands. The 800 m long beach is the longest beach on this rocky section of shore. It faces east across the bay and is backed by sandstone slopes, with a few patches of low foredune. It is bordered by rock platforms, with sand flats extending 400 m offshore.

Beaches NT 801-804 all occupy the next 1.5 km long bay and are all fronted by 300 m wide sand flats. Beach **NT 801** is a 400 m long south-facing sand beach, with a low foredune and some minor dune activity behind. It is bordered by low sandstone slopes, with mangroves fringing the eastern rocks and a rock outcrop toward the western end of the beach. Beach **NT 802** is located 200 m to the west and is an 80 m long strip of high tide sand wedged in between the low rocky shoreline. Three hundred metres further west is beach **NT 803**, a similar 60 m long strip of sand. The southernmost beach on the peninsula is beach **NT 804**, a 400 m long east-facing sandy beach backed by sloping sandstone, and a low foredune.

NT 805-814 BUCKINGHAM BAY (E)

No.	Beach	Rating HT LT	Type	Length
NT805	Warranyin (S)	1 1	R+sand flats	250 m
NT806	Warranyin	1 1	R+tidal flats+rocks	400 m
NT807	Warranyin (E)	1 1	R+tidal flats	700 m
NT808	Buckingham Bay (E4)	1 2	R+tidal flats+rocks	500 m
NT809	Buckingham Bay (E5)	1 2	R+tidal flats+rocks	150 m
NT810	Buckingham Bay (E6)	1 2	R+tidal flats+rocks	100 m
NT811	Buckingham Bay (E7)	1 2	R+tidal flats+rocks	100 m
NT812	Buckingham Bay (E8)	1 2	R+rock flats	50 m
NT813	Buckingham Bay (E9)	1 2	R+rock flats	250 m
NT814	Buckingham Bay (E10)	1 2	R+sand flats	300 m
Spring & neap tide range = 4.1 & 2.8 m				

Buckingham Bay is a large, deep, U-shaped bay bordered by the Napier and Flinders peninsulas. The bay is 11 km wide at the entrance and extends to the southwest for 20 km. The boundary arms of the peninsulas are dominated by sandstone bedrock shore and generally small pocket beaches, such as beaches NT 779-804 (above). The southern end of the bay however consists of a 25 km long section of mangrove-fringed tidal flats backed by approximately 23 000 ha of salt flats and wetlands. They are drained by the Buckingham River, a large tidal creek. There is very limited vehicle access to the bay shore, with the only communities nearby being the Bularring community on the Napier Peninsula and Garrata community on the Flinders Peninsula. The eastern shoreline is dominated by the generally rugged sandstone, which rises to heights of 30-60 m behind the shore. The sandstone has been heavily dissected and contains numerous drowned valleys and bays. The 15 km of shoreline between the Warrawuruwoi River, the easternmost tidal creek in the bay, and the Kurala River contain 15 beaches (NT 805-819).

Beach **NT 805** is located immediately east of the first small sandstone headland on the eastern side of the bay. It is a 250 m long northwest-facing sand beach, backed and bordered by sandstone slopes, with mangroves growing along the eastern end of the beach. It is fronted by 400 m wide sand flats. Immediately to the north is the small Warranyin headland, where the only access track on the eastern side of the bay terminates. A 1.5 km wide bay on the eastern side of the headland contains two beaches.

Beach **NT 806** is a 400 m long low energy beach, bordered by the headland and a low rocky point, with rocks and mangroves scattered along its eastern half. A storm-deposited coral-cobble ridge backs the sand beach, with vegetated slopes behind. On the eastern side of the rocks is 700 m long beach **NT 807**, the outermost ridge in a 400 m wide series of beach ridges that have filled the backing valley (Fig. 5.94). The beach is bounded in the east by a small tidal creek, draining a 100 ha of salt flats.

Figure 5.94 Low energy, mangrove-fringed beaches NT 806 & 807.

Beach **NT 808** is located along the rocky shore that extends north of the creek. It is a crenulate, low energy, northwest-facing high tide beach, backed by a coral-cobble ridge, and fronted by scattered mangroves and rock flats.

On the eastern side of the point is the next bay. It is 1 km wide, faces north with most of its shoreline fringed by mangroves. On the eastern headland are two pockets of sand (beaches **NT 809** & **NT 810**), which are 150 m and 100 m long respectively. They both face west, are bordered by low mangrove-covered rocky points and are fronted by tidal flats.

The next bay to the east is a similar 1.5 km wide mangrove-lined embayment, with beaches restricted to the west-facing shores of the eastern headland. Beaches **NT 811**, **NT 812** and **NT 813** are neighbouring 100 m, 50 m and 250 m long pockets of high tide sand, each bordered by low rocky points, with rocks scattered along the tidal flats and a few mangroves on the points.

Beach **NT 814** is located on the tip of the headland. It is a 300 m long north-facing sand beach bordered by low rocky points and backed by a coral-cobble ridge. It has a few rock outcropping along the beach and is fronted by an attached low tide bar.

NT 815-823 BUCKINGHAMS BAY-KURALA RIVER

No.	Beach	Rating HT LT	Type	Length
NT815	Buckingham Bay (E11)	1 1	R+sand flats	300 m
NT816	Buckingham Bay (E12)	1 1	R+sand flats	200 m
NT817	Buckingham Bay (E13)	1 1	R+sand flats	50 m
NT818	Buckingham Bay (E14)	1 2	R+rock flats	100 m
NT819	Buckingham Bay (E15)	1 2	R+rock flats	100 m
NT820	Kurala R (W1)	1 2	R+rock flats	100 m
NT821	Kurala R (W2)	1 1	R+sand flats	50 m
NT822	Kurala R (W3)	1 1	R+sand flats	50 m
NT823	Kurala R (W4)	1 1	R+sand flats	200 m
Spring & neap tide range = 4.1 & 2.8 m				

The north-south trending Kurala River forms the boundary of the first sandstone section on the eastern shores of Buckingham Bay. Beaches NT 805-823 lie to the west of the river mouth, with the equally rocky Flinders Peninsula shore and beaches to the east.

The western head of the river is a 20 m high sandstone headland, with beaches NT 815-817 located in the first bay west of the head. The 1 km wide northwest-facing bay is dominated by mangroves, with beach **NT 815** located in the southwestern corner. It is a 300 m long strip of high tide sand, bordered by dense mangroves, backed by low wooded slopes and fronted by 500 m wide sand flats. Beach **NT 816** is located in a small embayment on the eastern headland of the bay. It is a 200 m long sand beach, bordered by a rocky point in the west and a corner filled with mangroves in the east. Beach **NT 817** is located just past the mangroves and is a 50 m long pocket of sand wedged in between a low rocky point and the remainder of the headland. Coral-cobble ridges back both beaches (Fig. 5.95).

Figure 5.95 Beaches NT 816 & 817 are both sandy beaches backed by coarse ridges of coral debris.

Outside the bay the shoreline continues on to the northeast for 2 km to the river entrance. Beaches **NT 818** and **NT 819** are located halfway along and consist of adjoining 100 m long beaches, both bordered by low rocky points, with some rock scattered along both beaches. They are backed by wooded slopes rising to 20 m.

The Kurala River beaches are located in a 1 km long east-facing bay just inside the western entrance to the river. Beaches **NT 820**, **NT 821** and **NT 822** are three small pockets of sand, each bounded by sloping rocky points. They are 100 m, 50 m and 50 m long respectively, with the first also fronted by scattered rocks, the latter two by 200 m wide sand flats. At the southern end of the small bay is beach **NT 823**, a 200 m long northeast-facing sand beach, backed by a low foredune, then wooded slopes, and fronted by the sand flats.

NT 824-831 **KURALA RIVER-FLINDERS PENINSULA (N 1)**

No.	Beach	Rating HT	Rating LT	Type	Length
NT824	Kurala R (E1)	1	1	R+sand flats	200 m
NT825	Kurala R (E2)	1	1	R+sand flats	200 m
NT826	Kurala R (E3)	1	1	R+sand flats	100 m
NT827	Kurala R (E4)	1	1	R+sand flats	300 m
NT828	Kurala R (E5)	1	1	R+rock flats	1.4 km
NT829	Flinders Pens. (N1)	1	1	R+rock flats	50 m
NT830	Flinders Pens. (N2)	1	1	R+sand flats	150 m
NT831	Flinders Pens. (N3)	1	1	R+sand flats	200 m
Spring & neap tide range = 4.1 & 2.8 m					

The **Flinders Peninsula** is a 30 km long, north then northeast-trending sandstone peninsula that separates Buckingham and Ulundurwi bays. Ridges of sandstone up to 100 m high form the backbone of the heavily dissected peninsula. The Kurala River flows due north along its western boundary before entering Buckingham Bay. The peninsula is 7 km wide east of the river mouth, its widest point, narrowing to between 1-3 km for the northern 15 km. There is vehicle access to the Garrata community located 2 km south of beach NT 833, with a track to a mangrove-lined bay located 2 km southeast of the community.

The northern shore of the peninsula between the Kurala River and the tip at Flinders Point consists of 35 km of crenulate shoreline containing about 12 small bays, with intervening sandstone points and headlands and a total of 29 beaches (NT 824-852). There is no vehicle access to any of the beaches.

Beaches **NT 824**, **NT 825**, **NT 826** and **NT 827** are four adjoining pocket beaches located along a 1.2 km section of steep wooded slopes which rise to 65 m along the eastern entrance to the Kurala River. The beaches are 200 m, 200 m, 100 m and 300 m long respectively and each is fronted by sand flats, then the deeper river channel. Beaches NT 824 and 827 are backed by small foredunes.

Beach **NT 828** begins 200 m to the north and is a crenulate 1.4 km long strip of high tide sand fronted by a broken scattering of rocks and low mangroves. It wraps around the northern entrance to the bay, facing west then northwest at the rocky northern tip. On the eastern side of the tip are beaches NT 829 and 830, two neighbouring north-facing beaches. Beach **NT 829** is 50 m long, bordered by low sandstone rocks and fronted by a strip of rocks. Beach **NT 830** is 150 m long with 100 m wide sand flats and a prominent 20 m high eastern headland.

On the eastern side of the headland is a 1 km wide bay which is dominated by mangroves, apart from a 200 m long beach (**NT 831**) located in a small embayment on the western side of the bay. The beach faces north and is fronted by 200 m wide sand flats, while it is bordered by the headland and a lower vegetated rocky point, with steep vegetated slopes behind rising to 60 m behind the beach.

NT 832-842 **FLINDERS PENINSULA (N 2)**

No.	Beach	Rating HT	LT	Type	Length
NT832	Flinders Pens. (N4)	1	1	R+sand flats	400 m
NT833	Flinders Pens. (N5)	1	1	R+sand flats	200 m
NT834	Flinders Pens. (N6)	1	2	R+rock flats	50 m
NT835	Flinders Pens. (N7)	1	1	R+sand flats	200 m
NT836	Flinders Pens. (N8)	1	1	R+sand flats	1.7 km
NT837	Flinders Pens. (N9)	1	2	R+rock flats	700 m
NT838	Flinders Pens. (N10)	1	1	R+sand flats	1.3 km
NT839	Flinders Pens. (N11)	1	1	R+sand flats	400 m
NT840	Flinders Pens. (N12)	1	1	R+sand flats	300 m
NT841	Flinders Pens. (N13)	1	2	R+rock flats	100 m
NT842	Flinders Pens. (N14)	1	1	R+sand flats	150 m
Spring & neap tide range = 4.1 & 2.8 m					

The middle section of the northern side of Flinders Peninsula begins at beach NT 832, located 2 km north of the small Garrata community. The coast between here and beach NT 842 consists of 15 km of crenulate north- to northwest-facing bays and bordering headlands and rocky points, with mangroves filling the deeper bays. Wooded slopes rise behind most of the beaches, reaching 60 m in the south and 30 m in the north. There is no formed vehicle access to any of the beaches.

Beach **NT 832** is located on the western headland of the second bay east of the Kurala River mouth. The 400 m long beach faces north and is bordered by the headland to the west and a low wooded point to the east (Fig. 5.96). It is fronted by 300 m wide sand flats and backed by a low foredune then lightly wooded slopes. Beach **NT 833** is located on the eastern side of the bay, with mangroves occupying the protected central section. The beach is a 200 m long northwest-facing strip of sand, backed by a low foredune, with a small rocky point to the west, and a continuation of the eastern headland on the eastern side.

Figure 5.96 Headland-bound beach NT 832.

Beach **NT 834** lies out toward the tip of the headland, a 1.5 km long 200-300 m wide and up to 40 m high ridge of wooded sandstone. The beach is a 50 m pocket of sand located on the eastern side of the headland 200 m south of the northern tip. It is backed by 10 m high rocky bluffs and bordered and fronted by rocks.

The headland forms the western boundary of a 1 km wide 1.5 km deep bay. Beach **NT 835** is located in the southwest corner of the bay, the remainder of the shore dominated by rocks and mangroves. The beach is a 200 m long strip of sand, bordered by mangroves to the east, and fronted by 600 m wide tidal flats. This is the nearest beach to the **Garrata** community, which is located 1 km to the south.

To the east of the bay is a 4 km long more open northwest-facing embayment, containing six beaches (NT 836-841). Beach **NT 836** is a curving, northwest-facing, 1.7 km long beach, bordered by low rocky points. It is backed by a few beach ridges at the southern end, and wooded slopes elsewhere. It terminates at a protruding point, around which is 700 m long beach **NT 837**. This beach is fronted by near continuous rocks and rock flats, with rock reefs extending up to 1 km off the beach and linking it to two small wooded islands.

Beach **NT 838** is a 1.3 km long northwest-facing beach that begins on the northern side of the point. It is backed by one to two vegetated beach ridges and bordered by rocks to the south and mangrove-covered rocks to the north. Wooded slopes back the beach.

Beaches NT 839-841 occupy a 1 km wide embayment at the northern end of the bay. Beach **NT 839** is a 400 m long west-facing beach, bordered by mangroves at either end, backed by a small foredune and fronted by 400 m wide sand flats. Its neighbour, beach **NT 840**, is a similar 300 m long beach, with bordering mangroves at the southern end and a rocky slope at the northern end, which runs out to the northern point of the bay. Just inside the tip of the wooded 20 m high point is beach **NT 841**, a 100 m long pocket of sand that faces south down the bay, with rock and rock flats to either side and rocks off the beach.

Beach **NT 842** lies on the northern side of the point. It is a 150 m long northwest-facing sand beach fronted by 200 m wide sand flats. It is bordered by rock platforms and vegetated slopes, with a small foredune behind the beach.

NT 843-848 **FLINDERS PENINSULA (N 3)**

No.	Beach	Rating HT	LT	Type	Length
NT 843	Flinders Pens. (N15)	1	1	R+ridged sand flats	200 m
NT844	Flinders Pens. (N16)	1	1	R+sand flats+rocks	1.1 km
NT845	Flinders Pens. (N17)	1	1	R+sand flats+rocks	150 m
NT846	Flinders Pens. (N18)	1	1	R+sand flats	300 m
NT847	Flinders Pens. (N19)	1	1	R+sand flats	300 m
NT848	Flinders Pt (W)	1	1	R+sand flats	500 m
Spring & neap tide range = 4.1 & 2.8 m					

The northernmost 5 km of the peninsula up to Flinders Point consist of a series of small embayments containing a total of six generally north-facing beaches (NT 843-848), all fronted by sand flats. There is no vehicle access to this section of the coast.

Beach **NT 843** lies at the base of a 500 m wide, 500 m deep, U-shaped, northwest-facing bay. The bay is almost completely filled with ridged sand flats, with low waves breaking out on the edge of the flats at low tide (Fig. 5.97). The beach is a 200 m long strip of curving sand that lines the inner shore of the bay. Rocks impinge on each side, together with some mangroves lining parts of the bay sides. During higher waves a rip runs out along the southern boundary of the bay.

Figure 5.97 Beach NT 843 occupies a small bay partly filled with intertidal ridged sand flats, shown here at high tide.

On the northern side of the bay entrance is 1.1 km long beach **NT 844**. This north-facing beach lies along the base of vegetated slopes, which rise to 50 m. A low foredune separates the beach from the slopes, while rocks outcrop along much of the beach, and sand flats extend 100 m offshore. It terminates at a small rocky point, on the north side of which is a curving 1.5 km long embayment containing three small beaches (NT 845-847) all linked by 100-200 m wide sand flats. The first is beach **NT 845**, a 150 m long strip of sand, bordered and partly fronted by rocks, with a low foredune then vegetated slopes behind. Beach **NT 846** is a 300 m long sandy beach, bordered by rock platforms. It has two gullies cutting the backing slopes and a low active foredune. Beach **NT 847** is located at the northern end of the bay. It is a 300 m long west-facing beach backed by a 10 m high foredune, then slopes rising to 20 m. Rock flats border each end, the northern rocks continuing as a narrow 10 m high, 300 m long rocky point.

The rocky point forms the western arm of a 500 m wide, semicircular, north-facing embayment that contains beach **NT 848**. The 500 m long beach curves round the southern shore of the bay, with sand flats filling much of the bay, while it is backed by an active foredune, which reaches 15 m in height. The beach is bordered by irregular rock platforms, while the northern tip of the peninsula, Flinders Point, borders the eastern side of the bay.

NT 849-853 **FLINDERS POINT (E)**

No.	Beach	Rating HT	Rating LT	Type	Length
NT849	Flinders Pt (E1)	1	2	R+sand flats+rocks	100 m
NT850	Flinders Pt (E2)	1	1	R+sand flats	200 m
NT851	Flinders Pt (E3)	1	2	R+sand flats+rocks	80 m
NT852	Flinders Pt (E4)	1	1	R+sand flats	80 m
NT853	Flinders Pt (E5)	1	1	R+sand flats	60 m
Spring & neap tide range = 4.1 & 2.8 m					

Flinders Peninsula narrows to a few hundred metres towards its northern tip at **Flinders Point,** a 20 m high vegetated sandstone point. The point forms the western entrance to both Nalwarung Straits and further south the large protected Arnhem Bay. The eastern side of the peninsula trends back toward the southwest into Ulundurwi Bay. Only the first 3 km of shore are exposed to moderate waves as the coast to the south is protected by the linear Probable, Gwakura and Rekala islands.

Six beaches (NT 849-853) are located with 2 km of the tip of the point. Beach **NT 849** is a 100 m long strip of east-facing sand, bordered by rocky shore and backed by the narrowest part of the point, with beach NT 848, 200 m to the west. The rocky shore, with a scattering of low mangroves, continues for 200 m into a small east-facing bay, containing 200 m long beach **NT 850**. This is the most exposed beach on the point and is fronted by 300 m wide sand flats.

The second small bay south of the point contains the remaining three beaches (NT 851-853). Beach **NT 851** is an 80 m long strip of high tide sand fronted by a jumble of intertidal sandstone boulders, with a small foredune behind. One hundred metres further south is beach **NT 852**, a steep 80 m long open moderately sandy beach, with a few large pink sandstone slabs on the centre of the beach (Fig. 5.98) and then 200 m wide sand flats. Immediately south around a small rocky point is beach **NT 853**, a 60 m long pocket of steep sand, that grades into rock and rock flats at its southern end.

Figure 5.98 The steep beach face and sandstone slabs of beach NT 852.

NT 854-864 PROBABLE ISLAND (E)

No.	Beach	Rating HT	Rating LT	Type	Length
NT854	Probable Is (E1)	1	1	R+sand flats	900 m
NT855	Probable Is (E2)	1	1	R+sand flats	300 m
NT856	Probable Is (E3)	1	1	R+sand flats	150 m
NT857	Probable Is (E4)	1	1	R+sand flats+rocks	150 m
NT858	Probable Is (E5)	1	1	R+tidal flats	50 m
NT859	Probable Is (E6)	1	1	R+sand flats	300 m
NT860	Probable Is (E7)	1	1	R+sand flats	300 m
NT861	Probable Is (E8)	1	1	R+rock flats	50 m
NT862	Probable Is (E9)	1	1	R+sand flats	150 m
NT863	Probable Is (E10)	1	1	R+sand flats+rocks	100 m
NT864	Probable Is (E11)	1	1	R+sand flats+rocks	100 m
Spring & neap tide range = 4.1 & 2.8 m					

Probable Island is located 3 km due south of Flinders Point, and lies just 400 m off the peninsula along its northwestern shore. The uninhabited island is 7 km long and between 0.5-2 km wide. It trends north-south with a 60 m high sandstone ridge forming the backbone of the island. The western shore is relatively straight, with steep slopes and a fringe of mangroves. The more exposed eastern shore is dissected by several small streams and valleys, and contains eleven small headland-bound beaches (NT 854-864).

Beach **NT 854** occupies the northeast-facing northern shore of the island. The 900 m long beach is bordered by low rocky points, and backed by dense woodlands, with 300 m wide sand flats off the beach.

The eastern-side beaches commence in lee of the northeastern tip of the island at beach **NT 855**. This is a 300 m long east-facing beach, exposed to the trade winds blowing across the bay. The beach is moderately steep, backed by a low foredune, then wind-blasted vegetation, with 200 m wide sand flats off the beach. One hundred metres to the south is beach **NT 856**, a similar 150 m long beach, also bounded by sloping vegetated headlands.

One kilometre to the south is beach **NT 857**, a 150 m long strip of sand at the base of steep, densely vegetated slopes, with rock platforms to either side and 100 m wide sand and rock flats off the beach.

Beach **NT 858** is located another 1.5 km to the south and lies at the base of a 500 m deep V-shaped valley. The 50 m wide beach fills the back of the valley, with mangroves bordering the base of the boundary slopes, as well as extending several tens of metres up a small creek that drains across the southern side of the beach.

Beach **NT 859** occupies the next small bay to the south. It is a 300 m long east-facing beach, partially protected by an island 300 m off the beach. It is backed and bordered by steep vegetated slopes with mangroves fringing both ends. Sand flats extend out to the rear of the island.

Immediately to the south is a 1.2 km long more open bay containing three beaches (NT 860-862) all fronted by continuous 200 m wide sand flats. Beach **NT 860** is a 300 m long strip of sand at the base of 50 m high wooded slopes. Beach **NT 861** lies 200 m to the south and is a 50 m long pocket of sand occupying a small wooded valley, with 10 m high cliffs to either side and rock flats immediately off the beach. Beach **NT 862** lies in the southern corner of the 'bay' and is a 150 m long beach, bordered by a rock platform and slopes to the north, with mangroves growing on its southern end.

The southernmost 'bay' on the eastern side of the island contains the last two beaches. Beaches **NT 863** and **NT 864** are both 100 m in length, bordered by 10 m high red cliffs and wooded slopes and fronted by low tide rubble rock flats, together with rock reefs located in the centre of the bay (Fig. 5.99).

Figure 5.99 The embayed beaches NT 863 & 864.

NT 865-875 YALLIQUIN-EVERETT ISLAND-RAYMANGIRR

No.	Beach	Rating HT	Rating LT	Type	Length
NT865	Flinders Pens. (S)	1	2	R+rock flats	800 m
NT866	Yalliquin	1	2	R+tidal flats	1.6 km
NT867	Everett Is (N1)	1	2	R+sand flats	3.5 km
NT868	Everett Is (N2)	1	2	R+rock flats	200 m
NT869	Everett Is (N3)	1	2	R+rock flats	1.2 km
NT870	Everett Is (E1)	1	2	R+sand flats	800 m
NT871	Everett Is (E2)	1	2	R+sand flats	1.3 km
NT872	Everett Is (E3)	1	2	R+sand flats	800 m
NT873	Everett Is (E4)	1	2	R+sand flats	700 m
NT874	Raymangirr (N)	1	2	R+tidal flats	1.4 km
NT875	Raymangirr (N)	1	2	R+tidal flats	2 km
Spring & neap tide range = 4.1 & 2.8 m					

The coastline south of the Flinders Peninsula between Probable Island and Mallison Island is occupied by Ulundurwi Bay and the larger Arnhem Bay. Both bays are well protected by the peninsula, several islands and Cape Newbald, resulting in 180 km of a generally low energy mangrove-lined shoreline. The only beaches along this shore are one located on the southern side of the peninsula (NT 865), several around Everett Island (NT 867-873) and the two at Raymangirr (NT 874 & 875). Elsewhere mangroves, salt flats and a few rocky points dominate the shore. The only settlements in the area are at the Yalliquin community on the south side of Ulundurwi Bay, the Raymangirr community on the western shores of Arnhem Bay, the Burrum community located 5 km to the west, and the Rorruwuy community on the northern shore of the bay. The only vehicle access to this section of coast is at the small Yalliquin, Raymangirr and Rorruwuy communities.

Beach **NT 865** is located 11 km southwest of Flinders Point and 4 km west of the southern end of Probable Island. It is the only beach on the southern side of the peninsula south of the cluster just south of the point. The beach is 800 m long, faces southeast, lies at the base of steep 60 m high vegetated slopes and is fronted by 200 m wide rock flats.

Approximately 12 km due south on the eastern side of Ulundurwi Bay is the small **Yalliquin** community. The community is located toward the northern end of a 1.6 km long crenulate west-facing beach (**NT 866**), that is backed by 10 m high bluffs in the north, grading into a series of beach ridges and finally several recurved spits, that terminate amongst mangroves. The beach spits are the terminus for sand moving around Everett Island and south along the shore.

Two kilometres to the north on the other side of an open tidal creek is the western end of beach **NT 867**. This beach forms the northeastern side of Everett Island. This 'island' is separated from the mainland by a 500 m wide band of mangroves. There is an access track from Yalliquin, that runs due east though the narrowest section of mangroves to the 'island'. The beach is 3.5 km long, faces north and curves between an eastern bright red headland and a series of recurved spits in the west. Sand is being eroded from the 10 m high laterite bluffs in the east, moving along the beach around the western point and slowly down towards Yalliquin. Ridged sand flats extend 500 m off the beach.

On the eastern side of the bluffs is a 200 m long pocket of high tide sand (beach **NT 868**). Dense woodland backs the scarped bluffs (Fig. 5.100), with rock flats in the intertidal zone. The beach is bordered by more bluffs to the east, on the eastern side of which is curving, southeast-trending, 1.2 km long beach **NT 869**. It too is backed by the red bluffs and fronted by increasing rock flats to the south. At the eastern end of the beach the bluffs turn due south, with beach **NT 870** extending for 800 m along their base to a southern clump of mangroves and a 20 m high red point.

On the southern side of the point is the straight 1.3 km long east-facing beach **NT 871**. The beach is also backed by red 10 m high bluffs and fronted by rock and sand flats. It terminates at an inflection in the shore, on the south side of which is beach **NT 872**, an 800 m long beach that begins beneath the bluffs, then continues as a series of low beach ridges. This beach and its neighbour (NT 873) represents the terminus for sand transported down the eastern side of the island. Beach **NT 873** curves round as a recurved spit for 700 m, with the spit continuing west in amongst mangroves for another 1 km.

Figure 5.100 Beach NT 868 shown at high tide at the base of bright red bluffs.

Six kilometres to the south is the northern end of beach **NT 874**. This 1.4 km long beach faces southeast across 30 km wide Arnhem Bay. It consists of a low beach and beach ridge and backing salt flats and mangroves, which are drained by a creek which forms the southern boundary of the beach. On the south side of the 50 m wide creek is beach **NT 875**, a 2 km long low sand beach that is initially backed by a few beach ridges, then by low wooded bluffs. The **Raymangirr** community and landing ground is located on the low bluffs above the southern end of the beach.

NT 876-881 **ARNHEM BAY (S)**

No.	Beach	Rating HT	Rating LT	Type	Length
NT876	Arnhem Bay (S1)	1	2	R+rock flats	700 m
NT877	Arnhem Bay (S2)	1	2	R+rock flats	250 m
NT878	Arnhem Bay (S3)	1	2	R+rock flats	900 m
NT879	Arnhem Bay (S4)	1	2	R+rock flats	700 m
NT880	Arnhem Bay (S5)	1	1	R+tidal flats	600 m
NT881	Arnhem Bay (S6)	1	1	R+tidal flats	1.8 km
NT882	Cliffy Pt (N)	1	1	R+tidal flats	1.2 km
Spring & neap tide range = 4.1 & 2.8 m					

Arnhem Bay is one of the larger bays on the north coast of the Northern Territory. The roughly circular bay is up to 45 km across and 30 km deep, with the northern entrance between Everett and Mallison islands 7 km wide. The bay has 150 km of predominantly very low energy shoreline dominated by mangrove-lined tidal flats and only six low energy beaches (NT 876-881). The southern shore is dominated by the mangrove-fringed mouths of the Darwarunga, Habgood, Baralminar, Gobalpa and Goromuru rivers, with the Cato, Peter John and Burungbirinung rivers converging on the northeast corner of the bay. In between are low wooded headlands, with the six southern beaches all located on three of the headlands.

Beaches **NT 876**, **NT 877** and **NT 878** are three neighbouring east-facing beaches located on the northern side of the Habgood River mouth. They are 700 m, 250 m and 900 m long respectively, and are all backed by low wooded coastal plain and fronted by 200 m wide intertidal rock flats, with mangroves growing toward the northern ends of the latter two beaches. The beaches face across the 6 km wide mouths of the Habgood, Baralminar and Gobalpa rivers. On the eastern side of the mouths is a second headland, with beach **NT 879** located along its northern side. The beach is a crenulate 700 m long strip of high tide sand, fronted by 200-300 m wide rock flats and scattered mangroves.

A third headland begins 3 km to the east, with beach **NT 880** located in an open embayment dominated by mangroves. The 600 m long beach occupies a gap in the mangroves, which together with the mangroves fronts a 3.3 km long series of beach ridges and backing salt flats. The entire system is fronted by 300 m wide tidal flats.

Beach **NT 881** is located on the eastern side of the headland and is an outward curving 1.8 km long strip of high tide sand, fronted by tidal flats. Sediment moving down the beach has built a 700 m long spit, presently enclosed in mangroves.

To the east of beach NT 881 is an undulating mangrove-lined shoreline that extends for 7 km to the 500 m wide mouth of the Goromuru River. On the eastern side of the river and for 12 km to Cliffy Point is a 500 m wide band of mangroves, backed by salt flats up to 10 km wide. On the northern side of the 15 m high point is 1.2 km long beach **NT 882**, the only beach on the eastern shore of the bay. This is a very low energy beach, backed by a beach ridge and 50 ha mangrove-filled lagoon, while it is fronted by 200 m wide tidal flats, together with a scattering of mangroves along the shore.

NT 883-892 **RORRUWUY-GARADANDANBOI BAY**

No.	Beach	Rating HT LT	Type	Length
NT883	Rorruwuy	1 1	R+sand flats	2.6 km
NT884	Rorruwuy (W1)	1 1	R+sand flats	700 m
NT885	Rorruwuy (W2)	1 1	R+sand flats	2.8 km
NT886	Rhodes Pt (E)	1 1	R+sand flats	700 m
NT887	Rhodes Pt (N1)	1 1	R+sand flats	200 m
NT888	Rhodes Pt (N2)	1 1	R+tidal flats	1 km
NT889	Rhodes Pt (N3)	1 1	R+tidal flats	700 m
NT890	Garadandanboi Bay	1 1	R+tidal flats	600 m
NT891	Cape Newbald (S2)	1 1	R+sand flats	200 m
NT892	Cape Newbald (S1)	1 1	R+sand flats	200 m
Spring & neap tide range = 4.1 & 2.8 m				

The northeast corner of Arnhem Bay between Cliffy Point and the Rorruwuy community is occupied by 20 km of mangrove-lined salt flats associated with the Cato, Peter John and Burungbirinung rivers. At the mouth of the Burungbirinung River the shoreline turns and trends west for 10 km as a low straight bedrock-dominated shore to Rhodes Point. This section of coast faces south across the 30 km wide bay, and receives only waves generated by the trade winds blowing across the bay. The waves are sufficient to maintain a series of nine beaches (NT 883-891) between the river mouth and the northeastern end of the shore opposite Mallison Island.

Beach **NT 883** is a 2.6 km long south-facing beach, with **Rorruwuy** community located behind the centre of the beach. It is bordered by 10 m high bluffs in the east and mangroves in the west, with three small creeks also draining across the beach and linked to backing patches of mangroves, while sand flats up to 200 m wide front the beach. On the western side of the mangroves is beach **NT 884**, a dynamic 700 m long beach that is located across a creek mouth, and fronted by tidal shoals extending 300 m into the bay. A small 20 ha V-shaped mangrove swamp backs the beach.

On the western side of the creek 10 m high red bluffs dominate the shore for 4 km to Rhodes Point. Two beaches are located along the base of the bluffs. Beach **NT 885** is a straight 2.8 km long sand beach, fronted by 100 m wide sand flats, with rock outcropping toward the western end. Beach **NT 886** is a 700 m long strip of high tide sand, which grades west into the red bluffs. A blufftop vehicle track and landing ground parallel the beach.

Three beaches (NT 887-889) extend north of Rhodes Point. Beach **NT 887** begins on the eastern side of the point and trends due north for 2.5 km into Garadandanboi Bay. The 200 m long beach is backed by 10 m high bright red laterite bluffs, with a few mangroves and rock flats immediately north of the point and then the narrow high tide beach, which is fronted by 100 m wide sand flats. Besides the landing ground on the point, there is a house above the southern end of the beach and a vehicle track running along the top of the bluffs. The beach terminates at a small protrusion in the bluffs, beyond which is a clump of mangroves then 1 km long beach **NT 888**. This is a similar narrow high tide beach, backed by the bluffs and fronted by both tidal flats and some rock flats and reefs extending up to 500 m off the beach. Finally 700 m long beach **NT 889** begins on the north side of a second small protrusion in the bluffs. The beach is initially backed by the bluffs, which give way as it transforms into a series of low energy recurved spits that have partially filled the eastern side of the bay, which is backed by a 10 ha V-shaped mangrove-filled estuary. The spits terminate at a 100 m wide creek mouth, while the sand flats widen to 1 km wide tidal shoals.

On the western side of the creek the shore trends to the west, and after a 1 km long section of mangroves, beach **NT 890** is located between the mangroves and sparsely vegetated 60 m high sandstone cliffs that continue east behind the beach. The beach is fronted by tidal shoals extending up to 500 m off the beach. This beach forms part of a 2 km long series of crenulate recurved spits that have filled the western side of the bay. A 200 m wide belt of mangroves now fronts most of the spit (Fig. 5.101).

Figure 5.101 Beach NT 890 consists of an active spit, which grades right into an inactive mangrove-fronted older spit.

The cliffs extend west for 2.5 km and as these decrease in height, beach **NT 891** is located in a 200 m long gap, with 20 m high cliffs to either side and 5-10 m high vegetated bluffs behind. The straight beach faces south and is fronted by 100 m wide sand flats. One kilometre further west the shoreline turns and trends north toward Cape Newbald, with Mallison Island separated by a 1 km wide channel. Beach **NT 892** is located 1 km into the strait. It is a 200 m long southwest-facing beach, bordered by low rocky points with a few mangroves and a clump of rocks toward the centre of the beach. It is backed by lightly vegetated sandstone slopes.

NT 893-901 NALWARUNG STRAITS (1)

No.	Beach	Rating HT	Rating LT	Type	Length
NT893	Mudhamul	1	1	R+sand flats	600 m
NT894	Nalwarung Straits (2)	1	2	R+rock flats	100 m
NT895	Nalwarung Straits (3)	1	1	R+tidal flats	500 m
NT896	Nalwarung Straits (4)	1	1	R+tidal flats	600 m
NT897	Nalwarung Straits (5)	1	2	R+rock flats	200 m
NT898	Nalwarung Straits (6)	1	2	R+rock flats	150 m
NT899	Nalwarung Straits (7)	1	1	R+tidal flats	100 m
NT900	Nalwarung Straits (8)	1	2	R+rock flats	200 m
NT901	Nalwarung Straits (9)	1	2	R+rock flats	100 m
Spring & neap tide range = 4.1 & 2.8 m					

Mallison Island and Cape Newbald form the southwestern entrance to the 40 km long Nalwarung Straits, which run between the mainland and Inglis Island. Immediately east of Cape Newbald is the small **Mudhamul** community at Godijboi Point. The first 10 km of coast east of the cape contain a series of small bays, with the shoreline largely dominated by mangroves. There are nine small low energy beaches (NT 893-901) located in some of the bays and on the intervening headlands. The only access to the coast is at Mudhamul, which also has a landing ground.

Mudhamul beach (**NT 893**) is a steep, 600 m long, northwest-facing sand and cobble beach, fronted by sand to mud flats extending 200 m offshore. The beach is bordered by low sandstone boulders and a few mangroves, with a few rocks outcropping in the centre. The two community houses, water tower and landing ground are all located toward the northern end of the beach, with a low foredune backing the southern half.

Beach **NT 894** lies 3 km to the east. It is a 100 m long strip of sand located on a small headland. It is bordered by mangrove-covered rocks, backed by vegetated slopes, with rock flats off the beach. Three hundred metres to the east is a small bay containing 500 m long beach **NT 895**. The beach is bordered by vegetated 20 m high rounded headlands, with rocks and mangroves at each end and some mangroves scattered along the beach. Tidal and rock flats extend 300 m off the beach, filling the bay.

Beach **NT 896** is located on the eastern side of the next largely mangrove-filled bay. The beach begins at a low rocky point and trends due north for 600 m before being overrun with mangroves. One kilometre to the north are two beaches on the next headland. Beaches **NT 897** and **NT 898** are adjoining 200 m and 150 m long beaches. The first is backed by densely wooded slopes, the second by a low red scarp then the slopes, with mangrove-covered rocks separating the two. Rocks and rock flats dominate the intertidal of both beaches and extend up to 200 m into the strait.

The next headland lies 2 km to the west and has three small beaches (NT 899-901). Beach **NT 899** is a 100 m long pocket of sand bordered by mangrove-covered rocks to each end. Two hundred metres of mangroves separate it from beach **NT 900**, a 200 m long sand beach, also bordered by rocks and mangroves, while beach **NT 901** is a similar 100 m long beach. All three beaches are backed by wooded slopes slowly rising to 30 m and are fronted by 500 m wide sand flats and inner rock flats.

NT 902-906 NALWARUNG STRAITS (2)

No.	Beach	Rating HT	Rating LT	Type	Length
NT 902	Mata Mata	1	1	tidal flats	100 m
NT903	Gikal	1	1	R+sand flats	300 m
NT904	Gurundu (1)	1	1	R+sand flats	250 m
NT905	Gurundu (2)	1	1	R+sand flats	300 m
NT906	Gurundu (3)	1	2	R+rock flats	200 m
Spring & neap tide range = 4.1 & 2.8 m					

The **Mata Mata** community is located 15 km northeast of Cape Newbald in a section of the straits that narrows to 3 km. The community has a landing strip, with about six houses clustered on the northern side of the strip. The 'beach' (**NT 902**) has been produced by the clearing of the 100 m wide band of mangroves at the base of the northern landing ground (Fig. 5.102), both to assist the safe landing of aircraft and to provide easy access to the strait. The beach is used by the community to launch and moor small boats on the 100 m long beach. It is fronted by 1 km wide tidal flats and bordered on each side by the 100 m wide mangrove forests.

Figure 5.102 The Mata Mata community, landing ground and beach (NT 902).

Seven kilometres to the east on the eastern side of the narrow strait section is the **Gikal** community which is located on beach **NT 903**. The community consists of two houses in a sandy clearing just behind the western end of the 300 m long beach. It is a moderately steep high tide sand beach fronted by 400 m wide sand flats, then the deep waters of the strait. It is fringed by mangroves at each end and is linked by road to the Mata Mata community.

Four kilometres further east are a cluster of the three Gurundu beaches (NT 904-906). Beach **NT 904** is a 250 m long north-facing sand beach fronted by 100 m wide sand flats. It is bordered by low rocks, with some rocks extending along the eastern end of the beach. A solitary house is located in a cleared area behind the centre of the beach. Three hundred metres to the east is a small 800 m long embayment containing beaches NT 905 and 906. Beach **NT 905** is a moderately steep sandy beach with 300 m wide sand flats. It is 300 m long with rocks and a few mangroves bordering each end and a tin shed located behind the centre of the beach. Beach **NT 906** is located on the rocky eastern arm of the bay and consists of a 200 m long strip of high tide sand, fronted by sand and rock flats extending 400 m into the strait, with a few mangroves growing on the rocks and to either end.

NT 907-913 **NALWARUNG STRAITS-MALAY RD**

No.	Beach	Rating HT	Rating LT	Type	Length
NT907	Nalwarung Straits (14)	1	1	R+ridged sand flats	1.7 km
NT908	Nalwarung Straits (15)	1	1	R+ridged sand flats	600 m
NT909	Malay Road (1)	1	1	R+rock-sand flats	400 m
NT910	Malay Road (2)	1	1	R+sand-rock flats	350 m
NT911	Malay Road (3)	1	1	R+ridged sand flats	500 m
NT912	Malay Road (4)	1	1	R+ridged sand flats	250 m
NT913	Malay Road (5)	1	1	R+ridged sand flats	400 m
Spring & neap tide range = 4.1 & 2.8 m					

The northern end of the Nalwarung Straits terminates opposite Bosanquet Island where the straits are 6.5 km wide. The straits are replaced by a section of water between the mainland and the offshore islands known as Malay Road, with the narrowest section of the 'road' named after the famous Malay seafarer Pobassoo.

At the junction of the strait and road is a relatively straight 5.5 km long section of coast containing seven sandy beaches (NT 907-913). These beaches are more exposed to east through northwest waves and respond with the formation of ridged sand flats off the beaches.

Beach **NT 907** is a straight 1.7 km long northwest-facing sand beach. It is backed by low slowly rising woodlands, with two clumps of rocks spaced equally along the beach, together with low sandstone rocky points and a few mangroves at each end. Sand flats extend 200 m off the beach and contain two inner sand ridges (Fig. 5.103). A low wooded sandstone point protrudes 200 m at the end of the beach, on the northern side of which is beach **NT 908**. This is a more embayed beach, with the eastern headland extending 400 m into the strait and a wide fringe of mangroves along its side. The beach runs straight between the two headlands for 600 m, with sparsely vegetated sandstone slopes behind. Sand flats containing up to eight sand ridges extend 400 m off the beach.

Figure 5.103 The upper section of beach NT 907, with the sand flats largely covered by the tide.

On the northern side of the headlands is a narrow 400 m long high tide beach (**NT 909**), fronted by low tide rock flats, with backing sandstone slopes. Mangroves border the western end against the headland, with the rocks dominating for 300 m to the beginning of beach **NT 910**. This 350 m long beach occupies a small valley. The beach has blocked the creek, forming a 1 ha mangrove swamp in lee of the low beach and overwash flats. The creek drains across the centre of the beach forming a small tidal shoal, with sand flats extending 300 m into the small bay.

Immediately east of the bay is a relatively straight 1.5 km long stretch of coast containing beaches NT 911-913, each separated by small sandstone points. Beach **NT 911** is 500 m long, bordered by the low vegetated sandstone points, with a low foredune, then lightly wooded slopes behind. Beach **NT 912** is 250 m long with rocky points extending 100 m off each end, while beach **NT 913** is 400 m long, grading into a straight section of rocky shore at its northern end. All three beaches are fronted by ridged sand flats narrowing from 200 m in the west to 100 m in the east.

NT 914-920 **MALAY ROAD (2)**

No.	Beach	Rating HT	Rating LT	Type	Length
NT914	Malay Road (6)	1	2	R+rock-sand flats	250 m
NT915	Malay Road (7)	1	2	R+rock flats	50 m
NT916	Malay Road (8)	1	1	R+tidal flats	500 m
NT917	Malay Road (9)	1	1	R+tidal flats	1 km
NT918	Malay Road (10)	1	2	R+rock flats	200 m
NT919	Malay Road (11)	1	2	R+rock-sand flats	500 m
NT920	Dholtji (W)	1	2	R+rock flats	200 m
Spring & neap tide range = 4.1 & 2.8 m					

Beaches NT 914-918 are located around the perimeter of a 2 km wide northwest-facing embayment 1.5 km west of the Dholtji community, while beaches NT 919 and 920 lie along a north-facing section of sandstone coast immediately west of the community. There is no development or vehicle access to the beaches.

Beaches NT 914 and 915 lie on the western headland of the embayment, a 1 km wide 10 m high wooded sandstone point. Beach **NT 914** is a 250 m long north-facing strip of high tide sand fronted by low tide sandstone rocks and reefs. It is bordered by a low rocky point and mangroves in the west grading into rocks to the east, with wooded slopes behind. Beach **NT 915** is located at the eastern tip of the headland and is a 50 m long pocket of high tide sand, bordered and fronted by sandstone rocks and boulders, with a few mangroves growing on the rocks.

One kilometre to the east in the more protected waters of the bay is 500 m long northwest-facing beach **NT 916**. The beach is bordered by a low wooded point with fringing mangroves and fronted by 300 m wide sand flats, with a small upland creek draining across the eastern end of the beach. Beach **NT 917** begins 500 m to the northeast and lies in the more sheltered west-facing section of the bay. The 1 km long beach has substantial areas of mangroves at either end as well as two large clumps of mangroves along the beach, with ridged sand flats extending 800 m off the beach. The Dholtji community lies 1 km due east of the beach.

Beach **NT 918** is located on the northern point of the bay, a low wooded sandstone point that trends west into the strait. The beach is a 200 m long northwest-facing sand beach fronted by ridges of intertidal sandstone and bordered by low sandstone rocks, with some mangroves on the western rocks.

The sandstone-dominated shoreline continues for 2 km east of the point and contains two beaches. Beach **NT 919** is a 500 m long north-facing sand beach, with outcrops of sandstone dominating the lower beach and intertidal area. One kilometre further east is beach **NT 920**, a 200 m long strip of high tide sand, with sandstone ridges and boulders dominating the lower beach. It terminates at a low sandstone point composed of jointed pink Proterozoic sandstone. The point extends east into the next embayment as a submerged linear 500 m long rock reef.

NT 921-929 **DHOLTJI**

No.	Beach	Rating HT	Rating LT	Type	Length
NT921	Dholtji	1	1	R+sand flats+rocks	1.8 km
NT922	Dholtji (E1)	1	1	R+sand flats+rocks	900 m
NT923	Dholtji (E2)	1	1	R+sand flats+rocks	250 m
NT924	Dholtji (E3)	1	1	R+sand flats+rocks	1 km
NT925	Dholtji (E4)	1	1	R+sand flats+rocks	50 m
NT926	Dholtji (E5)	1	1	R+rock flats	150 m
NT927	Dholtji (E6)	1	1	R+rock flats	50 m
NT928	Dholtji (E7)	1	1	R+rock flats	250 m
NT929	Dholtji (E8)	1	1	R+sand flats	900 m
Spring & neap tide range = 4.1 & 2.8 m					

Immediately east of the sandstone point and in lee of the rock reef is **Dholtji** community. The community is located at the western end of an open north-facing bay containing nine beaches (NT 921-929). The only development in the bay is the small community, which contains about three houses, with no vehicle access. A sandstone peninsula begins at beach NT 922 and extends for 20 km to Point William. It varies in width from 200 m to 3 km wide and reaches up to 60 m in height.

Dholtji beach (**NT 921**) is a curving 1.8 km long northeast- to north-facing sand beach, fronted by 200-300 m wide sand flats (Fig. 5.104), as well as a linear rock reef off the western point and a scattering of rocks and reefs over the flats and on the beach. The western kilometre of beach is backed by a 300 m wide beach ridge plain, upon which the community is located, with sandstone slopes backing the eastern half of the beach. Two small creeks cross the beach, the westernmost in lee of the point and containing a small area of mangroves.

Beach **NT 922** is a more crenulate 900 m long north-facing beach, which is bordered by low sandstone points, with rocks outcropping on the beach and across the 500 m wide sand flats. Wooded slopes rise to over 20 m in lee of the beach. Beach **NT 923** lies immediately to the east, and is a curving 250 m long sand beach bordered by a linear rocky point which extends as a reef across the front of the beach, enclosing a protected 'lagoon'. The beach forms one side of a sand foreland at its eastern end, on the other side of which is the beginning of beach **NT 924.** This is a 1 km long sand beach, with a central foreland in lee of a rock reef located on the 400 m wide sand flats, with a creek also draining across the eastern end and a low boundary sandstone point. Vegetated beach ridges then wooded slopes back both beaches.

Figure 5.104 View east along the curving beach NT 921.

Beach **NT 925** lies 100 m east of the boundary point which marks the start of a 1.5 km long section of rock-dominated shore containing the next four small beaches, all backed by sparsely vegetated sandstone slopes and bordered by low sandstone points and boulders. The 50 m long beach occupies a small northwest-facing valley and partially blocks a creek. Two hundred metres further east is beach **NT 926**, a straight 150 m long north-facing sand beach, bordered by rock which partially extends along the low tide line, as well as occurring as rock reefs on the 200 m wide sand flats. Beach **NT 927** lies 500 m to the east and is a 50 m pocket of sand bordered by a sloping rocky point with scattered rock reefs off the beach. Beach **NT 928** lies another 500 m to the east and is a similar 250 m long beach, with rocks and boulders scattered the length of the beach.

The easternmost beach in the bay is beach **NT 929**, a slightly curving 900 m long sandy beach backed by a low foredune, then wooded slopes. It faces northwest across the bay and is fronted by 100 m wide sand flats, with low sandstone points to either end.

NT 930-936 **MALAY ROAD (E)**

No.	Beach	Rating HT LT	Type	Length
NT930	Malay Road (22)	1 2	R+rock flats	400 m
NT931	Malay Road (23)	1 1	R+tidal flats	1.8 km
NT932	Malay Road (24)	1 1	R+sand flats+rocks	400 m
NT933	Malay Road (25)	1 2	R+rock flats	250 m
NT934	Malay Road (26)	1 2	R+rock flats	800 m
NT935	Malay Road (27)	1 2	R+rock flats	60 m
NT936	Malay Road (28)	1 2	R+rock flats	80 m
Spring & neap tide range = 4.1 & 2.8 m				

Immediately east of beach NT 929 is a 10 m high wooded sandstone headland that forms the eastern boundary of the beach and bay. One beach is located on the north side of the headland (NT 930), which also forms the western boundary of the next bay containing beach NT 931.

Beach **NT 930** is a straight northwest-facing high tide sand beach, fronted by continuous 50-100 m wide intertidal sandstone strata, with a low foredune on the backing vegetated slopes. There is a circular clump of mangroves on the northern tip of the point, which forms the western boundary of a 1.2 km wide semicircular-shaped bay, containing beach **NT 931**. Mangroves extend for 500 m down the eastern side of the headland into the bay, before the curving 1.8 km long beach commences. The beach is backed by a series of low beach ridges, which widen to 300 m along the eastern half, with creeks crossing the centre and eastern end of the beach. The latter connects a 30 ha area of mangroves and salt flats into which drain three streams, which rise on the backing 60 m high sandstone ridge of the peninsula.

A 2.5 km wide headland commences at the eastern end of the beach, extending east to Elizabeth Bay. Beach **NT 932** is located at the western end of the sloping wooded headland. It consists of a 400 m long west-facing slightly curving sand beach, bordered by low rocky points, with mangroves on the northern point and sand flats extending 200 m off the beach. On the northern side of the mangroves is 250 m long beach **NT 933**, a crenulate beach located in lee of irregular 100-200 m wide rock reefs, and backed by some low beach ridges before the woods.

One hundred metres of red 10 m high bluffs separate beach NT 933 from beach **NT 934**. This is an 800 m long double crenulate sandy beach with a central foreland in lee of rock reefs, and a mixture of rock reefs and sand flats extending 200-300 m off the beach. Vegetated 10 m high bluffs back the beach. It terminates at the beginning of the narrow rocky tip of the headland, on whose western side are beaches **NT 935** and **NT 936**. The beaches are 60 m and 80 m long respectively, face west, and are bordered by low rocky points and partly fronted by rock reefs, with a 50-100 m wide vegetated 5-10 m high ridge behind.

NT 937-943 **ELIZABETH BAY**

No.	Beach	Rating HT LT	Type	Length
NT 937	Elizabeth Bay (1)	1 1	R+sand flats	1.6 km
NT938	Elizabeth Bay (2)	1 1	R+sand flats	1.3 km
NT939	Elizabeth Bay (3)	1 1	R+sand flats	800 m
NT940	Elizabeth Bay (4)	1 1	R+sand flats	1.6 km
NT941	Elizabeth Bay (5)	1 1	R+sand flats	60 m
NT942	Elizabeth Bay (6)	1 1	R+sand flats	70 m
NT943	Point William	1 2	R+coral flats	900 m
Spring & neap tide range = 4.1 & 2.8 m				

Elizabeth Bay is 4 km wide, 3 km deep, north to northwest-facing bay lying in lee of Point William, which forms its eastern boundary. The entire bay is protected from the trade winds by its orientation and the backing slopes, which rise to the 30-60 m high cliffs of Cape Wilberforce. The bay contains seven beaches (NT 937-943), which occupy most of the 10 km of shoreline. The only development in the bay area is a pearl farm whose racks are located off beach NT 937, with a shed to service the farm on the western point of the bay, and a few boats usually moored off the beach.

Beach **NT 937** extends immediately east of the narrow sandstone point that separates it from beach NT 936. The beach is 1.6 km long, curves slightly and faces due north across 300-400 m wide sand flats, with the pearl racks located seaward of the flats. The eastern end of the beach grades into a low mangrove-fringed point which curves round for 50 m to the small tidal creek, which forms the western boundary of beach **NT 938**. This beach is 1.3 km long, faces north to northwest and is backed in the west by a series of low beach ridges and a small mangrove-lined tidal creek and salt flats. Sand flats extend up to 800 m off the beach. The western end is bordered by a 50 m high, wooded sandstone point.

Beach **NT 939** lies on the eastern side of the 200 m wide point. The 800 m long beach is straight, faces northwest, is backed by a low foredune then rising wooded slopes and fronted by 600 m wide sand flats. Beach **NT 940** lies around a 200 m long low rocky point and has a few rocks outcropping along the first couple of hundred metres of the 1.6 km long curving northwest-facing beach (Fig. 5.105). Two small mangrove-lined creeks drain off the backing slopes at either end of the beach, while sand flats extend 300 m into the bay.

Figure 5.105 Beaches NT 940-942 lie in lee of the rugged rocky shoreline of Cape Wilberforce (foreground).

One hundred metres of low sandstone rocks separate it from beach **NT 941**, a 60 m long pocket of sand, separated from its neighbour beach NT 942 by a 50 m protrusion in the sandstone. Beach **NT 942** is 70 m long, relatively steep and bordered by sandstone rocks to either side, with a few mangroves adjacent to the southern rocks. The northern rock continues for 2 km to the next beach. Sand flats extend 300 m off both beaches, and the deeper water off these flats offers the best anchorage under southeast wind conditions. The peninsula is only 150 m wide behind beach NT 942, and it is well worth a walk up over the sandstone to view the steep exposed cliffs of Cape Wilberforce on the eastern side.

Point William is one of the northernmost points of the Northern Territory mainland at 11°53' S, and represents the end of the narrow elongate peninsula, which separates the Malay Road from Melville Bay. Beach **NT 943** is located on the protected western side of the point and extends 900 m south into Elizabeth Bay. The beach consists of a high tide sand beach backed by a low windswept foredune, with a few rocks along the beach and sand and coral reef flats extending 200 m off the beach.

NT 944 **CAPE WILBERFORCE**

No.	Beach	Rating HT	LT	Type	Length
NT944	Cape Wilberforce	2	5	R (boulder)	300 m
Spring & neap tide range = 2.2 & 1.2 m					

Cape Wilberforce is a prominent 40 m high sandstone cliff that lies 4 km south of Point William. Mid-way between the two points is a 300 m long boulder beach (**NT 944**), the only beach along the 20 km of steep rocky cliffs that form the southern shore of the Cape Wilberforce peninsula. This beach represents the first beach on the east coast of the Northern Territory. While the beach is difficult and dangerous to approach from the sea, it lies within 1 km of the protected beaches NT 942 and 943 in Elizabeth Bay and can be accessed on foot from these beaches.

NT 945-957 **MELVILLE BAY (N)**

No.	Beach	Rating HT	LT	Type	Length
NT945	Melville Bay (1)	2	3	R+LTT	1.7 km
NT946	Melville Bay (2)	2	3	R+sand flats	200 m
NT947	Melville Bay (3)	2	3	R+sand flats	2.1 km
NT948	Bakirra (E 1)	1	1	R+tidal sand flats	700 m
NT949	Bakirra (E 2)	1	1	R+sand flats	1.6 km
NT950	Melville Bay (5)	1	2	R+rock flats	500 m
NT951	Melville Bay (6)	1	2	R+sand flats+rocks	400 m
NT952	Melville Bay (7)	1	2	R+sand flats+rocks	600 m
NT953	Melville Bay (8)	1	2	R+rock flats	400 m
NT954	Melville Bay (9)	1	2	R+rock flats	800 m
NT955	Melville Bay (10)	1	2	R+LTT	600 m
NT956	Melville Bay (11)	1	2	R+LTT	300 m
NT957	Melville Bay (12)	1	2	R+LTT+rocks	900 m
Spring & neap tide range = 2.2 & 1.2 m					

Melville Bay is a 30 km wide, open, east-facing bay bordered by Cape Wilberforce in the north and the Gove Peninsula at Cape Wirawawoi and Bremer Island to the south. The bay has 180 km of shoreline containing 66 beaches (NT 945-1011). Its southern more protected section is also the site of Gove Harbour.

The northern section of the bay immediately south of Cape Wilberforce contains an east-facing, 4 km wide, 6 km deep bay, which terminates at a 1 km wide inlet. On the protected shores of the inlet is the small Bakirra community and landing ground. Three beaches (NT 945-947) are located in the bay north of the inlet, and ten beaches (NT 948-957) extend from the south side of the inlet for 6 km out along the southern rocky tip of the bay.

As the narrow Cape Wilberforce peninsula joins the mainland, the coast trends due south for 4 km to the tidal inlet and contains the three exposed east-facing beaches. Beach **NT 945** is a 1.7 km long sand beach fronted by a continuous 100 m wide low tide bar, with normal easterly waves averaging about 1 m and breaking across a 100 m wide surf zone. Permanent rips are located against the rocks at each end and a small creek drains across the middle of the beach. The southern rocky point forms the northern side of a 200 m long 'bay' that contains beach **NT 946**. The beach is 200 m long, moderately protected by the point and adjacent rock reefs, with only a 20 m wide surf at the beach. An abandoned landing ground is located just behind the beach.

Beach **NT 947** extends for 2.1 km south of the southern rocks. It is the outermost spit of a 300-600 m wide series of recurved spits that have extended south to partly fill and enclose a 400 ha mangrove-filled lagoon (Fig. 5.106). The 300 m wide tidal creek forms the southern boundary of the beach. Its inlet tidal shoals extend up to 1.2 km off the creek mouth and up the beach, lowering waves at the shore.

Figure 5.106 A series of prograded spits form the southern end of beach NT 947.

Beaches NT 947 and 948 are separated by the 1 km wide inlet with the creek channel occupying the northern side. The **Bakirra** community is located inside the inlet on the western shore. Beach **NT 948** is part of a 2.6 km long series of recurved spits that have partly filled the southern side of the bay. The beach is 700 m long, faces due north and extends from some eroding clumps of mangroves as a 200 m wide series of spits. It is fronted by tidal sand shoals extending up to 2 km to the north. Beach **NT 949** is located on the eastern side of the mangroves and is a curving north- to northeast-facing 1.6 km long beach that terminates at a low sandstone point. It is backed by older spits of variable width, with a small creek, draining a 20 ha area of mangroves, crossing the eastern end of the beach.

Between the rocks at the northern end of beach NT 949 and a major point at beach NT 957 are 4.5 km of generally north-facing shore dominated by low rocky points and reefs. The points divide the shoreline into eight beaches (NT 950-957), all influenced to some degree by the points and bedrock reefs. All are backed by gently sloping wooded terrain, with no vehicle access to the beaches.

Beach **NT 950** is a relatively straight, northwest-facing, rock-controlled, 500 m long beach, with scattered intertidal rocks and reefs extending up to 200 m off the beach, resulting in low waves at the shore. Immediately east is beach **NT 951**, a curving 400 m long beach, bordered by intertidal rocky points and fronted by 300 m wide irregular sand flats, which cause waves to break well off the beach. The eastern corner of the beach is experiencing erosion and is backed by slumping 10 m high bluffs.

Beach **NT 952** is a double crenulate 600 m long beach dominated by 50 m wide intertidal rock reefs along the first 400 m, with a second more protected pocket of sand immediately beyond the rocks. Waves break across the rocks at high tide producing a right-hand break at the western end. The bluffs are being eroded at the more protected western end and are slumping down onto the beach.

Beach **NT 953** is a 400 m long west-facing strip of high tide sand fronted by continuous 50 m wide intertidal rocks, then 300 m wide sand flats. The shoreline turns to face northwest along its neighbouring beach, **NT 954**, a similar 800 m long crenulate beach also fronted by continuous rocks, with some rock reefs lying a few hundred metres off the beach.

The final three beaches (NT 955-957) in the 'bay' are slightly more exposed and with fewer rocks, resulting in higher waves and more surf. Beach **NT 955** is 600 m long, faces northeast and, while bordered by low rocky points and reefs, has a central 400 m long sand beach section, fronted by an attached bar with waves averaging about 1 m and producing a 50 m wide surf zone. A rip runs out against the western rocks, with a couple of beach rips also occurring during higher wave conditions. There is a small protected pocket of sand at the eastern end of the beach, then a 200 m long sand-rock point, with a right-hand surf break along the rocks.

Beach **NT 956** lies immediately east of the point. It is a 300 m long sand beach, also bordered by low rock points, with waves breaking across a 50 m wide bar. Finally, beach **NT 957** is a curving 900 m long beach, which begins in lee of 50 m wide rock reefs, then grades into a sand bar, with wave height decreasing to the east in lee of the major rocky point, which forms the southern boundary of the bay. The beach curves round to face west in lee of the point, with rock reefs extending 200-300 m off the point.

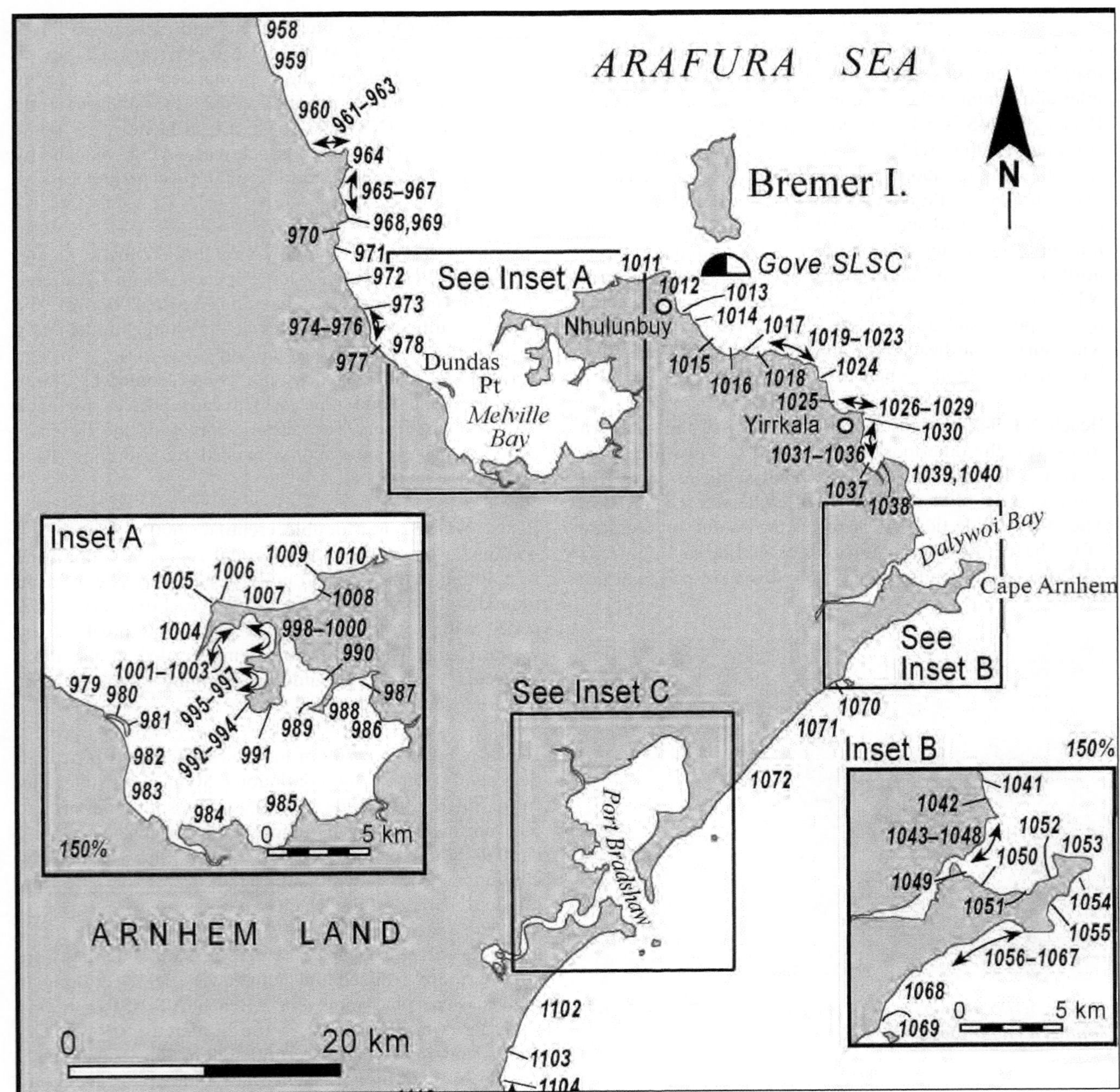

Figure 5.107 Regional map of the northeastern tip of Arnhem Land centred on the major town of Nhulunbuy. Beaches NT 958-1069.

NT 958-964 MELVILLE BAY (MOUNT BONNER)

No.	Beach	Rating HT	Rating LT	Type	Length
NT958	Melville Bay (13)	3	4	R+LTR	5.2 km
NT959	Melville Bay (14)	1	2	R+LTT	100 m
NT960	Melville Bay (15)	2	3	R+LTT+rocks	5.3 km
NT961	Mount Bonner (N5)	1	1	R+sand flats	800 m
NT962	Mount Bonner (N4)	1	2	R+rock flats	900 m
NT963	Mount Bonner (N3)	1	2	R+LTT	300 m
NT964	Mount Bonner (N2)	2	3	R+LTT	800 m
Spring & neap tide range = 2.2 & 1.2 m					

The low rocky point and reefs at the northern end of beach NT 958 mark a change in shoreline direction, with the coast turning to trend south-southeast for 40 km, deep into the protected southern waters of Melville Bay-Gove Harbour. The beaches south of the point generally face east into the trade winds and receive moderate waves which gradually decrease in lee of the Gove Peninsula. The first 13 km of coast between the point and the prominent Mount Bonner point contain seven beaches (NT 958-964). The only vehicle access on this section of coast is to the southern end of beach NT 958.

Beach **NT 958** is a straight 5.2 km long east-facing steep sandy beach composed of medium-coarse yellow sand. The beach receives moderate waves averaging about 1 m which break across a narrow attached bar (Fig. 5.108), with deep water off the beach at high tide and usually 20 beach rips cutting across the bar at low tide. The beach is backed by a mixture of low vegetated bedrock slopes, and two small tidal creeks and wetlands, the southern creek exiting at the southern end of the beach against some low rocks. Beach **NT 959** begins just beyond the rocks, and is a 100 m long pocket of sand bordered by low rocky points and reefs and backed by vegetated slopes.

Figure 5.108 Waves breaking across the narrow bar at the base of beach NT 958.

Beach **NT 960** is a 5.3 km long beach, backed at the northern end by a 30 m high vegetated knoll. The southern 4 km of beach are backed by low 200-300 m wide beach-foredune ridges, then a 100 ha mangrove-dominated wetland which is drained by a 50 m wide creek at the southern end of the beach. The beach is steep with a narrow bar, then irregular deeper rock reefs extending up to 1 km off the beach. The reefs lower waves to about 0.5 m at the beach, resulting in no rips. Beach **NT 961** begins on the southern side of the creek mouth and is an 800 m long northeast-facing sand beach, fronted by 500 m wide tidal shoals associated with the creek mouth. The shoals together with a headland to the east lower waves to less than 0.5 m at the shore. The beach is backed by a 200 m wide series of beach ridges and terminates at the massive sandstone boulders of Mount Bonner point, with a few mangroves growing along the protected rocky shore.

The point area contains four beaches. Beach **NT 962** is a crenulate 900 m long rock-dominated beach. It consists of a high tide sand beach fronted by near continuous inter to subtidal rocks and reefs, with vegetated slopes behind. Beach **NT 963** lies on the northern side of the point and faces due north. The beach is 300 m long and bordered by low rocks to the west and a more prominent sandstone boulder-point to the east. The sandy beach impounds a 3 ha wetland drained by a creek toward the eastern end of the beach. It is fronted by a 50 m wide bar, with one to two rips crossing the bar and along the boundary rocks when waves exceed 0.5 m.

Beach **NT 964** is located at the eastern end of the point and consists of an 800 m long east-facing sand beach and 50 m wide attached bar. It is bordered by a 20 m high sandstone headland to the north and lower sandstone point to the south, both capped by prominent large sandstone boulders. A low foredune backs the beach.

NT 965-971 **MELVILLE BAY (MOUNT BONNER S)**

No.	Beach	Rating HT LT	Type	Length
NT965	Mt Bonner (N 1)	1 1	R+LTT	1.4 km
NT966	Mt Bonner (S 1)	2 3	R+LTT	1.3 km
NT967	Mt Bonner (S 2)	2 3	R+LTT	800 m
NT968	Mt Bonner (S 3)	1 1	R+LTT	50 m
NT969	Mt Bonner (S 4)	1 2	R+rock flats	300 m
NT970	Melville Bay (25)	2 3	R+LTT	1.2 km
NT971	Melville Bay (26)	1 1	R+sand flats	200 m
Spring & neap tide range = 2.2 & 1.2 m				

On the south side of Mount Bonner point the coast continues to trend south with six beaches (NT 965-970) located within the next 5 km. Beach **NT 965** is located on the southern side of the 500 m long sandstone reef that forms the eastern extremity of the point. The reef affords protection to the 1.4 km long east-facing beach, lowering waves to about 0.5 m, resulting in a steep reflective beach and narrow surf zone. A line of casuarina trees backs the beach, with wooded slopes behind (Fig. 5.109). A low narrow 100 m long rocky point forms the southern boundary. Sixty metre high Mount Bonner is located 1.5 km due west of this point.

Figure 5.109 The reef-sheltered, steep beach NT 965.

Beach **NT 966** lies immediately south of the rocky point, with a 50 m wide tidal creek flowing out next to the rocks and forming the northern boundary of the 1.3 km long east-facing beach. The beach is moderately steep with waves averaging about 1 m and breaking across a 50 m wide surf zone. The trade winds have generated some minor dune activity with some dunes encroaching on the backing 15 ha mangrove wetland. The beach terminates at some low rocks and bluffs, beyond which is a curving 5-10 m high bluff, fronted by 800 m long beach **NT 967**. The narrow beach is wedged in below the bluffs, with some rocks on the beach and in the surf. At the end of the beach the bluffs turn and trend west to beach **NT 968**. This is a 50 m long pocket of sand bordered by 5 m high bluffs.

One hundred metres to the south the shoreline again turns into a curving south-facing section of shore, initially occupied by 300 m long beach **NT 969**. The beach is backed by densely wooded 5 m high bluffs, with rocks scattered along the beach, and reefs located across the 300 m wide sand flats which front the low energy beach. Beach **NT 970** begins immediately to the south. This is a relatively straight 1.2 km long beach, backed for the most part by a low foredune and a small salt flat-mangrove wetland, with low vegetated bedrock bordering the southern end. The shoreline trends west for 400 m at the southern boundary as a series of three bedrock-controlled crescents, with boulders dominating the shore, apart from the 200 m long sandy central crescent (beach **NT 971**). Rock and sand flats extend 300 m off the beach.

NT 972-983 **MELVILLE BAY (S)**

No.	Beach	Rating HT LT	Type	Length
NT972	Melville Bay (27)	1 2	R+LTT	4.1 km
NT973	Melville Bay (28)	1 2	R+LTT	1 km
NT974	Melville Bay (29)	1 2	R+LTT	500 m
NT975	Melville Bay (30)	1 2	R+LTT	400 m
NT976	Shepherd Bluff (N)	1 2	R+LTT	400 m
NT977	Shepherd Bluff (S)	1 1	R+sand flats	300 m
NT978	Melville Bay (33)	1 2	R+LTT	2.6 km
NT979	Melville Bay (34)	1 2	R+LTT	2.6 km
NT980	Yudu Yudu	1 1	R+sand flats	1.2 km
NT981	Parfitt Pt	1 1	R+sand flats	800 m
NT982	Parfitt Pt (S)	1 1	R+sand flats	600 m
NT983	Yanungbi	1 1	R+sand flats	300 m
Spring & neap tide range = 2.2 & 1.2 m				

The southwestern 15 km long section of Melville Bay lies opposite Gove Harbour and the Gove Peninsula. The peninsula affords increasing protection to the southern beaches, and while all the beaches lie within 10 km of Gove, access to and development on the coast is limited to two access tracks and two houses.

Beach **NT 972** is an exposed 4.1 km long east-facing beach that is bordered by small tidal creeks at either end, with generally low wooded slopes in between. The two creeks drain 20 ha and 100 ha mangrove wetlands respectively. The beach receives waves averaging about 1 m which break across a continuous 50 m wide bar. Beach **NT 973** lies on the southern side of the south creek and is a 1 km long casuarina-covered beach ridge, with the larger wetland behind. It terminates at a low rocky point.

Beyond the point are three exposed beaches each bordered by low rocky points. Beach **NT 974** is 500 m long with the rocky points to each end and dense woodlands behind. It has a central rock reef which results in a slight protrusion of the beach with waves usually breaking across a narrow low tide bar. Beaches **NT 975** and **NT 976** are similar 400 m long beaches with narrow bars, both located between low rock outcrops and backed by woodlands. The southern rocks of beach NT 976 are more substantial and extend for 100 m west as 16 m high Shepherd Bluff, to the beginning of beach **NT 977**. This is a 300 m long beach located between the rocks and a tidal creek, with tidal sand shoals extending up to 200 m off the beach.

Beach **NT 978** begins on the south side of the 50 m wide creek, which drains a 100 ha area of mangroves. The beach faces northeast and trends relatively straight for 2.6 km to a second tidal creek. It receives waves averaging about 1 m which maintain a continuous 50 m wide bar. Beach **NT 979** is located between the second and third tidal creeks and has similar exposure and beach conditions. It is also 2.6 km in length and is backed by continuous mangrove swamp up to 1 km wide. The two creeks interconnect to drain the 250 ha area of mangroves.

Beach **NT 980** begins on the eastern side of the 50 m wide creek. There is vehicle access to the beach and the two houses of **Yudu Yudu** community located close to the creek mouth. Beyond the creek mouth the beach extends east as a narrow north-facing spit, fronted by 200 m wide tidal sand shoals and backed by a 100-200 m wide band of mangroves. The beach extends for 1.2 km to where the shoreline turns and trends due south as an 800 m long active sand spit called **Parfitt Point** (beach **NT 981**). This beach consists of a series of recurved spits, backed by mangroves and fronted by 300-400 m wide sand flats. These spits and the following two beaches (NT 982 and 983) represent the terminus for sand moving south along the western shore of Melville Bay (Fig. 5.110).

Figure 5.110 Parfitt Point consists of beaches NT 980 (foregound) and NT 981 extending south as an active sand spit.

Beach **NT 982** is a slowly migrating 600 m long sand spit, backed by older recurved spits and mangroves, with tidal creeks to either end. Finally beach **NT 983** is a 300 m long terminal strip of sand deposited by the southerly movement of sand into the bay. The spit beaches are increasingly protected from the ocean swell with waves decreasing into the bay, and calm conditions common along the shore. A vehicle track reaches the beach which is known as Yanungbi.

NT 984-990 **MELVILLE BAY (SE)**

No.	Beach	Rating HT LT	Type	Length
NT984	Melville Bay (39)	1 1	R+tidal flats	300 m
NT985	Dolphin Rocks	1 1	R+tidal flats	500 m
NT986	Melville Bay (41)	1 1	R+tidal flats	1.4 km
NT987	Melville Bay (42)	1 1	R+tidal flats	1 km
NT988	Melville Bay (43)	1 1	R+tidal flats	400 m
NT989	Melville Bay (44)	1 1	R+tidal flats	600 m
NT990	Melville Bay (45)	1 1	R+tidal flats	700 m
Spring & neap tide range = 2.2 & 1.2 m				

The southeastern corner of Melville Bay is partly enclosed by the Gove Peninsula. It consists of a low energy crenulate shoreline alternating between protruding laterite peninsulas, separated by mangrove-filled bays, with a few beaches fronted by tidal flats on the more exposed section of the shore. The 30 km of shoreline between beaches NT983 and 990 contains seven beaches (NT 984-990) totalling 5 km in length, the remaining shoreline fringed by mangroves. Beaches NT 986-990 are all accessible by vehicle tracks off the main Gove Peninsula road.

Beach **NT 984** is located on the southern shore of the bay, along the north side of a 200 ha island that partly blocks the Giddy River mouth. The 300 m long beach is part of a 1.1 km long recurved spit, the majority of the spit fronted by mangroves, with a series of older spits extending up to 400 m inland. The spits have been developed by easterly waves generated within the bay. Four kilometres to the east in lee of the small Dolphin Rocks is a small peninsula, along the western shore of which is located 500 m long beach **NT 985**. The beach is backed by wooded laterite plain and fronted by intertidal sand to rock flats.

Beach **NT 986** is located 5 km to the northeast at the eastern extremity of the bay. The beach is sheltered from easterly winds though exposed to westerly winds and waves blowing across the bay. The 1.4 km long beach faces southwest across the bay and wraps around a 1 km wide peninsula, with the extremities of the beach forming multiple spits that have built 1.2 km and 2 km respectively into the adjacent bays, the majority of the spits fronted by mangroves. A vehicle track parallels the back of the beach.

Beach **NT 987** lies opposite the northern side of beach NT 986 and consists of a 1 km long beach fronting the wooded coastal plain, with mangrove-fringed spits also trailing for 1 km to the east of the beach. Beach **NT 988** is located on the eastern side of a narrow peninsula 1 km to the southwest. The beach is 400 m long, faces southeast with mangroves bordering each end. Beach **NT 989** is located on the east side of the 200 m wide peninsula. It is a narrow 600 m long beach, with mangroves fringing either end. Beach **NT 990** is located at the northern junction of the peninsula and curves round for 700 m, facing southwest down the 2 km wide bay. A continuous vehicle track commencing at beach NT 990 links the four beaches to the main road.

NT 991-1000 **DRIMMIE PENINSULA**

No.	Beach	Rating HT LT	Type	Length
NT991	Drimmie Hd (W)	1 1	R+tidal flats	500 m
NT992	Drimmie Hd	1 1	R+tidal flats	600 m
NT993	Butjumurru	1 1	R+tidal flats	250 m
NT994	McIntyre Pt (1)	1 1	R+tidal flats	50 m
NT995	McIntyre Pt (2)	1 1	R+tidal flats	200 m
NT996	Middle Bay	1 1	R+tidal flats	800 m
NT997	Half Tide Pt	1 1	R+tidal flats	100 m
NT998	Inverell Bay (S)	1 1	R+tidal flats	100 m
NT999	Gurrukpuy Beach	1 1	R+tidal flats	1.3 km
NT1000	Wanaka Bay	1 1	R+tidal flats	800 m
Spring & neap tide range = 2.2 & 1.2 m				

The **Drimmie Peninsula** is low wooded peninsula that extends from the Nhulunbuy-Gove road 4 km south into Melville Bay. The peninsula is the site of the Gove Yacht Club and further south the Gunyangara community, consisting of a number of houses and boating facilities. There are ten low energy beaches along the southern and western shores of the peninsula, all accessible by formed roads.

Beach **NT 991** is located on the southeast side of the peninsula. It is a straight 500 m long sand beach fronted by 200 m wide sand and rock flats, with two clumps of mangroves and a few rocks along the beach. Beach **NT 992** occupies the southern tip of the peninsula and is a crenulate 600 m long beach, which wraps around the point and faces south then southwest. Mangroves form the eastern boundary, with low rock bluffs to the west, and two houses located behind the beach and northern point.

Beach **NT 993** is located on the north side of the rock bluffs and is a curving 250 m long west-facing beach known as Butjumurru. It is the site of boating facilities, with a concrete boat ramp across the middle of the beach and boat storage facilities behind, while several boats are usually moored off the beach (Fig. 5.111).

Beach **NT 994** is located on McIntyre Point and is a 50 m long pocket of sand bordered by large granite boulders. Immediately to the north is beach **NT 995**, a 200 m long west-facing beach bordered by vegetated granite points, with two houses located at the southern end. The backing wooded slopes rise to 42 m high Drimmie Hill located just behind the beach.

Figure 5.111 Butjumurru beach (NT 993) is the site of boating facilities with moorings off the beach.

On the north side of the beach is 1 km wide Middle Bay containing beach **NT 996**, a curving west-facing 800 m long beach. The beach is backed by the **Gunyangara** community, which consists of several beachfront houses, and fronted by 400 m wide tidal flats that partially fill the bay. Half Tide Point forms the northern arm of the bay and on its southern side contains the pocket 100 m long beach **NT 997**. The beach faces southwest and is bordered by low vegetated granite points, including a few mangroves. A vehicle track reaches the centre of the beach.

The northern side of Half Tide Point forms the southern arm of Inverell Bay, a circular 2 km wide bay containing three beaches (NT 998-1000). Beach **NT 998** is located in a 100 m gap in the mangroves that line the northern side of the point, with a vehicle track leading to the back of the beach. **Gurrukpuy Beach (NT 999)** is a curving 1.3 km long west-facing beach, which is the site of the Gove Yacht Club and is the most popular and heavily utilised beach in the Gove region. Up to 100 boats are usually moored off the beach and numerous dinghies usually litter the beach, with a public boat ramp next to the yacht club. The road along the back of the beach links the southern half of the peninsula to the mainland via a 50 m long causeway.

Wanaka Bay is located at the northern end of the bay. It contains a curving 800 m long south- to southeast-facing low energy beach and tidal flats (**NT 1000**) fringed by mangroves and fronted by tidal flats. It is backed by the main Gove Harbour road, then bauxite storage, but there is no development on the beach.

NT 1001-1003 GOVE HARBOUR

No.	Beach	Rating HT	LT	Type	Length
NT1001	Galupa	1	1	R+tidal flats	400 m
NT1002	Gove Harbour jetty	1	1	R+tidal flats	100 m
NT1003	Dundas Pt (E)	1	1	R+sand flats	2 km
Spring & neap tide range = 2.2 & 1.2 m					

Beaches NT 1001-1003 were originally one beach that has been truncated and cut into three by the port development in Gove Harbour. All three are backed by a continuous gravel road with the small **Galupa** community in the east, then various port and plant facilities associated with the bauxite plant, including fuel storage tanks in lee of Dundas Point. Beach **NT 1001** begins immediately west of Wanaka Bay, commencing at a mangrove-lined rocky point and extending west for 400 m to a seawall and groyne. The sand and gravel beach faces south and is crossed toward the western end by a 100 m long fuel jetty. The Galupa community is located east of the jetty. Beach **NT 1002** is wedged in between two groynes. The eastern groyne is the site of the barge landing and a jetty. The beach is 100 m long and also backed by a rubble seawall, then part of the bauxite plant. Beach **NT 1003** begins on the western side of the groynes and is a curving 2 km long beach that finally faces east where it terminates at sandy **Dundas Point** (Fig. 5.112). Much of the beach has been backed by a rubble seawall, with a gravel road on top and a car park at the point. A 300 m long jetty extends across the 200 m wide sand flats into the bay from the car park area. Only the southern 100 m of the sandy point remain in a natural state.

Figure 5.112 Dundas Point is a long sand spit occupied by Gove port facilities and bordered by beaches NT 1003 & 1004.

NT 1004-1011 GOVE PENINSULA (N)

No.	Beach	Rating HT	LT	Type	Length
NT1004	Dundas Pt (W)	1	1	R+LTT	2.1 km
NT1005	Wargarpunda Pt (W)	1	2	R+LTT+rocks	200 m
NT1006	Wargarpunda Pt (E)	1	1	R+LTT	300 m
NT1007	Birritjimi/Wallaby Bch	1	1	R+LTT	5 km
NT1008	Lombuy Ck	1	1	R+sand flats	400 m
NT1009	Lombuy Ck (E1)	1	1	R+LTT	300 m
NT1010	Lombuy Ck (E2)	1	1	R+LTT	2.9 km
NT1011	East Woody Beach	1	2	R+LTT	3.6 km
Spring & neap tide range = 2.2 & 1.2 m					

The northern shore of the Gove Peninsula extends from Dundas Point in the west for 15 km to Cape Wirawawoi in the east. The cape also marks the boundary of the Arafura Sea and Gulf of Carpentaria. The near continuous sandy shoreline contains eight beaches (NT 1004-1011), with the Gove bauxite plant located in lee of the westernmost beaches. The beaches are moderately exposed, generally receiving wind waves less than 1 m in height.

Beach **NT 1004** is located along the western side of Dundas Point. It faces west across Melville Bay and receives low refracted trade wind waves. The 2.1 km long beach has several crenulations caused by beachrock reefs which sub-parallel the beach. At the southern end of the point is the major 500 m long T-jetty for the loading of bauxite, with the conveyor-belt running the length of the point. There is a car park adjacent to the jetty shared with beach NT 1003.

Beach **NT 1005** is located at the top of the point, a series of rounded granite rocks and boulders called **Wargarpunda Point**. The 200 m long beach is wedged in on the west side of the point. It is cut in places by the rocks and fronted by scattered rock reefs and shallow seafloor. The fenced bauxite plants lies immediately behind the beach. On the eastern side of the point is 300 m long north-facing beach **NT 1006**, which extends from a sandy foreland at the point to a small red bauxite-covered rocky point in the east. It has a steep reflective high tide beach grading into a low tide terrace.

Birritjimi-Wallaby Beach (**NT 1007**) extends from the eastern side of the rocky point for 5 km to Lombuy Creek. The beach initially faces north where it is called Birritjimi Beach, and curves slowly round to face northwest at the sandy creek mouth. The main Gove road runs along behind the beach with a few beachfront houses located toward the centre, where it is called Wallaby Beach. A narrow bar parallels the beach, widening to sand flats within 1 km of the creek mouth. Beach **NT 1008** is located between the creek mouth and a sand-covered, rock-controlled point. The beach is 400 m long and consists of a low partly vegetated sand spit grading into sand flats, which are backed by mangroves toward the creek mouth.

Beach **NT 1009** begins at the western tip of the low rock-controlled point. It extends east for 300 m as a series of four small sand scallops between the granite boulders, offshore rock reefs and intervening sand. Beach **NT 1010** begins as the rocks disappear and continues on to the east for 2.9 km to the creek mouth in lee of East Woody Point as a relatively steep high tide beach and narrow bar. Both beaches are backed by a continuous 200-300 m wide band of mangroves, which are drained by the creeks at either end. As a consequence there is no vehicle access to either beach and no development.

East Woody Island (also known as Dhamitjinya) is a conical 25 m high granite peak, which lies at the western end of **East Woody Beach** (**NT 1011**), and is permanently connected by the beach to the mainland (Fig. 5.113). The beach faces north and slowly curves round for 3.6 km to **Cape Wirawawoi**, the northern tip of the Gove Peninsula. The beach is backed by a series of beach ridges and is accessible by road from Nhulunbuy, located just 2 km to the south. A gravel road parallels the back of the beach, with car parks located at either end and in the centre. The small **Galaru** community is located toward the western end of the beach and the Nhulunbuy sewage treatment plant is located in the wetland behind the eastern end of the beach. Shoals moving around East Woody Point can form irregular holes and gutters along the more popular western end of the beach, with rips draining the gutters when waves are breaking.

Figure 5.113 East Woody Island with beach NT 1010 to right and NT 1011 to left.

NT 1012-1018 **GOVE PENINSULA (NE)**

No.	Beach	Rating HT	Rating LT	Type	Length
NT1012	Gadalathami/Town	2	2	R+LTT	3.3 km
	GOVE SLSC				
	Patrols: May-October - Sunday 0800-1400				
NT1013	Town (S 1)	2	3	R+rock flats	600 m
NT1014	Town (S 2)	2	2	R+LTT	1 km
NT1015	Town (S 3)	2	2	R+sand-rock flats	2.7 km
NT1016	Rainbow Cliff (N)	1	2	R+rock flats	1.3 km
NT1017	Rainbow Cliff	2	2	R+rock-sand flats	800 m
NT1018	Rainbow Cliff (S1)	1	2	R+rock-sand flats	1.4 km
Spring & neap tide range = 2.2 & 1.2 m					

At Cape Wirawawoi the coast turns and begins the long trend south into the Gulf of Carpentaria. The beaches are again exposed directly to the southeast trades, and the Northern Territory's highest energy beaches and barriers are located to the south, particularly south of Cape Arnhem. The first 10 km of coast trend to the southeast and contain seven beaches (NT 1012-1018) between the cape and the Rainbow Cliff, including the main Town Beach.

Town Beach (also known as Gadalathami, Beach **NT 1012**) is the main surfing beach for the Nhulunbuy-Gove region and site of the **Gove Surf Life Saving Club**. The club was established in 1974 and patrols the beach between May and October. The clubhouse is located on grassy bluffs at the very southern end of the beach, where it offers views up the beach and is backed by a large car park (Fig. 5.114). The beach is 3.3 km long and faces east, and is afforded some protection from the waves by Cape Arnhem, with waves averaging about 1 m in height. They maintain a steep coarse sand beach fronted by a flat 50 m wide low tide bar (Fig. 5.115), together with some rocks in the surf in front of the club house.

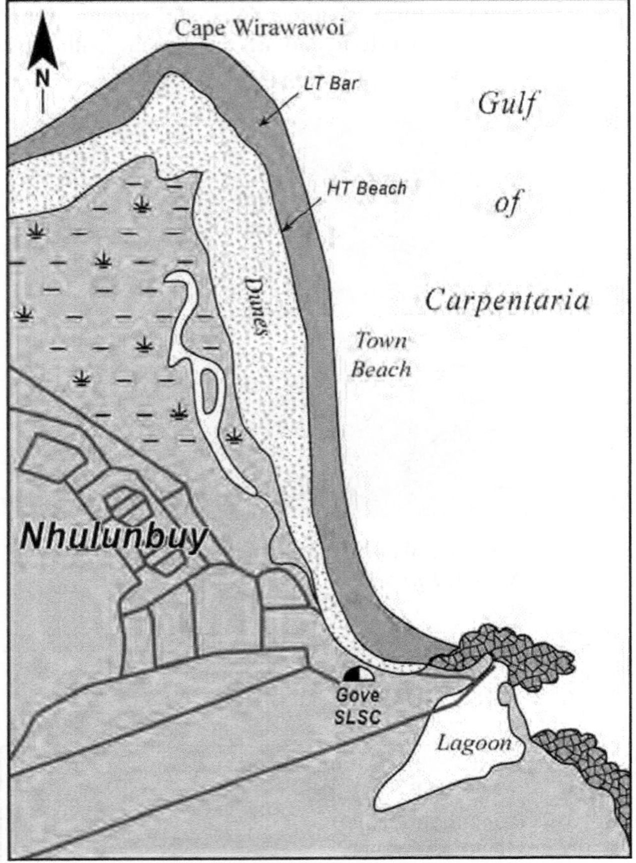

Figure 5.114 Town beach lies immediately east of Nhulunbuy and is the site of Gove Surf Life Saving Club the northernmost in Australia.

Figure 5.115 Town Beach (NT 1012) is patrolled by the Gove SLSC.

The Gove lifesavers also provide patrols at Cape Wirrawawoi at the northern end of the beach, Shady-Yirrakala (NT 1025) and Little Bondi (NT 1041) on some weekends depending on tide and wave conditions and community requirements.

Immediately east of the club is beach **NT 1013**. It extends out along a crenulate rocky point for 600 m, with waves running down along the side of the beach, and rock flats exposed at low tide. A vehicle track from the clubhouse car park runs along the back of the beach to the point.

At the point the shore turns south and the rocks impound a 50 ha V-shaped area of salt flats and mangroves, which are fronted by the narrow beach **NT 1014**. The 1 km long beach partially blocks the lagoon as well as continuing on for a few more hundred metres in front of low laterite bluffs. A vehicle track follows the southern side of the lagoon out to the bluffs and along behind beach NT 1015. Beach **NT 1015** extends south along the bluffs for 1.5 km before impounding a 100 ha mangrove-filled lagoon, with the total beach length 2.7 km, the beach terminating at a migratory creek that drains the lagoon. The beach is fronted by a 200-300 m wide mixture of sand and rock flats.

Beach **NT 1016** is located on the south side of the 50 m wide creek. It trends south then east along a low laterite point for 1.3 km, with rock rubble and flats paralleling the beach and extending up to 300 m offshore. A vehicle track reaches the southern side of the creek mouth and follows the back of the beach out to the point.

Beach **NT 1017** begins on the eastern side of the point and consists of a narrow 800 m long beach which becomes increasingly dominated by vegetated and scarped bluffs which reach a height of 40 m, with bluff debris spilling onto and across the beach. The exposed red and white bluffs are the 'rainbow cliffs', also known as Banambarrna. The beach is fronted by intertidal sand and rock flats up to 300 m wide and terminates at a small protrusion in the bluffs, beyond which is beach **NT 1018.** This is a narrow 1.4 km long curving beach, backed by a few grassy beach ridges then wooded coastal plain, with the eastern 300 m of beach trending north to a low rocky point. Rock flats border the beach, while rock and sand flats extend up to 500 m offshore, resulting in usually calm conditions at the shore.

NT 1019-1024 **MOUNT DUNDAS**

No.	Beach	Rating HT	Rating LT	Type	Length
NT1019	Rainbow Cliff (S 2)	1	2	R+rock flats	1.2 km
NT1020	Mt Dundas (N)	1	2	R+rock flats	600 m
NT1021	Mt Dundas	1	2	R+rock flats	1 km
NT1022	Mt Dundas (S1)	1	2	R+rock flats	200 m
NT1023	Mt Dundas (S2)	2	3	R+LT bar+rips	600m
NT1024	Mt Dundas (S3)	2	3	R+LT bar+rips	1 km
Spring & neap tide range = 2.2 & 1.2 m					

Between Rainbow Cliff and the Mount Dundas region is a 4 km long section of generally north-facing coast containing six beaches (NT 1019-1024). The beaches are initially dominated by shore-parallel rock reefs, which diminish to the east permitting higher easterly waves to reach the shore. The only vehicle access to the coast is at the easternmost beaches (NT 1023 & 1024).

Beach **NT 1019** is a relatively straight 1.2 km long north-northeast-facing high tide sand beach, fronted by low tide rock flats that widen from 100 m in the west to 500 m in the east. Waves break across the outer edge of the reefs, resulting in usually calm conditions at the shore, particularly at low tide (Fig. 5.116). The beach is backed by a small beach ridge plain in the east then scarped 5-10 m high bluffs, all capped by woodlands. It terminates at a low protruding rocky point, on the eastern side of which is beach **NT 1020**. This beach is a curving 600 m long north-facing high tide beach, backed by a low foredune and a 10 m high vegetated bluff, with bordering rock flats, and the edge of the rock reef lying 300 m off the beach. Beach **NT 1021** lies immediately to the east and is a more crenulate 1 km long north-facing high tide sand beach, with the rock flats narrowing to 100 m, but still usually calm conditions at the shore. It is backed by a 30 m high vegetated ridge in the centre that rises rapidly to 73 m high Mount Dundas (also known as Djawulpawuy).

Figure 5.116 Rainbow Cliff (foregound) backing beach NT 1019, which is fringed by rock reefs.

Beach **NT 1022** lies 500 m east of Mount Dundas. It is a curving 200 m long pocket beach, bordered by shallow rock flats that extend out 100 m to either side. The reefs link at the outer extremity, forming a calmer 'lagoon' between the reefs and the beach. Densely wooded slopes that rise slowly to Mount Dundas back the beach. Immediately east is beach **NT 1023**, which is located on a low curving 600 m long rocky point backed by a high tide strip of sand. The point and beach terminate in lee of a rock reef that extends 500 m off the beach.

Beach **NT 1024** is a 1 km long northeast-facing beach largely free of reef, which receives waves averaging about 1 m. These break across a 50 m wide surf zone and generate a few beach rips particularly at low tide. It has a low partly cleared headland and surrounding rock flats at the southern end, with a vehicle track from nearby Yirrkala crossing the headland to reach the southern end of the beach.

NT 1025-1029 **YIRRKALA**

No.	Beach	Rating HT LT	Type	Length
NT1025	Yirrkala	2 3	R+LT rips	2.5 km
NT1026	Yirrkala (E1)	1 2	R+rock flats	150 m
NT1027	Yirrkala (E2)	1 2	R+rock flats	100 m
NT1028	Yirrkala (E3)	1 2	R+rock flats	150 m
NT1029	Yirrkala (E4)	1 2	R+rock flats	200 m
Spring & neap tide range = 2.2 & 1.2 m				

The **Yirrkala** community is located at the southern end of Yirrkala beach (**NT 1025**) in lee of a laterite headland that extends 2.5 km to the east (Fig. 5.117). The beach begins on the southern side of the 100 m wide headland that separates it from beach NT 1024 and trends south before curving round at the southern end to face north. The northernmost 1 km of beach has offshore rock reefs at either end, and some rocks outcropping along the beach. The next 1 km of beach is exposed to moderate waves averaging just over 1 m, which maintain a 50-100 m wide surf zone, with up to 12 rips occurring south of the reefs. Waves decrease in the southern corner and calm conditions often prevail. A small often blocked creek, that drains a 200 ha largely infilled lagoon, flows across the southern end of the beach and against the small rocky point that forms the southern boundary. The beach is backed by foredune ridges in the south, grading to largely vegetated transgressive dunes in the north and extending up to 500 m inland. The 1000 strong Yirrkala community is largely located immediately south of the southern end of the beach, and a track from the community runs behind the beach. The beach is occasionally patrolled by the Gove SLSC.

Figure 5.117 Yirrkala community is located at the southern end of beach NT 1025, with four smaller beaches (NT 1026-1029) located along the base of the headland.

The northern side of the Yirrkala headland contains four small protected beaches (NT 1026-1029). Beach **NT**

1026 is located on the eastern side of the small boundary point. It consists of a 150 m long high tide sand beach, bordered by low rocky points, fronted by 100 m wide rock flats, and backed by a low wooded bluff containing two houses. Immediately east is beach **NT 1027**, a 100 m long high tide sand beach, bordered by a curving rocky spit at its eastern end. The Yirrkala boat ramp crosses the centre of the beach, where some of the rocks have been cleared from the 100 m wide rock flats that front the beach to improve boat access. Three houses also back the eastern half of the beach.

On the eastern side of the curving point is a small 'lagoon' beyond which is 150 m long beach **NT 1028**. This beach is fronted by continuous rock flats, with waves breaking across the rocks up to 100 m offshore. A solitary blufftop house overlooks the western end of the beach. Immediately east is 200 m long beach **NT 1029**, a double crenulate sand beach, bordered and centred by rock reefs, with two sand patches in between. This beach has no vehicle access. It extends south to the northern side of Yirrkala point, with a small transgressive dune from its neighbouring southern beach crossing the back of the point to spill onto the beach.

NT 1030-1038 **YIRRKALA POINT-ROCKY BAY**

No.	Beach	Rating HT	Rating LT	Type	Length
NT1030	Yirrkala Pt (S1)	2	3	R+LTT+rocks	600 m
NT1031	Yirrkala Pt (S2)	1	2	R+LTT+rocks	300 m
NT1032	Yirrkala Pt (S3)	1	2	R+LTT+rocks	100 m
NT1033	Miles Is (S1)	1	2	R+LTT+rocks	700 m
NT1034	Miles Is (S2)	1	2	R+LTT+rocks	600 m
NT1035	Rocky Bay (N2)	1	2	R+rock flats	200 m
NT1036	Rocky Bay (N1)	1	2	R+rock flats	100 m
NT1037	Garrirri Beach	1	2	R+LTT	1.6 km
NT1038	Rocky Bay (S1)	1	2	R+sand flats	200 m
Spring & neap tide range = 2.2 & 1.2 m					

On the southern side of Yirrkala point the coast trends due south for 4 km to the shores of the protected Rocky Bay, and contains nine generally small rock-controlled beaches (NT 1030-1038), apart from the longer Rocky Bay beach (NT 1037). There is vehicle access to the coast south of the point at beach NT 1031 and at either end of Rocky Bay.

Beach **NT 1030** is located on the southern side of the 10 m high massive sandstone of Yirrkala point. The 600 m long beach initially faces south in the protected lee of the point, then curves round to face east, exposing it to waves averaging about 1 m. These break across a 50 m wide low tide bar, as well as across some rocks scattered along the beach and particularly at either end, with a permanent rip against the southern rocks. The beach is backed by a 200 m wide early Holocene grassy dune field that extends across to the northern side of the point (Fig. 5.118).

Figure 5.118 Beaches NT 1030 & 1031 curve south of Yirrkala headland.

A low rocky point forms the southern boundary of the beach, on the southern side of which is the curving 300 m long beach **NT 1031**. This beach occupies a small valley, with rocks and reefs protruding out either end, and sandier conditions in the centre. A gravel road runs out to the northern point, with some farmland located behind the beach. Beach **NT 1032** is located on the southern wooded point, just past the rocks. It is a straight northeast-facing 100 m long beach, backed by wooded bluffs and fronted by a mixture of sand and rock reefs.

One hundred metres to the south the coast trends more southerly, with 10 ha Miles Island lying 500 m off beach **NT 1033**. This 700 m long beach faces due east and consists of three continuous crenulate sections each tied to rock reefs. It terminates at the southernmost cuspate foreland. Beach **NT 1034** is slightly more exposed and continues on south of the foreland for 600 m to a 200 m long section of low laterite bluffs. A vehicle track reaches the northern end of the beach.

Beaches **NT 1035** and **NT 1036** are two near continuous 200 m and 100 m long strips of sand, located on the southern side of the bluffs. They are backed by wooded slopes and fronted by 100 m wide rock flats and reefs. They terminate just north of the northern side of Rocky Bay beach.

Garrirri Beach (**NT 1037**) forms the western shore of Rocky Bay. The 1.6 km long beach initially faces east, curving around in the south to face north. The beach is the outermost of up to ten beach ridge-spits in a 1-1.5 km wide series that have filled the outer portion of a 4 km deep embayment. A creek borders the southern end of the beach and connects a backing 200 ha mangrove-filled lagoon with the bay. The beach has a continuous bar in the north, but receives increasing protection to the south from a prominent headland. This permits tidal sand shoals to extend up to 500 m off the creek mouth. There is vehicle access to the creek mouth which is used for launching small boats.

Beach **NT 1038** is located on the eastern side of the 100 m wide creek, and consists of a protected north-facing 200 m long sand spit, backed in the east by densely vegetated slopes rising rapidly to 30 m.

NT 1039-1048 **ROCKY-DALYWOI BAYS**

No.	Beach	Rating H	LT	Type	Length
NT1039	Rocky Bay (S2)	3	4	R+LT rips	250 m
NT1040	Rocky Bay (S3)	3	4	R+LT rips	200 m
NT1041	Little Bondi	2	3	R+LTT	200 m
NT1042	Turtle Beach	3	4	R+LT rips	100 m
NT1043	Needle Pt (1)	1	2	R+rock flats	50 m
NT1044	Needle Pt (2)	1	2	R+rock flats	100 m
NT1045	Macassan Beach	1	2	R+LTT	200 m
NT1046	Dalywoi Bay (N2)	1	2	R+sand flats	200 m
NT1047	Dalywoi Bay (N1)	1	2	R+sand flats	300 m
NT1048	Dalywoi Bay	1	2	R+sand flats	500 m
Spring & neap tide range = 2.2 & 1.2 m					

Between Rocky and Dalywoi bays are 12 km of generally more exposed east-facing shore dominated by red laterite and jagged calcarenite cliffs, with ten small beaches (NT 1039-1048) occupying a total of 2.1 km of the shore. This section of coast is readily accessible by 4WD vehicle from Nhulunbuy and Yirrkala and is a relatively popular area for day tripping and camping, and tidal and beach fishing, as well as surfing on the exposed beaches. There are no facilities at any of the beaches.

Beaches **NT 1039** and **NT 1040** are located 2 km south of Rocky Bay in an indentation in a 20-30 m high section of jagged red calcarenite bluffs. The two adjoining beaches are 250 m and 200 m long respectively, both facing east into waves averaging over 1 m. These maintain a 50 m wide bar, with rips common at low tide and against the boundary rocks. A vehicle track winds along the top of the backing cliffs.

One kilometre to the south is the more embayed **Baringura** or **Little Bondi** beach (**NT 1041**), a 200 m long white sand beach, bordered by 10-20 m high red laterite cliffs, extending 200-300 m off the beach. These reduce waves to just under 1 m and maintain a low tide bar, with rips usually against the points. There is a small foredune and both active and older dune transgression extending up to 200 m in lee of the beach (Fig. 5.119), with informal camping in the deflation hollow. This is a relatively popular surfing beach, and at times the entire Gove surf life saving patrol moves to the beach, so people can enjoy the surf in greater safety.

Figure 5.119 Little Bondi (Baringura) beach (NT 1041).

Another kilometre to the south is the more exposed **Ngumuy** or **Turtle Beach** (**NT 1042**), a 100 m long strip of sand wedged in between laterite bluffs to the south and calcarenite bluffs to the north. Waves average over 1 m and produce a 50 m wide surf zone with a permanent rip flowing out against the southern rocks (Fig. 5.120). However if swimming or surfing be careful of the rip and the rugged rocky shore to either side.

Figure 5.120 Turtle Beach (Ngumuy) (NT 1042) is a popular surfing beach.

The Macassan beaches are located in the next small embayment 1.5 km to the south. The 1 km wide southeast-facing bay contains three small beaches, which become increasing exposed to the south. They are all accessible by 4WD. Beach **NT 1043** is located in lee of the northern protruding narrow **Needle Point**, and consists of a 50 m pocket of sand backed by 10 m high grassy slopes, with rock flats extending 100 m off the beach. Immediately to the east is beach **NT 1044**, a similar 100 m long strip of sand at the base of steeper 10 m high bluffs. The southern 200 m of the bay are occupied by **Garanhan** or **Macassan** beach (**NT 1045**), site of Macassan camps in the past. The beach receives waves averaging less than 1 m which break across a 50 m wide bar, with rips flowing out against the rocks at either end. There is informal camping in lee of the beach.

South of Macassan Beach the coast turns and trends southwest into the more protected Dalywoi Bay. Beach **NT 1046** is located in a small 200 m wide southeast-facing bay. It is a low energy beach fronted by 300-500 m wide sand flats, which extend out to the ends of the eroding laterite headlands, with vehicle tracks running out to the tip of both headlands. The next small bay south contains 300 m long beach **NT 1047**, which also faces southeast across tidal sand shoals that extend up to 1 km off the beach. The trade winds have caused some minor dune activity in lee of the beach.

Beach **NT 1048** is located along the northern side of the Dalywoi Bay tidal channel. The 500 m long estuarine beach represents the outer portion of a series of spits that have moved sediment from the open bay into the tidal channel. This is the most popular beach in the region, and is used for camping, boat launching and fishing, as well as being the closest to Nhulunbuy. Swimming is not advised owing to the presence of crocodiles, as well as the deep tidal channel just off the beach.

NT 1049-1053 **DALYWOI BAY**

No.	Beach	Rating HT	Rating LT	Type	v Length
NT1049	Dalywoi Bay (1)	1	2	R+tidal shoals	900 m
NT1050	Dalywoi Bay (2)	1	2	R+ridged tidal flats	2.2 km
NT1051	Dalywoi Bay (3)	1	2	R+ridged tidal flats	900 m
NT1052	Dalywoi Bay (4)	1	2	R+ridged tidal flats	1 km
NT1053	Nanydjaka	2	4	R+rock flats	1.4 km
Spring & neap tide range = 2.2 & 1.2 m					

Dalywoi Bay is a 3 km wide northeast-facing bay located between Macassan Beach and the lee of Cape Arnhem. The bay is 3 km deep and is backed by 350 ha double Holocene and inner Pleistocene barrier systems, as well as a 1200 ha estuarine system. The northerly orientation of the bay beaches and Cape Arnhem provides considerable protection to the shoreline, which is dominated by four low energy beaches (NT 1049-1052), and extensive tidal and ridged sand flats. The entire bay has consequently acted as a major sink for both Pleistocene and Holocene marine sediment, transported around Cape Arnhem by both wave and tide-driven currents. In addition high energy beaches south of Cape Arnhem developed early Holocene transgressive dunes that have extended up to 2 km across the cape and into the bay. There is rough to sandy 4WD vehicle access from the south to all the beach systems.

Beaches NT 1049 and 1050 are parts of the main bay barrier system. They are 900 m and 2.2 km long respectively, face northeast and are separated by a 200 m long section of low rocks (Fig. 5.121). Beach **NT 1049** abuts the main tidal channel and has tidally driven sand shoals extending up to 1 km off the beach. The tidal shoals rapidly diminish along beach **NT 1050** and are replaced by 1.5 km wide ridged sand flats containing up to eight low amplitude ridges that fill much of the southern half of the bay. Both beaches are backed by 0.5-1 km wide beach ridge plain containing several ridges, then an older Holocene barrier and finally a Pleistocene barrier system.

Figure 5.121 Beaches NT 1049 & 1050 (left) converge at a low rocky point.

Beach **NT 1051** commences past a second low rocky inflection in the bay. It faces north and is 900 m long with ridged sand flats extending well into the bay. It is bordered in the east by more low rocky shore, which is capped by stable transgressive dunes that have crossed the cape from the south. Beach **NT 1052** runs along the back of more stable transgressive dunes for 1 km to the rocky rear shore of Cape Arnhem. The narrow, low energy beach faces northwest across 1 km wide ridged sand flats.

The northern shore of **Cape Arnhem** (also known as Nanydjaka) is a straight 1.4 km long rocky shore, backed by a continuous strip of high tide sand (beach **NT 1053**). There is one 200 m long gap in the rocks toward the western end of the beach. It receives waves lowered by refraction around the cape which break across the 50 m wide rock flats.

Region 4: East Arnhem Land-Gulf Coast: Cape Arnhem-NT/QLD Border

Length of coast: 1302 km (3727-1288 km)
Number of beaches: 435 (NT 1054-1488)

Aboriginal Land: Arnhem Land (including Bickerton and Groote islands), Marra
National Parks: none

Regional Figures:			
Figure 5.122	page 291	beaches NT	958-1117
Figure 5.128	page 298	beaches NT	1073-1275
Figure 5.141	page 313	beaches NT	1275-1414
Figure 5.154	page 329	beaches NT	1415-1488
Figure 5.166	page 341	Bickerton Island	BI 1-53
Figure 5.168	page 342	Groote Eylandt	GE 1-284

Towns/communities: Milyakburra (Bickerton Is), Alyangula (Groote Eylandt), Numbulwar, Borroloola

The eastern Arnhem Land and western Gulf coast initially trends 350 km south into Limmen Bight, then trends southwest to the border. Offshore of the coast are the large Groote Eylandt and Bickerton Island, and in the south the Sir Edward Pellew Group of islands. The coast ranges from the most exposed and highest energy in the gulf at Cape Arnhem, to the low energy tidal-flat-dominated shores of much of the southern gulf.

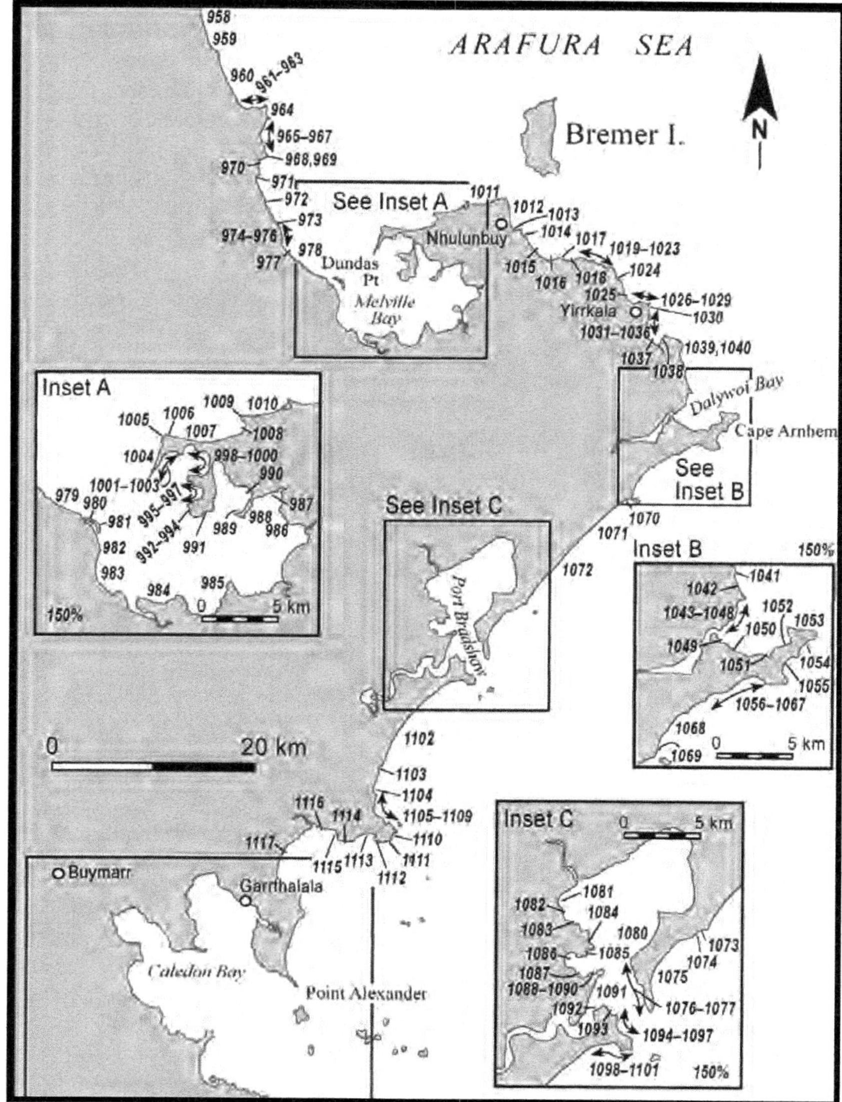

Figure 5.122 Regional map of northeastern Arnhem Land. Region 4 commences at Cape Arnhem and includes the eastern Arnhem Land and Gulf coast down to the Queensland border, beaches NT 958-1117. See Figures 5.128, 5.141, 5.154, 5.166 and 5.168 for full coverage of coast and beaches.

NT 1054-1061 **CAPE ARNHEM (S1)**

No.	Beach	Rating HT	Rating LT	Type	L ength
NT1054	Cape Arnhem (S1)	2	3	R+LTT+reef	200 m
NT1055	Cape Arnhem (S2)	2	3	R+rock flats	1.8 km
NT1056	Cape Arnhem (S3)	2	3	R+rock flats	150 m
NT1057	Cape Arnhem (S4)	1	3	R+LTT+reef	300 m
NT1058	Cape Arnhem (S5)	1	3	R+LT bar+rips	200 m
NT1059	Cape Arnhem (S6)	3	4	R+LT bar+rips	300 m
NT1060	Cape Arnhem (S7)	3	4	R+LT bar+rips	1.4 km
NT1061	Cape Arnhem (S8)	3	4	R+LT bar+rips	100 m
Spring & neap tide range = 2.2 & 1.2 m					

Cape Arnhem is the easternmost tip of the northern Northern Territory coast. The coast from the cape to Cape Shield, 130 km to the south, is the most exposed and highest energy section of shoreline in the Territory and the Gulf of Carpentaria. It is exposed to the full force of the southeast trade winds, as well as to waves generated across the full fetch of the gulf. While no accurate wave data are available, these waves average over 1 m with a period of 5-6 s, arriving from the southeast in winter and more easterly in summer. During strong southeasterlies the waves can reach 2-3 m, though wave period remain short. The shoreline reflects the high level of energy with rip-dominated beaches backed by the most extensive transgressive dunes systems in the Territory, extending in places up to 5 km inland and to elevations of 50 m.

The first 6.5 km of coast south of the cape trend to the southwest, facing squarely into the trades. The shore is a mixture of initially laterite then calcarenite headlands and bluffs, with eight carbonate-rich beaches (NT1054-1061) occupying 4.5 km of the shore. Rough and in places long sandy 4WD tracks link all the beaches to the main track off the Central Arnhem road.

Beach **NT 1054** is located on the southern tip of the cape and is a 200 m southeast-facing high tide beach, fronted by patchy rock flats largely covered by a veneer of sand. It is bordered and backed by10 m high red bluffs, with a vehicle track running along the top of the sparsely vegetated bluff tops. Crenulate laterite bluffs extend for 1.5 km to the south, to the beginning of beach **NT 1055**. This is a curving southeast to east-facing high tide sand beach, fronted by rock reefs that lower waves at the shore to less than 1 m, maintaining a steep reflective usually cusped beach. The beach is backed by older partially destabilised Holocene dunes extending 1 km across the cape to Dalywoi Bay and beaches NT 1051 and 1052. The beach terminates at a prominent laterite headland, 1 km west of which is beach NT 1056.

Beach **NT 1056** is a 150 m long semicircular southwest-facing pocket beach, bordered by 100 m wide rock flats, with a small partly rock-filled 'lagoon', and the beach in between. Vehicle tracks run right to the back of the beach. Beach **NT 1057** commences immediately to the west and also faces southwest down the coast. It is a relatively straight 300 m long beach, that has deeper rock reefs off the eastern half, with a permanent rip draining the western half of the beach. To the lee a foredune increases in height to the west as it climbs the bluffs, which reach 20 m at the western end of the beach.

At the bluffs the shoreline turns and trends southwest, with the shoreline dominated by Pleistocene calcarenite bluffs and intervening beaches facing directly into the trades. All the beaches and bluffs are backed by both lithified Pleistocene and vegetated to, in places, active Holocene dunes extending up to 1 km inland. Beach **NT 1058** occupies a 200 m gap in the 15 m high bluffs. It is fronted by a 100 m wide surf zone, consisting of low tide rips spaced approximately every 150 m, which continues on linking the beaches and bluffs for 2.5 km down to beach NT 1061. The beach is backed by two active blowouts that have climbed the backing bluffs and extend 100 m inland.

Beach **NT 1059** is located 200 m to the south and is a straight 300 m long beach, bordered by calcarenite bluffs, with the southern bluffs forming a small headland. The surf and rips continue past the bluffs and along the beach, with generally stable dunes behind the beach. Around the small headland is the beginning of beach **NT 1060**, a straight 1.4 km long beach with a surf zone containing up to eight rips, and with an active foredune behind, then the older dunes. It terminates at a groyne-like protrusion in the calcarenite. On the southern side of the 'groyne' is a small indentation in the shore containing 100 m long beach **NT 1061**. This is an open exposed beach backed by 10-15 m high calcarenite bluffs. There is a vehicle track to the back of the 'groyne', with a 100 m walk around the rocks required to reach the beach.

NT 1062-1069 **CAPE ARNHEM (S2)**

No.	Beach	Rating HT	Rating LT	Type	Length
NT1062	Cape Arnhem (S9)	2	3	R+LTT	50 m
NT1063	Cape Arnhem (S10)	2	3	R+LTT	150 m
NT1064	Cape Arnhem (S11)	2	3	R+LTT	200 m
NT1065	Cape Arnhem (S12)	1	2	R+LTT	150 m
NT1066	Cape Arnhem (S13)	3	4	R+LT bar+rips	800 m
NT1067	Cape Arnhem (S14)	1	2	R+LTT	200 m
NT1068	Cape Arnhem (S15)	3	4	R+LT bar+rips	3.5 km
NT1069	Cape Arnhem (S16)	2	3	R+LT bar+rips	200 m
Spring & neap tide range = 2.2 & 1.2 m					

On the southern side of beach NT 1061 is a 500 m long south-facing section of 20-40 m high calcarenite bluffs and clifftop dunes, with a predominantly rocky shoreline. Beach **NT 1062** is located in the centre of the cliffs. It is a 50 m pocket of sand backed by steep 40 m high vegetated slopes and bordered by 10-20 m high grassy bluffs. Two hundred metres to the west at the end of the bluffs the shoreline turns and faces southeast with seven sandy beaches (NT 1063-1069) dominating the shore down to beach NT 1069.

The first two beaches are separated by a small outcrop of calcarenite rock. Beach **NT 1063** is 150 m long and backed by a climbing blowout that extends 500 m inland, while its neighbour, beach **NT 1064**, is 200 m long with grassy slopes behind. It terminates at a 20 m high 100 m wide calcarenite headland. On the southern side of the headland is beach **NT 1065**, a curving 150 m long pocket beach with a larger headland on its southern side. The two calcarenite points converge to provide an 80 m opening for the protected beach. The second headland extends for 200 m to the northern end of beach NT 1066.

Beach **NT 1066** commences in lee of the headland which lowers waves to about 1 m and provides a continuous attached bar. Once clear of the headland wave height increases to over 1 m and low tide rips, spaced approximately every 150 m, dominate the remainder of the 800 m long beach. The beach is backed by active blowouts, climbing up over older vegetated dunes, the latter extending up to 2 km inland. The southern end of the beach is tied to an isolated 100 m wide headland-knoll, with two rocks lying off the headland (Fig. 5.123). This knoll forms the northern boundary of beach **NT 1067**, a curving 200 m long beach impounded by a slightly larger knoll on its southern side, the two headlands affording moderate protection to the beach. An active foredune backs the beach.

Figure 5.123 Beach NT 1066 is backed by a 2 km wide dune field, with the smaller beaches NT 1063-1067 to either side.

Beach **NT 1068** is the first of the longer beaches south of the cape. The 3.5 km long beach extends from the southern side of beach NT 1067 down to a foreland formed in lee of a 30 ha island. The 40 m high island is composed of Archaean granite and lies 300 m off the tip of the foreland. This is an exposed southeast to east-facing beach, which is dominated by up to 20 low tide beach rips. Generally stable dunes, with patchy blowouts, extend up to 2.5 km in from the beach. At the southern tip of the beach, right on the foreland, is beach **NT 1069**, a 200 m long convex strip of sand backed and bounded by granite boulders, and lying directly in lee of the island. This beach receives refracted waves from both sides, as well as wave and tidal currents flowing between the island and the foreland, and consequently has lower energy though a dynamic beach and surf zone.

NT 1070-1076 **CAPE ARNHEM (S3)**

No.	Beach	Rating HT	LT	Type	Length
NT1070	Caves Beach	5	5	TBR	800 m
NT1071	Cape Arnhem (S18)	6	6	TBR	150 m
NT1072	Oyster Beach	6	6	TBR+LBT	11 km
NT1073	Cape Arnhem (S20)	5	5	TBR	1.4 km
NT1074	Cape Arnhem (S21)	2	2	R	100 m
NT1075	Gwapilina Pt (N)	6	6	TBR	2.2 km
NT1076	Gwapilina Pt	2	2	R	200 m
Spring & neap tide range = 1.4 & 0.0 m					

Thirteen kilometres southwest of Cape Arnhem the mangrove-dominated wetlands of Dalywoi Bay almost reach the coast in lee of the granite island. The coast trends away to the southwest from the lee of the island for 20 km down to Gwapilina Point. This section contains Oyster or Lurrupukurru Beach (NT 1072), the highest energy beach in the Territory and gulf, all backed by one of the more massive Pleistocene and Holocene dune systems on the coast, totalling 3200 ha in area. Vehicle tracks link the northern beaches and run behind the massive dune system to the southern end of Oyster Beach.

Caves or **Rangura** beach (**NT 1070**) commences immediately south of the foreland in lee of the island. The straight 800 m long beach is bordered by the granite to the north, and calcarenite-capped granite boulders to the south. It is well exposed to easterly waves and has a continuous 100 m wide bar usually cut by rips every 150 m. A 4WD track and large parking area are located in lee of the northern rocks, with the remainder of the beach backed by an active foredune that is transgressing over older stable Holocene dunes that extend up to 1.5 km inland. A 500 m long calcarenite bluff separates it from beach **NT 1071**. This is a 150 m long pocket of sand wedged in between 20 m high jagged bluffs. The gap has provided a conduit for dune sand, with an active blowout extending 300 m in from the beach. A vehicle track runs down the northern side of the dunes, with a second track following the top of the southern bluffs.

Oyster or **Lurrupukuttu** Beach **NT 1072** commences 200 m to the south as the calcarenite bluffs are replaced by vegetated Pleistocene calcarenite and overriding stable and active Holocene dunes. This 11 km long, southeast-facing beach is probably the highest energy of the entire Territory and gulf coast. It is composed of fine sand, rich in carbonate, which in combination with wind waves averaging about 1.5 m maintains a 300 m wide double bar system. The 100 m wide inner bar is cut by rips approximately every 150 m, with up to 70 rips along the beach. The rips can produce pronounced shoreline rhythms, with the rips exiting the embayments. A shore-parallel trough lies off the inner bar, with a more parallel outer bar extending up to 300 m offshore. The whole beach system is backed by multiple episodes of Holocene parabolic dune transgression, riding over earlier Pleistocene dunes. The dunes extend up to 4 km inland and to heights of 30 m. A main vehicle track parallels the beach about 1 km inland. The beach terminates in the south as it narrows and is replaced by 10 m high calcarenite bluffs fronted by shallow rock reefs.

Beach **NT 1073** commences 1 km to the south on the south side of a low calcarenite foreland, in lee of scattered granite islets and reefs extending 2 km offshore. The beach is 1.4 km long, facing initially south then curving around to face southeast, terminating at a bulging sandy foreland in lee of more offshore islets and reefs and amongst a collection of large granite boulders at the end of the beach. The beach receives waves averaging over 1 m which maintain a single bar usually cut by several rips. Beach **NT 1074** is a 100 m pocket of sand located amongst the granite boulders, which extend over 100 m seaward from either end. A vehicle track follows a backing dune ridge to the rear of the picturesque beach.

Beach **NT 1075** commences on the southern side of the boulders and is a slightly curving 2.2 km long southeast- to east-facing beach that terminates 2 km north of Gwapilina Point. The beach is composed of coarser, carbonate-rich sand (30-50% carbonate), which maintains a steeper beach gradient and a single 100 m wide bar, cut by up to 14 rips. The beach is backed by near continuous active blowouts, which are transgressing over the 40-80 m high, 1 km wide, earlier Holocene dunes (Fig. 5.124).

Figure 5.124 Beach NT 1075 is backed by an active dune field.

Gwapilina Point is a calcarenite-capped granite point that forms the northern entrance to Port Bradshaw. Right on the tip of the point is a small island, tied to the point by a tombolo, the western side of which is beach **NT 1076**. The 200 m long beach faces south across the bay entrance and is protected by the island from most waves. The vehicle track terminates on the cleared slopes above the beach.

NT 1077-1080 **PORT BRADSHAW (E)**

No.	Beach	Rating HT	Rating LT	Type	Length
NT1077	Port Bradshaw (E1)	1	2	R+reef flats	300 m
NT1078	Port Bradshaw (E2)	1	1	R+sand flats	200 m
NT1079	Bawaka	1	1	R+sand flats	400 m
NT1080	Yalangbara	1	1	R+sand flats	4 km

Spring & neap tide range = 1.4 & 0.0 m

Port Bradshaw is a 5000 ha embayment formed from partial drowning of the sloping hinterland and blocked along much of its eastern side by the large barrier system that extends south of Dalywoi Bay-Cape Arnhem to Gwapilina Point. The bay has a 2 km wide bedrock-controlled entrance bordered by the granite Gwapilina and Binanangoi points. Several small upland streams flow to its western shore, with most circulation-driven by tidal flow through the entrance. The bay has 40 km of low energy shoreline, much dominated by a narrow fringe of mangroves and sand flats. There are 18 low energy beaches (NT 1077-1094) within the bay, all located in the southern half and within 9 km of the entrance. While the outermost beaches receive low refracted Gulf waves, most are maintained by wind waves generated in the bay.

The eastern shore of the bay commences at Gwapilina Point and extends roughly north for 12 km. Four beaches (NT 1077-1080) are located along the first 6 km of the shore, the remainder of the shoreline dominated by mangroves. Beach **NT 1077** is located 1.5 km north of the tip of the point. It is a bent 300 m long south-facing beach fringed by sand and coral reef flats up to 100 m wide, and backed by grassy slopes rising to 15 m. Granite boulders border the beach, with a few located on the beach. Beach **NT 1078** lies around the northern low granite point. It is a 200 m long west-facing strip of high tide sand fronted by sand flats and backed by lightly wooded slopes. The next small embayment, called Bawaka, contains beach **NT 1079**, a crenulate 400 m long west-facing beach with numerous granite boulders outcropping on the beach and sand flats.

The main eastern beach (**NT 1080**) is a curving 4 km west-facing beach that extends from the granite boulders north of its neighbour, deep into the bay to a protruding sand foreland-spit, beyond which mangroves dominate the shore. The water is relatively deep off the southern half of the beach, widening as sand flats toward the foreland-spit. The small **Yalangbara** community is located on the tip of the point.

NT 1081-1090 **PORT BRADSHAW (W)**

No.	Beach	Rating HT	Rating LT	Type	Length
NT1081	Port Bradshaw (W1)	1	1	R+sand flats	800 m
NT1082	Port Bradshaw (W2)	1	1	R+sand flats	300 m
NT1083	Port Bradshaw (W3)	1	1	R+sand flats	150 m
NT1084	Port Bradshaw (W4)	1	1	R+sand flats	600 m
NT1085	Port Bradshaw (W6)	1	1	R+sand flats	600 m
NT1086	Dhulmulmiya	1	1	R+tidal flats	500 m
NT1087	Port Bradshaw (W8)	1	1	R+sand flats	150 m
NT1088	Port Bradshaw (W9)	1	1	R+sand flats	300 m
NT1089	Port Bradshaw (W10)	1	1	R+sand flats	150 m
NT1090	Port Bradshaw (W11)	1	1	R+sand flats	100 m

Spring & neap tide range = 1.4 & 0.0 m

The 20 km long western shores of Port Bradshaw are dominated by mangroves in the north, including a beach ridge plain associated with two creek systems. Along the southern half granite bedrock dominates the shore, resulting in a series of small embayments, most occupied by the ten generally small, low energy estuarine beaches (NT1081-1090) that occupy 10 km of the shoreline.

Beach **NT 1081** is an east-facing 800 m long strip of high tide sand, backed by low wooded slopes and fronted by sand flats. A clump of mangroves occupies the southern corner of the beach. Just around the small southern point is beach **NT 1082**, a 300 m long northeast-facing sand beach, bordered by a more prominent southern headland extending 500 m into the bay. The next embayment to the south contains a northeast-facing 150 m long pocket beach (**NT 1083**) at its head, with a small upland creek draining out against the southern end of the beach.

Beach **NT 1084** occupies the next embayment and faces west across the bay toward beach NT 1080. The beach is 600 m long and bounded by low vegetated headlands. Beach **NT 1085** faces down the bay toward the entrance, 5 km to the southeast. It is bordered by vegetated granite points, and has a small foredune toward its southern end.

Beach **NT 1086** is the most sheltered of the bay beaches, being located deep inside a semicircular bay, with the area behind the beach called Dhulmulmiya. The 500 m long east-facing beach is fringed by mangroves at either end and crossed by two upland creeks, which have combined with the sand flats to produce tidal shoals along the beach. One and a half kilometres across the embayment is a 2 km long bedrock point containing four small beaches (NT 1087-1090). Beach **NT 1087** is a 150 m long north-facing pocket beach, bordered by small vegetated points, together with a few mangroves against the eastern end. Its eastern neighbour, beach **NT 1088**, is a curving 300 m long north-facing beach, backed by a grassy beach ridge. Dunes from beach NT 1091 are transgressing onto the northern end of the beach. Beach **NT 1089** is located on the northern tip of the point, and is a 150 m long strip of sand, bounded and partly cut by granite boulders. Finally on the eastern tip of the point is the pocket beach **NT 1090**, a 100 m long strip of sand that faces east across the bay toward beach NT 1079.

NT 1091-1097 **HOLLY INLET & BINANANGOI POINT**

No.	Beach	Rating HT	Rating LT	Type	Length
NT1091	Dhaniya	1	1	R+sand flats	2.8 km
NT1092	Holly Inlet (E)	1	1	R+sand flats	700 m
NT1093	Binanangoi Pt (N5)	1	2	R+rock flats	800 m
NT1094	Binanangoi Pt (N4)	2	2	R	400 m
NT1095	Binanangoi Pt (N3)	2	2	R	200 m
NT1096	Binanangoi Pt (N2)	2	2	R	300 m
NT1097	Binanangoi Pt (N1)	2	2	R	100 m
Spring & neap tide range = 1.4 & 0.0 m					

The southwestern shore of Port Bradshaw contains two beaches (NT 1091 & 1092), located either side of the entrance to Holly Inlet, a southern tributary of the port, and five beaches (NT 1093-1097) along the northern side of Binanangoi Point, the southern entrance to the port. All seven beaches receive low refracted swell entering the port and become increasingly wave-dominated toward the entrance.

Beach **NT 1091** is located along the western entrance to Holly Inlet. It faces due east toward the port entrance and receives low refracted waves, particularly along the northern half of the 2.8 km long beach. The waves have moved marine sands into Holly Inlet, building a low beach ridge plain up to 200 m wide that grades to a spit toward the southern half. The deep inlet channel parallels the southern half of the beach, with tidal sand shoals off the northern half. The 1500 ha inlet extends for another 12 km to the south, backing the long beach-barrier system (beaches NT 1102-1103). A vehicle track runs out to the centre of the beach, where the small **Dhaniya** community, consisting of four fishing shacks, is located.

Beach **NT 1092** is located on the eastern side of the inlet. The beach is tied at its eastern end to a low granite point and extends to the southwest for 700 m, terminating in a small recurved spit. A small salt lake backs the eastern end of the beach, while sand flats up to 300 m wide lie between the beach and the deeper inlet channel.

The **Binanangoi Point** beaches begin on the eastern side of the 35 m high granite point, and extend for 3 km down to the point. All tend to face northeast across the port entrance toward Gwapilina Point. Beach **NT 1093** is a curving 800 m long beach, bounded by low gradient boulders, as well as having granite outcrops on the beach and dominating the intertidal flats. Beaches **NT 1094** and **NT 1095** are adjoining crescentic reflective beaches 400 m and 200 m long respectively, connected by a small tombolo in lee of a granite outcrop. The beaches receive waves averaging 0.5 m and are backed by a 50 m wide grassy beach ridge.

Beach **NT 1096** is a curving 300 m long northeast-facing beach, bounded by low granite points, with rocks outcropping on the centre of the beach. Finally, beach **NT 1097** is located at the northeastern tip of the point. It is a 100 m long pocket of sand, facing east out into the gulf, but still only receiving low refracted waves, which maintain a steep reflective beach.

NT 1098-1103 **BINANANGOI POINT (S)**

No.	Beach	Rating HT	Rating LT	Type	Length
NT1098	Binanangoi Pt (S1)	3	3	R/LTT	200 m
NT1099	Binanangoi Pt (S2)	3	3	R/LTT	200 m
NT1100	Binanangoi Pt (S3)	3	3	R/LTT	300 m
NT1101	Binanangoi Pt (S4)	4	4	LTT	50 m
NT1102	Holly Inlet	6	6	TBR	11.6 km
NT1103	Holly Inlet (S)	4	4	LTT	2 km
Spring & neap tide range = 1.4 & 0.0 m					

Binanangoi Point forms the northern boundary of an open 15 km wide east-facing embayment, dominated by the near continuous 14 km long Holly Inlet beach and dune system (NT 1102 & 1103), with eleven smaller bedrock-controlled beaches to either end (NT 1098-1101, 1104-1110).

Beaches NT 1098-1101 are all located around the point. Beaches **NT 1098** and **NT 1099** are located at the southern extremity of the point. They are two adjoining 200 m long beaches that face due south. They are however partly protected by two calcarenite-capped granite islands and several reefs off the point, resulting in waves averaging less than 1 m at the shore. The waves and medium sand produce a reflective to at times low tide terrace beach.

Inside the western lee of the point are beaches **NT 1100** and **NT 1101**, two straight 300 m and 50 m long southeast-facing beaches bordered by large granite bounders and slopes and backed by steep partly vegetated sandy slopes. They are also partly protected by the island, with waves averaging just under a metre which maintain a narrow low tide bar.

The main beach (**NT 1102**) begins immediately south of the granite boulders. It extends to the southwest, then south for 11.6 km to a low 50 m wide granite outcrop, which separates it from beach NT 1103. Wave energy increases rapidly down the beach, averaging about 1.5 m as far as a small cuspate foreland formed in lee of the 11 m high Rigel Kent Rocks. From here down to the point waves average about 1 m. The higher waves maintain a 100 m wide rip-dominated bar system, with rips spaced approximately every 150 m, resulting in up to 60 rips along the beach. South of the foreland the bar is more continuous. The entire beach is backed by massive, 2300 ha, largely vegetated, multiple parabolics extending up to 3.5 km inland (Fig. 5.125), with Holly Inlet backing the entire dune system.

Figure 5.125 Part of beach NT 1102 with its backing multiple parabolic dunes.

Beach **NT 1103** extends due south of the granite point for 2 km to the northern granite slopes of 100 m high Mount Alexander. The beach receives protection from Wanyanmera Point and scattered offshore rock reefs, resulting in waves averaging 1 m or less, and a continuous bar along the beach, usually free of rips. The beach is backed by a 200-300 m wide barrier then the salt flats of the southern reaches of Holly Inlet.

NT 1104-1110 **MOUNT ALEXANDER**

No.	Beach	Rating HT	Rating LT	Type	Length
NT1104	Mt Alexander (N6)	3	3	LTT	100 m
NT1105	Mt Alexander (N5)	2	2	R	50 m
NT1106	Mt Alexander (N4)	2	2	R	1.5 km
NT1107	Mt Alexander (N3)	2	2	R	100 m
NT1108	Mt Alexander (N2)	2	2	R	800 m
NT1109	Mt Alexander (N1)	2	2	R	150 m
NT1110	Wanyanmera Pt	4.5	4.5	LTT/TBR	600 m
Spring & neap tide range = 1.4 & 0.0 m					

Mount Alexander is a 100 m high lightly wooded granite tor that dominates the southern section of the embayment. The surrounding shoreline for 3 km to the north and south contains several smaller granite points and seven intervening beaches (NT 1104-1110).

Beach **NT 1104** is a 100 m long east-facing pocket of sand located at the southern end of beach NT 1103, and separated from it by a small granite point. It receives waves averaging less than 1 m which maintain a continuous low tide bar. Granite slopes dominate the next 200 m, with beach **NT 1105** a 50 m pocket of sand located between a granite outcrop and a southerly inflection in the bedrock. The beach faces east and is reflective.

One hundred metres to the south the rocks form the northern boundary of 1.5 km long beach **NT 1106**, a relatively straight east-facing reflective beach, protected by the point and several islands and reefs, and backed by low grassy slopes. Beach **NT 1107** is located 100 m around the southern granite point. It is a 100 m long northeast-facing pocket beach, partly surrounded by grassy slopes that rise steeply to 100 m high Mount Alexander, located 500 m south of the beach. Several rocks also outcrop off the beach.

Beaches **NT 1108** and **NT 1109** are adjoining steep reflective beaches, 800 m and 150 m long respectively, joined by a small tombolo in lee of a linear granite reef. The beaches are usually cusped and are backed by a 50-100 m wide beach ridge, with steeper slopes to either end.

Beach **NT 1110** is located 1 km to the east, immediately south of 30 m high **Wanyanmera Point**, which extends east as a 400 m long grassy headland. The beach faces east and receives waves averaging about 1 m, which maintain a 70 m wide bar, cut by three to four rips during higher wave conditions, including permanent rips against the boundary rocks. A partly vegetated 10 m high bluff runs the length of the beach. The southern end is bordered by an eroding section of bluffs and bluff debris.

NT 1111-1117 WANYANMERA POINT (W)

No.	Beach	Rating HT	Rating LT	Type	Length
NT1111	Wanyanmera Pt (W1)	1	2	R+rocks	200 m
NT1112	Wanyanmera Pt (W2)	1	2	R+rocks	300 m
NT1113	Wanyanmera Pt (W3)	2	3	R/LTT	2.3 km
NT1114	Wanyanmera Pt (W4)	2	3	R+rock flats	1.2 km
NT1115	Buymarr (E)	1	2	R+rock flats	1.3 km
NT1116	Buymarr	2	3	R+sand flats	600 m
NT1117	Buymarr (S)	3	4	LTT/TBR	5.6 km
Spring & neap tide range = 1.4 & 0.0 m					

At the southern end of beach NT 1110 the shoreline turns and trends west for 7 km, forming the northern shore of an open 6 km wide east-facing bay, with **Wanyanmera Point** forming the northern boundary and the small Buymarr community located in the northwest corner of the bay. Beaches NT 1111-1115 are located along the northern shore of the bay, with beaches NT 1116 and 1117 part of a 6 km long barrier-lagoonal system that has infilled the western end of the bay. Beaches NT 1118-1122 lie along the southern arm of the bay. Several granite reefs, islets and small islands, including White Rock and the Three Hummocks, lie off the entrance to the bay, slightly lowering waves to their lee. Vehicle tracks reach the western end of beach NT 1113 and through to beach NT 1117.

Beach **NT 1111** is a 200 m long southeast-facing reflective beach, protected by a group of reefs and islets 1 km off the shore. Waves average about 0.5 m and maintain a steep narrow cusped beach, backed by scarped 10-20 m high red bluffs, capped by older vegetated clifftop dunes. Deeper rock reefs dominate the nearshore. It is separated from beach NT 1112 by a narrow 100 m long laterite point. Beach **NT 1112** is 300 m long, faces south and is also protected by the reefs. It is backed by steep grassy slopes and fronted by a mixture of sand and rock reefs.

Beach **NT 1113** occupies a 2 km wide south-facing embayment located 1 km south of Mount Alexander. The beach curves round within the embayment, facing initially west, then south and finally southeast at the western low rocky point. The beach is cut by a periodically open tidal inlet toward the west and is backed by both foredune ridges and older stable transgressive dune systems extending up to 1.5 km inland, with a largely infilled lagoon extending up to 4.5 km in from the beach (Fig. 5.126). The beach receives waves averaging about 1 m, which maintain a reflective to occasionally low tide terrace system.

Beach **NT 1114** commences on the western rocky point and is a relatively straight 1.2 km long south-facing strip of high tide sand, backed by a low foredune and fronted by rocks and rock reefs extending out into the bay. At its western end the shoreline turns to the north and beach **NT 1115** commences. This is a crenulate 1.3 km long lower energy beach also dominated by shallow rock reefs and backed by a low beach ridge. It terminates at a conical 40 m high granite knoll.

Figure 5.126 Beach NT 1113 (right) is backed by a mangrove-filled lagoon, with beach NT 1114 to left.

Beach **NT 1116** commences on the north side of the knoll and extends for 600 m to a 100 m wide tidal inlet. The small **Buymarr** community, consisting of several houses, is located on the 400 m wide beach ridge plain behind the beach, with a landing ground 1 km inland. The beach is relatively sheltered and fronted by 200 m wide tidal sand shoals associated with the inlet.

Beach **NT 1117** commences on the southern side of the inlet and curves to the south for 5.6 km, terminating at a 50 m long rock-controlled inflection on the shore. The beach faces into the trade winds and receives waves averaging about 1 m, which break across a 50-100 m wide surf zone (Fig. 5.127), with small rips spaced every 100-150 m common during trade wind conditions. The beach is backed by a 1-2 km wide sand plain, consisting of inner vegetated transgressive dunes up to 20 m high and an outer and more southerly series of ten or more foredune ridges. These in turn are backed by a 1500 ha mangrove and salt-flat-dominated wetland.

Figure 5.127 Beach NT 1117 with waves breaking across the low tide bar.

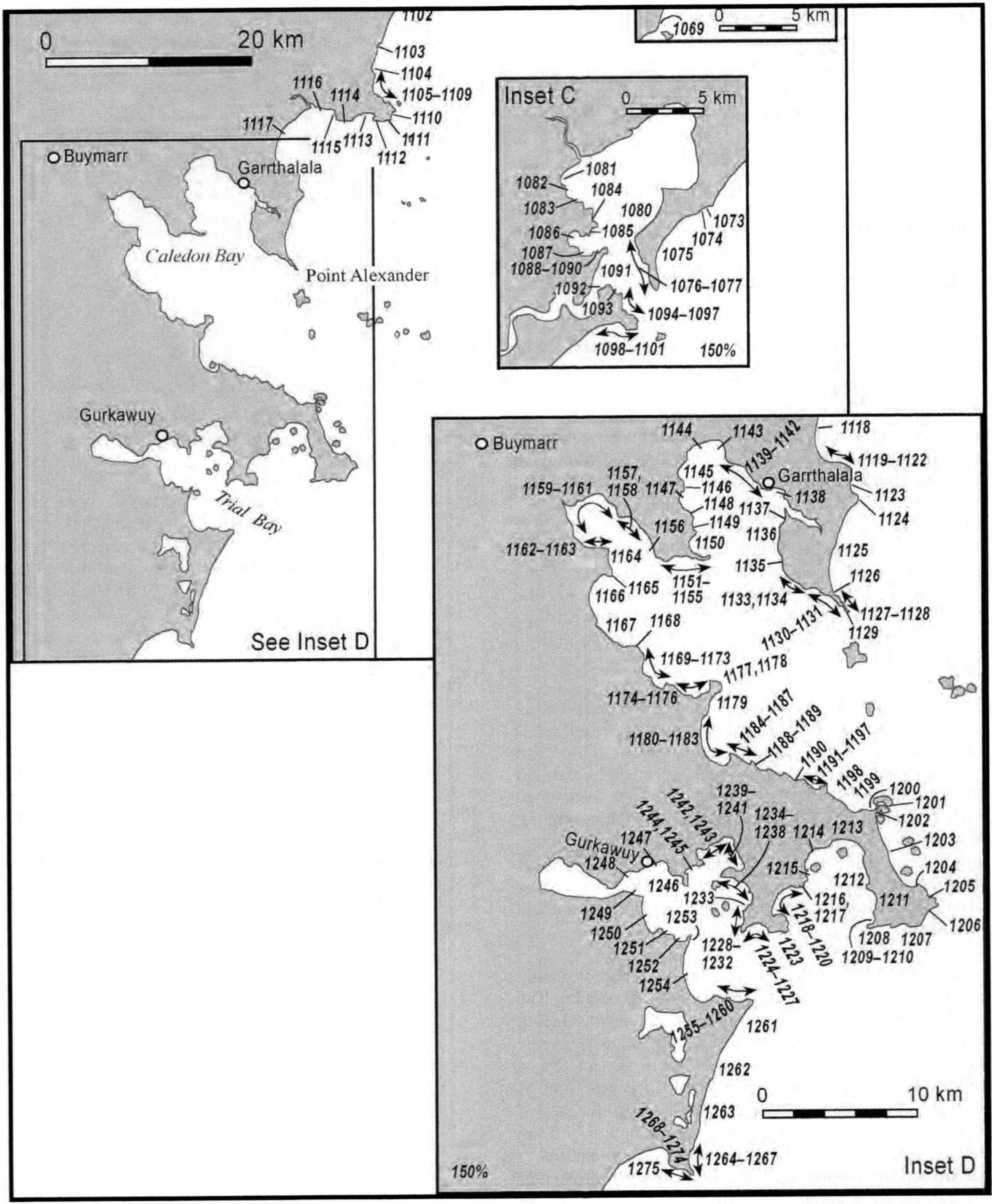

Figure 5.128 Regional map showing the numerous beaches in Caledon and Trial bays, beaches NT 1073-1275.

NT 1118-1128 POINT ALEXANDER

No.	Beach	Rating HT LT	Type	Length
NT1118	Pt Alexander (N11)	2 3	R/LTT	800 m
NT1119	Pt Alexander (N10)	2 3	R/LTT+rocks	150 m
NT1120	Pt Alexander (N9)	2 2	R+rock flats	400 m
NT1121	Pt Alexander (N8)	2 2	R+rock flats	500 m
NT1122	Pt Alexander (N7)	2 2	R+rock flats	1.1 km
NT1123	Pt Alexander (N6)	2 2	LTT-R	1.7 km
NT1124	Pt Alexander (N5)	2 3	LTT	300 m
NT1125	Pt Alexander (N4)	3 3	LTT	5.3 km
NT1126	Pt Alexander (N3)	2 3	R+rock flats	1.4 km
NT1127	Pt Alexander (N2)	2 2	R+rock flats	500 m
NT1128	Pt Alexander (N1)	2 2	R+rock flats	700 m
Spring & neap tide range= 1.4 & 0.0 m				

Point Alexander is a low granite headland capped by laterite that forms the northern entrance to Caledon Bay. Between the southern end of Buymarr beach (NT 1117) and Point Alexander are 13 km of rock and reef-dominated beaches linked by the 5 km long main beach (NT 1125). There is vehicle access to the northern beaches (NT 1118 & 1119), but none south of beach NT 1120.

Beach **NT 1118** is a slightly curving 800 m long east-facing beach exposed to waves averaging less than 1 m that decrease to the south. It is backed by a few low grassy foredunes and a vehicle track, and is normally fronted by a narrow bar, usually devoid of rips. It terminates at a narrow groyne-like strip of laterite surrounded by rock reef. Beach **NT 1119** is located on the eastern side of the 'groyne' and consists of a 150 m long pocket beach, bordered by the groyne and low sand-covered rocks to the east. It receives waves averaging between 0.5-1 m and has a narrow attached bar and a central sandy seafloor between the two bordering reefs. A low foredune backs the beach.

Beaches NT 1120-1122 are three adjoining beaches, each bordered and fronted by irregular rock reefs. Beach **NT 1120** is a straight 400 m long high tide beach bordered by low rocks and fronted by patchy rock reefs extending up to 300 m off the beach. A wooded coastal plain backs the beach, with the 40 m high vegetated transgressive sand ridge, originating from beach NT 1125, reaching the eastern end of the beach. Beach **NT 1121** begins immediately to the east and is a more crenulate north-facing 500 m long sand beach, with patchy rock reef extending 300 m off the beach, and the sand ridge 50-100 m behind the beach. Beach **NT 1122** is a continuous 1.1 km long north to northeast-facing sand beach, initially fronted by rock reefs, then clear sand, and finally 500 m wide mangrove-covered rock and reef flats forming the eastern boundary. Low woodlands back the western end, while low sand dunes originating from beach NT 1123 back the eastern half.

Beach **NT 1123** is a curving 1.7 km long east-facing beach, bordered by prominent rock reefs. It is exposed to waves averaging about 1 m at the northern end, which decrease to the south as the southern reefs extend up to 1 km off the beach and form a second 300 m long sandy crescent. The northern end of the beach has a narrow low tide bar, with more reflective conditions and some rock flats toward the southern end. Northwest-trending vegetated transgressive dunes, probably early Holocene, extend up to 1 km inland. Beach **NT 1124** is an east-facing 300 m long pocket beach wedged in between low rocky points, that extend 500 m seaward as reefs and scattered rock islets.

Beach **NT 1125** is the main beach. It is 5.3 km long, faces east-southeast and curves slightly toward the southern end in lee of the northern rocks and reefs of Point Alexander. It is well exposed to the southeast winds and waves, with waves averaging about 1 m, which maintain a steep beach face fronted by a continuous low tide bar, cut by small rips during higher wave conditions. Some active blowouts back the northern half of the beach with generally stable transgressive dunes behind. The dunes are 500 m wide in the south, extending up to 3 km inland and to heights of 40 m in the north. A strip of eroding beachrock occupies the centre of the beach, with eroded slabs of the rock thrown up onto the back of the beach.

The remaining 2 km of shoreline out to the tip of the point contain three increasingly rock and reef-dominated beaches. Beach **NT 1126** is a northeasterly-facing, 1.4 km long, steep, slightly crenulate sand beach fronted by scattered reefs extending up to 500 m off the beach, which lower waves to about 0.5 m at the shore. Beach **NT 1127** is a curving 500 m long beach that has prominent reefs extending 500 m off each end, together with low tide rocks along the southern half of the beach. Waves break over the reefs, with low wave to calm conditions at the shore. Finally, beach **NT 1128** runs down to the foot-like tip of Point Alexander. It is a double crenulate 700 m long high tide sand beach, fronted by rocks and reef flats extending from 100-500 m offshore. Both left and right-hand surf breaks peel off the reefs on the north side of the 'foot' (Fig. 5.129).

Figure 5.129 Point Alexander with beach NT 1128 fronted by rocks and reef flats, with waves breaking over the reefs.

NT 1129-1137 CALEDON BAY (1)-POINT ALEXANDER-NANJIWOI CREEK

No.	Beach	Rating HT	Rating LT	Type	Length
NT1129	Pt Alexander (W1)	1	2	R+rock flats	200 m
NT1130	Pt Alexander (W2)	1	2	R+rock flats	500 m
NT1131	Pt Alexander (W3)	1	2	R+sand/rock flats	800 m
NT1132	Pt Alexander (W4)	1	2	R+sand/rock flats	1.2 km
NT1133	Pt Alexander (W5)	1	2	R+sand/rock flats	1.1 km
NT1134	Pt Alexander (W6)	1	2	R+sand/rock flats	300 m
NT1135	Pt Alexander (W7)	1	2	R+sand flats	1.8 km
NT1136	Pt Alexander (W8)	1	1	R+sand flats	2 km
NT1137	Nanjiwoi Ck (S)	1	1	R+sand flats	1.2 km
Spring & neap tide range = 1.4 & 0.0 m					

Point Alexander forms the northern entrance to **Caledon Bay**, a 13 000 ha moderately protected bay, bordered by Point Alexander and Caledon Point, 7 km to the southwest, with 65 km of shoreline containing 49 beaches (NT1129-1177). The small communities of Garrthalala and Birany Birany are located on the northern and western shores respectively.

Inside the point the shoreline trends northwest for 5 km, then north into the smaller Grays Bay, with the prominent Nanjiwoi Creek located on the eastern shore. Between the point and the creek are 9 km of southwest to west-facing bay shore containing nine protected low energy beaches (NT 1129-1137), though some are exposed to the strong trade winds and have dune activity. There is no vehicle access to these beaches.

Beach **NT 1129** is located on the 'sole' of Point Alexander. It is a slightly curving, south-facing, 200 m long strip of sand, largely fronted by low tide rocks and 50 m wide rock reefs, with only one clear patch of sand toward the western end of the beach, backed by a solitary small blowout. At the 'heel' of the point the shoreline turns and trends north as curving, 500 m long beach **NT 1130**. This steep, narrow beach is backed in the north by six blowouts, and fronted by low tide rocks and patchy rock reefs.

Beach **NT 1131** is exposed to sand waves moving into the bay, and varies in size depending on the location of the sand waves. It faces southwest across the bay entrance and can be up to 800 m long. It is backed by 5-10 m high dune-capped laterite bluffs. The bluffs continue along the back of beach **NT 1132**, a curving, south-facing, 1.2 km long beach fronted by 100-200 m wide sand flats. It also contains two blowouts in reactivated early Holocene sands, located on top of the predominantly densely vegetated blufftops.

Beach **NT 1133** is a crenulate 1.1 km long beach, that faces generally southwest and is dominated by rock and reef, with the reefs and sand flats extending up to 500 m offshore. The entire beach is backed by active sand dunes extending up to 500 m inland and to heights of 20 m, with older densely vegetated dunes behind. At the western end of the beach is a 300 m long inflection in the shore containing beach **NT 1134**. This beach is bordered by low rocks and backed by the wooded laterite coastal plain. At its western end the shoreline turns and trends due north.

Beach **NT 1135** faces essentially due west across the entrance to Grays Bay. It is a crenulate 1.8 km long beach, dominated by low laterite bluffs, rocks and reefs, with the vegetated dune and coastal plain behind and sand flats widening to 500 m off the beach. It terminates at a protruding low point and on the north side of the point is 2 km long beach **NT 1136**. This is a low energy, west-facing series of rock and reef-controlled crenulations, fronted by 1 km wide sand flats. Sand is accumulating at the northern end of the beach, forming a slowly migrating 300 m long, 200 m wide recurved spit.

The terminus of the sand moving around Point Alexander and up the eastern side of the bay is the increasingly wide sand flats, around the recurved spits of beach NT 1136 and beach **NT 1137**. The latter is a 1.2 km long beach attached to the coastal plain in the south, and extending north as a narrow, 1 km long sand spit, which forms the southern entrance to 200 m wide Nanjiwoi Creek (Fig. 5.130). This mangrove-lined creek extends 2 km to the east to the back of beach NT 1125. It has a total wetland area of approximately 700 ha.

Figure 5.130 Beach NT 1137 terminates as a narrow spit across the mouth of Nanjiwoi Creek.

NT 1138-1144 GRAYS BAY-GARRTHALALA

No.	Beach	Rating HT	Rating LT	Type	Length
NT1138	Nanjiwoi Ck (N)	1	1	R+sand flats	800 m
NT1139	Garrthalala (S)	1	1	R+sand flats	400 m
NT1140	Garrthalala	1	1	R+sand flats	1 km
NT1141	Garrthalala (N1)	1	1	R+sand flats	700 m
NT1142	Garrthalala (N2)	1	1	R+sand flats	1.2 km
NT1143	Garrthalala (N3)	1	1	R+sand flats	900 m
NT1144	Grays Bay	1	1	R+sand flats	2.3 km
Spring & neap tide range = 1.4 & 0.0 m					

Grays Bay is a 5 km wide, 6 km deep, south-facing bay, bounded by the Nanjiwoi Creek area in the east and Middle Point to the west, and faces into the large Caledon Bay. The bay has 20 km of sandy, predominantly low energy, bluff-backed shoreline, containing 13 beaches (NT 1138-1150), with the small Garrthalala community located on the eastern shore, just above Nanjiwoi Creek. The community is accessible by vehicle.

Immediately north of the mouth of Nanjiwoi Creek is an 800 m wide southwest-facing embayment containing beach **NT 1138**. The beach is also 800 m long and links low mangrove-fringed bluffs to either side of the embayment. It blocks a small upland creek, that connects to a central 10 ha area of salt flats. The beach is relatively straight, usually calm and fronted by 700 m wide sand flats, including the tidal shoals and channel of the large creek that extends to the northwest off the beach.

The northern side of the bay turns and faces southwest, with beach **NT 1139** replacing the mangroves. The 400 m long beach is bordered by a low tree-covered point to the south, scattered rocks and mangroves to the north, with some rocks and sand flats extending off the beach. Trees behind the northern end of the beach have been cleared to provide aircraft access to the Garrthalala landing ground.

The small **Garrthalala** community consists of several houses located between a 500 m long landing ground and 1 km long beach **NT 1140**. This is a crenulate beach consisting of five sandy crenulations each bordered by low rocks and rock reefs, with both rocks and sand flats extending into the bay. The beach terminates at a small usually dry creek, on the north side of which is 700 m long beach **NT 1141**, which trends northwest to the next small creek mouth. This is also a crenulate beach backed by a few beach ridges in the south and low vegetated bluffs in the north, while rock and sand flats parallel the shore.

Beach **NT 1142** begins on the north side of the creek and runs roughly west for 1.2 km to a low point where the shore turns north. The beach has four rock-controlled crenulations and intervening sand flats, and is backed by a mixture of low beach ridges and wooded coastal plain. On the north side of the inflection is beach **NT 1143**, a crenulate 900 m long west-facing low energy beach, fronted by near continuous rocks and rock reefs, as well as sand flats.

Beach **NT 1144** occupies the northern apex of Grays Bay and is the ultimate sink for sand moving into the bay. The beach is 2.3 km long, faces southeast into the trade winds, and is crossed by two upland creeks which merge with the extensive sand flats. The wind waves have moved sand into the bay to form a low beach ridge plain up to 1 km wide and containing at least 15 ridges, totalling about 100 ha, which are truncated in the centre and west by the creeks. Both creeks are lined by mangroves and backing salt flats totalling about 200 ha in area.

NT 1145-1150 **GRAYS BAY (W)**

No.	Beach	Rating HT	Rating LT	Type	Length
NT1145	Grays Bay (W1)	1	1	R+sand flats	1.2 km
NT1146	Grays Bay (W2)	1	1	R+sand flats	900 m
NT1147	Grays Bay (W3)	1	1	R+sand+rock flats	1.6 km
NT1148	Grays Bay (W4)	1	1	R+rock+sand flats	600 m
NT1149	Grays Bay (W5)	1	1	R+rock+sand flats	1.1 km
NT1150	Middle Pt (N)	1	1	R+sand flats	1.1 km
Spring & neap tide range = 1.4 & 0.0 m					

The western shore of Grays Bay trends south for 6 km to the narrow Middle Point and contains six low energy beaches (NT 1145-1150), most backed by vegetated to scarped laterite bluffs, with rocks and reefs bordering the central four beaches. There is no vehicle access to the beaches.

Beach **NT 1145** commences on the southern side of the western boundary creek of beach NT 1144. The 1.2 km long beach faces southeast down the bay. It is bordered by low vegetated laterite bluffs and blocks a small upland creek forming a 50 ha salt flat-wetland behind the southern end of the beach, with a few mangroves near the mouth. Beach **NT 1146** is located below the bluffs which form the southern boundary. It trends south for 900 m to a more southerly inflection in the bluffs. The beach is crenulate between rock outcrops with wooded slopes behind the beach rising to a 34 m high crest.

Beach **NT 1147** is a curving 1.6 km long, double crenulate, east-facing beach, that is fronted by scattered rock flats and sand flats. A small usually dry upland creek drains across the centre of the beach, with wooded 20 m high laterite slopes behind. Beach **NT 1148** is located along a protruding section of shore. The 600 m long beach is fronted by near continuous rocks and rock reefs, and sand flats, with wooded slopes rising behind.

Beach **NT 1149** is located on the south side of the protrusion. It is a double crenulate 1.1 km long beach, with prominent rock reefs at either end, and sand flats in between, while dense woodlands back the beach. Finally, beach **NT 1150** occupies the 1.5 km wide embayment on the north side of Middle Point. The 1.1 km long beach extends south along the northern shore of the embayment. It is backed by vegetated, then scarped, red 10 m high laterite bluffs, which replace the beach along the southern half of the bay as they rise to 20 m at the point.

NT 1151-1155 **CALEDON BAY (N 1)**

No.	Beach	Rating HT	Rating LT	Type	Length
NT1151	Middle Pt (W1)	1	2	R+rock flats	200 m
NT1152	Middle Pt (W2)	1	2	R+rock flats	80 m
NT1153	Middle Pt (W3)	1	2	R+rock flats	50 m
NT1154	Middle Pt (W4)	1	1	R+rock flats	1 km
NT1155	Middle Pt (W5)	1	1	R+rock flats	900 m
Spring & neap tide range = 1.4 & 0.0 m					

The northern side of Caledon Bay consists of three embayments, Grays Bay to the east, an unnamed 2 km wide, 5 km deep bay to the west of Middle Point, and the third which contains the Birany Birany community. From Middle Point the shore trends west for 3.5 km and contains five crenulate south-facing bluff and rock-controlled beaches (NT 1151-1155). There is no vehicle access to any of the beaches.

Beach **NT 1151** is located on the southern tip of Middle Point. It is a curving, 200 m long, south-facing high tide beach, backed by 10 m high red bluffs, and fronted by continuous rock reefs linking the boundary point and forming a calm 'lagoon' to their lee. Immediately to the west is 80 m long beach **NT 1152**, a strip of high tide sand at the base of vegetated 15 m high bluffs, fronted by continuous low tide rocks and rock flats. The third point beach (**NT 1153**) is located in the next indentation and consists of a 50 m strip of high tide sand, also backed by vegetated 20 m high bluffs, with continuous rock flats.

At the junction of the point the shoreline turns and trends southwest for 1 km. It consists of vegetated bluffs, decreasing from 20 m to 5 m in height, which are fronted by beach **NT 1154**. The beach is a strip of high tide sand with shallow rock reefs extending up to 400 m off the beach. At the southern end of the beach the shoreline turns and trends northwest into the bay. Beach **NT 1155** is located along the first double crenulate 900 m of shore, with rocks exposed on the foreland in lee of rock reefs. A small sandy foreland backs the central foreland. At the end of the beach, the shore turns and trends northerly.

NT 1156-1164 **CALEDON BAY (N 2)**

No.	Beach	Rating HT	Rating LT	Type	Length
NT1156	Caledon Bay (N1)	1	1	R+rock+sand flats	1.2 km
NT1157	Caledon Bay (N2)	1	1	R+sand flats+rock	700 m
NT1158	Caledon Bay (N3)	1	1	R+sand flats+rocks	700 m
NT1159	Caledon Bay (N4)	1	1	R+sand flats+rock	50 m
NT1160	Caledon Bay (N5)	1	1	R+tidal shoals	1.4 km
NT1161	Caledon Bay (N6)	1	1	R+tidal shoals	600 m
NT1162	Caledon Bay (N7)	1	1	R+sand flats	300 m
NT1163	Caledon Bay (N8)	1	1	R+rock flats	200 m
NT1164	Caledon Bay (N9)	1	1	R+rock+sand flats	1.2 km
Spring & neap tide range = 1.4 & 0.0 m					

The northwestern corner of Caledon Bay consists of a 2-3 km wide, 6 km deep embayment, which faces toward the entrance, 15 km to the southeast. This protected bay receives sufficient wave energy to maintain nine beaches around the perimeter of the bay shore, most dominated by the low laterite bluffs and reefs. Beaches NT 1160 and 1161 are the ultimate sink for sand moving into the bay, while beaches NT 1162 and 1161 are also active recurved spits and indicate that sand is still moving into and prograding the bay shoreline.

Four south to southwest-facing beaches (NT 1156-1159) occupy the northern shore of the bay. Beach **NT 1156** begins immediately west of beach NT 1155 and occupies the next 1.2 km of shore. The beach curves round to face southwest at its northern end, where it terminates amongst a 200 m long strip of mangroves, growing in lee of shallow rock reefs. A 20 ha mangrove-dominated wetland backs the northern half of the beach.

Beach **NT 1157** commences at the western end of the mangroves and is a 700 m long southwest-facing beach, backed by a mixture of low vegetated bluffs and beach ridges, and fronted by rock and sand flats. A small foreland at its western end separates it from beach **NT 1158**, a curving 700 m long south-facing beach, bordered by low rock points and reefs, with mangroves growing on the western rocks. It is backed by a narrow strip of beach ridges. Beach **NT 1159** is located on the northern side of the western point, and consists of a 50 m long pocket of sand, located below vegetated 10 m high bluffs, with rock reefs extending 100 m offshore and mangroves dominating the shoreline to the north.

Beaches NT 1160 and 1161 are located at the northern apex of the bay (Fig. 5.131), which has been the sink for sand moving along both sides of the bay. Two creeks drain into the northern end and separate the beaches. Beach **NT 1160** is 1.4 km long and faces southeast into the trade winds, though only low wind waves reach the beach. Mangroves fringe either end as well as growing along sections of the beach, with seagrass-covered sand flats extending off the beach. It is backed by a series of up to ten recurved spits, and a 300 ha salt flat-dominated wetland, which drains via a creek to the southern end of the beach. Beach **NT 1161** is located on the southern side of the 100 m tide creek mouth. It is an east-facing 600 m long strip of low sand, formed by recurved spits migrating from the south across the second creek mouth. The migratory nature of the low sand spits and the presence of the creek result in a mobile beach system, which varies in size, presence and length, with the 1969 aerial photographs showing a mangrove-fringed shoreline in lee of a beach-spit still present in 1997. The spits originate from the vicinity of beach NT 1162.

Figure 5.131 Beaches NT 1160 & 1161 (left) are separated by a tidal creek and fronted by extensive tidal sand flats.

The two creek mouths lie 500 m apart and their combined tidal channels and shoals dominate the southern half of the bay apex, with tidal shoals extending 400 m into the bay. Between the southern creek mouth and beach NT 1162 are 3 km of mangrove-fringed shoreline, backed by older recurved spits toward beach NT 1163. Beach **NT 1162** is a dynamic recurved spit and represents the conduit for transporting sand from the updrift beaches deeper into the bay. It varies in length and shape and can detach and migrate deeper into the bay, attaching to parts of the mangrove shoreline, as far as beach NT 1161. In 1997 it was 300 m long. It is usually backed by mangroves, while fronted by 100-200 m wide sand flats and attached at the updrift end to the densely wooded coastal plain.

Beach **NT 1163** occupies the next 200 m of mainland shore and consists of a narrow, crenulate, north-facing strip of sand, which acts as a transition from the spit to the mainland shore. At the eastern end of the beach, the shoreline turns and trends south for 1.2 km as beach **NT 1164**. This is a continuous beach containing three segments, each fronted by rocks and reefs, with a mixture of rock and sand flats extending 50-100 m off the beach. Dense woodlands back the beach.

NT 1165-1168 **BIRANY BIRANY**

No.	Beach	Rating HT	Rating LT	Type	Length
NT1165	Birany Birany (N)	1	2	R+rock flats	200 m
NT1166	Birany Birany	1	1	R+LTT-tidal shoals	3.8 km
NT1167	Birany Birany (S1)	1	1	R+tidal shoals	600 m
NT1168	Birany Birany (S2)	1	1	R+sand flats	1.2 km
Spring & neap tide range = 1.4 & 0.0 m					

Birany Birany is a small community located behind the northern end of beach NT 1166. It consists of several houses, a landing ground and radio tower and is connected to the central Arnhem Road, 14 km to the west.

Beach **NT 1165** is located at the tip of the northern headland, 1 km east of the community. It is a low energy, 200 m long, east-facing beach bordered by a low 100 m long rock reef, with rock flats filling the area between. As a result mangroves fringe either end of the beach, together with a few growing in lee of some central rocks. Mangroves dominate the shore between the point and the community beach.

Birany Birany beach (**NT 1166**) is a curving, east-facing, 3.8 km long beach that receives sufficient wave energy along its northern 2 km to maintain a low energy beach system (Fig. 5.132). The southern half is fronted by the tidal shoals and sand flats associated with a tidal creek that forms the southern boundary. The beach is the outermost of a 500 m wide series of approximately ten beach ridges in the north, with older (possibly Pleistocene) beach ridges extending up to 1.5 km inland in the southern creek area. It is bordered by low wooded laterite headlands to either end.

Figure 5.132 Birany Birany beach (NT 1166) with the landing ground behind.

Beach **NT 1167** commences on the southern side of the creek and is a very low energy 600 m long beach ridge, fronted by 500 m wide tidal shoals, with active recurved spits migrating from east to west along the beach. The spits originate from beach **NT 1168**, a 1.2 km long north-facing beach, tied to the laterite coastal plain. Sand migrates along the beach as migratory recurved spits, which vary in location and size as they move toward beach NT 1167. In lee of both beaches is a 1 km wide splay of up to ten recurved spits, backed by a 500 ha area of salt flats.

NT 1169-1178 **CALEDON POINT (N)**

No.	Beach	Rating HT	Rating LT	Type	Length
NT1169	Caledon Pt (N9)	1	1	R+rock/sand flats	2.2 km
NT1170	Caledon Pt (N8)	1	2	R+rock/sand flats	150 m
NT1171	Caledon Pt (N7)	1	1	R+sand flats	1.6 km
NT1172	Caledon Pt (N6)	1	2	R+rocks	200 m
NT1173	Caledon Pt (N5)	1	2	R+rocks	150 m
NT1174	Caledon Pt (N4)	1	2	R	1.2 km
NT1175	Caledon Pt (N3)	1	2	R	500 m
NT1176	Caledon Pt (N2)	1	1	R+sand flats	700 m
NT1177	Caledon Pt (N1)	1	2	R+rock flats	500 m
NT1178	Caledon Pt	2	3	R+rock flats	300 m
Spring & neap tide range = 1.4 & 0.0 m					

Caledon Point forms the southern entrance to Caledon Bay. The 18 m high point is composed of Archaean granite with a capping of densely vegetated dune sands. To the west of the point the coast trends west then northwest for 6 km to the beginning of beach NT 1169, with nine near continuous beaches (NT 1169-1177) in between, all backed by dense woodlands, with no vehicle access to the point or any of the beaches.

Beach **NT 1169** commences as the shore turns and trends south at the end of beach NT 1168. The crenulate beach is 2.2 km long, faces east and runs along the front of wooded, then scarped red laterite bluffs, which rise to 15 m in the south. Rock flats and patchy sand flats extend 200-300 m off the beach. Beach **NT 1170** lies at the southern end of the beach. It is a 150 m long pocket of northeast-facing sand, bounded by rocks and rock reefs to either end, and backed by low vegetated bluffs. At the eastern end of the beach the shoreline turns and trends southeast, as beach **NT 1171**. This is a curving 1.6 km long beach, consisting of a low 200-300 m wide barrier occupying the first 1 km, then a north-facing section backed by a low beach ridge and wooded slopes, and finally a small upland creek crossing the southern end of the beach.

The creek exits in lee of a linear reef formed of granite boulders, with beach **NT 1172** commencing on the eastern side of the rocks. This is a straight 200 m long strip of sand, fronted by near continuous intertidal granite boulders, with a 30 m clear beach section, then a clump of mangroves at the eastern end. A 100 m wide low granite point, separates it from beach **NT 1173**, another 150 m long pocket of sand, which while bordered by rocks, has a central reflective sandy beach section.

Beach **NT 1174** commences on the south side of the granite boulders, as the shore again turns and trends southeast. The beach is 1.2 km long and consists of a narrow vegetated barrier, that decreases from 100 m to 50 m in width to the south, with a small usually dry creek crossing the southern end. The beach, while bordered by granite boulders, is free of rocks and reflective. The southern 100 m of the beach turns and faces north, with boulders scattered along the beach. Beach **NT 1175** commences at the end of the boulders. It faces due north and curves round for 500 m as a low energy reflective beach, backed by a low grassy beach ridge.

Beach **NT 1176** is located in lee of Caledon Point. It extends from a small protrusion that separates it from beach NT 1175, for 700 m to the rocks of the point. It faces northwest into Caledon Bay and is backed by low woodlands, with the higher energy beach NT 1179 located 500 m behind the beach on the south side of the point.

Two beaches are located on the point. Beach **NT 1177** runs along the north side of the point for 500 m. It is a crenulate high tide beach fronted by continuous irregular rocks and rock flats which extend up to 200 m off the beach. Finally beach **NT 1178** lies immediately to the east on the northeast side of the point. It is 300 m long and fronted by rock flats that widen to 500 m off the northern end, with a few mangroves located past the southern end of the beach.

NT 1179-1183 **CALEDON POINT (S 1)**

No.	Beach	Rating HT	Rating LT	Type	Length
NT1179	Caledon Pt (S1)	1	2	R+rock flats	1.1 km
NT1180	Caledon Pt (S2)	1	2	R+rock flats	250 m
NT1181	Caledon Pt (S3)	2	3	LTT	1 km
NT1182	Caledon Pt (S4)	2	3	LTT-R-sand flats	3.2 km
NT1183	Caledon Pt (S5)	1	1	R+tidal shoals	200 m
Spring & neap tide range = 1.4 & 0.0 m					

On the south side of Caledon Point the coast trends south for 4 km, before turning southeast for 15 km down to Cape Grey. Within the first 4 km is an open east-facing embayment containing five beaches (NT 1179-1183).

Beach **NT 1179** commences on the southern side of the point and trends southwest for 1.1 km. While exposed to the trade winds it is protected by rock reefs to either end, with only a central reflective sandy section, backed by a few small active blowouts. It terminates at a 100 m wide granite point, on the south side of which is 250 m long beach **NT 1180**. This beach is bordered by granite boulder points, with rocks also fronting its southern half, and only the northern reflective section clear of rocks. Partly vegetated granite slopes back the beach, rising to 60 m, 500 m inland.

Beach **NT 1181** begins on the south side of the 400 m wide granite headland. The beach trends due south for 1 km to the lee of two sets of prominent granite boulders located in the surf zone. Waves average about 1 m and maintain a low gradient beach with a 50 m wide bar, which is cut by rips during higher wave conditions. Large vegetated transgressive dunes extend from the beach up to 700 m inland to the base of the granite slopes of 100 m high Mount Caledon. Beach **NT 1182** commences on the southern side of the boulders and trends south before curving around to finally face north in lee of the southern headland (Fig. 5.133), with a total beach length of 3.2 km. The northern-central section of beach has low surf and a continuous bar, with a chance of rips along the northern half, while in lee of the headland a creek exists against the rocks and maintains 400 m wide tidal shoals. The creek drains a 400 ha wetland, dominated by salt flats, with mangroves lining the creek near the mouth. Both beaches are backed by a continuous barrier system, consisting of beach ridges in the south, grading into foredunes and then largely vegetated transgressive dunes, which reach a height of 30 m. Turtles nest in the foredunes.

Figure 5.133 Beach NT 1182 (right) terminates at a sheltered creek mouth in lee of a low rocky headland, with smaller beaches NT 1183 & 1184 located either side of the headland.

Beach **NT 1183** is located on the inside of the 14 m high granite headland that forms the southern boundary. It faces west across the creek mouth and shoals, with granite boulders bordering each end as well as outcropping on the low energy beach.

NT 1184-1197 **CALEDON POINT (S 2)**

No.	Beach	Rating HTLT	Type	Length
NT1184	Caledon Pt (S6)	1 2	R+rocks	400 m
NT1185	Caledon Pt (S7)	1 2	R	700 m
NT1186	Caledon Pt (S8)	1 2	R+rocks	30 m
NT1187	Caledon Pt (S9)	1 2	R+rocks	40 m
NT1188	Caledon Pt (S10)	2 3	R/LTT	600 m
NT1189	Caledon Pt (S11)	2 3	R/LTT	250 m
NT1190	Caledon Pt (S12)	2 3	R/LTT	500 m
NT1191	Caledon Pt (S13)	2 3	LTT	100 m
NT1192	Caledon Pt (S14)	2 3	LTT	250 m
NT1193	Caledon Pt (S15)	1 2	R	300 m
NT1194	Caledon Pt (S16)	1 2	R	50 m
NT1195	Caledon Pt (S17)	2 3	LTT	800 m
NT1196	Caledon Pt (S18)	1 2	R	150 m
NT1197	Caledon Pt (S19)	1 2	R	200 m
Spring & neap tide range = 1.4 & 0.0 m				

The 6 km of shoreline between beaches NT 1184 and 1197 consists of five north-facing embayments, each bordered by 20-40 m high granite points, with some smaller beaches occupying parts of the headlands, resulting in a total of 14 beaches (NT 1184-1197). There is a vehicle track from the Bukudal community to beach NT 1195 and via beach NT 1198 to beach NT 1197.

Beach **NT 1184** is located along the northern side of the first headland. It is 400 m long, faces north, and is a steep and reflective beach, fronted by a near continuous field of partly submerged granite boulders, with grassy to wooded slopes behind. Beach **NT 1185** commences at the southern end of the boulder field and curves round for 700 m to the southern end of the first small embayment. It receives waves averaging less than 1 m in the north, which decrease to the south in lee of a boulder reef and the southern 20 m high headland, with reflective conditions the length of the beach. It is backed by a 50-100 m wide grassy beach ridge then woodlands, with a small usually blocked creek located toward the southern end of the beach.

The next headland contains two pockets of sand 30 m and 40 m long respectively (**NT 1186** & **NT 1187**), both bordered by large granite boulders, with only the southern beach providing sandy access to the shore. Beach **NT 1188** lies 100 m to the south, and is a relatively straight 600 m long northeast-facing sand beach, which receives sufficient waves to maintain a narrow bar at the base of the usually steep cusped beach face. It is backed by a grassy to wooded 100-200 m wide beach ridge plain. Beach **NT 1189** lies 200 m to the south beyond a wooded granite headland. It is a 250 m long northeast-facing low gradient beach, bordered by granite boulders, with a usually dry creek crossing the southern end of the beach, and a 500 m long granite headland beyond.

The next embayment contains the solitary beach **NT 1190**, a slightly curving, 500 m long, northeast-facing sand beach. It is bordered by granite headlands and boulders and backed by a 200-300 m wide grassy to wooded beach ridge plain that has infilled the small V-shaped valley. A mangrove-lined creek runs along the southern side of the plain and valley side, and periodically flows across the beach. The beach itself has a narrow bar in the north grading, as waves decrease, to reflective in the south.

The fourth embayment contains four beaches (NT 1191-1194). Beach **NT 1191** is located on the eastern side of the low granite boundary point. It is a low 100 m pocket of sand bordered by granite points. It faces east into the trades and receives waves averaging about 1 m which maintain a 50 m wide bar, with an active blowout crossing the back of the 100 m wide point to link with beach NT 1192. Beach **NT 1192** is located 200 m to the south, and consists of a 250 m long northeast-facing beach bordered and backed by vegetated granite slopes, with an incised creek at the southern end. The beach receives sufficient waves to maintain a narrow continuous bar. Beach **NT 1193** is located 100 m further on at the southern end of the embayment. It is 300 m long, faces due north and is sheltered by a granite point and small island to the east, resulting in low waves and a reflective beach. A 50 m wide grassy beach ridge backs the beach, then steeper wooded slopes. Finally, immediately east of the beach is beach **NT 1194**, a 50 m pocket of north-facing sand, that has formed on the point as a small sand foreland in lee of the island.

The fifth and southernmost of the embayments contains beaches NT 1195-1197. Beach **NT 1195** commences 300 m south of the island-point. It is 800 m long, faces northeast and receives waves averaging about 1 m, which maintain a continuous 50 m wide bar. It is backed by a 200-300 m wide beach ridge plain and is accessible by vehicle toward the southern end. A small creek also periodically breaks out across the southern end. Granite boulders border each end, with a few large boulders separating it from beach **NT 1196**. This is a 150 m long, more protected, north-facing beach, bordered by the boulders to the west and a 200 m wide sloping granite point to the east. Beach **NT 1197** is located on the eastern side of the point. It is a 200 m long north-facing reflective beach, with a conical boulder-strewn 100 m wide point forming the eastern boundary. It abuts the dunes of beach NT 1198, with a vehicle track running along the back of the dunes to the beach.

NT 1198-1204 **DU PRE BAY & POINT**

No.	Beach	Rating HT LT	Type	Length
NT1198	Du Pre Bay	2 3	LTT	2.5 km
NT1199	Du Pre Bay (S1)	1 2	R	50 m
NT1200	Du Pre Pt (N)	1 2	R	500 m
NT1201	Du Pre Pt (S1)	1 2	R	80 m
NT1202	Du Pre Pt (S2)	5 6	TBR	600 m
NT1203	Du Pre Pt (S3)	3 4	LTT-R	4.1 km
NT1204	Du Pre Pt (S4)	1 2	R	400 m
Spring & neap tide range = 1.4 & 0.0 m				

Du Pre Bay (also known as Six Pack Bay) is an open northeast-facing bay that extends for 3 km northwest of Du Pre Point. While open to the northeast the bay is sheltered by several granite rocks and reefs off the centre and to the south and east by Du Pre Point and the large Du Pre Rocks. As a result waves average less than 1 m. All the beaches are accessible by vehicle track or along the beach from the nearby Bukudal community.

The main beach (**NT 1198**) faces northeast and extends for 2.5 km from the northern conical granite point to the inner rocks on the north side of Du Pre Point. It has a central sandy foreland formed in lee of the offshore rock reefs. The beach is composed of medium to coarse sand and has a steep beach face, with a narrow bar in the north and more reflective conditions to the south. It is backed by a foredune and low grassy dune area, with a vehicle track running along behind the foredune. A small creek breaks out across the middle of the beach during the wet season. A low granite point forms the southern boundary, with 50 m long beach **NT 1199** located between this point and a 300 m long section of granite. It receives low waves, faces north and is reflective.

Beach **NT 1200** is located between the granite and the north side of Du Pre Point. It is a 500 m long north-facing reflective beach, linking with Du Pre Point to form a tombolo. The point is a conical 18 m high granite headland. Dune sand from the adjoining beach NT 1202 is blowing across the tombolo and spilling onto the beach (Fig. 5.134). Beach **NT 1201** is located in a 50 m wide gap in the granite, which has permitted a curving 80 m long reflective beach to link the boulders. It faces east but is protected by the boulders and the offshore rocks.

On the southern side of the point and south of the rocks is the more exposed beach **NT 1202**. This curving 600 m long beach faces east and, while slightly protected at either end, receives waves averaging over 1 m along the central section which maintain a 100 m wide surf zone cut by three to four beach rips. Active dunes blow from the beach across the tombolo to beach NT 1200. It is bordered in the south by a 100 m wide, 40 m high granite head, beyond which is the main beach.

Figure 5.134 Du Pre Point, with beaches NT 1200 (right), 1201 on the point, curving NT 1202 (centre) and the longer NT 1203 (top), all backed by active dunes.

The main beach (**NT 1203**) trends south for 4.1 km to the northern rocks of Cape Grey. Despite its exposure to the trades, it is sheltered by offshore granite reefs, rocks and islets, as well as Cape Grey to the south. Waves average less than 1 m and decrease to the south, as well as forming a central cuspate foreland. The beach has a few rips in the north, but for the most part has a continuous bar to reflective conditions. The trade winds maintain an active dune in lee of the beach, with dunes blowing up to 300 m inland, while older vegetated dunes extend up to 700 m inland and to heights of 36 m.

Beach **NT 1204** is separated from the main beach by a small granite outcrop. It continues on for 400 m to the more continuous rocks of the point. It faces north and is protected by the point, with reflective conditions prevailing.

NT 1205-1208 **CAPE GREY**

No.	Beach	Rating HT LT	Type	Length
NT1205	Cape Grey	3 4	R+rocks&reefs	500 m
NT1206	Cape Grey (S1)	6 6	TBR+rocks	1.1 km
NT1207	Cape Grey (S2)	5 5	TBR-LTT	800 m
NT1208	Cape Grey (S3)	2 3	R+rocks&reefs	900 m
Spring & neap tide range = 1.4 & 0.0 m				

Cape Grey is located on the eastern tip of an exposed 600 ha headland. The 20-30 m high headland has a granite basement, capped by Tertiary laterite and blanketed by Holocene dunes, which have originated from the three southern cape beaches (NT 1206-1208).

Beach **NT 1205** is located at the eastern tip of the cape, with Cape Grey forming the southern boundary of the slightly curving 500 m long east-facing beach. The beach is however sheltered by the cape and rock reefs, with waves breaking off the beach resulting in low wave to calm conditions at the shore. Rocks, reefs and rock flats border and fringe the beach, with only the central section relatively free of rocks.

Beach **NT 1206** is located on the more exposed south side of the cape. The 1.1 km long beach faces southeast into the trades, receiving waves averaging between 1 and 1.5 m toward the eastern end, one of the higher energy beaches on the Territory coast. These break across a 100 m wide surf zone cut by a permanent rip against the cape, and two to three beach rips, with wave height decreasing to the west in lee of rock reefs. Vegetated transgressive dunes extend up to 1 km in from the beach.

Beach **NT 1207** is located 500 m to the west. It is a curving 800 m long south-facing beach exposed to waves averaging just over 1 m, which maintain a permanent rip against the eastern rocks and three to four beach rips, with rocks lining the western 100 m of beach. Immediately to the west is beach **NT 1208**, a double crenulate 900 m long beach, with a central rock outcrop. It is partially protected by rock reefs, resulting in lower waves at the shore and reflective beach conditions. It is backed by some active blowouts (Fig. 5.135) and together with beach NT 1207 has largely vegetated transgressive dunes, including clifftop dunes, extending up to 2.5 km inland and to 30 m in height.

Figure 5.135 Beach NT 1208 is backed by multiple episodes of dune transgression.

NT 1209-1215 WONGA BAY/BUKUDAL

No.	Beach	Rating HT	Rating LT	Type	Length
NT1209	Wonga Bay (E1)	1	1	R+sand flats	1 km
NT1210	Wonga Bay (E2)	1	1	R+sand flats	600 m
NT1211	Wonga Bay (E3)	1	1	R+sand flats	1.1 km
NT1212	Wonga Bay (E4)	1	2	R+sand flats+rocks	1.9 km
NT1213	Bukudal	1	1	R+sand flats	2.2 km
NT1214	Wonga Bay (W1)	1	2	R+sand flats+rocks	2.6 km
NT1215	Wonga Bay (W2)	1	2	R+sand flats+rocks	100 m
Spring & neap tide range = 1.4 & 0.0 m					

Wonga Bay is a 5 km wide south-facing bay located on the western side of Cape Grey and bounded by Guyuwiri Point to the west. It is 5 km deep, with the small Bukudal community located to the northern apex of the bay. It has 18 km of shoreline, dominated by 12 sandy beaches. The only vehicle access to the beaches is in the vicinity of the community.

The southwest tip of Cape Grey consists of a 500 m long, 12 m high, laterite-capped granite point, on the north side of which the shoreline trends north into Wonga Bay, with granite boulders and reefs scattered off the beaches. The boulders also form the tips of the cuspate forelands that separate the four eastern shore beaches. The first of the bay beaches (**NT 1209**) commences on the north side of the point and curves round for 1 km to the first granite-tipped foreland. It is backed by dunes blowing across the cape from the southeast, and fronted by sand flats and scattered boulders extending 1 km into the bay.

Beach **NT 1210** is a curving 600 m long beach located between the next two forelands, with a few boulders scattered along the beach. It is backed by vegetated transgressive dunes and fronted by sand flats and boulders. Beach **NT 1211** extends to the northwest for 1.1 km from the foreland to a 28 m high, 300 m wide vegetated granite point. It is backed by a low foredune, with a partly vegetated blowout backing the more southerly-facing northern end of the beach. Sand flats, boulders and granite islets extend up to 1.5 km off the beach.

Beach **NT 1212** extends from the north side of the point for 1.9 km. It is a continuous north-trending beach consisting of three boulder-tied crenulations, together with sand flats, boulders and islets scattered up to 1 km off the beach. It is backed by low, shore-parallel, partly stablised transgressive dunes.

The **Bukudal** community is located at the very northern tip of the bay, at the western end of 2.2 km long beach **NT 1213**. The beach extends from a granite-tipped foreland in the east, curving around to face south against the western 40 m high wooded granite knoll. The community consists of several houses, with a landing ground located 1.5 km northwest of the houses. Sand flats and a 15 m high granite islet front the low energy beach. The southerly orientation of the beach into the trades has produced a partially active foredune and blowouts extending up to 300 m inland, with older vegetated dunes extending up to 1 km in from the beach.

To the west of the community is the granite knoll fronted by a small islet, then 2.6 km long beach **NT 1214**. This is a crenulate beach that is exposed along its southeast-facing eastern half and backed by moderately active foredune and small blowouts. The western half curves round to face east in lee of a prominent granite islet and backing granite boulders, reefs and sand flats which extend to the beach. The beach terminates at a low granite boulder point, on the south side of which is 100 m long east-facing beach **NT 1215**. The beach is bordered and partly fronted by granite boulders and backed by a grassy beach ridge plain which links it with the southern end of beach NT 1214.

NT 1216-1220 WONGA BAY (W)

No.	Beach	Rating HT LT	Type	Length
NT1216	Wonga Bay (W3)	2 2	R	250 m
NT1217	Wonga Bay (W4)	2 2	R	250 m
NT1218	Wonga Bay (W5)	2 2	R	1.7 km
NT1219	Wonga Bay (W6)	2 2	R	600 m
NT1220	Wonga Bay (W7)	2 2	R+rocks	200 m
Spring & neap tide range = 1.4 & 0.0 m				

The southwestern side of Wonga Bay contains five more exposed east to southeast-facing beaches (NT 1216-1220), which while receiving lower waves are sufficiently exposed to the trade winds to be backed by some moderately active dune transgression and more extensive older dunes.

Beaches **NT 1216** and **NT 1217** are similar neighbouring beaches. They are both 250 m long, bordered by boulder-strewn 20 m high granite points, face southeast and are backed by active blowouts extending up to 400 m inland and to heights of 30 m, with older vegetated dunes extending 1 km inland (Fig. 5.136). Waves are however reduced within the bay and average less than 1 m, maintaining a steep reflective beach.

Figure 5.136 Beaches NT 1216 & 1217 are similar headland–bound beaches backed by active dunes.

Beach **NT 1218** commences on the western side of the western point and curves round for 1.7 km to a collection of low boulders lying just off the beach. The northern section of the beach is exposed to the southeast winds and is backed by extensive Holocene transgressive dunes extending up to 1.5 km inland, which are largely vegetated today by grasses, with casuarina trees lining the back of the beach. Beach **NT 1219** is located on the south side of the boulders, and is a curving 600 m long reflective beach backed by a casuarina-capped foredune and two older grassy foredunes. It terminates at the northern granite rocks of Guyuwiri Point, with beach **NT 1220** located 300 m to the east in the first embayment in the rocks. It is a 200 m long north-facing strip of high tide sand, bordered and largely fronted by granite boulders.

NT 1221-1227 GUYUWIRI POINT

No.	Beach	Rating HT LT	Type	Length
NT1221	Guyuwiri Pt (N2)	3 3	LTT	400 m
NT1222	Guyuwiri Pt (N1)	2 2	R+rock flats	200 m
NT1223	Guyuwiri Pt	2 2	R+rocks	600 m
NT1224	Guyuwiri Pt (W1)	2 2	R+rocks	150 m
NT1225	Guyuwiri Pt (W2)	2 2	R+rocks	250 m
NT1226	Guyuwiri Pt (W3)	3 3	LTT	200 m
NT1227	Guyuwiri Pt (W4)	4 4	LTT/TBR	900 m
Spring & neap tide range = 1.4 & 0.0 m				

Guyuwiri Point is a 37 m high vegetated headland, consisting of a granite core, capped by Tertiary laterite and some Holocene dunes. A sandy foreland (beaches NT 1222 & 1223) surrounds the point, together with large granite boulders. It forms the western entrance to Wonga Bay and the northeastern boundary of Trial Bay.

Beaches NT 1221 and 1222 lie immediately north of the point. Beach **NT 1221** is bordered in the north by a 500 m wide granite point and an attached granite islet. The beach faces east with some sheltering from the point and reefs, but receives sufficient wave energy to maintain a continuously attached low tide bar, with a stable foredune behind that covers the backing bedrock slopes. It is bordered by 200 m of scarped red bluffs, beyond which is beach **NT 1222**, a 200 m long east-facing high tide beach, fronted by continuous sand and rock flats, with the rocks extending up to 300 m off the beach. It forms the northern side of the cuspate foreland at is Guyuwiri Point. Beach **NT 1223** forms the southern side of the foreland. It is 600 m long with two large clumps of boulders located along the beach and rock flats off the beach.

Immediately west of the point is a 1.5 km wide south-facing embayment containing beaches NT 1224-1227. Beach **NT 1224** is a reflective 150 m long cuspate foreland attached to a boulder field, as well as bordered by boulders around the point. One hundred metres to the west is beach **NT 1225**, a curving 250 m long pocket beach surrounded by sloping 20 m high vegetated granite slopes, with some older dunes covering the backing slopes. Boulder points and some boulder reefs lie to either end of the reflective beach.

Beach **NT 1226** lies past the next granite point, 200 m to the west. It is a similar 200 m long pocket beach, free of boulders, with a narrow low tide bar paralleling the beach and more gently vegetated slope behind. Finally beach **NT 1227** occupies the western end of the embayment. It is a slightly curving, 900 m long, southeast-facing beach bordered by red granite boulders. It is exposed to lower southeast waves and winds, with the waves producing several beach rips during strong southerly conditions, the rip channels remaining as holes in the bar during calmer conditions. A solitary blowout at the western end of the beach and up to 500 m wide low grassy dune flats back the beach (Fig. 5.137).

Figure 5.137 Beach NT 1227 (right), with beach NT 1228-1232 extending to left foreground.

NT 1228-1232 **TRIAL BAY (E 1)**

No.	Beach	Rating HT LT	Type	Length
NT1228	Trial Bay (E1)	2 2	R+rocks	100 m
NT1229	Trial Bay (E2)	2 2	R+rocks	150 m
NT1230	Trial Bay (E3)	2 2	R	150 m
NT1231	Trial Bay (E4)	1 1	R+sand flats	200 m
NT1232	Trial Bay (E5)	1 1	R+sand flats	200 m
Spring & neap tide range = 1.4 & 0.0 m				

Trial Bay is a 3500 ha southeast-facing bedrock-controlled embayment that has a relatively narrow 2 km wide entrance and 31 km of low energy shoreline, and the Gurkawuy community located on the northwestern side of the bay. Proterozoic granite dominates the bay entrance, as well as all the high points in the bay, several small bay islands and numerous reefs. The eastern shore of the bay trends north from the western end of Guyuwiri Point, for 6 km to the top of the subsidiary St Davids Bay. It contains eleven low energy generally west-facing beaches, exposed to local wind waves only.

The first five beaches north of the point (NT 1228-1332), face west and represent a transition from beaches exposed to low gulf waves, to those exposed to bay wind waves only. They each occupy small embayments in 20-40 m high vegetated granite slopes. Beach **NT 1228** is a 100 m long pocket of sand bordered and fronted by granite boulders, with boulders also littering the western half of the beach. Beach **NT 1229** occupies the second embayment. It is a curving 150 m long beach, with boulder points to either end, as well as some submerged off the points, with only the centre relatively free of rock. Older dunes from beach NT 1127 have transgressed across the point and over the backing slopes.

Beach **NT 1230** lies in the next embayment. It is a 150 m long reflective beach and, while bordered by bedrock points, is free of rocks. Vegetated dunes from beach NT 1127 drape the southern half of the backing slopes, with a 200 m high flattop point bordering the northern half of the beach. On the northern side of the point is beach **NT 1231**, a curving 200 m long sand beach, bordered and backed by steep vegetated slopes, with tidally controlled sand waves extending 500 m off the beach. Beach **NT 1232** occupies the fifth embayment and is a 200 m long southwest-facing sand beach, bordered by wooded slopes, with irregular sand flats off the beach.

NT 1233-1238 **TRIAL BAY (E 2)**

No.	Beach	Rating HT LT	Type	Length
NT1233	Trial Bay (E6)	1 1	R+sand flats	200 m
NT1234	Trial Bay (E7)	1 1	R+sand flats	900 m
NT1235	Trial Bay (E8)	1 1	R+sand flats	250 m
NT1236	Trial Bay (E9)	1 1	R+sand flats	1.4 km
NT1237	Trial Bay (E10)	1 1	R+sand flats	150 m
NT1238	Trial Bay (E11)	1 1	R+sand flats	100 m
Spring & neap tide range = 1.4 & 0.0 m				

The second section of Trial Bay's eastern shore consists of 4 km of west to southwest-facing shore containing two embayments with five beaches (NT 1233-1237), while beach NT 1238 is located on the western tip of the northern point. All beaches receive only low wind waves generated within the bay and are all fronted by irregular sand flats and some scattered rocks.

Beach **NT 1233** lies just inside the southern point of the first embayment. It is a 200 m long north-facing strip of sand, bordered by a couple of granite boulders in the west and protruding 20 m high vegetated slopes to the east fronted by scattered mangroves. Beach **NT 1234** begins on the north side of the slopes and is a curving 900 m long west-facing beach, bordered at the northern end by a small vegetated granite outcrop, with mangroves around its base. Sand moving into the bay has been deposited in up to 10 beach ridges that have resulted in up to 500 m of beach progradation. Sand flats now fill the embayment between the two beaches and extend over 1 km into the bay. A meandering tidal creek flows out on the north side of the point, which also forms the southern boundary of beach **NT 1235**. This 250 m long northwest-facing high tide beach forms the northern side of the tombolo that ties the granite outcrop to the shore.

Beach **NT 1236** begins on the north side of a vegetated spur and curves round with a double crenulate shoreline for 1.4 km to the northern point. It has a central cuspate foreland lying in lee of offshore rocks, with sand flats and scattered rocks filling this section of the bay. A 200 m section of vegetated rocky slopes separates it from beach **NT 1237**, a 150 m long strip of sand at the base of densely vegetated 10-20 m high slopes. On the western tip of the point is 100 m long beach **NT 1238**, a pocket beach bordered and backed by steep, vegetated 20 m high slopes.

NT 1239-1243 **ST DAVIDS BAY**

No.	Beach	Rating HT	Rating LT	Type	Length
NT1239	St Davids Bay (1)	1	1	R+tidal flats	500 m
NT1240	St Davids Bay (2)	1	1	R+tidal flats	300 m
NT1241	St Davids Bay (3)	1	1	R+tidal flats	1 km
NT1242	St Davids Bay (4)	1	1	R+tidal flats	200 m
NT1243	St Davids Bay (5)	1	1	R+tidal flats	400 m
Spring & neap tide range = 1.4 & 0.0 m					

St Davids Bay is a protected bay located on the northern side of Trial Bay. The 300 ha bay has a 1.2 km wide entrance between prominent vegetated granite headlands, and 5 km of low energy shoreline, consisting of five beaches (NT 1239-1243) and tidal flats separated by mangrove-lined rocky shores.

Beach **NT 1239** is located 1 km inside the eastern headland. It is a protected northwest-facing 500 m long beach ridge, bordered by 300 m and 500 m long mangrove-fringed bedrock slopes. It is backed by a narrow strip of beach ridges, which are cut by two small streams at either end. On the northern side of the eastern mangroves is 300 m long beach **NT 1240**. This is a relatively straight high tide sand beach, fronted by 200 m wide sand flats, and backed by a small densely wooded valley.

The northern end of the bay is occupied by beach **NT 1241**, a curving 1 km long south-facing beach. It is backed by a 400 m long section of beach ridges at its eastern end, that extend up to 600 m inland, with small creeks draining across the beach at either end. The western half of the beach is backed by wooded slopes, with two clumps of mangroves and rocks dominating this more crenulate section. Beach **NT 1242** is located on the western side of the boundary rocks and mangroves. It is a 200 m long south-facing strip of sand, fronted by sand, bordered by rocks and rock flats and backed by wooded slopes.

Beach **NT 1243** is located at the western extremity of the bay in a 500 m wide embayment. The beach blocks a small upland stream that drains out at the southern end of the beach, with mangroves lining the southwestern corner of the bay. Tidal shoals lie off the creek mouth and sand flats partly fill the bay.

NT 1244-1248 **TRIAL BAY-GURKAWUY**

No.	Beach	Rating HT	Rating LT	Type	Length
NT1244	Trial Bay (N1)	1	1	R+sand flats+rocks	250 m
NT1245	Trial Bay (N2)	1	1	R+sand flats	1 km
NT1246	Trial Bay (N3)	1	1	R+sand flats+rocks	450 m
NT1247	Gurkawuy	1	1	R+sand flats+rocks	2.2 km
NT1248	Gurkawuy (W)	1	1	R+tidal shoals&channel	1.6 km
Spring & neap tide range = 1.4 & 0.0 m					

To the west of St Davids Bay the northern Trial Bay shoreline continues for 4 km to the Gurkawuy community, to the west of which is the 300 m wide funnel-shaped entrance to the westernmost section of the bay. There are five beaches (NT 1244-1248) located between St Davids Bay and the entrance.

Beach **NT 1244** is located immediately west of the St Davids Bay entrance. It is a 250 m long southeast-facing beach, wedged in between 10 m high boulder points, with boulders also strewn along the eastern half of the beach. It is backed by densely wooded slopes rising to 20 m. Beach **NT 1245** is located on the western side of the rocks, and is a relatively straight 1 km long southeast-facing beach, bordered by the granite points, with mangroves growing along the shores of the larger southern headland. It is backed by a solitary stable foredune, and fronted by sand flats.

Beach **NT 1246** is located on the western side of the larger headland. It is a curving, 450 m long, south-facing strip of sand, bordered by granite points, with boulders also dotting the beach and intertidal area.

The small **Gurkawuy community** is located along the southern half of beach **NT 1247**, a 2.2 km long southeast-facing beach, fronted by sand flats that widen to 600 m in the centre, in lee of a cluster of granite reefs. The tidal shoals and the deep channel of the westernmost bay extend off the western end of the beach. The community consists of several houses and a landing ground (Fig. 5.138) and is linked by road to the central Arnhem Road, 26 km to the northwest. The beach terminates amongst a cluster of granite boulders, beyond which is beach **NT 1248**. This beach forms the northern side of the entrance into the western funnel-shaped bay. It is a low energy tide-dominated beach, with the tidal channel and shoals paralleling the entire 1.6 km long strip of sand.

Figure 5.138 Gurkawuy beach (NT 1247) is backed by the small community and landing ground.

NT 1249-1252 **TRIAL BAY (W)**

No.	Beach	Rating HT LT	Type	Length
NT1249	Trial Bay (W1)	1 1	R+sand flats	500 m
NT1250	Trial Bay (W2)	1 1	R+sand flats	2.8 km
NT1251	Trial Bay (W3)	1 1	R+sand flats	400 m
NT1252	Trial Bay (W4)	1 1	R+sand flats	1.6 km
Spring & neap tide range = 1.4 & 0.0 m				

The southwestern shores of Trial Bay commence at the southern entrance to the western funnel-shaped bay and trend southeast toward Bald Point, the southern boundary of the bay. The 15 km of shoreline contain 12 beaches (NT 1249-1260), contained in two large sweeping bays. The first bay contains four beaches (NT 1249-1252).

Beach **NT 1249** is a dynamic migratory spit that forms the southern entrance to the westernmost embayment of Trial Bay. The spit varies in occurrence, shape, and length, averaging about 500 m. It represents the movement of marine sand into the 5500 ha bay which is largely filled by a flood tide delta. The spit is tied to the shoreline at its eastern end where there are two clusters of granite boulders.

Beach **NT 1250** commences on the southern side of the second group of boulders and trends south for 2.8 km, finally spiralling round to face northwest. The northern section is exposed to the trades and is backed by older vegetated transgressive dunes extending up to 1.5 km inland, with active dunes extending up to 300 m inland and to heights of 20 m. Sand and tidal shoals widen from a few hundred metres to over 1 km in the south, and lower waves to less than 0.5 m, resulting in reflective conditions at high tide. The southern end is bordered by a cluster of granite boulders, to the south of which is curving 400 m long beach **NT 1251**, which terminates at a small 100 m wide vegetated granite headland. The beach is fronted by 500 m wide sand flats, and backed by largely vegetated transgressive dunes up to 400 m wide and 20 m high (Fig. 5.139).

Figure 5.139 Beach NT 1251 is backed by active and vegetated dunes.

Beach **NT 1252** is a curving 1.6 km long beach that forms the southern shore of the embayment. It is tied to the granite point in the west and a narrow granite point in the east. The beach is sheltered by its orientation, the eastern point and sand shoals extending up to 500 m offshore, resulting in usually calm conditions at the shore. Sand from beach NT 1253 blows across the tip of the point onto the beach.

NT 1253-1260 **TRIAL BAY-BALD POINT**

No.	Beach	Rating HT LT	Type	Length
NT1253	Trial Bay (W5)	1 1	R+sand flats	500 m
NT1254	Trial Bay (W6)	1 1	R+sand/rock flats	5.4 km
NT1255	Bald Pt (W5)	1 1	R+sand flats+rocks	300 m
NT1256	Bald Pt (W4)	1 1	R+sand flats+rocks	500 m
NT1257	Bald Pt (W3)	1 1	R+sand flats+rocks	700 m
NT1258	Bald Pt (W2)	1 1	R+sand flats+rocks	400 m
NT1259	Bald Pt (W1)	2 3	R+rock flats	200 m
NT1260	Bald Pt	2 3	R+rock flats	100 m
Spring & neap tide range = 1.4 & 0.0 m				

The southernmost section of Trial Bay consists of a 5 km wide northeast-facing embayment, bordered in the east by the grassy Bald Point. The embayment contains 10 km of shoreline and seven generally protected beaches (NT 1253-1259).

Beach **1253** is a 500 m long east-facing beach located between the tip of the sand-covered granite point and a sand foreland formed in lee of a 20 m high island located 200 m off the southern end of the beach. The beach is sheltered by the island and rock reefs, resulting in waves averaging less than 0.5 m, which maintain reflective beach conditions.

On the south side of the foreland and island is 5.4 km long beach **NT 1254**, which trends south then curves round in lee of Bald Point to finally face northwest against a low rocky shore. It has reflective conditions the length of the beach, with irregular rock and sand shoals filling the southern end of the spiral, while it narrows toward the northern. The exposed northern section is backed by vegetated older dunes that extend up to 2 km inland and to 42 m in height, with a 2 km long section of active blowouts moving 200 m inland and to heights of 15 m. A solitary creek flows out during the wet season toward the centre of the beach. It drains an interconnected series of dammed wetlands, including the 400 ha Lake Peterjohn.

Between the southern end of beach NT 1254 and Bald Point is a 2 km section of north-facing shoreline containing four arcuate beaches (NT 1255-1258), then the two point beaches (NT 1259 & 1260). All are backed by low-lying grassy to shrubby terrain. Beach **NT 1255** is a curving 300 m long north-facing strip of high tide sand, tied to low rocky points and flats, with irregular rock flats extending up to 200 m off the beach. Beach **NT 1256** extends for 500 m from the eastern rocky point, to low bluffs, where it continues below to a low rock point. It is fronted by a mixture of rock and sand flats.

Beach **NT 1257** is tied to two low rocky points and is largely free of rocks in the centre. It is 700 m long and faces north across 300 m wide sand shoals. Beach **NT 1258** occupies the final arc, curving for 400 m to the lee of Bald Point. The eastern section is very sheltered by the point and shallow rock flats and has scattered mangroves along the shore.

Beach **NT 1259** is located on the north side of the 10 m high grassy **Bald Point**. The irregular high tide beach is 200 m long and fronted by 50-100 m wide intertidal rock flats. Beach **NT 1260** is located on the northeastern tip of the point and is a 100 m long east-facing strip of high tide sand, fronted by gently sloping 100 wide rock flats, with waves breaking over the rocks, resulting in low waves at the shore. It is backed by the scarped red laterite bluffs of the point.

NT 1261-1267 BALD-BAGBIRINGULA POINTS

No.	Beach	Rating HT	Rating LT	Type	Length
NT1261	Bald Pt (S1)	6	6	TBR	4.2 km
NT1262	Bald Pt (S2)	6	6	TBR	250 m
NT1263	Bald Pt (S3)	6	6	TBR-LTT	4.6 km
NT1264	Bagbiringula Pt (N4)	4	4	LTT-R	1.1 km
NT1265	Bagbiringula Pt (N3)	3	4	R+rock flats	100 m
NT1266	Bagbiringula Pt (N2)	3	4	R+rock flats	100 m
NT1267	Bagbiringula Pt (N1)	3	4	R+rock flats	200 m
Spring & neap tide range = 1.4 & 0.0 m					

The 12 km of coast between Bald and Bagbiringula points is one of the more exposed, high energy sections of the Territory coast. A near continuous beach links the two laterite points, with a few rock outcrops dividing the near continuous sand into four high energy beaches (NT 1261-1264), and three rock-controlled beaches (NT 1265-1267) on Bagbiringula Point, most backed by extensive active and older vegetated dune transgression.

Beach **NT 1261** commences on the southern side of **Bald Point** and trends south for 4.2 km to a 700 m long section of laterite bluffs. The beach is fully exposed to the southeast waves and winds and receives waves averaging about 1-1.5 m. These maintain about 20 well developed rips cutting across the 100 m wide bar, with at times rhythmic megacusps formed along the beach in their lee. Older transgressive dunes extend up to 3.5 km inland and to heights of 45 m, with active dunes presently extending up to1.5 km inland as a series of multiple nested blowouts (Fig. 5.140). In the north the older dunes have crossed the rear of Bald Point and spilled onto the Trial Bay beaches (NT 1250-1258).

Figure 5.140 Beach NT 1261 (top) is backed by a series of multiple-nested parabolic dunes.

Beach **NT 1262** is located between a small rock outcrop and the red laterite bluffs. The beach is 250 m long and usually has two permanent rips flowing out each end of the beach. The dunes are more stable to the lee of the beach and bluffs.

Beach **NT 1263** commences on the southern side of the bluffs and continues on for 4.6 km to a slightly protruding sandy section in lee of scattered reefs. The beach receives high waves along the central-northern section with rips dominating the surf, decreasing closer to the reefs. It is also backed by vegetated and active transgressive dunes, the vegetated dunes linking with those of beach NT 1261. The dunes of both systems are transgressing into the 400 ha Lake Peterjohn in the north and four freshwater wetlands in the south.

South of the reefs the shore continues as a lower energy beach (**NT 1264**). The curving 1.1 km long east-facing beach is backed by an active foredune in the north and red laterite bluffs rising to 10 m in the south. Reefs dominate the offshore, lowering waves to about 1 m and maintaining a low tide terrace to reflective beach in the south, usually free of rips.

The three southern beaches (NT 1265-1267) are located on the southeast tip of Bagbiringula Point. Beach **NT 1265** is a 100 m long pocket of sand wedged in between small laterite points and backing bluffs, with reefs offshore. Immediately south is beach **NT 1266**, a similar 100 m long strip of sand fronted by intertidal rocks and reefs. Beach **NT 1267** is located at the tip of the point and is a curving 200 m long beach bordered by laterite-capped granite boulders, with dunes climbing up over the backing bluffs of the point. All three are backed by older grassy blufftop dunes.

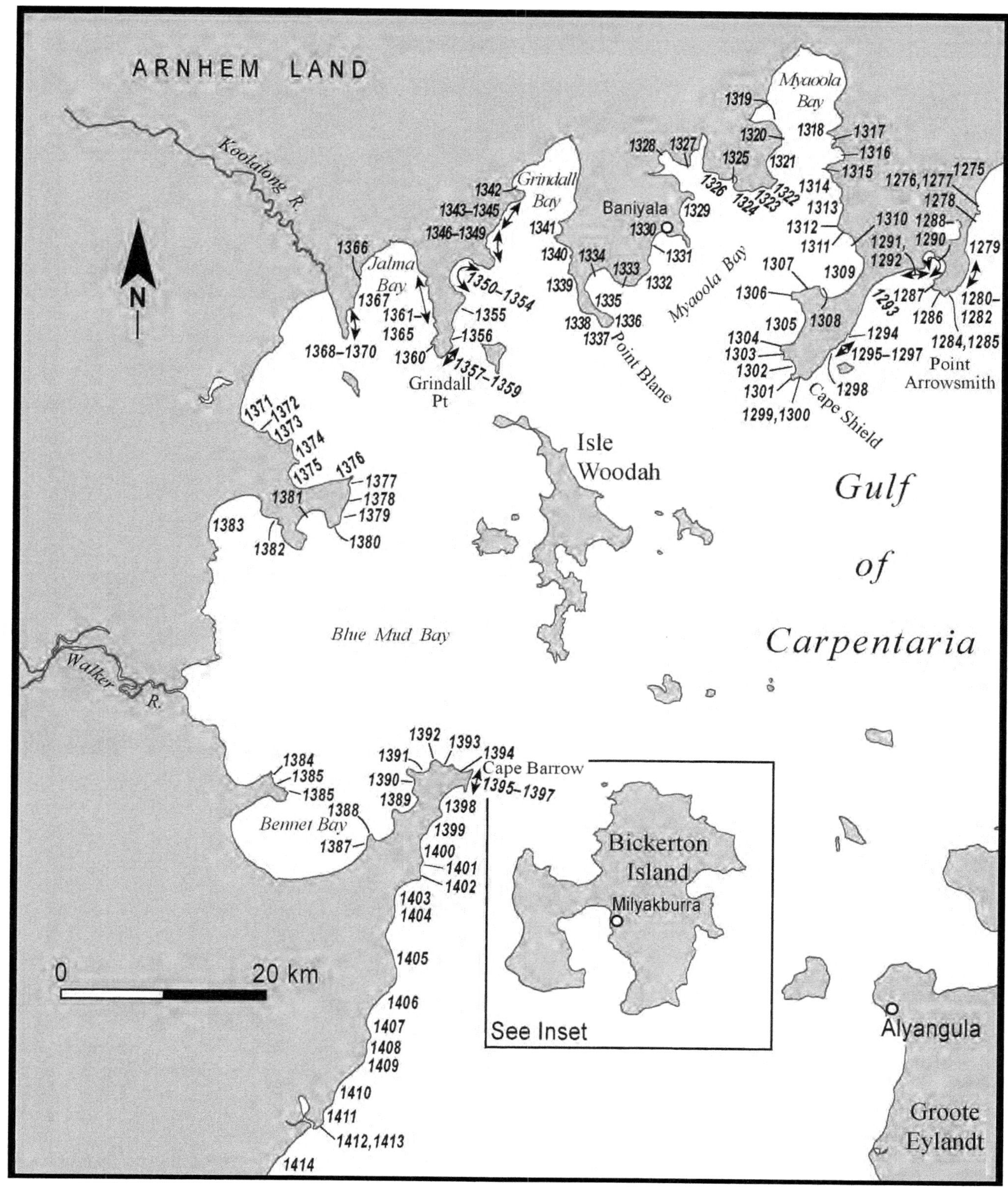

Figure 5.141 Regional map of Blue Mud Bay in eastern Arnhem Land and the adjacent coast and beaches NT 1275-1414. See Figure 5.166 for Bickerton Island and Figure 5.168 for Groote Eylandt.

NT 1268-1274 BAGBIRINGULA POINT (W)

No.	Beach	Rating HT	Rating LT	Type	Length
NT1268	Bagbiringula Pt (W1)	2	3	R+rock flats	30 m
NT1269	Bagbiringula Pt (W2)	2	3	R+rock flats	50 m
NT1270	Bagbiringula Pt (W3)	2	3	R+rock flats	30 m
NT1271	Bagbiringula Pt (W4)	2	3	R+rock flats	30 m
NT1272	Bagbiringula Pt (W5)	2	3	R+rock flats	150 m
NT1273	Bagbiringula Pt (W6)	2	3	R+rock flats	300 m
NT1274	Bagbiringula Pt (W7)	2	3	R+rock flats	250 m
Spring & neap tide range = 1.4 & 0.0 m					

Immediately west of Bagbiringula Point the shore trends northwest for 1.5 km as a series of small crenulations bordered by laterite reefs and backed by red laterite bluffs, with rock flats extending off the shore. Seven small high tide beaches (NT 1268-1274) are located in each of the crenulations and exposed to waves averaging about 0.5 m, with each fronted by intertidal rock flats. The first four (**NT 1268-1271**) extend along the western side of the narrow tip of the point and are all less than 50 m in length. Beach **NT 1272** is 150 m long and backed by 10-20 m high bluffs. Only beach **NT 1273** has a backing foredune that has blown up onto the bluffs, while beach **NT 1274** is again backed by 10 m high bluffs, with bluff debris partly covering the ends of the beach.

NT 1275-1279 WARDARLEA BAY

No.	Beach	Rating HT	Rating LT	Type	Length
NT1275	Wardarlea Bay (1)	5	5	LTT-TBR-LTT	5.8 km
NT1276	Wardarlea Bay (2)	3	4	LTT	400 m
NT1277	Wardarlea Bay (3)	3	3	LTT	400 m
NT1278	Wardarlea Bay (4)	4	4	LTT-R	4 km
NT1279	Wardarlea Bay (5)	3	4	R+rock flats-R	900 m
Spring & neap tide range = 1.4 & 0.0 m					

Wardarlea Bay is a spiralling, relatively open, south- then east-facing bay located to the lee of Bagbiringula Point. The point provides some protection to the northeast corner of the bay, with most of the shore exposed to moderate waves from the east. There is no vehicle access to the bay.

The main northern beach (**NT 1275**) begins in lee of the point as a south-facing lower energy beach fronted by a continuous bar. Once clear of the point waves pick up to about 1 m and maintain usually small rips crossing the bar every 100-150 m, with up to 30 rips located along the main section of beach. The northern 2 km of beach is backed by a massive 3 km wide series of up to 30 beach-foredune ridges, with the earliest ridge transgressing inland. Two small creeks transect the ridges and a larger third creek forms their western boundary. The southern 4 km face southeast and is backed by reactivated foredunes that are presently transgressing up to 700 m inland. They are backed in the south by two wetlands. The beach terminates at a section of granite boulders and backing partly vegetated slopes.

Beach **NT 1276** is located in amongst the granite section. It is 400 m long and faces east out into the gulf. The granite extends offshore as reefs and lowers waves at the shore, resulting in a low tide terrace fronting the clear pockets of sand between the rocks.

Beach **NT 1277** is located at the southern end of the granite. It curves round to face north for 400 m as a tombolo that ties a small granite outcrop to the shore (Fig. 5.142). The beach is moderately protected with waves averaging about 0.5 m and breaking across a narrow bar, with a few boulders littering the beach and nearshore.

Figure 5.142 Beaches NT 1277 & 1278 form either side of a well developed tombolo.

Beach **NT 1278** extends from the south side of the tombolo for 4 km to the first of the red laterite bluffs, which dominate all the way to Point Arrowsmith. The beach faces due east and is slightly crenulate in lee of deeper offshore reefs. The reefs lower waves to less than 1 m, maintaining a continuous bar and occasional rips in the north, with steep reflective conditions dominating most of the beach to the south. The back beach is well exposed to the trade winds, which have resulted in multiple episodes of dune transgression, moving up to 1 km inland and to 30 m in height with a 150 ha lake backing the dunes.

Beach **NT 1279** commences below the laterite bluffs and trends south for 900 m as a narrow high tide beach fronted by 100-200 m wide intertidal laterite flats. It terminates at an inflection in the bluffs, which marks the beginning of the Point Arrowsmith beaches.

NT 1280-1286 **POINT ARROWSMITH**

No.	Beach	Rating HT	Rating LT	Type	Length
NT1280	Pt Arrowsmith (N5)	3	4	R+rock flats	1 km
NT1281	Pt Arrowsmith (N4)	4	4	R+rock flats/reef	400 m
NT1282	Pt Arrowsmith (N3)	4	4	R+rock flats/reef	1.3 km
NT1283	Pt Arrowsmith (N2)	4	4	R-LTT-TBR	600 m
NT1284	Pt Arrowsmith (N1)	4	4	R+reef	300 m
NT1285	Pt Arrowsmith (E)	4	4	R-R+rock flats	700 m
NT1286	Pt Arrowsmith (W1)	3	3	R+rock flats	1 km
Spring & neap tide range = 1.4 & 0.0 m					

Point Arrowsmith lies at the southern tip of a 4 km long, 2 km wide laterite peninsula that separates Wardarlea and Marajella bays. The point is 8 m high with the backing terrain rising to a peak of 29 m. The seven more exposed beaches to the east of the point (NT 1280-1286) are all influenced to some degree by laterite bluffs, rock flats and reefs. There is no vehicle access to the point area.

Beach **NT 1280** commences at a low laterite bluff, with fronting rock flats and reefs, which marks a more southerly trend in the shore. It extends south for 1 km, fronted initially by 50 m wide rock flats and reefs further out, then is a little more open but backed by 10 high laterite bluffs, partly draped by dunes, with reefs extending up to 1 km off the southern sandy point.

Beach **NT 1281** commences immediately to the south and consists of a relatively straight 400 m long sandy beach, bordered by rocks and reefs to either end, with deeper reefs off the beach resulting in lower waves at the shore and a steep reflective beach face. It is backed by three coalescing parabolic dunes that extend up to 700 m inland and rise to 30 m (Fig. 5.143). The southern point continues on for 200 m as a 10 m high scarped laterite bluff to the beginning of beach **NT 1282**. This 1.3 km long beach is also fronted by deeper reefs extending up to 500 m offshore, which permit lowered waves to maintain a steep reflective beach. It is backed by a mixture of bluffs, older red probably Pleistocene dunes and a central active dune section extending 100-200 m inland. It terminates at a 16 m high bluff.

The bluff forms a small headland, with beach **NT 1283** commencing inside the south side of the bluff. It initially faces south where it is protected by the headland resulting in low waves and reflective beach conditions. It then curves to face southeast for most of its 600 m length, with wave height increasing down the beach to maintain a continuous low tide terrace and a few beach rips toward the southern end. The beach is backed by bluffs which grade from scarped in the south to vegetated in the north, where they rise to 20 m and are capped by a small clifftop dune. Beach **NT 1284** commences on the south side of a sandy foreland, which separates it from beach NT 1283. It is a curving 300 m long beach tied to low laterite rocks at either end, with reefs extending across the beach, resulting in low waves and reflective conditions at the shore. It is backed by 100 m of active dunes, then more stable dunes extending 500 m inland, which link with the dunes behind beach NT 1285.

Figure 5.143 Beach NT 1281 (left) is backed by three large parabolic dunes, extending up to 700 m inland.

Beach **NT 1285** extends southwest of beach NT 1284 as a crenulate 700 m long beach. It has an eastern section relatively clear of reef containing a reflective to low tide terrace beach, while the western section is fronted by continuous rock flats and reefs. Transgressive dunes back the western end, with small stable blufftop dunes along the western half. The beach terminates at the tip of Point Arrowsmith, an 8 m high dune-capped laterite point. To the west of the point red bluffs curve to the north for 200 m to the beginning of beach **NT 1286**. The 1 km long beach is narrow in the east and backed by the red bluffs. These diminish to the west where the beach widens and is backed by low vegetated dunes. It terminates at Point Arrowsmith, a sandy reef-controlled foreland.

NT 1287-1291 **POINT ARROWSMITH (W)**

No.	Beach	Rating HT	Rating LT	Type	Length
NT1287	Pt Arrowsmith (W2)	1	2	R+rock-sand flats	1.3 km
NT1288	Pt Arrowsmith (W3)	1	2	R+sand flats	2.2 km
NT1289	Marajella Bay (1)	1	2	R+sand flats	1.9 km
NT1290	Marajella Bay (2)	1	2	R+rock flats	600 m
NT1291	Marajella Bay (3)	1	2	R+rock flats	500 m
Spring & neap tide range = 1.4 & 0.0 m					

To the west of **Point Arrowsmith** the coast trends north for 4 km into the upper reaches of Marajella Bay, before turning and running southwest for 16 km to Cape Shield. Five relatively protected beaches (NT 1287-1291) are located along the western shore of the point and in the upper reaches of the bay.

Beach **NT 1287** commences 1 km west of Point Arrowsmith as the shore turns at a low laterite point and extends north for 1.3 km as a prograding series of several recurved spits separated by mangroves. The present spit terminates in a splay of three smaller spits, the size and shape of which can be expected to vary as sand moves along the spits into the bay. The spit is initially fronted by 200 m wide rock flats which grade into sand flats. A left-hand surf break is located off the southern tip of the point, though it requires strong southeasterly conditions to have sufficiently high waves to break.

The northern ends of the spits are slowly building out across the southern end of beach **NT 1288**, producing a sheltered mangrove-lined shore along this curving, 2.2 km long, west to finally south-facing beach. The beach represents the ultimate terminus for much of the sand moving into the bay. The sand is manifest as a series of spits in the south and up to ten beach ridges in the north which have prograded the shoreline up to 500 m to the present beach. The beach is now backed by a beach ridge and fronted by sand flats. It terminates at a low vegetated section of laterite, beyond which is beach **NT 1289**. This beach curves round for 1.9 km between the laterite and the beginning of a rock-dominated section of shore and faces southeast toward Point Arrowsmith. This beach has also prograded 200-300 m as a series of low beach ridges, with a freshwater swamp behind the northern end of the beach. A small creek occasionally breaks out across the beach to drain the swamp.

The southern end of the beach marks the beginning of a 3 km long section of shore fronted by laterite rock flats and reefs. Beach **NT 1290** occupied the first 600 m. It is a crenulate northeast to east-facing high tide sand beach fronted by irregular rock flats up to 300 m wide. Beach **NT 1291** commences as the shore turns and faces southeast. The beach extends for 500 m as a similar crenulate rock- and reef-fringed beach.

NT 1292-1300 **MARAJELLA BAY-CAPE SHIELD**

No.	Beach	Rating HT	Rating LT	Type	Length
NT1292	Marajella Bay (4)	2	3	R+rock flats	400 m
NT1293	Marajella Bay (5)	4	4	R-LTT-R	8.8 km
NT1294	Djarrakpi	4	4	LTT+reefs	1.3 km
NT1295	Djarrakpi	3	3	R	300 m
NT1296	Djarrakpi	3	3	R+rock flats	300 m
NT1297	Djarrakpi	3	3	R+rock flats	100 m
NT1298	Djarrakpi	3	3	LTT	3.7 km
NT1299	Cape Shield (E 2)	3	3	R+rock flats	200 m
NT1300	Cape Shield (E 1)	4	4	R+rock flats	200 m
Spring & neap tide range = 1.4 & 0.0 m					

The western shore of **Marajella Bay** is a relatively open southeast-facing stretch of shore that extends down to Cape Shield. While all the beaches face into the southeast trades, Gooninnah Island in the south, extensive inshore laterite reefs and, in places, rock flats all lower waves to generally less than 1 m at the shore. The small Djarrakpi community and landing ground are located 2 km in from the coast to the west of beach NT 1294, with vehicle access to the community and from there by 4WD to many of the bay and cape beaches.

Beach **NT 1292** marks the beginning of one of the longer sections of beach and dunes on this section of coast. The beach is 400 m long, faces south and is bordered by low laterite rocks to the east and a sandy reef-controlled foreland to the west, with deeper reefs paralleling most of the beach. It is backed by active dunes extending up to 300 m inland. On the southern side of the foreland begins the 8.8 km long main beach (**NT 1293**). It initially faces south-southeast, curving round to face east-southeast in the south. Despite its exposed orientation, nearshore and inshore reefs, including coral reefs off the southern end, lower waves particularly towards either end. It has reflective conditions in these locations, with a continuous bar along the central 6 km, which is only cut by rips during higher wave conditions. The entire beach is backed by moderately active dunes that average about 500 m in width. Older vegetated parabolic dunes have crossed the 2 km wide cape to reach the shores of Blue Mud Bay. Two small, usually closed, streams cross toward either end of the beach.

Beach **NT 1294** commences on the south side of a 20 m high laterite bluff that forms the southern boundary of beach NT 1293. The beach trends south-southeast for 1.3 km to a bluff and rock-reef-dominated section of shore. It is backed by dune-capped, scarped red, 15-20 m high bluffs along the northern half, with the vegetated dunes then extending up to 500 m inland. In the south lower dunes are being reactivated and moving up to 300 m inland. Deeper reefs and rocks in the south front the beach, lowering waves to less than 1 m which maintain a narrow continuous bar along the open section.

On the south side of the southern rocks and reefs is a bluff-backed, zeta-curved 800 m long open embayment, containing the next two beaches. Beach **NT 1295** is a 300 m long strip of sand located at the base of 10 m high red bluffs, with usually reflective conditions along the beach. Two hundred metres to the south is beach **NT 1296**, a similar 300 m long sand strip, backed by bluffs, with intertidal rock flats fronting the beach. Just around the corner of the bluffs is beach **NT 1297**, a 100 m long strip of south-facing sand wedged in between the bluffs and rock flats.

Beach **NT 1298** commences 200 m east of beach NT 1297. It trends south then southwest, and is initially backed by red bluffs which soon give way to active dunes extending up to 1.5 km inland. The dunes reach an inner height of 30 m and are actively transgressing over a 30 ha wetland. The **Djarrakpi** community is located 500 m north of the northern end of the dunes and a vehicle track runs from the community down to the wetland. The beach lies in the lee of **Gooninnah Island**, located just over 1 km offshore, with wave refraction around the island producing a cuspate foreland (Fig. 5.144). Reflective beach conditions dominate each end of the beach with a 100 m wide low tide terrace along the foreland. The beach terminates in the south at a sloping laterite bluff and rocky shore.

Figure 5.144 Beach NT 1298 is part of a curving cuspate foreland and is backed by massive transgressive dunes extending up to 1.5 km inland.

The bluffs curve round to face south, with beach **NT 1299** tucked in their lee. The 200 m long beach faces south and is backed by the sloping sand-draped bluffs and fronted by continuous rocks and reefs. Five hundred metres to the west and in the lee of Cape Shield is beach **NT 1300**. This is a 200 m long east-facing beach, moderately exposed to waves which break across a 100 m wide bar and some rock reefs, with permanent rips maintained at either end of the beach. There is the possibility of a right-hand surf break off the southern point during higher wave conditions. A 100 m wide dune field backs the beach then the 10 m high slopes of Cape Shield.

Blue Mud Bay

Blue Mud Bay is a large, 90 km long, 35 km deep, shallow bay bordered by Cape Shield in the north and Cape Barrow 46 km to the south. The two projecting points together with Isle Woodah, which extends for 25 km between the two points, and several smaller islands, maintain a well to very well protected bay shoreline. The bay has a total shoreline length of 339 km containing 94 low energy beaches (NT 1301-1394). Several medium-sized rivers and creeks drain the coastal range to the west, delivering sediment to the bay and forming the extensive mangrove-fringed deltaic systems, with tidal flats up to 10 km in width and more than 80 km of mangrove-fringed shoreline. The small communities of Barraratjpi, Baniyala and Yilila are located on the shores of the bay, with several other communities located within a few kilometres of the bay.

NT 1301-1307 **CAPE SHIELD (W)**

No.	Beach	Rating HT	Rating LT	Type	Length
NT1301	Cape Shield	2	3	R+rock flats	500 m
NT1302	Cape Shield (W1)	1	2	R+rock-sand flats	2 km
NT1303	Cape Shield (W2)	1	2	R+rock-sand flats	1.5 km
NT1304	Cape Shield (W3)	1	2	R+rock flats	600 m
NT1305	Cape Shield (W4)	1	1	R+rock sand	5.8 km
NT1306	Cape Shield (W5)	1	2	R+rock flats	700 m
NT1307	Cape Shield (W6)	1	2	R+sand flats+rocks	2.4 km
Spring & neap tide range = 1.4 & 0.0 m					

At **Cape Shield** the shoreline turns and trends north for 8.5 km and contains seven generally west-facing beaches (NT 1301-1307) located in four embayments. Beach **NT 1301** is located on the southern tip of the cape, wrapping round the cape for 500 m. It consists of a high tide sand beach fringed by continuous 100 m wide intertidal rock flats. Low vegetated dunes originating from beach NT 1300 back the beach. At its western end the shoreline trends due north into an embayment that contains 2 km long beach **NT 1302**. This beach initially faces west then curves round to face due south at the northern end. Rock flats fringed by coral reefs front the southern half of the beach with sand flats along the northern half. It is backed by a foredune abutting 10 m high vegetated bluffs, with minor dune transgression behind the northern end of the beach.

The northern end of the embayment is bordered by rock flats with a small curving embayment on the northern side containing 1.5 km long beach **NT 1303**. The curving beach faces west and is tied to more rock flats at its northern end, with sand flats in between. The beach is backed by 50-100 m wide active and vegetated dunes. Beach **NT 1304** is located in lee of the northern rock flats. The crenulate beach faces due west and extends for 600 m to the beginning of the next embayment. Beach **NT 1305** is a 5.8 km long west-facing beach, which curves to face northwest against the low rocky northern point. It is fronted by sand flats between the rocks and is backed by a narrow band of beach ridges and a 50 ha freshwater swamp toward the centre of the beach, with some minor dune transgression on the northern point. The **Djarrakpi** community and landing ground is located 2.5 km east of the southern end of the beach.

The northern point trends relatively straight for 700 m as west-facing beach **NT 1306**. The beach is tied to two low rocky points, with scattered rock flats and coral patches in between and a low foredune backing the beach. Beach **NT 1307** extends east of the northern point for 2.4 km to the beginning of the next embayment. The beach has sand flats and scattered rock outcrops along the shore, and is backed by the low wooded plain.

NT 1308-1312 **BARRARATJPI**

No.	Beach	Rating HT	Rating LT	Type	Length
NT1308	Barraratjpi (S)	1	2	R+mud flats	700 m
NT1309	Barraratjpi	1	2	R+mud flats	5.6 km
NT1310	Barraratjpi (N1)	1	2	R+mud flats	1.5 km
NT1311	Barraratjpi (N2)	1	2	R+mud flats+rocks	300 m
NT1312	Barraratjpi (N3)	1	2	R+rock/mud flats	500 m
Spring & neap tide range = 1.4 & 0.0 m					

The small **Barraratjpi** community and landing ground are located adjacent to the northern end of beach NT 1309. This beach and its two neighbours occupy a protected 5 km wide northwest-facing embayment. A vehicle track from the community backs the main beach and continues on to the Djarrakpi community, located 8 km to the south.

Beach **NT 1308** lies in the southwestern corner of the embayment. The 700 m long beach faces east to northeast into the bay, resulting in a very sheltered location. It is bordered by rock flats in the west and low scarped red laterite bluffs to the east, with a narrow beach in between, together with a few clumps of mangroves along the shore. The main beach (**NT 1309**) commences on the eastern side of 100 m long bluffs. It is a curving 5.6 km long beach that faces north, northwest and finally west at the community. It is backed by a low beach ridge, with vegetated transgressive dunes crossing the 1.5 km wide peninsula from beach NT 1293 to back the northern half of the beach. The northern end, nearest the community, terminates at a 500 m long section of mangrove-covered rock flats.

At the end of the mangroves the shoreline turns and trends north as 1.5 km long beach **NT 1310**, which curves slightly to face southwest. A small tidal creek flows across the northern end, with sand and tidal flats widening to 300 m off the creek mouth. The beach terminates at a southwest-facing 300 m long section of low laterite bluffs. The bluffs continue on for 800 m and are fronted by two narrow beaches. Beach **NT 1311** extends for 300 m and is largely fronted by tidal flats and scattered rocks, while crenulate beach **NT 1312** continues on for 500 m with more rock flats than sand flats off the beach. It terminates at a 1.5 km long section of wooded laterite shoreline.

NT 1313-1318 **MYAOOLA BAY (E)**

No.	Beach	Rating HT	Rating LT	Type	Length
NT1313	Barraratjpi (N4)	1	2	R+tidal flats	1.3 km
NT1314	Barraratjpi (N5)	1	2	R+tidal flats	3.1 km
NT1315	Myaoola Bay (E1)	1	2	R+tidal flats	1 km
NT1316	Myaoola Bay (E2)	1	2	R+mud flats	500 m
NT1317	Myaoola Bay (E3)	1	2	R+tidal flats	1.1 km
NT1318	Myaoola Bay (E4)	1	2	R+tidal flats	1.5 km
Spring & neap tide range = 1.4 & 0.0 m					

Myaoola Bay is located in the northeast corner of Blue Mud Bay. The 12 km deep bay faces due south to a 3.5 km wide entrance and has an area of 7500 ha. Nine beaches (NT1313-1321) line the outer eastern and western shores, with the inner 20 km of shoreline dominated by mangrove forests up to 2.5 km wide and some extensive salt flats up to 4 km wide. The only vehicle access to the area is the Barraratjpi track, which runs 2-3 km in from the eastern shore of the bay.

Beach **NT 1313** commences 4 km north of the Barraratjpi community at the northern end of a section of low laterite bluffs. The beach faces west across the bay and consists of a 1.3 km long strip of high tide sand cut by two small rock outcrops. It is backed by a few beach ridges in the south and low wooded bluffs to the north, and fronted by tidal mud flats. It terminates at a low rocky point, on the north side of which begins 3.1 km long beach **NT 1314**. This is a more crenulate beach with a few rock outcrops, rock flats and mangroves, together with more continuous tidal flats. Wooded laterite plains back most of the beach apart from a few wetland-backed beach ridges in the south.

Beach **NT 1315** begins at a low rock point that separates it from beach NT 1314 (Fig. 5.145). The shoreline turns and faces northwest, with the beach extending for 1 km into a mangrove-filled apex of a 1.5 km wide embayment. The beach consists of two sandy crenulations, then a lower energy rock and partly mangrove-fronted section that grades into extensive mangroves at its eastern end. On the north side of the mangroves is 500 m long west-facing beach **NT 1316**, a very low energy strip of high tide sand fronted by mud flats.

Figure 5.145 Beaches NT 1314 (right) & 1315 (left).

Beaches NT 1317 and 1318 are located either side of a V-shaped point and are the last beaches on the eastern side of the bay. Beach **NT 1317** is a 1.1 km long southwest-facing strip of sand backed by a few beach ridges and fronted by tidal flats and rocks toward the western point. Beach **NT 1218** commences on the northern side of the point and is a 1.5 km long northwest-facing strip of sand fronted by mud flats, which widen from 100 m at the point to 300 m in the east. Mangroves dominate most of the next 20 km of shore, which include the entire top end of the bay.

NT 1319-1326 **MYAOOLA BAY-BLUE MUD BAY**

No.	Beach	Rating HT	Rating LT	Type	Length
NT1319	Myaoola Bay (W1)	1	2	R+rock flats	2.2 km
NT1320	Myaoola Bay (W2)	1	2	R+rock flats	1.1 km
NT1321	Myaoola Bay (W3)	1	2	R+rock flats	3.5 km
NT1322	Blue Mud Bay (N1)	1	2	R+rock/coral flats	600 m
NT1323	Blue Mud Bay (N2)	1	2	R+rock/coral flats	1.5 km

NT1324	Blue Mud Bay (N3)	1	2	R+rock/coral flats	2.8 km
NT1325	Blue Mud Bay (N4)	1	2	R+rock/coral flats	300 m
NT1326	Blue Mud Bay (N5)	1	2	R+rock/coral flats	800 m
Spring & neap tide range = 1.4 & 0.0 m					

The western shore of **Myaoola Bay** contains 10 km of mangrove-dominated shore, before wave energy becomes sufficient to maintain low energy beaches between the low laterite bluffs and intertidal rock flats. The southern half of the bay contains three beaches (NT 1319-1321) before the shore turns and trends west within the main Blue Mud Bay, with another five rock flat and coral reef-fringed beaches (NT 1322-1326) occupying the first 8 km of shore. There is no vehicle access to the northern and western sides of the bay.

The western bay beaches commence at beach **NT 1319**, which emerges from a mangrove-lined shore as a high tide sand beach located between vegetated bluffs and irregular rock outcrops and 100 m wide rock flats. The crenulate 2.2 km long beach initially faces north before curving round to face northeast. It terminates at a southerly inflection in the shore, which marks the beginning of beach **NT 1320**. This beach faces due east, is 1.1 km long and is backed by a few beach ridges then the bluffs, with irregular rock flats averaging 100 m in width.

Beach **NT 1321** is located in an open east-facing embayment, bordered by the rocks of beach NT 1320 in the north and red laterite bluffs to the south. The beach is 3.5 km long, faces generally east, gradually curving round within the embayment. There are several rock-controlled inflections along the shore, some capped by mangroves, with 100 m wide rock flats and some coral reefs paralleling the beach.

Beach **NT 1322** begins 500 m south of beach NT 1321. The beach wraps round a low laterite headland for 200 m before a straight 400 m long southeast-facing section. The entire beach is fronted by 50-100 m wide rock flats then more uniform coral reefs extending up to 300 m off the beach. It is backed by dense woodlands on the headland, and bordered in the west by a smaller 500 m wide low wooded headland.

Beach **NT 1323** commences on the western side of the headland and curves to the west for 1.5 km, terminating at a rock-controlled sandy foreland. The beach is bordered by rock flats and fronted by continuous 300 m wide coral reef flats. It is backed by a low foredune then woodlands (Fig. 5.146).

Beach **NT 1324** commences on the western side of the foreland. It is a 2.8 km long beach that trends west then curves round into an inlet mouth as a series of recurved spits to finally face west. The inlet drains a 400 ha area of mangroves and salt flats. Some active dunes back the exposed southerly section and extend up to 100 m inland. The shoreline however is fringed by 30 m wide coral reef flats which are replaced by some rocks, mangroves and tidal sand shoals in the inlet mouth.

Figure 5.146 Beach NT 1322 (right) and the longer beach NT 1323 (right) are both dominated by intertidal rock flats and fringing coral reef.

Beach **NT 1325** is located on the western shore of the inlet and consists of a 300 m long strip of sand bordered by low rocks and mangroves and fronted by 100 m wide coral reef flats. One kilometre to the west is beach **NT 1326**, a crenulate 800 m strip of sand bordered and interspersed with rocks and mangroves, and fronted by an irregular 500 m wide section of rocks and coral reef flats.

NT 1327-1333 **BANIYALA**

No.	Beach	Rating HT	LT	Type	Length
NT1327	Baniyala (N3)	1	2	R+rock flats	2.2 km
NT1328	Baniyala (N2)	1	2	R+tidal flats	900 m
NT1329	Baniyala (N1)	1	2	R+sand-rock flats	3.5 km
NT1330	Baniyala	1	1	R+sand flats	2.2 km
NT1331	Baniyala (S1)	1	2	R+rock-coral flats	2.4 km
NT1332	Baniyala (S2)	1	2	R+rock-coral flats	2.4 km
NT1333	Baniyala (S3)	1	2	R+rock-coral flats	3.2 km
Spring & neap tide range = 1.4 & 0.0 m					

The **Baniyala** community consists of about 20 houses and a landing ground located at the eastern end of a curving beach (NT1330) 10 km northeast of Point Blane and 9 km from Cape Shield, the northern entrance to Blue Mud Bay. The surrounding coastline consists of low energy beaches to the immediate north and south (NT 1329-1333), with two additional beaches (NT 1327 and 1328) located in a three-armed bay to the north. Only beach NT 1329 and the main Baniyala beach (NT 1330) are accessible by vehicle.

Beach **NT 1327** is located on a low laterite promontory 6 km north of the community. The beach wraps around the shoreline for 2.2 km, initially facing east and then curving to the south. It is only impacted by low wind waves generated within the 1-2 km wide bay and is fronted by continuous irregular rock flats. Two kilometres to the west is beach **NT 1328**, a 900 m long southeast-facing beach formed in the apex of the central of the three arms. It consists of a low beach backed by a grassy beach ridge and fronted by 200 m wide mud flats. A small upland stream and tidal creek drains across the northern and southern ends respectively, with a 50 ha wetland behind the southern end.

Beach **NT 1329** lies immediately north of the community. It is a 3.5 km long east-facing beach consisting of a curving 2 km long section generally clear of rocks and a southern 1.5 km long crenulate section in lee of 0.5-1 km wide rock flats and fringing coral reefs. The beach grades into mangroves at the southern end, with a vehicle track along the southern portion of the beach.

Baniyala beach (**NT 1330**) is located to the lee of a mangrove, rock and reef-fringed eastern point. The community is located at the northern end of the curving 2.2 km long south to southeast-facing beach. Rock flats and reefs fringe both ends, with a sandy beach and 100 m wide sand flats along the central section. Two to three hundred metre wide low beach ridges back the beach, then a 200 ha wetland that links to a 100 ha freshwater swamp 3 km west of the community.

Beach **NT 1331** commences beyond the rocks at the southern end of the main beach. It trends due south for 2.4 km as a crenulate high tide sand beach fronted by rock then coral reef flats extending up to 500 m off the beach. It terminates at a southwest inflection in the shore, which marks the beginning of 2.4 km long beach **NT 1332**. This is a similar crenulate southeast-facing beach with rocks and coral flats extending out to 300 m off the beach. It terminates at another major inflection in the shoreline, where beach **NT 1333** trends west, then northwest, as a highly crenulate 3.2 km rock- and reef-fronted beach, with the reefs narrowing from 400 m in the east to 200 m in the west.

NT 1334-1341 **POINT BLANE-GRINDALL BAY**

No.	Beach	Rating HT	Rating LT	Type	Length
NT1334	Pt Blane (N2)	1	1	R	3.8 km
NT1335	Pt Blane (N1)	1	1	R+rock-coral flats	3.6 km
NT1336	Pt Blane	1	1	R+rock-coral flats	800 m
NT1337	Pt Blane (W1)	1	1	R+rock-coral flats	900 m
NT1338	Pt Blane (W2)	1	1	R+rock-coral flats	2.3 km
NT1339	Pt Blane (W3)	1	1	R+rock flats	3.9 km
NT1340	Grindall Bay (E1)	1	2	R+rock flats	4.2 km
NT1341	Grindall Bay (E2)	1	1	R+sand flats	1.2 km
Spring & neap tide range = 1.4 & 0.0 m					

Point Blane is located at the southern tip of a 7 km long, 0.5-1 km wide, low wooded laterite peninsula that protrudes into Blue Mud Bay and forms the eastern boundary of Grindall Bay. A vehicle track runs to the tip of the point where a 100 m high communication tower is located. Apart from the tower there is no other development in the area.

Beach **NT 1334** is located at the eastern junction of the peninsula where it forms a 2.5 km wide southeast-facing bay, with the 3.8 km long beach facing down the bay into the trades. Much of the beach forms a low 100-300 m wide barrier, which dams two 100 ha freshwater wetlands at either end. The first 2 km of beach have a sandy shore and receive waves averaging about 0.5 m. The southern half of the beach is fronted by coral reef flats that widen to 500 m, resulting in calm conditions at the shore.

Beach **NT 1335** begins at a southeasterly inflection in the shore and continues on for 3.6 km to the northern tip of Point Blane. The highly crenulate beach is fronted by irregular rock, then smoother coral reef flats averaging 500 m in width (Fig. 5.147). The point road runs along the back of beaches NT 1334 and 1335 to reach the point.

Figure 5.147 The crenulate beach NT 1335 (top) and beach NT 1336 (lower) are fringed by a continuous coral reef. Note tower clearing to left.

Beach **NT 1336** is located on the southeast-facing side of the point, with the tower and four associated buildings located in a circular clearing 200 m in from the southern side of the beach. The 800 m long high tide beach is fronted by irregular rock flats then 300 m wide coral reef flats. A low point is located 200 m south of the beach, beyond which the shore trends west for 900 m as beach **NT 1337**. This is another highly crenulate rock-controlled beach fronted by more irregular 100-200 m wide coral reef flats. It terminates at a 200 m long section of low red bluffs.

Beach **NT 1338** commences on the northern side of the bluffs and trends north then west as a curving 2.3 km long beach. The first 1 km is rock-controlled and crenulate, while the second south-facing section is more continuous and freer of rocks. However the entire beach is fronted by 50-100 m wide rock flats, together with some coral reef in the south. It terminates at a northerly inflection in the shore, beyond which beach **NT 1339** commences. This beach trends roughly north for 3.9 km, as a crenulate rock-fronted beach backed by dense woodlands.

Beach **NT 1340** commences at a northerly inflection in the shore and trends north for 2.5 km before curving round to the tip of a sandy foreland, with a total length of

4.2 km. The rock outcrops diminish along the beach, but gentler crenulations dominate the entire shoreline. Two small streams drain 10 ha and 40 ha freshwater wetlands behind the beach. Beach **NT 1341** is located on the northern side of the foreland. This is a low energy northwest- to west-facing beach, partly fringed by mangroves and fronted by tidal flats. It faces into Grindall Bay and receives very low wind waves which have built a low, splaying, grassy, 100-300 m wide beach ridge plain, containing at least 10 ridges.

The northern 20 km of **Grindall Bay** are dominated by intertidal mangroves and high tide salt flats, with the northernmost flats extending 8 km inland.

NT 1342-1349 **GRINDALL BAY (W)**

No.	Beach	Rating HT	Rating LT	Type	Length
NT1342	Grindall Bay (W1)	1	2	R+rock flats	300 m
NT1343	Grindall Bay (W2)	1	2	R+sand-rock flats	1 km
NT1344	Grindall Bay (W3)	1	2	R+rock-coral flats	500 m
NT1345	Grindall Bay (W4)	1	2	R+rock-coral flats	600 m
NT1346	Grindall Bay (W5)	1	2	R+rock-sand flats	2.2 km
NT1347	Grindall Bay (W6)	1	2	R+sand-rock flats	300 m
NT1348	Grindall Bay (W7)	1	2	R+rock-coral flats	600 m
NT1349	Grindall Bay (W8)	1	2	R+rock flats	1.6 km

Spring & neap tide range = 1.4 & 0.0 m

The western shore of Grindall Bay commences in dense mangroves at the northern apex. Mangroves line the first 10 km of shore until a low laterite point with beach NT 1342 located on its northern shore. South of the point low energy sandy beaches dominate the shore with a mixture of rocks, coral reef, sand flats and a few mangroves fronting the east-facing beaches. There is no vehicle access to any of the bay shoreline, the nearest track terminating 7 km northwest of beach NT 1342.

Beach **NT 1342** is a 300 m long very low energy strip of high tide sand fronted by rocks and scattered mangroves that face northeast deep into Grindall Bay. It terminates at a low laterite point, with laterite dominating the next 1 km of shore to the beginning of beach NT 1343.

Beach **NT 1343** trends southwest for 1 km as a relatively straight sand beach fronted by narrow sand flats that widen to 200 m at the southern end. It is backed by a grassy 100 m wide beach ridge plain, with rocks, a few mangroves and patchy coral reef forming the southern point. Beach **NT 1344** is located 100 m to the south and is a 500 m long straight southeast-facing sand beach fronted by sand flats that widen to 100 m in the south. It terminates at a slight projection in the low laterite bluffs.

Beach **NT 1345** commences on the southern side of the bluffs and trends south for 600 m to the lee of sand flats and coral reefs. The latter lie up to 300 m off the beach and connect to the low sandy point that forms the southern boundary of the beach. Beach **NT 1346** commences on the southern side of the point and trends south to southwest for 2.2 km. The slightly crenulate beach is fronted by three patches of rock and reef flats, with sand flats in between. Two small creeks drain across the beach, which is backed by a low 100 m wide band of wooded foredunes.

Beach **NT 1347** is located on the south side of a small mangrove-covered laterite point. The beach is 300 m long and fronted by a continuous 200 m wide coral reef flat, with mangroves bordering the southern laterite point. Beach **NT 1348** is located 500 m to the east on the other side of the point. It is a double crenulate, 600 m long, low energy, east-facing beach that has two rock outcrops capped by mangroves on the shore and patchy coral reef off the beach.

Beach **NT 1349** is located along the eastern side of a low laterite headland that protrudes 2 km into the bay. The 1.6 km long beach consists of a strip of high tide sand fronted by irregular mangrove-capped patchy rock, rock flats, patches of coral reef, and toward the south two areas of sand flats fronted by coral flats up to 200 m wide (Fig. 5.148).

Figure 5.148 The crenulate beach NT 1349 is dominated by rock flats and some patches of fringing coral reef.

NT 1350-1359 MOUNT GRINDALL-GRINDALL POINT

No.	Beach	Rating HT	Rating LT	Type	Length
NT1350	Mt Grindall (N5)	1	2	R+coral flats	200 m
NT1351	Mt Grindall (N4)	1	2	R+coral flats	400 m
NT1352	Mt Grindall (N3)	1	2	R-coral flats	2.7 km
NT1353	Mt Grindall (N2)	1	2	R+rock-coral flats	300 m
NT1354	Mt Grindall (N1)	1	2	R+rock-coral flats	200 m
NT1355	Grindall Pt (N4)	1	2	R+coral flats	3 km
NT1356	Grindall Pt (N3)	1	2	R+coral flats	1 km
NT1357	Grindall Pt (N2)	1	2	R+coral flats	300 m
NT1358	Grindall Pt (N1)	1	2	R+rock-coral flats	100 m
NT1359	Grindall Pt	1	2	R+rock-coral flats	100 m
Spring & neap tide range = 1.4 & 0.0 m					

The southwestern shores of **Grindall Bay** commence in an open southeast-facing embayment containing beaches NT 1350-1354, the latter at the foot of 55 m high Mount Grindall. The shore then trends due south for 6 km to Point Grindall, the southern tip of a 1-2 km wide, 10 km long peninsula. The core of the peninsula is a north-south-trending ridge of Proterozoic coarse quartz-rich sandstones and conglomerate, with the higher eastern section reaching 76 m and composed of metasedimentary rocks including slates, schists, phyllites, siltstones and volcanics. The slopes are lightly wooded, with bedrock exposed along some of the steeper eastern sections above the beaches. There is no vehicle access to the peninsula.

Beach **NT 1350** is located toward the western end of a 2 km long headland that forms the northern boundary of the 3 km wide southeast-facing embayment. The beach faces west into the bay and consists of a 200 m long beach backed by a low beach ridge then 10 m high laterite bluffs. It is tied via a low gravel ridge to a small rock-reef outcrop in the east, while abutting the bluffs in the west. Beach **NT 1351** is located 100 m to the west and commences below the bluffs. It curves to the west for 400 m to link with a linear mangrove-capped rock outcrop, forming a small tombolo with the northern end of beach NT 1352. It is backed by minor foredune activity and fronted by 300 m wide patchy coral reef flats which link with the rocks off beach NT 1350.

Beach **NT 1352** is the main beach, occupying the base of the embayment. The 2.7 km long beach faces southeast into the trade winds, which provide sufficient energy deep into Blue Mud Bay to maintain a 2 km long reflective beach with the southern section fringed by coral flats that widen to 300 m. The beach is backed by an 800 m wide beach to foredune ridge plain, and three freshwater wetlands totalling 70 ha in area, the southernmost connected by a small stream to the southern end of the beach.

The southern shore of the embayment is dominated by the metasedimentary rocks and contains two pocket beaches at the base of Mount Grindall. Beach **NT 1353** is located on the south side of 100 m long mangrove-capped rocks. It is a 300 m long east-facing beach with mangroves at each end, and coral flats widening from 100 m to 300 m off the beach. Beach **NT 1354** is located 500 m to the east and is a 200 m pocket of sand at the base of 70 m high wooded bedrock slopes. It is fronted by 300-400 m wide coral flats.

One kilometre to the east the shoreline turns and trends south for 6 km to Grindall Point. Beach **NT 1355** occupies the first 3 km. This is a slightly curving sand beach, backed by 50-200 m wide wooded foredunes and fronted by coral flats which range from 100 m to 200 m in width. Beach **NT 1356** is located on the southern side of a 300 m long, low, mangrove-capped rocky point. It is a relatively straight 1 km long east-facing beach fronted by straight 200 m wide coral flats, with some rock outcropping on the beach. Low mangroves and rock flats border the southern end.

Beach **NT 1357** is located on the southern side of a low protruding wooded point. The beach is a 300 m long pocket of sand bordered by linear rock flats and fronted by a central 50 m wide coral flat. Beach **NT 1358** is located immediately to the south and consists of a 100 m long pocket of sand bordered by 100 m wide rock flats with a small central coral section. Five hundred metres to the south at the tip of Grindall Point is 100 m long beach **NT 1359**, a high tide beach fronted by continuous 50 m intertidal rock flats, then 100 m wide coral flats.

NT 1360-1365 GRINDALL POINT-JALMA BAY (E)

No.	Beach	Rating HT	Rating LT	Type	Length
NT1360	Grindall Pt (W)	1	1	R+sand flats	300 m
NT1361	Jalma Bay (E1)	1	1	R+sand flats	300 m
NT1362	Jalma Bay (E2)	1	1	R+sand flats	150 m
NT1363	Jalma Bay (E3)	1	2	R+mud flats	200 m
NT1364	Jalma Bay (E4)	1	3	R+mud flats	200 m
NT1365	Jalma Bay (E5)	1	4	R+mud flats	400 m
Spring & neap tide range = 1.4 & 0.0 m					

Grindall Point forms the eastern boundary of **Jalma Bay**, a U-shaped, south-facing, 8 km wide, 11 km deep bay, that is bordered in the west by Kapui Point. The Koolatong River flows into the northwestern side of the bay, depositing extensive 5-10 km wide mangrove-fringed tidal flats. Mangroves dominate 8 km of the 30 km long bay shore, with the remainder consisting of bedrock and six low energy pocket beaches (NT1361-1366), most fringed by mangroves.

At **Grindall Point** the shoreline turns and trends north into Jalma Bay, with the six small beaches located along the next 8 km of predominantly mangrove-fringed rocky shore, followed by mangrove-tidal flats commencing at 9 km. Beach **NT 1360** is located 1.5 km northwest of the point. It is a 300 m long south-facing sand beach, bordered by mangrove-fringed bedrock points and fronted by sand flats. A grassy beach ridge backs the beach. Beach **NT 1361** is located in the next embayment

1 km to the north. It is a curving 300 m long east-facing beach, also bordered by dense mangroves on the bedrock points.

Beach **NT 1362** is located in the next embayment 600 m to the north. It consists of two 100 m and 50 m long pockets of sand separated by a central 100 m long fringe of mangroves, with mangrove-lined headlands to either side. Beach **NT 1363** lies 1 km to the north and is a 200 m long strip of high tide sand fronted by 100 m wide mud flats and bordered by mangroves.

Beach **NT 1364** lies 1.5 km further north and is a 200 m pocket of sand wedged in between near continuous mangroves and fronted by 200 m wide mud flats. Finally, beach **NT 1365** is located 500 m to the north in the last bedrock-controlled embayment before the tidal flats. It is a 400 m long southwest-facing high tide sand beach, with 100-200 m wide mud flats off the southern half, then a small rock outcrop and a rock-flat-fronted northern half. The tidal flats of the Koolatong River delta commence 1.5 km to the north.

NT 1366-1370 **JALMA BAY-KAPUI POINT**

No.	Beach	Rating HT	Rating LT	Type	Length
NT1366	Jalma Bay (W)	1	2	R+rock flats	1.2 km
NT1367	Kapui Pt (N3)	1	2	R+rock flats	800 m
NT1368	Kapui Pt (N2)	1	2	R+rock flats	300 m
NT1369	Kapui Pt (N1)	1	2	R+rock flats	600 m
NT1370	Kapui Pt	1	2	R+rock flats	150 m

Spring & neap tide range = 1.4 & 0.0 m

Five kilometre long **Kapui Point** forms the western boundary to Jalma Bay. The point narrows from 2 km to 200 m at the point and is composed of a low 15 m high ridge of Proterozoic sandstone and conglomerate. The bedrock dominates the more exposed eastern shoreline, which contains five small beaches (NT 1366-1370) all fronted by rock flats. The western shore faces across St Nicholas Inlet, whose shoreline is entirely dominated by mangrove-fringed salt flats extending several kilometres inland.

Beach **NT 1366** is located 5.5 km north of the point emerging from the southern tidal flats of the Koolatong River delta. The low energy beach consists of a crenulate 1.2 km of 50-100 m wide irregular rock flats and scattered mangroves. Beach **NT 1367** lies 1.5 km to the south and consists of a relatively straight southeast-facing beach, bordered by low rocky points and fronted by continuous 200 m wide rock flats. Beach **NT 1368** is 300 m to the south. It is a slightly curving 300 m long east-facing beach bordered by mangrove-fringed rock points, with continuous rock flats exposed at low tide.

Beach **NT 1369** is located 400 m to the south in the next small embayment. It is a 600 m long east-facing beach, also fronted by rock flats, as well as scattered mangroves along the shore. A few beach ridges back the northern half of the beach. Beach **NT 1370** lies at the southern tip of Kapui Point. It consists of a 150 m long strip of south-facing sand wedged in between two fingers of protruding rock flats, with rock flats also off the beach.

NT 1371-1381 **HADDON HEAD**

No.	Beach	Rating HT	Rating LT	Type	Length
NT1371	Haddon Hd (N6)	1	2	R+tidal flats	600 m
NT1372	Haddon Hd (N5)	1	2	R+tidal flats	300 m
NT1373	Haddon Hd (N4)	1	2	R+tidal flats	1.6 km
NT1374	Haddon Hd (N3)	1	2	R+tidal flats	2.2 km
NT1375	Haddon Hd (N2)	1	2	R+mud flats	2 km
NT1376	Haddon Hd (N1)	1	2	R+rock-coral flats	2 km
NT1377	Haddon Hd	1	2	R+rock-coral flats	1.5 km
NT1378	Haddon Hd (S1)	1	2	R+rock-coral flats	800 m
NT1379	Haddon Hd (S2)	1	2	R+rock-coral flats	1.5 km
NT1380	Haddon Hd (S3)	1	2	R+rock-coral flats	1.5 km
NT1381	Haddon Hd (S4)	1	1	R+tidal flats	3.6 km

Spring & neap tide range = 1.4 & 0.0 m

Haddon Head is located on the central western shores of Blue Mud Bay. The low wooded head is composed of Tertiary laterite and protrudes 11 km into the bay with 40 km of shoreline. Wide tidal flats and mangroves lie north and south of the head, with eleven beaches (NT 1371-1381) located on the more exposed north and eastern-facing shores of the head, while coral fringes the easternmost shoreline. There is a solitary 60 km access track from the Central Arnhem Road, which reaches the southern shore of beach NT 1374.

Six beaches (NT 1371-1376) are located along the 16 km long northern side of the head. Beach **NT 1371** commences at the southern end of 6 km long mangrove-fringed tidal flats. The beach faces north and consists of a crenulate 600 m long strip of high tide sand bordered in the west by mangroves and fronted by a mixture of rocks and tidal flats. It terminates at a low laterite outcrop, on the eastern side of which is 300 m long beach **NT 1372**. This beach faces northeast and is bordered to the east by a mangrove-fringed low rocky point.

Beach **NT 1373** is located 1 km to the east. It is a roughly curving 1.6 km long beach that faces northeast then north, terminating at a mangrove-lined shoreline. It has some rock outcrops along the beach and is fronted by tidal flats. The mangroves extend for 1 km to a low V-shaped rocky point on the southern side of which is beach **NT 1374**. The beach curves to the south for 2.2 km terminating at a rounded rocky point. The beach is backed by a 50-100 m wide beach ridge plain that has impounded a 10 ha freshwater swamp behind the northern half of the beach. A small stream also crosses toward the centre of the beach, just south of which the solitary vehicle track reaches the shore.

On the southern side of the round point is a 1 km long section of mangroves, which gives way to 2 km long beach **NT 1375**. This is a curving east to northeast-facing beach, also bordered by mangroves in the south. Mud flats front the beach which is backed by a 200-300 m wide beach ridge plain and a southern 30 ha freshwater wetland, connected by a small stream to the southern corner of the beach. The shoreline trends east of the beach and is dominated by mangroves for 2.5 km to beach **NT 1376**. This is a 2 km long north-facing beach which terminates at a V-shaped sandy foreland tied to low laterite rocks and rock flats (Fig. 5.149). Patchy rock flats and more continuous coral flats widen from 100 m in the west to 200 m in the east, and 800 m off the point.

Figure 5.149 Beaches NT 1376 (right) & NT 1377 (left) form a low sandy foreland tied to laterite rocks and reefs.

At the point the shore turns and trends south with three eastern shore beaches (NT 1377-1379) occupying 6 km of shore. Beach **NT 1377** curves south of the point for 1.5 km to a section of laterite rocks and flats and scattered mangroves, while coral flats extend 500 m offshore. Beach **NT 1378** is located on the south side of the rocks and is a narrow 800 m long east-facing beach, backed by scarped 5 m high red laterite bluffs, with rock flats fringing the base of the beach and coral flats extending 400 m offshore. In addition a clump of mangroves is located in lee of rock flats which form the southern boundary.

Beach **NT 1379** commences on the southern side of the rocks and is a 1.5 km long east-southeast-facing slightly curving beach that terminates against mangrove-covered rock flats. Some bordering rock flats and coral flats extend 400 m off the beach and rock flats dominate the next 2 km of shore, with beach **NT 1380** located to the lee of the flats. This is a convex crenulate 1.5 km long beach that is dominated by the rocks and flats. It faces southeast, south, then curves round into an embayment to finally face west. The western end of the beach consists of an active spit backed by some older mangrove-fronted spits that have been deposited along the eastern side of the embayment.

Beach **NT 1381** is the only beach on the southern side of the head. It is located within a semicircular bay that is 2 km wide at the entrance, where both low heads are fringed by mangroves and 100-300 m wide coral flats. The beach is 3.6 km long and curves round to face southeast out of the bay. It is backed by a low 200 m wide beach ridge plain and two freshwater wetlands totalling 30 ha in area, with tidal flats fronting the beach.

NT 1382-1383 **HADDON HEAD (W)**

No.	Beach	Rating HT LT	Type	Length
NT1382	Haddon Hd (W1)	1 2	R+mud flats	200 m
NT1383	Haddon Hd (W2)	1 2	R+mud flats	300 m
Spring & neap tide range = 1.4 & 0.0 m				

The shoreline of Blue Mud Bay west and south of Haddon Head is dominated by mangrove-fringed tidal flats and low coastal plain. Between beach NT 1381 and Yilila head are 50 km of continuous mangrove-lined shore apart from two small pockets of sand (beaches NT 1382 & 1383).

Beach **NT 1382** is located 2 km due west of beach NT 1381 and consists of a 200 m long south-facing strip of sand bordered by mangroves, fronted by mud flats and backed by the low wooded coastal plain. Five kilometres to the east across a mangrove-lined embayment is beach **NT 1383**, a 300 m long east-facing strip of high tide sand, also bordered by mangroves and mud flats, with the low coastal plain behind.

NT 1384-1386 **YILILA**

No.	Beach	Rating HT LT	Type	Length
NT1384	Yilila (N3)	1 2	R+tidal flats	300 m
NT1385	Yilila (N2)	1 1	R	1.2 km
NT1386	Yilila (N1)	1 2	R+rock flats	250 m
Spring & neap tide range = 1.4 & 0.0 m				

The **Yilila** community consists of two houses and is located on the southern side of a low wooded 2.5 km long laterite headland, that is connected to the mainland by 4 km wide salt flats. A vehicle track runs along the back of the headland and across the tidal flats, when dry, to connect with formed track 12 km east of the community. Along the north side of the headland are three adjoining low energy beaches (NT 1386-1389), while to the north and south mangroves dominate the entire southwest shores of Blue Mud Bay.

Beach **NT 1384** is located at the north end of the headland and consists of a 300 m long northeast-facing beach, backed by a low foredune and fronted by 100 m wide sand and rock tidal flats. Three hundred metre long rock flats border its northern end and 5 m high red laterite bluffs the southern end. Beach **NT 1385** is located on the south side of the 50 m wide bluffs. It is a relatively straight 1.2 km long steep reflective beach, backed by a few low beach ridges and a 50 ha freshwater wetland. Rock flats border the southern end, on the south side of which is beach **NT 1386**. This is a narrow 250 m long strip of sand wedged in at the base of scarped 5 m high laterite bluffs, with some rock and coral flats extending 100 m off the beach. The community is located 1 km southwest of this beach.

NT 1387-1394 **CAPE BARROW (W)**

No.	Beach	Rating HT	Rating LT	Type	Length
NT1387	Cape Barrow (W8)	1	2	R+rock flats	200 m
NT1388	Cape Barrow (W7)	1	2	R+rock/tidal flats	100 m
NT1389	Cape Barrow (W6)	1	1	R+sand flats	3.7 km
NT1390	Cape Barrow (W5)	1	1	R+sand flats	2.8 km
NT1391	Cape Barrow (W4)	1	1	R+sand flats	2.5 km
NT1392	Cape Barrow (W3)	1	2	R+rock flats	100 m
NT1393	Cape Barrow (W2)	1	2	R+sand flats	1.7 km
NT1394	Cape Barrow (W1)	1	3	LTT-R	2 km
Spring & neap tide range = 1.4 & 0.0 m					

Cape Barrow forms the southern boundary of Blue Mud Bay. The low, wooded, northeast-trending cape is composed of Tertiary laterite, which outcrops as occasional low bluffs and rock flats, with most of the cape shoreline occupied by beaches and sand flats. There is no vehicle access to the cape area. To the west of the cape the shoreline trends west, then southwest into **Bennet Bay**. The first 12 km of shore contain eight increasingly lower energy beaches (NT 1387-1394), beyond which is a sweeping 20 km long section of mangroves fronting the 6-10 km wide Wurindi Swamp. The mangroves terminate at Yilila headland.

Beaches NT 1387 and 1388 are two strips of high tide sand located on the west and northern sides of a low wooded 500 m wide headland that forms the eastern boundary of the swamp. Beach **NT 1387** is a 200 m long strip of sand fronted by 50-100 m wide rock flats and scattered mangroves. Beach **NT 1388** is located on the northern side of the headland below scarped 5 m high laterite bluffs. It is a 100 m long pocket of sand bordered by rock flats, with tidal flats in the centre.

Three kilometres to the east is a similar sized headland, with the bay between occupied by mangroves. On the western side of the headland is a fringe of mangroves then the curving 3.7 km long beach **NT 1389**. This is a low energy north to northwest-facing beach and sand flats exposed to westerly waves generated within the bay. It is backed by a 500 m wide series of more than eight low grassy beach ridges and bordered in the east by a rock-controlled sandy foreland.

Beach **NT 1390** is a continuous high tide beach that commences on the north side of the foreland and trends north for 2.8 km, curving round in lee of a low rounded laterite point. The beach is largely backed by a narrow foredune and fronted by 300 m wide sand flats, with rock flats to either end. As the point curves round to face north, beach **NT 1391** commences and trends west (Fig. 5.150) in a curving arc for 2.5 km to a crenulate 1 km wide section of low laterite bluffs. It is backed by a 100-200 m wide fringe of beach ridge-foredunes, with a 10 ha freshwater swamp behind the centre of the beach. Sand flats extend up to 1 km off the centre of the beach.

Figure 5.150 Continuous beaches NT 1389 (top), with NT 1390 curving round the point, and NT 1391 extending to the left.

Beach **NT 1392** is a 100 m long sliver of high tide sand located on the point below the laterite bluffs. It faces north and is fronted by two mangroves and 100 m wide rock flats.

Beach **NT 1393** commences immediately west of the point and is a more exposed north-facing 1.7 km long slightly curving beach. It receives low refracted waves along the western end, with 100-200 m wide sand flats and scattered reef patches off the beach, particularly toward the more protected eastern end. Active and stable foredunes climb to 21 m behind the beach and extend up to 200 m inland, with a linear 30 ha freshwater swamp behind the foredunes. The beach is bordered in the east by a protruding section of rock flats, with beach **NT 1394** commencing on the eastern side. This is a curving, north-facing 2 km long beach which receives waves averaging about 0.5 km at the western end, which decrease to the east in lee of a 500 m long reef that runs off the northern tip of Cape Barrow. The centre of the beach is backed by 200 m wide active and stable foredunes that rise to 13 m. The rock flats and reefs of Cape Barrow form the eastern boundary.

NT 1395-1402 **CAPE BARROW (S)**

No.	Beach	Rating HT	Rating LT	Type	Length
NT1395	Cape Barrow (S1)	1	2	R+rock flats	600 m
NT1396	Cape Barrow (S2)	1	2	R+rock flats	400 m
NT1397	Cape Barrow (S3)	1	2	R+rock flats	100 m
NT1398	Cape Barrow (S4)	1	1	R+ridged sand flats	4.4 km
NT1399	Cape Barrow (S5)	1	2	R+rock flats-R	4.3 km
NT1400	Cape Barrow (S6)	1	2	R+rock flats	800 m
NT1401	Cape Barrow (S7)	1	2	R+rock flats	1.3 km
NT1402	Cape Barrow (S8)	1	2	R+rock flats	500 m
Spring & neap tide range = 2.2 & 1.0 m					

At **Cape Barrow** the coast turns and trends generally southwest for 130 km to the mouth of the Roper River. While the shore is exposed to the southeast winds, Bickerton and Groote islands afford considerable protection from the higher 'deepwater' gulf waves for approximately 22 km south of the cape to the northern end of Almarlangij beach (NT 1406).

The first three beaches south of the cape (NT 1395-1397) are located on the rock flats that extend south for 2 km. Beach **NT 1395** commences at the northern tip of the cape and curves roughly south for 600 m. It is fronted by 300 m wide intertidal rock flats, with scattered mangroves and fringing coral reefs and backed by a beach ridge and then wooded coastal plain. It terminates at a sandy foreland, on the south side of which is 400 m long beach **NT 1396**, a double crenulate strip of high tide sand fronted by scattered low mangroves, rock flats and then fringing coral reefs. One kilometre further south is 100 m long beach **NT 1397**, a curving strip of sand located toward the southern end of the rock flats.

Beach **NT 1398** commences 1.5 km south of the cape in lee of a low rocky point. It initially faces southeast, curving round for 4.4 km to face east against the low southern rocky point. It is fronted by 400 m wide ridged sand flats in the north which narrow to 200 m in the south. It is backed by a narrow casuarina-fringed beach ridge plain, which widens to 500 m in the south. The beach has a moderately steep narrow high tide beach, with the low gradient sand ridges exposed at low tide (Fig. 5.151). Beach **NT 1399** begins on the south side of the point and trends southwest for 4.3 km to another small rocky point, with a sand foreland located in lee of rock reefs in the centre of the beach. It is backed by a 200-300 m wide beach ridge plain north of the foreland and a narrower system to the south. Waves average less than 0.5 m at the beach.

Figure 5.151 Casuarina fringed beach NT 1398 with the low tide exposing a ridge and runnel bar system.

Beach **NT 1400** is a crenulate east-facing 800 m long beach located between two low rocky points with 50-100 m wide rock flats paralleling the beach, and wooded coastal plain behind. Beach **NT 1401** extends south of the southern point for 1.3 km. It is fronted by shallow 100-200 m wide sand and rock flats, with a 50-100 m wide foredune systems backing the beach. Beach **NT 1402** is a 500 m long double crenulate rock-controlled beach located on the southern rocks. It is fronted by 400 m wide rock and reef flats, and backed by a 400 m wide beach ridge system which links the beaches to the north and south.

NT 1403-1405 **AMAYA-DHARRNI**

No.	Beach	Rating HT LT	Type	Length
NT1403	Amaya	1 1	R+sand flats	3.6 km
NT1404	Amaya (S)	1 2	R+rock-coral flats	1.2 km
NT1405	Dharrni	1 2	R+rock-coral flats	7.2 km
Spring & neap tide range = 2.2 & 1.0 m				

Amaya beach (**NT 1403**) commences on the southern side of the reef that projects seaward of beach NT 1402. The beach initially faces south, curving round for 3.6 km to face east against the southern low rocks. The beach is composed of coarse shell-rich sand and has a steep beach face backed by a 300 m wide grassy beach ridge plain, with two interconnected 40 ha freshwater wetlands behind the southern half of the beach. A creek diagonally crosses the beach ridges to flow out toward the centre of the beach, while sand flats front the beach, grading into patchy rock and coral flats toward the south.

At the southern end of the beach the shoreline trends due south as a crenulate 1.2 km long beach (**NT 1404**) fronted by 300-400 m wide rock and coral flats. It is backed by a 100 m wide series of 3-4 grassy beach ridges. A sandy inflection at the southern end of the beach marks the beginning of 7.2 km long **Dharrni** beach (**NT 1405**). This beach initially curves to the south, then extends round a slightly protruding laterite foreland, to terminate at a 200 m wide laterite outcrop. The northern half of the beach is backed by 200-400 m wide grassy beach ridges (Fig. 5.152), with a backing elongate freshwater swamp, connected by a small stream to the centre of the beach. The beach ridges narrow along the foreland, with the southernmost section containing a 500 m section of stable transgressive dunes, which marks the beginning of the higher energy conditions to the south. A 4WD vehicle track skirts the northern end of the wetland and runs out to the beach to the small **Dharrni** community.

NT 1406-1414 **ALMARLANGIJ-ARNDANI-MONNIE CREEKS**

No.	Beach	Rating HT LT	Type	Length
NT1406	Almarlangij Ck	4 5	UD	3.6 km
NT1407	Almarlangij Ck(S1)	1 2	R+rock-coral flats	1.2 km
NT1408	Almarlangij Ck (S2)	1 2	R+rock-coral flats	1.5 km
NT1409	Aryillarlarg	1 2	R+rock-coral flats	800 m
NT1410	Aryillarlarg (S1)	1 2	R+rock-coral flats	5.1 km
NT1411	Arndani (N2)	1 2	R+rock-coral flats	1.7 km
NT1412	Arndani (N1)	1 2	R+rock-coral flats	400 m
Spring & neap tide range = 2.2 & 1.0 m				

Figure 5.152 View south along Dharrni beach (NT 1405).

The low laterite outcrop that marks the beginning of beach NT1406 also marks the start of a 60 km long section of east to southeast-facing shore that is clear of the protection of Bickerton and Groote islands and fully exposed to the southeast winds and waves. This is the last higher energy section on the Arnhem Land coast and extends as far south as Amamarrity Island. South of the island larger rivers such as the Rose and Roper and extensive tidal flats dominate the coast, as waves are also reduced by the decreasing fetch and water depth in the lower gulf. While this section is well exposed to the gulf waves, producing some higher energy beaches and dune systems, there are also long sections of rock and coral flats, which lower waves at the shore.

Beach **NT 1406** commences at the rock outcrops and trends southwest for 3.6 km to the small mouth of **Almarlangij Creek**. Apart from two small rock outcrops in the north the beach is an exposed low gradient ultradissipative system. During stronger winds waves up to 1-1.5 m maintain a 100 m wide surf. In the north it is backed by an older area of vegetated transgressive dunes that have blown 2 km inland and to heights of 18 m, while the southern half of the beach is backed by a 300 m wide foredune-blowout system and the 100 ha salt flats of **Malgaryungu Lagoon**. These in turn are backed by a 200-400 m wide inner (possibly Pleistocene) barrier, which blocks the 100 ha freshwater **Amunngale Swamp**. The entire barrier-wetland system totals about 600 ha and Almarlangij Creek drains the entire system.

Beach **NT 1407** commences on the southern side of the creek and trends south for 1.2 km to a low red laterite outcrop. The low energy beach is protected by the inlet tidal shoals, with coral reef flats increasing in width from 200 m to 500 m. It is backed by a 200 m wide grassy beach ridge plain.

Beach **NT 1408** commences on the southern side of the 100 m long laterite outcrop. It is a curving 1.5 km long low energy beach fronted by patchy coral reef flats extending 1 km offshore, with some ribbons of sand across the flats. A 300 m wide wooded section (possibly Pleistocene barrier?) backs the beach with a circular 20 ha shallow lagoon behind. It terminates at a 200 m long section of low rock outcrops, beyond which is beach **NT 1409**. This is a narrow 800 m long beach fronted by narrow rock flats, then sand flats with patchy reef flats extending 1 km offshore. It terminates at a low 200 m long rocky section.

Beach **NT 1410** begins on the south side of the rocks and trends generally southwest for 5.1 km to a rock-controlled inflection in the shore. The entire beach is fronted by shallow coral reef flats averaging 1 km in width, which lower waves at the shore to less than 0.5 m at high tide. The 5 ha **Aryillarlarg Billabong**, together with several other wetlands, backs the northern end of the beach. Between the wetlands and the shore is a 0.5-1 km wide series of possibly Pleistocene, then onlapping Holocene beach ridges, including some minor dune transgression. It is possible this and some of the adjoining barrier systems were more active during the mid-Holocene following the sea level stillstand. At this time the present reef flats were deeper or non-existent permitting higher waves to reach the shore and possibly produced the more active dune systems. Subsequent shoaling by the reefs decreased wave height and consequently beach and dune type.

Beach **NT 1411** commences on the south side of the inflection and curves to the south for 1.7 km before terminating at a low convex headland and rock flats. The beach is fronted by sand patches and coral reef flats up to 1.5 km wide. It is backed by 200-300 m wide grassy foredunes and a 30 ha freshwater wetland. Beach **NT 1412** is located on the south side of the headland, and is a 400 m long strip of high tide sand, backed by calcarenite transgressive dunes, and fronted by coral reef flats which narrow from 500 m to 200 m in the south. The now vegetated dune has extended up to 800 m inland and to heights of 16 m.

The southern end of the beach is bordered by a 400 m long section of dune calcarenite, which protects the entrance to **Arndani inlet** and backing swamp. The 10 m high calcarenite bluffs are also paralleled by a submerged section of calcarenite (beachrock?) which provides additional protection for the inlet.

NT 1413-1415 ARNDANI-MINNIE CREEK

No.	Beach	Rating HT	Rating LT	Type	Length
NT1413	Arndani	1	4	R+rock-coral flats	600 m
NT1414	Minnie Ck	4	5	Ultradissipative	6.6 km
NT1415	Minnie Ck (S)	1	2	R+sand-coral flats	900 m
Spring & neap tide range = 2.2 & 1.0 m					

Arndani inlet is a partly sheltered 100 m wide entrance in lee of the calcarenite bluffs and reefs. The reefs deflect the tidal channels along the front of beach NT 1413, and permit a relatively easy entry to the inlet for small boats, at high tide and during normal 1 m high easterly waves. Once inside the inlet there is an excellent anchorage behind the bluffs and adjacent to a good camping area. The channel bifurcates and leads east to the 100 ha **Arndani Swamp** (actually a shallow lagoon) (Fig. 5.153) and south to a 6 km long tidal channel, which backs the massive dune systems of beach NT 1414. Mangroves fringe both the swamp and tidal creek.

The inlet beach (**NT 1413**) is a curving 600 m long, southeast-facing, moderately steep beach that is relatively sheltered from waves by the reef and inlet shoals, but which is paralleled by the deep channel and strong tidal currents. It is backed by a 500 m wide partly active transgressive dune system. Dune calcarenite outcrops at the southern end of the beach and forms the southern boundary. Numerous colourful fishing nets are usually found caught on the jagged calcarenite.

Beach **NT 1414** commences at the rocks and trends straight southwest for 6.6 km to the small mouth of Minnie Creek. This is a low gradient fine sand ultradissipative beach, with a 200 m wide surf during strong southeasterly winds and seas. It is backed by a largely stable, 1 km wide in the south to 2.5 km in the north, transgressive dune system, which reaches up to 35 m in height. A section of the northern dunes is actively spilling into the creek. The dunes are almost encircled by the Arndani tidal creek and the southern Minnie Creek and associated wetlands.

Beach **NT 1415** is located on the southern side of the 50 m wide Minnie Creek mouth. It is a protected, crenulate, northeast-facing, 900 m long beach fronted by 600 m wide sand then coral flats, with patches of rocks both on the beach and flats, and forming the southern boundary. It is backed by a 200-300 m wide band of about five beach ridges, then the low woodlands.

Figure 5.153 Arndani Swamp contains the lagoon fringed by mangroves and salt flat. The inlet connects it to the Gulf and flows past beach NT 1413 (left). Beach NT 1412 is located in right foreground.

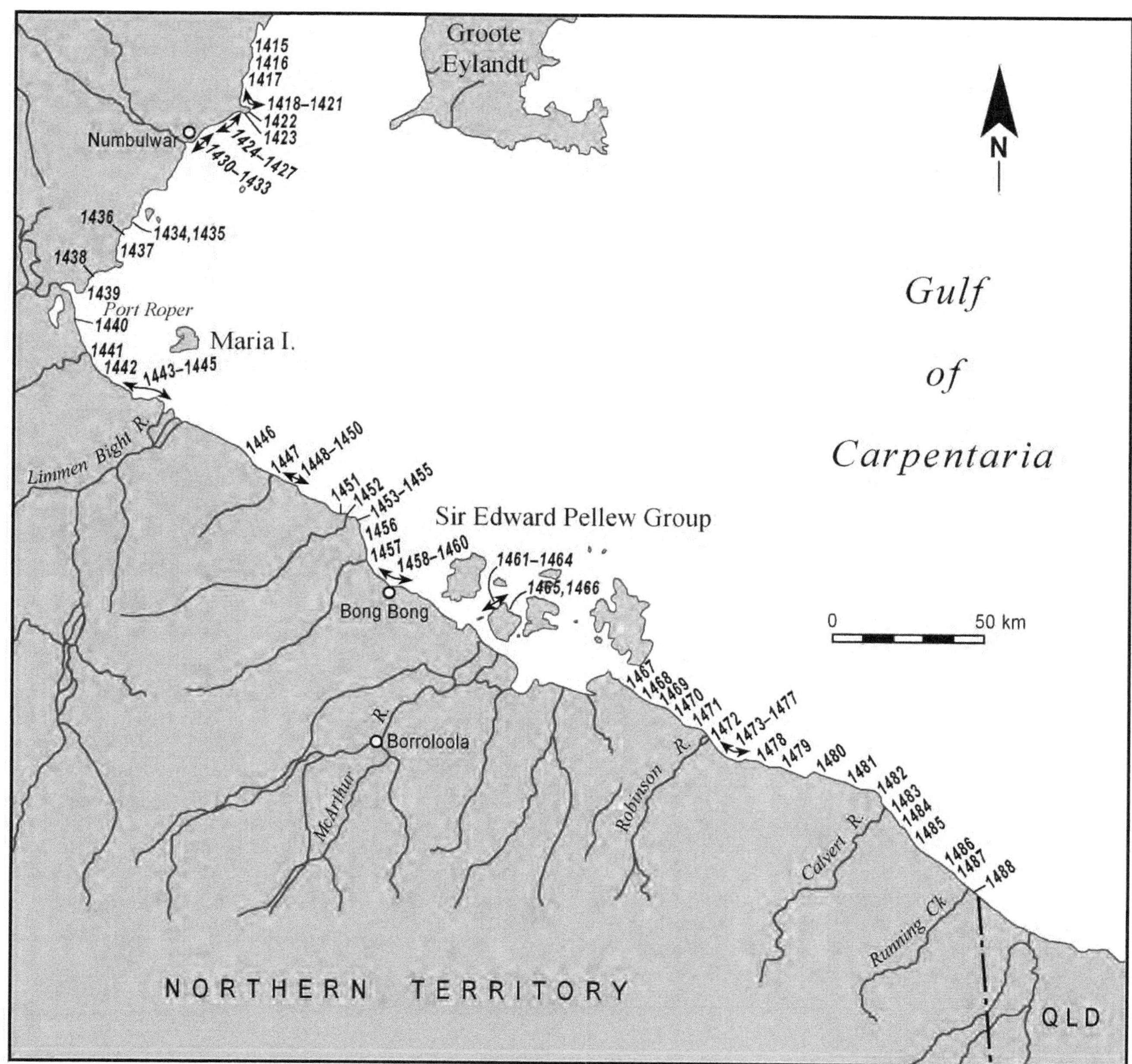

Figure 5.154 The southern western Gulf coast between Numbulwar and the Northern Territory-Queensland border contains the low energy tide dominated beaches NT 1415-1488.

NT 1416-1420 **MIWUL-HART R-MUNTAK CK**

No.	Beach	Rating HT	LT	Type	Length
NT1416	Miwul	1	2	R+rock-coral flats	5.1 km
NT1417	Miwal	1	2	R+rock-coral flats	4.3 km
NT1418	Hart R	1	2	R+tidal shoals	1.5 km
NT1419	Muntak Ck (N)	1	2	R+ridged sands flats	4.9 km
NT1420	Muntak Ck	1	2	R+ridged sands flats	700 m
Spring & neap tide range = 2.2 & 1.0 m					

South of the rocks that form the southern boundary of beach NT 1415, the coast trends roughly due south for 1 km to the Hart River mouth, then another 7 km to the mouth of Muntak Creek. The entire section of exposed east-facing shore is fronted by rock, sand and coral flats between 0.2-1.5 km wide as well as the river mouth shoals. As a consequence waves are low at the shore and calm conditions prevail at low tide. There is abundant evidence of higher energy wave conditions, probably at the sea level stillstand, with numerous vegetated (mid-Holocene?) parabolics along this section of shore. The small Miwul community is located on the shores of beach NT 1416, with 4WD access also to the two neighbouring beaches.

Beach **NT 1416** commences on the southern side of the small rock outcrop. It is a south-trending slightly protruding beach that extends for 5.1 km to a small rocky corner. The few **Miwul** community houses are located halfway down the beach, immediately north of a usually closed creek mouth. The beach is fronted by near continuous narrow rock flats, then patchy sand and coral flats widening from 200 m in the north to 500 m in the south. A few beach ridges, then stable, probably mid-Holocene parabolic dunes extend between 0.3 km and 1.7 km in from the beach and to heights of 31 m. A vehicle track parallels the length of the beach.

Beach **NT 1417** commences on the south side of the rocky corner and curves slowly to the south and southwest for 4.3 km terminating amongst the mangroves on the northern side of the 150 m wide **Hart River** mouth. The beach is fronted by rock, then reef and finally tidal sand shoals, which increase in width to the south from 400 m to 3 km off the river mouth. A narrow band of beach ridges, then stable parabolics extending between 0.8 km and 3 km inland back the beach. The two 15 ha Miwal Billabongs are located behind the central parabolics. There is vehicle access from Miwul to the northern end of the beach.

The southern side of the river mouth is a reasonably dynamic system, as a result of a northward-migrating sand spit (**NT 1418**) which, depending on its length, affords varying protection to the backing shoreline. In 1999 the spit was 1.5 km long almost reaching to the river mouth, and backed by 10 ha of mangroves. Over time several similar spits have attached to the southern side of the river prograding the shoreline 600 m. Three kilometre wide sand then reef flats extend seaward of the beach.

Beach **NT 1419** commences at a protrusion in the spit and trends almost due south as a relatively straight beach for 4.9 km to the northern mouth of **Muntak Creek**. This beach acts as a conduit for sand moving north along the coast, with the 0.5-1.5 km wide reef flats covered by multiple transverse sand ridges. The beach itself is composed of northern and southern spits, as well as a few older spits and beach ridges, then still earlier dune transgression, including one large parabolic extending 2.5 km inland.

Beach **NT 1420** is a sand island located in the middle of the bifurcating creek mouth. It consists of a 700 m long beach ridge, backed by a 500 m wide band of mangroves, then the main southern creek channel. A 0.5-1.5 km wide band of vegetated parabolics extends west of the creek and south for 2 km to the rear of Rantyirrity Point.

NT 1421-1423 **RANTYIRRITY POINT**

No.	Beach	Rating HT	Rating LT	Type	Length
NT1421	Rantyirrity Pt (N)	1	2	R+rock-coral flats	1.9 km
NT1422	Rantyirrity Pt	1	2	R+rock-coral flats	700 m
NT1423	Rantyirrity Pt (S1)	1	2	R+sand flats	400 m
Spring & neap tide range = 2.2 & 1.0 m					

Rantyirrity Point is a curving 500 m wide 10-20 m high laterite point that protrudes 3 km into the gulf. The entire point is surrounded by rock and coral flats and contains four low energy beaches (NT 1421-1424) (Fig. 5.155). Muntak Creek lies 2 km north of the point with a curving section of mangroves between the southern creek mouth and the beginning of beach **NT 1421**. The 1.9 km long high tide beach is convex and faces northeast as it curves round to the eastern tip of the point. It is fronted by continuous 50-100 m wide rock flats, then a fringing 200 m wide reef flat.

Figure 5.155 Rantyirrity Point, with beaches NT 1421 to right, NT 1422 on the point, and NT 1423 & 1424 to the south.

At the eastern tip of the point the shoreline turns and faces southeast, with 700 m long beach **NT 1422** occupying this section of the point. Rock flats, then patchy reef, extend 500 m offshore. Immediately west of the southern tip of the point is a small embayment containing 400 m long beach **NT 1423**. The slightly curving south-facing beach is bordered by 200 m long mangrove-fringed low rocky points, with a sand flat in between, and reef flats extending up to 400 m offshore.

NT 1424-1429 **RANTYIRRITY POINT-AMAMARRITY ISLAND**

No.	Beach	Rating HT	Rating LT	Type	Length
NT1424	Rantyirrity Pt (S2)	1	2	R+rock-coral flats	2.5 km
NT1425	Rantyirrity Pt (S3)	1	2	R+sand-rock-coral flats	2 km
NT1426	Minintirri (1)	1	2	R+rock-coral flats	400 m
NT1427	Minintirri (2)	1	2	R+rock-coral flats	3.3 km
NT1428	Amamarrity (1)	1	2	R+sand-rock-coral flats	2 km
NT1429	Amamarrity (2)	1	2	R+rock-coral flats	50 m
Spring & neap tide range = 2.2 & 1.0 m					

South of Rantyirrity Point the shoreline trends southwest for 20 km to Numbulwar at the mouth of the Rose River. The first 10 km are occupied by six beaches (NT 1424-1429) all fronted by extensive sand, rock and coral flats.

Beach **NT 1424** begins on the south side of Rantyirrity Point, with 1 km of mangroves separating it from beach NT 1423. As the beach begins the shoreline curves round to face south and finally southeast at the southern end of the 2.5 km long beach. It terminates at a small creek mouth, which drains a 100 ha area of salt flats and a few mangroves along the creek. The beach is fronted by patchy sand and reef flats extending 1-1.5 km into the gulf. It is backed by a narrow band of beach ridges, and then older vegetated parabolics extending up to 1 km inland. A 4WD track follows the northern side of the creek to reach the southern end of the beach.

Beach **NT 1425** commences on the south side of the creek. It faces southeast and curves round slightly for 2 km to a 500 m long section of 10 m high bluffs. The beach is fronted initially by the sand flats of the creek mouth, then rock and reefs flats that extend 1-1.5 km offshore. It is backed by a 200-500 m wide barrier consisting of inner older vegetated transgressive dunes and a strip of active dunes behind the beach, with the active dunes reaching 21 m in height. These in turn are backed by the 150 ha salt flat-dominated lagoon that is drained by the small creek. A 4WD vehicle track reaches the bluffs at the southern end of the beach.

Beach **NT 1426** is a 400 m long strip of high tide sand located at the base of the bluffs. It is fronted by a few rock outcrops, then the 500 m wide reef flats, with the small **Minintirri Island** located 2 km offshore. Beach **NT 1427** commences on the southern side of the bluffs. It faces southeast and curves round slightly for 3.3 km to a sand foreland formed in lee of **Amamarrity Island**, located 1 km offshore. Beach **NT 1428** extends from the south side of the foreland for another 2 km to the northern end of a 500 m long section of low rock bluffs. Reef and patchy sand flats connect the island with the beach, with the flats narrowing to either end of the beaches. They are backed by a largely active transgressive dune system that contains active parabolics extending up to 1.5 km inland and to heights of 24 m. These in turn are backed by a 100 ha salt lake (Fig. 5.156).

Figure 5.156 Beaches NT 1427 & 1428 lie either side of the sandy foreland, with extensive transgressive dunes actively migrating inland.

Beach **NT 1429** is a 50 m long pocket of sand located at the southern end of the rock bluffs. It is fronted by rocks and fringing reef, which extends 200 m offshore.

NT 1430-1433 **KULARRUTY CK-NUMBULWAR**

No.	Beach	Rating HT LT	Type	Length
NT1430	Kularruty Ck (N)	1 2	R+ridged sand flats	2.4 km
NT1431	Kularruty Ck (S1)	1 2	R+ridged sand flats	3.5 km
NT1432	Kularruty Ck (S2)	1 2	R+ sand flats	1.3 km
NT1433	Numbulwar	2 3	R+sand flats-tidal channel	5.2 km
Spring & neap tide range = 2.2 & 1.0 m				

Immediately south of beach NT 1429 the shoreline trends initially west then curves round to trend southwest for 10 km to the Rose River mouth. This section of coast is dominated by three recurved spits, two creek mouths, the river mouth and a backing beach-foredune ridge plain that reaches 2 km in width in lee of Numbulwar, with a transgressive dune system forming the inner boundary. The entire system totals approximately 2000 ha in area. It represents three phases of deposition. First, an initial mid-Holocene phase probably accompanying the sea level stillstand associated with deeper water and high waves leading to the inner transgressive dunes. This was followed by onshore sand transport to be deposited as prograding beach-foredune ridges possibly accompanying the growing reefs and shoaling nearshore. More recently sediment is moving longshore toward the river mouth, a migratory spit phase. This may be a result of infilling of the 'embayment' which previously trapped the sand as beach-foredune ridges. The sand now moves longshore in the form of the migrating spit-island and sand pulses along the Numbulwar beach.

Beach **NT 1430** commences at the southern tip of the bluffs and trends initially west then southwest for 2.4 km to the northern entrance of **Kularruty Creek**. The southern tip of the beach is a sand spit that is slowly migrating south and deflecting the creek mouth. It is backed by four earlier spits, while sand and then reef flats extend 1 km offshore. Beach **NT 1431** commences on the south side of the 100 m wide creek mouth and is a migratory spit 3.5 km in length. It is backed by one earlier spit, then some stable transgressive dunes that extend up to 2.5 km inland and to the south the beginning of the 2 km wide beach to foredune ridge plain. The southern side of the spit splays to a series of smaller recurved spits at the southern entrance to Kularruty Creek. An arm of the creek parallels the entire back of the beach.

Beach **NT 1432** is the third spit in the series. It is a 1.3 km long collection of five recurved spits, separated from the mainland by a narrow southern tributary of Kularruty Creek. This is in turn backed by the 2 km wide beach ridge plain. **Numbulwar Beach NT 1433** commences immediately south of the small boundary creek and curves round in a convex shape for 5.2 km, facing initially southeast and finally south into the funnel-shaped 500 m wide mouth of the Rose River. The small

Wungumana Creek flows along behind the northern 2 km of beach before crossing at a slight inflection in the shore. Pulses of sand moving southward along the beach and into the river mouth produce a series of crenulations in the shoreline. The **Numbulwar** community is located behind the centre of the beach. The community has a fluctuating population of up to several hundred people and has an administrative centre, store and school. A road runs from the community to the barge landing toward the western end of the beach, with the landing ground 1 km further west. Numbulwar beach is initially bordered by the deep Rose River channel along the western 1 km. The channel then swings to the northeast and moves 100-500 m offshore, with sand flats to either side, the outer flats extending 3 km offshore.

Numbulwar is the largest community in the southwest section of the gulf, with the nearest comparable communities located 250 km to the north at Nhulunbuy-Yirrkala, 90 km to the east at Alyangula on Groote Eylandt, and 200 km to the southeast at Borroloola.

NT 1434-1438 **WIYAKIPA BEACH**

No.	Beach	Rating HT	Rating LT	Type	Length
NT1434	Wiyakipa (N2)	1	1	R+ridged sand flats	1.2 km
NT1435	Wiyakipa (N1)	1	1	R+ridged sand flats	500 m
NT1436	Wiyakipa Beach	1	1	R+ridged sand flats	10.7 km
NT1437	Wuyagiba	1	1	R+sand flats	700 m
NT1438	Yikikukunyiyanga	1	1	R+ridged sand flats	800 m
Spring & neap tide range = 2.2 & 1.0 m					

South of Numbulwar the shoreline trends southwest for 60 km to the mouth of the Roper River and Limmen Bight. This is an increasingly low wave energy coast with tidal flats and mangroves dominating the shore. The reduction in waves is due to a combination of factors including decreasing fetch into the southern gulf, decreasing water depth including offshore islands and shoals, and the decreasing inshore gradients resulting from the supply of river sediments to the coast, particularly from the Roper River. In addition tide range increases slightly into the southwest gulf with a spring range of 2.2 m at Numbulwar and 2.3 m at Centre Island. There are only five beaches (NT 1434-1438) totalling 14 km in length along this section, all located just south of Edward Island, with most of the remaining shoreline dominated by mangroves and wide tidal flats.

Beaches NT 1434 and NT 1435 are two contiguous beaches located 4 km west of Edward Island. Beach **NT 1434** is 1.2 km long, with a 200 m long clump of mangroves separating it from 500 m long beach **NT 1435**. Both beaches share common ridged sand flats that have up to 14 ridges extending over 500 m off the beaches, with tidal flats continuing out for another 1-2 km. Low mangrove-fringed rock flats border each end, with a 300-500 m wide beach ridge plain backing both beaches, which is in turn backed by a 100 m wide band of 10-15 m high, probably mid-Holocene transgressive dunes. A 4WD track runs to the rear of the dunes.

Wiyakipa Beach (NT 1436) is located 1 km to the south. This is a 10.7 km long east-facing beach that is fronted by a 200-500 m band of ridged sand flats. The low waves break across a wide low gradient ultradissipative surf (Fig. 5.157). The beach is crossed by two creeks, with a third northern boundary creek. A 500 m wide beach-foredune ridge plain backs the beach, with older vegetated transgressive dunes behind. In the south these extend another 1 km inland and to heights of 32 m, and represent a mid-Holocene higher energy phase of deposition.

Figure 5.157 Wiyakipa Beach (NT 1436), shown here at high tide as low waves break across the submerged multiple sand ridges.

The small **Wuyagiba** community is located at the very southern end of the beach. It consists of several houses located in lee of the low rock bluff and outcrop which form the southern boundary. It is connected by vehicle track to the Numbulwar road, and has tracks running north along the back of the coast for 20 km to the mouth of Miyangkala Creek.

Beach **NT 1437** is located on the southern side of the bluffs and curves to the south for 700 m to the 50 m wide mouth of Ardurrurru Creek. One kilometre wide tidal sand shoals front the beach, with the mangrove-fringed creek and salt flats extending 4 km inland.

Warrakunta Point lies 2.5 km southeast of the community, with beach **NT 1438** located 4.5 km southwest of the point. This is an 800 m long strip of high tide sand bordered in the north by a mangrove-fringed tidal creek, and low rock flats in the south. It is fronted by sand and rock flats that extend 1 km southeast to **Yikikukunyiyanga Islet**. A 1 km wide splay of cuspate beach ridges and recurved spits has formed in lee of the islet and rock flats, with the present beach forming the eastern boundary, while dense mangroves border the southern side.

NT 1439-1445 ROPER-LIMMEN BIGHT RIVERS

No.	Beach	Rating HT	Rating LT	Type	Length
NT1439	Roper River	1	2	R+ridged sand flats	3.1 km
NT1440	Towns River (N)	1	2	R+ridged sand flats	700 m
NT1441	Towns River	1	2	R+ridged sand flats	1.2 km
NT1442	Towns River (S1)	1	2	R+ridged sand flats	8 km
NT1443	Towns River (S2)	1	2	R+ridged sand flats	8.5 km
NT1444	Spillen Ck (W)	1	2	R+ridged sand flats	5.8 km
NT1445	Spillen Ck (E)	1	2	R+ridged sand flats	3.6 km
Spring & neap tide range = 2.2 & 1.0 m					

The **Roper River** is located in the southwestern corner of the Gulf of Carpentaria, an area known as Limmen Bight. The Arnhem Land coast extends north of the river for 300 km to Cape Arnhem, while to the west the coast curves round along the southern shores of the Gulf to Queensland and beyond. The river also marks the boundary between the large Arnhem Land Aboriginal Land and the smaller Marra Aboriginal Land to the west. The Roper is a major river system with a 1 km wide mouth and a massive deltaic system dominated by salt flats and fringing mangroves (Fig. 5.158). The delta has 30 km of shoreline and extends for up to 40 km inland. The nearest coastal access is to the Roper River No. 1 Landing, located 5 km west of the mouth on the south side of the river.

Figure 5.158 The large Roper River mouth.

The **Limmen Bight River** mouth is located 55 km southeast of the Roper and in between are seven low energy beach systems (NT 1439-1455), each fronted by ridged sand and tidal flats, and some in places up to 10 km of river mouth shoals. Both rivers are delivering sediment to the coast and have built extensive low gradient deltas dominated by salt flats, with both Pleistocene and Holocene beach ridge-spit barrier systems. In addition longshore sediment transport is converging on the Roper River from the north and south and contributing to the extensive tidal deposits.

Beach **NT 1439** extends for 3.1 km north of the northern Roper River entrance. It is the only beach system associated with the river and its deltaic plain, with the remaining gulf shore fringed by mangroves. The beach consists of a low beach ridge that faces southwest as it curves round into the river mouth. It is backed by two older ridges and fronted by low gradient ridged sand flats grading into 5 km wide river mouth tidal flats. The northern end of the beach is eroding, exposing mangroves, while the southern end is prograding into the river mouth as a recurved spit, which terminates in a small mangrove woodland.

Mangroves dominate the next 15 km of shore south of the river mouth to beach **NT 1440**. This is a 700 m long strip of east-facing sand located on the north side of a 100 m wide tidal creek. Ridged sand flats extend 500 m off the beach, grading into 2.5 km wide tidal shoals. The beach consists of overwash flats in the north, grading into a vegetated beach ridge to the south, with salt flats extending 2 km to the west.

The Holocene beach ridges and recurved spits of beaches NT 1441-1445 are backed by an inner Pleistocene beach ridge to recurved spit system that extends longshore from the northern side of the Limmen River mouth for approximately 50 km to the southern side of the Roper River. The barrier ranges from 1-4 km in width and totals approximately 6000 ha. The inner barrier is in turn backed by 4-10 km wide salt flats.

Beach **NT 1441** commences 4 km to the south. It is a 1.2 km long strip of sand located on the north side of the 200 m wide **Towns River** mouth (Fig. 5.159). The beach faces east and curves into the river mouth, with mangroves to either end and some scattered along the tidal flats. Five hundred metre wide ridged sand flats converge on the river mouth, where the deep tidal channel lies immediately off the beach. Tidal shoals extend 2 km seaward of the sand ridges.

Figure 5.159 The Towns River mouth with beaches NT 1441 & 1442 to either side.

Beach **NT 1442** commences on the south side of the river mouth and is an 8 km long relatively straight northeast-facing beach. The beach is part of a prograding series of beach ridges, which grade into a 3 km wide splay of recurved spits towards the river mouth. Up to 20 ridges extend between 0.5-3 km inland. The beach formerly extended for at least another 8 km, but has been gradually overlapped by the migrating beach NT 1443.

Beach **NT 1443** onlaps the southern end of beach NT 1443, and is an active northward-migrating recurved spit. The entire 8.5 km long beach is part of a migrating barrier island, which is eroding along its southeast end, truncating and exposing older spits, mangroves and tidal creeks, and accreting as recurved spits at its northwest end. The remains of more than 30 recurved spits compose the beach.

Beach **NT 1444** is a narrow, eroding beach ridge cutting into older recurved spits and mangroves. It splays to a series of three recurved spits at its western end where it onlaps with beach NT 1443, the two separated by a 50 m wide creek mouth. Ridged sand flats then tidal flats extend more than 1 km off the beach.

Spillen Creek separates beaches NT 1444 and 1445. The relatively straight 5 km long creek channel drains a 1500 ha area of salt flats and cuts through a 3 km wide section of Pleistocene beach ridges and swales. The beach (**NT 1445**) is eroding into Pleistocene (?) barrier deposits that are being transported westward via the Holocene barriers toward the Roper River. The eastern end of the beach is bordered by the mangrove-fringed salt flats of the **Limmen River Delta**. The delta and river mouth occupy the next 10 km of shore. The river bifurcates into three deep channels 200 m, 300 m and 400 m wide, which extend across the tidal shoals and up to 5 km offshore.

NT 1446-1450 **WURALIWUNTYA CREEK**

No.	Beach	Rating HT	Rating LT	Type	Length
NT1446	Wuraliwuntya Ck (W)	1	2	R+ridged sand flats	2.1 km
NT1447	Wuraliwuntya Ck (E1)	1	2	R+ridged sand flats	700 m
NT1448	Wuraliwuntya Ck (E2)	1	2	R+ridged sand flats	800 m
NT1449	Wuraliwuntya Ck (E3)	1	2	R+ridged sand flats	200 m
NT1450	Wuraliwuntya Ck (E4)	1	2	R+ridged sand flats	700 m
Spring & neap tide range = 2.3 & 0.7 m					

The 70 km of coast between the larger Limmen Bight River and Rosie Creek are a low wave energy, low gradient, northeast-facing coastal plain, with an offshore zone dominated by the coral reef flats of the Labyrinth Shoals, which extend up to 20 km seaward. The Holocene deposits are dominated by mangrove-fringed shoreline fronted by 1-2 km wide sand flats, with inner Pleistocene beach ridge barriers toward the two boundary rivers. The only Holocene beach-barrier deposits are located along a 10 km long section either side of Wuraliwuntya Creek (NT 1446-1450), located midway between the systems, and around Rosie Creek. Salt flats between 1 km and 5 km wide back the entire shoreline and prevent vehicle access to the coast.

Beach **NT 1446** is located immediately west of Wuraliwuntya Creek mouth. It consists of a low 2.1 km long north-facing beach, backed by a partly vegetated beach ridge, 1 km wide salt flats and a 1.5 km wide Pleistocene barrier. It is fronted by ridged sand flats, which widen to 2 km off the creek mouth (Fig. 5.160), and then shallow coral reef flats. The beach terminates in the east at the mangrove-lined mouth of 100 m wide Wuraliwuntya Creek.

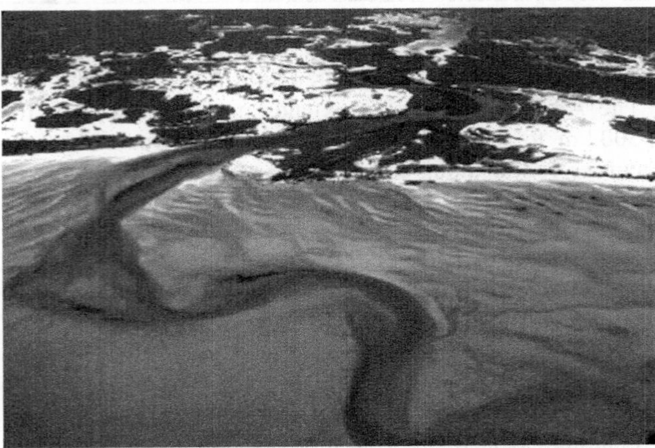

Figure 5.160 Wuraliwuntya Creek mouth with beaches NT 1446 & 1447 to either side, and extensive ridged sand flats and the tidal channel extending into the gulf.

Beach **NT 1447** commences on the east side of the creek mouth and trends due east for 700 m to a 50 m wide tidal creek and channel. It is backed by salt flats and fronted by 2 km wide sand flats then reef flats. Beach **NT 1448** commences 200 m to the east and is a relatively straight north-facing 100-300 m wide vegetated beach ridge, also backed by salt flats and fronted by 2 km wide sand flats. Small creeks and channels cross the inner sand flats and border each end.

Beach **NT 1449** is a 200 m long pocket of sand wedged in between the boundary creek with beach NT 1448 and a second creek, with both creeks linking behind the beach to enclose the small beach and beach ridge in mangrove-fringed tidal channels. Ridged sand flats extend 2 km off the beach.

Beach **NT 1450** is located 2 km to the east and consists of a narrow 700 m long beach ridge, backed by an earlier beach ridge and salt flats. It has a small creek on its western end and a more substantial though diffuse creek at the eastern boundary, with 2 km wide ridged sand flats off the beach.

NT 1451-1454 **ROSIE CREEK**

No.	Beach	Rating HT	Rating LT	Type	Length
NT1451	Rosie Ck (W)	1	2	R+ridged sand flats	3.2 km
NT1452	Rosie Ck	1	2	R+sand flats	400 m
NT1453	Rosie Ck (E1)	1	2	R+sand flats	3 km
NT1454	Rosie Ck (E2)	1	2	R+sand-rock flats	1.3 km
Spring & neap tide range = 2.3 & 0.7 m					

Rosie Creek is a medium sized upland stream that reaches the coast in lee of 5 ha, 400 m long, low laterite bluffs (Fig. 5.161), with low laterite outcrops also extending 4 km to the west and then 3 km seaward as intertidal rock flats. The coast faces north between the rock flats and Rosie Creek and contains four beaches (NT1451-1454). At the rock flats the shoreline turns and trends southeast toward Bing Bong. Extensive sand flats up to 5 km wide back the creek region and there is no vehicle access to the coast. The laterite rocks are the most southern bedrock outcrop on the western gulf shore. The next exposure is located 60 km to the east at Sharker Point, then it is another 890 km to the low laterite bluffs of Worbody Point on the western shores of Cape York Peninsula.

Figure 5.161 Rosie Creek with beach NT 1451 to right, beach NT 1452 in creek mouth and the longer beach NT 1453 extending to east.

Beach **NT 1451** commences at the laterite bluffs and extends to the west for 3.2 km to the mouth of a small mangrove-lined creek. It is a relatively straight north-facing beach backed by a narrow to in places bare overwashed beach ridge, then a 1 km wide stand of mangroves, older beach ridges and finally 3 km wide salt flats. A 500 m wide mixture of ridges and flats, tidal shoals and rock flats fronts the beach. The bluffs extend 200 m off the eastern end of the beach providing a sheltered, if shallow, corner.

Beach **NT 1452** is attached to the southern side of the laterite bluffs on the western side of Rosie Creek mouth. The 400 m long beach faces east across the creek channel and tidal shoals. It is backed by two truncated beach ridges and intervening salt flats.

Beach **NT 1453** commences on the eastern side of the creek mouth and initially curves round to trend southeast for 3 km to a low linear rock outcrop that forms an inflection in the shore. A 200 m wide zone of active foredune backs the beach, then the first in a 7 km wide sequence of wooded Pleistocene beach ridges and intervening salt flats. The western end of the beach lies in lee of the 1 km wide creek mouth tidal channel and shoals, with sand flats toward the centre and sand and rock flats to the east. Beach **NT 1454** begins immediately east of the linear rock flats and trends southeast for 1.3 km to the lee of the extensive rock flats off the point. This beach is also backed by the Pleistocene barrier and fronted by the 1-3 km wide sand and rock flats.

NT 1455-1460 **ROSIE CREEK (S)**

No.	Beach	Rating HT	Rating LT	Type	Length
NT1455	Rosie Ck (S1)	1	2	R+sand-rock flats	1.2 km
NT1456	Rosie Ck (S2)	1	2	R+ridged sand flats	8.3 km
NT1457	Rosie Ck (S3)	1	2	R+sand flats	5.8 km
NT1458	Rosie Ck (S4)	1	2	R+tidal flats	2.8 km
NT1459	Rosie Ck (S5)	1	2	R+tidal flats	1.2 km
NT1460	Rosie Ck (S6)	1	2	R+tidal flats	3.2 km
Spring & neap tide range = 2.3 & 0.7 m					

The main **Rosie Creek** mouth is located 4 km west of a major inflection in the shore in lee of low laterite rock flats. The shoreline turns and trends south to southeast for 25 km to Bing Bong Creek. A second channel of the creek reaches the coast 15 km south of the point. To either side and between the two channels is an extensive Holocene and Pleistocene (?) barrier system, which begins at the point and runs for 25 km to the mouth of Bing Bong Creek, with sediment moving south and converging on Bing Bong. Six beaches (NT 1455-1460) are located along the barrier each separated by tidal creeks. The barrier system ranges from 1 km wide in the south to 3 km in the north and is backed by salt flats, which also widen from 1-5 km. While there are vehicle tracks behind the southern salt flats, there is no formed access to the coast.

Beach **NT 1455** commences at the rock flat-controlled point and trends south for 1.2 km to a sandy-cobble foreland formed in lee of a low rock outcrop. The beach is composed of coarse sand and cobbles. It is fronted by a band of rock flats, then patchy ridged sand flats and more rock flats and coral reefs, which extend up to 3 km off the northern end of the beach.

Beach **NT 1456** commences on the southern side of the foreland and trends south-southwest for 8.3 km to a sand foreland formed in lee of a reef located 2.5 km off the beach. The beach is backed by a continuous 2-3 km wide series of overwashed vegetated beach ridges, then 3-5 km wide salt flats. It has a moderately steep beach face fronted by 1 km wide ridged sand flats, with a patch of reef off the centre and off the southern foreland. Beach

NT 1457 extends from the southern side of the foreland to the southern 300 m wide mouth of Rosie Creek. The backing barrier is 2 km wide in the north and consists of inner recurved spits fronted by continuous beach ridges, which in the south have prograded for 2 km past an earlier extension of beach NT 1458. The creek trends southeast and separates these two sections of barriers. The beach is fronted by sand flats, which widen from 500 m in the north to 1.5 km off the creek mouth.

South of the creek mouth wave energy decreases as tidal flats increase in dominance and the shore becomes increasing protected by its orientation and the Sir Edward Pellew islands 20 km to the east. The three beaches (NT 1458-1460), are backed by 1 km wide beach ridge plains, then a salt flat-interbarrier depression and probably Pleistocene inner beach ridge barrier extending up to 4 km inland. Beach **NT 1458** commences inside the southern side of the creek mouth and trends southeast for 2.8 km to a smaller 50 m wide creek mouth and channels. It is fronted by 2 km wide low gradient sandy tidal flats.

Beach **NT 1459** commences on the south side of the creek mouth and continues on for 1.2 km to another small creek mouth, with mangroves lining the channel that extends out across the inner section of the tidal flats. Finally, beach **NT 1460** is a transition from beach to tidal flats. It is a convex 3.2 km long strip of sand, which curves round into the mouth of **Bing Bong Creek**, where it is fronted by irregular 2 km wide tidal flats. It marks the southern end of the inner 3 km wide beach ridge barriers.

Bing Bong

Bing Bong loading facility provides one of the few locations on the entire Gulf of Carpentaria where a sealed road runs to the coast and provides public access to the low gradient very low energy shoreline. The road from Borroloola, located 50 km to the south, was built by the McArthur River Mine in the mid 1990s, to provide access for the zinc ore loading facility. The mine is located about 120 km to the south on the Borroloola road. The public is permitted to use the road but not enter the facility. The facility consists of a dredged basin (Fig. 5.162) where barges are loaded with ore then towed out to ore ships anchored offshore in deeper water.

The shoreline to either side of the facility consists of tidal sand flats and scattered mangroves. The former 9 km long beach system is backed by a 1-2 km wide series of up to 12 beach ridges with recurved spits to either end. A vehicle track leads from the entrance to the facility for 5 km to the east to the end of one of the recurves spits on Mule Creek, where there is a concrete boat ramp used by local fishermen. The ramp and a few channel markers were constructed during the building of the loading facility and are deteoriating.

Figure 5.162 The Bing Bong channel and port facility.

NT 1461-1466 **SOUTH WEST ISLAND (N)**

No.	Beach	Rating HT	Rating LT	Type	Length
NT1461	South West Island (N1)	1	2	R+rock flats	300 m
NT1462	South West Island (N2)	1	2	R+sand-rock flats	200 m
NT1463	South West Island (N3)	1	2	R+ridged sand flats	250 m
NT1464	South West Island (N4)	1	2	R+rock-sand flats	300 m
NT1465	South West Island (N5)	1	2	R+ridged sand flats	1.1 km
NT1466	South West Island (N6)	1	2	R+rock-sand flats	150 m
Spring & neap tide range = 2.3 & 0.7 m					

South West Island is a 9200 ha island composed of a core of Proterozoic sandstone rising to a maximum height of 93 m, with a mantling of Tertiary laterite. The island has been linked to the coast by the progradation of salt flats and mangrove shoreline in front of the Carrington Channel, a northern arm of the McArthur River delta. The island is one of the Sir Edward Pellew Group which includes several other substantial islands of similar or larger size and numerous smaller islands. The entire south and western and most of the eastern shorelines are fringed by mangroves. The only beaches (NT 1461-1466) are located on the more open and exposed 14 km long rocky northern shore.

Beach **NT 1461** commences just inside the rocky northwestern tip of the island. The 300 m long beach is bordered by low mangrove-fringed sandstone headlands, and fronted by a 400 m wide mixture of sand and rock flats. Beach **NT 1462** is located 200 m to the east past mangrove-covered rock flats. This is a similar 200 m long beach with sand and rock flats, as well as rock reefs off the beach. A sandstone ridge rises behind the beach and a small creek reaches the shore 300 m to the east.

Beach **NT 1463** is located 1 km to the east and occupies the southern 250 m of the 2 km long northward protrusion of the sandtone. The slightly curving west-facing beach is fringed by rock flats and mangroves, with a few scattered mangroves also on the 500 m wide ridged sand flats. A few beach ridges, then a sandstone ridge, back the beach. Immediately north is beach **NT 1464**, a crenulate 300 m long beach that extends up to the northern tip of the sandstone point. Several rock outcrops with scattered mangroves lie along the beach with a mixture of sand and rock flats off the beach. It is backed by a 50 ha area of bare dissected sandstone.

Beach **NT 1465** is located 2 km southeast of the point in the next embayment. It is a 1.1 km long slightly curving north-facing beach fronted by 1 km wide ridged sand flats, and bordered by the sandstone point to the west and a small rock outcrop to the east. Beach **NT 1466** is located on the eastern side of the 200 m wide outcrop. It is a 150 m long pocket beach, bordered by rocks and fronted by rock flats including a row of small mangroves. A vehicle track connects the beach to the few houses of the small **South West Island** community which is located a few hundred metres in behind the beach.

Mcarthur and Wearyan deltas

The McArthur River delta is one of the largest deltaic systems on the Territory coast. It has formed in the protected lee of the Sir Edward Pellew Group of islands. The main river branches into several channels spreading the delta across a 40 km wide delta front bordered by Sharker Point in the east. The entire delta is dominated by mangrove-fringed salt flats, fronted by intertidal mud flats up to 5 km in width.

The mangrove-fronted salt flats continue on east of Sharker Point for another 25 km across the mouth of the Wearyan River to Pelican Spit. Combined, the two rivers have a 60 km long delta and a total deltaic plain consisting primarily of salt flats and tidal creeks of approximately 80 000 ha.

Vanderlin Island

Vanderlin Island is the largest of the Sir Edward Pellew Group with an area of approximately 26 500 ha. It is also the easternmost island in the group and its 30 km long east coast is fully exposed to the southeast trade winds and waves. A near continuous coral reef fringes 23 km of the east coast, the largest reef system in the Territory. Along the now protected coast in lee of the 1-3 km wide reef systems are extensive mid-Holocene transgressive dunes, which blanket the eastern half of the island, with an area of approximately 15 000 ha. They extend up to 8 km inland and to a height of 50 m, producing one of the largest dune fields in the Territory.

NT 1467-1471 **PELICAN POINT-ROBINSON RIVER**

No.	Beach	Rating HT	Rating LT	Type	Length
NT1467	Pelican Pt	1	2	R+ridged sand flats	7.8 km
NT1468	Robinson R (N3)	1	2	R+ridged sand flats	4.8 km
NT1469	Robinson R (N2)	1	2	R+ridged sand flats	3 km
NT1470	Robinson R (N1)	1	2	R+ridged sand flats	12.5 km
NT1471	Robinson R	1	2	R+ridged sand flats	5.2 km

Spring & neap tide range = 2.3 & 0.7 m

The **Robinson River** is a medium-sized river that reaches the coast 33 km east of Pelican Point. Between the river mouth and the point are a series of five barrier island-spits (beaches NT 1467-1471) that represent the westward movement of sand toward the point. The five north to northeast-facing spits form the outer shoreline of a 1-2 km wide Holocene barrier-spit plain, backed by salt flats and a 3-5 km wide Pleistocene beach ridge plain, backed in turn by 3-5 km wide salt flats. The entire two barrier systems have an area of approximately 12 000 ha and represent in part the delta of the Robinson River.

Pelican Point (**NT 1467**) is a crenulate 7.8 km long sand spit, with the crenulations induced by a smaller series of recurved spits that are migrating westward along the shoreline, resulting in alternating shoreline accretion of 100-200 m, with erosion in between, in places exposing the backing mangroves (Fig. 5.163). As a consequence the total length, width and shape of the spit changes on an annual basis. In addition up to 1 km wide ridged sand flats parallel the beach. It is backed by a mixture of beach ridges, truncated recurved spits, mangroves and a small tidal creek and the larger Fat Fellows Creek, which has been deflected 10 km to the west to its present mouth at Pelican Point.

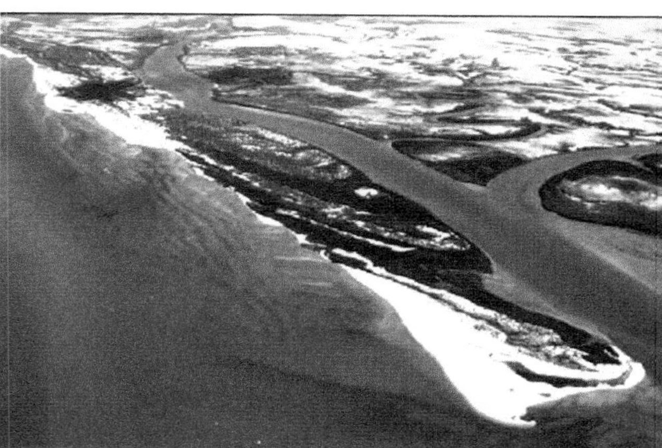

Figure 5.163 Pelican Point (NT 1467) is part of a migratory series of recurved spits that have deflected Fat Fellows Creek mouth 10 km to the west.

Beach **NT 1468** onlaps with Pelican Point and is a 4.8 km long narrow migrating sand spit of a similar form to Pelican Point. The spit is bounded by two deflected creek mouths, and fronted by 300 m wide ridged sand flats. It is backed by a narrow meandering west-flowing tidal creek, then a 5 km wide series of Pleistocene spits and beach ridges.

Beach **NT 1469** is located between the next two creek mouths. It is a narrow, eroding 3 km long spit backed by a small meandering tidal creek which drains to the east and then Pleistocene spits and ridges. Ridged sand flats widen off both creek mouths to more than 1 km off the eastern creek, narrowing in the centre of the beach.

Beach **NT 1470** is 12.5 km long and the longest and largest of the migrating spits. It trends east for 5.5 km to an inflection then for another 7 km to the southeast. Sand is moving westward along the beach in the form of a 200-300 m wide band of shore-transverse and parallel sand ridges. The beach is prone to overwashing and largely unvegetated, with a series of active foredunes and blowouts and vegetated transgressive dunes overlaying the earlier inner spits sequences and in places the Pleistocene barrier. The Holocene barrier widens to the centre to reach a maximum width of 2.5 km. It onlaps an outer Pleistocene barrier, which is backed by 1 km wide salt flats and an inner 2 km wide Pleistocene barrier.

Beach **NT 1471** commences on the west side of the 500 m wide Robinson River mouth, and extends east for 5.2 km to onlap with beach NT 1470. The beach is a curving convex-shaped strip of sand, lying in lee of the 1 km wide river mouth shoals in the west, which narrow to 200 m at the eastern boundary creek. It is backed by a 500 m wide series of Holocene spits, the Pleistocene barrier deposits.

NT 1472-1475 **ROBINSON R-STOCKYARD CK**

No.	Beach	Rating HT	Rating LT	Type	Length
NT1472	Robinson R (E)	1	2	R+ridged sand flats	7.5 km
NT1473	Shark Ck	1	2	R+ridged sand flats	6 km
NT1474	Seven Emu Ck	1	2	R+ridged sand flats	3.2 km
NT1475	Stockyard Ck	1	2	R+ridged sand flats	3.4 km
Spring & neap tide range = 2.3 & 0.7 m					

To the east of the Robinson River the shoreline continues southeast for 50 km to the mouth of the Calvert River. In between are eleven beach systems (NT 1472-1482), all associated with the westerly migrating barrier spits, each separated by usually deflected tidal creeks, and backed by extensive Pleistocene barriers and salt flats. The only vehicle access to the coast is in the vicinity of the Robinson River via the Seven Emu Homestead (permit required from homestead, phone (077) 42 8258).

The first 20 km of coast contain four beach systems between the river and Stockyard Creek. Beach **NT 1472** commences on the eastern side of the river mouth and extends southeast for 7.5 km to the mouth of **Shark Creek**. The beach is initially well protected by the 1 km wide river mouth shoals, and a curving sand island located on the shoal. Once away from the shoals the beach is fronted by 500 m wide ridged sand flats, which narrow further to the east. It terminates at a 1.5 km long eastern deflection in the creek mouth. The beach is backed by a 2 km wide beach ridge plain containing about 10 ridges.

Beach **NT 1473** commences on the eastern side of Shark Creek and is initially protected by the creek, which cuts a channel across the 800 m wide sand flats. The beach continues for 6 km to the southeast to the mouth of Seven Emu Creek, which is deflected 3.5 km to the west. Five hundred metre wide ridged sand flats parallel the beach, increasing to 1 km off the Shark Creek mouth. This beach is also backed by a 2-3 km wide beach ridge plain, containing approximately 15 ridges which splay to the east as recurved spits.

Beaches NT 1474 and 1475 represent a boundary between sediment moving east from the vicinity of the Robinson River and west from the Calvert River. The two narrow beaches are 3.2 and 3.4 km long respectively and are separated by a small tidal creek. Whereas beach **NT 1474** is composed of westward-trending recurved spits, beach **NT 1475** has eastward-trending recurves. Both beaches are backed by 5 km wide salt flats with no Pleistocene barrier deposits. Beach NT 1475 terminates at the mouth of **Stockyard Creek**.

NT 1476-1482 **STOCKYARD CK-CALVERT R**

No.	Beach	Rating HT	Rating LT	Type	Length
NT1476	Stockyard Ck	1	2	R+ridged sand flats	8.1 km
NT1477	Skeleton Ck	1	2	R+ridged sand flats	900 m
NT1478	Skeleton Ck (E)	1	2	R+sand flats	3.2 km
NT1479	Calvert R (W3)	1	2	R+sand flats	5.1 km
NT1480	Calvert R (W2)	1	2	R+LTT	6.7 km
NT1481	Calvert R (W1)	1	2	R+ridged sand flats	8.6 km
NT1482	Calvert R	1	2	R+tidal sand flats	5 km
Spring & neap tide range = 2.3 & 0.7 m					

The **Calvert River** is the most eastern major river on the Territory coast and the river mouth marks a change in shoreline orientation as it swings round more to the southeast. River sediment delivered to the gulf has been reworked to the west by easterly waves. This sediment probably combined with shelf supply has deposited a massive series of westward-trending Holocene and Pleistocene beach ridge-recurved spit barrier systems between the river mouth and **Stockyard Creek**, 38 km to the west. The system is crossed by the four tidal creeks. The Holocene barrier ranges between 100 m and 3 km in width and totals about 2000 ha in area. The Pleistocene barriers extend up to 5 km inland and total about 8000 ha in area, producing a total system of about 10 000 ha, which is backed by an equivalent area of salt flats averaging about 5 km in width. As a result of the salt flats and tidal creeks there is no formed vehicle access to this section of coast.

Beach **NT 1476** is an 8.1 km long series of beach ridge-recurved spits that is actively moving sand westward across the mouth of Stockyard Creek. The beach is paralleled by inner shore-transverse to outer shore-parallel sand ridges averaging about 500 m in width. The transverse ridges are indicative of longshore sand transport and induce crenulation in the shoreline. The barrier is composed of multiple recurved spits and spit splays to the west and is separated from inner spits by shore-parallel tidal creeks and mangroves. In 1968 the beach was breached by a creek which was closed by 1997. The three Holocene barrier spits are 2 km wide in the west, all backed by Pleistocene spits which widen from 1 km in the west to 4 km in the east, and then by 4-6 km wide salt flats.

Beach **NT 1477** is located on the western side of a 200 m wide tidal channel of Skeleton Creek. The narrow 900 m long beach is fronted by 1 km wide tidal shoals and backed by a 3 km wide series of Pleistocene beach ridge-spits, then 3 km of salt flats. Beach NT 1476 overlaps the western end of the beach. Beach **NT 1478** commences on the eastern side of Skeleton Creek and is a narrow 3.2 km long beach that terminates at a large 500 m wide creek mouth. A 2 km wide series of four Pleistocene beach ridge-spits backs the beach, then 5-6 km wide salt flats. The beach is fronted by the tidal shoals of both creeks which extend over 1 km offshore. The eastern end of the beach is being overlapped by the migrating western end of beach NT 1479.

Beach **NT 1479** is a 5.1 km long barrier spit that extends between two creek mouths. It has migrated 2 km across the western creek mouth overlapping beach NT 1478. Sediment is moving along the beach in the form of 1-2 shore-parallel bars, a product of slightly higher wave energy, with wider tidal shoals off each creek mouth. A 6 km wide sequence of approximately 12 Holocene beach ridge-spits backs the beach, which are in turn backed by the inner reaches of the tidal creek and another 6 km of salt flats.

Beach **NT 1480** extends for 6.7 km between two creek mouths and like its western neighbour has wide shoals off the creek mouths, with 1-2 shore-parallel ridges along the beach. It is backed by a 5 km sequence of Pleistocene barrier-spits and then 4 km wide salt flats.

Beaches NT 1481 and 1482 are two converging beaches that form the western side of the Calvert River mouth and are backed by a Holocene beach ridge plain that contains up to 15 ridges and reaches 2 km in width in lee of beach NT 1482. Beach **NT 1481** commences on the eastern side of the 100 m wide creek mouth that separates it from beach NT 1480 and trends east-southeast for 8.6 km. It initially abuts eroding Pleistocene (?) barrier deposits, then is backed by an increasingly wide Holocene beach ridge series that terminates in a low 2 km long sand spit backed by a 1 km wide mangrove forest. River mouth shoals increase substantially in width off this spit and continue around into the river entrance. The beach is paralleled by a 500 m wide band of 2-3 shore-parallel bars. Beach **NT 1482** commences 500 m east of the spit as a low continuous convex beach ridge that curves round into the Calvert River mouth, finally facing east across the 700 m wide river channel (Fig. 5.164). The beach is initially backed by a continuation of the mangrove forest, then the eastern end of the Holocene beach ridge plain, which reaches a maximum of 2 km behind the mangroves. These are in turn backed by a 2 km wide Pleistocene barrier plain, then up to 8 km of salt flats.

Figure 5.164 Beach NT 1482 curves east into the Calvert River mouth (top) and is backed by extensive mangroves and beach ridges.

NT 1483-1488 **CALVERT R-NT/QLD BORDER**

No.	Beach	Rating HT	LT	Type	Length
NT1483	Calvert R (E1)	1	2	R+ridged sand flats	4.9 km
NT1484	Calvert R (E2)	1	2	R+ridged sand flats	7.6 km
NT1485	Sandy Ck	1	2	R+ridged sand flats	4.2 km
NT1486	Dingo Ck	1	2	R+ridged sand flats	8.9 km
NT1487	Mountain Ck	1	2	R+sand flats	6.2 km
NT1488	Mountain Ck (E)	1	2	R+ridged sand flats	9.1 km
Spring & neap tide range = 2.3 & 0.7 m					

To the east of the Calvert River mouth the coast trends southeast for 41 km to the Northern Territory-Queensland border. In between are six beach systems (NT 1483-1488) each separated by tidal creeks and backed by a 1-2 km wide Holocene beach ridge system totalling about 5000 ha in area, then salt flats which widen from 1 km in the east to 5 km near the river mouth. All the beaches are fronted by usually ridged sand flats, which widen off the creek mouths. Because of the salt flats and creeks there is no formal vehicle access to the coast.

Beach **NT 1483** commences on the eastern side of the Calvert River mouth. The 4.9 km long beach is the outermost in a series of several spits that have prograded the shoreline 2 km into the gulf. The eastern end of the beach is also backed by older spits that prograded the shoreline of the boundary creek 1.5 km. The creek is a distributary of the river, which forms a delta mouth island, with the beach forming the northern boundary. The creek was blocked by a partly vegetated berm in 1997, and may break out during wet season floods. The beach has extensive river mouth shoals off the western end which narrow to 500 m wide ridged sand flats to the east, widening again at the creek mouth.

Beach **NT 1484** commences on the eastern side of the creek and extends southeast for 7.6 km to the deflected mouth of **Sandy Creek**. The beach is fronted by ridged sand flats, which widen to the east, reaching over 1 km off the creek mouth. It is backed from west to east by a 2 km wide zone of vegetated spits, then beach ridges, and then a 4 km long spit that has deflected Sandy Creek to the east. Some minor foredune blowouts and dune activity are occurring on the spit. The entire system is backed by 4-5 km wide salt flats.

Beach **NT 1485** emerges from the rear of the spit and deflected creek mouth to continue on to the southeast for 4.2 km to **Dingo Creek**. Extensive ridged sand flats up to 2 km wide parallel the beach, together with the creek channels and tidal shoals to either end (Fig. 5.165). The beach is backed by a more massive 700 ha, 2 km wide, vegetated barrier which contains boundary spits, and a central section that has experienced blowouts and minor dune transgression to largely mask the original beach-foredune ridges. Two to three kilometre wide salt flats back the barrier.

Beach **NT 1486** commences on the eastern aside of 100 m wide Dingo Creek and trends to the southeast for 8.9 km to **Mountain Creek**. The beach is fronted by 1-2 km wide ridged sand flats and backed by a 1-2 km wide barrier system. The barrier has boundary recurved spits, with vegetated and active dune transgression in between, particularly to the east, with reactivated dunes reaching 20 m in height and transgressing over the vegetated barrier. Salt flats extend 2-3 km inland from the barrier.

Beach **NT 1487** extends from Mountain Creek for 6.2 km to the southeast to the next creek mouth, the last on the Territory coast. The beach is relatively straight with permanent creek mouths to either end. It is fronted by 1-2 km wide sand flats, and backed by a low 1 km wide vegetated barrier, containing some remnants of earlier dune activity. This is in turn backed by variable 1-2 km wide salt flats, then a narrow section of the backing coastal plain that just reaches the back of the barrier.

Beach **NT 1488** is the easternmost beach in the Northern Territory and extends from the creek mouth for 9.1 km southeast to the Queensland border, the beach continues on the other side of the border for 3.9 km to a creek mouth, a total length of 13 km. The beach is paralleled by 1-2 km wide ridged sand flats, and backed by a 1-1.5 km wide series of approximately 10 beach-foredune ridges, then 1-2 km wide salt flats.

Figure 5.165 Dingo Creek with beaches NT 1485 to west and NT 1486 to east (left). The backing low barriers and extensive salt flats are typical of the southern Gulf coast.

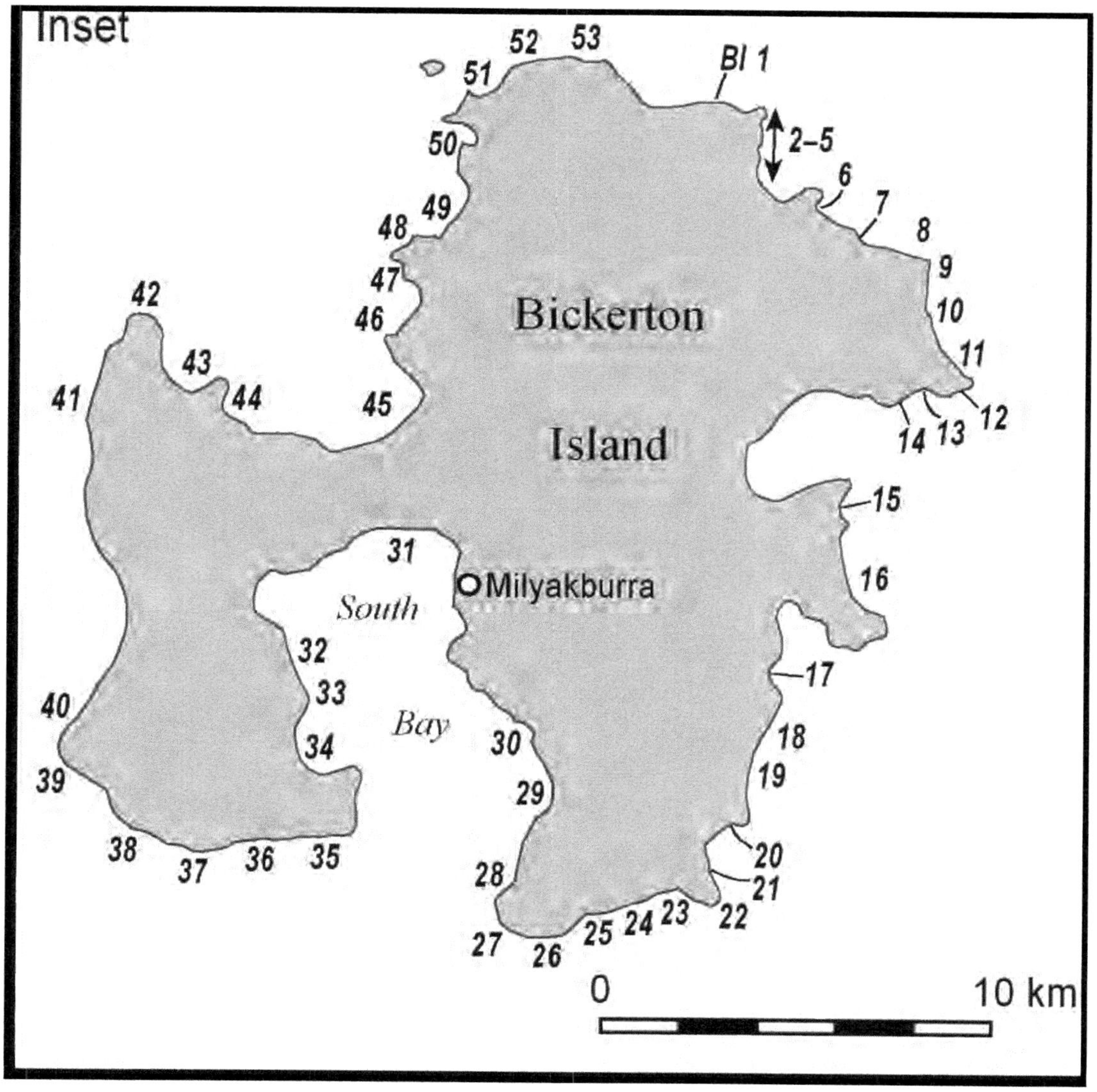

Figure 5.166 Bickerton Island, see Figure 5.151 for location.

Bickerton Island is a 23 000 ha island with 117 km of shore containing 53 beaches (Figs. 5.166). The beaches surround the irregular shaped island and range from the most exposed along the southeastern shore to those on the sheltered lee side beaches and in South Bay (Fig. 5.167). The island has one main community at Milyakburra and an island population of approximately 200. The barge landing for the island is in South Bay at beach BI 29.

Figure 5.167 Beach BI 25 (foreground) with South Bay to rear. The main community Milyakburra is located on the northern side of the bay.

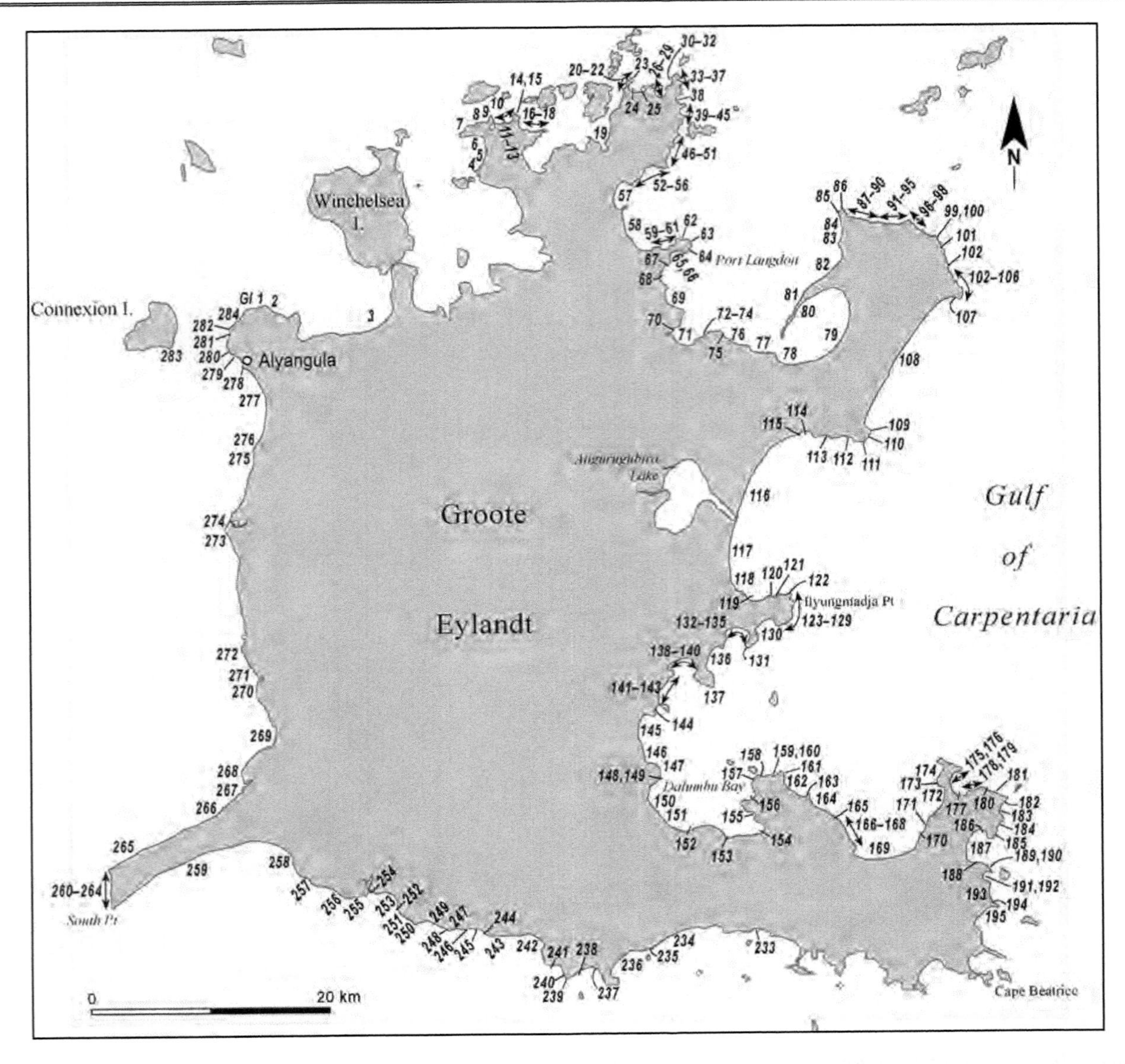

Figure 5.168 Groote Eylandt.

Groote Eylandt is a 246 100 island with 480 km of coast containing 284 beaches. The beaches range from sheltered along the west and irregular north coast (Fig. 5.169) to exposed higher energy on parts of the east coast, where extensive dunes back some beaches (Fig. 5.170). The island has one main town at Alyangula, which also services the islands extensive open cut bauxite and manganese mines, and a total population of about 1200.

Figure 5.169 Beaches GI 20-24 are located on the irregular rock-dominated northern tip of the island.

Figure 5.170 Angurugubira Inlet on the exposed eastern shore is bordered by extensive transgressive sand dunes. The inlet and its sand clogged tidal delta separated beaches GI 116 and 117.

6 CAPE YORK

The Cape York coast consists of three coastal regions: the 380 km long southeast section of the Gulf of Carpentaria between the Northern Territory border and Karumba: the 1089 km long eastern gulf shore along the western side of the Cape up to Cape York; and the 1032 km long eastern side of the Cape down to Cooktown (Table 6.1 & Fig. 6.1).

Table 6.1 Cape York coastal regions

	Region	Boundaries	km	Net km	Beaches	No. beaches	Beach km & (%)
1	Southeast Gulf	NT/Qld border-Norman R	0-380	380	Q 1-29	30	140 (39%)
2	West Cape York	Karumba-Cape York	380-1469	1089	Q 30-213	184	798 (73%)
3	East Cape York	Cape York-Cooktown	1469-2501	1032	Q 214-640	427	571 (55%)
	Total					641	1509 (60%)

Table 6.2 Maps of coastal regions and beaches

	Region	Regional maps	Beaches	Figure/s	Page
1	Southeast Gulf	1	Q 1-29	6.2	344
2	West Cape York	1	Q 30-73	6.2	344
		2	Q 74-231	6.14	357
3	East Cape York	3	Q 214-479	6.37	377
		4	Q 478-640	6.70	411

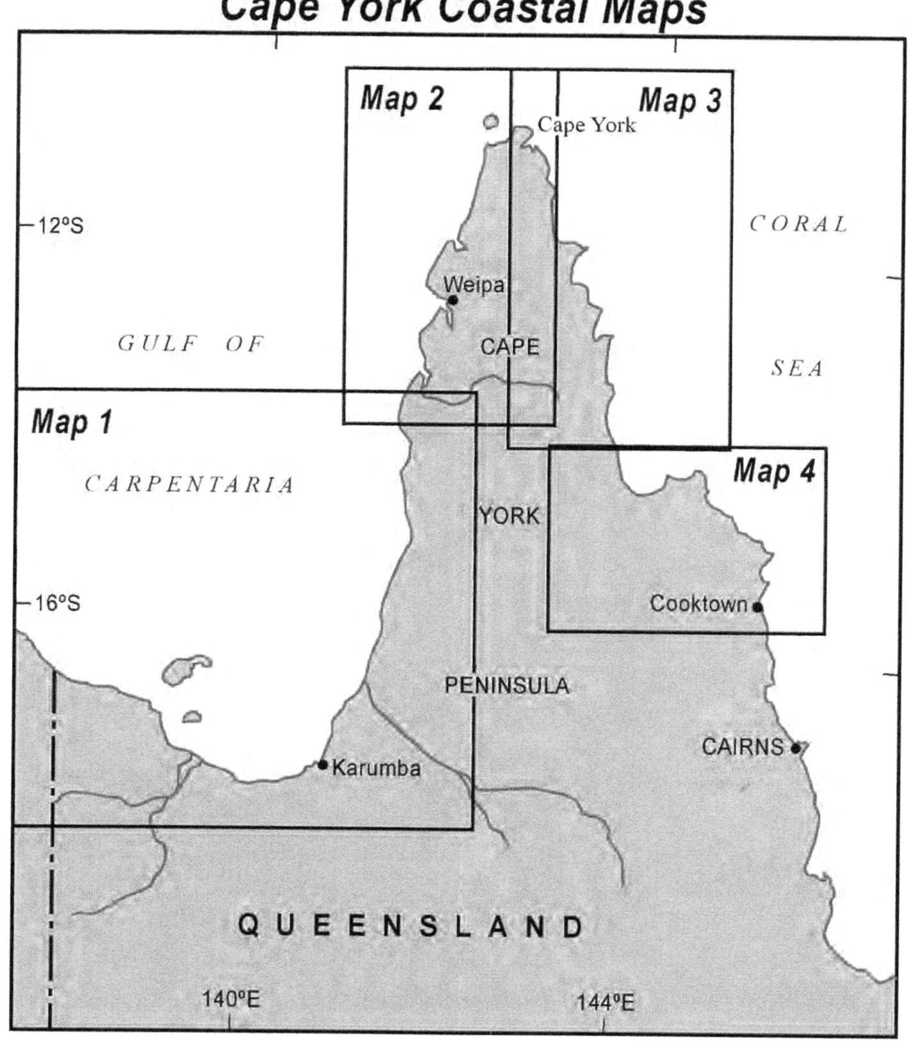

Figure 6.1 Cape York region and coverage of regional maps

Region 1 Southeast Gulf: NT/Qld border to Norman River

Length of coast: 380 km (0-380 km)
Number of beaches: 30 (Q 1-30)
Figure 6.2

Aboriginal Land: 115-126 km
National Parks: Finuncane Island

Towns/communities: Burketown, Karumba

The 380 km of shoreline between the Northern Territory border and the Norman River mouth at Karumba are prograding, very low gradient, low energy shore, receiving abundant fine sediments from approximately 30 creeks and rivers during the Wet season. These sediments contribute to the extensive tidal flats, with the coarse material and shells being deposited on 30 beach systems and in the extensive creek and river mouth tidal shoals. The beaches occupy 140 km (39%) of the shore, with much of the remaining 240 km dominated by wide bands of mangroves. The Nicholson, Leichhardt, Flinders, Bynoe and Norman rivers all deliver sediments to the southernmost gulf shore, where very low shoreline gradient and mangroves dominate. The prevailing southeast trades blow off to alongshore, resulting in generally calm to low wave conditions at the shore, and little dune activity. Only the occasional tropical cyclone produces high seas and storm surges, which can extend up to tens of kilometres in across the wide supratidal flats. Apart from a few vehicle tracks across the salt flats to the coast, the only development along this section is at Karumba.

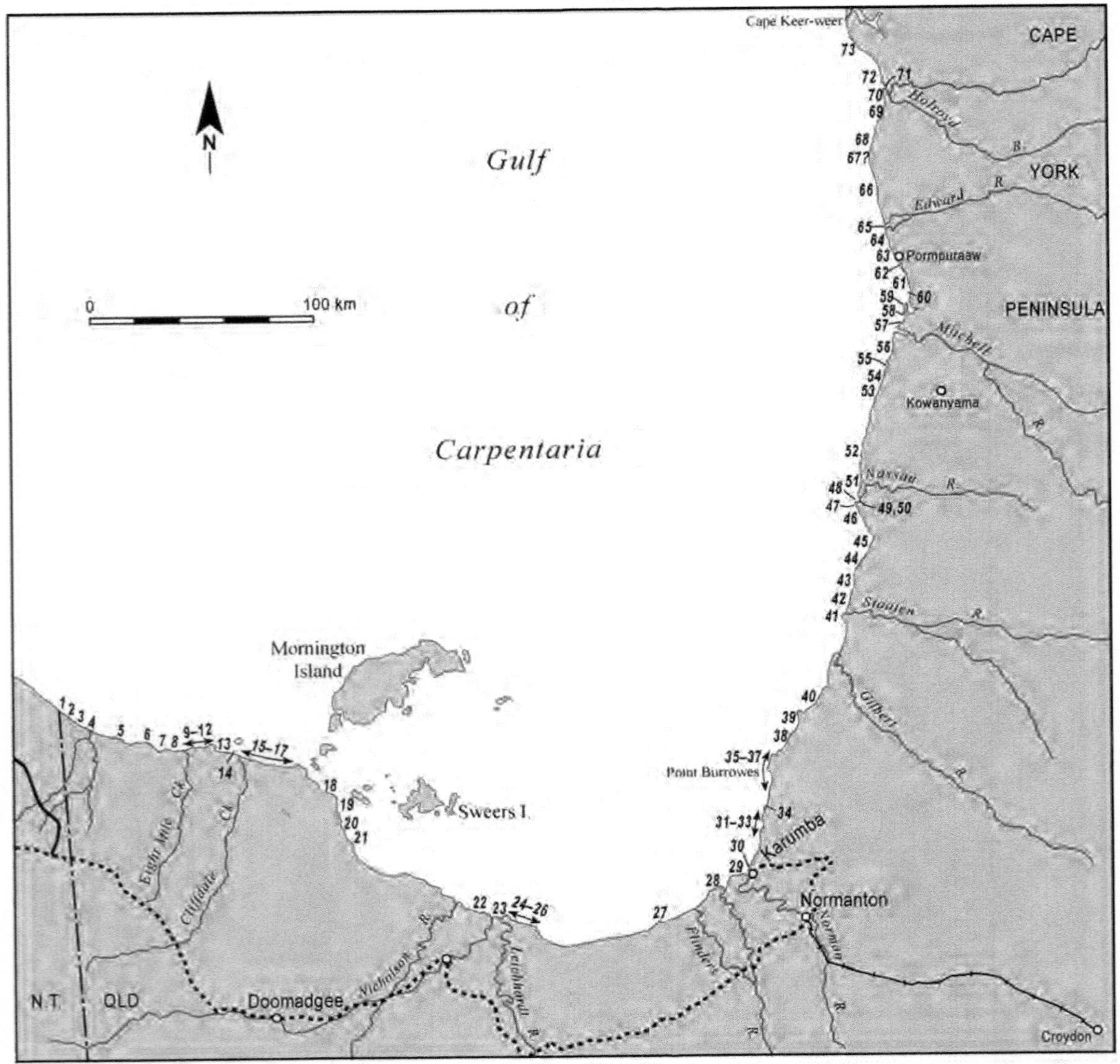

Figure 6.2 Cape York regional map 1: the southeast Gulf and lower western Cape (beaches Q 1-73).

Q 1-5 NT BORDER-MASSACRE INLET

No.	Beach	Rating HT	LT	Type	Length
Q1	Gum Ck (W 2)	1	2	R+ridged sand flats	3.9 km
Q2	Gum Ck (W 1)	1	2	R+ridged sand flats	7.8 km
Q3	Gum Ck (E)	1	2	R+ridged sand flats	4.9 km
Q4	Tully Inlet	1	2	R+ridged sand flats	5 km
Q5	Tully-Massacre inlets	1	2	R+ridged sand flats	20.2 km
Spring & neap tidal range = 2.3 & 0.7 m					

The Queensland border and coast begin at 138°E latitude. At the coast this is located towards the eastern end of a 13 km long beach. This beach commences in the Northern Territory as 9.1 km long beach NT 1488, and continues into Queensland as 3.9 km long beach Q 1.

The 42 km of coast between the border and Massacre Inlet are made up of five northeast to north-facing beaches (Q 1-5), all fronted by ridged intertidal sand flats, and separated by creeks and inlets. All the beaches are backed by 1-3 km wide beach ridge plains, then wider salt flats. The barriers total 8000 ha in area, and widen to the east where they finally splay into a 4 km wide series of up to 30 beach ridges to recurved spits. The barrier is backed by more than 10 000 ha of salt flats. The only access to the coast is via tracks from the Wollogorang and Westmoreland homesteads. The tracks follow high ground on the east side of Camel Creek leading to the coast between Tully and Massacre inlets, with an outstation and landing ground located 5 km west of Massacre Inlet.

Beach **Q 1** commences at the border and extends to the southeast for 3.9 km to a 100 m wide creek mouth. The creek borders a 1.5 km wide beach-foredune ridge plain before dispersing amongst 3 km wide salt flats. Sand ridges extend up to 500 m off the beach, with the shoals widening off the creek mouth (Fig. 6.3). Beach **Q 2** commences on the eastern side of the creek mouth and continues on for 7.8 km, past a usually closed shallow creek, to the mouth of 200 m wide **Gum Creek**. The beach is backed by more than 20 beach-foredune ridges, which extend up to 2 km inland. Ridged sand flats parallel the shore, with wider shoals off the boundary creeks.

Beach **Q 3** is a 4.9 km long northeast-facing beach bordered by Gum Creek to the west and an unnamed 100 m wide creek to the east. The beach is backed by a 2 km wide barrier consisting of inner beach-foredune ridges, and outer active dune transgression, including a number of low coalescing blowouts. Five hundred metre wide ridged sand flats front the beach, widening off the creek mouths. Beach **Q 4** commences on the eastern side of the creek and continues for 5 km to **Tully Inlet** at the mouth of **Camel Creek**. The beach is backed by up to 500 m wide dune transgression, then 1.5-2.5 km wide beach-foredune ridges, then the salt flats.

Figure 6.3 Beach Q 1 extends across the border and 4 km into Queensland, and is fronted by wide ridged sand flats.

Beach **Q 5** is the longest beach in the area, and extends for 20.2 km between Tully and Massacre inlets. A vehicle track reaches this beach and runs along behind to a landing ground located in lee of an inflection in the beach, that marks the beginning of the large **Massacre Inlet**. The western section of beach is fronted by ridged sand flats, while east of the inflection wider tidal sand flats extend 2-4 km off the beach and inlet. The western 12 km of beach are backed by active and stable dune transgression, then 1-2 km wide beach-foredune ridges and more irregular salt flats. To the east the ridges widen to 3-4 km on the western side of the inlet, as a series of up to 30 splaying recurved spits, which onlap and wrap around abandoned inlet channels. The 300 m wide inlet is bordered by a wide low gradient shoreline, with salt flats reaching the coast where they grade into the wide sand to mud flats.

Q 6-12 HORSE ISLAND-EIGHT MILE CREEK

No.	Beach	Rating HT	LT	Type	Length
Q6	Horse Is	1	2	R+mud flats	11.2 km
Q7	Horse Is (E1)	1	2	R+ridged sand flats	2.8 km
Q8	Hann Ck (W)	1	2	R+ridged sand flats	2.2 km
Q9	Hann Ck (E)	1	2	R+ridged sand flats	5.2 km
Q10	Eight Mile Ck (E1)	1	2	R+ridged sand flats	4.5 km
Q11	Eight Mile Ck (E2)	1	2	R+ridged sand flats	3.5 km
Q12	Eight Mile Ck (E3)	1	2	R+ridged sand flats	5.0 km
Spring & neap tidal range = 3.5 & 2.4 m					

Horse Island is bounded by Massacre Inlet to the west and a smaller inlet to the southeast. It forms the western boundary of a 34 km wide east-trending curve in the shoreline occupied by seven near continuous beach-barrier systems (Q 6-12), each separated by tidal inlets, fronted by extensive tidal flats, particularly at the creek mouths and backed by extensive beach-foredune ridges between 2 km and 7 km wide, then 2-5 km of high tide salt flats. There is no vehicle access to the coast.

Horse Island (**Q 6**) is an 11 km long, slightly convex, north to northeast-facing beach backed by a westward-growing series of over 30 multiple recurved spits that extend up to 2 km inland, with more than 20 inner ridges and salt flats extending another 6 km to the south. The ends of the outer systems are bordered by the tidal inlets and their 1-2 km wide sand flats, with mud flats paralleling the central 4 km of shore.

Beach **Q 7** is a narrow 2.8 km long casuarina-capped beach ridge that is fronted by ridged sand flats up to 1 km wide, and backed by 500 m wide salt flats, then a 6 km wide series of creek-truncated inner beach ridges and salt flats. Beach **Q 8** is a similar 2.2 km long narrow barrier, that has a new low 1 km long beach forming a few hundred metres off the inner beach. Mangroves are also growing on the eastern tidal flats adjacent to the inlet, shielding this end of the beach from direct wave action. It is backed by a 2 km wide series of 10 beach ridges, then 5 km of salt flats.

Hann Creek forms the boundary between beaches Q 8 and Q 9 and its inlet sand shoals protrude over 2 km into the gulf. Beach **Q 9** is a relatively straight, north-facing, 5.2 km long, westward-migrating series of up to 20 recurved spits fronted by up to 15 low intertidal sand ridges. The 500 m wide spit complex is backed by a continuous meandering tidal creek, then a 2 km wide series of approximately 15 beach ridges and finally 3 km wide salt flats. The backing creek has been deflected a total of 7 km, and now also crosses the shore between beaches Q 9 and Q 10, as the mouth of **Eight Mile Creek**. Beach **Q 10** commences on the eastern side of the 50 m wide creek mouth. The beach consists of an inner 4.5 km long, 1 km wide beach ridge system, with a new low narrow beach ridge forming off the central 3 km of beach. It is fronted by ridged sand flats that widen to 1 km off the boundary inlets. The inner system is backed by two more beach ridge systems containing up to 17 ridges and extending 3 km to the south, then 1 km wide inner salt flats. There is an access track from Hells Gate Roadhouse across the salt flats and ridges to the eastern end of the beach.

A 50 m wide meandering tidal creek separates beaches Q 10 and Q 11. Beach **Q 11** is a 3.5 km long low energy beach that is afforded increasing protection toward the east from a slight northerly protrusion in the shoreline, with mangroves covering the easternmost 1 km of the tidal flats. One kilometre wide ridged sand flats form the remainder of the beach. The beach is backed by a 4 km wide discontinuous series of beach ridges and salt flats, then 1-2 km wide salt flats.

Beach **Q 12** is located on a slight protrusion in the shoreline, and forms the eastern end of the arc that extends from Massacre Inlet. The 5 km long beach faces north and consists of a 2.5 km long central section, with recurved spits extending 1 km to the east and west, giving the entire system a convex shape (Fig. 6.4). The system is fronted by 1-2 km wide irregular sand flats. It is backed by 4-5 km wide salt flats, then an inner converging 1 km wide beach ridge system, and then another 4-5 km of salt flats.

Figure 6.4 Beach Q 12 is a strip of sand backed by 5 km wide salt flats and fronted by 2 km wide sand flats.

Q 13-16 CLIFFDALE-BERYL CREEKS

No.	Beach	Rating HT	LT	Type	Length
Q13	Cliffdale Ck (W 2)	1	2	R+mud flats	8.9 km
Q14	Cliffdale Ck (W 1)	1	2	R+sand flats	3.1 km
Q15	Cliffdale Ck (E)	1	2	R+mud flats	8.2 km
Q16	Beryl Ck (E)	1	2	R+mud flats	6 km
Q16A	Passmore Ck	1	2	R+tidal flats	1.8 km
Spring & neap tidal range = 3.5 & 2.4 m					

Beaches Q 13-16A occupy 26 km of a relatively straight 36 km long section of north-northeast-facing low energy shore centred on Cliffdale Creek and terminating at Passmore Creek in the east. Each of the beaches is part of 2 km wide, increasing to 4 km in the south, beach ridge-recurved spit systems, backed by 3-5 km of high tide salt flats, then low wooded coastal plains. Each beach is bordered and separated by tidal creeks with tidal sand shoals extending up to 1 km into the gulf, while the beaches in between are fronted by intertidal mud flats. There is vehicle access across the salt flats to the mouths of Cliffdale and Beryl creeks.

Beach **Q 13** is a slightly curving 8.9 km long beach which is part of a series of more than 25 recurved spits that have extended the beach-barrier at least 5 km eastward to the mouth of a small creek. The beach ridge-spits widen from 200 m in the west to 1 km in the east. The system is backed by 0.1-2 km wide salt flats, then a 1-2 km wide inner beach ridge barrier and then 1-4 km wide high tide salt flats. It is bordered by tidal sand flats associated with the creeks and with mud flats between the creeks.

Beach **Q 14** is 3.1 km long and bordered by the larger Cliffdale Creek in the east and the smaller boundary creek to the west. The two creeks combine to deposit extensive sand flats off the beach, resulting in very low energy conditions at the shore. It is backed by a 100-500 m wide low barrier, then a 1-2 km wide inner beach ridge barrier, then 4 km wide salt flats.

Beach **Q 15** is a slightly curving 8.2 km long beach fronted by mud flats with boundary creek mouth sand shoals. It is backed by up to 15 beach ridges which grade into recurved spits toward the boundary Cliffdale and Beryl creeks, indicating sediment transport both onshore to build the ridges and alongshore in both directions to build the spits. There is vehicle access to the eastern side of Cliffdale Creek and the neighbouring elongate lagoon, located between two of the outer spits.

Beach **Q 16** extends east of **Beryl Creek** for 6 km as a low, increasingly narrow, strip of high tide sand and backing casuarinas, backed by salt flats up to 1 km wide, then an inner 3 km wide series of discontinuous beach ridges and salt flats, and finally the inner 1-2 km wide salt flats. The eastern tip of the beach grades into open tidal flats and mangroves formed in lee of low nearshore gradients and shoals that protect a 20 km long section of mangrove-fringed shoreline to Bayley Point.

Passmore Creek is located 12 km east of beach Q 16, with beach **Q 16A** extending west of the creek mouth. In between is a relatively straight section of very low energy predominantly mangrove-fringed shoreline. The beach is 1.8 km long, faces due north and is bordered by mangroves in the west and the 100 m wide mangrove-fringed creek mouth in the east. It is fronted by extensive tidal flats and shoals associated with the creek mouth, and is backed by a 50 m wide vegetated beach ridge, then a 7-8 km wide series of recurved spits and salt flats. There is vehicle access across the salt flats to the eastern end of the beach.

Q 17-18 BAYLEY POINT

No.	Beach	Rating HT	Rating LT	Type	Length
Q17	Bayley Pt	1	2	R+rock flats	600 m
Q18	Syrell Ck (N)	1	2	R+tidal flats	1.7 km
Spring & neap tidal range = 3.5 & 2.4 m					

Bayley Point is a low right angle inflection in the shore in lee of Bayley Island with the larger Forsyth and Mornington islands located 10 and 30 km offshore respectively. The point is composed of Holocene calcarenite that extends up to 500 m off the beach. It forms the eastern boundary of east-facing 600 m long beach **Q 17**, whose barrier extends 3 km to the west, much of it fronted by a wide band of mangroves and subtidal mudflats.

Mangroves dominate the shoreline south of the point, extending to the southeast for 8 km to beach **Q 18**. This beach faces northeast and trends to the southeast for another 1.7 km to within 1 km of the mangrove-fringed mouth of Syrell Creek. The beach is backed by a 200-300 m wide vegetated beach ridge, then 4-7 km of salt flats.

Q 19-21 POINT PARKER

No.	Beach	Rating HT	Rating LT	Type	Length
Q19	Pt Parker	1	2	R+ridged sand flats	6.5 km
Q20	Pt Parker (S1)	1	2	R+sand-mud flats	3.2 km
Q21	James Ck	1	2	R+sand-mud flats	1 km
Spring & neap tidal range = 3.5 & 2.4 m					

Point Parker is a low but prominent shoreline inflection (Fig. 6.5) located immediately west of Allen Island and 25 km west of the larger Bentinck Island. The point marks the western boundary of a 60 km long curving east to northeast-facing section of coast that extends to the next major inflection at Gore Point. In between is 130 km of predominantly mangrove-lined shoreline, interrupted by numerous tidal creeks and the Nicholson, Leichhardt and Albert rivers. The river section is backed by up to 25 km of high tide salt flats. In the east there are three beaches located at and immediately south of Point Parker.

Figure 6.5 Point Parker consists of a series of cheniers, with east-facing beach Q 19 in foreground.

Beach **Q 19** commences on the northern side of Point Parker, extends for 1 km to the point, and then continues on to the south for another 5.5 km. In the vicinity of the point the beach is fronted by intertidal flats of Cretaceous siltstone. South of the point 500 m wide ridged sand flats face east into the more open gulf waters, with mangroves forming the southern boundary. Both sides of the point area are backed by a series of up to ten well developed beach ridges extending 2 km inland. The ridges thin to the south and some dune instability has resulted from the easterly orientation. There is vehicle access to the point.

Beach **Q 20** commences 13 km due south of the point and is a 3.2 km long east-facing beach fronted by 500 m wide sand to mud flats, and backed by a 100-200 m wide barrier consisting of some minor dune instability and vegetated beach ridges. It is bordered by mangroves to either end.

Beach **Q 21** lies a further 4 km to the south and is a 1 km long, very low energy beach that is bordered by James Creek to the south. The straight beach has mangroves growing toward either end, with only a few hundred metres clear of mangroves toward the southern end. Beyond the creek mangroves dominate the next 70 km of coast to the Albert River.

Q 22-26 ALBERT RIVER-GORE POINT

No.	Beach	Rating HT	Rating LT	Type	Length
Q22	Albert R (W)	1	2	R+ridged sand flats	1.6 km
Q23	Albert R (E)	1	2	R+tidal flats	2.4 km
Q24	Gore Pt (W3)	1	2	R+ridged sand flats	6.7 km
Q25	Gore Pt (W2)	1	2	R+sand flats	800 m
Q26	Gore Pt (W1)	1	2	R+sand flats	2.6 km
Spring & neap tidal range = 3.5 & 2.4 m					

The **Albert** and **Leichhardt rivers** enter the gulf within 4 km of each other and together contribute considerable sediments to the coast. This sediment, combined with the low gradient shoreface and mid-Holocene uplift of 1-2 m, has resulted in some of the most extensive Holocene shoreline progradation in Australia, with high tide salt flats extending up to 30 km inland from the shoreline. Burketown, located on the banks of the Albert River, lies 30 km southwest of the river mouth. Scattered across the tidal flats are discontinuous cheniers and beach ridges, with five north-facing beaches (Q 22-26) forming part of the prograding shoreline.

Beach **Q 22** is located on the western side of the Albert River mouth. It is an irregular, 1.6 km long, high tide beach fronted by 1 km wide intertidal ridged sand flats, and then another 2-3 km of tidal shoals associated with the river mouth. The narrow beach ridge is in turn backed by the extensive salt flats. Beach **Q 23** lies on the eastern side of the 500 m wide river mouth and is a 2.4 km long beach that is fronted by sand flats and the river channel in the west, grading into extensive mangroves to the east and then the mouth of the Leichhardt River, with 4 km wide river mouth shoals off the beach.

Finuncane Island National Park

Beach: Q 23
Coast length: 4 km (251-255 km)

Finuncane Island National Park is a small park located on the 'island' between the Albert and Leichhardt river mouths. The park preserves a section of the low gradient mangrove fringed shore and backing cheniers and salt flats. It has no facilities and is best accessed by boat.

Beach **Q 24** commences 5 km east of the Leichhardt River mouth. It is a relatively straight 6.7 km long northeast-facing beach, bordered by deflected creek mouths and backed by a 100-300 m wide casuarina-covered beach ridge. Beach **Q 25** is located immediately to the east and is also wedged between tidal creeks. The convex 800 m long beach is fronted by 1 km wide ridged to irregular tidal shoals and some scattered mangroves. Beach **Q 26** borders the eastern creek and is a straight 2.6 km long beach, which terminates as a series of recurved spits that form Gore Point to the east. All three beaches are backed by the salt flats, cheniers, tidal creeks and Disaster Inlet (Fig. 6.6) with the nearest bedrock 15 km inland.

Figure 6.6 Beach Q 24 is backed by a series of cheniers and intervening salt flats.

To the east of Disaster Creek mangrove-fringed salt flats dominate the low shoreline for the next 60 km to beach Q 27 at the mouth of Spring Creek.

Q 27 SPRING CREEK

No.	Beach	Rating HT	Rating LT	Type	Length
Q 27	Spring Ck	1	2	R+sand-mud flats	1.2 km
Spring & neap tidal range = 3.5 & 2.4 m					

Beach **Q 27** is located on the eastern side of the protruding mouth of Spring Creek. Mangrove-lined shores extend for 60 km to the west and 25 km to the east. The beach is composed of coarser sands and shells reworked from the creek mouth to form the relatively straight northwest-facing strip of sand. It is a low beach prone to overwashing and backed by overwash flats and then salt flats, with cheniers 1-2 km inland and then 20-30 km of supratidal salt flats. The beach is fronted by 300-500 m of ridged sand flats, then mud flats extending 1 km offshore. Mangroves border each end of the beach with some scattered along the beach.

Q 28-30 BYNOE RIVER-KARUMBA

No.	Beach	Rating HT	Rating LT	Type	Length
Q28	Bynoe R (W)	1	2	R+sand-mud flats	3 km
Q29	Bynoe R (E)	1	2	R+sand-mud flats	4.5 km
Q30	Karumba	1	2	R+sand-mud flats	3 km
Spring & neap tidal range = 3.5 & 2.4 m					

The Flinders, Bynoe and Norman rivers reach the southeastern corner of the gulf within 35 km of each other and like most of the gulf rivers have delivered substantial amounts of fine sediments to the coast resulting in Holocene shoreline progradation. Much of the prograding shoreline is composed of mangrove-fringed mud flats, with only occasional low energy beaches. While the Flinders River mouth is bordered by mangroves, there are beaches either side of the Bynoe (Q 28 and 29) and on the east side of the Norman (Q 30).

Beach **Q 28** is located 4 km west of **Bynoe Inlet**. It is a narrow, 3 km long, low energy, northwest-facing strip of high tide sand fronted by 500 m wide ridged sand flats, then low tide mud flats. It is low and prone to overwashing and sits atop low tide mud forming a chenier, with mangroves bordering each end. A series of several earlier cheniers and intervening salt flats extend 15 km inland. Beach **Q 29** commences on the eastern side of the 500 m wide river mouth. It trends northeast for 4.5 km, finally merging in behind the mangroves that then run for another 6 km to the Norman River mouth. The beach is fronted by irregular 500 m wide sand flats with scattered mangroves, then wider low tide mud flats. It is backed by a 12 km wide series of cheniers and salt flats, with recurved spits to either end. These have been eroded in places by the meandering of the Norman River, which has exposed the cheniers on its banks.

Karumba Beach (**Q 30**) is one of the few publicly accessible beaches on the gulf and probably the most popular and most frequented. The beach itself begins inside the eastern side of the river mouth running for 1 km up to the point, where it turns and trends northeast for 2 km (Fig. 6.7), gradually running in behind dense mangroves that then continue for another 9 km. The steep, shell-rich beach also has shell-rich beachrock exposed along this section. The western river mouth section of the beach has tidal sand flats, and the Elbow Bank, that extends 6 km offshore. To the east these give way to 1 km wide mud flats. The Karumba Point settlement backs the western section of the beach. It includes a beachfront tavern and beer garden, which provide a view of sunsets over the gulf, a caravan park, boat ramp and a few shops. The airport parallels the central-eastern section of the beach. The main town and port of Karumba is located 3 km to the south on the banks of the Norman River (Fig. 6.8).

Swimming: This is the Gulf's most accessible beach, and like many beaches around the Gulf, while wave are usually low, it is hazardous for bathing owing to stingers in summer, and crocodiles year round. In addition the western tip by the camping area drops off into the deep water of the Norman River with its strong tidal currents and crocodiles. So if you do intend to get wet here, check the water is too shallow for crocodiles and stay close to the beach.

Summary: For most travellers this is the only Gulf beach they will see. It is worth the drive out to the Point to view the beach or camp, however it's not a suitable bathing beach owing to the presence of crocodiles.

Figure 6.7 Karumba beach (Q 30) at the mouth of the Norman River.

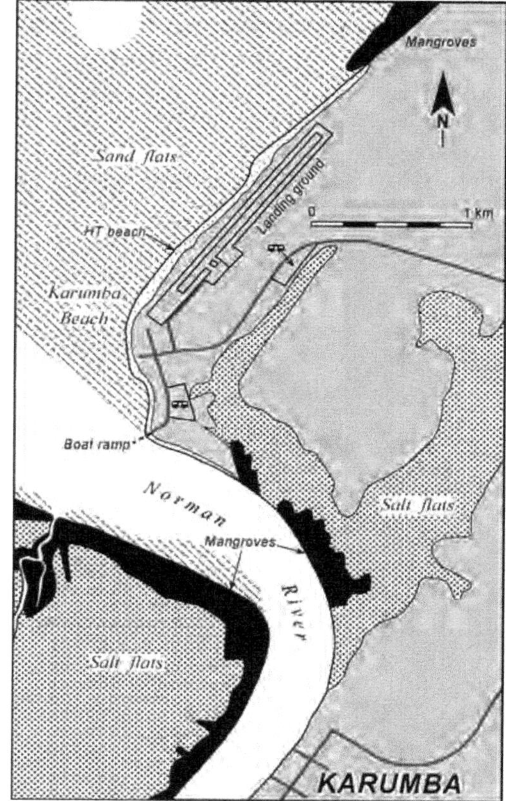

Figure 6.8 Karumba is located near the mouth of the Norman River, with Karumba beach located on the eastern side of the mouth.

Region 2 Western Cape York: Karumba to Cape York

Length of coast:	1089 km	(380-1469 km)		
Number of beaches:	183	(Q 31-213)		

Regional map 1	Beaches Q 30-73	Figure 6.2	Page 344
Regional map 2	Beaches Q 74-231	Figure 6.14	Page 357

Aboriginal Land:	Kowanyama, Pormpuraaw, Mapoon, Weipa, Bamaga			
	Coast length	632-668 km	Beaches:	Q 54-56
		685-888 km		Q 60-75
		892-902 km		Q 78
		1152-1240 km		Q 140-148
		1261-1363 km		Q 149-164
		1367-1398 km		Q 170-176
		1422-1452 km		Q 180-204

National Parks: none

Towns/communities: Karumba, Kowanyama, Pormpuraaw, Aurukun, Weipa, Bamaga, Seisia

At Karumba the shoreline turns and begins the 1000 km long trend north to Cape York. This entire section of coast faces west into the gulf with the prevailing southeast trades blowing offshore, while the lighter summer northwesterlies blow onshore producing low seas. Only the occasional tropical cyclone produces high seas. As a consequence this is a low energy coast, with the spring tides ranging between 3-4 m. Numerous rivers and creeks deliver sediment to the coast during the Wet, which is deposited on the extensive creek and river mouth tidal sand shoals and gradually reworked longshore into the beach systems. Apart from the first 20 km north of Karumba, which is dominated by mangroves fringed mudflats, generally long beaches separated by creeks and river mouths dominate the entire shore up to the cape. The first 500 km consist solely of sedimentary deposits, with bauxite deposits outcropping along the shore north of Worbody Point. In all, this section has 184 beaches (Q 30-213), which occupy 798 km or 73% of the shore, together with 63 creeks and 17 rivers. The beaches average 4.5 km in length (σ = 6.7 km), with creeks and or rivers located on average every 10 km along the shore.

Q 31-36 BRANNIGAN-ACCIDENT CREEKS

No.	Beach	Rating HT	LT	Type	Length
Q31	Brannigan Ck (S3)	1	2	R+ridged sand flats	800 m
Q32	Brannigan Ck (S2)	1	2	R+ridged sand flats	2 km
Q33	Brannigan Ck (S1)	1	2	R+sand flats	1.6 km
Q34	Accident Ck (S)	1	2	R+sand flats	4 km
Q35	Accident Ck (N1)	1	2	R+sand flats	3.7 km
Q36	Accident Ck (N2)	1	2	R+sand flats	4.6 km
Spring & neap tidal range = 3.5 & 2.4 m					

Brannigan Creek is located 20 km north of Karumba with most of the shoreline in between consisting of mangrove-lined tidal flats. Beach **Q 31,** 5 km south of the creek mouth, represents the first stretch of sand north of Karumba. The 800 m long beach faces west and consists of a low generally bare beach ridge, prone to overwashing, that has developed 100-300 m off the old trend of the shore, and represents the ongoing progradation of this section of coast. It is fronted by 50 m wide ridged sand flats, while it is separated from the older shore by mangrove-covered intertidal flats. Tidal creeks border either end of the beach. Beach **Q 32** commences 2 km to the north and is a 2 km long west-northwest-facing strip of high tide sand fronted by 500 m wide ridged sand flats, containing a few scattered mangroves. It terminates in the north at the mouth of a tidal creek, which has tidal sand shoals extending up to 2 km offshore. As a consequence waves are low at the beach and mangroves fringe the northern end. Five kilometre wide salt flats back both beaches.

Beach **Q 33** is part of a 1.6 km long, low northwest-facing, 100 m wide barrier, that extends north of the creek to its northern boundary at Branigan Creek. The creek also backs the barrier together with salt flats up to 5 km in width. The narrow high tide beach is fronted by 200-300 m wide intertidal sand flats.

Beach **Q 34** is located between Brannigan and Accident creeks. It is a convex-shaped 4 km long beach dominated at both ends by the tidal channels and extensive tidal shoal deposits that extend up to 2 km offshore. The beach is backed by a 2 km wide sequence of beach ridges and barrier deposits, then several kilometres of salt flats and some inner barriers. There is a vehicle track across the flats to the creek mouth where a rough fishing camp is located.

Beach **Q 35** extends from the north side of Accident Creek at Fitzmaurice Point north for 3.7 km to a small tidal creek. The beach is part of a 'drumstick' barrier that has prograded south to the creek mouth as a series of more than 15 recurved spits. Beach **Q 36** commences on the north side of the small creek and trends north for 4.6 km, finally grading into mangroves on the open shore in lee of the extensive tidal shoals of the Smithburne River. It is a narrow barrier also composed of south-trending recurved spits. Both beaches are fronted by 200 m wide sand flats, then low tide mud flats. They are backed by salt flats and inner Holocene and Pleistocene barriers extending more than 10 km inland.

Q 37-40 SMITHBURNE RIVER-SNAKE CREEK

No.	Beach	Rating HT	LT	Type	Length
Q37	Bold Pt	1	2	R+ridged sand flats	9.6 km
Q38	Van Diemen Inlet	1	2	R+ridged sand flats	18.5 km
Q39	Duck Ck	1	2	R+ridged sand flats	3.8 km
Q40	Kelso Ck	1	2	R+ridged sand flats	17.8 km
Spring & neap tidal range = 3.1 & 1.6 m					

Beaches Q 37-40 occupy most of the 53 km of coast between the Smithburne River and Snake Creek. The four generally low beaches are each bounded by prominent tidal creeks and associated tidal sand shoals. They form the outer boundary of a 10 km wide sequence of outer Holocene and Pleistocene barriers backed by extensive supratidal salt flats. There is vehicle access across the tidal flats to beaches Q 37, 38 and 40.

Beach **Q 37** commences on the northern side of the Smithburne River at Point Austin and trends north for 9.6 km past the slight curvature at Bold Point (Fig. 6.9) to Point Burrowes at Van Diemen Inlet. There is vehicle access to fishing camps on the banks of the tidal creeks at both ends of the beach. The beach has extensive tidal shoals off each boundary inlet, with 300 m wide ridged sand flats along most of the beach, backed by low casuarina-topped foredunes. The beach is part of a series of some 20 elongate recurved spits that narrow to the north and have migrated southward, in the process deflecting the river mouth southward.

Figure 6.9 Bold Point marks the southern end of beach Q 37.

Beach **Q 38** commences at 300 m wide Van Diemen Inlet and trends northeast for 18.5 km, past Top Dam stockyard, to 4 km south of Duck Creek, where beach Q 39 has been building southward along the former northern end of the beach. The long beach is part of a low barrier that narrows to a few hundred metres in the centre, widening to either end as 2-3 km wide splays of beach ridges terminating in spits. The beach is fronted by ridged sand flats that also widen towards the creek and narrow to 200 m in the centre. There is vehicle access to the coast at Top Dam. Beach **Q 39** is a low 3.8 km long beach ridge that extends from the mouth of Duck Creek southwards to where it overlaps the northern end of beach Q 38. The beach is backed by a meandering mangrove-lined deflected tidal creek, and fronted by migratory sand waves that are moving sediment to the south.

Beach **Q 40** commences on the northern side of 500 m wide Duck Creek and trends northeast for 17.8 km, past one small creek to the mouth of Snake Creek, where it curves round into the creek mouth. The beach appears to be eroding along the major central portion where mud flats front the coarse steep high tide beach. The beach is backed by older Holocene deposits that extend 2-4 km inland. There is vehicle access to the coast in the south along Kelso Creek and to the northern end at Snake Creek where a landing ground is also located.

Gilbert River

North of Snake Creek low gradient, low energy, mangrove-lined shores dominate the coast for 20 km to the mouth of the Gilbert River, then another 15 km to the mouth of the Staaten River. This 35 km long section of coast is backed by Holocene and Pleistocene beach ridge-chenier plains and intervening salt flats, extending between 10-15 km inland.

Q 41-45 STAATEN RIVER-SALT ARM CREEK

No.	Beach	Rating HT	Rating LT	Type	Length
Q41	Staaten R (S)	1	2	R+mud flats	2.7 km
Q42	Staaten R (N1)	1	2	R+mud flats	10.3 km
Q43	Staaten R (N2)	1	2	R+mud flats	2.5 km
Q44	Salt Arm Ck (S)	1	2	R-ridged sand flats	9.3 km
Q45	Salt Arm Ck (N)	1	2	R+ridged sand flats	10.4 km
Spring & neap tidal range = 3.1 & 1.6 m					

An extensive coastal plain extends north of the Staaten River mouth for 30 km to the Coolaman Swamp region, the plain representing part of the Staaten River delta. It consists of outer Holocene beach ridges and cheniers, inner Pleistocene ridges, and extensive salt flats, the entire system averaging 20 km in width. Five beaches (Q 41-45) are located between the southern side of the river mouth and the swamp.

Beach **Q 41** is a very low energy, west-facing, low gradient, 2.7 km long, high tide beach. It commences on the north side of the mangroves that extend 10 km north from the Gilbert River and terminates at the mangrove-lined mouth of the Staaten River. It is fronted by mud flats extending several hundred metres into the gulf, with mangroves scattered along the intertidal zone. The beach and backing 1-2 km of salt flats are a fauna reserve, with the salt flats extending another 3 km east to Pleistocene beach ridges.

Beach **Q 42** commences 1 km north of the 300 m wide Staaten River mouth. The 10.3 km long beach trends north past one tidal creek to a mangrove-lined tidal creek mouth that separates it from beach Q 43. The beach-barrier consists of a 100-200 m wide beach ridge, which is eroding in the north exposing mangrove peats while it grades into elongate recurved spits to the south, backed by the 20 km wide coastal plain. The beach is rich in shells, moderately steep, capped by casuarina and fronted by 1 km wide intertidal mud flats.

Beach **Q 43** commences on the northern side of the creek, which drains an extensive area of salt flats. It is a west-facing, low gradient, 2.5 km long, low energy beach, prone to overwashing and fronted by scattered low mangroves growing on the wide intertidal mud flats.

Beach **Q 44** commences on the northern side of a 3 km stretch of mangroves. It then trends north-northeast for 9.3 km to the 50 m wide mouth of Salt Arm Creek. The straight beach is backed by older south-trending beach ridge spits, with the outer beach-barrier consisting of a 100-500 m wide series of beach ridges splaying to wider ridges and elongate spits to the south. Two fence lines reach either end of the beach and the Inkerman prawn farm is located on the salt flats behind the northern junction to Salt Arm Creek. Beach **Q 45** commences on the northern side of the creek mouth and continues north-northeast for 10.4 km where it terminates at a mangrove-lined shore. The 50 m wide beach is moderately steep, has numerous shells winnowed from the low amplitude sand ridges that extend several hundred metres into the gulf. A solitary fence line reaches the middle of the beach. The beach is backed by several casuarina-capped beach ridges extending 400-500 m inland, with older Holocene beach ridges and recurved spit deposits widening to 1.5 km in the south. Salt flats associated with the Coolaman Swamp and inner barriers extend up to 10 km inland.

Q 46-49 ROCKY CREEK

No.	Beach	Rating HT	Rating LT	Type	Length
Q46	Jacks Pocket (S)	1	2	R+sand ridges	7.3 km
Q47	Jacks Pocket (N)	1	2	R+sand ridges	5.2 km
Q48	Rocky Ck (S2)	1	2	R+tidal channel	2.2 km
Q49	Rocky Ck (S1)	1	2	R+tidal channel	100 m
Spring & neap tidal range = 1.8 & 1.6 m					

Rocky Creek is a 20 km long tidal creek that has during the Holocene been deflected 12 km to the north to reach the gulf just 1 km inside the mouth of the Nassau River, which has been deflected 10 km to the south. The two systems flow across a 15-20 km wide coastal plain composed of extensive salt flats and at least four sequences of Holocene beach ridge-recurved spits, as well as an inner Pleistocene beach ridge barrier. There are four beaches (Q 46-49) extending for 12 km south of the Rocky Creek mouth.

Beach **Q 46** is a low, 7.3 km long beach ridge-spit, which commences on the south side of the 50 m wide creek mouth which drains Jacks Pocket. The spit represents the outermost sequence in a 4 km wide series between the present coast and Rocky Creek. It is capped by casuarina trees, and consists of a 50 m wide moderate gradient shell-rich beach fronted by a 300 m wide series of 3-4 low sand ridges, which represent the southward-moving sediments toward the terminal linear spit. It terminates at the next small creek mouth (Fig. 6.10), with beach **Q 47** extending from the north side of the creek mouth for 5.2 km to the southern side of the Nassau River mouth. The beach faces west-southwest and forms the western part of a beach ridge-spit that is deflected to the east at the river mouth as beach Q 48. The low beach is also fronted by a 300 m wide series of low migratory sand ridges. It is backed by a northward-deflected mangrove-lined tidal creek, then inner spits and beach ridges and intervening salt flats.

Figure 6.10 A creek draining extensive salt flats separates beaches Q 46 & 47.

Beach **Q 48** commences at the southern edge of the river mouth where the spit is deflected eastward into the river mouth. This dynamic spit grew east more than 1 km between 1969 and 1983, enclosing an earlier spit of similar dimensions. Over time it will change considerably in length and shape, and in 1998 a creek had breached the western section of the spit. It fronts the deep tidal channel of the Nassau River. Beach **Q 49** was a 1.5 km long inner spit in 1969, which had by 1994 been overlapped by the newer spit (beach Q 48), and in 1998 was only 100 m long, terminating near the entrance to Rocky Creek. Just inside the creek mouth is a jetty and fishing camp, with boats moored in the creek during the fishing season. The camp is connected by a vehicle track across the ridges and salt flats to Inkerman station.

Q 50-53 NASSAU RIVER-TOPSY CREEK

No.	Beach	Rating HT LT	Type	Length
Q50	Nassau R mouth	1 2	R+sand flats	1.3 km
Q51	Nassau R (N)	2 3	R+LT rips	13.5 km
Q52	Horse Ck (N)	2 3	R+LT rips	14 km
Q53	Horse-Topsy Cks	2 3	R+LT rips	18 km
Spring & neap tidal range = 1.8 & 1.6 m				

To the north of the **Nassau River** is a 45 km long, relatively straight, west-facing section of coast comprising four beaches (Q 50-53). The central 20 km of shoreline are eroding, while the terminal ends at the Nassau River and Topsy Creek both consist of splays of 10-12 beach ridges, which widen to 1 km at the river mouths. The entire system is backed by a 10-15 km wide low coastal plain containing outer Holocene beach ridges and salt flats and an inner Pleistocene beach ridge sequence. To the north of the river mouth the beaches receive sufficient wave energy to maintain a low tide bar cut by northward-skewed rips.

Beach **Q 50** is a 1.3 km long south-facing beach that forms the northern shore of the Nassau River mouth and links the ends of the terminal recurved spits. It is a relatively narrow, steep beach fronted in part by the deep 200 m wide tidal channel. Beach **Q 51** commences at the northern river mouth entrance and curves slightly as it trends north for 13.5 km to a small tidal creek that separates the southern accreting barrier from the eroding section to the north. It is backed by a beach ridge plain that widens from 100 m in the north to 1 km at the river mouth. The beach is moderately steep, with shellrock debris on the shore, and fronted by a 100 m wide north-skewed series of low tide bars and rips.

Beach **Q 51** commences on the northern side of the small creek and trends almost due north for 14 km to 100 m wide Horse Creek. This is an eroding section of shore with the beach truncating Holocene beach ridges and salt flats and some small usually blocked creek mouths. It has a moderately steep crenulate beach fronted by north-skewed low tide bars and rips.

Beach **Q 52** commences on the northern side of Horse Creek and extends north for 18 km to **Topsy Creek**. The southern half of the beach is eroding into earlier Holocene beach ridges and salt flats, while the northern half widens from a 0.1-1 km wide series of beach ridges. The moderately sloping sandy beach is fronted by continuous low tide bars and rips skewed to the north and spaced about every 200-300 m.

Kowanyama and Pormpuraaw Aboriginal Lands

Beaches: 10 (Q54-56, 60-75 & 78)
Coast length: 249 km (632-668 km, 685-888 km & 892-902 km)

Kowanyama and Pormpuraaw Aboriginal Lands extends from Topsy Creek to Aurukun at the mouth of the Archer River, encompassing 250 km of shoreline and 10 beaches systems. Kowanyama is located just north of Topsy Creek but 28 km inland, with Pormpuraaw 70 km to the north at the mouth of the Chapman River, and Aurukun another 170 km further north.

Q 54-61 MITCHELL RIVER DELTA

No.	Beach	Rating HT	Rating LT	Type	Length
Q54	South Mitchell R (S)	2	3	R+LT rips	17 km
Q55	South Mitchell R	2	3	R+tidal channel	1 km
Q56	South Mitchell R (N)	2	3	R+LT rips	17 km
Q57	Mitchell R (N 1)	1	2	R+mud flats	6 km
Q58	Mitchell R (N 2)	1	2	R+mud flats	6 km
Q59	North Arm	1	2	R+mud flats	2.5 km
Q60	Coleman River	1	2	R+mud flats	1.5 km
Q61	Malaman Ck	1	2	R+mud flats	1.5 km

Spring & neap tidal range = 1.8 & 1.6 m

The **Mitchell River delta** extends along approximately 40 km of coast with Holocene and Pleistocene deltaic beach ridges, cheniers, salt flats, distributaries and abandoned channels extending up to 22 km inland. In all the delta covers an area of approximately 50 000 ha. The river itself bifurcates into a series of active channels including the South Mitchell, Mitchell, North Arm and Coleman. Eight beaches (Q 54-61) occupy 52 km of deltaic shoreline and grade from relatively higher energy in the south with low surf and rips to extensive mud flats north of the Mitchell River mouth.

Beach **Q 54** is a relatively straight, 17 km long, west-northwest-facing beach bordered by Topsy Creek and the South Mitchell River. The beach is backed by a 3 km wide beach-foredune ridge plain containing over 20 casuarina-capped ridges. It has a moderate gradient and is fronted by a 50 m wide low tide bar cut by rips skewed to the north every 200-300 m. There is vehicle access to either end of the beach via tracks from Kowanyama, which is located 25 km to the east.

Beach **Q 55** lies in the curving river mouth channel of the South Mitchell River and faces west into the gulf across the river channel and river mouth shoals. While the 1 km long beach receives low wave to calm conditions, it is paralleled by the strong tidal currents of the river channel. Beach **Q 56** commences on the northern side of the river mouth where a series of south-migrating recurved spits episodically extend south and connect to beach Q 55, before they are eroded by floods (Fig. 6.11). The beach initially trends northeast then more northerly for 17 km to the mouth of the Mitchell River, where it also recurves into the protruding river mouth. The beach is backed by a series of beach-foredune ridges, which widen from 200 m in the centre to 500 m in the south and 1.5 km in the north, then extensive salt flats and inner older ridge systems. It is fronted by a low tide bar cut by northward-trending rips every 200-300 m.

Figure 6.11 The Mitchell River mouth with beach Q 56 curving round into the mouth.

Beach **Q 57** commences on the north side of the 500 m wide river mouth and trends northeast for 6 km. The beach represents a more recent barrier island that has developed 1 km westward of the former shoreline as well as extending northward 1-2 km, with mangroves and salt flats separating it from the inner shoreline, which is in turn backed by up to 13 km of Holocene ridges and salt flats. The beach is fronted by extensive intertidal mud flats. At its northern end it overlaps beach **Q 58**, a curving 6 km long west-facing low energy beach, backed by several beach ridge-foredunes, then the 5 km wide tidal flats of North Arm. It is fronted by wide intertidal mud flats and terminates at the mouth of North Arm.

Beach **Q 59** is located between the mouths of North Arm and Coleman River, both large tidal creeks connected to the Mitchell River 15 km to the south. The beach is a 2.5 km long west-northwest-facing barrier island, backed by overwash flats that widen to 200 m in the north, then extensive mangroves, salt flats and tidal channels. It is fronted by wide mud flats with the tidal channels to either end.

Beach **Q 60** commences north of the mangroves that fringe the northern side of the Coleman River and trends north for 1.5 km to the smaller Malaman Creek. It is a very low energy, low beach system, prone to overwashing and fronted by scattered mangroves and extensive mud flats. Beach **Q 61** on the north side of the creek is a similar 1.5 km long west-facing beach fronted by 2 km wide mud flats. The beach grades into mangroves to the north which extend north for another 10 km along the Melamen Plain.

Q 62-65 PORMPURAAW

No.	Beach	Rating HT LT	Type	Length
Q62	Melamen Plains	2 3	R+ridged sand flats	3.5 km
Q63	Pormpuraaw	2 3	R+ridged sand flats	8 km
Q64	Moonkan Ck	2 3	R+LT rips	8.6 km
Q65	Edward R	2 3	R+LTT	4 km
Spring & neap tidal range = 1.8 & 1.6 m				

Pormpuraaw is largest coastal Aboriginal community on western Cape York. The township is located 1 km in from the shoreline, on a 3 km wide outer Holocene foredune ridge plain, with the landing ground paralleling the rear of the beach. A series of 14 near continuous sandy beaches (Q 62-75), separated by tidal creeks, extend from just south of the township for 175 km north to the Archer River mouth.

Beach **Q 62** emerges from the 10 km of mangroves that front the Melamen Plain to trend north-northwest for 3.5 km to the mouth of the 50 m wide Chapman River. The beach has a moderately low gradient and is fronted by a 200 m wide subdued intertidal sand ridge. It is backed by a 4 km wide densely vegetated beach ridge plain, then the southern salt flats of the Chapman River and the inner Nine Mile Ridge Pleistocene barrier.

Pormpuraaw beach (Q 63) commences on the north side of the Chapman River and trends northwest for 8 km to the small mouth of Moonkan Creek. There is a landing and boat ramp just inside the creek mouth, and a crocodile farm next to the creek, with vehicle tracks leading up the beach and to the township, 2 km north of the creek (Fig. 6.12). The beach is moderately steep and fronted by low ridged sand flats extending 200 m off the beach.

Figure 6.12 Pormpuraaw is located on higher ground 1 km in from its main beach (Q 63).

Beach **Q 64** is an 8.6 km long west-facing beach that trends relatively straight between Moonkan Creek and the small Edward River mouth. Wave energy increases gradually north of the Melamen Plain, and the beach receives sufficiently high waves to maintain a 100 m wide low tide bar cut by northward-deflected enechelon rips, spaced every 400-500 m. It is backed by a 1-2 km wide Holocene outer beach ridge barrier, a 3 km wide interbarrier depression and the inner Pleistocene barrier.

Beach **Q 65** is a 4 km long beach bordered by the interconnected Edward River and Balurga Creek. It is backed by a 500 m wide beach ridge plain, then the linked salt flats and tidal channels. The beach has tidal sand shoals off either end and a low tide bar in between.

Q 66-69 CHRISTMAS CREEK-KENDALL RIVER

No.	Beach	Rating HT LT	Type	Length
Q66	Christmas Ck	2 3	R+LTT	27 km
Q67	Hersey Ck	2 3	R+LTT	13 km
Q68	Hersey-King Ck	2 3	R+LTT	3.8 km
Q69	Kulinchin	2 3	R+LTT	19 km
Spring & neap tidal range = 1.8 & 1.6 m				

To the north of Balurga Creek-Edward River are 62 km of near continuous, relatively straight, west-facing, sandy beaches, cut by the small Christmas, Hersey and King creeks and terminating at the larger Holroyd River.

Beach **Q 66** is one of the longer beaches on the cape at 27 km in length. It trends north from the mouth of Balurga Creek, past the small deflected mouth of **Christmas Creek** to an inflection in the shore adjacent to Christmas Creek. The entire system is backed by a massive beach-foredune ridge plain, containing up to 70 ridges, extending between 1-5 km inland, which are in turn backed by a 1-3 km wide interbarrier depression and a 4 km wide inner Pleistocene barrier system, making it one of the larger beach ridge systems in Australia. The modern beach has a moderate slope and is paralleled by a continuous 100 m wide low tide bar. It is backed by casuarina-capped beach to low foredune ridges. Mangrove-lined Christmas Creek has been deflected up to 6 km to the south with several recent beach ridges to low spits between the creek and the shore. There is vehicle access to and along the beach and a two house community at the creek mouth.

The beach ridges paralleling the deflected Christmas Creek form an inflection in the shoreline, with beach **Q 67** commencing at the inflection and trending north-northeast for 13 km to the small Hersey Creek mouth. The Christmas Creek beach ridges pinch out at the coast 9 km north of the inflection, with the northern half of the beach backed by eroding 4 km wide uplifted interbarrier deposits, with shell-rich beachrock exposed along the shore, and in places eroded to form storm boulder ridges. To the lee of the interbarrier depression is the 1-2 km wide inner Pleistocene barrier. The beach is relatively steep, shelly, interrupted by the beachrock and prone to overwashing. Beach **Q 68** trends north-northeast for 3.8 km between Hersey and King creeks. The entire beach is being eroded with interbarrier depression deposits exposed at the shore. The beach is similar in form to beach Q 67.

Beach **Q 69** commences at King Creek and curves gradually to trend due north where it terminates as a series of recurved spits at the 1 km wide **Holroyd River** mouth. The 19 km long beach is backed by the eroding interbarrier depression and exposed beachrock for the first 7.5 km, then a widening band of beach-foredune ridges that terminate in the 3 km wide splay of more than 20 recurved spits at the river. There is a semi-permanent fishing camp located on a recurved spit on the southern arm of the river. For much of its length the beach has a moderately steep gradient fronted by a 50 m wide low tide bar, which begins widening 5 km south of the river mouth to a 1.5 km wide series of more than 20 intertidal low sand ridges.

Q 70-73 **HOLROYD RIVER-CAPE KEERWEER**

No.	Beach	Rating HT	Rating LT	Type	Length
Q70	Holroyd R bar	1	2	R+sand flats	2 km
Q71	Holroyd R mouth	2	3	R+tidal channel	200 m
Q72	Holroyd R (N)	1	2	R+ridged sand flats	5 km
Q73	Cape Keerweer	2	3	R+LTT	24 km
Spring & neap tidal range = 1.7 & 0.8 m					

The **Holroyd River** and its tributaries drain a substantial catchment and reach the coast as a funnel-shaped 1 km wide river mouth, with a dynamic barrier island lying off the mouth. The river protrudes 5 km into the gulf, with extensive deltaic beach ridge plains spreading to the north and south.

Beach **Q 70** is a low shelly barrier island that forms off the river mouth and episodically extends southward to link with the northern end of beach Q 69. In 1969 the beach was 1 km long and largely unvegetated and surrounded by bare sand flats. By 1989 it had extended another 1 km to the south and linked to the beach, and become partially vegetated and backed by extensive mangroves on the sand flats, a position it maintained in 1994 (Fig. 6.13). Over time however it is expected that major floods will rework the entire beach system, and the process starts again.

Beach **Q 71** is located just inside the northern side of the river mouth. It is a narrow, curving, southwest-facing beach that truncates the backing beach ridges, and is fronted by the deep river channel.

Beach **Q 72** commences on the northern side of the river mouth and curves slightly to the north for 5 km to the 50 m wide mouth of Knox Creek. The creek meanders across a 9 km wide beach ridge plain, before it is deflected 3 km to the south to its present position. The beach is backed by the beach ridges and fronted by 1 km wide ridged sand flats, which originate from the extensive river mouth shoals.

Figure 6.13 The dynamic beach Q 70, which forms across the mouth of the Holroyd River.

Beach **Q 73** commences on the north side of the creek and curves round to the northwest and then back to the north for a total of 24 km, terminating at the mouth of the Kirke River at **Cape Keerweer**. This is one of the longer beaches on the cape and marks the outer boundary of one of the most extensive beach ridge plains on the Australian coast. The near continuous ridges extend from south of the Holroyd River to the cape region, a distance of 40 km, and up to 70 ridges are spread up to 10 km inland. They have formed from southerly sand transport as a series of both beach ridges and terminal spits to recurved spits. In all the plain totals about 23 000 ha in area. It is undeveloped apart from two abandoned landing grounds, one near the coast, 8 km north of Knox Creek. The beach is exposed to moderate gulf waves and characterised by moderate slopes and a continuous 50-100 m wide low tide bar. The cape is famous as the location where, in 1606, the Dutch yacht *Duyfken* turned about on its exploratory trip down the cape, hence the name *keerweer* or 'turn again'.

Q 74-75 **KIRKE-LOVE RIVERS**

No.	Beach	Rating HT	Rating LT	Type	Length
Q74	Kirke R	2	3	R+LTT	50 km
Q75	Love R	2	3	R+LTT	15 km
Spring & neap tidal range = 1.7 & 0.8 m					

At **Cape Keerweer** the shoreline turns gradually and trends north-northeast, then north for 50 km to the mouth of the Love River. Beach **Q 74** occupies the 50 km of shoreline and is the longest beach on Cape York. For its entire length it is backed by a series of beach to foredune ridges, which reach a maximum width of 2 km, and which narrow to each end in front of deflected river mouths. The **Kirke River** has been deflected 10 km to the south, while the Love is deflected 8 km to the north (Fig. 6.15). The rivers drain extensive wetlands that interlink behind the beach and range from 1-10 km in width, with Pleistocene barriers to the east. The beach consists of a moderate gradient sandy shore, fronted by a continuous 50-100 m wide low tide bar, with waves averaging less than 0.5 m. Shell-rich beachrock is exposed in sections, suggesting beach erosion. There is vehicle access across the narrower section of wetlands toward the northern end of the beach.

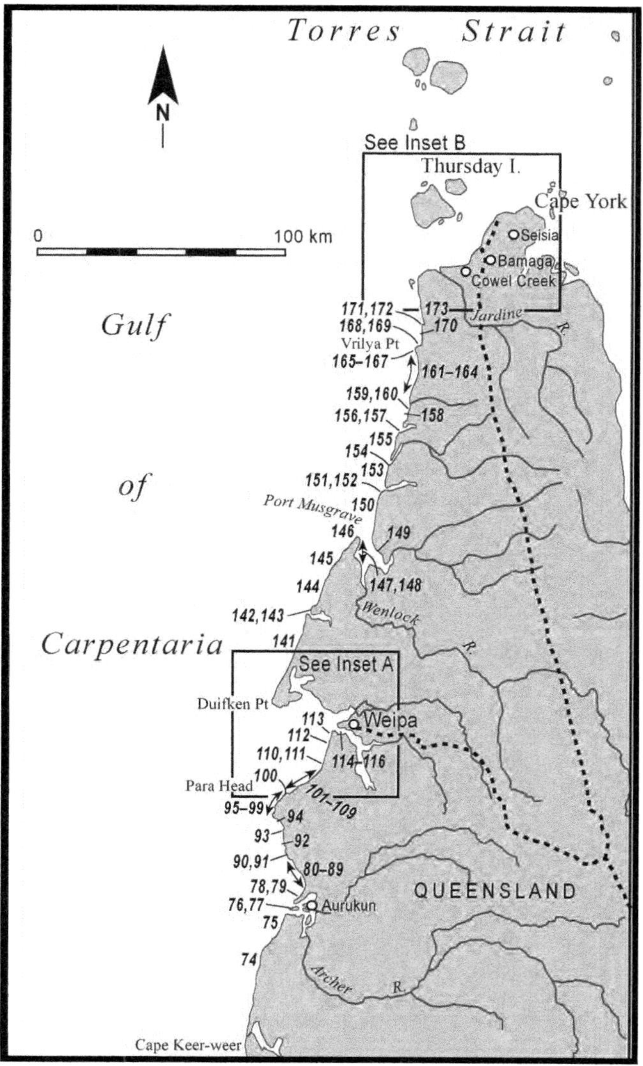

Figure 6.14 Cape York regional map 2: the western gulf (beaches Q 74-231). See Figure 6.18 for Weipa insert and Figure 6.35 for Cape York insert.

Figure 6.15 The deflected mouth of the Love River and beach Q 74.

Beach **Q 75** commences on the northern side of the 200 m wide **Love River**, and curves slightly to the north-northeast for 15 km to the southern entrance to Archer Bay. The beach is backed by four major splays of north-trending beach ridge-spits that terminate at the Archer River. They increase in width from 4 km in the south to 9 km in the north. The combined Kirke and Lowe rivers barrier totals 20 000 ha in area. Mangrove-lined Nundah Creek flows south for 12 km between the outer two ridge systems, restricting vehicles access to a southern fishing camp and landing ground just inside the entrance to the Love River. There is a second fishing camp just inside **Archer Bay** at the very northern end of the beach. The beach consists of a moderate gradient sandy beach fronted by a continuous 50-100 m wide low tide bar, in addition to the river mouth channel and shoals at both ends.

Q 76-79 WORBODY POINT

No.	Beach	Rating HT LT	Type	Length
Q76	Wallaby Is	1 2	R+sand flats	4 km
Q77	Worbody Pt (S)	1 2	R+tidal channel	1.5 km
Q78	Worbody Pt (N1)	1 2	R+sand flats	12.5 km
Q79	Worbody Pt (N2)	1 2	R+sand flats	200 m
Spring & neap tidal range = 1.7 & 0.8 m				

Wallaby Island occupies much of the mouth of 4 km wide **Archer Bay**. The bay is located at the confluence of the Ward, Watson and Archer rivers and a few smaller creeks, with the Aurukun community located 8 km inside the mouth adjacent to the central Watson River mouth. There are two 500 m wide entrances to the bay either side of Wallaby Island, with tidal sand shoals extending up to 3 km into the gulf.

Wallaby Island is 600 ha in area and consists of two sets of beach ridge-recurved spits, with the 4 km long beach (**Q 76**) bordering the western set. The beach is double crenulate in shape and fronted by the 2-3 km wide intertidal sand flats. Beach **Q 77** is located opposite the northern end of the beach and borders the northern channel for the bay, where it truncates a 2 km wide series of recurved spits. The beach is 1.5 km long and faces south across the tidal channel to the island and sand flats. It terminates in the west at low sandy Worbody Point.

Beach **Q 78** commences at Worbody Point and curves slightly to the north-northeast to the first outcrop of laterite on the southern gulf shores, and then for another 3 km along the base of the laterite bluffs, a total length of 12.5 km. The beach is backed by the 2 km wide series of up to 20 beach ridges, in total shaped like a sperm whale. The northern end of the beach and ridges abuts the 10-15 m high laterite bluffs. The steep shell-rich beach (Fig. 6.16) is fronted by 2 km wide sand flats off the point, narrowing to a few hundred metres toward the northern end. There is vehicle access that runs south for 6 km along the top of the bluffs to reach the beach.

Figure 6.16 The steep shelly beach face of beach Q 78 grades abruptly into sand flats at its base.

Beach **Q 79** is located between a protrusion in the laterite bluffs and a major 500 m west-trending section of bluffs. The 200 m long beach is backed by scarped red laterite bluffs, with the vehicle track on top. There are some laterite boulders and debris on the beach and a combination of laterite and sand flats off the beach.

Q 80-87 INA CREEK

No.	Beach	Rating HT	LT	Type	Length
Q80	Ina Ck (S5)	1	2	R+LTT	800 m
Q81	Ina Ck (S4)	1	2	R+LTT	300 m
Q82	Ina Ck (S3)	1	2	R+LTT	1.7 km
Q83	Ina Ck (S2)	1	2	R+LTT+rock flats	500 m
Q84	Ina Ck (S1)	1	2	R+ridged sand flats	1 km
Q85	Ina Ck	1	2	R+sand flats	1.8 km
Q86	Ina Ck (N1)	1	2	R+sand flats	3.2 km
Q87	Ina Ck (N2)	1	2	R+sand/rock flats	1.2 km
Spring & neap tidal range = 1.7 & 0.8 m					

To the north of the Worbody Point beach ridge plain (Q 78) 10-20 m high eroded Tertiary laterite bluffs dominate the coast to Weipa, 70 km to the north. Sandy beaches lie at the base of most of the bluffed shoreline, with foredunes only accumulating in lee of lower sections in the bluffs, often associated with drainage depressions and creek mouths. Between beaches Q 80-87 is a 12 km long section of generally curving northwest-trending bluffs, cut by only two creeks, one at beach Q 85 and the larger Ina Creek at beach Q 86.

Beaches Q 80-82 are part of a curving, 3 km long, west-southwest-facing, bluffed embayment, with two protrusions in the bluffs cutting the sandy beach into three. Beach **Q 80** occupies the first 800 m long section, beach **Q 81** a 300 m long central section and beach **Q 82** the northern 1.7 km long section which curves round to face south in lee of the northern bluffs. The bluffs average 10 m in height and are scarped for the most part, with some sections fronted by vegetated bluff debris. The beaches all consist of a moderately steep high tide beach, some parts littered in red laterite boulders, fronted by a continuous low tide bar, which widens in lee of the northern point. A straight vehicle track from North Camp reaches the top of the bluffs in lee of beach Q 82, with tracks then running along the top of the bluffs the length of the three beaches.

Beach **Q 83** is a crenulate 500 m long southwest-facing beach located on a low vegetated bluff-headland, and fronted by a mixture of sand and rock flats, the latter extending over 500 m off the beach. Beach **Q 84** lies immediately to the north and is a curving 1 km long west to south-facing beach. It is more exposed in the south with a low tide bar, while to the north three ridged sand flats have formed in lee of the headland and adjacent rock flats.

Beach **Q 85** occupies the next west-facing embayment. It is 1.8 km long and bordered by intertidal rock flats at either end, with a southern bar grading into 200 m wide sand flats in the north. A 10 ha freshwater swamp lies 1 km in from the beach and is connected by a small creek that drains across the centre of the beach. Low foredunes parallel much of the beach.

Beaches Q 86 and Q 87 occupy a curving 4.5 km long west to south-facing curve in the bluffs. Beach **Q 86** is a 3.2 km long west-facing beach that curves southwest in the north. It is bordered by low vegetated bluffs to either end linked by a series of foredunes that widen from 200 m to 800 m in the north and are in turn backed by a 250 ha freshwater swamp. The swamp is drained by **Ina Creek**, which runs behind much of the beach to cross at its southern end, while the beach is fronted by 200-300 m wide sand flats. Beach **Q 87** is a 1.2 km long curving south-facing lower energy beach fronted by sand and rock flats.

Q 88-94 FALSE PERA HEAD-THUD POINT

No.	Beach	Rating HT	LT	Type	Length
Q88	False Pera Hd (S3)	2	3	R+LT rips	2.8 km
Q89	False Pera Hd (S2)	1	2	R+sand/rock flats	500 m
Q90	False Pera Hd (S1)	1	2	R+sand/rock flats	3.6 km
Q91	False Pera Hd	1	2	R+sand/rock flats	1.3 km
Q92	Norman Ck (S)	2	3	R+LT rips	5.3 km
Q93	Norman Ck (N)	1	2	R+sand/rock flats	4.8 km
Q94	Thud Pt (S)	1	2	R+sand/rock flats	2.4 km
Spring & neap tidal range = 1.7 & 0.8 m					

False Pera Head and Thud Point are both low protrusions in the 10 m high Tertiary lateritic bluffs. To either side of both points are curves in the bluffs fronted by continuous sandy beaches and varying degrees of intertidal rock flats.

Beach **Q 88** commences 8 km south of False Pera Head and is a slightly curving, 2.8 km long, west-facing beach. It is well exposed along most of its length, resulting in minor dune activity along the southern half of the beach, with some dunes reaching 20 m in height. The beach is fronted by a low tide bar which is usually cut by rip channels skewed to the north and spaced approximately every 150-200 m. Vegetated bluffs back the northern half of the beach. The bluffs increasingly impinge on the northern end, and separate it from 500 m long beach **Q 89**. This is a curving, lower energy beach bordered by the protruding vegetated bluffs and fronting rock flats, with a combination of sand and rocks flats extending 500 m off the beach.

Beach **Q 90** trends due north from the point for 3.6 km, past a small creek mouth to a sandy foreland in lee of intertidal rocks and rock flats. Beach **Q 91** continues on the other side of the foreland for 1.3 km to **False Pera Head**, another protruding low sand foreland formed in lee of rock flats. Beach Q 90 is backed by 300 m wide partly unstable foredunes, then a linear 2.5 km long freshwater swamp which parallels the rear of the foredunes, while beach Q 91 has some minor dune activity covering the backing 5 m high bluffs. Both beaches are fronted by sand and rock flats up to 500 m wide.

Beach **Q 92** extends from False Pera Head north to the mouth of **Norman Creek**, a distance of 5.3 km. It faces due west, is backed by 10-25 m high bluffs, and fronted by a continuous low tide bar, cut by rips skewed to the north every 150-200 m (Fig. 6.17). There is a landing ground paralleling the northern half of the beach, which is connected by a straight vehicle track to North Camp. Beach **Q 93** commences on the north side of the 50 m wide mouth of Norman Creek, and curves round for 4.8 km to finally face south in lee of a rock-bound sandy foreland. It is backed by about six foredune ridges, then the salt flats of Norman Creek, which widen to 1 km behind the mouth.

Figure 6.17 The rhythmic beach face of beach Q 92, caused by the alternating rips and bars along the beach.

Beach **Q 94** is a 2.4 km long west-facing beach that continues up to Thud Point, the entire beach fronted by 500 m wide rock flats, and backed by some minor foredune activity.

Q 95-100 THUD PT-PERA HEAD-BOYD PT

No.	Beach	Rating HT LT	Type	Length
Q95	Thud Pt (N)	2 3	R+LT rips	3.2 km
Q96	Pera Hd (S2)	1 2	R+LTT/reef	1 km
Q97	Pera Hd (S1)	1 2	R+LTT	1.3 km
Q98	Pera Hd	1 2	R+LTT	300 m
Q99	Pera Hd (N)	2 3	R+LT rips	5 km
Q100	Boyd Pt (W)	1 2	R+LTT/reef	1 km
Spring & neap tidal range = 2.2 & 0.8 m				

At 10 m high **Thud Point** the bluffed shoreline turns and trends relatively straight to the north-northwest for 12 km to Boyd Point, with Pera Head located midway between the two. In between are six bluff-backed beaches (Q 95-100), with varying degrees of rock reef off some of the beaches.

Beach **Q 95** commences on the north side of low sandy crenulate Thud Point and, once past the rocks of the point, trends north-northwest for a total of 3.2 km, all the while backed by 10-15 m high scarped red bluffs, with the southernmost section capped by early Holocene yellow dune sand. The steep sandy beach face is fronted by a low tide bar cut by rips every 200 m. It terminates at a small protrusion in the bluffs, beyond which beach **Q 96** continues straight for another 1 km to the next protrusion. This marks the commencement of beach **Q 97** which continues for another 1.3 km to the south side of Pera Head. Both beaches are backed by 15 m high scarped red bluffs and fronted by a mixture of low tide bar and boundary rock reefs.

Pera Head consists of two long arms of laterite that extend a few hundred metres into the gulf, with the 300 m long beach **Q 98** curving in between. The beach is bordered by vegetated headlands, with considerable rock debris at the southern end of the beach, and a mixture of low tide bar and rock reef paralleling the shore.

Beach **Q 99** commences on the northern side of the linear rock reef and trends relatively straight for 5 km toward Boyd Point. The beach is backed by a mixture of slumped and scarped 15 m high bluffs, with a vehicle track running along the top of the bluffs. The beach is sufficiently exposed to have a low tide bar usually cut by about 20 rips. A small lagoon backs the southern end of the beach and flows out against the northern rocks of Pera Head.

Boyd Point also consists of two rock reefs extending a few hundred metres to the northwest, with 1 km long beach **Q 100** curving in between and facing northwest. The beach is moderately exposed and has a mixture of low tide bar and rock reef off the beach. It is backed by active dunes and blowouts extending up to 200 m inland and to heights of 15 m.

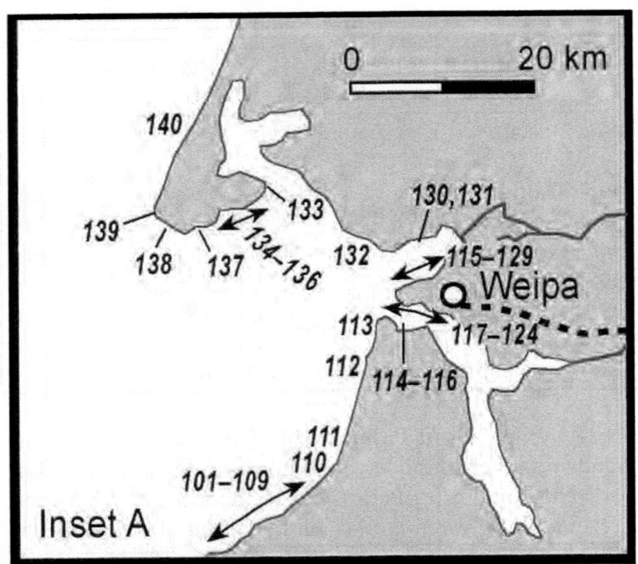

Figure 6.18 The Weipa region and beaches Q 101-140. See Figure 6.14 for regional location.

101-106 BOYD POINT (E)

No.	Beach	Rating HT LT	Type	Length
Q101	Boyd Pt (E1)	1 2	R+rocks	500 m
Q102	Boyd Pt (E2)	1 2	R+rocks	500 m
Q103	Boyd Pt (E3)	1 2	R+sand flats-LT rips	4 km
Q104	Boyd Pt (E4)	1 2	R+sand/rock flats	500 m
Q105	Boyd Pt (E5)	1 2	R+sand/rock flats	2.6 km
Q106	Boyd Pt (E6)	1 2	R+sand/rock flats	500 m
Spring & neap tidal range = 2.2 & 0.8 m				

Boyd Point protrudes 1 km north of the trend of the backing bluffed coastline. At the point the shoreline turns south for 1 km, then turns and curves round to trend toward the northeast. There are two beaches on the eastern side of the point (Q 101 & 102) and the beginning of a series of beaches that continue the run of the coast with beaches Q 103-105 occupying the next 7 km.

Beach **Q 101** is located along the first 500 m of crenulate shoreline on the eastern side of Boyd Point. It consists of a moderately steep high tide beach, fronted by protected irregular intertidal laterite rocks and flats of variable width. The dunes from beach Q 100 spill across the 200 m wide point onto the beach. Beach **Q 102** continues south along the eastern side of the point for another 500 m. It is a protected curving low energy beach, backed by 10 m high bluffs and fronted by sand flats (Fig. 6.19). The area in lee of the point offers a reasonably well protected anchorage, particularly for the southeasterlies.

Figure 6.19 The sheltered beach Q 102 located to the lee of Boyd Point.

Beach **Q 103** commences where Boyd Point links with the main trend of the shoreline, and trends initially east before curving to the northeast for a total of 4 km, to the beginning of a 3 km long series of laterite rocks and reefs. The beach is protected in the west, with wave height increasing sufficiently to the east to maintain a low tide bar and rip sequence, with up to 20 rips occupying the eastern 3 km of the beach. It is backed by scarped red laterite bluffs which reach 25 m in height along the central-northern section of the beach.

Beach **Q 104** continues the trend of beach Q 103, only it is located in lee of the 600 m wide rock reefs, resulting in low waves at the shore and no rips, while it is backed by a 300 m wide sequence of low grassy beach ridges. At the northern end of the beach the shoreline turns more to the northeast and continues as beach **Q 105** for another 2.6 km. The reefs widen to 1 km, and the entire beach is backed by a 100-200 m wide series of grassy beach ridges. Beach **Q 106** lies immediately to the east. It faces north and consists of a narrow high tide beach located along the base of 10 m high red bluffs, with a mixture of sand, rocks and laterite reef flats off the beach.

Q 107-111 BOYD POINT (E)

No.	Beach	Rating HT LT	Type	Length
Q107	Boyd Pt (E7)	1 2	R+sand/rock flats	1.6 km
Q108	Boyd Pt (E8)	1 2	R+rock flats	250 m
Q109	Boyd Pt (E9)	1 2	R+rock flats	1.4 km
Q110	Boyd Pt (E10)	1 2	R+rock flats	4.4 km
Q111	Boyd Pt (E11)	1 2	R+rock flats	5.8 km
Spring & neap tidal range = 2.2 & 0.8 m				

To the east of beach Q 106 is a 1.5 km long section of eroding 15-20 m high red bluffs, which mark the only non-sandy section of shoreline between Worbody Point, 50 km to the south, and Weipa 20 km to the north. Beach **Q 107** commences at the eastern end of the bluffs and consists of a northwest-facing, 1.6 km long beach located at the base of scarped and slumped 15-25 m high red laterite bluffs, with laterite reefs extending off either end.

Beach **Q 108** lies immediately to the north and is a curving, northwest-facing, 250 m long beach backed by wooded bluffs which decrease in height to the north, and fronted by extensive laterite reefs.

Beach **Q 109** commences to the east and is a curving 1.4 km long northwest-facing sandy beach, backed by a series of grassy beach ridges which widen to 300 m in the north, and fronted by 1 km wide laterite reefs, covered by sand flats off the beach.

Beach **Q 110** continues on from the reef-controlled point at the northern end of beach Q 109. It is a curving-crenulate, 4.4 km long, northwest-facing beach backed by a 100-200 m wide grassy beach ridge plain. It lies in lee of an irregular laterite reef that extends up to 1 km off the beach, including a straight 2 km long reef-breakwater, lying 300-400 m offshore that parallels the northern end of the beach. Beach **Q 111** commences on the northern side of the inflection induced by the breakwater. It continues north-northeast for 5.8 km to the next reef-induced inflection. It is backed by a mixture of patchy beach ridges up to 100 m wide and some areas of low red bluffs, with the southern arm of Triluck Creek lying a few hundred metres to the east. It is fronted by sand flats, apart from a central reef section.

Q 112-115 WOOLDRUM POINT

No.	Beach	Rating HT	Rating LT	Type	Length
Q112	Trillick Ck (S1)	1	2	R+sand flats	12.5 km
Q113	Wooldrum Pt (S)	1	2	R+tidal flats	500 m
Q114	Wooldrum Pt (E)	2	3	R+tidal channel	1.7 km
Q115	Urquhart Pt (E)	1	2	R+tidal flats	1.5 km

Spring & neap tidal range = 2.2 & 0.8 m

Wooldrum and Urquhart points mark the southern entrance to the Embley River, on whose northern banks is the town and port of Weipa. The two points are part of a 12 km long Holocene barrier-spit that commences at the base of beach Q 112 and includes beaches Q 113-115.

Beach **Q 112** is a 12.5 km long north-trending sandy beach that commences at a reef-induced inflection in the shore where it is backed by low wooded bluffs. Five hundred metres north of the point Trillick Creek is deflected and parallels the back of the beach to its mouth 12 km to the north (Fig. 6.20). The mangrove-lined creek runs 50-200 m in behind the beach, with low grassy beach ridges and the mangroves in between. An earlier inner barrier spit lies to the east, and is backed by the north-trending Leithen Creek. The beach consists of a moderately steep beach face fronted by low gradient sand flats extending a few hundred metres off the beach, and widening to several hundred metres toward the creek mouth.

Figure 6.20 The deflected mouth of Trillick Creek marks the northern end of beach Q 112.

Beach **Q 113** commences on the northern side of the 50 m wide Triluck Creek mouth and trends north for 500 m to the low sandy Wooldrum Point. It is bordered by tidal channels and fronted by tidal sand shoals extending up to 1 km off the beach. At the point the shoreline turns and trends northeast for 1.7 km to sandy Urquhart Point as beach **Q 114**, which has beachrock exposed along the shore. This beach faces north across the South Channel of Embley River, whose deep waters and strong tidal currents parallel the beach. Both beaches are backed by a series of 10-15 recurved beach ridge-spits, which curve round into the river entrance.

Beach **Q 115** commences at Urquhart Point where it faces into the South Channel, with channel markers located off the beach. It then trends southeast then south into the Embley River mouth. It runs along the truncated end of the ridges, in places exposing beachrock. It is a narrow, very low wave energy, east-facing beach, fronted by tidal flats that widen to 500 m to the south. On its southern side it is bordered by the tidal channel and flats of Leithen Creek.

Q 116-121 WEIPA (S)

No.	Beach	Rating HT	Rating LT	Type	Length
Q116	Franjum Pt	1	2	R+tidal flats	800 m
Q117	Weipa South	1	2	R+tidal flats	600 m
Q118	Napranum	1	2	R+tidal flats	1.6 km
Q119	Lorim Pt	1	2	R+tidal flats	500 m
Q120	Evans Landing	1	2	R+tidal flats	300 m
Q121	Gonbung Pt (W)	1	2	R+tidal flats	400 m

Spring & neap tidal range = 2.2 & 0.8 m

Weipa is located on a 10-20 m high relatively flat laterite peninsula bordered by the Mission River to the north and the Embley River to the south. The entire shoreline of the peninsula is low energy and is dominated by mangroves along the river channels with several low energy beaches (Q 117-129) in amongst the mangroves and on the more exposed western shores that face into Albatross Bay. The port facilities are located along the southern Embley shoreline.

Beach **Q 116** is located opposite Weipa South, at Franjum Point on the southern side of the Embley River. The 800 m long, low, narrow beach faces northeast across the 3 km wide channel to Weipa. The beach represents the innermost wave-deposited beach ridges on the southern river banks. They have built a 100-300 m wide low beach ridge series, backed by 20 ha salt flats and fronted by 500 m wide tidal flats. A small creek from the salt flats drains across the ridges. A small community is located at the southern end of the beach.

Beach **Q 117** lies 3 km northeast of Franjum Point and represents the easternmost wave-deposited ridges on the northern side of the river channel. The 600 m long beach commences at cuspate Jessica Point and curves round to face southwest across the river channel. It is fronted by tidal flats that narrow towards the point, and is backed by a stand of coconut trees then the houses of the large **Napranum** community (Fig. 6.21). There is a boat ramp at the tip of Jessica Point, with beach **Q 118** commencing on its northern side. This 1.6 km long beach faces towards the river mouth 10 km to the west. It is fronted by 200-300 m wide tidal flats and backed by a reserve, then the community's oval in the south and houses in the north. A small creek flows across the northern half of the beach and is fronted by a shallow tidal delta.

Figure 6.21 The Napranum community and beach Q 117.

Beach **Q 119** is located adjacent to the main Weipa port loading facility and has been modified by the port development. It consists of a curving 500 m long southwest-facing, narrow high tide beach, fronted by 300 m wide intertidal flats, and bordered and backed by port facilities. Beach **Q 120** is located 1 km to the west at Evans Landing. It is a 300 m long, narrow, curving, south-facing beach fronted by 200 m wide tidal flats, with boating facilities located along the shore to the west of the beach.

Beach **Q 121** lies to the west of Evans Landing and is a low, 400 m long, 100 m wide, crenulate sandy spit tied to the laterite reefs of Gonbung Point. Navigation lights are located on the point which marks the entrance to Weipa harbour. Strong tidal currents sweep past the point.

Q 122-129 **WEIPA (N)**

No.	Beach	Rating HT	Rating LT	Type	Length
Q122	Gonbung Pt (W)	1	2	R+tidal flats	300 m
Q123	Kerr Pt (S)	1	2	R+tidal flats	500 m
Q124	Kerr Pt (N)	1	2	R+tidal flats	50 m
Q125	Nghanambonna Pt (S)	1	2	R+tidal flats	50 m
Q126	Nanum Beach (S)	1	2	R+tidal flats	900 m
Q127	Nanum Beach (N)	1	2	R+tidal flats	100 m
Q128	Kumrunja Beach	1	2	R+tidal flats	1 km
Q129	Rocky Pt	1	2	R+tidal flats	400 m
Spring & neap tidal range = 2.2 & 0.8 m					

Gonbung Point marks the southern entrance to Weipa harbour, with the shoreline trending to the northwest for 3 km to Nghanambonna Point, then northeast for 8 km to Awonga Point, the mangrove-lined entrance to Mission River. The entire shore between the three points is very low energy and dominated by mangroves together with eight very low energy beaches and associated tidal flats (Q 122-129). All beaches are accessible by vehicle from a road which parallels the shoreline, while the main center of Weipa lies immediately south of beaches Q 126-129. Beach **Q 122** is a 300 m long southwest-facing strip of high tide sand, bordered by the laterite reefs of Gonbung Point to the south, and mangroves to the north. It is fronted by 1 km wide tidal flats, with trees and then the road 200 m behind the beach. Beach **Q 123** lies a few hundred metres to the northwest and is a slightly curving 500 m long low beach, backed by scattered trees, then the road, and fronted by 600 wide sand flats. It is bordered by mangroves to either end, with the northern mangroves continuing west for 500 m to Kerr Point.

At **Kerr Point** the shoreline turns and trends northeast for 500 m to beach **Q 124**. This is a 50 m long pocket of sand bordered by mangroves and laterite rocks and fronted by 500 m wide ridged sand flats. The mangroves then continue east past Nghanambonna Point for 2 km to beach **Q 125**, a 50 m long pocket of high tide sand bordered by mangroves. It is fronted by low energy sand and rock flats.

Nanum Beach is the first of the two main Weipa beaches and is located 1.5 km west of the town centre. The beach consists of two parts, the southern 900 m long beach **Q 126,** and the northern 100 m long beach **Q 127,** with the small mouth of Trunding Creek forming the boundary in between. Both beaches are narrow, with melaleuca trees growing right to the high tide line, and are fronted by

500 m wide tidal flats with scattered rocks (Fig. 6.22). The main beach has a caravan park in the trees at the southern end, a vehicle track paralleling the remainder of the beach, and a few houses behind the northern beach.

Figure 6.22 The low energy Nanum Beach (Q 126) is fronted by wide sand flats.

Kumrunja Beach (**Q 128**) lies to the east ands abuts Weipa town centre. It is a very low energy 1 km long northwest-facing high tide beach, fronted by 300 m wide tidal flats that narrow to the east. There is a reserve behind the southern half of the beach, with a small creek crossing at the northern end, then houses and town facilities along the northern half. It terminates at Rocky Point where a boat ramp is located. Beach **Q 129** is located on the eastern side of the point and continues east for 400 m as a very low energy strip of high tide sand and tidal flats which widen from 100 m at the point to 500 m to the east.

Q 130-132 **RED BEACH**

No.	Beach	Rating HT	LT	Type	Length
Q130	Red Beach (E)	2	3	R+tidalchannel	500 m
Q131	Red Beach (W)	1	2	R+tidal flats	350 m
Q132	Andoomajettie Pt(W)	1	2	R+ridged sand flats	1.6 km
Spring & neap tidal range = 2.2 & 0.8 m					

The Weipa foreshore faces north across the 4-5 km wide funnel-shaped entrance to Mission River to the northern shore where Red Beach and Andoomajettie Point are located. While both beaches are accessible by vehicle, the adjoining shoreline is dominated by mangroves and undeveloped, with the backing hinterland an area of extensive open cut bauxite mining.

Red Beach is a low cuspate foreland located 3 km due north of Weipa's Rocky Point. The beach has two parts. The east-facing beach (**Q 130**) is 500 m long and fronted by narrow sand flats, then the deep tidal channel of Andoom Creek and then more tidal flats. It links at the foreland with the southwest-facing western side (beach **Q 131**), a 350 m long straight section of beach fronted by 1 km wide tidal flats. Both beaches consist of red sands and grade into mangroves at their base. There is vehicle access to both beaches.

Andoomajettie Point lies 5 km to the east, with beach **Q 132** commencing in lee of the mangrove-covered point and curving to the northwest for 1.6 km, to mangroves at the western end. The beach is backed by a 400 m wide series of up to eight grassy beach ridges, and fronted by 500 m wide ridged sand flats. Mangroves dominate the shoreline for the next 15 km to the entrance to Pine River Bay. There is vehicle access across the ridges to the centre of the beach.

Q 133-139 **DUYFKEN-LANDFALL POINTS**

No.	Beach	Rating HT	LT	Type	Length
Q133	Bagley Channel (1)	1	2	R+tidal sand flats	2.4 km
Q134	Bagley Channel (2	1	2	R+tidal sand flats	700 m
Q135	Bagley Channel (3)	1	2	R+tidal sand flats	900 m
Q136	Landfall Pt	1	2	R+ridged sand flats	6 km
Q137	Hatchman Pt	1	2	R+ridged sand flats	2.5 km
Q138	Duyfken Pt	1	2	R+ridged sand flats	3 km
Q139	Jantz Pt	1	2	R+LTT	1.6 km
Spring & neap tidal range = 2.2 & 0.8 m					

The northern entrance to Albatross Bay is marked by Jantz Point, at which point the low sandy shoreline turns and trends west into the bay for 12 km to the **Bagley Channel** in Pine River Bay. Between the point and channel entrance are seven generally south-facing beaches (Q 133-139), along which considerable sediment has been transported eastward into the bay and the channel. There is no vehicle access to the beaches.

Beach **Q 133** is the westernmost and lowest energy of the seven beaches and forms the western side of Bagley Channel. It is the terminal part of a 7 km long series of large recurved spits that are attached at Landfall Point and are the product of sand transport into the channel. The entire system contains multiple spits and recurves and totals about 1000 ha in area. The beach is 2.4 km long, and faces east across 100-500 m wide sand flats then the 3-4 km wide channel. Sand has been transported along the beach by both refracted southerly waves and flood tidal currents. Beaches Q 134 and 135 are part of an active recurved spit moving more sand into the channel. Beach **Q 134** is a 700 m long recurve that has all but enclosed a 15 ha lagoon between it and the former southern shore of beach Q 133 (Fig. 6.23). Beach **Q 135** is the 900 m long spit along which sediment is moving toward the recurve. The entire system is active and is expected to change considerably in size, shape and location over periods of years to decades. In 1998 the western end of beach Q 135 was eroding, with casuarina trees falling onto the beach. A navigation tower is located at the southern end of the beach.

Beach **Q 136** forms the base of the spit. It commences at **Landfall Point**, a sandy point tied to laterite reefs and curving to the north, then east for 6 km. It is exposed to southerly waves which maintain a series of migratory sand ridges, which narrow to the east in lee of offshore laterite reefs. The spit is narrow in the west where it is backed by a 4 km wide Crawford Creek mangrove forest, and widens to the east as the older recurved spit splays northward into the former channel.

Figure 6.23 Beach Q 134 is a dynamic recurved spit that has enclosed a small lagoon.

Beach **Q 137** is a curving 2.5 km long south to southwest-facing beach located between Landfall and Hatchman points. Both points are capped by 20 m high vegetated dunes and are tied to laterite reefs. The beach receives southerly waves and has a series of migratory sand ridges in lee of **Hatchman Point** grading east, as wave energy increases, to a low tide bar.

Beach **Q 138** commences at Hatchman Point and curves slightly west for 3 km to **Duyfken Point**. It faces south-southwest into the southerly waves and also has the migratory sand ridges in the west grading to a low tide bar to the east, in response to increasing wave energy. Beach **Q 139** extends from Duyfken Point for 1.6 km to **Jantz Point**. It faces southwest into the gulf, but is afforded some protection from the laterite reef that extends 1 km south of Jantz Point. The beach has a moderate slope and a continuous low tide bar, with the rock reefs off the points. It is backed by a mixture of vegetated dunes and laterite bluffs, with a navigation tower located on the 20 m high eastern bluffs.

Jantz-Vrilya Points

The 220 km of coast between the laterite bluffs of Jantz and Vrilya points are dominated by nine relatively long (5-35 km) Holocene barrier systems containing 28 beaches (Q 140-167) separated by the Pennefather, Ducie, Skardon, Jackson, MacDonald, Doughboy and Cotterell rivers, as well as a few creeks, and all backed by extensive interbarrier depression and inner Pleistocene barrier systems. The whole system forms a 200 km long Quaternary coastal plain with an area of over 20 000 ha.

Q 140-141 **JANTZ PT-PENNEFATHER RIVER**

No.	Beach	Rating HT LT	Type	Length
Q140	Jantz Pt (N1)	2 3	R+LT rips	35.9 km
Q141	Jantz Pt (N2)	2 3	R+LT rips	3.7 km
Spring & neap tidal range = 2.4 & 0.8 m				

At **Jantz Point** the shoreline turns and trends north-northeast for 40 km to the mouth of the Pennefather River. The relatively straight beach in between consists of two systems, the long beach Q 140, and beach Q 141 which terminates at the river mouth. Both beaches are well exposed to the highest gulf waves and are dominated by rip-controlled beach morphology. There is vehicle access from the Mapoon Road to the northern ends of both beaches.

Beach **Q 140** is a 36 km long, relatively straight, continuous sand beach. It is backed by a 0.5-1 km wide series of Holocene foredunes, then the extensive wetlands of Nomenade Creek. The beach has a moderate to steep slope and is well exposed to the gulf waves, with periodic high waves maintaining a series of well developed rhythmic low tide bars and rips (Fig. 6.24), and backing highly crenulate shoreline topography (Fig. 6.25), with the bars and rips spaced about every 200 m. This results in up to 180 bar-rip systems located along the beach.

The beach terminates in lee of a 3 km long section of laterite reefs where the beach protrudes a few hundred metres to the west. Beyond the foreland beach **Q 141** continues on for 3.7 km to the dynamic mouth of the **Pennefather River**. Toward the river mouth the foredunes splay into the mouth, widening to 1.5 km. The beach is backed by partially active foredune, through which vehicle tracks run up to the river mouth. Rhythmic bars and rips parallel the beach, widening into 500 m wide river mouth shoals adjacent to the river. At times a low recurved sand spit extends from the northern tip of the beach into the river mouth.

Q 142-143 **PENNEFATHER RIVER MOUTH**

No.	Beach	Rating HT LT	Type	Length
Q142	Pennefather R (S)	2 3	R+tidal channe	2 km
Q143	Pennefather R (N)	2 3	R+tidal channel	1.2 km
Spring & neap tidal range = 2.4 & 0.8 m				

The **Pennefather River** occupies a drowned valley that extends 10 km inland and is fed by four creeks. The shoreline is dominated by mangroves, while tidal sand shoals occupy much of the bay. Beaches Q 142 and 143 are located either side of the 300 m wide 1.5 km long entrance to the river.

Figure 6.24 Well developed rhythmic bar and beach topography along beach Q 140. When active rips flow out of each embayment and across the bar.

Figure 6.25 The highly rhythmic shoreline associated with the bars and rips along beach Q 140.

Beach **Q 142** extends east for 2 km inside the southern entrance to the river. It is a narrow, low wave energy, north-facing strip of high tide sand, backed by the truncated end of the foredune ridges. Periodically a sand spit extends from the southern entrance of the river along the outer section of the beach, enclosing it behind a small lagoon. Most of the beach faces across the deep tidal channel to beach **Q 143** on the northern side of the entrance. This is a 1.2 km long beach that runs along the inner side of the spit that forms the southern end of beach Q 144.

Q 144-146 **FLINDERS & SAMMY'S CAMPS**

No.	Beach	Rating HT LT	Type	Length
Q144	Flinders Camp	2 3	R+LT rips	23.6 km
Q145	Janie Ck	2 3	R+tidal channel	900 m
Q146	Sammy's Camp	2 3	R+LT rips	12.8 km
Spring & neap tidal range = 2.4 & 0.8 m				

On the northern side of the Pennefather River the shoreline trends straight for 24 km to Janie Creek, then gradually curves round to the east for another 13 km to Cullen Point, the southern entrance to Port Musgrave. Beach **Q 144** commences at the Pennefather River mouth as a sandy spit which is deflected by the river mouth shoals for 500 m to the southwest. It then trends straight to Janie Creek where it also terminates in a slightly curving sand spit. In between is a 23.6 km long beach, for most of its length facing the west-northwest and backed by a 200-500 m wide foredune system up to 23 m high, then a 1 km wide interbarrier depression which grades from a freshwater swamp in the south to salt flats then mangroves in the north, all of which are backed by a 1-2 km wide, 10-20 m high Pleistocene barrier. The beach system is well exposed and dominated by a highly rhythmic shoreline with transverse bars and rips spaced about every 200 m, resulting in up to 100 rips along the beach. Flinders Camp is located in the foredunes toward the southern end of the beach and is accessible by 4WD.

Janie Creek has been deflected 4 km to the north by the northern end of beach Q 144. Beach **Q 145** is located in the 100 m wide creek mouth and runs along in lee of beach Q 144 for 900 m to the northern side of the creek and the beginning of beach Q 146. It is a low wave energy beach fronted by the deep tidal channel, then sand flats and the spit and creek mouth shoals which extend up to 500 m off the beach.

Beach **Q 146** commences on the north side of the creek and trends northwest and finally west for 12.8 km, terminating at **Cullen Point**, the southern entrance to the large Port Musgrave. The beach is backed by low foredunes, an interbarrier depression and inner Pleistocene barrier totalling up to 4 km in width, but narrowing toward the point. There is a landing ground at the point, and vehicle tracks the length of the beach down to Janie Creek. Sammy's Camp is located in the foredune behind the centre of the beach. The beach itself is open to waves and dominated by low tide rips for the first 4 km north of the creek mouth. It then runs in lee of laterite reefs that fringe the beach in places and extend up to 1 km offshore. As a consequence the beach receives low waves along the 3 km long north-facing point section, where patches of beachrock are exposed along the tip of the point.

Q 147-149 **PORT MUSGRAVE**

No.	Beach	Rating HT LT	Type	Length
Q147	Cullen Pt (S)	1 2	R+sand flats	4 km
Q148	Red Beach	1 2	R+tidal flats	800 m
Q149	Namaleta Ck	1 2	R+mud flats	900 m
Spring & neap tidal range = 2.7 & 0.5 m				

Port Musgrave is a 10 000 ha embayment fed by the Wenlock and Ducie rivers. It has 50 km of shoreline largely dominated by mangroves, including some very extensive forests along the Wenlock River. The mangroves are fronted by intertidal sand and mud flats, in places 1-2 km wide, together with tidal shoals in the bay, all of which have considerably filled and shoaled the bay floor. The Mapoon community is located just to the north of the Wenlock River mouth and is the only settlement in the area. There are three beaches(Q 147-149) located toward the outer entrance to the port.

Beach **Q 147** commences at **Cullen Point**, the southern entrance to the port, and curves to the southwest for 4 km. It faces southeast across the 15 km wide bay and receives sufficiently low wind waves to have prograded 500 m into the bay as a series of low wooded beach ridges fronted by 200-300 m wide tidal flats. There are a few houses scattered along the back of the beach all connected by a vehicle track from Mapoon.

Red Beach (**Q 148**) is located at **Mapoon**. It is a narrow 800 m long northeast-facing sand beach, fronted by 500 m wide sand to mud tidal flats, with mangroves fringing either end. Beachfront houses are spread along the beach.

Beach **Q 149** is located on the north side of the port immediately to the west of Namaleta Creek mouth. It is a 900 m long southeast-facing strip of high tide sand, backed by a 30 ha series of cheniers. The beach is fronted by 2 km wide mud flats, and grades into mangroves to the east.

Q 150-152 **SKARDON RIVER (S)**

No.	Beach	Rating HT LT	Type	Length
Q150	Skardon R (S2)	2 3	R+LT rips	23 km
Q151	Skardon R (S1)	1 2	R+sand flats	4.5 km
Q152	Skardon R	1 2	R+tidal flats	800 m
Spring & neap tidal range = 2.7 & 0.5 m				

The **Skardon River** occupies a meandering 15 km long drowned valley fed by the small river and several creeks. Most of the shoreline is bordered by salt flats and mangroves, all flushed by the deep 300 m wide river mouth. Beaches Q 150-152 are part of a continuous 2500 ha barrier that extends 25 km north to the river mouth.

Beach **Q 150** commences at the mouth of Namaleta Creek and trends north for 23 km to the beginning of the Skardon River mouth shoals, where it beings to curve northeast toward the river mouth. The beach is backed by a 0.5-2.5 km wide series of up to 20 foredune ridges which terminate at each end in recurved spits at the creek and river mouth. The outer barrier is backed by an interbarrier depression drained by Namaleta Creek to the south and a tributary of the Skardon to the north, with inner Pleistocene ridges backing the northern section of the barrier. There is vehicle access across the barrier to the centre of the beach. The beach is well exposed to gulf waves for most of its length, with a finer sand resulting in a low gradient beach fronted by a low tide bar, usually attached to the beach, but at times cut by rips every 200 m.

Beach **Q 151** is a continuation of the beach in lee of the Skardon River tidal shoals. It curves round to the northeast then east for 4.5 km, terminating at the river mouth (Fig. 6.26). It is backed by a continuation of the outer and inner barrier, and fronted by initially 1 km wide tidal sand shoals, which narrow and give way to the deep tidal channel at the mouth. Beach **Q 152** commences immediately inside the river mouth and trends to the southeast for 800 m, terminating at mangroves. It is initially fronted by the tidal channel, then tidal flats which widen to 300 m at the mangroves.

Figure 6.26 The Skardon River mouth with beaches Q 151 & 152 to either side.

Q 153-154 **SKARDON-JACKSON RIVERS**

No.	Beach	Rating HT LT	Type	Length
Q153	Skardon-Jackson R	2 3	R+LT rips	10 km
Q154	Jackson R	2 3	R+tidal channel	1.3 km
Spring & neap tidal range = 2.7 & 0.5 m				

Beach **Q 153** is located between the mouths of the **Skardon** and **Jackson rivers**. At each end the beach curves into the river mouths in lee of extensive tidal sand shoals that extend up to 1 km into the gulf. In between, the relatively straight 10 km long beach faces west-northwest and receives relatively high gulf waves, which maintain a moderately steep beach fronted by a low tide bar, cut by rips every 200 m during and following periods of higher waves. The beach is backed by series of foredunes which widen from 200 m in the south to 500 m in the north, and which recurve into both river mouths. The outer barrier is backed by a 1 km wide interbarrier depression and a 1-4 km wide inner Pleistocene barrier that continues for 7 km to the north on the eastern side of the Jackson River channel, and in places extends up to 7 km in from the present coast.

Beach **Q 154** is a continuation of the northern spit that follows the 300 m wide tidal channel of the Jackson River for 1.3 km. It curves round to face northwest across the channel to the southern end of beach Q 155. While it receives very low waves it is paralleled by the deep channel and swept by the strong tidal currents.

Q 155-156 JACKSON-MACDONALD RIVERS

No.	Beach	Rating HT	Rating LT	Type	Length
Q155	Jackson-MacDonald R	2	3	R+LT rips	15.5 km
Q156	MacDonald R	1	3	R+sand flats	2.2 km
Spring & neap tidal range = 2.7 & 0.5 m					

Beach **Q 155** is a relatively straight north-northwest-trending beach that extends for 15.5 km between the Jackson and MacDonald rivers. The beach is backed by a 1 km wide series of low foredunes in the north that terminate in a low 2 km long sand spit at the river mouth, while recurved spits form the southern 9 km of beach where they have deflected the Jackson River. The inner Pleistocene barrier lies to the east of the river and extends up to 6 km inland. Away from the terminal spits the beach has a moderate slope, is well exposed to gulf waves and is dominated by a low tide bar cut by rips every 150-200 m. Low active foredunes extend up to 100 m in from the beach.

At the **MacDonald River** the sand spit is fronted by tidal sand shoals extending up to 1 km seaward and cut by the 300 m wide tidal channel. The funnel-shaped river is fed by three small creeks and extends 7 km inland. Beach **Q 156** is located on the southern shore inside the river mouth and curves around to face northwest. The 2.2 km long beach is well protected from waves, fronted by tidal sand flats and paralleled by strong tidal currents. There is no vehicle access to either beach.

Q 157-15 9 MACDONALD-DOUGHBOY RIVERS

No.	Beach	Rating HT	Rating LT	Type	Length
Q157	MacDonald R (N)	2	3	R+tidal channel	1.5 km
Q158	MacDonald-DoughboyR	2	3	R+LT rips	7.5 km
Q159	Doughboy R	2	3	R+tidal channel	1.8 km
Spring & neap tidal range = 2.9 & 0.5 m					

The MacDonald and Doughboy rivers are separated by 7.5 km long beach Q 158, which interlinks in the backing mangrove-dominated wetlands. Both rivers have extensive river mouth shoals which extend 1-2 km into the gulf as well as beaches located inside their mouths (beaches Q 157-159).

Beach **Q 157** is a curving 1.5 km long partly vegetated sand spit that extends from the 500 m wide river channel northwards in lee of the deflected channel and river mouth shoals, to the open coast where beach Q 158 commences. While the beach receives low waves it is swept by the strong tidal currents (Fig. 6.27).

Figure 6.27 The MacDonald River mouth, with the elongate spit of beach Q 157 to the left.

Beach **Q 158** commences as the shoreline straightens and trends straight for 7.5 km to the mouth of the Doughboy River. The river mouth shoals dominate each end of the beach, with only the central 5.5 km dominated by the gulf waves. This is characterised by a relatively steep beach face fronted by a continuous low tide bar cut by rip channels every 200 m. The beach is backed by some moderately active foredunes, then a 1-1.5 km wide series of up to 12 foredunes, then a 2 km wide interbarrier depression dominated by mangrove forests, with an inner Pleistocene barrier widening from 3 km on the north to 5 km in the south, which is in turn backed by a 1-2 km wide swampy backbarrier depression. Because of the extensive mangrove swamps there is no vehicle access to the beaches.

Beach **Q 159** is located inside the **Doughboy River**. It curves around the southern shoreline for 1.8 km and is backed by some of the inner foredune ridges. While the beach faces west it is fronted by the river channel and extensive sand shoals, receiving low waves, but swept by strong tidal currents. Scattered mangroves are located along the sand flats. The river has been deflected to the south for 4 km by beach Q 161, and is fed by the two low gradient swampy creeks.

Q 160-163 DOUGHBOY-COTTERELL RIVERS

No.	Beach	Rating HT	Rating LT	Type	Length
Q160	Doughboy R (N)	2	3	R+tidal channel	600 m
Q161	Doughboy-Cotterell R	2	3	R+LT rips	10 km
Q162	Cotterell R mouth	2	3	R+tidal channel	1.7 km
Q163	Cotterell R (N)	2	3	R+LT rips	6 km
Spring & neap tidal range = 2.9 & 0.5 m					

To the north of the Doughboy River is another straight 10 km long barrier system extending up to the Cotterell River, then a barrier continuing on the north side for another 6 km to the first bedrock shoreline since Duyfken Point, 215 km to the south. Beaches Q 161 and 163 form the two barriers, while beaches Q 160 and 162 occupy the river mouths.

Beach **Q 160** is located on the north side of the Doughboy River inside the sand spit that forms the northern entrance. It faces southeast across the 400 m wide river to beach Q 159. It is a low narrow beach, backed by truncated foredunes and paralleled by the deep river channel.

Beach **Q 161** commences at the Doughboy River mouth and trends north-northwest for 10 km to the Cotterell River mouth. River mouth sand shoals dominate the southern 2 km and northern 1 km of beach, with an exposed beach in between. The central section has a moderately steep slope fronted by a low tide ridge and runnel, which is cut by rips during and following periods of higher waves. The beach is backed by a 700 m wide series of foredunes, then an earlier series of recurved spits that migrated south the length of the barrier, deflecting the Doughboy River 5 km to the south. A 2 km wide mangrove wetland backs the barrier, with the Doughboy and Cotterell rivers interlinked by a tidal creek. This is then backed by a low, inner, 2 km wide Pleistocene inner barrier then an old Pleistocene shoreline consisting of scarped laterite bluffs.

Beach **Q 162** forms the northern spit of the Cotterell River. It runs south-southeast for 1.7 km in lee of the river channel and 500 m wide tidal shoals. It is a narrow, low wave energy beach swept by strong tidal currents, and backed by truncated foredunes.

Beach **Q 163** commences as the tidal shoals give way to a straight 6 km long north-trending beach, that terminates at the first laterite bluffs since Jantz Point. It is fronted by a 200 m wide low tide ridge and runnel cut by occasional rips, and backed by a 1 km wide Holocene barrier, then the northern end of a 1.5 km wide mangrove system, drained by the meandering river, and finally the inner Pleistocene barrier and older laterite shoreline.

Q 164-169 VRILYA POINT

No.	Beach	Rating HT	Rating LT	Type	Length
Q164	Vrilya Pt (S3)	2	3	R+LTT/rocks	6.5 km
Q165	Vrilya Pt (S2)	2	3	R+LTT/rocks	600 m
Q166	Vrilya Pt (S1)	2	3	R+sand/rock flats	250 m
Q167	Vrilya Pt	2	3	R+rock flats	1.2 km
Q168	Vrilya Pt (N1)	2	3	R+rock flats	200 m
Q169	Vrilya Pt (N2)	2	3	R+sand/rock flats	800 m
Spring & neap tidal range = 2.9 & 0.5 m					

Vrilya Point is the most prominent bedrock headland between Boyd Point 200 km to the south and Mutee Head 40 km to the north. The 20-30 m high 2 km long laterite point protrudes 1.5 km west of the general trend of the coast and is surrounded by laterite rocks and reefs, as well as several rock-controlled beaches (Q 164-169). There is vehicle access to the point and some of the adjacent beaches.

Beach **Q 164** commences 8 km south of the point as the first laterite rocks and reefs begin to outcrop at the shore and mark the boundary with beach Q 163. The beach continues due north for 6.5 km to a more substantial reef outcrop and cuspate foreland that separates it from beach Q 165. The beach is backed by partially active foredunes that have blanketed the backing 10 m high laterite bluffs, while it is fronted by a continuous low tide bar, together with scattered rocks and reefs.

Beach **Q 165** is a 600 m long, slightly curving, west-facing beach, backed by partly vegetated dunes extending up to 200 m inland to the laterite coastal plain. It is fronted by a low tide terrace, bordered by the reefs to the south and the westward protrusion of Vrilya Point to the north. Beach **Q 166** lies immediately to the west on the southern side of the point. The 250 m long beach faces southwest, is bordered by laterite points and reefs, and backed by densely vegetated laterite bluffs. It is fronted by sand flats and scattered rock reefs.

Beach **Q 167** occupies the western side of the point. The 1.2 km long beach faces due west and is bordered by rocks and rock reefs to the south and the 15 km high wooded point to the north, with rocks and reefs dominating much of the intertidal zone in between. It is backed by a mixture of scarped and vegetated 10 m high laterite bluffs.

Vrilya Point consists of a wooded 3 ha 'island', connected by a 100 m long tombolo and rock flats to the mainland bluffs. At the point the shoreline turns to the east, with beach **Q 168** tucked in on the eastern side of the point. It is a protected 200 m long northeast-facing narrow beach with sand flats and intertidal laterite reefs scattered off the beach. It is bordered and backed to the east by densely vegetated 15 m high laterite bluffs. The bluffs continue to the east for 1 km to the beginning of beach **Q 169**, an 800 m long west-facing beach with a prominent mangrove-topped reef off the northern section. This beach is backed by a foredune ridge plain that widens to 200 m in the north, then vegetated bluffs that rise to 30 m. A mixture of sand and rock flats front the beach.

Q 170-174 **CRYSTAL CREEK**

No.	Beach	Rating HT LT	Type	Length
Q170	Vrilya Pt (N3)	1 2	R+sand flats	10 m
Q171	Crystal Ck (S2)	1 2	R+sand flats	1 km
Q172	Crystal Ck (S1)	1 2	R+sand flats	2 km
Q173	Crystal Ck (N1)	1 2	R+sand flats	2.5 km
Q174	Crystal Ck (N2)	1 2	R+sand flats	9.3 km
Spring & neap tidal range = 2.9 & 0.5 m				

Between Vrilya and Slade points is a 25 km section of gently curving sandy shoreline, backed by discontinuous Holocene and a backing Pleistocene barrier, the two barriers separated by a 20 km long mangrove swamp, drained in the centre by Crystal Creek.

Beach **Q 170** commences at the northern rock reefs of Vrilya Point, adjacent to beach Q 169, and curves slightly as it trends almost due north for 10 km to a 500 m wide creek mouth. The beach has a moderate gradient and is fronted by 400 m wide intertidal sand flats. It is backed by low foredunes that extend 300-600 m inland, then a 1-2 km wide mangrove-filled interbarrier depression, and finally a 1 km wide Pleistocene barrier and backing 2 km wide swamp-filled backbarrier depression. There is vehicle access to the southern bedrock-connected end of the beach, and in 1994 a wrecked fishing boat was stranded toward the northern end of the beach.

Beaches **Q 171** and **Q 172** are located between the creek mouth and Crystal Creek 3 km to the north. The beaches are 1 km and 2 km long respectively, but are subject to considerable change owing to the migration of the creek mouth, tidal channels and flats, the latter extending up to 2 km off each beach. Beach Q 171 in particular is a low curving tide-dominated beach and prone to overwashing (Fig. 6.28). Beach Q 172 is eroding into earlier Holocene foredunes as Crystal Creek migrates to the south. The foredunes are in turn backed by the mangroves, then the Pleistocene barrier.

Beach **Q 173** commences on the north side of **Crystal Creek** and curves to the north for 2.5 km, all the while in lee of the tidal shoals, which are 2 km wide off the creek mouth and narrow to the north. The beach consists of a 0.5-1 km series of splaying foredune ridges, backed by earlier southward-migrating recurved spits, which deflect the creek mouth to the south. The creek drains a 4 km wide mangrove swamp, with the Pleistocene barrier extending 4-6 km to the east.

Figure 6.28 The low beach Q 171 (right) and the eroding beach Q 172 (left) are located at the mouth of Crystal Creek, with tidal flats extending 2 km into the gulf.

Beach **Q 174** continues north of the tidal shoals for 9.3 km as a curving barrier system that terminates at low sandy **Slade Point**, in lee of Crab Island. The beach is fronted by sand flats, which widen from 300 m in the south to several hundred metres toward the point. The entire system is backed by a complex series of multiple beach-foredune ridges and inner recurved spits, which widen to 2.5 km to the north. An interbarrier depression is occupied by a freshwater swamp in the north grading into the mangroves of Crystal Creek to the south, then the Pleistocene barrier to the east, the entire system up to 9 km wide.

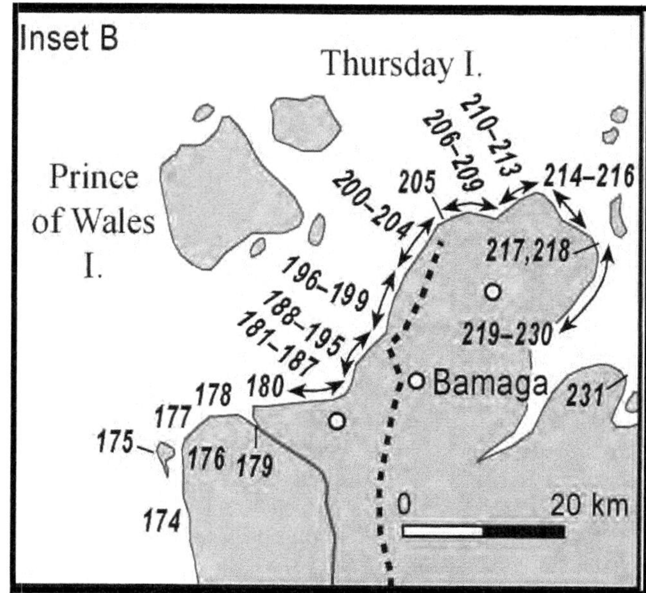

Figure 6.29 The tip of Cape York showing location of beaches Q 174-231. See Figure 6.14 for regional location.

Q 175-178 **SLADE-VAN SPEULT POINTS**

No.	Beach	Rating HT LT	Type	Length
Q175	Crab Island	1 2	R+sand flats	6.5 km
Q176	Slade Pt (N1)	1 2	R+sand flats	2.5 km
Q177	Slade Pt (N2)	1 2	R+sand flats	1.7 km
Q178	Van Spoult Pt	1 2	R+sand flats	9.2 km
Spring & neap tidal range = 2.9 & 0.5 m				

At Slade Point the shoreline begins to curve round to the northeast then east, terminating 12 km to the northeast at Van Speult Point, at the mouth of the Jardine River. Extensive tidal sand shoals lie off this section of coast, the shoals part of the massive flood tide delta of Torres Strait. Part of the shoals are manifest as sandy Crab Island which is located adjacent to Slade Point. The massive amounts of sand that have been deposited by the tidal currents and delivered by the Jardine River in this area have been reworked onshore to form the 10 000 ha Holocene and Pleistocene barrier systems that back the point. The coast and hinterland between Slade and Van Speult points is a coastal recreation reserve.

Crab Island (beach **Q 175**) is a 6.5km long, semicircular convex west-facing sand island, made up of two arms composed of multiple spits both recurving to the east, with the linking centre of the island low and narrow. The outer edge of the island is swept by the strong tidal currents of Endeavour Strait, and sand shoals, banks and coral reefs extend right across the western entrance of the 18 km wide strait.

Beach **Q 176** commences on the north side of Slade Point, 1 km to the lee of the island. It is a curving 2.5 km long west-facing beach terminating at a low cuspate foreland. It is fronted by 2 km wide tidal shoals and the island, with a tidal channel paralleling the shoreline. A 300 m wide series of low grassy beach ridges backs the beach, then older tree-lined ridges extending up to 3 km inland.

Beach **Q 177** commences on the north side of the foreland and curves around for 1.7 km to the next sandy foreland, which is composed of a series of low migratory sand spits. The beach is fronted by 1-1.5 km wide tidal shoals and a central channel, and backed by densely vegetated beach to foredune ridges.

Beach **Q 178** commences at the foreland and trends round to the east for 9.2 km to the 1 km wide mouth of the **Jardine River**. Much of the beach has formed as a 200-300 m wide eastward-migrating series of spits that terminate at low sandy Van Speult Point. There are some areas of dune activity along the spit, which is largely covered by casuarina. It is fronted by sand shoals, which narrow from 2 km in the west to 300 m at the river mouth.

Q 179-180 **JARDINE RIVER**

No.	Beach	Rating HT LT	Type	Length
Q179	Jardine R	1 2	R+tidal sand shoals	500 m
Q180	Jardine R (E)	1 2	R+sand flats	3.8 km
Spring & neap tidal range = 2.1 & 0.9 m				

The **Jardine River** is the northernmost river on Cape York and drains much of the northern tip of the cape, with its tributaries reaching 90 km to the southeast, almost to the eastern side of the cape. The river has deposited a 12 000 ha deltaic plain consisting of Pleistocene and Holocene beach ridges and recurved spits, together with extensive swamp and mangrove forests. The mouth is wide but filled with tidal shoals (Fig. 6.30), which extend well into the gulf. The western side of the river mouth, marked by Van Speult Point, is part of a camp and recreational reserve, while the eastern side which commences at Sunshine Point is aboriginal land.

Figure 6.30 The funnel-shaped, sand chocked Jardine River mouth, with beach Q 180 on its northern bank.

Beach **Q 179** is located just inside the eastern side of the mouth and curves south from the low lateritic Sunshine Point for 500 m into the 1 km wide sediment-choked river mouth. It is a low, narrow crenulate beach, fronted by the myriad of sand shoals and channels of the shallow mouth, and backed by densely vegetated beach ridges.

Beach **Q 180** commences at Sunshine Point and curves away to the northeast for 3.8 km to Mutee Head where it attaches to bedrock of Carboniferous rhyolite. The beach is backed by a series of low foredunes, which narrow from 500 m at the mouth to 300 m at the head, and then older recurved spits and ridges, which extend up to 4 km inland. River mouth shoals extend 1 km off Sunshine Point and the western end of the beach, narrowing to the east as 200-300 m wide sand flats. There is vehicle access along the rear of the Holocene ridges to beach Q 179 and Sunshine Point.

Q 181-187 MUTEE HEAD-COWAL CREEK

No.	Beach	Rating HT	Rating LT	Type	Length
Q181	Mutee Head	1	2	R+sand/rock flats	300 m
Q182	Ichera Hd (W)	1	2	R+sand flats	1.5 km
Q183	Ichera Hd (E1)	1	2	R+sand/rock flats	300 m
Q184	Ichera Hd (E2)	1	2	R+sand/rock flats	300 m
Q185	Ichera Hd (E3)	1	2	R+sand/rock flats	2.1 km
Q186	Cowal Ck (W)	1	2	R+rock flats	800 m
Q187	Cowal Ck mouth	1	2	R+sand/rock flats	600 m
Spring & neap tidal range = 2.1 & 0.9 m					

Mutee Head is the first outcrop of ancient bedrock on the western side of Cape York and is composed of 300 million year old carboniferous rhyolite, part of the Cape York Inlier, which includes the high islands of Torres Strait as well as the headlands between Mutee Head and the cape. Between 30 m high Mutee Head and Cowal Creek, 7 km to the east, is a series of seven north-facing beaches (Q 181-187), each bordered by generally low wooded rhyolite headlands.

Beach **Q 181** is bordered on the west by Mutee Head, backed by wooded bedrock slopes and curves to the east for 300 m. It is a relatively steep high tide beach, fronted by 200 m wide sand and rock flats. The remains of a World War 2 jetty are located toward the eastern end. The area behind the beach was used as a wartime storage area, and later an islander community, but now is only used for camping and fishing off the beach. There is vehicle access to the beach.

Beach **Q 182** lies in the next embayment and is a curving 1.5 km long beach bordered in the east by the low rhyolite Ichera Head. Bedrock outcrops at either end forming tree-covered headlands, with the sweeping golden sand beach in between. The beach is relatively steep and fronted by sand flats which widen from 100 m in the east to 300 m to the west. Beach **Q 183** is located on the eastern side of Ichera Head, and is a curving 300 m long north-facing beach, backed by tree-covered bedrock slopes. It is fronted by a 200 m wide mixture of rock and sand flats.

Beach **Q 184** commences 500 m to the east and is a 300 m long sandy beach, backed by a 200 m wide lightly wooded plain, then slopes. It terminates at a rock-reef-controlled cuspate foreland, with beach **Q 185** extending to the east for 2.1 km to the next low wooded bedrock headland. It is a north to northwest-facing sand beach, backed by a few wooded foredunes, with a small creek draining out across the centre, then wooded slopes rising to 50 m in the east.

Beach **Q 186** commences on the eastern side of the small headland and trends to the northeast for 800 m, as a crenulate sandy high tide beach, fronted by 200 m wide rock flats, which narrow to the east. Wooded slopes rise to 40 m behind the beach, the same slopes forming the western boundary of **Cowal Creek**, a 500 m wide funnel-shaped tidal creek that connects to upland streams. Beach **Q 187** extends to the southeast for approximately 600 m into the western side of the creek mouth. The beach runs along the base of the wooded slopes for a few hundred metres, then splays into a series of recurved spits, which have narrowed the creek to 150 m at the eastern end of the beach.

Q 188-195 COWAL CREEK-SEISIA

No.	Beach	Rating HT	Rating LT	Type	Length
Q188	Cowal Ck (N1)	2	3	R+rock flats	800 m
Q189	Cowal Ck (N2)	2	3	R+rock flats	1 km
Q190	Cowal Ck (N3)	2	3	R+rock flats	700 m
Q191	Aloa Beach (1)	2	3	R+rock flats/coral	300 m
Q192	Aloa Beach (2)	2	3	R+rock flats/coral	250 m
Q193	Aloa Beach (3)	2	3	R+rock flats/coral	250 m
Q194	Seisia (S)	2	3	R+rock flats/coral	3.2 km
Q195	Seisia	2	3	R+rock/sand flats	2 km
Spring & neap tidal range = 2.1 & 0.9 m					

Between Cowal Creek and Seisia are 8 km of northeast-trending shoreline, containing eight beaches (Q 188-195), all tied to low rhyolite headlands, and dominated by intertidal rock flats, with patchy coral reef increasing to the north. The aboriginal and islander communities of Injinoo and Seisia are located at Cowal Creek and Seisia respectively, with the main township of Bamaga located 5 km inland southeast of Seisia, while the communities of Umagico and New Mapoon are located between Bamaga and the coast. The five communities of both Torres Strait islanders and aborigines are the largest concentration of native people on the cape.

Beach **Q 188** commences on the northern side of the low red bedrock headland that forms the northern entrance to Cowal Creek (Fig. 6.31) and trends relatively straight to the northeast for 800 m, all the while fronted by 500 m wide sand and irregular rock flats. There is a rock jetty at the point, backed by a parking area, toilets and picnic area. A road runs along the back of the beach which is also used as a camping area. The **Injinoo** community is located 500 m in from the southern end of the beach. Beach **Q 189** commences on the north side of a rock-controlled inflection at the northern end of beach Q 188. The beach continues to the northeast for another 1 km, to a more substantial rock-controlled inflection. This beach is backed by wooded 10 m high bedrock slopes, and also fronted by the irregular 400 m wide sand and rock flats. A road runs along the top of the slopes.

Beach **Q 190** is a curving 700 m long northwest-facing beach, also tied at either end to bedrock slopes, the northern slopes rising to 55 m. Rocks outcrop at either end, and it is fronted by 300 m wide sand and rock flats.

Figure 6.31 Cowal Creek, with Injinoo community located behind, and beach Q 188 extending to the north.

Beaches **Q 191, Q 192** and **Q 193** are three 250 m long scallops of sand running along the base of low wooded slopes. Each beach faces northwest and is bordered by protruding bedrock slopes, and fronted by a 200 m wide mixture of sand and rock flats. There is vehicle access to the three beaches, with the **Umagico** community located 500 m south of beach Q 193. A road runs straight from the community to the beach where there is a boat ramp, with camping permitted behind the beach.

Beach **Q 194** is the longest of the eight beaches. It is a slightly curving 3.2 km long northwest-facing sand beach, tied at either end to low bedrock points and scattered rocks and rock flats, as well as beachrock exposed along the northern end, while coral reef flats parallel much of the beach. The beach is backed by a 1.5 km wide Quaternary coastal plain, containing an outer Holocene, and possibly inner Pleistocene barrier, separated by a 500 m wide wetland. There is vehicle access to either end of the beach.

Seisia beach (**Q 195**) commences at a 10 m high bedrock headland with a small creek draining across the southern end of the beach. It curves round to the north for 2 km, terminating at the cuspate foreland (Red Island Point) formed in lee of Red Island. The settlement of **Seisia** is located on the low foreland, together with the jetty and landing ramp (Fig. 6.32). The southern half of the beach is dominated by sand and rock flats, as well as patchy reef, while the steep northern half fronts the deep 300 m wide tidal channel between the foreland and the island. The Seisia community extends halfway down the beach, and a number of facilities are located on the foreland including a beachfront camping and caravan park, motel, restaurant, fuel station and store.

Q 196-204 **SEISIA-LARADEENYA CREEK**

No.	Beach	Rating HT	Rating LT	Type	Length
Q196	Seisia (N1)	2	3	R+sand+rock flats	800 m
Q197	Seisia (N2)	2	3	R+sand+rock flats	1.8 km
Q198	Paterson Ck (S)	2	3	R+sand+rock flats	4 km
Q199	Paterson Ck	1	2	R+sand flats	200 m
Q200	Paterson Hill	1	2	R+sand flats	500 m
Q201	Laradeenya Ck (S2)	1	2	R+sand flats	700 m
Q202	Laradeenya Ck (S1)	1	2	R+sand flats	2.3 km
Q203	Laradeenya Ck (N1)	1	2	R+sand flats	700 m
Q204	Laradeenya Ck (N2)	1	2	R+sand flats	2.3 km

Spring & neap tidal range = 2.1 & 0.9 m

To the north of Red Island Point the shoreline trends north-northwest, past 83 m high **Paterson Hill**, the highest point on the entire gulf coast, to Laradeenya Creek. This 13 km long section of coast parallels Endeavour Strait which contains a series of linear tidal-dominated banks, coral reefs and bedrock islets and islands, all reducing waves at the shoreline. At the same time the strong tidal currents of the Strait flow parallel to the shore. There are nine near continuous low energy beaches along this section (Q 196-204), with vehicle access to most of the beaches and a few camps scattered along the coast.

Figure 6.32 Seisia is located in lee of a sandy foreland bordered by beaches Q 195 & 196.

Beach **Q 196** commences at the tip of Red Island Point and curves slightly to the northeast for 800 m, to a low bedrock point. The southern half of the beach is backed by the beachfront caravan park, while trees back the northern half. It is fronted by 200-300 m wide sand and rock flats, then the deep 200 m wide channel between the shore and Red Island. Beach **Q 197** commences on the northern side of the rocks and continues on to the northeast for another 1.8 km, to a rock-controlled inflection in the shore. Rocks and 200 m wide rock flats dominate much of the beach, with wooded bedrock slopes behind. There is vehicle access to both beaches and a few beachfront houses and camps behind each.

Beach **Q 198** at 4 km is the longest beach along this section. It commences in the south tied to the rocks, but then extends to the north as a series of wooded beach-foredune ridges, backed by a 3 km long creek that feeds in the northern 2 km into the mangrove-lined Paterson Creek. The beach is fronted by sand and rock flats in the south, grading to 200 m wide sand flats to the north, with occasional beachrock on the shore. A number of fishing shacks and campsites are spread along the beach. Beach **Q 199** is a recurved spit that begins at the northern end of beach Q 198 where beachrock outcrops at the shore, and curves round to the east for 200 m to the mangrove-lined mouth of Paterson Creek. Sand flats extend 500 m off the beach.

Beach **Q 200** commences on the northern side of the small creek mouth and runs north for 500 m along the base of the conical Paterson Hill to some rocks outcropping at the base of the hill. Beach **Q 201** continues on the northern side of the rocks for another 700 m to the next set of low rocks. Both beaches are backed by wooded slopes and fronted by sand and some rock flats extending 500 m off the beaches.

Beach **Q 202** is tied to low rocks in the south, then extends for 2.3 km to the north as a low wooded barrier, backed by a freshwater swamp and then mangroves that flow into **Laradeenya Creek**, which reaches the shore at the northern end of the beach. Sand flats extend up to 500 m off the beach. There is vehicle access to the southern half of the beach where a number of fishing camps are located.

Beach **Q 203** commences on the northern side of the small creek mouth and trends north for 700 m to a smaller creek, beyond which beach **Q 204** continues on for another 2.3 km before it merges into mangroves in lee of Possession Island (Fig. 6.33). Both beaches are part of a beach ridge plain, with intervening mangrove-filled swales, that has migrated up to 3 km to the north and reaches 500 m in width. The low gradient, low energy beaches are fronted by 500 m wide sand flats and patchy coral reefs. There is vehicle access to the northern side of Laradeenya Creek.

Figure 6.33 View north along beaches Q 203 & 204.

Q 205-213 **PEAK POINT-CAPE YORK**

No.	Beach	Rating HT	LT	Type	Length
Q205	Roonga Pt	1	2	R+sand/rock flats	250 m
Q206	Peak Pt	2	3	R+rock flats	150 m
Q207	Smiley Pt	1	2	R+sand/rock flats	200 m
Q208	Smiley Pt (E)	1	2	R+ridged sand flats	1.6 km
Q209	Battery Pt	1	2	R+ridged sand flats	3.8 km
Q210	Big Ck	1	2	R+tidal sand shoals	700 m
Q211	Big Ck (E)	1	2	R+ridged sand flats	1.5 km
Q212	Bay Pt	1	2	R+sand flats	150 m
Q213	Frangipani Bch	1	2	R+ridged sand flats	2.2 km
Spring & neap tidal range = 2.1 & 0.9 m					

Roonga and Peak points mark the northwestern tip of Cape York Peninsula, with Cape York located 11 km to the east, the northern tip of the peninsula and the Australian mainland. The entire northern tip of the peninsula is dominated by Carboniferous rhyolite rocks reaching a maximum elevation of 150 m at Peak Hill, and 50 m at Bay Point and Cape York. At the base of and in between the hills are nine beaches (Q 205-213), most facing north into Endeavour Strait. There is vehicle access to most of the beaches, with the northernmost Frangipani Beach being one of the most visited in Australia as people walk across its sands on the way to the tip of the Cape.

Beach **Q 205** is located between Roonga and Peak points. It is a 250 m long, low energy, west-facing, sandy beach, backed by a 50-100 m wide wooded beach ridge and small area of salt flats, while it is bordered by mangrove-fringed bedrock points. It is fronted by narrow sand flats then a mixture of coral reef and rock flats.

Beach **Q 206** lies on the eastern side of Peak Point at the base of Peak Hill and occupies the base of a steep narrow valley. The 150 m long beach is composed of cobble and boulders, as well as coral debris, and fronted by narrow intertidal rock flats. Beach **Q 207** lies 500 m to the southeast in the next valley, and is a 200 m long sand beach fronted by a narrow band of sand and rock flats. It is bordered by rock flats and bedrock points including the eastern Smiley Point, and backed by densely vegetated slopes.

Beach **Q 208** commences at Smiley Point and curves to the east for 1.6 km to low rocky Battery Point. It is a steep sandy beach fronted by ridged sand flats to the west, with rocks and rock outcrops dominating the eastern half. The beach is backed by a 100-200 m wide beach-foredune ridge plain, with vehicle access to the rear of the beach and a few houses located in lee of Battery Point. Beach **Q 209** commences on the eastern side of Battery Point and curves gently east for 3.8 km, to a cuspate foreland formed in lee of Mona Rock. This is a relatively steep sandy beach, with some rocks outcropping toward Battery Point, but otherwise dominated by a 300 m wide series of ridged sand flats, and backed by a 200-300 m wide band of wooded beach-foredune ridge. Punsand Bay Private Reserve is located behind the beach and offers accommodation and camping facilities.

Beach **Q 210** is a very low energy strip of sand that partially blocks, and deflects to the east, the mouth of Big Creek. The 700 m long beach commences at the foreland and extends to the creek mouth. It is fronted in places by mangroves, then 600 m wide tidal shoals, and backed by the mangrove-lined meandering creek. Beach **Q 211** commences at the creek mouth and trends to the east-northeast for 1.5 km to 6 m high Bay Point. This is a relatively low narrow sand beach backed by a 200-300 m wide bank of densely vegetated beach ridges, and fronted by 400 m wide ridged sand flats.

Beach **Q 212** is located on the north-facing tip of Bay Point. It is a 150 m long strip of sand with steep bedrock slopes to either side.

Australia's northernmost (mainland) beach runs from Bay Point for 2.2 km all the way to Cape York itself (Fig. 6.34). It is located in Punsand Bay and is known as **Frangipani Beach** (**Q 213**). The 10-40 m high Cape forms a prominent northern boundary to the beach, protruding 500 m out from the beach, with a band of mangroves running out along the side of the cape. The beach faces northwest and consists of a 20-50 m wide dry high tide beach, fronted by low gradient sand flats up to 400-500 m wide at low tide (Fig. 6.35). The sand flats consist of four to five low undulating bars and swales paralleling the beach. The beach is backed by a 200-400 m wide, densely vegetated, low sand barrier, which in turn is backed by salt flats and a tidal creek, which crosses the beach beside the western boundary, at Bay Point.

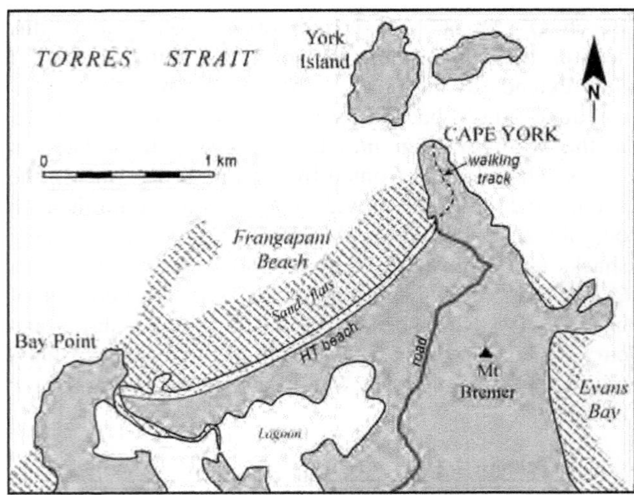

Figure 6.35 Frangapani Beach is located adjacent to Cape York, with the wide intertidal beach extending west to Bay Point.

There is a car park located behind the eastern end of the beach, with a track leading down to the beach. People also park here to make the 500 m walk to the tip of the Cape and Australia. Cape York Pajinka Wilderness Lodge is located just before the car park and provides both lodge accommodation and camp sites overlooking the beach.

Swimming: The beach is relatively safe for bathing, particularly below the car park and at high tide. At low tide it can be a long walk across the shallow to bare sand flats to reach deeper water. The western Bay Point end is more hazardous with a deep tidal inlet and tidal currents, and also the greater possibility of encountering crocodiles associated with the creek.

Figure 6.34 Cape York and the eastern end of Frangipani Beach (Q 213), Australia's northernmost mainland beach.

THURSDAY ISLAND

No.	Beach	Rating HT	Rating LT	Type	Length
TI 1	Jetty Beach	1	2	R+sand/rock flats	600 m
Spring & neap tidal range = 2.4 & 0.9 m					

Thursday Island is located 37 km west of Cape York. The island is about 300 ha in area, and reaches a maximum elevation of 100 m. It is surrounded on all sides by deep tidal channels between the island and the adjoining Horn, Prince of Wales, Friday and Hammond islands. The township of Thursday Island is the major town and administrative centre for the Torres Strait Islands. It occupies the southern half of the island and is bordered along its southern shore by the main island beach (**TI 1**) (Fig. 6.36). The beach faces southeast across Ellis Channel, and extends for 600 m from near the main jetty in the east to Vivien Point. The main road runs along the back of the beach, with a narrow shady reserve between the road and the top of the beach, which is steep and sandy. There are usually a few small boats moored high on the beach, with larger boats anchored off the beach. Low tide exposes 200 m wide tidal and rock flats.

Figure 6.36 Thursday Island town and main beach (left of jetty)

Region 3 East Cape York: Cape York to Endeavour River (Cooktown)

Length of coast: 1032 km (1469-2501 km)
Number of beaches: 427 (Q 214-640)

Regional map 3	Beaches Q 214-479	Figure 6.37	Page 377
Regional map 4	Beaches Q 478-640	Figure 6.70	Page 411

Aboriginal Land:	Lockhardt	Coast length	1885-2080 km	Beaches:	Q 417-486
			2083-2090 km		Q 487-488
	Hope Vale		2373-2499 km		Q 586-640
National Parks:	Jardine River	Coast length	1610-1655 km	Beaches:	Q 272-293
	Iron Range		1839-1858; 1870-1886		Q 394-402; 411-416
	Lakefield		2128-2134; 2143-2172km		none
	Cape Melville		2221-2348 km		Q 515-594
	Endeavour River		2499-2501 km		Q 640

Towns/communities: Portland Roads, Lockhart River (Cooktown)

The east coast of Cape York extends for 1032 km between the Cape and Cooktown. A number of features dominate the coast and its beaches. First, is the Great Barrier Reef, which protects the coast from ocean waves and swell resulting in a lower energy wind wave dominated shore. Second, is the southeast Trades which blow onshore for much of the year delivering the waves to form the beaches and move sediment northward, and winds to from the numerous and in places massive dune systems. Third, is the geology and its structure with from north to south the resilient Cape York inlier, the low Carpentaria Basin (Cape York to Cape Grenville), the Coen Inlier (Temple Bay to Claremont Point), Laura Basin occupied by the large Princess Charlotte Bay, and the uplifted Hodgkinson basin forming the high capes between Cape Melville and Hinchinbrook Island. Fourth, is the humid tropical climate which brings high rainfall to the eastern cape resulting in numerous creeks and a few rivers that deliver abundant sand to the coast that is reworked northward into the beach and shoal systems. Finally, is the 2-3 m tide which floods to the north, inducing both tide-modified and tide-dominated beaches, as well as assisting the waves in moving sediment northward.

Depending on orientation and in some cases protection from reefs, the beaches range from very low energy tide-dominated to some of the most exposed on the Cape, with numerous rips cutting across the higher energy beaches. The only settlements on the east coast are the small fishing community at Portland Roads, the aboriginal community at Lockhart River, and Cooktown.

Q 214-218 **EVANS-SOMERSET BAY**

No.	Beach	Rating HT LT	Type	Length
Q214	Evans Pt	1 2	R+sand flats	200 m
Q215	Evans Bay	1 2	R+ridged sand flats	2.3 km
Q216	Muddy Bay	1 2	R+sand/mud flats	300 m
Q217	Somerset Bay	1 2	R+sand flats/reef	300 m
Q218	Lyons Pt	1 2	R+sand flats	100 m
Spring & neap tidal range = 2.6 & 0.2 m				

At Cape York the shoreline trends away to the east-southeast for 11 km to Sheridan Point. In between are several north-trending 100 m high spurs of the Carnegie Range, while the southern half of the shore is paralleled by Albany Island, with 0.5-1 km wide Albany Passage between the island and mainland. The combination of the northerly orientation, prominent headland and protecting Albany Island results in a very low energy shoreline, with some of the bays filled with mud flats and ringed by mangroves. There are however five beaches, two in the north (Q 214-215) near the cape, one in Muddy Bay (Q 216), one at Somerset (Q 217) and a small pocket of sand to the south at Lyons Point (Q 218).

Beach **Q 214** is located 1 km southeast of Cape York, essentially at the eastern base of the cape. The cape and 128 m high Mount Bremer back the western side of the beach, with 15 m high Evans Point to the east. The 200 m long beach faces north across 200 m wide sand flats rimmed by deeper reefs, with a low narrow vegetated bedrock spur separating it from Evans Bay beach. Rhyolite boulders litter the base of the beach.

Evans Bay is a 1.5 km wide northeast-facing bay bordered by Evans Point and 21 m high Ida Point. The bay contains a curving 2.3 km long white sand beach (**Q 215**) which receives sufficient wave energy to maintain a 300-500 m wide ridged sand flat (Fig. 6.38), with the number of ridges increasing for three to five in the south. It is a low narrow beach backed by a densely vegetated narrow band of beach ridges then the steeply rising bedrock slopes.

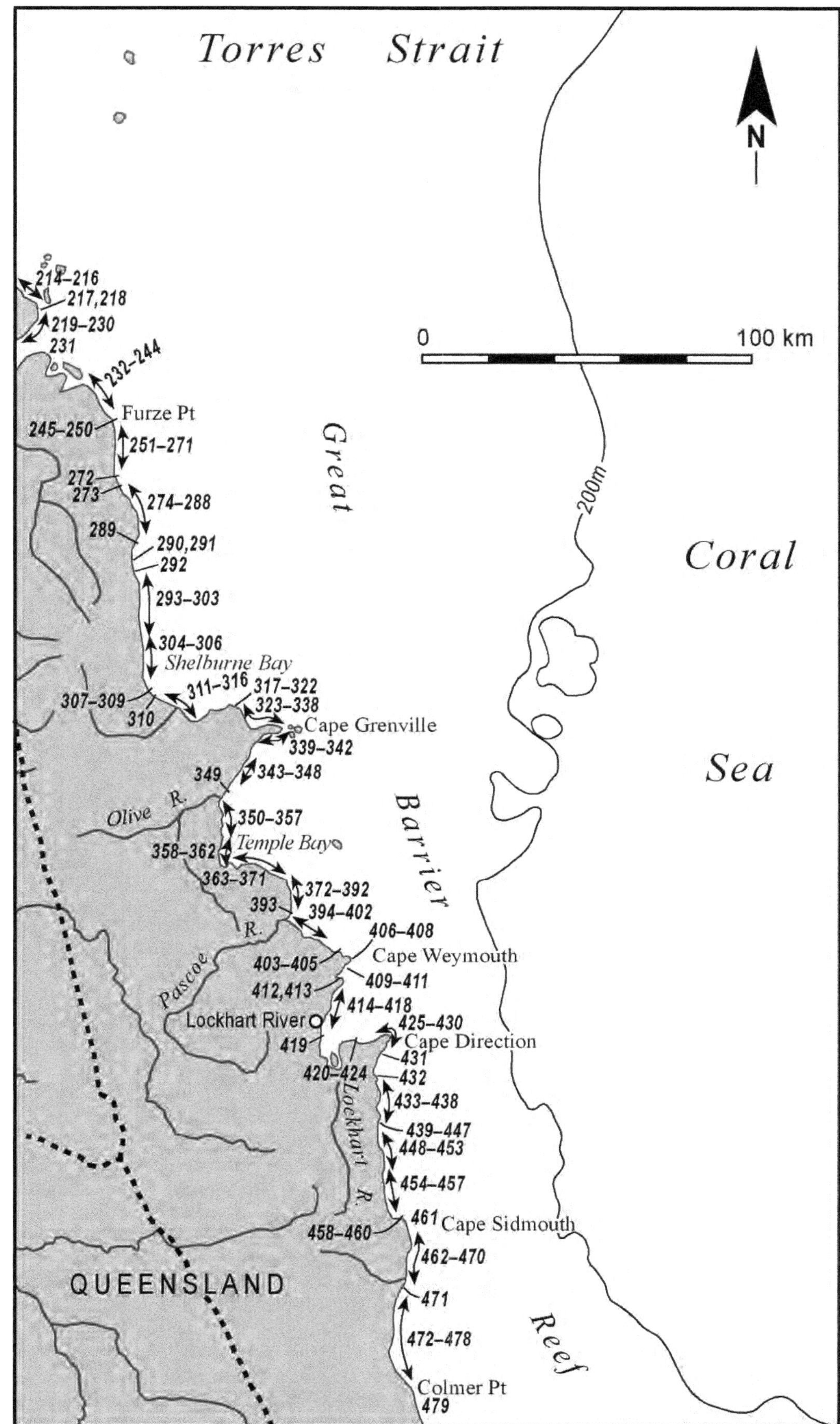

Figure 6.37 Cape York regional map 3 showing the northern part of the eastern Cape, beaches Q 214-479.

Q 214-218 **EVANS-SOMERSET BAY**

No.	Beach	Rating HT LT	Type	Length
Q214	Evans Pt	1 2	R+sand flats	200 m
Q215	Evans Bay	1 2	R+ridged sand flats	2.3 km
Q216	Muddy Bay	1 2	R+sand/mud flats	300 m
Q217	Somerset Bay	1 2	R+sand flats/reef	300 m
Q218	Lyons Pt	1 2	R+sand flats	100 m
Spring & neap tidal range = 2.6 & 0.2 m				

At Cape York the shoreline trends away to the east-southeast for 11 km to Sheridan Point. In between are several north-trending 100 m high spurs of the Carnegie Range, while the southern half of the shore is paralleled by Albany Island, with 0.5-1 km wide Albany Passage between the island and mainland. The combination of the northerly orientation, prominent headland and protecting Albany Island results in a very low energy shoreline, with some of the bays filled with mud flats and ringed by mangroves. There are however five beaches, two in the north (Q 214-215) near the cape, one in Muddy Bay (Q 216), one at Somerset (Q 217) and a small pocket of sand to the south at Lyons Point (Q 218).

Beach **Q 214** is located 1 km southeast of Cape York, essentially at the eastern base of the cape. The cape and 128 m high Mount Bremer back the western side of the beach, with 15 m high Evans Point to the east. The 200 m long beach faces north across 200 m wide sand flats rimmed by deeper reefs, with a low narrow vegetated bedrock spur separating it from Evans Bay beach. Rhyolite boulders litter the base of the beach.

Evans Bay is a 1.5 km wide northeast-facing bay bordered by Evans Point and 21 m high Ida Point. The bay contains a curving 2.3 km long white sand beach (**Q 215**) which receives sufficient wave energy to maintain a 300-500 m wide ridged sand flat (Fig. 6.38), with the number of ridges increasing for three to five in the south. It is a low narrow high tide beach backed by a densely vegetated narrow band of beach ridges then the steeply rising bedrock slopes.

Figure 6.38 View across the wide intertidal ridges and flats of Evans Bay (Q 215) at mid tide.

Muddy Bay lies 2 km to the southeast and is a 1 km deep bay containing 1 km wide sand-mud flats and a mangrove-dominated shoreline. Along the base of the bay is a 700 m long chenier, the central 300 m of which are exposed to waves at high tide and form beach **Q 216**. The small Mew River drains out across the northern end of the beach which is fronted by 200 m wide ridged sand flats, then the mud flats.

Somerset Bay is the site of the historic Somerset settlement, which was abandoned in 1942. Today only the ruins and some gravestones remain. There is vehicle access to the bay and camping is permitted behind the beach. The beach (**Q 217**) is 300 m long, faces east across Albany Passage, and consists of a relatively steep high tide beach, with 200 m wide intertidal sand flats grading into deeper patch reefs. This is a popular beach visited by most who travel to the Cape, and suitable for swimming toward high tide, so long as there are no crocodiles about.

Beach **Q 218** lies 1 km southeast of Somerset in a small indentation bordered by Lyons Point on the southern side. It is just 100 m long, bordered by low rocky points and backed by densely vegetated slopes.

Q 219-224 **PUTTA PUTTA-NARAU BEACHES**

No.	Beach	Rating HT LT	Type	Length
Q219	Putta Putta Beach	2 3	UD	200 m
Q220	Putta Putta Beach (S)	2 3	UD	150 m
Q221	Freshwater Beach	2 3	UD	700 m
Q222	Nanthau Beach (N)	2 3	UD	250 m
Q223	Nanthau Beach	3 4	R+LT rips	750 m
Q224	Narau Beach	3 4	R+LT rips	1.3 km
Spring & neap tidal range = 2.6 & 0.2 m				

Fly Point marks the southern entrance to Albany Passage and the northern tip of the funnel-shaped Newcastle Bay, which is 15 km wide at the entrance narrowing to the 2 km at the mouth of Kennedy Inlet and Jackey Jackey Creek, 20 km to the south. The first 14 km south of the point are dominated by low laterite headland and intervening sandy beaches, initially well exposed to the southeast trades. The first six beaches (Q 219-224) receive moderately high waves and maintain wide beaches. They are also all accessible by a vehicle track which runs behind the beaches and over each of the low headlands and which link up with the Somerset road.

Putta Putta Beach (**Q 219**) is the northernmost of the beaches and is located 500 m southwest of Fly Point. It is a 200 m long southeast-facing beach, with a 300 m wide low gradient ultradissipative intertidal bar, while partially active transgressive dunes extend 400 m inland. It is bordered by a sloping headland to the north and a boulder headland to the south, with low mangroves in amongst the boulders. The 200 m wide southern point separates it from beach **Q 220**, a similar 150 m long low gradient beach, bordered by low vegetated headlands to either side. It is also backed by dunes extending 300 m inland including an active blowout.

Freshwater Beach (**Q 221**) occupies Freshwater Bay 300 m south of the Putta Putta beaches. It has a low sloping headland to the north and the more prominent Vallack Point protruding 300 m east to the south. The beach curves round to the lee of the point and consists of a very low gradient intertidal bar backed by a low foredune (Fig. 6.39), then older vegetated transgressive dunes, which extend up to 500 m inland.

Figure 6.39 View across Freshwater Beach (Q 221) at low tide.

The rocky shore of **Vallack Point** continues for 1 km southwest to **Nanthau Beach**, which consists of a northern 250 m long ultradissipative beach (**Q 222**), with a grassy 200 m wide low headland separating it from the main 750 m long beach (**Q 223**). This beach receives slightly higher waves and commonly has low tide rips, particularly toward the southern end. Both beaches are backed by a partially active foredune, backed by 200-500 m wide vegetated transgressive dunes.

Narau Beach (**Q 224**) lies 500 m to the south and is a well exposed 1.3 km long east-facing beach, which has a 200 m wide intertidal bar, with several small rips commonly forming along the base of the bar. It is backed by densely vegetated transgressive dunes up to a few hundred metres wide, with a small creek running out of the centre of the dunes and across the beach.

Q 225-231 **KENNEDY INLET**

No.	Beach	Rating HT	Rating LT	Type	Length
Q225	Chandogoo Pt (W1	1	2	R+ridged sand flats/reef	50 m
Q226	Chandogoo Pt (W2)	1	2	R+ridged sand flats/reef	200 m
Q227	Saldogoo Beach	1	2	UD	2.7 km
Q228	Kilbie Beach	1	2	R+ridged sand flats	1.1 km
Q229	Congora Beach	1	2	R+ridged sand flats	800 m
Q230	Kennedy Inlet (W)	1	2	R+tidal sand flats	1.5 km
Q231	Kennedy Inlet (E)	1	2	R+tidal sand flats	1.2 km
Spring & neap tidal range = 2.6 & 0.2 m					

From **Chandogoo Point** the bedrock-controlled shoreline curves to the southwest for 7 km to the beginning of the wide band of mangroves that dominate the sides of Jackey Jackey Creek and the larger Kennedy Inlet and much of the southern shore of Newcastle Bay. There are five beaches located along the bedrock section (Q 225-229) and two in amongst the mangroves either side of the 5 km wide inlet (Q 230 & 231).

Beaches Q 225 and 226 are two pockets of sand located in indentations in lee of 10 m high Chandogoo Point. Beach **Q 225** is a 50 m long pocket of sand wedged in between laterite points, and fronted by a patch of sand flats then coral reef flats extending 200 m offshore. The reef continues west for 300 m to beach **Q 226**. This 200 m long beach is fronted by three intertidal sand ridges, then reef off the eastern half of the sand flats. Densely vegetated gentle bedrock slopes back both beaches.

Chaldogoo Beach (**Q 227**) is a 2.7 km long east-facing beach, bordered by 30 m high Cliffy Point to the south. The beach protrudes slightly in lee of inlet shoals and receives sufficient wave energy to maintain a 300 m wide ultradissipative beach. Vegetated early Holocene and possibly Pleistocene transgressive dunes extend up 4 km in from the beach, together with an active blowout near Cliffy Point (Fig. 6.40). A creek drains out of the dunes and across the centre of the beach.

Figure 6.40 Chaldogoo Beach (Q 227) at low tide, with the active parabolic dune at the southern end.

Kilbie Beach (**Q 228**) commences on the southern side of 300 m wide Cliffy Point, and trends south for 1.1 km. It is also backed by the same 3-4 km wide vegetated dunes, with a few minor blowouts. It receives lower waves and is fronted by five sand ridges, then patchy coral reef.

Congora Beach (**Q 229**) lies 500 m further south and into the inlet. It is an 800 m long east-facing beach also fronted by ridged sand flats, and backed by largely vegetated transgressive dunes. A bedrock islet marks the southern end of the beach, beyond which the shoreline curves to the west for 1.5 km as beach **Q 230**, followed by the mangroves of Jackey Jackey Creek. This is a very low energy inlet beach fronted by an 800 m wide mixture of intertidal sand ridges and tidal sand shoals. It is backed by densely vegetated dunes that have blown over from neighbouring Congora Beach.

Beach **Q 231** lies on the other side of Kennedy Inlet 7 km to the east. It is a low beach ridge deposited on a predominantly mangrove-dominated shoreline, where the backing mangroves extend for up to 12 km inland. It is fronted by a 600 m wide mixture of shore parallel and transverse tidal sand ridges.

Turtle Head Island

Turtle Head Island lies across the entrance to the Escape River, with the river mouth flowing out between the southern tip of the island and Sharp Point, 1 km to the southeast. The island is 7.5 km long, 1-2 km wide and about 1,100 ha in area. It trends to the north then west from the river mouth. It has 5 km of exposed east-facing shore with a 1.2 km long beach backed by dunes that have transgressed up to 2 km across the island. Along the northern shore are nine lower energy beaches. The dunes reach 36 m in height and form the highest topography on the generally low, undulating, densely vegetated island. The western shore faces across 1-2 km wide Middle River tidal channel to the mangrove-covered delta of the Escape River and is dominated by continuous band of mangroves. On the southwestern tip of the island is a landing ground and the small OK Village community.

Much of the Escape River mouth is part of the Escape River Fish Habitat Reserve.

Jardine River Resources Reserve

Coast length:	45 km	(1565-1610 km)
Beaches:	41	(Q 232-272)

Q 232-237 **SHARP-SHADWELL POINT**

No.	Beach	Rating HT	LT	Type	Length
Q232	Sharp Pt (S1)	2	3	UD+rocks	900 m
Q233	Sharp Pt (S2)	1	2	R+LTT	150 m
Q234	Sharp Pt (S3)	2	3	UD+rocks	1.5 km
Q235	Shadwell Pt (N3)	2	3	UD+rocks	200 m
Q236	Shadwell Pt (N2)	2	3	UD+rocks	400 m
Q237	Shadwell Pt (N1)	2	3	UD+rocks	600 m
Spring & neap tidal range = 2.8 & 0.2 m					

Sharp Point marks the northern tip of a 100 km long exposed section of coast that trends south to Double Point in Shelburne Bay. Much of the coast in between is well exposed to the southeast trades and dominated by higher energy beaches, and backed in many places by extensive Holocene and Pleistocene long-walled transgressive dune systems. These dune systems between beaches Q 232-272 are contained within the Jardine River Resources Reserve.

Sharp Point is a round, flat 12 m high laterite point partially covered with clifftop dunes (Fig. 6.41). It is backed by the mouth of the Escape River and its intricate river mouth shoals. Beach **Q 232** commences at the point and curves to the south then east for 900 m to the lee of the next point, a dune-capped series of red laterite rocks. The beach is well exposed in the north and consists of a 200 m wide ultradissipative beach with scattered rocks across the intertidal zone, grading into 300 m wide migratory sand shoals in the more protected southern end, where trees grow to the back of the beach. It is backed by largely vegetated dunes that have transgressed right across the 1 km wide back of the point into the Escape River. Beach **Q 233** is located on the southern point. It is a narrow 150 m long north-facing lower energy beach, with low rocks to either end.

Figure 6.41 Sharp Point, with the mangrove-fringed shore to the west, and the beginning of beach Q 232 in foreground.

Beach **Q 234** commences on the south side of the rocks and trends due south for 1 km before curving round to the east in lee of the next low rocky point. The east-facing section is well exposed and has a low gradient 200 m wide ultradissipative beach, backed by densely vegetated scarped transgressive dunes extending 2 km to the Escape River, with a couple of large blowouts right behind the beach. Groundwater seeping out of the dunes helps maintain a low, wet intertidal beach. The southern end curves round and has an intertidal dominated by sand shoals that have migrated round the point and are part of the overall northward sand transport along this coast.

Beach **Q 235** lies on the east side of the rocks, and is a 200 m long east-facing beach, bordered by low rocks to either end, with a vegetated 20 m high bluff behind the southern rocks. The beach is ultradissipative and 200 m wide. Beach **Q 236** continues on the south side of the bluffs for 400 m. It is backed by densely vegetated dunes, and protected in the south by a rocky point that extends 400 m to the east. The northern 100 m of beach are ultradissipative, with the remainder fronted by two migratory sand shoals extending 300 m offshore. There is vehicle access to the southern end of the beach.

Beach **Q 237** begins on the south side of the point, faces northeast and curves round for 600 m to **Shadwell Point**, a low rocky laterite point that extends 200 m to the north. The beach has a 200 m wide ultradissipative northern section, with migratory sand shoals off the more protected southern half, where trees back the beach. A vehicle track terminates on top of the point.

Q 238-244 **TERN CLIFFS**

No.	Beach	Rating HT	Rating LT	Type	Length
Q238	Shadwell Pt	1	2	R+LTT	1.5 km
Q239	Tern Cliffs	2	3	R+LT rips	250 m
Q240	Tern Cliffs (S1)	2	3	R+LT rips	200 m
Q241	Tern Cliffs (S2)	2	3	R+rock flats	100 m
Q242	Tern Cliffs (S3)	2	3	R+LTT	100 m
Q243	Tern Cliffs (S4)	2	3	R+LTT	200 m
Q244	Tern Cliffs (S5)	2	3	R+LTT	300 m
Spring & neap tidal range = 2.8 & 0.2 m					

At **Shadwell Point** the shoreline turns and curves southeast for 1.5 km to Tern Cliffs, then due south for 2 km to the lee of Thomson Islet. Beach Q 238 lies between the point and the cliffs, while beaches Q 239-244 begin immediately south of the cliffs. A vehicle track reaches the coast at beach Q 244 and runs about 300-400 m in from the beaches north to Shadwell Point, with some informal vehicle access to the 20 m high laterite bluffs that back all the beaches. All the beaches have in the past supplied sand to the backing clifftop transgressive dune sheet that extends up to 10 km northwest to the Escape River.

Beach **Q 238** is a curving 1.5 km long white sand beach, bordered by extensive rocks and reef of Shadwell Point and Tern Cliffs to either end, and backed by the densely vegetated transgressive dunes extending 6 km inland as longwalled parabolics. The northern half of the beach is exposed to the southeast waves and has a couple of low tide rips, while the southern half curves round in lee of Tern Cliffs and is fronted by migratory sand ridges. The vehicle track runs through the backing frontal dunes to the point.

The **Tern Cliffs** are 40 m high scarped east-facing laterite bluffs, that extend for 500 m, before becoming fronted by six pocket beaches (Q 239-244). The first two beaches are well exposed and dominated by low tide rips (Fig. 6.42). Beach **Q 239** occupies the first 250 m long curve in the cliff-line, with the 30-40 m high cliffs backing the beach. The beach receives moderate southeast waves and consists of a 100 m wide low gradient bar with usually four rips along its base. To the rear the foredune has climbed part way up the backing bluffs, and in the past has deposited clifftop dunes that extend 8 km inland to the shores of the Escape River. Beach **Q 240** lies 50 m to the south in the next 200 m long curve in the cliff-line. It is a similar low gradient beach with 3-4 low tide rips, backed by exposed and periodically eroded rocks and no foredune.

Figure 6.42 Beaches Q 239 & 240 are backed by the 40 m high bluffs of the Tern Cliffs.

Beach **Q 241** lies at the southern end of the beach and consists of a 100 m long pocket of high tide sand, located in a semicircular amphitheatre in the backing 40 m high red laterite cliffs. Cliff debris backs the beach, while rocks and rock flats extend over 100 m offshore. The beach is only accessible by boat, through a small gap in the rocks at high tide.

Beaches Q 242-244 are three adjoining beaches also backed by the bluffs, but afforded some protection from Thomson Islet, a 20 m high 10 ha laterite island located 500 m due east of beach Q 243. Beach **Q 242** is a 100 m long high tide pocket of sand fronted by a 100 m wide intertidal bar that extends for 150 m along the base of the 40 m high cliffs. At low tide it links with beach **Q 243**. This is a 200 m long, 100 m wide, low gradient beach, backed by lightly vegetated 30 m high slopes, and is the most accessible of the beaches. Finally, beach **Q 244** lies immediately to the south and is a 300 m long, 100 m wide, intertidal beach, that extends to the backing bluffs at high tide, with some rocks outcropping on the beach.

Q 245-250 **SADD-FURZE POINTS**

No.	Beach	Rating HT	Rating LT	Type	Length
Q245	Sadd Pt (N4)	1	2	R+LTT	700 m
Q246	Sadd Pt (N3)	1	2	R+LTT	250 m
Q247	Sadd Pt (N2)	2	3	R+rock flats	100 m
Q248	Sadd Pt (N1)	2	3	R+LT rips	150 m
Q249	Sadd Pt	2	3	R+LT rips	200 m
Q250	Furze Pt	1	2	R+ridged sand flats	1.8 km
Spring & neap tidal range = 2.8 & 0.2 m					

In lee of Thomson Islet the laterite bluffs turn and trend southeast for 5 km past Sadd Point to Furze Point. A combination of 20-30 m high red laterite bluffs and six generally small pocket beaches (Q 245-250) occupy this section of coast, with vehicle access to the southern side of Sadd Point. The point is the southern end of the 11 km long series of 30-70 m high Pleistocene and early Holocene transgressive dune sheets that extend as densely vegetated longwalled parabolic dunes all the way to the banks of the Escape River, a distance of up to 10 km, covering a total area of 4000 ha.

Beach **Q 245** commences at the southern end of a 1.5 km line of 30 m high cliffs. The 700 m long beach curves around to face northeast, with a protruding 20 m high point bordering the southern end. The northern end of the beach is exposed to the southeast trades and backed by a series of small blowouts, then vegetated older transgressive dunes. The entire beach has a low gradient intertidal bar that extends at high tide to the dunes. Beach **Q 246** is located on the eastern side of the 200 m wide point and is a 250 m long northeast-facing pocket beach bordered by a 300 m long laterite point to the east. It consists of a 100 m wide intertidal beach, with some rock outcropping in the intertidal and with subtidal migratory sand ridges moving around the eastern point and lying off the eastern half of the beach.

Beach **Q 247** is a 100 m long high tide pocket of reddish sand, backed by 20 m high vegetated bluffs and wedged between protruding laterite arms at the northern end of Sadd Point. It is fronted by a continuous 50-100 m wide band of intertidal rocks, with some sand in the subtidal. A 100 m wide flat-topped laterite headland separates it from beach **Q 248**. This is a 150 m long white sand pocket beach, also bounded by 20 m high laterite headlands. It is backed by an unstable foredune climbing up the backing bluffs and fronted by a 100 m wide low gradient intertidal beach with low tide rips flowing out against either headland.

Twenty metre high, 200 m wide **Sadd Point** forms the boundary between beaches Q 248 and 249. Beach **Q 249** is a 200 m long, east-facing, exposed beach backed by several active blowouts extending up to 150 m inland climbing up and over the backing older transgressive dunes and bluffs. It is a 100 m wide low gradient intertidal beach, fronted by boundary rips that flow out against Sadd Point, a low band of southern boundary rocks.

Beach **Q 250** is bordered in the north by a 50 m wide tidal creek that flows out against the southern side of Sadd Point. From here the beach trends to the southeast for 1.8 km to the lee of 40 m high **Furze Point**. The point affords moderate protection for the beach, which consists of a low gradient 100 m wide intertidal beach, fronted in the north by 200 m wide tidal shoals of the creek mouth, while in the south an intricate 500 m wide series of five rhythmic, migratory sand ridges lies between the point and the beach. A 15 m high densely vegetated foredune backs the beach, with some higher dune instability at the creek mouth (Fig. 6.43).

Figure 6.43 Furze Point (foreground) and beach Q 250 with its intricate intertidal sand ridges.

Q 251-258 **FURZE-REID-NUMBER TWO POINTS**

No.	Beach	Rating HT LT	Type	Length
Q251	Furze Pt (S1)	2 3	R+sand/rocks	100 m
Q252	Furze Pt (S2)	2 3	R+LT rips	200 m
Q253	Furze Pt (S3)	2 3	R+LT rips	80 m
Q254	Reid Pt	2 3	R+LT rips	250 m
Q255	Reid Pt (S1)	3 4	R+LT rips/rocks	150 m
Q256	Reid Pt (S2)	3 4	R+LT rips/rocks	300 m
Q257	Reid Pt (S3)	3 4	R+rock flats/rips	200 m
Q258	No 2 Point	3 4	R+rock flats/rips	200 m
Spring & neap tidal range = 2.6 & 0.6 m				

At **Furze Point** the shoreline trends to the south for 25 km to Orford Ness and includes Orford Bay. Much of the shore is dominated by 20 m to, in places, 60 m high laterite-capped horizontally-bedded Cretaceous sandstones. Along the base of the bluffs and cliffs, and in indentations in the bedrock, are several small generally exposed beaches. Beaches Q 251-258 occupy the first 4 km between Furze and Number Two points.

Beach **Q 251** is located in a small grassy valley, surrounded by 40 m high vegetated bluffs on the southern side of Furze Point. It consists of an exposed, 100 m long, east-facing beach composed of medium yellow sand. It forms a moderately steep beach fronted by a continuous intertidal bank of rocks, then a low tide bar cut by a rip. Beach **Q 252** lies 100 m to the south and is a 200 m long high tide beach, composed of coarser yellow sand, and fronted by a 50 m wide low tide bar composed of finer white sand. A rip drains out at low tide between the bordering rock flats. Beach **Q 253** lies another 200 m to the south and is bordered by steep 40 m high vegetated bluffs. It consists of an 80 m long, low gradient, 100 m wide, white sand intertidal beach, with the high tide waves reaching the base of the bluffs. The beach is fronted by a skewed migratory sand ridge.

Beach **Q 254** is a 250 m long, east-facing, 100 m wide, low gradient intertidal beach located on the southern side of 40 m high **Reid Point**. It backs onto low scarped bluffs, and is fronted by a low tide rip and rocks off the northern half of the beach. Beach **Q 255** lies 500 m to the south, past a boulder-strewn 40 m high headland. The beach is 150 m long, backed by steep vegetated 40 m high bluffs, with bluff debris covering both ends of the beach and only a central 100 m central section free of debris. The 100 m wide beach has two low tide rips flowing out against the bordering boulders.

Beach **Q 256** is backed by more dissected bluffs, with a small creek draining out across the northern end. It has a 200 m wide boulder field off the northern headland and scattered rocks off the southern head. In between is the 300 m long sand beach, which abuts vegetated slopes in the rear, and is fronted by a bar with four low tide rips.

Beaches Q 257 and Q 258 are two crenulate sand beaches wedged in between backing steep, vegetated, 60 m high laterite bluffs, and bordered and largely fronted by rocks and intertidal rock reefs. Beach **Q 257** is 200 m long and has a northern half fronted by continuous rocks, while the southern half has a low tide bar cut by two rips. The backing grassy slopes contain a number of prominent termite mounds. Beach **Q 258** is also a 200 m long double crescent of high tide sand, paralleled by a band of intertidal rock reefs, with only a small southern section free of rocks, and one reef-controlled rip draining the low tide beach. It terminates on the northern side of Number Two Point.

Q 259-266 NUMBER TWO-USSHER POINTS

No.	Beach	Rating HT	Rating LT	Type	Length
Q259	No 2 Point (S1)	2	3	R+LT rips	700 m
Q260	No 2 Point (S2)	2	3	R+LTT+rocks	400 m
Q261	No 2 Point (S3)	2	3	R+LT rips+rocks	300 m
Q262	No 2 Point (S4)	2	3	R+LT rips	300 m
Q263	No 2 Point (S5)	1	2	R+LT rips-LTT	1.8 km
Q264	No 2 Point (S6)	1	2	R+LTT	400 m
Q265	Ussher Pt (N 2)	1	2	R+LTT-sand ridges	1.8 km
Q266	Ussher Pt (N 1)	1	2	R+LTT	600 m
Spring & neap tidal range = 2.6 & 0.6 m					

Between Number Two and Ussher points are 5.5 km of generally south-trending coast backed by some of the highest cliffs on the cape, reaching to 82 m, 1 km south of Number Two Point. In the small valleys and indentations carved in the bluffs are eight small sandy beaches (Q 259-266). They are well exposed in the north, becoming slightly protected toward Ussher Point by the more prominent headland. There is vehicle access to Ussher Point but not to any of the beaches.

Beach **Q 259** is a straight, 700 m long, east-facing, moderate gradient, sandy beach, fronted by five low tide rips, and bordered by intertidal rock flats. The backing steep vegetated bluffs rise to 60 m. Beach **Q 260** is a 400 m long intertidal sand beach at the base of 80 m high cliffs and steep bluffs. It has a low tide bar and rock reefs off the beach, with a rip flowing between the beach and reefs.

Beach **Q 261** commences on the southern side of the high cliffs and is a 300 m long strip of sand and rocks, backed by steep vegetated bluffs, and fronted by a rock-controlled rip. Beach **Q 262** commences immediately to the south and is a 300 m long beach, backed by lower moderately sloping bluffs, and fronted by scattered rocks off the northern end and two low tide rips crossing the 80 m wide bar.

Beach **Q 263** is a slightly curving, 1.8 km long, east-facing beach which is backed by low scarped bluffs, then an area of active dune transgression extending up to 400 m inland. It curves round at the southern end in lee of a sand-capped point that extends 300 m to the east (Fig. 6.44). A small upland creek drains out in the southern corner of the beach. It has a few low tide rips along the northern half of the beach, grading into a 100 m wide low tide bar towards the more protected southern end. Beach **Q 264** commences on the southern side of the point and extends to the south for 400 m to the next 200 m long point. The beach is moderately protected by the southern point and rocks protruding off each point, and consists of a 150 m wide low tide bar, with a rip against the northern rocks.

Figure 6.44 Beaches Q 263, 264 & 265 are backed by unstable climbing dunes.

Beach **Q 265** commences at the tip of the point and trends straight south for 1.8 km. It is backed by irregular vegetated bluffs and older dunes. The beach is fronted by a 100 m wide low tide bar and some rocks in the north, grading into 200 m wide sand ridges in the southern corner. An upland creek flows across the southern end of the beach. Beach **Q 266** extends for 600 m between the point and the base of Ussher Point. It is protected from southeast waves by **Ussher Point** and rock reefs, resulting in a narrow high tide beach fronted by a 100 m wide intertidal bar, then shallow subtidal sands out to the reefs. It is backed by partly vegetated, steeply climbing older dunes which rise to 50 m.

The older transgressive dunes that commence in lee of beach Q 263 continue south for 2.5 km to the southern end of beach Q 265 and extend up to 3 km inland as longwalled parabolic dunes, climbing to a maximum height of 94 m. Offshore of these beaches are shore transverse sand waves which probably the product of waves and tidally induced northward sediment transport.

Q 267-270 **USSHER POINT (S)**

No.	Beach	Rating HT LT	Type	Length
Q267	Ussher Pt (S1)	1 2	R+LTT+rocks	500 m
Q268	Ussher Pt (S2)	1 2	R+LTT+rocks	500 m
Q269	Ussher Pt (S3)	1 2	R+LTT+rocks	100 m
Q270	Ussher Pt (S4)	1 2	R+LTT+rocks	300 m
Spring & neap tidal range = 2.6 & 0.6 m				

Ussher Point is a 50 m high deeply laterised headland, with steep red cliffs extending for 2 km to the south. Along the base of the cliffs are four east-facing, bluff-bound and bedrock-controlled beaches (Q 267-270). There is a vehicle track that runs along the top of the bluffs behind the beaches and terminates at the point.

Beach **Q 267** commences at the tip of the point and is a 500 m long, narrow, high tide beach, wedged in at the base of the steep vegetated bluffs, and fronted by rock flats off the northern half and a 100 m wide low tide bar and some rocks off the southern half. Beach **Q 268** is a more crenulate 500 m long beach wedged in between the bluffs and a near continuous band of intertidal rock reef. The backing 30-40 m high bluffs have slumped onto the back of the beach, partially burying the southern end.

Beach **Q 269** is a 100 m long pocket of sand contained in a small amphitheatre of slumping 30 m high red bluffs. It is fronted by a 50 m wide band of continuous rocks, then a mixture of low tide sand and rocks. Beach **Q 270** is a straight 300 m long sand beach, backed by bluffs decreasing in height from 30 m to 15 m, and fronted by a narrow low tide bar, then a mixture of sand and rock reef.

The high bluffs behind all four beaches are backed by now well vegetated longwalled parabolic dunes extending up to 2 km inland and to heights of 81 m.

Q 271-275 **ORFORD BAY**

No.	Beach	Rating HT LT	Type	Length
Q271	Left Hill	2 3	R+LT rips	2.4 km
Q272	Logan Jack Ck (N)	1 2	R+LTT	3.9 km
Q273	Logan Jack Ck (S)	1 2	Ridged sand flats	2.5 km
Q274	Orford Bay (S 1)	1 2	Ridged sand flats	2.6 km
Q275	Orford Bay (S 2)	1 2	Ridged sand flats	2 km
Spring & neap tidal range = 2.6 & 0.6 m				

The southern 13 km of **Orford Bay** are occupied by five longer, but increasingly protected and lower energy beaches (Q 271-275). All five are backed by the massive Pleistocene and early Holocene transgressive dune field that begins 4 km to the north. The laterite bluffs that dominate the coast south of Slade Point continue to beach Q 272, giving way to prograding barrier islands south of Logan Jack Creek, as the southern end of the bay fills with shallow sand deposits.

Beach **Q 271** commences on the southern side of a 500 m long section of steep 40-50 m high vegetated bluffs, with vehicle access to the top of the bluffs. The low tide bar continues south from beach Q 270, beneath the bluffs, to this beach. The beach continues south for 2.4 km as a relatively exposed beach, terminating in lee of an offshore reef that has enabled a slight protrusion in the backing shoreline. It is a moderately sloping, yellow sand, high tide beach, fronted by a 100 m wide low gradient bar containing some scattered intertidal rock flats and low tide rips. It is backed by low vegetated bluffs and the vegetated dune field, which reaches a maximum height of 108 m at Left Hill, 1.5 km west of the beach. Beach **Q 272** commences immediately to the south and marks a transition from wave to tide-dominated beaches. The beach is 3.9 km long, trending south then deflecting southwest to the mouth of Logan Jack Creek. It is fronted by a 100 m wide low gradient bar until the inflection, where tidal shoals of the creek extend up to 500 m offshore. The northern bluffs give way to moderately active blowouts, with several larger parabolics migrating slowly across the backing 6 km wide dune field.

Jardine River National Park

Area:	237 000 ha	
Coast length:	45 km	(1610-1655 km)
Beaches:	22	(Q 272-293)

The Jardine River National Park encompasses the headwaters of the Jardine River, the so-called 'wet desert' encountered by the Jardine brothers in 1864. It extends for 45 km along the coast between Orford Bay and Hunter Point and encompasses 22 beaches (Q 272-293). These lie along a section of well exposed, east-facing coast, that is for the most part backed by massive Pleistocene and Holocene transgressive dunes, in the form of longwalled parabolic dunes. These are generally well vegetated, through there are some sections of active blowouts behind the more exposed beaches, as well as some larger parabolic dunes slowly moving northwest. All the dunes are aligned to the dominant southeast trade winds. They extend up to 6 km inland and in places rise to over 100 m, and represent a massive deposit of wind-blown marine sands. While there is some activity today, most dune formation probably accompanied Pleistocene and Holocene sea level transgressions.

Logan Jack Creek is an upland creek that meanders across the middle of the dune field and whose meandering 50 m wide mouth forms the boundary between the two beaches as well as the northern boundary of Jardine River National Park (Fig. 6.45). Beach **Q 273** commences on the southern side of the creek mouth and extends to the south for 2.5 km to a second creek mouth. The entire beach is backed by a 10-20 m high, generally vegetated foredune, then the transgressive dunes. The southern creek meanders between the narrow beach and 500 m wide tidal shoals, then deeper sand waves extending 1.5 km offshore.

Figure 6.45 Logan Jack Creek with beaches Q 273 & 274 to either side, and the tidal shoals and ridges extending 500 m offshore.

Beach **Q 274** commences at the southern creek mouth and continues due south for 2.6 km to a third creek. This beach is a barrier island backed by a 400-800 m wide mangrove-filled interbarrier depression, then the older transgressive dunes. The mangroves are drained by tidal creeks to either end. In total the shoreline has prograded up to 1 km into the bay during the mid to late Holocene. It is fronted by more than 10 low intertidal sand ridges that extend 1.5 km into the bay. Beach **Q 275** commences on the southern side of the tidal creek and trends to the southeast for 2 km, gradually merging into the protected mangrove-lined southern shore of the bay. It forms a barrier spit attached to the shore in the south, with the mangrove-lined interbarrier depression widening to 300 m at the creek mouth, and older dunes to the west. It is fronted by up to 20 intertidal sand ridges extending 2 km into the bay.

Q 276-284 **ORFORD NESS**

No.	Beach	Rating HT	Rating LT	Type	Length
Q276	Orford Ness	2	3	R+rock-sand ridges	1 km
Q277	Orford Ness (S1)	2	3	R+rock-sand ridges	800 m
Q278	Orford Ness (S2)	1	2	R+LTT	800 m
Q279	Orford Ness (S3)	1	2	R+LTT-sand flats	1.3 km
Q280	Orford Ness (S4)	2	3	R+rock-sand ridges	700 m
Q281	Orford Ness (S5)	2	3	R+rock-sand ridges	700 m
Q282	Orford Ness (S6)	1	2	R+ridged sand flats	600 m
Q283	Orford Ness (S7)	2	3	R+rock-sand ridges	600 m
Q284	Orford Ness (S8)	2	3	R+rock-sand ridges	800 m
Spring & neap tidal range = 2.6 & 0.6 m					

Orford Ness is a dune-capped laterite point, with the densely vegetated dunes reaching 30 m in height. Low laterite rocks and reef surround the point and dominate most of the coast down to False Orford Ness, 11 km to the southeast. In amongst the laterite reefs and points are 13 beaches (Q 276-288). Beaches Q 276-284 occupy the first 7 km south of the Ness. All the beaches are backed by massive longwalled parabolic dunes, extending in places up to 5 km inland and to heights of 76 m. Sand is actively moving northward along the beaches, and is usually manifest as migratory sand ridges running transverse to parallel to the low tide beach, and filling the protected southern end of some embayed beaches. This sand has contributed to the massive sand shoals and ridges that have filled the southern end of Orford Bay and led to the shoreline progradation of beaches Q 275 and Q 274. There is no vehicle access to this section of coast and no development.

Beach **Q 276** runs along the top of Orford Ness for 600 m in lee of scattered rock flats, then extends for another 400 m into Orford Bay as a low sandy recurved spit. It is backed by the dune-capped point, with the rock flats and reefs extending up to 200 m off the point, then sand ridges beyond (Fig. 6.46).

Figure 6.46 Dune-capped Orford Ness, with crenulate beach Q 275 (right) and beach Q 276 curving to the south.

Beach **Q 277** commences at the eastern tip of the point and curves roughly to the southeast for 800 m. The beach is backed by generally vegetated transgressive dunes rising to 30 m, with laterite flats and boulders scattered along much of the beach and intertidal, then a migratory sand ridge paralleling the beach, with seagrass beyond. Beach **Q 278** occupies the next 800 m long southward-curving indentation. It is bordered by low laterite points to either end, but is otherwise free of rocks, apart from an outcrop on the southern section. It is fronted by a near continuous 50 m wide low tide bar, then seagrass. Several active blowouts and parabolics back the beach, the largest climbing to 68 m.

Beach **Q 279** commences immediately south of the rocks and trends to the southeast for 1.3 km. The northern third is exposed and contains a 50 m wide low tide bar, backed by some blowouts and parabolics, while the southern half is well protected by the next point and the beach is fronted by sand and some rock flats widening to 400 m with seagrass offshore. A creek drains through the well vegetated backing dunes and crosses the southern corner of the beach. Beach **Q 280** is a 700 m long easterly extension of this beach. It faces north and runs out to a low, rocky, dune-capped, 10 m high laterite point. The beach is well protected from southerly waves and is fronted by a 200 m wide mixture of sand and rock flats.

Beach **Q 281** commences at the point and trends almost due south for 700 m to a low rock-controlled inflection. The beach is backed by several active climbing blowouts, reaching 20-30 m. It is fronted by 100 m wide sand, then rock flats, on the point, grading south into sand ridges. Beach **Q 282** commences at the inflection and continues southeast for another 300 m to the northern end of a 500 m long section of red laterite bluffs, then along the base of the bluffs for another 300 m. The beach is fronted by a 100 m wide low tide bar edged in places by rock reef.

Beach **Q 283** commences beneath the 20 m high red bluffs and continues south for 600 m to the lee of 200 m long east-trending laterite point. The beach is backed by stable bluffs and dunes, apart from one central blowout, and fronted by a 150 m wide mixture of sand and rock flats. Beach **Q 284** extends from the eastern tip of the point for 800 m to the southeast, as a double crenulate high tide beach fronted by 100 m wide rock flats, fringed by sandy seabed then seagrass meadows. It is backed by some minor foredune activity then vegetated transgressive dunes, with one longwalled parabolic dune extending 4.5 km inland.

Q 285-288 **FALSE ORFORD NESS**

No. Beach	Rating HT LT	Type	Length
Q285 False Orford Ness (N4)	2 3	R+LT rips	1.5 km
Q286 False Orford Ness (N3)	1 2	R+sand ridges	500 m
Q287 False Orford Ness(N2)	2 3	R+LT rips+rock flats	1.6 km
Q288 False Orford Ness (N1)	1 2	R+rock flats/sand ridges	1 km
Spring & neap tidal range = 2.6 & 0.6 m			

False Orford Ness is a prominent inflection to the coast between Sharp Point and Shelburne Bay. It consists of a 14 m high bluffed headland composed of deeply weathered Mesozoic sandstone, with vegetated transgressive dunes behind. To the north the shoreline trends for 12 km northeast to Orford Ness, with beaches Q 277-288 in between. All are bordered, and to a large degree dominated, by the sandstone and laterite outcrops and reefs. Beaches Q 285-288 occupy the first two embayments north of the Ness.

Beach **Q 285** is a relatively straight, south-trending, 1.5 km long beach bordered by low rock flats in the north. The northern half is well exposed to the southeast waves and has a 100 m wide low gradient intertidal bar, with northward skewed rips spaced about every 100 m and cutting across the low tide bar. The southern half of the beach is protected by a low dune-capped point that extends 1 km to the east. While the low bar continues to the south it is fronted by two migratory sand ridges that widen to 500 m in the south, grading into seagrass meadows off the ridges (Fig. 6.47). Several active blowouts and parabolics back the central to southern section of the beach. They are climbing up and into 30-40 m high vegetated transgressive dunes, which extend up to 4.5 km inland. A small creek runs through the dunes and along behind the northern half of the beach. Beach **Q 286** extends along the north-facing side of the point. It is a narrow protected, crenulate, lower energy beach, backed by densely vegetated 40 m high transgressive dunes, that have originated from beach Q 287. It is fronted by the 100 m wide rock flats, then the eastern end of the sand ridges that extend west to attach to beach Q 285.

Figure 6.47 Curving beach Q 285 (top) grades south into extensive intertidal sand ridges. It is backed by several active blowouts. Crenulate beach Q 286 (lower).

Beach **Q 287** commences on the east side of the point and trends almost due south for 1.6 km to a low rock outcrop, where a small creek exits next to the rocks. The northern half is backed by 10 m high laterite bluffs and fronted by a mixture of 100 m wide sand and rock flats. The southern half is free of rocks with the 100 m wide intertidal bar fronted by low tide rips spaced every 100 m. The central-southern section of beach is backed by moderately active blowouts and parabolics extending up to 500 m inland, with vegetated transgressive dunes extending another 2 km. Beach **Q 288** commences on the southern side of the rocks and curves to the southeast as a narrow, crenulate beach for 1 km toward False Orford Ness. It is backed by vegetated transgressive dunes, and terminates at the 10 m high red bluffs that continue onto the Ness. It is also fronted by an irregular band of intertidal rocks and sand, then a 1.5 km long migratory subtidal sand ridge extending from the Ness to connect with the southern end of beach Q 287. The ridge lies up to 300 m off the beach.

Q 289-291 **FALSE ORFORD NESS**

No.	Beach	Rating HT	Rating LT	Type	Length
Q289	False Orford Ness (S)	2	3	R+LT rips	11.8 km
Q290	Hunter Ck	1	2	R+ridged sand flats	1.5 km
Q291	Hunter Pt (N)	1	2	R+ridged sand flats	700 m
Spring & neap tidal range = 2.6 & 0.6 m					

False Orford Ness forms the northern bedrock point of an open 12 km wide embayment, bordered to the south by Hunter Point. In between are three beaches (Q 289-291), all backed by massive vegetated Pleistocene and Holocene longwalled parabolic dunes.

Beach **Q 289** is one of the longer beaches on the eastern cape. It extends south-southwest, then south for 11.8 km from False Orford Ness to Hunter Creek. It consists of two halves. The northern 7 km is straight and faces directly into the prevailing southeast winds and waves. These maintain a medium gradient high tide beach, fronted by a continuous 100 m wide low tide bar, usually cut by rips every 80-100 m, the rips skewed to the north. The exposed beach is backed by the transgressive dunes extending up to 6 km inland and to heights of 94 m, with some of the parabolics containing perched lakes. Numerous streams drain the dune field and flow out onto the beach. The southern half trends more due south for 5 km to Hunter Creek. This section consists of a narrow high tide beach, fronted by a 50 m wide intertidal bar containing skewed sand ridges, then a fringing coral reef between 100-300 m wide, which lowers waves at the shore. Close to Hunter Creek the ridges widen to 300 m and partially bury the reef.

Beach **Q 290** commences on the southern side of Hunter Creek. The upland creek originates at the Hunter Falls located in a gorge 3 km inland and drains through the southern dune field to the beach, where it is deflected north for 500 m along the base of the dunes. The narrow high tide beach runs along the base of the backing vegetated dunes for 1.5 km, curving round in the south to terminate at a low rock outcrop. It is fronted by a series of five sand ridges extending 500 m offshore. Beach **Q 291** commences on the eastern side of the rocks and continues as a crenulate low energy beach to Hunter Point, 700 m to the east. It is fronted by a 500 m wide mixture of sand ridges and rock flats, then two migratory sand ridges that link with beach Q 290 (Fig. 6.48).

Heathland Resources Reserve

Coast length: 18 km (1655-1673 km)
Beaches: 11 (Q 295-304)

Figure 6.48 Hunter Point (foreground) provides shelter for the extensive sand ridges that migrate along beaches Q 290 & 291.

Q 292-296 **HUNTER POINT**

No.	Beach	Rating HT	Rating LT	Type	Length
Q292	Hunter Pt (S1)	1	3	R+sand flats/coral reef	2.2 km
Q293	Hunter Pt (S2)	1	3	R+LTT/coral reef	5 km
Q294	Hunter Pt (S3)	1	3	R+sand flats/coral reef	1 km
Q295	Hunter Pt (S4)	1	3	R+sand flats/coral reef	1 km
Q296	Hunter Pt (S5)	1	3	R+rock flats	600 m
Spring & neap tidal range = 2.1 & 0.5 m					

Hunter Point is a dune-capped 20 m high headland, underlain by a basement of Mesozoic sandstone. The point marks the beginning of a relatively straight south-trending section of bedrock-dominated coast that extends for 35 km to Shelburne Bay. While this section of coast faces east into the prevailing trades, and is in places backed by transgressive dunes, for the most part it is a moderate to low energy section of coast, owing to the presence of both fringing and offshore coral reefs, which substantially lower waves at the shore. As a consequence of the lower waves, tidal currents become more dominant in the nearshore between the reefs and shore and in places subtidal shore transverse tidal sand ridges are slowly moving northward along the coast. Beaches Q 292-296 occupy the first 10 km of coast south of the point.

Beach **Q 292** commences on the southern side of the point and trends due south for 2.2 km to a low rocky outcrop and shoreline inflection. It has a narrow crenulate high tide beach fronted by 100 m wide shallow ridged sand flats, then 100-200 m wide fringing coral reef. It is backed by densely vegetated transgressive dunes extending 1 km inland. Beach **Q 293** begins on the southern side of the rocks and trends south for 5 km to the mouth of **Camisade Creek**. The beach is slightly crenulate and consists of a narrow high tide beach, fronted by a generally narrow (50 m wide) low tide bar, fronted by a fringing reef which widens from 100 m in the north to 500 m in the south. The entire beach is backed by densely vegetated 1-2 km wide transgressive dunes. Transverse tidal ridges lie off the reef along both beaches (Fig. 6.49).

Figure 6.49 Beach Q 293 is fronted by a fringing coral reef and the shore transverse tidally driven subtidal sand waves.

Beach **Q 294** commences at the northward-deflected mouth of Camisade Creek. It trends for 1 km to the south-southeast as a narrow crenulate high tide beach, fronted by ridged sand flats widening to 500 m off the southern end, with fringing reef beyond. The reef and flats lower waves sufficiently at the shore to permit clumps of mangroves to grow at the base of the high tide beach. Beach **Q 295** occupies the next southeast-trending embayment. It is a similar low energy 1 km long crenulate beach, fronted by a mixture of 500 m wide sand and rock flats, then fringing reef. Finally beach **Q 296** occupies the last embayment before the coast trends south. It is a narrow curving 600 m long high tide beach, fronted by 200 m wide rock flats, then a narrow strip of fringing reef. All three beaches are backed by 20-30 m high, vegetated, transgressive dunes.

Q 297-303 **CAPTAIN BILLY LANDING**

No.	Beach	Rating HT	Rating LT	Type	Length
Q297	Captain Billy Landing (N3)	2	3	R+rock flats/coral	800 m
Q298	Captain Billy Landing (N 2)	1	3	R+LTT/coral	2.5 km
Q299	Captain Billy Landing (N1)	1	3	R+sand/rock/coral	500 m
Q300	Captain Billy Landing	1	3	R+sand ridges/coral	600 m
Q301	Captain Billy Landing (S1)	1	3	R+sand ridges/coral	800 m
Q302	Captain Billy Landing (S2)	1	2	R+LTT	700 m
Q303	Captain Billy Landing (S3)	1	3	R+LTT/coral	2 km

Spring & neap tidal range = 2.1 & 0.5 m

Captain Billing Landing is an old jetty site that is connected by a 70 km long vehicle track to the main cape Telegraph Road. Today it consists of an old open shed, a deteriorating ramp and a sign warning of crocodiles. It was chosen as a landing as it is one of the few beaches along this section of coast free of fringing reef and backing dunes. Beaches Q 297-303 occupy a 10 km long east-facing section of coast, generally backed by horizontally-bedded laterised, Mesozoic sandstone-siltstone. The shore is fringed by scattered rock flats and coral reefs, with the seven beaches (Q 297-303) wedged in between.

Beach **Q 297** commences 5 km north of the landing and extends to the south for 800 m. It is backed by lightly vegetated sandstone slopes rising to 40 m, with the narrow high tide beach lying along the base of the bluffs, and 100 m wide rock flats off the beach, then a narrow band of fringing reef. Beach **Q 298** commences against the 30 m high bluffs, then trends south for 2.5 km, for most of the way backed by minor blowouts, then densely vegetated dunes extending 3 km inland as two longwalled parabolics. Rocks again outcrop to the south. It is fronted by a 50 m wide low tide bar, then 100-200 m wide fringing reef. Beach **Q 299** lies immediately south of the southern rocks, and is a slightly curving 500 m long beach, bordered by rock flats, and backed by a few blowouts that have partly climbed the backing 40 m high bluffs.

Captain Billy Landing (**Q 300**) is a 600 m long sand beach, backed by 40 m high laterised horizontally bedded sandstone. Several blowouts have partly climbed the bluffs behind the northern half of the beach, while the landing area is clear of dunes, and afforded slight protection from the southern headland that extends 300 m to the southeast. The road winds down from the bluffs to the beach and ramp. The beach is only accessible by small boat at high tide, as it is fronted by two to three intertidal sand ridges, scattered laterite boulders, then a low tide fringing reef.

Beach **Q 301** commences on the southern side of the headland, and runs along the base of the 40 m high vegetated headland for 800 m. The narrow beach is fronted by 100 m wide sand flats, scattered rock flats, then deeper fringing reef. Beach **Q 302** occupies a 600 m long infilled valley between two headlands and their fringing rock flats and reef. It is a slightly curving 700 m long beach fronted by a 100 m wide low tide bar, the centre of which is free of rocks and reef, owing to the deeper valley floor. Captain Billy Creek drains the valley and flows across the northern end of the beach. Beach **Q 303** commences on the south side of the southern head, and trends due south for 2 km to the next eastern inflection in the shore. The beach is initially backed by 20 m high, vegetated bluffs, which give way to small blowouts backed by 1 km wide vegetated transgressive dunes. The beach is fronted by a 100 m wide low tide bar and some sand ridges, then 100-200 m wide fringing reef.

Q 304-309 **MESSUM HILL-RED CLIFFS**

No.	Beach	Rating HT LT	Type	Length
Q304	Messum Hill (N)	2 3	R+LT rips/rock flats	4.5 km
Q305	Messum Hill	1 2	R+LT rips	4.9 km
Q306	Messum Hill (S)	1 2	R+sand flats/coral reef	6.2 km
Q307	Red Cliffs (N)	1 2	R+ridged sand flats/coral reef	1.2 km
Q308	Red Cliffs	1 2	R+ridged sand flats/coral reef	1.2 km
Q309	Red Cliffs (S)	1 2	R+ridged sand flats/coral reef	2.5 km

Spring & neap tidal range = 2.1 & 0.5 m

Messum Hill is a prominent 85 m high active parabolic dune located 500 m behind beach Q 305. The hill is part of a 19 km long series of massive Pleistocene and Holocene longwalled parabolics that have prograded up to 9 km inland. Today most of the dunes are densely vegetated, though about 10% of the 9500 ha field is presently active. There is no vehicle access through the dunes to the six beaches (Q 304-309) along this section of coast, and no development in this southeastern section of Shelburne Bay.

Beach **Q 304** commences on the southern side of a 500 m long, 20 m high, partly vegetated laterite bluff. It trends due south for 4.5 km to a medium sized upland creek, which emerges from the backing dunes and has a meandering 50 m wide mouth and tidal shoals extending 200 m off the beach. The beach consists of two halves. The northern 2 km have a 100 m wide low tide bar fronted by 100-200 m wide rock flats composed of sloping sandstone beds. The rocks give way in the south to a continuous sand beach, with a more exposed low tide bar cut by well developed rips every 80-100 m, fronted by subtidal, shore transverse, tide driven sand ridges extending another 500 m into the bay. These ridges parallel the outer edge of the flats and reefs to the southern end of beach Q 309. The beach is backed by transgressive dunes averaging 1-2 km in width and up to 56 m in height.

Beach **Q 305** commences on the southern side of the creek mouth and continues south for 4.9 km to the next creek mouth, which flows into a 200 m deep indentation in the shoreline. The beach is bordered at each end by the creek mouth tidal shoals, with a continuous 100 m wide low tide bar in between, cut by rips every 80-100 m, with the rip channels skewed to the north (Fig. 6.50). The beach is backed by several active climbing blowouts, then a series of active parabolics including the prominent Messum Hill. The edges of the dunes are eroding and coffeerock is exposed along the shore of beaches Q 304 and Q 305.

Beach **Q 306** commences at the creek mouth and curves outwards to the south for 6.2 km to a smaller creek mouth. The beach is protected by fringing reefs, as well as atolls scattered across the southern end of the bay. At the shore it consists of a lower energy sand beach with a low tide bar, 100-200 m wide in places, which widens into ridged tidal flats off the southern creek mouth. It is backed by several active blowouts and the widest section of the transgressive dunes, including some active parabolics located 2-5 km in from the beach. Freshwater swamps occupy some of the dune deflation hollows.

Figure 6.50 Well developed low tide transverse bars and rips formed along beach Q 305.

Beach **Q 307** extends from the creek mouth south along the rocks to the north of Red Cliffs. It consists of a 1.2 km long, narrow, high tide beach, fronted by 500 m wide ridged sand flats at the creek mouth, which narrow to the south as fringing reef and intertidal rocks become more prominent. Beach **Q 308** initially runs along the base of **Red Cliffs**, and is then backed by vegetated dunes for a total of 1.2 km, terminating at the mouth of a small linear creek, which emerges from the densely vegetated dunes. The beach is fronted initially by 500 m wide rocks then fringing reef, then ridged sand flats toward the creek mouth. Beach **Q 309** commences at the creek mouth and trends due south for 2.5 km where it merges with the mangrove-fringed shoreline that continues southeast for 6 km to Etatapuma Point. This beach becomes increasingly protected by sand and reef flats that widen from 500 m in the north to 1.5 km in the south. The three lower energy southern beaches are all backed by stable foredunes and generally densely vegetated transgressive dunes extending 9 km to the lee of Red Cliffs.

Q 310-314 **ETATAPUMA (DOUBLE) POINT**

No.	Beach	Rating HT LT	Type	Length
Q310	Etatapuma Pt (N2)	1 2	R+sand flats/coral reef	200 m
Q311	Etatapuma Pt (N1)	1 2	R+ridged sand flats	150 m
Q312	Etatapuma Pt	1 2	R+sand/rock flats	250 m
Q313	Etatapuma Pt (S1)	1 2	R+ridged sand flats	500 m
Q314	Etatapuma Pt (S2)	1 2	R+ridged sand flats	900 m

Spring & neap tidal range = 2.1 & 0.5 m

Etatapuma (or **Double**) **Point** is a low sandstone point capped by 54 m high dunes, located in the southern corner of Shelburne Bay. The point protrudes to the north of the mangrove-lined shore to either side, sufficiently to receive enough wave energy to maintain five low energy beaches (Q 310-314). The dunes are remnants of the massive Cape Grenville Pleistocene dune field.

Beach **Q 310** is located 3 km to the west of the point, and is an isolated, 200 m long, east-facing, very low energy strip of high tide sand. It is fringed by mangroves and fronted by 1 km wide sand flats, then scattered reefs further out.

Beach **Q 311** is located on the northern tip of Etatapuma Point. It is a northeast-facing 150 m long pocket of sand, backed by densely vegetated 20 m high bluffs, bordered by mangrove-covered rock debris, and fronted by a 300 m wide series of ridged sand flats. The eastern rocks terminate at the point, with beach **Q 312** commencing on the eastern side of the point and trending southeast for 250 m. The beach is bounded by low boulder-strewn points, and consists of a narrow high tide beach, backed by climbing vegetated dunes and fronted by 200 m wide sand flats, that link with the small rock reef called Kennedy Island to the south.

Beach **Q 313** extends from the southern side of the rocks for 500 m to a second rock outcrop. It is backed by scarped, partly vegetated, white Pleistocene dunes and fronted by 300-400 m wide ridged sand flats, with Kennedy Island off the northern end. Beach **Q 314** commences on the southern side of the rocks and continues southeast for 900 m to the first mangroves of the Harmer Creek delta. It is backed by scarped Pleistocene dunes and fronted by 300-200 m wide sand flats, then 2-5 km wide subtidal sand shoals that front the entire 7 km long delta front. A vehicle track reaches the northern end of the delta, and runs along behind the mangroves to the southern end of the beach.

Cape Grenville Dune Field

To the south and west of Cape Grenville is a massive Pleistocene and Holocene transgressive dune field, containing scores of longwalled parabolic dunes, averaging 40 m in height and reaching 70 m. The prevailing southeast trades have blown the dunes from the Temple Bay beaches up to 29 km inland, and in places onto the southern shore of Shelburne Bay, where they have contributed to the white beaches and 2-5 km wide sand flats that dominate this section of the bay. In total the dunes cover an area of 425 km^2. While most are vegetated, there are numerous active parabolics, including Wolona Point that is actively migrating into Shelburne Bay. In addition there are scores of perched freshwater lakes and swamps in many of the dune deflation hollows.

Q 315-316 **WOLONA (WHITE) POINT**

No.	Beach	Rating HT	Rating LT	Type	Length
Q315	Wolona Pt (W)	1	2	R+sand/tidal flats	600 m
Q316	Wolona Pt (E)	1	2	R+ridged sand flats	900 m

Spring & neap tidal range = 2.1 & 0.5 m

Wolona Point is the northwestern tip of a large active longwalled parabolic dune, which has migrated 22 km from its origin on Temple Bay beach. The dune has migrated across the base of Cape Grenville to the southern shores of Shelburne Bay. It is also known as White Point, after the bright white, leached Pleistocene dune sands, which are still actively migrating over the 54 m high dune crest and spilling into the bay (Fig. 6.51).

Figure 6.51 Wolona (White) Point is the leading edge of a white parabolic dune that is actively migrating (left to right) into the shallow waters of Temple Bay.

The western side of Wolona Point consists of 600 m long northwest-facing beach **Q 315**. This is a very low energy beach, bordered by the mangroves of the Harmer Creek delta to the west, and with sand, then tidal flats, extending 2 km off the beach. The beach has been formed by easterly waves reworking the dune sand of the point, as a series of low west-trending beach ridges. Beach **Q 316** commences at the tip of the point and runs southeast for 900 m, where it merges into mangroves that continue east for 13 km to Round Point. This beach is part of the sidewall of the 8 km long, up to 1 km wide, longwalled parabolic, whose northern tip forms the mobile point. The narrow high tide beach is fronted by 2 km wide sand flats containing up to 14 intertidal sand ridges, grading into subtidal tide-driven transverse ridges further out.

Q 317-322 **ROUND-THORPE POINTS**

No.	Beach	Rating HT	LT	Type	Length
Q317	Round Pt (S1)	1	2	R+sand flats	1.2 km
Q318	Round Pt (S2)	1	2	R+ridged sand flats	1.4 km
Q319	Round Pt (S3)	1	2	R+ridged sand flats	1 km
Q320	Round Pt (S4)	1	2	R+ridged sand flats	2 km
Q321	Thorpe Pt (N)	1	2	R+LTT/sand flats	1.4 km
Q322	Thorpe Pt	1	2	R+sand flats	150 m
Spring & neap tidal range = 2.2 & 0.4 m					

Round Point forms the southern boundary of Shelburne Bay, while Thorpe Point, located 6 km to the southeast, is the northwestern boundary of Margaret Bay. The two points are formed of Mesozoic sandstone and together with some smaller bedrock outcrops form the boundaries of the six beaches (Q 317-322) located between the two points. All the beaches face east-northeast and are afforded protection from direct southerly waves by Cape Grenville which is located 10 km east of Thorpe Point.

Beach **Q 317** commences at the tip of low sandy Round Point and curves to the southeast for 1.2 km, terminating at a sand foreland in lee of low intertidal rocks. The beach has prograded seaward as a series of low foredune ridges, and protrudes as a central foreland formed from wave refraction around Rodney Island, located 1 km off the beach. Sand flats extend from the beach to connect with the island. Beach **Q 318** curves from the southern side of the rocks for 1.4 km to the south, then southeast. The low sandy beach has a central small mangrove-covered rock outcrop, and terminates amidst mangroves at the southern end of the small embayment, with a small creek draining out through the mangroves. The beach is backed by a couple of low vegetated foredunes, and fronted by ridged sand flats that extend 500 m offshore where they are edged by fringing coral reef.

Beach **Q 319** occupies the next embayment and is a curving 1 km long north-facing white sand beach, tied to two low sandstone headlands. The intervening embayment is almost filled with sand flats, which are 400 m wide in the centre. Beach **Q 320** extends from the eastern side of the rocks for 2 km as a double crenulate white sand beach, with a central foreland formed in lee of a small rock islet located 500 m off the beach. In addition a creek drains out from the centre of the foreland, with a few stunted mangroves off the mouth. Both sides of the foreland contain 400-500 m wide ridged sand flats.

Beach **Q 321** commences at the eastern rocks and curves for 1.4 km to the south then east, to terminate in lee of 34 m high **Thorpe Point**. The northern end of the beach is more exposed and has a low tide bar, while the more protected southern half of the embayment is filled with ridged sand flats that widen to 400 m at the point (Fig. 6.52). Finally beach **Q 322** is a 150 m long pocket of sand located on the northeast side of the point. The red laterite of the point is exposed above the northern end of the beach, with steep vegetated bluffs behind the remainder. It is fronted by 200 m wide sand flats containing some scattered rocks, then deeper seagrass meadows.

Figure 6.52 The sheltered southern corner of beach Q 321 is filled with ridged sand flats and fringed by seagrass meadows.

Q 323-327 **THORPE POINT (S)**

No.	Beach	Rating HT	LT	Type	Length
Q323	Thorpe Pt (S1)	1	2	R+LTT	2.3 km
Q324	Thorpe Pt (S2)	1	2	R+tidal sand flats	350 m
Q235	Thorpe Pt (S3)	1	2	R+tidal sand flats	500 m
Q326	Thorpe Pt (S4)	1	2	R+ridged sand flats	300 m
Q327	MacMillan R (N)	1	2	R+ridged sand flats	1.8 km
Spring & neap tidal range = 2.2 & 04 m					

Thorpe Point marks the northwestern boundary of the open, north-facing, 9 km wide Margaret Bay. The eastern boundary is the top side of Cape Grenville. The bay has beaches extending along each of its arms, with the central 5 km dominated by mangroves spreading to the east of the MacMillan River mouth. Beaches Q 323-327 extend south of Thorpe Point for 6 km to the western side of the MacMillan River. All the beaches are afforded increasing protection to the south by Cape Grenville, and all are backed by older Pleistocene transgressive dunes, and some younger vegetated Holocene dunes.

Beach **Q 323** commences on the south side of the 30 m high red cliffs of Thorpe Point. It trends essentially due south, with a central bulge in lee of a reef 700 m off the beach. The beach consists of a narrow high tide beach fronted by a 100 m wide low tide beach north of the foreland, with decreasing wave energy leading to some sand ridges in the south, all fronted by seagrass meadows. It is backed by a few small blowouts then the densely vegetated transgressive dunes. Beach **Q 324** is a 350 m long pocket of sand wedged between two low wooded headlands, with a creek draining out across the south end of the beach. The result is a low energy beach fronted by irregular tidal sand flats that fill the embayment.

Beach **Q 325** commences on the southern side of the point and is a 500 m long white sand beach fronted by sand ridges that widen to 200 m in the south, then deeper transverse sand ridges and then seagrass meadows. Beach **Q 326** occupies the next 300 m long embayment, and is a similar beach, with the sand ridges widening into the more protected southern end, then the shore parallel transverse ridges.

Beach **Q 327** commences on the southern side of a small anvil-shaped wooded headland. It trends almost due south for 1.8 km to the 500 m wide funnel-shaped mouth of the **MacMillan River**. The river is a tidal creek that drains the 1-2 km wide interbarrier depression between the older Pleistocene transgressive dunes and the Holocene Indian Bay barrier. The beach is eroding into the backing 30 m high Pleistocene dunes, with high tide waves reaching the dunes. A low gradient 100 m wide bar parallels the dunes, then deeper transverse ridges extend 500 m offshore in the north, widening to 1.5 km in the south.

Q 328-332 **MARGARET BAY (E)**

No.	Beach	Rating HT	Rating LT	Type	Length
Q328	Margaret Bay (E1)	1	3	R+coral reef	50 m
Q329	Margaret Bay (E2)	1	3	R+coral reef	100 m
Q330	Waterhole Bay	1	2	R+ridged sand flats	100 m
Q331	Waterhole Bay (E1)	1	2	R+rock/coral flats	100 m
Q332	Waterhole Bay (E2)	1	2	R+rock/coral flats	200 m
Spring & neap tidal range = 2.2 & 0.4 m					

The eastern side of **Margaret Bay** is bordered by the back of Cape Grenville. The cape is a 700 ha peninsula, formed of resilient Palaeozoic rhyolite, which rises to a crest of 62 m at the central Hillgate Hill. The eastern end of the bay is a popular anchorage with the fishing fleet and passing yachts. Beaches Q 328-332 are five small, low energy, bedrock-dominated beaches located on the northwestern side of the cape.

Beach **Q 328** is a 50 m long pocket of intertidal sand, backed by sloping bare then vegetated rhyolite. It is fronted by a 100 m wide fringing reef. Beach **Q 329** lies 200 m to the north and is a similar 100 m long high tide beach, with a 100 m wide fringing reef. A straight track has been cut through the vegetation above the northern end of the beach, leading up and over the 30 m high headland to beach Q 330.

Beach **Q 330** is located on the western side of Waterhole Bay, a 600 m wide semicircular bay. The bay is filled with 600 m wide ridged sand flats, with mangroves bordering most of the shoreline. The beach is a 100 m long strip of high tide sand fronted by rock flats, then the sand ridges.

Beach **Q 331** is located immediately east of the bay mouth and is a 100 m long pocket of sand, bordered by steep rhyolite bluffs and backing vegetated slopes. The moderate gradient 100 m wide intertidal beach is fronted by low tide rock then reef flats (Fig. 6.53). Beach **Q 332** lies 100 m to the east and is a similar 200 m long beach backed by a steeply sloping vegetated valley, and fronted by the rock then reef flats.

Figure 6.53 The small beach Q 331 has a wide moderately sloping intertidal beach.

Q 333-338 **CAPE GRENVILLE (NE)**

No.	Beach	Rating HT	Rating LT	Type	Length
Q333	Cape Grenville (NE1)	1	3	R+coral reef	150 m
Q334	Cape Grenville (NE2)	1	3	R+coral reef	300 m
Q335	Cape Grenville (NE3)	1	3	R+rock flats/coral reef	150 m
Q336	Cape Grenville (NE4)	1	3	R+sand flats/coral reef	300 m
Q337	Cape Grenville (NE5)	1	3	R+rock flats/coral reef	50 m
Q338	Cape Grenville (NE6)	1	3	R+coral reef	250 m
Spring & neap tidal range = 2.2 & 0.4 m					

The northeastern side of Cape Grenville consists of a 3 km long rocky shoreline containing a series of small embayments occupied in part by six small sand beaches, all backed by lightly vegetated sloping bedrock and fronted ultimately by a near continuous fringing coral reef. The subtidal Bremner Shoal is also attached to this section of the Cape, before it extends to the northwest for 3 km toward the western side of Margaret Bay.

Beaches Q 333 and Q 334 are adjoining 150 m and 300 m long beaches on the northeastern tip of the cape. Beach **Q 333** has a moderate gradient high tide beach, fronted by a 50 m wide strip of sand, then 200 m wide fringing reef, and separated from beach **Q 334** by a 50 m wide clump of rocks. This southern beach is similar, with low tide sand and rock flats widening to 100 m, before the reef.

Beach **Q 335** is located at the western end of the next 800 m long small embayment. It is a 150 m long sloping high tide beach bordered by a few mangroves and fronted by 100 m wide sand and rock flats, then the fringing reef. A mangrove-lined creek enters the centre of the embayment, while beach **Q 336** is located along the southern side of the bay. It is a narrow slightly curving 300 m long high tide beach, fronted by a 200 m wide mixture of sand, rock and finally reef flats.

Beach **Q 337** lies 500 m to the east and is a 50 m pocket of sand, bordered by low rocky points and fronted by a 100 m wide mixture of rock and reef flats. Beach **Q 338** commences 100 m to the south and is a straight, east-facing, 250 m long, relatively steep, high tide sand beach fronted by fringing reef that widens from 50 m in the north to 200 m in the south.

Q 339-342 **CAPE GRENVILLE-INDIAN BAY**

No.	Beach	Rating HT LT	Type	Length
Q339	Cape Grenville (S1)	1 2	R+ridged sand flats/coral reef	800 m
Q340	Cape Grenville (S2)	1 2	R+ridged sand flats/coral reef	800 m
Q341	Cape Grenville (S3)	1 2	R+ridged sand flats/coral reef	200 m
Q342	Indian Bay	1 2	R+ridged sand flats	4 km
Spring & neap tidal range = 2.2 & 0.4 m				

The southern side of **Cape Grenville** consists of 4 km of southeast then southwest-facing bedrock shore, the southeastern section protected by the extensive Home Islands, a series of reef-fringed, high rhyolite islands. There is a disused landing strip on the cape, but otherwise no development or access. The cape is tied to the mainland by Indian Bay beach (Q 342) which links with the Temple Bay beaches to the south.

Beach **Q 339** is an 800 m long southeast-facing beach that lies in lee of Gore Island and is also protected by the larger Hicks Island. Waves are therefore low at the beach, which consists of a narrow high tide beach backed by a veneer of vegetated dunes over the bedrock, and fronted by a 200 m wide series of ridged sand flats, then a continuous fringing reef which widens to 400 m in the south. Beach **Q 340** is located 300 m to the west and is an 800 m long south-facing beach also protected by Gore and Orton islands. It is backed by vegetated dunes blanketing the backing bedrock, and fronted by a narrow high tide beach and a 200 m wide series of ridged sand flats, then a more irregular fringing coral reef, with one clear sandy patch through the reef toward the southern end of the beach (Fig. 6.54).

Beach **Q 341** is a very low energy, south-facing, 200 m long strip of high tide sand tucked in at the western base of the cape. It is fronted by 500 m wide ridged sand flats that extend south to a small 5 m high bedrock islet. The sand flats also continue west into Indian Bay.

Indian Bay beach (**Q 342**) commences against the western side of the Cape and curves slightly to the southwest for 4 km to where reef-fronted mangroves dominate the shore. While the beach faces southeast it is protected by fringing coral reefs and the larger Mason and Nomad reefs located 10-25 km offshore. A few mangroves grow along the back of the fine sand, low gradient beach and its 300 m wide series of ridged sand flats, fringed by irregular fringing reefs. The beach is backed by a 0.5-1 km wide series of densely vegetated blowouts and small parabolics, then the salt flats and mangroves of the southern shore of Margaret Bay and MacMillan River to the west. A well defined walking track leads from the eastern end of the beach along the base of the cape to the eastern end of Margaret Bay.

Figure 6.54 Beach Q 340 is fronted by ridged sand flats, then a patchy strip of fringing coral reef.

Q 343-349 **TEMPLE BAY (N)-OLIVE RIVER**

No.	Beach	Rating HT LT	Type	Length
Q343	Temple Bay (N1)	1 3	R+ridged sand flats/coral reef	500 m
Q344	Temple Bay (N2)	1 3	R+ridged sand flats/coral reef	300 m
Q345	Temple Bay (N3)	1 3	R+ridged sand flats/coral reef	2.8 km
Q346	Temple Bay (N4)	1 3	R+ridged sand flats/coral reef	800 m
Q347	Temple Bay (N5)	1 3	R+LTT/coral reef	250 m
Q348	Temple Bay (N6)	2 3	R+LT rips/coral reef	2.8 km
Q349	Olive R (N)	1 3	R+ridged sand flats/coral reef	6.7 km
Spring & neap tidal range = 2.2 & 0.4 m				

The northern shoreline of **Temple Bay**, between the mangroves of Indian Bay and the mouth of the Olive River, consists of 15 km of exposed southeast-facing shoreline. It is however protected to a significant degree by continuous fringing coral reef, with extensive reefs and atolls lying between 10-40 km offshore, while the Great Barrier Reef is located 80 km offshore. As a consequence both waves within the Great Barrier Reef lagoon and at the shore are constrained and further reduced by presence of the reefs. This section of coast did receive higher wave energy during the early Holocene 'high-energy' window, when the reefs were lower, water deeper and waves consequently higher. During this time the massive backing Pleistocene and early Holocene dunes were activated and blown up to 29 km inland. Today the dunes are largely vegetated and lakes occupy many of the deflation hollows in the longwalled parabolic dunes. There is no vehicle access to or development along this section of the coast. Because of their orientation into the southwest winds and waves the beaches are littered with flotsam and jetsam, particularly fishing floats and all manner of floatable boating debris.

Beach **Q 343** is the first beach south of the 6 km long band of mangroves and their fringing reef that trends southwest from Indian Bay. The narrow 500 m long southeast-facing beach is fronted by 200-300 m wide ridged sand flats, then the fringing reef extending up to 1 km offshore. It is bordered by mangroves with some low mangroves on the beach. A small mangrove-lined creek separates it from beach **Q 344**. This is a straight 300 m long southeast-facing beach, bordered in the south by a prominent, grassy, 40 m high, laterite-capped, sandstone headland, that extends 200 m to the southeast. The beach consists of a narrow strip of high tide sand, backed by a line of casuarina trees, and fronted by 300 m wide ridged sand flats, then the reef.

Beach **Q 345** commences on the southern side of the headland and trends southwest for 2.8 km to a 20 m high slight bedrock protrusion. The beach is backed by moderately steep, grassy Pleistocene dunes rising to 60 m. The narrow high tide beach is paralleled by 100 m wide ridged sand flats in the north that narrow to less than 50 m in the south, with the fringing reef dominating the intertidal and extending up to 500 m offshore. Beach **Q 346** extends from the southern side of the rocks for 800 m to a protruding 25 m high laterite-capped headland. The beach curves around to the lee of the headland, as the sand flats widen from 50-300 m and extend out over the reef to the tip of the head. The reef parallels the shore and extends up to 500 m offshore.

Beach **Q 347** is wedged between the lower northern rocks and the 40 m high laterite cliffs to the south. The beach has prograded about 100 m out to the rocks and is backed by a low grassy plain. It consists of a 50 m wide sloping high tide beach, then a 100 m wide low tide bar, fronted by deeper sand flats, with scattered rocks to either end, and fringing reef further out. Beach **Q 348** commences on the southern side of the headland and trends southwest for 2.8 km to a prominent 25 m high laterite-capped headland. While the beach is paralleled by fringing reef, it receives sufficient wave energy to maintain an irregular 100-200 m wide low tide bar, which in places is cut by rip channels, particularly to the south. This is paralleled by a 100-200 m wide band of subtidal sand then the fringing reef. The beach is backed by generally vegetated transgressive dunes, with two adjacent prominent active parabolic dunes rising to 60 m behind the northern section of the beach.

Beach **Q 349** begins inside the headland and trends to the southwest for 6.7 km to the 1 km wide mouth of the **Olive River**. For the first 5 km the beach has a 100 m wide low tide bar, which is initially cut by low tide rips, then it runs as a continuous ultradissipative beach to the river mouth shoals, with deeper fringing reef further out. As the shoals widen to 1 km the beach becomes increasingly protected and curves round into the river mouth, terminating as a series of recurved spits. The entire beach is backed by massive parabolic dune systems, with some active dunes behind the northern half, while the dunes are densely vegetated behind the more protected southern half.

Q 350-355 **OLIVE RIVER (S)**

No.	Beach	Rating HT	Rating LT	Type	Length
Q350	Olive R (S1)	1	2	UD	1.2 km
Q351	Olive R (S2)	1	3	R+LTT/coral reef	1.6 km
Q352	Olive R (S3)	1	3	UD/coral reef	200 m
Q353	Olive R (S4)	1	3	R+LTT/coral reef	800 m
Q354	Olive R (S5)	1	3	R+rock flats	50 m
Q355	Olive R (S6)	1	3	R+rock flats	100 m
Spring & neap tidal range = 2.2 & 0.4 m					

The **Olive River** is the first significant east coast river south of Cape York. It enters the sea in lee of a prominent 56 m high laterite-capped headland, composed of horizontally bedded sandstone and conglomerate and surrounded by sandstone boulder debris. The deep river channel hugs the southern side of the mouth, with the river mouth shoals extending over 1 km seaward of the headland. The river mouth provides an excellent anchorage for fishing boats and passing yachts. A solitary shack is located just inside the mouth to the lee of the headland.

Beach **Q 350** is a 1.2 km long east-facing ultradissipative beach that extends from the base of the headland to the southern 53 m high headland. It is free of inshore reef and receives sufficient wave energy to maintain a 300 m wide intertidal beach, with a 50-100 m wide surf zone during strong southeasterlies. The beach is backed by a series of active climbing blowouts reaching 40 m toward the northern end (Fig. 6.55), and backed in turn by a band of densely vegetated transgressive dunes, the entire system 300 m wide, and all backed by a rainforest-lined creek. A walking track leads from the shack through the dunes to the northern end of the beach.

Beach **Q 351** commences on the southern side of the south headland and curves to the south for 1.6 km to the lee of a headland that extends 500 m to the east. The northern end of the beach is exposed but protected by 200 m wide fringing reef. To the south the beach becomes increasingly protected by the reef, which widens to 600 m, and the headland, with mangroves growing along the base of the rocks. The entire beach is backed by vegetated laterite slopes.

Beach **Q 352** is located at the eastern end of the southern point. It is a 200 m long ultradissipative beach, wedged in at the base of 20 m high, steep, vegetated laterite bluffs. It is bordered and almost enclosed by bedrock points, with rock reef in the subtidal. The 200 m wide intertidal beach narrows to 50 m at high tide. Beach **Q 353** commences immediately to the south and is an 800 m long high tide beach, fronted by a 200-300 m wide low tide beach, which to the south grades into ridged sand flats as it becomes increasingly protected by widening fringing reef and a southern headland. Sloping bedrock backs the beach, becoming densely vegetated to the south, with one patch of dunes activity in the centre.

Figure 6.55 Beach Q 350 with its backing active blowouts and wide ultradissipative beach and surf zone.

Beaches Q 354 and Q 355 are located on the southern headland. Beach **Q 354** is a 50 m long pocket of sand located at the base of a 50 m high cliff, with rocks to either side and in the intertidal. It is separated from beach **Q 355** by a 200 m long headland. This beach is a 100 m long pocket of sand with 40 m high headlands to either side and a grassy valley behind. It consists of a moderately steep high tide beach, then intertidal boulder and rock flats, with sand again the lower intertidal.

356-362BOLT HEAD (TEMPLE BAY- S)

No.	Beach	Rating HT	LT	Type	Length
Q356	Bolt Head (N2)	1	2	R+LTT/coral reef	1.5 km
Q357	Bolt Head (N1)	1	2	R+LTT/coral reef	1.4 km
Q358	Bolt Head	2	3	R+rock flats	200 m
Q359	Bolt Head (S1)	1	3	R+sand/rock flats	400 m
Q360	Bolt Head (S2)	1	3	R+LTT/coral reef	1.9 km
Q361	Bolt Head (S3)	1	3	R+LTT/coral reef	4.2 km
Q362	Bolt Head (S4)	1	2	R+ridged sand flats	2.6 km
Spring & neap tidal range = 2.2 & 0.4 m					

The southeastern corner of **Temple Bay** consists of a south-trending 14 km long section from beach Q 356, past Bolt Head to the junction with the mangrove-lined southern shore, which turns and trends due east for 10 km to Mosquito Point. Beaches Q 356-362 occupy the east-facing bedrock-backed section of coast, with most of the beaches accessible via a vehicle track that originates at Bramwell Station, about 60 km to the west.

Beach **Q 356** is a 1.5 km long east-southeast-facing beach bordered by 30-40 m high steep bluffs, with a lower central section. The beach begins below the steep, vegetated bluffs, initially fronted by rock flats, then by a 100 m wide low tide bar extending to the southern headland, with reef fringing the outer edge of the bar and widening to 200 m in the south. There is a prominent blowout in the centre of the beach exposing red Pleistocene sand and extending 400 m inland. Beach **Q 357** commences on the south side of the 100 m long, 20 m high headland. It continues south for 1.4 km to a protected mangrove-fringed embayment to the lee of Bolt Head, which extends 1 km to the east. The beach is backed by vegetated sloping bedrock, and fronted by a 50 m wide low tide bar, then fringing reef that extends across the bay to Bolt Head.

Beach **Q 358** is located on the northeastern section of **Bolt Head**. It is a 200 m long east-facing strip of high tide sand, backed by steeply sloping 40 m high bluffs of the head and fronted by 50-100 m wide intertidal rock flats. Beach **Q 359** is located 300 m to the south on the southern side of Bolt Head. It consists of a 400 m long, 50 m wide, high tide beach, fronted by irregular 50-200 m wide rock flats and reefs, with some rock patches on intertidal. The backing headland and bedrock slopes are crisscrossed by mining tracks, and a 4WD track to the head and beaches.

Beach **Q 360** commences on the southern side of a 100 m long section of bedrock shore, and continues to the south for 1.9 km to the next bedrock outcrop. For the most part the beach has a narrow high tide beach fronted by a 100 m wide low tide bar, with patches of rock flats to either end, and fringing reef becoming more prominent and up to 300 m wide in the south. It is backed by a veneer of vegetated dune sands over the backing gently sloping bedrock, with some minor dune activity towards the southern end of the beach.

Beach **Q 361** commences on the southern side of the 200 m long section of rocks and trends to the south for 4.2 km to a small creek mouth. The beach is backed by largely vegetated sloping foredunes, with a few small blowouts, then a narrow high tide beach, and a 100 m wide low tide bar, paralleled by a near continuous fringing reef extending up to 200-300 m offshore. As wave energy drops to the south the bar is replaced by ridged sand flats. Beach **Q 362** extends from the creek mouth south for 2.6 km to the mangroves of the southwestern corner of Temple Bay. The beach is fronted initially by ridges and flats and patchy fringing reef, then by continuous sand flats that widen to 400 m in the south and contain up to 10 ridges. In lee of the mangroves is a 2 km long, up to 500 m wide series of recurved spits, that represent the earlier bay shoreline. Between these ridges and Mosquito Point is a 3000 ha mangrove forest drained by the Kangaroo River and Glennie and Hunter creeks, with the mangroves extending up to 5 km inland. The system and its 500 m wide mud flats are still actively prograding into the bay.

Q 363-370 **MOSQUITO POINT**

No.	Beach	Rating HT	Rating LT	Type	Length
Q363	Mosquito Pt (W3)	1	2	R+tidal sand flats	150 m
Q364	Mosquito Pt (W2)	1	2	R+tidal sand flats	300 m
Q365	Mosquito Pt (W1)	1	2	R+tidal sand flats	50 m
Q366	Mosquito Pt (E1)	1	2	R+rock flats	50 m
Q367	Mosquito Pt (E2)	1	2	R+tidal sand flats	800 m
Q368	Mosquito Pt (E3)	1	2	R+rock/tidal flats	50 m
Q369	Mosquito Pt (E4)	1	2	R+tidal sand flats	200 m
Q370	Mosquito Pt (E5)	1	2	R+tidal sand flats	50 m
Spring & neap tidal range = 1.9 & 0.4 m					

Mosquito Point is part of a series of granite outcrops which form the southeastern shoreline of Temple Bay. To the west the southern shore of the bay is dominated by mangroves, while to the southeast resilient rhyolite dominates the shore. The shoreline either side of the point tends to face north and is protected by its orientation, the small bays and 500 m wide tidal flats, resulting in a series of eight very low energy pocket beaches (Q 363-370) extending for 3.5 km either side of the point.

Beach **Q 363** lies 2 km west of the point and is a 150 m long strip of north-facing sand, backed and bordered by sloping granite rising to 20 m, and fronted by 500 m wide very low energy tidal flats.

Beach **Q 364** is located in the first small embayment west of Mosquito Point and is a low curving 300 m long high tide beach, which together with beach Q 367 has tied Mosquito Point to the mainland in a tombolo-like formation. The beach is backed by a 100 m wide grassy beach ridge, then salt flats, and fronted by 300 m wide tidal flats that fill the embayment. A clump of mangroves is located at the eastern end of the beach in lee of a small outcrop of rocks which are tied to Mosquito Point by beach Q 364. Beach **Q 365** is a 50 m long pocket of sand wedged in between the rocks, with mangroves at its eastern end, almost obscuring the beach from view. It is also fronted by the tidal flats.

Beach **Q 366** is located on the northern tip of 300 m wide Mosquito Point. It consists of a 50 m long pocket of high tide sand, backed by sloping granite, and fronted by rock flats. Beach **Q 367** links the southern side of Mosquito Point to the mainland with a curving, 800 m long, embayed, very low energy beach. The beach is backed by a 200 m wide grassy beach ridge, then the salt flats it shares with beach Q 364. The beach ridge narrows to the south as bedrock encroaches on the back of the beach. It is fronted by 300-400 m wide tidal flats, with mangroves bordering each end and a large clump in the centre of the beach.

Beach **Q 368** lies 700 m to the east and is a 50 m long pocket of high tide sand located at the mouth of a V-shaped valley. It is fronted by a narrow band of rock flats then 400 m wide tidal flats. Beach **Q 369** is located 200 m to the east and shares the same embayment and tidal flats. This is a curving 200 m long high tide sand beach, bordered and backed by granite slopes, while fronted by 300 m wide tidal flats. Mangroves line the eastern end of the beach and the boundary headland. Whitish termite mounds dot the slopes behind the two beaches.

Beach **Q 370** is located 200 m to the east on the eastern side of the same headland, with a low saddle between it and beach Q 369. It consists of a narrow high tide strip of sand fronted by 500 m wide tidal flats.

Q 371-375 **SECOND-FIRST STONY POINTS**

No.	Beach	Rating HT	Rating LT	Type	Length
Q371	Second Stony Pt (W2)	1	2	R+tidal sand flats	100 m
Q372	Second Stony Pt (W1)	1	2	R+tidal sand flats	300 m
Q373	Second Stony Pt	1	2	R+tidal sand flats	200 m
Q374	First Stony Pt (W2)	1	2	R+tidal sand flats/reef	300 m
Q375	First Stony Pt (W1)	1	2	R+ridged sand flats	500 m
Spring & neap tidal range = 1.9 & 0.4 m					

Second and First Stony points are composed of resilient reddish, Carboniferous rhyolite which extends inland as ranges rising over 200 m, with Huxley Hill, 3 km south of Second Stony Point, peaking at 288 m. The 4 km of shoreline in between the two points consist of alternately rocky outcrops and mangrove-lined shores protected by their northerly orientation and extensive tidal flats, with reefs further out. In between are five low energy beaches (Q 371-375).

Beach **Q 371** lies 1 km west of **Second Stony Point** and is a 100 m long pocket of sand bordered by a 20 m high headland to the west and low mangrove-covered rock flats to the east. Rock, then 600 m wide tidal flats, front the low energy beach. Beaches Q 372 and Q 373 are neighbouring low energy beaches on the eastern side of the point and lie across the base of a steep valley that rises to 123 m, 500 m in from shore. The creek draining the valley has deposited a mangrove-covered boulder delta at its mouth, with beach **Q 372** extending 300 m to the north, and beach **Q 373** 200 m to the south. Both beaches are fronted by 500-700 m wide tidal flats. These tidal flats widen to over 1 km and fill the remainder of the embayment to First Stony Point.

Beach **Q 374** is located at the base of the northeast-facing embayment, 1.5 km west of First Stony Point. It consists of a 300 m long strip of narrow, high tide sand fronted by 1.2 km wide tidal flats, then patchy reef, while it is bordered by mangroves. Huxley Hill rises 1.5 km west of the beach, providing a dramatic background. Beach **Q 375** commences at the rocky northern tip of **First Stony Point** and extends to the west for approximately 500 m, hugging the base of the point, then extending across the embayment as a series of older and at times active low recurved spits. It is fronted by ridged sand flats that widen to 600 m at the western end of the beach.

Q 376-380 **FIRST STONY POINT-FAIR CAPE**

No.	Beach	Rating HT	LT	Type	Length
Q376	First Stony Pt	1	2	R+sand flats	1.5 km
Q377	First Stony Pt (S1)	1	2	R+rock/ridged sand flats	50 m
Q378	First Stony Pt (S2)	1	2	R+ridged sand flats	1 00 m
Q379	Fair Cape (N2)	1	2	R+sand flats	800 m
Q380	Fair Cape (N1)	1	2	R+sand flats	350 m
Spring & neap tidal range = 1.9 & 0.4 m					

The 5 km of shoreline between First Stony Point and Fair Cape contain five beach systems, three longer sandy beaches (Q 376, Q 379, Q 380) and two pockets at the mouths of steep creeks (Q 377 & Q 378). Much of the shoreline is backed by steeply rising rhyolite slopes reaching over 200 m, 700 m south of beach Q 377, and 150 m at Fair Cape. The beaches while exposed to the southeast trades are protected by sand flats, then a 1-2 km wide field of transverse sand waves which parallel the coast, then extensive inner reefs 10 km offshore, all of which lower waves at the shore.

Beach **Q 376** is a 1.5 km sandy beach that extends south of First Stony Point. It is backed by a 200 m wide overwash plain, then moderately active blowouts moving into vegetated transgressive dunes, the entire system up to 500 m wide, which is backed in turn by the salt flats and mangroves it shares with the recurved spits of beach Q 375. The northern end of the beach receives low waves, which rapidly decrease to the south as the sand flats widen from 300-600 m, and some patchy reef grows off the middle of the beach. A small mangrove-backed creek crosses the southern end of the beach.

Beaches Q 377 and Q 378 are two pockets of sand at the base of a creek mouth descending from the 200 m high ridge that backs the southern 1.5 km long headland. The sand flats from beach Q 376 extend out along the base of the headland as a 300 m wide series of ridged flats, indicative of sand moving from east to west along the base of the headland and past the two small beaches (Fig. 6.56). Beach **Q 377** is a 50 m pocket of sand bordered and fronted by rocks, then the sand flats. Beach **Q 378** lies 200 m to the east and is a similar 100 m long strip of sand, bordered by the rocks and backed by a steep rocky ridge, with the sand flats continuing past the beach to the point.

Beaches Q 379 and Q 380 are located immediately south of the point, and trend south for a total of 1.2 km, separated by a small 20 m high rock islet, tied to the shore by a sandy tombolo. Beach **Q 379** is 800 m long, faces east and is backed by some small climbing blowouts, then densely vegetated 70 m high slopes. At the northern end of the beach some of the dunes have climbed over the 20 m high headland and spilled down onto the sand flats between the point and beach Q 378. It is fronted by subtidal sand flats, then some scattered reef, with the sand waves further out. Beach **Q 380** is 350 m long and backed by more active and larger blowouts climbing 20-30 m up the backing vegetated slopes. It is fronted by intertidal sand flats and sand ridges associated with sand moving around **Fair Cape** and into and along the embayment.

Figure 6.56 A series of migratory sand ridges passes the two small pocket beaches Q 377 and 378, en route to beach Q 376.

Q 381-387 **FAIR CAPE (S)**

No.	Beach	Rating HT	LT	Type	Length
Q381	Fair Cape (S1)	1	2	R+LTT	250 m
Q382	Fair Cape (S2)	1	2	R+LTT	200 m
Q383	Fair Cape (S3)	1	2	R+sand flats/reef	500 m
Q384	Fair Cape (S4)	1	2	R+sand flats/reef	500 m
Q385	Fair Cape (S5)	1	2	R+ridged sand flats/reef	800 m
Q386	Fair Cape (S6)	1	2	R+ridged sand flats/reef	50 m
Q387	Fair Cape (S7)	2	3	R+rock/sand flats/reef	300 m
Spring & neap tidal range = 1.9 & 0.4 m					

Fair Cape is a conical granite headland that rises steeply to 150 m. The steep rock-dominated shoreline trends to the south for 3.5 km, before turning and trending east for 1 km to the next headland. In between are six east-facing beaches (Q 381-385), with beach Q 386 in the corner of the embayment, and beach Q 387 strung out along the base of the embayment. Sand flats and fringing reef fill much of the bay in between the two points, lowering waves at the shoreline.

Beach **Q 381** commences 300 m south of the cape and is an exposed east-facing 250 m long steeply sloping beach, backed by a 50 m wide overwash apron, then the steeply rising partly vegetated slopes of the cape (Fig. 6.57). Large rhyolite boulders outcrop to either end of the beach, and a small usually blocked creek is located in the southern corner. It is fronted by a low tide bar then deeper sand waves. Beach **Q 382** commences 100 m to the south and is a similar 200 m long steep beach, also backed by steeply vegetated slopes.

Figure 6.57 Beach Q 381 towards high tide. It is bordered by large granite boulders.

Beach **Q 383** is located another 100 m to the south and is a straight 500 m long beach located at the base of the steep slopes. It is however fronted by northward migrating shallow sand waves, which shoal and widen to 200 m in the south, then a 200 m wide fringing reef. Beach **Q 384** runs south for 500 m to the mouth of a small valley. The straight sandy beach becomes increasingly littered with boulder debris to the south, with a sandy pocket of the beach partly cut off the by rocks. It is backed by the steep slopes and fronted by 300 m wide migratory sand waves, then the reef which extends 800 m offshore.

Beach **Q 385** commences on the southern side of the valley and trends south for 800 m, terminating at a 400 m long band of mangroves that continue to the corner of the embayment. The beach is backed by steep vegetated slopes and fronted by 800 m wide patchy sand waves and reef in the north grading into intertidal ridged sand flats in the south. Beach **Q 386** occupies the corner of the embayment. It is a 50 m long patch of sand wedged in between the base of the mangroves and rocks to the east with ridged sand flats extending 500 m off the beach. Beach **Q 387** commences immediately to the east and is a crenulate, 300 m long, north-facing, sand beach running along the base of steep vegetated slopes, with rocks and boulder debris littering the beach, and sand flats narrowing to 200 m in the east.

Q 388-393 **STANLEY HILL-PASCOE RIVER**

No.	Beach	Rating HT	Rating LT	Type	Length
Q388	Stanley Hill (1)	1	2	R+LTT	80 m
Q389	Stanley Hill (2)	1	2	R+sand flats/reef	200 m
Q390	Stanley Hill (3)	1	2	R+sand flats/reef	1.2 km
Q391	Pascoe R (N 3)	1	2	R+sand flats/reef	300 m
Q392	Pascoe R (N 2)	1	2	R+sand waves	700 m
Q393	Pascoe R (N 1)	2	3	R+sand waves/shoals	1.2 km
Spring & neap tidal range = 1.9 & 0.4 m					

Stanley Hill is a prominent 363 m high granite hill located 4.5 km due south of Fair Cape. The surrounding steep slopes dominate the shoreline between beaches Q 387 and Q 389. The granite-backed shoreline continues to the north side of the Pascoe River, which has carved a deep valley through the granite batholith. Between the base of the hill and the river mouth are six beaches (Q 388-393) all backed by steep vegetated granite slopes. Offshore sand waves emanating from the sand deposited by the river mouth form a 1-1.5 km wide band of shoaler water, affording moderate protection to the generally east-facing beaches.

Beach **Q 388** is an 80 m long pocket of sand located at the base of a narrow scree valley descending off Stanley Hill, which crests 1 km to the west. The beach is bordered by large granite boulders and fronted by the subtidal sand waves.

Beach **Q 389** commences 500 m to the south and is a straight east-facing 200 m long sand beach, located at the base of 300 m high slopes. It has coral reef lying 100-200 m offshore, with migratory intertidal sand waves between the beach and reef, and deeper sand waves seaward. Beach **Q 390** lies 100 m to the south and is a relatively straight 1.2 km long beach, backed in places by a narrow densely vegetated overwash plain, which is eroding in the south. It is fronted by subtidal sand waves, which extend 1.5 km offshore.

Beach **Q 391** lies 200 m to the south in a 300 m long indentation in the bedrock slopes. The southern headland has afforded some additional protection which has permitted sand flats to extend 150 m out to the point, narrowing to the northern end of the beach. These are in turn bordered by the deeper sand wave field. Beach **Q 392** commences on the south side of the point and trends south for 700 m to another 30 m high point composed of massive granite boulders. The beach has a low tide bar in the north-centre, widening to ridged sand flats in the southern corner, with several large boulders lying along the beach. A vehicle track from Wattle Hill homestead reaches the southern end of the beach, and the northern end of beach Q 393.

Beach **Q 393** begins on the southern side of the point, with a pocket of high tide sand connected at low tide to the remainder of the 1.2 km long beach which trends south to the 300 m wide **Pascoe River** mouth. The beach is moderately steep and backed by low flat overwash plain, which widens to 300 m toward the river mouth, then breaks up into a series of recurved spits at the mouth. It has a low tide bar in the north, then the sand waves, which in the south are replaced by the sandy river mouth shoals extending up to 1 km seaward.

Pascoe River sand and sand waves

The Pascoe River is incised in steep granite slopes right to the river mouth. The granite batholith which it drains supplies the coarse yellow sand to the river, while also constraining its flow and accommodation space. As a consequence most sand transported by the river is deposited initially on the 1 km wide river mouth shoals (Fig. 6.58). These sands are then reworked northward by the prevailing southeast trade wind waves and flood tide currents along the base of the steep bedrock shore to Fair Cape and beyond to the Stony points and Mosquito Point, probably terminating at and contributing to the sedimentation off the Kangaroo River mouth 30 km to the north. The sediment moves northward in the form of subtidal sand waves, arranged transverse to the shore and parallel to the waves and currents. The sand waves are up to 1.5 km across, with a spacing averaging 150 m, and probably an amplitude of 1-2 m, with each wave containing on the order of 100 000 m^3 of sand.

Figure 6.58 The Pascoe River mouth and its extensive sand shoals, with beaches Q 393 & 394 to either side.

Iron Range National Park

Area: 34 600 ha
Coast length: 35 km (1839-1858 km, 1870-1886 km)
Beaches: 15 beaches (Q 394-402, Q 411-416)

The Iron Range National Park is centred on the Iron Range, a north-south-trending band of deeply dissected Proterozoic metasedimentary rock, which averages between 100-300 m in height. To the east of the range is a board dome of Permian granite and dolerite that extends to the coast at Cape Weymouth. Bedrock dominates the 35 km of park shoreline, which extends from the mouth of the Pascoe River in the north to 5 km south of Cape Griffith in the south. The park is renowned for its extensive areas of tropical lowland rain forest which extend right to the shore along much of the park coastline.
The park is accessible via the Portland Roads road, with a spur off the Pascoe River in the north and Cape Weymouth to the east.

Q 394-399 **PASCOE RIVER-WEYMOUTH BAY**

No.	Beach	Rating HT	Rating LT	Type	Length
Q394	Pascoe R (S 1)	2	3	R+tidal sand shoals	500 m
Q395	Pascoe R (S 2)	1	2	R+rock/sand flats	100 m
Q396	Pascoe R (S 3)	1	2	R+sand/coral reef	800 m
Q397	Pigeon Is (S)	1	2	R+sand/coral reef	500 m
Q398	Weymouth Bay (1)	1	3	R+fringing reef	600 m
Q399	Weymouth Bay (2)	1	3	R+fringing reef	800 m

Spring & neap tidal range = 1.9 & 0.4 m

On the south side of the **Pascoe River** the coast begins to trend away to the southeast for the next 22 km to Cape Weymouth and Restoration Island. The intervening southern shore of Weymouth Bay faces north to northeast and is a generally low to very low wave energy shoreline, dominated for the most part by mangroves and fringing reef. Beaches Q 395-399 are located in the first 6 km southeast of the river mouth. There is vehicle access off the Portland Roads road to the river mouth and beach Q 394, otherwise there is no access or development along the coast.

Beach **Q 394** commences at the river mouth and trends due south for 500 m to low granite boulders scattered on the shore and intertidal. A well vegetated 500 m wide sand spit backs the beach, while river mouth tidal sand shoals extend up to 1 km seaward, narrowing to the south. Beach **Q 395** is a 100 m long strip of sand bordered by granite boulders to either end, and backed by vegetated granite slopes. It is fronted by a continuation of the sand flats and patchy coral reef. Beach **Q 396** continues on south of the rocks for 800 m to a mangrove-lined shore. The beach is backed by vegetated slopes and fronted by low energy sand and reef flats extending 1 km off the beach. It is also protected by Pigeon Island, located 2 km to the southeast, and its surrounding reefs. The reefs, island and northern orientation of the shore to the north lower waves sufficiently for a 1.5 km long band of mangroves to dominate the shore.

Beach **Q 397** is located in the next embayment to the south, a northeast-facing 1 km wide bay, with the beach occupying a 500 m long central section at the back of the bay. This is a low energy strip of high tide sand, with scattered mangroves along the shore, and ridge-sand flats extending out 400 m, and reef further out.

Beach **Q 398** commences at the base of a 50 m high conical hill and trends due south for 600 m to a small mangrove-covered rock outcrop, beyond which beach **Q 399** continues on for another 800 m to where it merges into the mangroves that dominate the next 6 km of shore. Both beaches are backed by low overwash flats, then vegetated bedrock slopes. They are fronted by fringing reef, growing right to the shore and extending 100-200 m offshore, with deeper reef further out.

Q 400-402 **WEYMOUTH BAY**

No.	Beach	Rating HT	Rating LT	Type	Length
Q400	Weymouth Bay (3)	1	3	R+tidal flats/reef	150 m
Q401	Weymouth Bay (4)	1	3	R+rock flats/reef	200 m
Q402	Weymouth Bay (5)	2	3	R+fringing reef	700 m
Spring & neap tidal range = 1.9 & 0.4 m					

Midway between the Pascoe River mouth and Portland Roads is a 100 m high densely vegetated spur of the backing granite range. The spur protrudes 500 m north of the line of mangroves and fringing reef that dominate the shore to either side. There are three beaches located either side of the spur (Q 400-402), each backed by steeply rising, densely vegetated, bedrock slopes.

Beach **Q 400** is located on the western side and is a 150 m long recurved spit that has formed at the western end of the bedrock, between the bedrock and the dense mangroves to the west. It is backed by the grassy 50 m wide spit which continues west behind the mangroves and attaches to the granite slopes to the east. Mangroves are scattered along the beach, which is fronted by 200 m wide tidal sand flats, then a narrow band of fringing reef. Beach **Q 401** lies 400 m to the east on the western tip of the spur. It is a curving 200 m long north-facing narrow strip of high tide sand, backed by steep wooded slopes, and fronted by beachrock and bedrock rock flats, with some outer sand flats and then the fringing reef.

Beach **Q 402** is located on the slightly more exposed eastern side of the spur and is a straight, northeast-facing, high tide sand beach, fronted by fringing reef which widens from 50 m in the north to 300 m in the south. Low rocks with scattered boulders border each end of the beach, with a narrow grassy beach ridge, then densely vegetated slopes rising to 100 m behind the beach.

Q 403-405 **PORTLAND ROADS**

No.	Beach	Rating HT	Rating LT	Type	Length
Q403	Portland Roads	1	2	R+sand/reef flats	100 m
Q404	Portland Roads (E1)	1	2	R+fringing reef	200 m
Q405	Portland Roads (E2)	1	2	R+fringing reef	200 m
Spring & neap tidal range = 1.9 & 0.4 m					

Portland Roads is the only freehold settlement between Cape York, 230 km to the north, and Cooktown, 370 km to the south. The small fishing community consists of several houses, together with other houses and shacks at Cape Weymouth, 4 km to the southeast, and on adjacent Restoration Island. The population fluctuates with the fishing season, with up to 30 boats anchored off the main beach during the season. The Roads is also a popular destination for recreational fishers, some of whom tow their boats over the rough 130 km track in from the Telegraph Road. Besides the score of houses the only facilities at the Roads are two pay telephones and freshwater. The area was the focus of mining activity in the 1930s and a military base during World War 2. The eroding stumps of a once 250 m long jetty are located at the end of the road at the eastern end of the small beach (Q 403). Camping is not permitted at the beach.

Portland Roads beach (**Q 403**) is a 100 m long strip of moderately steep high tide sand, wedged in between mangroves to the west and a granite point rising to 50 m to the east. The low energy beach faces northwest and is fronted by a veneer of sand over reef flats that extend 200 m off the beach, with the boats anchored beyond the reef. The beach is usually littered with small dinghies from the boats, and is also used for launching small boats. It is backed by a shady coconut grove, then the road. The few houses occupy the backing slopes.

Beaches Q 404 and Q 405 are two similar beaches located on the eastern side of the 120 m high Aylen Hills that back the Roads. Beach **Q 404** is a 200 m long, moderately steep high tide beach, backed by vegetated slopes, and fronted by a 300 m wide fringing reef (Fig. 6.59). Beach **Q 405** lies 200 m to the south, and is also 200 m long and fronted by the continuous fringing reef. The remains of a boulder-based shack is located in the rocks at the northern end of the beach.

Figure 6.59 The moderately steep beach Q 404 is fronted by fringing coral reef.

Q 406-411 **CAPE WEYMOUTH-CHILLI BEACH**

No.	Beach	Rating HT	Rating LT	Type	Length
Q406	Cape Weymouth (S1)	1	3	R+fringing reef	800 m
Q407	Cape Weymouth (S2)	1	3	R+rocks/fringing reef	60 m
Q408	Cape Weymouth (S3)	1	3	R+rocks/fringing reef	40 m
Q409	Chilli Beach (N2)	3	3	R+LTT/fringing reef	200 m
Q410	Chilli Beach (N1)	1	2	R+LTT/fringing reef	350 m
Q411	Chilli Beach	1	3	R+fringing reef	7 km
Spring & neap tidal range = 2.1 & 0.1 m					

Cape Weymouth is a prominent 50 m high granite headland located to the lee of 110 m high Restoration Island, the two marking one of the major inflections on the eastern cape coast. There is an access track to the cape and a scattering of houses and shacks behind the beaches and in the hills around the cape and on the island. Five beaches are located around the cape (Q 406-410) with Chilli Beach (Q 411) extending for 7 km to the south. There is vehicle access to all the beaches and camping is permitted, though no facilities are provided.

Beach **Q 406** commences on the eastern side of the cape and trends south as a slight cuspate foreland for 800 m. The foreland has formed from wave refraction around Restoration Island, which is located 500 m off the beach. The beach is backed by a low 50 m wide grassy foreland, then vegetated bedrock slopes. It is fronted by a 50 m wide fringing reef north of the foreland, which widens to 300 m on the southern side. Fishing boats sometimes anchor in the channel between the cape and the island.

Restoration Island

Restoration Island is famous as the site where Captain Bligh landed in 1792 with his boatload of *Bounty* sailors. It was here they 'restored' themselves before setting out for Batavia. Today the 80 ha island has changed little. It consists of a 111 m high densely vegetated central peak of granite, with a sandy cuspate foreland formed to the rear western side of the island, opposite beach Q 406 (Fig. 6.60). The foreland has a northern 300 m long beach and a southern 200 m long beach, both steep sandy beaches fronted by fringing reef, with beachrock exposed. Only the tip of the foreland is free of reef. A few shacks are located on the grassy plain of the foreland, shaded by scattered casuarina and coconut trees.

Figure 6.60 Restoration Island (right) and Cape Weymouth, with beaches Q 406 and 409 (left).

Beach **Q 407** lies on the southern side of a 50 m long clump of large granite boulders, with a more substantial low granite point forming the southern boundary of the 60 m long beach. It is composed of moderately sloping white sand, and backed by a veneer of dunes, then bedrock. It is fronted by the continuation of the 300 m wide fringing reef, together with the scattered granite rocks and reefs. Beach **Q 408** lies immediately to the south and is a similar 40 m long sandy beach, backed by an eroding early Holocene foredune and aboriginal midden, with a shack located behind the southern end of the beach. It is also fronted by the same fringing reef and rocks.

Beach **Q 409** is located on the southern side of a 300 m long, 150 m wide, 20 m high granite headland. The beach trends southwest for 200 m. It consists of a wider lower gradient high tide beach, fronted by a 200 m wide mixture of sand and rock flats, then fringing reef extending 500 m offshore. At high tide a rock- and reef-controlled rip flows out over the bar. Beach **Q 410** lies 100 m to the south and is a similar 350 m long lower sand beach, fronted by a more continuous low tide bar, then the fringing reef, with some granite rocks outcropping on the reef, and lying up to 500 m off the southern end (Fig. 6.61). Both beaches are backed by low grassy climbing foredunes.

Figure 6.61 Chilli Beach includes beach Q 410 (foreground) and the longer beach Q 411.

Chilli Beach (**Q 411**) is one of the better known and visited cape beaches. There is vehicle access to the northern end and the wide low gradient beach is suitable for vehicle traffic. It trends to the southwest, then curves to the south for 7 km into the protected Albatross Bay. For the most part it consists of a narrow high tide beach, fronted by a 100 m wide low tide bar, then a continuous fringing reef extending up to 500 m off the beach, together with granite boulders scattered on and off the northern end of the beach. The beach is backed by a 300-400 m wide series of early Holocene foredunes. A creek mouth forms the southern end of the beach, which together with the protection from Cape Griffith has formed 500 m wide tidal sand flats in the southern corner.

Q 412-415 **ALBATROSS BAY-CAPE GRIFFITH**

No.	Beach	Rating HT	Rating LT	Type	Length
Q412	Albatross Bay (1)	1	2	R+tidal sand flats	300 m
Q413	Albatross Bay (2)	1	2	R+tidal sand flats	50 m
Q414	Cape Griffith (S 1)	3	4	R+LT rips/rocks	500 m
Q415	Cape Griffith (S 2)	3	4	R+LT rips	800 m

Spring & neap tidal range = 2.1 & 0.1 m

Albatross Bay is a sheltered north-facing bay lying in lee of Cape Griffith. Between the creek mouth at the end of Chilli Beach and the Cape are 1.5 km of mangroves and bedrock shore fronted by tidal sand flats and two small beaches (Q 412 and Q 413).

Beach **Q 412** is located on the southern side of the creek mouth and continues for 300 m to the southeast. It is a narrow high tide beach, bordered at both ends by mangroves, and fronted by irregular creek mouth sand flats extending up to 600 m into the bay. Beach **Q 413** is located 1 km to the east in lee of the cape, and is a 50 m long pocket of sand located at the base of a steep grassy 120 m high cape. It is bordered by rocks on the western side, with mangroves to the east, and vegetated tidal flats extending 200 m into the bay.

Cape Griffith is a prominent 120 m high grassy headland, with wind-blasted shrubs on its windward side. Once around Cape Griffith the first 5 km of shoreline to the south are free of reefs, and wave energy reaches the shore unimpeded. The result is two of the higher energy beaches (Q 414 & Q 415) on this section of the cape.

Beach **Q 414** commences 3 km southwest of the cape and is a 500 m long east-facing beach, bordered by rocky granite points to either end, and backed by a grassy overwash plain, then vegetated slopes rising to 60 m. The beach is fronted by a 100 m wide low tide bar, as well as patchy rock flats, the two combining with the higher waves to form three to four low tide- and rock-controlled rips. Beach **Q 415** lies 100 m to the south, separated by a low grassy granite headland. This is a straight 800 m long, east-facing beach, backed by a few active blowouts climbing up the backing 40 m high bedrock slopes. The beach has a continuous low tide bar usually cut by four beach rips, with permanent rips against the boundary headlands.

Lockhardt Aboriginal Land

Coast length: 104 km (1885-2080 km, 2083-2090 km)
Beaches: 72 (Q 417-486, 487-488)

Q 416-419 **LOCKHART RIVER**

No.	Beach	Rating HT	Rating LT	Type	Length
Q416	Cape Griffith (S 3)	3	4	UD	1.6 km
Q417	Spring Ck	3	4	UD	4.4 km
Q418	Lockhart R (N)	1	2	R+ridged sand flats	2.2 km
Q419	Quintell Beach	1	2	R+ridged sand flats	6.5 km

Spring & neap tidal range = 2.1 & 0.1 m

South of Cape Griffith the coastline trends south for 20 km to the mouth of the Lockhart River, located in the shelter of Cape Direction and deep in Lloyd Bay. The headland at the southern end of beach Q 415 marks the end of the bedrock shoreline, and a near continuous string of four beaches (Q 416-419) extends south for 15 km to the mangrove-lined Lockhart River mouth. The beaches are initially well exposed with higher wave energy, however south of Lloyd Island they are afforded increasing protection by the island, reefs and the shelter of Cape Direction.

Beach **Q 416** commences on the southern side of the 400 m long granite headland, and trends due south for 1.6 km to a small creek mouth, which also marks the southern boundary of the Iron Range National Park. The exposed beach is backed by moderately active blowouts (Fig. 6.62) climbing 100-200 m inland over sloping bedrock terrain. The beach has a low gradient 200 m wide low tide bar, with deeper reefs further out into the bay. Beach **Q 417** commences on the southern side of the 50 m wide creek mouth and is a 4.4 km long wide sandy beach, that contains a subdued cuspate foreland in lee of Lloyd Island, located 1.5 km off the centre of the beach. North of the foreland the beach is more exposed and has an ultradissipative 200 m wide system with a few blowouts and is accessible by vehicle. South of the foreland the island plus some fringing reef lowers wave energy and transforms the bar into a continuous series of 200 m wide ridged sand flats and outer seagrass meadows, with a low stable foredune. The beach terminates 1 km south of Spring Creek at the next small creek mouth.

Beach **Q 418** commences at Spring Creek and continues south for 2.2 km to the next major creek. It is a lower energy beach fronted by 300 m wide ridged sand flats, then seagrass meadows and patchy fringing reefs, while it is backed by dense tropical rain forests. **Quintell Beach (Q 419)** commences on the south side of the 50 m wide creek and continues south for 6.5 km to the mouth of the Claudie River, which enters Lloyd Bay adjacent to the Lockhart River. The beach includes the Lockhart River barge landing, with some granite boulders scattered along the beach south of the landing. Continuous 200 m wide ridged sand flats front the beach, widening to 500 m at the river mouth. The beach is backed by low vegetated terrain, with a series of elongate recurved spits forming the southern 2 km, then older mangrove-lined, spit-remnants continuing on for another 6 km south into the river mouth. **Lockhart River** community is located 2 km west of the barge landing and the airfield another 4 km further inland. There are basic facilities at the beach landing including a parking area, beach boat ramp and toilets.

Figure 6.62 Typical trade wind wave breaking across the ultradissipative beach Q 416, which is backed by active blowouts.

Q 420-424 **ORCHID POINT**

No.	Beach	Rating HT	Rating LT	Type	Length
Q420	Orchid Pt (W)	2	3	R+sand flats/rocks/reef	200 m
Q421	Orchid Pt (E1)	2	3	R+sand flats/rocks/reef	500 m
Q422	Orchid Pt (E2)	2	3	R+sand flats/rocks/reef	400 m
Q423	Orchid Pt (E3)	2	3	R+sand flats/rocks/reef	150 m
Q424	Orchid Pt (E4)	2	3	R+sand flats/rocks/reef	400 m
Spring & neap tidal range = 2.1 & 0.1 m					

Orchid Point is a sloping tree-covered granite point that lies 1 km north of 136 m high Orchid Hill. Granite boulders dominate the shoreline either side of the point with five beaches (Q 420-424) located in amongst the boulders, all backed by steeply rising vegetated slopes. To the west of the point are the extensive mangroves of the Lockhart River mouth, while to the east a 500 m wide band of mangroves extends for 8 km to Chisholm Point.
Beach **Q 420** is located on the western tip of the point, and is a double crenulate 200 m long white sandy beach, with boulders and a few mangroves to either end and in the centre. It is protected by its northerly orientation, and ridged sand flats extend 150 m off the beach to a narrow band of fringing reef. A series of four chenier-recurved spits and intervening mangroves extend for 4 km to the west of the point, where they widen to 1.3 km.

Beach **Q 421** commences on the eastern side of the point and trends to the southeast for 500 m to a boulder-induced inflection in the shore. A few large boulders lie along and off the beach, including a series of small crenulations, together with 100 m wide sand flats, then a narrow band of reef. A fishing shack is located behind the centre of the beach.

Beach **Q 422** commences on the southern side of the boulders and is a similar 400 m long beach, with boulders and sand flats widening to 300 m, and an outer 100 m wide band of reef. A small creek drains across the centre of the beach and maintains a small sandy intertidal delta, while the southern corner of the beach is used as a camp site. Beach **Q 423** is a 150 m long pocket of sand bordered by boulders lying up to 50 m off the beach, as well as boulders on the centre of the beach with seagrass-covered sand flats extending out 200 m to patchy reef. Beach **Q 424** begins at the southern boulders and continues southeast for 400 m, finally merging into the mangroves as wave energy decreases. It is fronted by 200 m wide sand flats, then another 200 m of fringing reef.

Q 425-428 **CAPE DIRECTION**

No.	Beach	Rating HT	Rating LT	Type	Length
Q425	Chisholm Pt (W)	1	2	R+ridged sand flats/waves	400 m
Q426	Chisholm Pt (E)	1	2	R+ridged sand flats/waves	300 m
Q427	Cape Direction (W2)	1	2	R+ridged sand flats/waves	100 m
Q428	Cape Direction (W1)	1	2	R +ridged sand flats/waves	500 m
Spring & neap tidal range = 1.8 & 0.2 m					

Cape Direction marks the southeastern tip of Lloyd Bay and the beginning of a 180 km long section of exposed east-facing coast that trends south toward the protected shores of Princess Charlotte Bay. Chisholm Point is located 1 km west of the cape, and between the two capes are four small beaches (Q 425-428) (Fig. 6.63). The cape also marks a major inflection for the northward sediment transport along the coast. This transport is most manifest as a band of shore transverse sand waves which reach 5 km in width just south of the cape. They narrow to 3 km at the cape, and then accumulate up to 5 km north of the Cape, as well as moving westward around the cape as a shallow, elongate, 9 km long, up to 1.5 km wide, sand spit. The spit and northern shoals would conservatively contain over 30 000 000 m^3 of sand, and probably more than 100 000 000 m^3, all delivered to the cape by the northward transport during the Holocene. This would represent between 5 000 and 15 000 m^3 per year, a reasonable rate for the environment.

Beach **Q 425** commences on the western side of 40 m high **Chisholm Point** and trends southwest along the base of the granite point for 400 m, with some boulders outcropping on the beach. It faces northwest across the long ridged sand spit, which is 1 km wide off the beach. To the west of the beach is an inactive, 8 km long, curving series of several mangrove-lined recurved spits-cheniers, that represent early Holocene sedimentation in lee of the cape.

Figure 6.63 Cape Direction with the three pocket beaches Q 426-428, and associated ridged sand flats.

Beach **Q 426** lies to the east of the point and is a 300 m long, slightly curving, east-facing sand beach, backed by the rocky vegetated point, and fronted by a 300 m wide series of ridged sand flats, then the deeper sand shoals of the cape.

Beach **Q 427** lies 200 m to the east in the next indentation in the rocky shore. It is a curving 100 m long pocket of sand, bordered to the west by steeply rising massive granite rocks, with a lower, sand-draped, vegetated rocky point to the west. One hundred metre wide bare sand flats lie off the beach then deeper seagrass-covered sand shoals. Beach **Q 428** extends from the rocks to the lee of **Cape Direction**, a 20 m high dome of partly vegetated sloping granite. The beach has scattered large boulders at either end and ridged sand flats widening from 100 m in the west to 300 m in lee of the cape, then the seagrass-covered sand shoals.

Q 429-431 **VILLIS POINT**

No.	Beach	Rating HT	LT	Type	Length
Q429	Cape Direction (S)	2	3	R+LT rips	2.8 km
Q430	Villis Pt (N)	1	2	R+sand flats/rocks	150 m
Q431A	Villis Pt	2	3	R+rip	50 m
Q431	Villis Pt (S)	2	3	R+LT rips⇒sand flats	10.7 km
Spring & neap tidal range = 1.8 & 0.2 m					

Villis Point is a 60 m high granite point located 3 km south of Cape Direction. Between the two headlands are two beaches (Q 429 and Q 430), while beach Q 431 extends south of the point toward the Old Lockhart River Mission site. While this is an exposed east-facing section of coast, the wide sand wave field and in places capping coral reefs, plus granite Rocky Island, all substantially lower waves along much of the shore.

Beach **Q 429** commences on the southern side of Cape Direction and trends south before curving round in lee of Villis Point, with a total length of 2.8 km. The northern end of the beach is well exposed with rips cutting across the low tide bar. To the south these give way to a continuous low tide bar fronted by deeper seagrass meadows, with sand waves extending 3 km off the cape. The beach is backed by both Holocene and red Pleistocene transgressive dunes that have climbed 60 m up the sides of the backing 146 km high steep slopes of Direction Hill. The vegetated dunes are scarped in the south revealing brilliant coloured sands and palaesols. An upturned, partly buried wooden boat is also located close to the southern end of the beach.

Beach **Q 430** is located in amongst the northern rocks of **Villis Point**. It is a curving 150 m long northeast-facing moderately protected beach, fronted by a 300 m wide mixture of sand flats, seagrass and granite boulders. Apart from the boulders it provides a reasonable anchorage.

On the southern side of the Villis Point is beach **Q 431A** is a 50 m wide pocket of south-facing sand at the northern end, wedged in between large granite boulders, with a large slab of granite also crossing the beach. The main beach commencing immediately to the south with its surf zone imprinting on the beach and generating a topographic rips against the central rocks.

Beach **Q 431** commences adjacent to the beach and trends to the south for 10.7 km as a triple crenulate beach, owing to subdued cuspate forelands formed in lee of Rocky Island and reefs lying 1-4 km offshore. The northernmost 1-2 km receive sufficient wave energy to maintain low tide rips. However by the first foreland waves are lowered and a continuous low tide bar dominates for the next few kilometres. South of the southern foreland, increasing protection from Second Rocky Point and its associated reefs to the north have permitted a 1 km wide band of ridged sand flats to accumulate off the beach, with deeper sand waves extending up to 5 km offshore, the length of the beach. The beach is backed by a grassy to shrubland, wind-swept, Pleistocene and Holocene dune sheet, that has spread up and over the backing granite slopes (Fig. 6.64), reaching in places 2 km inland and to elevations of 60 m. Dinner Creek drains some of the backing slopes and flows through the stable dunes to reach the centre of the beach where it has deposited a small sandy delta. There is vehicle access to the southern end of the beach.

Figure 6.64 The exposed windswept rip prone northern end of beach Q 431.

Q 432-437 **SECOND RED ROCKY-ROUND POINTS**

No.	Beach	Rating HT	LT	Type	Length
Q432	Old Lockhart R	1	2	R+ridged sand flats/waves	1.3 km
Q433	Cutter Ck	1	2	R+sand flats/reef/waves	2.6 km
Q434	Cutter Ck (S)	1	2	R+rock-ridged sand flats/waves	1.8 km
Q435	Round Pt (N2)	1	2	R+sand-rock flats/waves	150 m
Q436	Round Pt (N1)	1	2	R+LTT/sand waves	1.9 km
Q437	Round Pt	1	2	R+ sand-rock flats/waves	600 m
Spring & neap tidal range = 1.8 & 0.2 m					

Between Second Red Rocky and Round points is a 5 km long section of generally east-facing shoreline containing two curving embayments and six beach systems (Q 432-437). The **Old Lockhart River Mission** is located on the western side of Second Rocky Point, while Cutter Creek enters the centre of the section and separates beaches Q 433 and Q 434. There is vehicle access from Cutter Creek north to the old mission site.

Beach **Q 432** commences on the eastern side of a small creek mouth that separates it from the long beach Q 431. The beach trends east for 1.3 km to the tip of Second Red Rocky Point. It is a crenulate beach with numerous granite rocks outcropping along the shore and in the intertidal towards the point, while wide sand flats in the west grade into ridged sand flats toward the point. Several houses and fishing shacks are scattered along the eastern end of the beach, with wooded slopes rising to 71 m high Mission Hill behind.

Beach **Q 433** commences at the tip of Second Red Rocky Point and trends south for 2.6 km to the mouth of Cutter Creek. The beach is initially backed by the grassy slopes of Mission Hill, then lower wooded terrain toward the creek mouth. While it faces east it is protected from larger waves by a 2 km wide band of sand flats and fringing reef, with sand waves extending up to 5 km offshore. Beach **Q 434** commences on the southern side of the creek mouth. The northern section of the 1.8 km long beach is a dynamic migratory spit, which deflects the creek mouth up to 500 m to the north. The low grassy spit is littered with flotsam primarily from fishing boats and floats. The position of the mouth and consequently length of the spit change over time. The base of the spit attaches to the bedrock slopes where the beach turns and trends southeast, then east to the next headland as a crenulate rock-controlled beach. For most of its length it is fronted by 300 m wide ridged sand flats, with rock flats increasing toward the point, and patchy reef and seagrass offshore (Fig. 6.65), and then sand waves further out. There is a small community located 1.5 km up Cutter Creek, with vehicle access down to the back of the mangroves adjacent to the creek mouth.

Figure 6.65 The southern end of beach Q 434, with rocks and ridged sand flats off the beach, and beach Q 435 on the point.

Beach **Q 435** is located on the tip of the headland. It is a 150 m long patch of northeast-facing sand, backed by grassy slopes, and fronted by a narrow patch of sand and numerous granite boulders, then the reefs. Beach **Q 436** extends south from the point for 1.9 km. It is backed by low grassy slopes, and fronted by 50 m wide sand flats widening to 200 m in the south, then a continuous 100 m wide fringing reef. Beach **Q 437** commences at the southern end of the beach where the shoreline turns and trends southeast for 600 m to **Round Point**. This is another rock-controlled beach with several crenulations, together with rocks and rock reefs scattered across the 100-200 m wide sand flats. It is backed by a mixture of wooded and grassy bedrock slopes.

Q 438-446 **ROUND-FIRST RED ROCKY POINTS**

No.	Beach	Rating HT	LT	Type	Length
Q438	Round Pt (S1)	1	3	R+sand/coral flats	3.8 km
Q439	Round Pt (S2)	2	3	R+sand/rock flats	150 m
Q440	Round Pt (S3)	2	3	R+sand/rock flats	150 m
Q441	Round Pt (S 4)	2	3	R+sand/rock flats	200 m
Q442	First Red Rocky Pt (N5)	2	3	R+sand/rock flats	100 m
Q443	First Red Rocky Pt (N4)	2	3	R+sand/rock flats	100 m
Q444	First Red Rocky Pt (N3)	2	3	R+sand/rock flats	50 m
Q445	First Red Rocky Pt (N2)	2	3	R+sand/rock flats	150 m
Q446	First Red Rocky Pt (N1)	1	3	R+coral reef/sand waves	600 m
Spring & neap tidal range = 1.8 & 0.2 m					

South of **Round Point** the shoreline trends to the south for 5.5 km to First Red Rocky Point, with nine beaches in between (Q 438-446). The two end beaches are longer sand systems, while beaches Q 439-445 are all small pockets of sand located along a rocky section of coast. All the beaches are backed by vegetated slopes rising up to 100 m, in places covered by tropical rain forest. While the coast faces east it is moderately protected by both patchy fringing and inshore reefs lying within 4 km of the shore, including Stork Reef. The relatively shallow waters between the reefs and shore are occupied by a field of shore transverse sand waves that commences more than 20 km to the south, south of Bobardt Point. There is no vehicle access to the coast.

Beach **Q 438** commences on the southern side of conical, 40 m high Round Point and trends due south for 2 km before curving into a slight embayment for 1.8 km to terminate at the beginning of the rocky granite section of shore. The beach is fronted by patchy sand and reef flats extending over 1 km offshore, which lower waves at the shore. It is backed by a northern beach ridge and southern foredune section of shoreline progradation, then the steeper bedrock slopes, with a small creek crossing the beach in the southern embayed section.

Beach **Q 439** is the first in the series of rock-dominated pockets of sand. It is 150 m long and bordered by low rocky points that extend out across the intertidal, together with a mixture of rock and sand flats extending up to 500 m offshore. Beach **Q 440** is a similar 150 m long beach, while beach **Q 441** is a more crenulate 200 m long strip of high tide sand and rocks, fronted by continuous intertidal rock flats that extend out to the point.

At the point the shoreline turns and trends south with beaches Q 442-445 occupying the next 500 m of rocky shore. All are backed by grassy slopes rising into shrubs and trees higher up. Beaches **Q 442** and **Q 443** are two neighbouring 100 m long pockets of sand separated by long granite points, with intertidal rocks extending 100 m off the beaches. Beach **Q 444** is a 50 m long pocket of sand almost encased in granite rocks. Beach **Q 445** is a 150 m long strip of rock-bordered sand, with a strip of intertidal rocks, then some sand flats off the beach, and then sand waves further out.

Beach **Q 446** lies immediately to the south and is a curving, east-facing, 600 m long sandy beach. It terminates in lee of **First Red Rocky Point**, a grassy sloping headland that rises 700 m inland, to a crest of 86 m. The beach is paralleled by a 100 m wide band of rock and reef flats then subtidal transverse sand waves extending up to 1 km offshore.

Q 447-453 **FIRST RED ROCKY-BOBARDT POINTS**

No.	Beach	Rating HT	Rating LT	Type	Length
Q447	First Red Rocky Pt	1	3	R+rock/sand flats	200 m
Q448	First Red Rocky Pt (S1)	1	2	R+sand/reef flats	2 km
Q449	First Red Rocky Pt (S2)	1	2	R+sand/reef flats	3.5 km
Q450	First Red Rocky Pt (S3)	1	2	R+ridged sand/reef flats	1.5 km
Q451	First Red Rocky Pt (S4)	1	2	R+ridged sand/reef flats	2.2 km
Q452	Bobardt Pt (N1)	1	2	R+ridged sand/reef flats	3 km
Q453	Bobardt Pt (N2)	1	2	R+ridged sand/reef flats	1.5 km

Spring & neap tidal range = 1.8 & 0.2 m

At **First Red Rocky Point** the shoreline continues south toward Princess Charlotte Bay. The next bedrock outcrop south of the point is Cape Sidmouth, 40 m to the south. In between is low to moderate energy reef-protected shoreline, with substantial areas of shoreline progradation, dominated by northward migrating spits and recurved spits, and massive sand wave fields paralleling the coast up to 4 km offshore. Numerous reefs are scattered off the coast, while fringing reefs parallel many of the beaches.

Beach **Q 447** is located on the eastern side of First Red Rocky Point and is a 200 m long northeast-facing strip of sand bordered by the sloping grassy point to the west, and low granite rocks and reefs to the east. It is fronted by 50 m wide intertidal rock flats, then patchy sand and reef flats extending out to the sand wave field.

Beach **Q 448** commences at the low rocky tip of the point and trends south for 2 km to a northward-deflected creek mouth. The beach is backed by grassy slopes down as far as the creek mouth and its tidal sand shoals. It is fronted by a 50 m wide low tide bar, then a few hundred metres of subtidal sand and patchy reefs, with the sand waves extending 4 km offshore. Beach **Q 449** commences on the southern side of the small creek mouth and continues south as a low sparsely vegetated sand spit to the next creek mouth, approximately 3.5 km to the south. Both beaches vary in length depending on the location of the migratory creek mouths. This beach is backed by a 300-600 m wide series of four to five earlier spits and their intervening mangrove-dominated swales. The beach is paralleled by a 200-300 m wide band of sand flats, then fringing reefs extending up to 600 m seaward and finally the 4 km wide sand wave field.

Beach **Q 450** commences on the southern side of the migratory creek and continues south for approximately 1.5 km to the next creek. It consists of a narrow, lightly vegetated spit, backed by dense mangroves in the backing creek, then a 500 m wide series of earlier spits and swales. It is paralleled in the north by 200 m wide sand flats widening to tidal shoals at the creek mouths, then the fringing reef and sand waves. To the south the sand flats widen to 500 m and contain multiple low sand ridges. Beach **Q 451** continues south of the creek mouth for 2.2 km as the northernmost in a series of three recurved spits that emanate from Bobardt Point, 4 km to the south. It is paralleled by 500 m wide sand ridges and flats, then the reef and sand waves (Fig. 6.66).

Figure 6.66 Beaches Q 450-452 are part of a series of migratory low sand spits.

Beaches Q 452 and Q 543 are two northward migrating recurved spits. Beach **Q 452** is up to 3 km long and consists of a low lightly vegetated sand spit that migrates north along the inner side of the ridged sand flats. Beach **Q 453** is a similar 1.5 km long spit that is attached in the south to the low sandy **Bobardt Point**. The point has formed as a cuspate foreland in lee of Night Island reef, 5.5 km to the east. The shoreline has prograded 2.5 km as a series of up to 15 recurved spits and their intervening swales. Ridged sand flats extend 0.5-1 km off the two beaches, with fringing reef and sand waves extending out to within 1 km of the island.

Q 454-455 **BOBARDT-VOADEN POINTS**

No.	Beach	Rating HT	Rating LT	Type	Length
Q454	Bobardt Pt (S)	1	3	R+sand flats/reef/sand waves	8 km
Q455	Voaden Pt	1	3	R+sand flats/reef/sand waves	8 km
Spring & neap tidal range = 1.8 & 0.2 m					

Bobardt and Voaden Points are two cuspate forelands formed in lee of inshore reefs, and represent the southern part of the 27 km long beach ridge-spit barrier system that commences at Round Point and terminates 5 km south of Voaden Point. The entire 1500 ha system consists of multiple beach ridge-spits-recurved spits and intervening swales containing outer mangroves grading to inner salt flats. The beaches are paralleled by sand flats, then fringing reef, then shore transverse sand waves, and scattered inshore reefs.

Beach **Q 454** commences on the southern side of the first creek mouth just south of **Bobardt Point** and trends south for 8 km to Hayes Creek. The beach is initially a low migratory sand spit backed by the 1 km wide beach ridge swamp, then by more elevated foredunes beginning 3 km south of the point, where a vehicle track reaches the shore. The beach terminates at Hayes Creek, with beach **Q 455** continuing on south for another 8 km. This beach also starts as a low elongate sand spit backed by mangroves for 4 km around the slightly protruding low sandy **Voaden Point**. The point is composed of a 1.5 km wide series of well vegetated foredunes which extend for another 5 km to the southern end of the beach, where Tertiary deposits are exposed at the shore. Both beaches are paralleled by 50-100 m wide irregular sand flats, edged by fringing reef extending up to 500 m offshore, then the 4 km wide sand wave field.

Q 456-460 **FRIENDLY POINT (N)**

No.	Beach	Rating HT	Rating LT	Type	Length
Q456	Friendly Pt (N4)	1	3	R+sand flats/reef/sand waves	1.6 km
Q457	Friendly Pt (N3)	1	3	R+sand flats/reef/sand waves	800 m
Q458	Friendly Pt (N2)	1	2	R+sand flats/sand waves	4.5 km
Q459	Friendly Pt (N1)	1	2	R+sand flats/sand waves	1 km
Q460	Friendly Pt	1	2	R+sand flats/sand waves	500 m
Spring & neap tidal range = 1.8 & 0.2 m					

Friendly Point is a low recurved spit formed in lee of Sharland Reefs, with a marked inflection in the shoreline to the west occurring at the point. The point and its low surrounding barrier beach represent a 700 ha accumulation of approximately 10 north-northwest-migrating recurved spits. Beaches Q 456-460 extend from the point for 8 km to the northwest, with extensive sand shoals and sand waves also located up to 4 km north of the point.

Beach **Q 456** is located 7 km northwest of Friendly Point and lies along the base of a 3 km long section of gently sloping, well vegetated Tertiary sediments that continue southeast across the back of the spits, while it is bordered by a small upland creek to the north and a larger southern creek which is deflected a few hundred metres to the north. It is fronted by 1 km wide southeast-trending en echelon sand waves and patchy reefs. Beach **Q 457** commences on the southern side of the creek mouth and continues southeast for 800 m to a small creek and its tidal sand shoals, which mark the southern end of the

Tertiary deposits at the shore. The wide band of sand waves continues along the beach, widening from 1 km to 2 km. A vehicle track follows the southern side of the northern creek to the beach.

Beach **Q 458** marks the beginning of the increasingly wide Holocene beach ridge-spit-mangroves deposits. The irregular beach trends to the east-southeast for 4.5 km to the beginning of the recurved spits. This is a low, lower energy beach that faces north across the sand wave field that widens from 2 km to 3 km. It is backed by a discontinuous series of spits, beach ridges and intervening mangroves that widen from 0.1 km in the west to 2 km in the east.

Beaches Q 459 and Q 460 are both associated with migratory recurved spits that form at the Friendly Point inflection and slowly (years-decade) migrate to the west, eventually merging with beach Q 458 (Fig. 6.67). In 1997 beach **Q 459** was a 1 km long, low, discontinuous merging spit, while beach **Q 460** commenced at the point and was a continuous 500 m long spit, in the form of multiple crenulate recurves, with a deflected creek mouth exiting between the two beaches. Over a period of years both beaches vary considerably in position, length and shape. The 3-4 km wide sand wave field extends north of the beach and represents the subtidal accumulation of sand delivered to the point region from the south.

Figure 6.67 Friendly Point consists of a series of dynamic northward migrating sand spits, with beaches Q 459 and 460 shown in this view.

Q 461-463 **FRIENDLY POINT-CAPE SIDMOUTH**

No.	Beach	Rating HT	LT	Type	Length
Q461	Friendly Pt (S1)	1	2	R+sand flats/sand waves	1.8 km
Q462	Friendly Pt (S2)	1	2	R+sand flats/sand waves	2.1 km
Q463	Cape Sidmouth (N)	1	2	R+sand flats/sand waves	1.8 km
Spring & neap tidal range = 1.8 & 0.2 km					

At Friendly Point the shoreline turns and curves to the south then southeast for 5 km to the vegetated granite slopes of Cape Sidmouth. In between are three migratory beaches (Q 461-463) which act as conduits for the sand moving around the cape and north to Friendly Point and beyond. Sand flats and sand waves extend 1-2 km offshore out to the Sharland Reefs, with additional sand deposits extending a total of 7 km offshore.

Beaches **Q 461** and **Q 462** are two adjoining migratory recurved spits that in 1997 were 1.8 and 2.1 km long respectively and consist of more than 20 individual recurves. Each recurve represents a migration of the northern end of the beach to the north, while the southern end is eroded. Over a period of decades both beaches are completely reworked. The shoreline has also built up to 2 km eastward, achieving its greatest width at the point. The beaches are separated by a migratory tidal creek, and paralleled by irregular sand flats, then the subtidal sand waves, then reef and deeper sand waves.

Beach **Q 463** is attached to Cape Sidmouth at its eastern end and extends to the northwest as a series of migratory recurved spits, which terminate at a creek mouth that separates it from beach Q 462. The beach is attached to the bedrock slopes in the east, with three to four older backing spits widening the Holocene deposits to 800 m at the northern end of the beach. To the east of the beach are sand flats, then 2 km wide sand waves, followed by reefs and the outer sand waves.

Q 464-467 **CAPE SIDMOUTH (S)**

No.	Beach	Rating HT	LT	Type	Length
Q464	Cape Sidmouth	1	2	R+sand waves	1.2 km
Q465	Cape Sidmouth (S1)	1	3	R+LT rips	4.5 km
Q466	Cape Sidmouth (S2)	1	2	R+sand flats/sand waves	2 km
Q467	Cape Sidmouth (S3)	1	2	R+sand flats/sand waves	3 km
Spring & neap tidal range = 1.9 & 0.4 m					

Cape Sidmouth is the last outcrop of bedrock exposed at the shore before the 100 km long sweep of sedimentary shoreline south to Princess Charlotte Bay, with the next bedrock at Bathurst Head on the eastern side of the bay, 110 km southeast of the cape. Cape Sidmouth represents the southern end of the resilient rocks of the Coen Inlier, the broad bay occupies the northern end of the sedimentary Laura Basin, while Bathurst Head and nearby Cape Melville represent the beginning of the rocks of the Hodgkinson Basin. Four beaches (Q 464-467) extend 12.5 km south of Cape Sidmouth, terminating in lee of the recurved spits of the Nesbit River delta. The delta is supplying sand that is being transported northward both as shore attached recurved spits, and as an inner and outer transverse sand wave field up to 4 km wide.

Beach **Q 464** is a 1.2 km long beach that runs along the eastern base of the cape, curving round to the lee of a 15 m high granite point. It is a narrow high tide beach for most of its length, backed by some small blowouts in the southern corner climbing to 20 m up the backing slopes. The beach is fronted by several shore transverse sand waves, with smaller sand waves spiralling around the southern point, while more extensive sand waves extend another 4 km east of the cape. The whole system represents a northern movement of sand towards the cape.

Beach **Q 465** commences on the south side of the point, where a creek deflected 1.5 km to the north runs out against the rocks. The beach trends south for 4.5 km as a low mangrove-backed sand spit, that narrows from 200 m in the north to eroding mangroves in the south. There is a slight cuspate foreland half way along in lee of Bell Bank reef located 3 km offshore. The beach is backed by the mangrove-lined creek then a 1 km wide series of earlier spits. It is fronted in the north by a rhythmic 200 m wide triple bar system with rips forming on the inner bar, then subtidal sand waves beyond. Beach **Q 466** commences immediately to the south and is a low relatively straight 2 km long sand beach backed by a mangrove-lined creek, then the inner spit ridges, which narrow to 300 m in the south. A creek deflected 1 km to the north separates it from beach Q 467.

Beach **Q 467** commences as a narrow sand spit paralleling the creek mouth, and continues to the south for a total of 3 km, terminating against one of the northern Nesbit River recurved spits (beach Q 468). The beach is backed by a 400 m wide series of six densely vegetated beach ridges, while shore parallel sand waves extend up to 2 km off the beach. Beach Q 468 periodically extends north as far as the southern end of the beach as a migratory recurved spit.

Q 468-471 **NESBIT RIVER/CAMPBELL POINT**

No.	Beach	Rating HT	Rating LT	Type	Length
Q468	Campbell Pt (3)	1	2	R+sand flats/sand waves	1 km
Q469	Campbell Pt (2)	1	2	R+sand waves	2.2 km
Q470	Campbell Pt (1)	1	2	R+sand waves	2 km
Q471	Nesbit R delta	1	2	R+sand waves	2 km

Spring & neap tidal range = 1.9 & 0.4 m

The sediment rich Nesbit River drains the granite of the McIlwraith Range, located 25 km to the west, delivering substantial deposits of sand to the river mouth. The sand has built a V-shaped delta (beach Q 471) and supplied sand to two migratory recurved spits (beaches Q 469 & 470), the first called Campbell Point, that extend up to 5 km to the north, overlapping at times beach Q 468 and the southern end of beach Q 467. In addition subtidal sand waves extend up to 3 km off the delta mouth.

Beach **Q 468** is the northernmost of the delta deposits. It was active and exposed to waves in 1958, and completely overlapped by beach Q 469 in 1997, the spit having migrated at least 1 km during the intervening 40 years. The inner beach is 1 km long and backed by vegetated sloping sandstone rising to 20 m. In its protected state mangroves spread along the beach and are then eroded when the outer spit moves on, once again exposing the beach to low wave activity.

Beach **Q 469** is the northern of the two migratory recurved spits (Fig. 6.68). In 1997 it was 2.2 km long including a 500 m long series of terminal recurves, while the 1.5 km long trailing arm was a low narrow strip of sand. A 200-400 m wide 'lagoon' separates it from beach Q 468. Beach **Q 470** is the first of the recurved spits and in 1998 was a 2 km long detached curving spit, called **Campbell Point**. The main Nesbit River channel is deflected northward along the back of the spit, exiting between beaches Q 470 and Q 469. Both spits are bordered by a 2 km wide series of generally transverse sand waves.

Figure 6.68 The migratory beaches Q 469 & 470 consist of multiple recurved sand spits.

Beach **Q 471** lies across the **Nesbit River** mouth. It is a narrow 2 km long strip of high tide sand, backed by up to 1.5 km of dense mangroves and some pockets of sand, while it is bordered in the south by a smaller meandering southern river channel. This is a dynamic beach that is periodically eroded by floodwaters and part of the northerly sand transport along the western shore of the bay. It is fronted by a series of intertidal sand ridges, then the subtidal sand waves.

Q 472-474 **NESBIT-CHESTER-ROCKY RIVERS**

No.	Beach	Rating HT	Rating LT	Type	Length
Q472	Nesbit R (S)	1	3	R+LT rips/sand waves	9.5 km
Q473	Chester R (N)	1	2	R+LT rips-bars/sand waves	7.5 km
Q474	Chester R (S)	1	2	R+ridged sand flats/sand waves	4.5 km

Spring & neap tidal range = 1.9 & 0.4 m

To the south of the **Nesbit River** mouth the shoreline continues trending south to south-southwest past the Chester and Rocky rivers and other creeks draining the granites of the McIlwraith Range. The relatively straight shoreline is dominated by a low Holocene coastal plain composed of migratory spits, beach ridges, salt flats and mangroves. While it faces east into the trades, the shoreline is afforded moderate protection by the outer reef 40 km to the east, the mid-shelf Claremont Island and reefs 20 m offshore, as well as scattered inner reefs, some fringing reefs, nearshore sand waves and finally intertidal sand flats that range from 0.2 km to as much as 1 km in width off the river and creek mouths. As the dominant waves and winds are from the southeast all sediment transport is northwards along the shore and nearshore, with bars skewed to the north. Beaches Q 472-475 occupy the next 21 km of south-trending shoreline.

Beach **Q 472** commences on the southern side of the Nesbit River and curves slightly to the south for 9.5 km to the northward deflected mouth of a medium-sized creek. In between is a 200-400 m wide low barrier, cut by a small tidal creek in the centre and backed by 0.2-1 km wide salt flats and some mangroves, then densely vegetated slopes rising to 124 m high Collins Hill and the 200-300 m high Embley Range 2-3 km inland. The beaches are backed by debris-covered overwash deposits and some low foredunes. They are moderately exposed with the waves maintaining a series of northward skewed shore transverse bars and rips spaced about every 80 m. Subtidal tidally driven transverse sand waves extend another 3-4 km offshore.

Beach **Q 473** commences at the creek mouth and continues due south for 7.5 km to the 100 m wide mouth of Chester River (Fig. 6.69). Between the two streams is a series of discontinuous beach spits and beach ridges cut by a few small drainage creeks. The ridges are 1 km wide in the north narrowing to a narrow deflected spit in the south, and backed by salt flats in the south-centre and mangroves in the north, the entire barrier system ranging from 1-1.5 km wide. The salt flats are backed by gently sloping, lightly wooded terrain. There is a vehicle track along the north side of the river, which terminates 2 km from the mouth. The beach is paralleled by northward skewed transverse bars and rips, then two to three shore parallel bars, with the sand waves further out. The bars widen into river mouth shoals off each of the stream mouths, and at the Chester River mouth the shoals extend 1 km seaward as ridged sand flats. The river itself drains part of the backing range and flows across a 3 km wide, swampy, coastal plain to reach the shore.

Beach **Q 474** extends south from the Chester River for 4.5 km, curving southwest into a 5 km wide embayment to the north of Colmer Point, and finally terminating at a small tidal creek in the centre of the embayment. The beach is backed by a 1 km wide low beach ridge plain in the north, partly attached to the backing slopes, that narrows to 300 m at the creek mouth, where salt flats extend 2 km inland. It is paralleled in part by the northward skewed bars and rips, then outer shore parallel ridges, then the deeper sand waves, with Frenchman Reef located 2 km offshore.

Figure 6.69 The Chester River mouth with beaches Q 473 & 474 to either side, and extensive ridged sand flats extending offshore.

Q 475-476 **ROCKY RIVER DELTA**

No.	Beach	Rating HT	Rating LT	Type	Length
Q475	Rocky R (N)	1	2	R+LT rips-ridges/sand waves	4.6 km
Q476	Rocky R (S1)	1	2	R+ridged sand flats/sand waves	3.8 km
Spring & neap tidal range = 1.9 & 0.4 m					

Rocky River reaches the coast at the low sandy Colmer Point where it is deflected to the north for 1 km, depositing a 1 km wide series of tidal sand ridges. Behind either side of the river mouth is a 1 km wide series of abandoned spits and beach ridges and intervening river channels that comprise the 8 km long, 500 ha Rocky River delta and beaches (Q 475 and Q 476).

Beach **Q 475** commences at the northern mouth of Rocky River and curves to the south for 4.6 km as a low sandy overwash spit, widening to the 1 km wide series of abandoned spits and terminating at the southern mouth. The river mouth migration and switching probably occur over periods of years to decades. The beach is fronted in the north by northward skewed transverse bars and rips, which widen to the 1 km wide ridged river mouth shoals to the south, which are also protected by reefs lying 1-2 km offshore.

Beach **Q 476** commences on the southern side of the southern mouth and curves to the south for another 3.8 km to where the backing ridges pinch out and mangroves are exposed at the shore. The entire system is linked to the river in the centre, which has built a 5 km wide Tertiary-Quaternary coastal fan with embayments to either side. Within the embayments the ridge systems are backed by 2 km wide salt flats. The beach begins in lee of the wide river mouth shoals and narrows to the centre as transverse bars and rips dominate the shoreline, widening to 1.5 km wide ridge sand flats in the south.

Q 477-479 **ROBERTS POINT**

No.	Beach	Rating HT LT	Type	Length
Q477	Roberts Pt (N2)	1 2	R+sand waves	2.3 km
Q478	Roberts Pt (N1)	1 2	R+ridged sand flats/ sand waves	3.7 km
Q479	Roberts Pt	1 2	R+ridged sand flats/ sand waves	5.5 km
Spring & neap tidal range = 1.9 & 0.4 m				

Roberts Point is a low sandy curving migratory sand spit, and part of a series of three migratory spits (beaches Q 477-479) that occupy 12 km of shoreline to the north of Massy Creek. All three beaches are paralleled by 0.5-1 km wide ridged sand flats and sand waves, and some fringing reef, and backed by 1-2 km wide sand flats. The stockyards and paddocks of Silver Plains station back the southern half of beach Q 479.

Beach **Q 477** is a 2.3 km long, curving, migratory spit which changes in length and shape over time. It represents the transition between the northern Roberts Point spits and the southern ridges of the Rocky River delta. It consists of a low, curving, partly vegetated sand spit, backed by the extensive mangroves of two converging meandering tidal creeks, which drain the wide salt flat plain, the mangroves extending up to 1.5 km in across the plain. The two creeks converge to form the boundary between beaches Q 477 and Q 478.

Beach **Q 478** commences at the northern end of the second spit and is a convex south-trending 3.7 km long low narrow spit, backed by a 100-200 m wide band containing a few ridges in places, mangroves elsewhere, then 2 km wide salt flats. Older ridges west of the flats are either early high sea level Holocene or Pleistocene in age, and extend to the Tertiary deposits of the backing gently sloping coastal plain. The beach is paralleled by a series of transverse, in the north, to parallel sand ridges and sand waves to the south, extending up to 1.5 km offshore.

Beach **Q 479** is a convex, south-trending, 5.5 km long, discontinuous barrier spit, that includes the low sandy **Roberts Point** at its most easterly protrusion. The beach is backed by a 100-200 m wide series of low former spits and mangrove-lined deflected creeks, then 2 km wide salt flats that narrow to 1 km at Massy Creek, then older ridges between the salt flats and the backing coastal plain. Ridged sand flats parallel the beach, extending 400-600 m offshore where they are fringed by 100-200 m wide reefs.

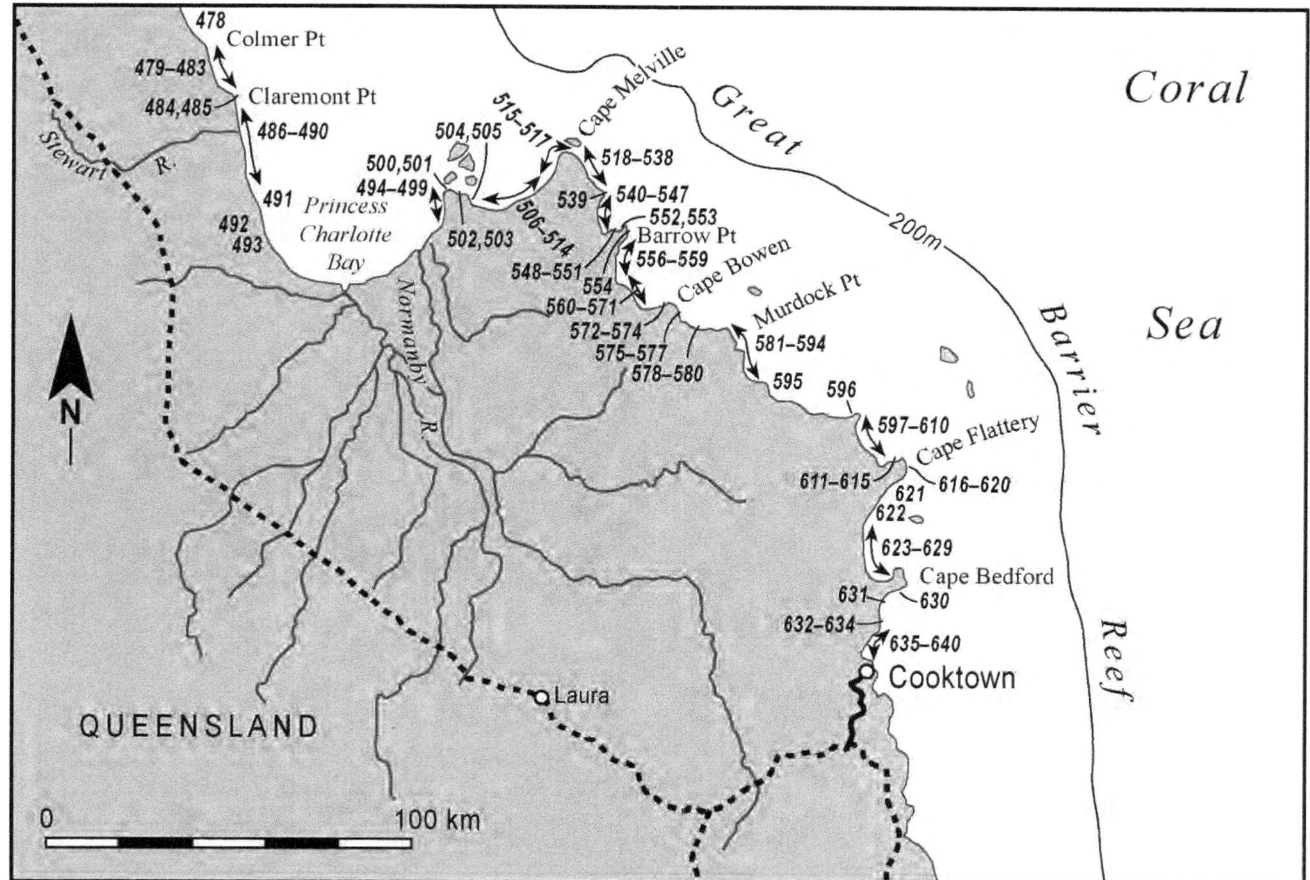

Figure 6.70 Cape York regional map 4 of the southern section of the eastern Cape, beaches Q 478-640.

Q 480-481 **MASSY-BREAKFAST CREEK**

No.	Beach	Rating HT	LT	Type	Length
Q480	Massy Ck (S)	1	2	R+ridged sand flats/ sand waves	3.5 km
Q481	Breakfast Ck (N)	1	2	R+ridged sand flats/ sand waves	2.5 km
Spring & neap tidal range = 1.9 & 0.4 m					

Massy and Breakfast creeks are part of a series of medium sized rivers and creeks that drain the McIlwraith Ranges and flow across the Tertiary outwash plain to the lower energy shore of Princess Charlotte Bay. The two creeks parallel each other 8 km apart, with Silver Plains station located to the north of Breakfast Creek 8 km in from the coast. Beaches Q 480 and Q 481 occupy the 6 km of shore between the two river mouths.

Beach **Q 480** is a low 3.5 km long east-northeast-facing migratory spit broken by three to four small tidal creeks, and backed by a 600 m wide series of earlier spits and 1-2 km of salt flats. The beach is paralleled by 400-600 m wide alternating creek mouth shoals and intervening ridged sand flats and some fringing reef. Beach **Q 481** is a similar 2.5 km long migratory barrier spit that terminates at the mouth of **Breakfast Creek**. It is also backed by outer and inner spit sequences and intervening and backing sand flats, the entire system up to 2 km wide. The tidal shoals extend over 1 km off the mouth, narrowing to 600 m wide ridged and transverse sand flats at the northern end of the beach.

Q 482-486 **CLAREMONT POINT-STEWART R**

No.	Beach	Rating HT	LT	Type	Length
Q482	Breakfast Ck (S1)	1	2	R+ridged sand flats/ sand waves	2 km
Q483	Claremont Pt (N2)	1	2	R+sand waves	2.7 km
Q484	Claremont Pt (N1)	1	2	R+sand waves	2.4 km
Q485	Claremont Pt	1	2	R+sand waves	700 m
Q486	Stewart R (N)	1	2	R+LT rips/sand waves	6.7 km
Spring & neap tidal range = 1.9 & 0.4 m					

The **Stewart River** is one of the larger rivers draining the backing ranges and flows across a 20 km wide braided outwash plain to reach the coast at Port Stewart. The sediment delivered by the river is reworked northward for 7 km to Claremont Point and then northwest along beaches Q 485-482, as a series of migratory recurved and barrier spits. This 12 km long system represents the Stewart River delta. In total the shoreline has prograded up to 2.5 km seaward and deposited multiple beach ridge, barrier and recurved spits totalling 3000 ha. Paralleling the dynamic shoreline deposits is a 1-2 km wide band of shore parallel and transverse sand ridges and deeper sand waves, as well as some fringing reef off Claremont Point. The wide low barrier system and its salt flats and mangrove-filled swales prevent access to the shore, apart from the small 'port' settlement on the northern banks of the Stewart River mouth.

Beach **Q 482** is a southeast-trending, 2 km long, low barrier spit, backed by a 1.5 km wide series of outer and inner ridges and intervening outer mangroves and inner salt flats. It is bordered by Breakfast Creek to the north and a smaller creek to the south. Creek mouth shoals dominate the intertidal, extending up to 1.5 km offshore.

Beaches Q 483, Q 484 and Q 485 are migratory recurved spits that originate at Claremont Point and move slowly westward, eventually merging with the shoreline along beach Q 482. All three vary considerably in location, shape and size over a period of years to decades. The following description is based on the 1960, 1970 and 1997 aerial photographs and the 1998 topographic maps. Beach **Q 483** is the westernmost, located up to 4 km from the point. It is a 2.7 km long migratory recurved spit lying up to 300 m off the main shoreline. It is backed by a shallow intertidal lagoon, then the mangrove-lined shoreline and backing older ridges totalling 1.5 km in width (Fig. 6.71). Beach **Q 484** is a 2.4 km long recurved spit, with a similar backing lagoon, mangrove-lined shoreline and 2 km wide outer and inner barrier ridges and swales. Migratory sand shoals extend up to 1.5 km off the beach. Beach **Q 485** commences at the tip of Claremont Point and periodically migrates as a recurved spit up to several hundred metres to the west, where it detaches and continues west as beach Q 484. It is backed by the widest section of older ridges totalling 2.5 km in width, while off the point sand ridges, sand shoals and sand waves extend up to 1 km seaward.

Figure 6.71 Beach Q 483 is the outermost in a 1.5 km wide series of prograding recurved spits and intervening mangrove-filled swales.

At **Claremont Point** the shoreline turns and trends south for 6.7 km to the Stewart River mouth as beach **Q 486**. This is a more exposed, low overwashed beach, with sufficient wave energy to maintain shore transverse, northward-skewed low tide bars and rips close to shore, together with inter to subtidal ridges extending 1 km offshore. The entire beach is backed by a 2-2.5 km wide series of up to 20 former beach ridge spit systems and intervening outer mangrove and inner salt flats. The beach terminates at the 200 m wide Stewart River mouth.

Q 487-489 **STEWART RIVER-EVANSON POINT**

No.	Beach	Rating HT	Rating LT	Type	Length
Q487	Stewart R (S)	1	2	R+ridged sand flats	2.4 km
Q488	Three Mile Lagoon	1	2	R+ridged sand flats/ shoals	1.5 km
Q489	Evanson Pt	1	2	R+ridged sand flats/ shoals	500 m
Spring & neap tidal range = 1.9 & 0.4 m					

At the **Stewart River** mouth the shoreline continues to trend south into the lower reaches of Princess Charlotte Bay. Beaches Q 487-489 occupy parts of the next 8 km of shoreline down to the creek at the southern end of Evanson Point. All three are fronted by 0.5-1 km ridged and transverse sand flats and tidal shoals, edged by fringing reefs, and backed by a 1-2 km wide outer and inner beach ridge plain. A vehicle track runs 2 km inland with no formal access across the swampy backshore.

Beach **Q 487** commences at the river mouth and trends due south for 2.4 km to a mangrove-lined tidal creek mouth. It is fronted by 500 m wide ridged sand flats and outer fringing reef, and backed by a 1.5 km wide densely vegetated beach ridge plain in the south, grading into spits recurving into the river mouth in the north.

Beach **Q 488** is located 1 km south of the creek mouth and consists of a narrow 1.5 km long strip of low sand bordered to either end by mangroves. Two small tidal creeks border the beach and one cuts across the centre, with all three depositing tidal sand shoals on the fronting 800 m wide tidal flats, which are in turn fringed by 200 m wide reefs. The beach is backed by a 1 km wide series of outer ridges and a 500 m wide band of more elevated inner ridges, with swales filled with mangroves that grade into inner salt flats, including Three Mile Lagoon.

Beach **Q 489** is located 2 km south of the southern side of the mangrove-lined **Evanson Point.** It lies in a 500 m long gap in the mangroves, which continue south for another 2 km. The beach is fronted by 1 km wide ridged and tidal sand flats and backed by a 1.5 km wide series of outer and inner ridges and intervening swales.

Q 490-492 **BALCLUTHA-RUNNING CREEKS**

No.	Beach	Rating HT	Rating LT	Type	Length
Q490	Balclutha Ck	1	2	R+ridged sand flats	1.8 km
Q491	Balclutha Ck (S)	1	2	R+ridged sand flats	8 km
Q492	Running Ck	1	2	R+ridged sand flats	1.8 km
Spring & neap tidal range = 2.2 & 0.4 m					

A distributary of Balclutha Creek forms the southern boundary of the Evanson Point beach (Q 489) with mangroves continuing south of the creek mouth for 2 km to the beginning of beach **Q 490**. This beach trends due south for 1.8 km to a small creek that separates it from the longer beach Q 491. There is vehicle access along the northern side of the creek to the southern end of the beach, which is backed by a 1.5 km wide series of vegetated beach ridges that continue to the south behind beaches Q 491 and Q 492, the entire system covering 2,000 ha. It is fronted by an 800 m wide series of ridged sand flats, then fringing reef.

Beach **Q 491** continues south of the creek mouth for 8 km to a 2.5 km long section of mangroves that separates it from beach Q 492. It is fronted by 1 km wide sand flats and patches of fringing reef, and backed by the 1-1.5 km wide beach ridge plain containing about 12 ridges, some separated by mangrove-filled swales.

Beach **Q 492** trends south of the mangroves for 1.8 km to the small indented mouth of Running Creek. The beach is backed by a series of inner spit-ridges which widen from 0.1 km to 1.5 km. Sand flats continue to parallel the beach.

Q 493 **GOOSE CREEK**

No.	Beach	Rating HT	Rating LT	Type	Length
Q493	Goose Ck	1	2	R+sand-mud flats	17.5 km
Spring & neap tidal range = 2.2 & 0.4 m					

Running Creek marks the northern end of 17.5 km long beach **Q 493**. The beach curves to the south then southeast into the bottom of Princess Charlotte Bay, where it gives way to mangrove-lined mud flats that continue on for 16 km to the mouth of the North Kennedy River. Behind the mangroves the former beach line splays as a series of cheniers, some of which almost reach the river mouth. The beach is fronted by 1 km wide sand flats in the north that grade into mud flats to the south. A few small tidal creeks including **Goose Creek** drain across the beach. The beach system is backed by a Holocene and Pleistocene series of inner ridges that hinge off the northern end of the beach. They widen to the south with the Holocene ridges reaching 2.5 km in width, while the Pleistocene lie up to 14 km inland, indicative of the very low gradient deposits and probably a combination of higher Pleistocene sea level and slight mid-Holocene uplift. A vehicle track parallels the back of the Holocene ridges along the southern half of the beach, with access to the shore via the tidal flats 5 km east of the southern end of the beach.

Princess Charlotte Bay cheniers

Princess Charlotte Bay is a 50 km wide, 30 km deep, U-shaped, north-facing bay, which occupies the northern end of the Laura Basin. In the south it is fed by the North Kennedy, Bizant, Normanby and Marrett rivers, all distributaries of the main North Kennedy and Normanby rivers. It is bordered by Evanson Point to the west and Bathurst Head 55 km to the southeast. The low bay gradient, the delivery of fine sediments by the rivers and the protected nature of the bay have all combined to develop a massive low deltaic system containing extensive salt flats, mangrove-lined creeks and rivers, and a series of both Pleistocene and Holocene chenier-beach ridges. The entire system has a curving shoreline length of 76 km, with deposits extending up to 25 km to the south. The entire delta covers about 90 000 ha, making it one of the larger deltaic plains outside of the southern Gulf of Carpentaria.

Mid-Holocene cheniers extend from the eastern end of beach Q 493 for another 15 km, almost reaching the Kennedy River. On the eastern side a 32 km long series of cheniers begins 3 km south of Bathurst Head and trends to the south then southwest for 32 km across the Marrett River to the Normanby. They are in places 3 km wide and cover an area of approximately 6000 ha.

Lakefield National Park

Area: 537 000 ha
Coast length : 37 km (2138-2134 km, 2142-2172 km)
Beaches: none (mangroves & mud flats)

Lakefield National Park incorporates much of the deltaic plain of the North Kennedy and Normanby rivers together with some of the backing low catchment of both rivers. The northern coastal boundary follows the curving southern shoreline of Princess Charlotte Bay, with two sections totalling 37 km. This shoreline is entirely dominated by mangroves and backed by many kilometres of low tidal flats and salt plains. Much of the upland terrain is covered by woodlands.

Q 494-499 **BATHURST HEAD**

No.	Beach	Rating HT	Rating LT	Type	Length
Q494	Bathurst Hd (S5)	1	3	R+mud flats	1.3 km
Q495	Bathurst Hd (S4)	1	2	R+sand flats	150 m
Q496	Bathurst Hd (S3)	1	2	R+sand flats	1.4 km
Q497	Bathurst Hd (S2)	1	2	R+sand flats	200 m
Q498	Bathurst Hd (S1)	1	2	R+sand flats	400 m
Q499	Bathurst Hd	1	2	R+sand flats	250 m
Spring & neap tidal range = 2.2 & 0.4 m					

Bathurst Head is a prominent, 74 m high, densely vegetated, sandstone headland that marks the northeastern boundary of Princess Charlotte Bay, and a return to the bedrock of the Hodgkinson Basin dominate the coast south to Hinchinbrook Island. The head lies at the northern end of the Bathurst Range, which is composed of horizontally bedded Cretaceous sandstone and shale. While the range has been deeply dissected, there are flat plateau remnants to the south. Six beaches (Q 494-499) are located on the western side of the head and extend 5 km to the south to the mangroves of the bay shoreline.

Beach **Q 494** is the outermost in a series of about twelve 1-2 km long spits that have been fed by sediment moving south along the base of the head. They have prograded the shoreline 1.5 km into the bay and represent the very northern hinge line of the massive beach ridge-chenier system that continues to the southwest for 32 km to the Normanby River. The present beach is tied to the bedrock in the north and consists of a low, partly vegetated, west-facing, 1.3 km long barrier spit, backed by dense mangroves of the backing swale, and fronted by 1 km wide mud flats. There is vehicle access along the base of the backing steep bedrock slopes to the northern end of the beach, and a landing ground built along the crest of one of the inner ridges. A 100 ha lagoon lies between the innermost ridge and backing sandstone slopes.

North of beach Q 494 steep vegetated sandstone slopes dominate the shore to Bathurst Head, with five beaches tucked in at the base of the slopes. Beach **Q 495** is located on the first headland and is a 150 m long pocket of southwest-facing sand, wedged into a steep valley, and fronted by 500 m wide tidal flats. Beach **Q 496** commences at the tip of the headland and trends north for 1.4 km to a slight rocky protrusion. It is backed by a 50-100 m wide band of grassy beach ridges, then the steep vegetated slopes. Tidal flats widen from 500 m off the southern end to 800 m in the north. Beach **Q 497** commences on the northern side of the rocks and continues north for 200 m to the base of a headland that extends 500 m to the west. Rocks outcrop at either end and in the centre of the beach, with sand flats extending 600 m into the bay.

Beach **Q 498** is located in a 400 m long west-facing embayment just below the head. The beach has filled two small backing valleys with 200-300 m wide series of low beach ridges. It is fronted by 200 m wide sand flats. Beach **Q 499** is located at the tip of Bathurst Head and is a 250 m long pocket of sand wedged between the two spurs of sandstone that form the head. It partly fills the small valley between the spurs and is fronted by 100 m wide sand flats, which are swept by strong tidal currents. There is a vehicle track along the crest of the backing ridges to the slopes above beaches Q 498 and Q 499.

Q 500-503 **COMBE POINT**

No.	Beach	Rating HT LT	Type	Length
Q500	Combe Pt (W)	1 2	R+sand flats	500 m
Q501	Combe Pt	1 2	R+sand flats	200 m
Q502	Combe Pt (E1)	1 2	R+sand flats	300 m
Q503	Combe Pt (E2)	1 2	R+sand flats	600 m
Spring & neap tidal range = 2.2 & 0.4 m				

Coombe Point is a 100 m high sandstone headland and the northernmost tip of the Bathurst Range. It has 12 km wide embayments to either side, with beaches Q 500 and Q 501 in the western embayment and beaches Q 502 and Q 503 to the east. A vehicle access track commences at the rear of beach Q 494 and winds up and along the crest of the backing ranges to beach Q 500.

Beach **Q 500** is a 500 m long north-facing pocket beach, with steep slopes extending out 300-400 m to either side. It has partly filled the backing valley with a 300 m wide series of low beach ridges, then a small lagoon, while sand flats extending 600 m offshore fill the remainder of the bay. The vehicle track descends the western slopes and runs along the back of the beach. Beach **Q 501** lies 700 m to the northeast and is a 200 m long pocket of sand also filling a small V-shaped valley with 150 m of beach deposits. Steep slopes rise up to 100 m behind the beach, which is fronted by 200 m wide sand flats.

Beach **Q 502** is located 1 km southeast of Combe Point. It is a narrow 300 m long strip of sand extending along the base of steep vegetated slopes that rise to 100 m just behind the beach. It is fronted by sand flats that widen from 100 m in the west to 300 m in the east. Beach **Q 503** commences 400 m to the southeast and shares the sand flats which widen to 400 m off the beach. It is a 600 m long beach that has built out a small 50-100 m wide series of beach ridges and terminates at a small upland creek mouth. To the east of the creek a 500 m long strip of mangroves fronts earlier beach deposits, terminating at the next steep headland behind.

Flinders Group

The Flinders Group of islands is located 7-18 km off Bathurst Head and between Princess Charlotte and Bathurst bays. The seven islands are all high islands formed of Cretaceous sandstone. The largest **Flinders Island** reaches 318 m at Flinders Peak and has an area of about 1500 ha. The 1 km wide Owen Channel separates it from the northern Stanley Island, while to the south Fly Channel separates it from the smaller Denham, Maclear and Blackwood islands. The western end of the Owen Channel is a well protected and favoured anchorage, and used as a refuelling site for the fishing fleet. There is also a small grassy cuspate foreland at the western end of the channel and several small beaches which permit easy landing. The entire island group is part of the 2962 ha **Flinders Group National Park**, and there is a landing and shelter on a small grassy foreland maintained by the parks midway along the southern side of the channel.

The channels and islands are also rimmed by fringing reefs, together with a 100 ha stand of mangroves on the southern side of Owen Channel.

Q 504-508 **BATHURST BAY (S)**

No.	Beach	Rating HT LT	Type	Length
Q504	Bathurst Bay (1)	1 2	R+sand flats	2.4 km
Q505	Bathurst Bay (2)	1 2	R+sand flats	1.4 km
Q506	Bathurst Bay (3)	1 2	R+sand flats	2.2 km
Q507	Bathurst Bay (4)	1 2	R+sand flats	2.3 km
Q508	Bathurst Bay (5)	1 2	R+sand flats	150 m
Spring & neap tidal range = 1.9 & 0.3 m				

Bathurst Bay is a broad 30 km wide north-facing bay located between Combe Point and Cape Melville. The Flinders Group of high islands lies between 7 km and 18 km off its western side, with more open conditions to the centre and east. While the coast is dominated by the north-trending Bathurst and Melville ranges, sandy beaches dominate the shore. Between the eastern side of the Bathurst Range and the western slopes of Bay Hill is an open 11 km wide north-facing embayment containing beaches Q 505-508. A vehicle track parallels the coast reaching to within 2 km of the shore in lee of beach Q 507, but with no formal access.

Beach **Q 504** commences 5 km east of Combe Point and is a straight more exposed, northeast-facing, 2.4 km long beach, that is backed by active foredune and blowouts extending up to 300 m inland, and fronted by narrow 100 m wide sand flats. It is bordered and backed by steep vegetated slopes rising in the west to a maximum of 200 m. Beach **Q 505** lies 500 m to the east and occupies an open 1.4 km wide V-shaped embayment, with sand flats widening from 100 m in the west to 500 m in the east and filling the embayment. The narrow beach is backed by grassy beach ridges widening to 100 m in the centre, then steep slopes rising to 200 m. It is separated from beach Q 506 by a 1 km wide 200 m high headland.

Beach **Q 506** commences on the eastern side of the headland and forms the eastern 2.2 km of a 7 km long curving stretch of beach and mangroves that has filled the bay between two prominent headlands, the eastern head at the base of Bay Hill. The beach faces north and is fronted by sand flats that widen from 100 m in the west to 600 m in the centre of the bay. It is backed by a series of beach ridge-cheniers that are hinged in the west and widen to 1.6 km behind the eastern end of the beach. These continue east as discontinuous beach ridges to the lee of beach Q 507. Beach **Q 507** is located on the western side of the bay and trends west for 2.3 km to the central band of mangroves. Tidal creeks border the ends of both beaches and drain the backing 1200 ha beach ridge system and its associated mangrove-lined creek and surrounding salt flats. The beach ridges narrow to 300 m behind the beach, while the sand flats narrow to 200 m.

Beach **Q 508** is located in a small V-shaped west-facing valley immediately east of beach Q 507. The 150 m long pocket of sand curves round the base of the backing slopes and faces west across 100 m wide sand flats.

Q 509-514 **BAY HILL**

No.	Beach	Rating HT LT	Type	Length
Q509	Bay Hill (1)	1 2	R+sand flats	1.5 km
Q510	Bay Hill (2)	1 2	R+sand flats	300 m
Q511	Bay Hill (3)	1 2	R+sand flats	800 m
Q512	Bay Hill (4)	1 2	R+sand flats	1.8 km
Q513	Bay Hill (5)	1 2	R+sand flats	2 km
Q514	Bay Hill (6)	1 2	R+sand flats	1.9 km
Spring & neap tidal range = 1.9 & 0.3 m				

Bay Hill is a prominent 430 m high hill located to the west of the Melville Range. Four beaches (Q 509-512) lie along the 5 km long northern base of the hill (Fig. 6.72), while beaches Q 513-515 link across the 5 km wide embayment between the hill and the Melville Range. There is vehicle access in the east to beaches Q 514 and Q 515, with the boundary of Cape Melville National Park located toward the western end of beach Q 514.

Figure 6.72 Beaches Q 509-512 form a near continuous beach below 430 m high Bay Hill (left).

Beach **Q 509** is a 1.5 km long north-facing beach that is backed by some minor beach deposits then the western slopes of Bay Hill that rise steeply to 200 m. A slight protrusion in the slopes separates it from beach **Q 510**, a curving 300 m strip of high tide sand at the base of steep slopes. Both beaches are paralleled by sand flats that widen from 100 m in the west to 200 m in the east.

Beach **Q 511** commences at the base of the slopes, then detaches and trends to the northeast as an 800 m long barrier spit to the mouth of a small migratory creek. It is backed by a 100 m wide band of mangroves then the basal slopes of the hill. Beach **Q 512** commences on the eastern side of the small creek and continues east for another 1.8 km to a 500 m wide mangrove-lined tidal creek. The beach is backed by mangroves that widen to 200 m, then an inner series of earlier barrier spits, that is hinged to the west and widens to 300 m in the east, adjacent to the creek. Both beaches are fronted by sand flats that widen to 400 m off the creek mouth, where they merge with the tidal sand shoals.

Beach **Q 513** commences on the eastern side of the creek and trends east for 2 km to a small tidal creek, which is deflected to the west. Beach **Q 514** continues on the other side of the creek for 1.9 km to the next small creek located at a protrusion in the beaches. The two beaches link across the shallow hill-range embayment and are backed by a 400-500 m wide series of discontinuous beach ridges, then the central meandering creek that trends south for 5 km and which is bordered by a 1 km wide band of mangroves and then marginal salt flats. Both beaches are fronted by 100-200 m wide sand flats that widen at each of the boundary creek mouths. A vehicle track runs along the base of the range along the eastern side of the embayment.

Cape Melville National Park

Area: 36 000 ha
Coast length: 127 km (2221-2348 km)
Beaches: 81 (Q 514-594)

Cape Melville National Park extends from the western side of Cape Melville to the cape and for 70 km southeast to Jeannie River, with a total shoreline length of 127 km. It includes the high granite relief of capes Melville and Bowen and the Melville Range, and a coastline containing 81 beaches ranging from short boulder-strewn beaches to long sand stretches and some backed by active and vegetated dunes, as well as sections of mangroves and fringing reef.

Q 515-516 **CAPE MELVILLE (W)**

No.	Beach	Rating HT LT	Type	Length
Q515	Cape Melville (W2)	1 2	R+sand flats	1.2 km
Q516	Cape Melville (W1)	1 2	R+sand flats	10 km
Spring & neap tidal range = 1.9 & 0.1 km				

Cape Melville is a prominent 176 m high pile of granite boulders and the northern tip of the Melville Range that extends for 19 km to the south and reaches heights of over 500 m. Many of the steep slopes are composed of bare granite boulders, which dominate the shoreline to the southeast of the cape. Beaches Q 515 and Q 516 are located along the western shoreline of the cape, and consist of a narrow beach-barrier system backed by the steeply rising boulder slopes.

Beach **Q 515** begins at the creek mouth on the slight protrusion in the sandy shore. It continues northeast for 1.2 km to the next small creek mouth. It is backed by a few low sandy beach ridges, a 500 m wide lagoon filled with mangroves, then salt flats, and finally the steep slopes. The eastern creek flows down the backing slopes to the shore where it is deflected to the west for 300 m.

Beach **Q 516** commences at the creek mouth and trends to the northeast for 10 km to the rocky outcrops of Cape Melville (Fig. 6.73). The beach is slightly crenulate and backed by a 100-300 m wide series of beach ridges, cut by several usually dry creeks that descend off the backing slopes and flow across the lower outwash plain to the shore. Both beaches are paralleled by 200-400 m wide sand flats. There is vehicle access to the southern section of the beach.

Figure 6.73 The eastern end of beach Q 516 abuts the granite boulders of Cape Melville.

Q 517-525 **CAPE MELVILLE**

No.	Beach	Rating HT	Rating LT	Type	Length
Q517	Cape Melville	1	2	R+sand/rock flats	100 m
Q518	Cape Melville (E1)	1	2	R+sand/rock flats	100 m
Q519	Cape Melville (E2)	1	2	R+sand/rock flats	50 m
Q520	Cape Melville (E3)	1	2	R+sand/rock flats	200 m
Q521	Cape Melville (E4)	1	2	R+sand/rock flats	150 m
Q522	Cape Melville (E5)	1	2	R+sand/rock flats	300 m
Q523	Cape Melville (E6)	1	2	R+sand/rock flats	300 m
Q524	Wedge Rocks	1	2	R+reef/sand/ rock flats	150 m
Q525	Wedge Rocks (S1)	1	2	R+reef/sand/ rock flats	150 m
Spring & neap tidal range = 1.9 0.1 m					

Cape Melville forms the corner of one of the major indentations on the east Cape coast, with Bathurst and Princess Charlotte bays extending up to 90 km to the west. The cape and its eastern shores are dominated by the large granite boulders that form much of the cape shore and backing slopes. At the cape the shoreline turns and trends due east for 3.5 km to Wedge Rocks, where it turns and continues to the south. Beaches Q 517-524 are located amongst the large granite boulders of this northern tip of the cape. All the beaches are backed by steeply rising boulder fields which reach over 200 m, 1 km to the south, with boulders also dominating the nearshore together with a mixture of sand patches and fringing reef and islets, such as the two Wedge Rocks.

Beaches Q 517 and Q 518 are adjoining strips of sand located at the tip of the cape. Beach **Q 517** is a 100 m long strip of high tide sand bordered and backed by large boulders, while a mixture of boulders and sand lies off the beach. There is a sign painted on the boulders indicating a source of freshwater emanating from the steep granite slopes. Beach **Q 518** is located 200 m to the east and is a similar 100 m long strip of sand, with fewer boulders, through also fronted by the boulder-rock flats.

Beaches **Q 519, Q 520 Q 521, Q 522** and **Q 523** occupy the next 1.5 km of shoreline. They are 50 m, 200 m, 150 m, 300 m and 300 m long respectively and all are bordered by large granite boulders and backed by steeply rising, partly vegetated slopes and boulders. Boulders also outcrop along the beach and lie in the offshore, together with patchy sand and reefs, all extending about 200 m offshore.

Beaches Q 524 and Q 525 lie in lee of the two rounded **Wedge Rocks**. Beach **Q 524** is a 150 m long pocket of sand fronted by fringing reef, then sand and rocks. It forms a foreland in the east in lee of a cluster of boulders, with beach **Q 525** extending for 150 m to the south of the foreland. The rocks and reefs extend out past the Wedge Rocks and up to 500 m offshore. The beach occupies a small valley which it has infilled with a low grassy overwash plain that extends 100 m behind the beach, to where two blowouts continue to climb another 100 m inland over the backing boulders.

Q 526-530 **WEDGE ROCKS (S)**

No.	Beach	Rating HT	Rating LT	Type	Length
Q526	Wedge Rocks (S2)	1	3	R+sand flats/reef	1 km
Q527	Wedge Rocks (S3)	1	3	R+sand flats/reef	200 m
Q528	Wedge Rocks (S4)	1	3	R+sand flats/reef	300 m
Q529	Wedge Rocks (S5)	1	3	R+sand flats/reef	250 m
Q530	Wedge Rocks (S6)	1	3	R+fringing reef	300 m
Spring & neap tidal range = 1.9 & 0.1 m					

Wedge Rocks lies 100 m off the eastern tip of Cape Melville, at which point the steep granite shoreline turns and trends to the southeast for 12 km to North Bay Point. In between is a series of 15 small sandy bedrock-dominated beaches (Q 526-539) all fronted by some patchy sand flats and near continuous fringing reef. Beaches Q 526-530 occupy the first 2.5 km between Wedge Rocks and Hales Island.

Beach **Q 526** is a 1 km long, more exposed, east-facing, moderately steep beach. It is fronted by a wedge of sand flats in the southern corner, with fringing reef attached to the northern half of the beach, then running along the outer end of the sand flats, with a gap in the reef providing access to the shore. Sand is actively blowing from the beach exposing a shell midden in the southern corner. A series of blowouts extends up to 300 m inland behind the northern end of the beach, and they are in turn backed by older vegetated dunes, extending another 100 m inland. The dunes are in stark contrast to the backing large granite boulders and bare slopes, which rise to 70 m (Fig. 6.74). A small, usually blocked creek flows through a steep valley and across the southern end of the beach. Beach **Q 527** trends east along the base of the granite rocks at the southern end of beach Q 526. It is a crenulate discontinuous 200 m long beach, fronted by 50 m wide sand flats then the 100 m wide fringing reef.

Figure 6.74 Beach Q 526 and its dunes are wedged in amongst the boulders of Cape Melville. A reef fringes the northern section of the beach.

Beach **Q 528** commences at the next small granite point at the end of beach Q 527. It is a 300 m long, more exposed, east-facing beach, which is also backed by active 100-200 m wide blowouts, and some backing older vegetated dunes, then the rocks. It is fronted by a 100 m wide band of sand flats, then reef flats extending up to 500 m offshore. Beaches **Q 529** and **Q 530** are two adjoining scallops of sand 250 m and 300 m long, both bordered and separated by large granite boulders, and backed by partly vegetated slopes rising to over 100 m. They face northeast toward **Hales Island**, located 500 m off beach Q 530. Beach Q 529 has a 50 m wide patch of sand, then reef flats, while beach Q 530 has fringing reef right to the shore, with both reefs extending to Hales Island.

Q 531-539 **HALES-ROCKY POINT ISLANDS**

No.	Beach	Rating HT	LT	Type	Length
Q531	Hales Is (S1)	1	3	R+fringing reef	150 m
Q532	Hales Is (S2)	1	3	R+fringing reef	200 m
Q533	Hales Is (S3)	1	3	R+sand flats/reef	700 m
Q534	Hales Is (S4)	1	3	R+sand flats/reef	2 km
Q535	Rocky Pt Is (N3)	1	3	R+sand flats/reef	300 m
Q536	Rocky Pt Is (N2)	1	3	R+sand flats/reef	200 m
Q537	Rocky Pt Is (N1)	1	3	R+sand flats/reef	300 m
Q538	Rocky Pt	1	3	R+sand flats/reef	700 m
Q539	Rocky Pt (S 1)	1	3	R+sand flats/reef	1.7 km
Spring & neap tidal range = 1.9 & 0.1 m					

Hales Island is a 16 m high, lightly vegetated, 5 ha granite outcrop located 500 m offshore, with fringing reefs connecting it to the mainland as well as extending another 500 m offshore. Between Hales Island and Rocky Point Island, 7 km to the southeast, is a continuous fringing reef, backed by sand flats, then two valleys bordered by 200 m high spurs, with eight beaches (Q 531-538) and some mangrove patches located along the shore.

Beach **Q 531** is located on the mainland 400 m south of Hales Island and around the low rocky corner from beach Q 530. The 150 m long beach faces east across 400 m wide fringing reef, together with some boulders strewn along the beach. It is backed by a central blowout-overwash section, which extends 100 m inland to the base of a 50 m high granite ridge. Beach **Q 532** commences 200 m to the south and is a 200 m long beach, also backed by some minor blowouts then rising granite slopes, and fronted by a patch of sand at the southern end, and the continuous fringing reef.

Beach **Q 533** commences immediately to the south and extends across the mouth of the first 1.5 km wide valley. It is a relatively straight 700 m long beach that is backed by some minor blowouts and densely vegetated slopes of valley floor. It is fronted by sand flats that widen to 250 m in the south and fringing reef that extends up to 700 m offshore of the beach, widening to 1.5 km off the southern end of the valley. Because of the increasing protection afforded by the reefs, mangroves grow along the southern few hundred metres of beach out across the sand flats. Beach **Q 534** commences as a low narrow spit that overlaps the southern 600 m of the mangroves, then connects to the rocky shore and continues south for a total of 2 km, terminating at a mangrove-covered inflection in the shore. It is backed by the mangroves and beach ridges in the south, then the steep rocky slopes rising 234 m to Twin Peaks, 1 km south of the beach. Sand flats then 1 km wide reef flats parallel the low energy beach.

Beach **Q 535** commences on the southern side of the mangrove-lined point and trends south for 300 m amongst scattered rocks and mangroves to terminate at a continuous band of mangroves that dominates the next 1 km of shoreline. Both the beach and mangroves lie across the 1.5 km wide mouth of the second valley, with inner beach ridges and mangrove-filled swales extending 500 m into the broad valley floor. Three hundred metre wide sand then reef flats extend 1 km offshore.

Beaches **Q 536** and **Q 537** commence 1 km east of the mangroves on the northern rocky side of Rocky Point. They are curving 200 m and 300 m long northeast-facing strips of high tide sand, bordered and backed by granite boulders, with backing slopes rising steeply to 200 m. They are fronted by narrow sand flats then deeper reefs.

Beaches Q 538 and Q 539 lie north and south of Rocky Point respectively (Fig. 6.75). Beach **Q 538** is a curving 700 m long north-facing beach bordered by a 60 m high dune-draped point to the north and the lower, flat Rocky Point and its offshore island to the south. The beach is fronted by 300 m wide fringing reef, which connects with the small 8 m high granite island. Sand has blown off the northern end of the beach onto the backing point, while at the southern end sand from beach Q 539 has blown across the point and spills onto the southern end of the beach. Beach **Q 539** commences on the southern side of Rocky Point and curves to the south for 1.7 km terminating in lee of 200 m high Nares Hill. Dense vegetation backs the southern half of the beach, while blowouts increase in size to the north, the northernmost actively blowing 300 m across the point to beach Q 538. The beach is paralleled by a 50 m wide low tide bar then a continuous 100 m wide fringing reef, then deeper reefs.

Figure 6.75 Rocky Point is bordered by beach Q 538 to the north and Q 539 to the south (left).

Q 540-547 **NORTH BAY POINT-NINIAN BAY**

No.	Beach	Rating HT	Rating LT	Type	Length
Q540	North Bay Pt (N)	1	3	R+rock-sand flats/reef	300 m
Q541	North Bay Pt (S1)	1	3	R+rock-sand flats/ sand waves	150 m
Q542	North Bay Pt (S2)	1	3	R+rock-sand flats/ sand waves	200 m
Q543	North Bay Pt (S3)	1	3	R+sand flats/sand waves	1 km
Q544	North Bay Pt (S4)	1	3	R+rock-sand flats	200 m
Q545	Temple Ck	1	3	R+rock/sand flats/ reef	3.5 km
Q546	Beabey Hill	1	3	R+sand flats/reef	1.6 km
Q547	Ninian Bay (W)	1	2	R+ridged sand flats/ reef	3.7 km
Spring & neap tidal range = 1.9 & 0.1 km					

North Bay Point is a prominent 120 km high headland composed of metasedimentary rocks, with slopes covered in grass and shrubs, in contrast to the exposed granite of the backing Melville Ranges. Streams have carved small steep valleys into the side of the point, with five beaches (Q 540-543) located across the base of the valleys, while beaches Q 545-547 are longer beaches sweeping south into Ninian Bay.

Beach **Q 540** is located on the northern side of the point and is a slightly curving 300 m long strip of sand at the base of shrubby slopes that rise steeply to over 100 m. It is bordered by rocky shores, has a low grassy foredune in the centre of the beach and is fronted by a partly sand-covered boulder delta, then a 50 m wide band of sand and fringing reef extending out 200 m.

Beaches Q 541 and Q 542 are two adjoining pocket beaches located in small valleys on the eastern face of the point. Beach **Q 541** is a 150 m long strip of high tide sand, with rocks to either end and in the centre, which is backed by steep grassy slopes rising initially to 50 m. A veneer of intertidal sand covers a small boulder delta, with subtidal sand further out. Beach **Q 542** lies 200 m to the south in the next small valley, and is a straight 200 m long beach at the base of a small, steep grassy valley, which has deposited a boulder delta at its base. The beach covers the top of the delta, with sand flats extending 200 m offshore to a subtidal sand wave field that continues around the point to feed a north-trending 500 m long subtidal sand spit.

Beach **Q 543** is a straight, 1 km long, east-facing beach, with debris-littered overwash flats grading into an outwash plain descending from the backing 100 m high slopes. The beach is composed of some coarse sand and gravel and is moderately steep. It is paralleled by 100 m wide sand flats in the north, widening to 150 in the south, then subtidal sand waves. Beach **Q 544** is located at the base of the southern headland and consists of a discontinuous 200 m long strip of high tide sand located in amongst a boulder debris field at the base of the sloping 50 m high headland. It is fringed by 100 m wide sand flats.

Beach **Q 545** commences against the southern side of the headland and trends south for 3.5 km to the next mangrove-lined headland. It is located across the mouth of a broad valley and has the main Temple Creek draining out across the northern section of the beach. The beach is backed by a low grassy overwash plain (Fig. 6.76), backed in turn by the mangrove-lined creek in the north, then a saline lagoon and finally low grassy valley slopes in the south. Beachrock is exposed along the southern half of the beach. Off the beach are sand flats north of the creek mouth, and fringing reef which widens from 50 m south of the creek to 300 m against the southern headland, with a patch of ridged sand flats in the southern corner, all indicative of a southward decrease in wave energy. A vehicle track crosses the base of the southern slopes to reach the southern section of the beach.

Figure 6.76 View south along beach Q 545 with its steep beach face and grassy overwash plain.

Beach **Q 546** runs along the eastern base of **Beabey Hill**, which rises to 141 m, 600 m west of the northern end of the beach. The straight east-facing beach is 1.6 km in length, with beachrock exposed its entire length. It is fronted by a 50 m wide sand flat, then continuous 100-200 m wide fringing reef. It is backed by a series of small blowouts, which climb the backing slopes.

Beach **Q 547** commences on the south side of the headland and trends south for 3.7 km to the mangrove-lined southern shore of **Ninian Bay**. It is backed by a low vegetated beach ridge plain up to 200 m wide, then 300-400 m wide mangroves in the north and freshwater swamps to the south, with a vehicle track terminating at the rear of the swamps. To the west of the wetlands are older Pleistocene dune deposits extending up to 3 km inland. A small creek draining the mangroves crosses the northern end of the beach where it maintains a small sandy tidal delta.. The beach is fronted by a mixture of narrow ridged sand flats and wider, though patchy, fringing reef, grading into mangroves in the south. The southern end of the beach was eroding in 1998 exposing dead trees along the shore.

Q 548-553 **NINIAN BAY-BARROW POINT**

No.	Beach	Rating HT	LT	Type	Length
Q548	Ninian Bay (S1)	1	2	R+tidal flats/reef	100 m
Q549	Ninian Bay (S2)	1	2	R+tidal flats/reef	200 m
Q550	Ninian Bay (S3)	1	2	R+tidal flats/reef	150 m
Q551	Ninian Bay (S4)	1	2	R+tidal flats/reef	1.2 km
Q552	Barrow Pt (W2)	1	2	R+tidal flats/reef	150 m
Q553	Barrow Pt (W1)	1	2	R+tidal flats/reef	300 m
Spring & neap tidal range = 1.9 & 0.1 m					

The southern shores of Ninian Bay are protected by their northern orientation, the eastern boundary Barrow Point and the two small Barrow islands located 500 m off the point. As a consequence sediment has accumulated along the southern shore which contains two 2 km long sections of mangroves, separated by a 1 km wide ridge of slatey-sandstone, along the base of which are four beaches (Q 548-551), with beaches Q 552 and Q 553 located to the lee of Barrow Point.

Beach **Q 548** is a 100 m long pocket of sand and cobbles bordered by slightly projecting, vertically-dipping metasedimentary rocks, including a low ridge in the centre of the beach. The moderately steep beach is backed by a steep vegetated V-shaped valley and composed of a mixture of sand and platy cobbles and boulders eroded from the bordering rocks. Beach **Q 549** lies 100 m to the east and is a similar 200 m long sand beach with cobbles and boulders increasing toward the eastern end, together with boulders also backing the beach. It is backed by moderately sloping partly cleared slopes, which are used as a campsite, with a vehicle track crossing over the eastern boundary ridge and running down to the beach. Both beaches are fronted by sand and rock flats with some patchy reef, which widens from 50 m off beach Q 548 to 200 m off the eastern end of beach Q 549.

Beach **Q 550** is located on the eastern side of the ridge and is a 150 m long cobble and boulder beach, backed by an open steeply sloping valley, and fronted by rocky flats, then fringing reef. Beach **Q 551** commences immediately to the east and extends east of the headland for 1.2 km to where it merges in with the southern bay mangroves, which continue for another 2 km to the lee of Barrow Point. Two small creeks drain across the centre and eastern end of the beach. The beach is moderately steep and backed by a series of vegetated beach ridges, which widen to 500 m in the centre and connect behind the mangroves with the inner rocks of Barrow Point. Beachrock outcrops along the lower half of the beach, which is paralleled by fringing reef and which widens to 1 km off the eastern end of the beach.

Beaches Q 552 and Q 553 are located on the western side of 50 m high Barrow Point. Beach **Q 552** is a 150 m long strip of sand that is attached to the rocks of the point in the east and extends west into mangroves. It represents the outermost of a series of recurved spits, which connect with the ridges of beach Q 551 in lee of the mangroves. Beach **Q 553** is a 300 m long steep rocky section of shore containing three patches of sand, rocks and beachrock, with the easternmost partly fronted by mangroves. Both beaches are fronted by 500 m wide tidal flats grading into fringing reef.

Figure 6.77 Beach Q 556 is an exposed beach backed by grassy overwash flats.

Q 554-556 **BARROW POINT (S)**

No.	Beach	Rating HT	Rating LT	Type	Length
Q554	Barrow Pt (S1)	1	3	R+sand flats/reef	300 m
Q555	Barrow Pt (S2)	2	3	R+rock/LT rips	250 m
Q556	Barrow Pt (S3)	2	3	R+LT rips	600 m
Spring & neap tidal range = 1.9 & 0.1 m					

Barrow Point is a rocky 50 m high ridge of Palaeozoic metasediments, with the 2 km wide ridge rising to 300 m, 2 km south of the point. Ninian Bay lies to the west of the point, while an initially more exposed shoreline trends to the south for 18 km toward Cape Bowen. There is vehicle access to the point and some of the beaches toward Cape Bowen. Beaches Q 554-556 are located on the east-facing side of the point.

Beach **Q 554** is located immediately east of Barrow Point and is a straight 300 m long beach, fronted by a 50 m wide strip of sand then fringing reef, extending 200 m offshore. It is an open valley, partly filled with grass-covered overwash deposits. A small creek runs along the northern side of the deposits to exit against the boundary rocks. Beach **Q 555** lies 2 km south of the point and is a 250 m long veneer of slightly curving high tide sand, the sand covering the rocks of a cobble-boulder delta, fed by a usually dry creek. Steep grassy slopes rise to 300 m, 500 m west of the shoreline. The beach has rocks to either end and in the intertidal, then a 100 m wide bar, cut by rips during higher waves.

Beach **Q 556** is located 1 km further south at the southern end of the ridge. It is an exposed 600 m long, east-southeast-facing beach, occupying a 500 m wide valley which it has partly filled with overwash deposits, backed by some tree-lined older dunes that have climbed up to 20 m over the backing slopes. The beach is fronted by a 100 m wide bar cut by several rip channels (Fig. 6.77).

Q 557-561 **SALTWATER CREEK**

No.	Beach	Rating HT	Rating LT	Type	Length
Q557	Barrow Pt (S4)	1	2	R+LTT/reef	4.7 km
Q558	Barrow Pt (S5)	1	2	R+ridged sand flats	1.5 km
Q559	Saltwater Ck	1	2	R+ridged sand flats	2.2 km
Q560	Wakooka Ck (N)	1	2	R+ridged sand flats	1.2 km
Q561	Wakooka Ck (S1)	1	2	R+tidal flats	700 m
Spring & neap tidal range = 1.9 & 0.1 m					

To the south of Barrow Point's rocky ridge is a 12 km long, 1-2 km wide barrier system. The barrier is comprised of beaches Q 557-561, which are bordered by small creeks and the larger Saltwater and Wakooka creeks. There is vehicle access along the southern base of the ridge to the northern beach Q 557, and in the south along the southern side of Wakooka Creek to beach Q 562 and beyond.

Beach **Q 557** commences against the steep relatively bare rocky slopes of the Barrow Point ridge, which rise to 130 m. It trends south for 4.7 km to a small tidal creek. The beach consists of an inner low tide bar, paralleled by a fringing reef lying between 100-300 m offshore and running the length of the beach. The beach is backed by a series of inner beach ridges up to 1 km wide, and then a 500 m wide backbarrier depression, occupied by inner salt flats, and mangroves toward the creek mouths. It terminates at a small creek mouth, which drains the backing wetland. Beach **Q 558** continues south for 1.5 km to the mouth of a 50 m wide creek. It is backed by a continuation of the beach ridges and wetlands, protected by 1 km wide ridged sand flats. Between the creek and Saltwater Creek is a 500 m long section of mangroves and tidal sand shoals.

Beach **Q 559** is a northward deflected recurved spit that terminates at the mouth of Saltwater Creek. It extends for 2.2 km south of the mouth to a small tidal creek, with beach **Q 560** continuing south for another 1.2 km to the mouth of Wakooka Creek. Both beaches are backed by a 1 km wide beach ridge plain, then a 500 m wide band of mangroves and salt flats, with mangroves bordering both of the boundary creeks. The beaches are fronted by 1 km wide ridged sand flats, with narrow fringing reef in the north and the Weigall Reefs lying 2-3 km offshore in the south.

Beach **Q 561** extends south from Wakooka Creek for 700 m south to link with the northern debris slopes of the Altanmoui Range. The beach is the outermost of several recurved spits that have migrated northward toward the creek mouth and prograded the shoreline up to 1 km. The present beach is a series of low recurved spits, backed and bordered in the north by mangroves and fronted by 1 km wide partly vegetated tidal flats.

Q 562-568 **CAPE BOWEN (N)**

No.	Beach	Rating HT	Rating LT	Type	Length
Q562	Wakooka Ck (S2)	2	2	R+rock/ridged sand flats	600 m
Q563	Wakooka Ck (S3)	2	2	R+rock/ridged sand flats	400 m
Q564	Cape Bowen (N5)	2	2	R+rock/ridged sand flats	200 m
Q565	Cape Bowen (N4)	2	2	R+rock/ridged sand flats	300 m
Q566	Cape Bowen (N3)	2	2	R+rock/ridged sand flats	200 m
Q567	Cape Bowen (N2)	2	2	R+rock/ridged sand flats	500 m
Q568	Cape Bowen (N1)	1	2	R+ridged sand flats	400 m
Spring & neap tidal range = 1.9 & 0.1 m					

To the east of Wakooka Creek mouth is a 5 km long southeast-trending bedrock-dominated coast continuing out to Cape Bowen. Seven low energy beaches (Q562-568) are located amongst the rocks and mangroves of the shore. A vehicle track runs from Wakooka Creek out to Cape Bowen, paralleling the coast about 200-300 m inland. The Altanmoui Range provides a dramatic backdrop to the beaches, with the steep slopes and escarpments rising to over 600 m within 2.5 km of the shore.

Beach **Q 562** is a 600 m long series of sand pockets linked by bordering rocks and mangroves. It is backed by low wooded slopes and fronted by inner rocks and rock flats and outer tidal flats extending 600 m offshore. Beach **Q 563** lies 500 m to the east and is a curving 400 m long stretch of high tide sand, bordered and fronted by rocks, with mangroves dominating the eastern section of the system. It is fronted by planar then ridged sand flats extending 600 m offshore.

Beach **Q 564** is a 200 m long strip of sand bordered by rocks and mangroves, with patchy rocks, then ridged sand flats extending out 500 m. Beach **Q 565** is a similar 300 m long, highly crenulate strip of high tide sand, with scallops formed by the rock flats, and scattered mangroves growing on the rocks, then the continuous ridged sand flats. Beach **Q 566** is a curving 200 m long beach, bordered by protruding rock flats, with a central clump of rocks and mangroves, then ridged sand flats which narrow to 200 m.

A narrow mangrove-lined, north-trending, joint-aligned creek separates beaches Q 566 and Q 567. Beach **Q 567** faces northeast and is a more exposed 500 m long beach, fronted by a continuous strip of low tide rock flats, with mangroves growing on the southern rocks. It is paralleled by ridged sand flats that widen from 100 m in the north to 300 m in the south, while it is backed by a few low blowouts. Beach **Q 568** occupies the next small embayment. The beach faces northeast and curves round for 400 m to terminate in lee of Cape Bowen. It is free of rock flats and fronted by ridged sand flats widening from 100 m in the north to 300 m at the cape. A small mangrove-lined creek runs out against the southern rocks. The beach is backed by active and vegetated blowouts climbing to 20 m and extending up to 200 m inland (Fig. 6.78, with the vehicle track running along the back of the dunes.

Figure 6.78 Beach Q 568 (foreground) curves to the lee of Cape Bowen, and has experienced moderate dune transgression.

Q 569-571 **CAPE BOWEN (S)**

No.	Beach	Rating HT	Rating LT	Type	Length
Q569	Cape Bowen (S 1)	1	3	R+boulder flats	150 m
Q570	Cape Bowen (S 2)	1	3	R+boulder flats	300 m
Q571	Cape Bowen (S 3)	1	3	R+LT rips-ridged sand flats	3.6 km
Spring & neap tidal range = 1.9 & 0.1 m					

To the south of **Cape Bowen** the shoreline trends south for 5 km before curving round to the lee of Red Point, located 13 km to the southeast. Immediately south of the cape are three more exposed beaches (Q 569-571) occupying the first 5 km of shore, then an 8 km long, more protected, mangrove-lined shoreline that extends to within 2 km of Red Point and includes the mouth of the Howick River.

Beach **Q 569** is a 150 m long sandy-cobble beach, wedged in a small valley and backed by a 50 m wide grassy overwash flat, then the valley sides. It is bordered by rocky points and fringing boulders and paralleled by 100 m wide intertidal boulder flats, the boulder derived from the backing slopes. Beach **Q 570** is located in the next valley and is a similar 300 m long beach with backing overwash plain, extensive high tide boulders toward the southern end and boulder flats paralleling the beach.

A protruding 300 m long boulder delta with stunted mangroves separates beaches Q 570 and Q 571. Beach **Q 571** extends south of the delta for 3.6 m terminating at the beginning of the mangroves of the **Howick River** mouth. The beach is more exposed at the northern end where it has a low gradient beach with a 100 m wide low tide bar cut by rips, as well as beachrock exposed in places across the bar and lower beach face. As wave energy decreases to the south the bar is replaced by ridged sand flats which widen to 400 m. It is backed by active blowouts in the north extending up to 200 m inland, while to the south it is backed by vegetated beach ridge-foredunes. The beach ridge deposits continue on in lee of the mangroves as a discontinuous series of beach ridge-cheniers that extend for another 6 km to the western rocks of Red Point.

Q 572-577 RED POINT

No.	Beach	Rating HT	Rating LT	Type	Length
Q572	Red Pt (W3)	1	2	R+ridged sand flats	300 m
Q573	Red Pt (W2)	1	2	R+ridged sand flats	1.1 km
Q574	Red Pt (W1)	1	2	R+ridged sand flats	300 m
Q575	Red Pt	1	2	R+LTT	150 m
Q576	Red Pt (S1)	1	2	R+LTT	800 m
Q577	Red Pt (S2)	1	2	R+LTT	4 km

Spring & neap tidal range = 1.9 & 0.3 m

Red Point and 123 m high Noble Island, 4 km to the north, are the northern end of a ridge of Palaeozoic metasediments. Beaches Q 572-574 extend for 2 km to the west of the point and beaches Q 575-577 for 5 km to the south, with mangroves to either side of the beaches. There is no vehicle access to the point region.

Beach **Q 572** emerges from the eastern mangroves of the Howick River mouth and is a 300 m long strip of high tide sand, backed by sloping densely vegetated bedrock, and fronted by scattered rocks, mangroves and 400 m wide ridged sand flats.

Beaches Q 573 and Q 574 share a 1.5 km wide north-facing embayment immediately west of Red Point. Ridged sand flat fills the embayment and links the two points (Fig. 6.79). Beach **Q 573** is a 1.1 km long north-northeast-facing beach, backed by a foredune that increases in height to the north where there is some minor dune transgression across the 200 m wide point. It is fronted by the ridged sand flats that widen from 50 m off the point to 400 m in the middle of the bay. Beach **Q 574** runs for 300 m along the base of 20-40 m high bluffs in the middle of the bay. It faces north across a narrow band of rocks and rock flats and the 300 m wide ridged sand flats.

Figure 6.79 Beaches Q 573 & 574 share a continuous field of ridged sand flats.

Beaches Q 575-577 are all more exposed east-facing beaches, however waves remain low to moderate owing to protection afforded by Murdoch Point, 15 km to the east, together with a series of coral reefs associated with the Cole Islands that extend east of Red Point and intervening tidal ridges. Beach **Q 575** is located in a 200 m wide indentation at the eastern tip of the point. It is a curving 150 m long beach, bordered by 20 m high grassy points, backed by a patch of active dunes climbing 20 m up the backing slopes, and fronted by a 100 m wide bar bordered by rock flats. Beach **Q 576** occupies the next longer embayment. It is a slightly curving 800 m long beach with a 50 m wide low tide bar, bordered by 20 m high rocky points and rock flats, and backed by largely stable dunes that have climbed up to 20 m onto the backing slopes.

Beach **Q 577** extends south from the southern rocks of the point for 4 km to merge with the mangroves of Rocky Creek. The beach faces due east and has a 50 m wide low tide bar for much of its length, grading into sand flats along the southern 500 m. It is backed by 100-200 m wide deflated surfaces, then both active and vegetated transgressive dunes extending up to 500 m inland. The dunes are backed by segmented freshwater swamps, with the northernmost swamp draining via a small creek across the beach, while the southernmost swamp connects to the southern mangroves. Earlier beach ridge-cheniers continue east for 3 km in lee of the mangroves to the next bedrock point.

Q 578-580 **BROWN PEAK**

No.	Beach	Rating HT	Rating LT	Type	Length
Q578	Brown Peak (1)	1	2	R+LTT	250 m
Q579	Brown Peak (2)	1	2	R+ridged sand flats	1.1 km
Q580	Brown Peak (3)	1	2	R+sand flats	300 m
Spring & neap tidal range = 1.9 & 0.1 m					

Brown Peak is part of a linear, north-south trending, 12 km long ridge of Palaeozoic metasediments that reaches the coast as two grassy points midway between Red and Murdoch points. The north-facing shoreline between, and to either side, of the Brown Peak points is dominated by the mangroves of Rocky Creek to the west and Dead Dog Creek to the east. Beaches Q 578 and Q 579 are located on the east-facing shore of the western point, and beach Q 580 is tucked in at the base of the eastern point, with continuous 1 km wide sand flats running between the points.

Beach **Q 578** commences at the tip of a 40 m high grassy bedrock ridge and faces northeast toward the Cole Islands. It trends south for 250 m to a slight protrusion in the shore. It is sufficiently exposed to be backed by three active blowouts which have climbed up onto the backing 20 m high ridge, with older vegetated dunes extending about 200 m inland. It is fronted by a 50 m wide low tide bar, then subtidal seagrass-covered sand flats. Beach **Q 579** continues immediately south of the protrusion for 1.1 km to a 1 km long band of mangroves, which connect to the eastern point. The beach faces northeast and is fronted by intertidal ridged sand flats that widen from 50 m in the north to 1 km in the south. To the lee of the beach are vegetated dunes that have climbed the 20-40 m high backing slopes. Both beaches lie in lee of a 5 km wide zone of linear tidal ridges that have been produced by tidal flows between the shoreline and the Cole Islands.

Beach **Q 580** is located at the base of the eastern point and is a 300 m long strip of high tide sand that is attached to the bedrock in the north and bordered by mangroves to the south, with a clump of mangroves also in the centre of the low energy beach. The beach represents the western tip of a 1.5-2.5 km long series of mid-Holocene inner beach ridge-spits and outer-cheniers that now lie in lee of the 500 m wide mangrove forest.

Q 581-588 **MURDOCH POINT-JEANNIE RIVER**

No.	Beach	Rating HT	Rating LT	Type	Length
Q581	Murdoch Pt	1	2	R+sand flats/ sand waves	500 m
Q582	Murdoch Pt (S1)	1	2	R+sand flats/ sand waves	400 m
Q583	Murdoch Pt (S2)	1	2	R+sand flats/ sand waves	800 m
Q584	Jeannie R (N2)	1	2	R+sand flats/ sand waves	1.5 km
Q585	Jeannie R (N1)	1	2	R+LT rips/sand waves	1.8 km
Q586	Jeannie R (S1)	1	2	R+tidal flats/ ridged sand flats	1 km
Q587	Jeannie R (S2)	1	2	R+rocks/ ridged sand flats	300 m
Q588	Jeannie R (S3)	1	2	R+rocks/ ridged sand flats	200 m
Spring & neap tidal range = 2.0 & 0.1 m					

Murdoch Point is the northern tip of a 6 km long series of multiple recurved spits that have been fed in the south by sediments from the Jeannie River. The river drains the backing sandstone range and escarpment and delivers quartz sand to the 1 km wide river mouth shoals, which are then reworked northward by both the waves and flood tide currents. Sand is actively moving northward along the intervening beaches (Q 581-585) as well as manifest offshore as subtidal ridges between the point and Murdoch Island. These ridges expand to the west as a 5000 ha fan of linear tidal shoals that extend between the shoreline and the Cole Islands and reach almost to Red Point, 16 km to the west. Furthermore, the exposed east-facing shore immediately north of the river mouth is capped by transgressive, longwalled parabolic dunes, stretching up to 1 km inland. In total, the Murdoch Point system represents a range of subtidal, intertidal and subaerial tide, wave and wind-dominated systems that compose the active delta of the Jeannie River and occupy a total area of at least 8000 ha.

Beach **Q 581** is located 5 km north of the river mouth and is the northernmost of the active recurved spits. Inactive older spits extend another 3 km to the west behind a 1 km wide band of mangroves. The present beach is 500 m long, faces northeast and consists of a low bare crenulate migratory sand spit, fronted by migratory tidal shoals between the shore and Murdoch Island, and backed by a 200-300 m wide band of mangroves, then older spits and mangrove-filled swales, extending up to 600 m inland (Fig. 6.80). Beach **Q 582** connects with the spit and trends south for approximately 400 m to where it is overlapped by the next beach-spit. In 1998 this beach was eroding into older spits and swales, resulting in some overwash aprons in amongst the mangroves as well as exposing some mangroves at the shore. It is also fronted by the migratory 1 km wide subtidal sand shoals.

Figure 6.80 Beaches Q 582 and 583 form a 2 km long recurved spit, which is backed by a mangrove-filled swale.

Beach **Q 583** is an 800 m long northeast-facing migratory spit, that is tied to dune deposits at its southern boundary. As it moves to the west sand is being eroded from the dunes and transported along the crenulate spit, gradually overlapping beach Q 582 and impounding a 200-300 m wide lagoon that is filling with mangroves. Beach **Q 584** extends southeast from where the spit attaches to a sandy point, the point formed in lee of a fringing reef located 500 m off the beach. The crenulate beach is 1.5 km long and runs along the side of eroding and vegetated transgressive dunes. The beach and its 500 m wide sand wave field are an active conduit for sediment moving north between the reef and shore and toward the northern spits. The eroding dune systems are also supplying sand to active blowouts that are transgressing over older 20 m high vegetated dunes, which extend up to 1 km to the northwest. Beaches Q 581-584 are three very unstable dynamic beach systems, which change in nature, shape, and length over periods of years to decades.

Beach **Q 585** is an east-facing, 1.8 km long beach that commences at the Jeannie River mouth and terminates at the point it shares with beach Q 584. The exposed beach is backed by active blowouts and parabolic dunes extending up to few hundred metres inland, with older vegetated parabolics up to 1.5 km inland. This system together with the three northern beaches in turn forms the eastern boundary of the 2000 ha mangroves and salt flats of Dead Dog Creek. The beach is paralleled in the centre by a 50 m wide bar cut by rips every 100 m, with sand flats widening to either end.

Beach **Q 586** commences on the southern side of the 300 m wide river mouth and curves to the southeast for 1 km, grading into a 1 km long band of mangroves. The beach is a low energy sand spit backed by a 500 m wide splay of older river mouth spits and mangrove-filled swales. It is fronted by 800 m wide tidal and outer ridged sand flats, with active river mouth shoals and bars located across the river mouth.

Beach **Q 587** is located 2.5 km east of the river mouth and is a 300 m long crenulate strip of high tide sand attached in the west to the rocks of a low dune-draped point. The beach faces northwest and grades into mangroves to the west. It is fronted by inner scattered coffeerock and mangroves, and a 500 m wide series of ridged sand flats. Beach **Q 588** is located immediately to the east, in a small indentation at the tip of the point. It is a narrow 200 m long northeast-facing strip of high tide sand, backed by 20 m high dune-draped rock bluffs and fronted by rocks then 300 m wide ridged sand flats. There is a vehicle track from the south to the back of both beaches.

Q 589-594 **JEANNIE RIVER (S)**

No.	Beach	Rating HT	LT	Type	Length
Q589	Jeannie R (S4)	1	3	R+rock/sand flats/reef	200 m
Q590	Jeannie R (S5)	1	3	R+rock/sand flats/reef	600 m
Q591	Jeannie R (S6)	1	3	R+rock/sand flats/reef	700 m
Q592	Jeannie R (S7)	1	3	R+fringing reef	800 m
Q593	Jeannie R (S8)	1	2	R+LTT	1.8 km
Q594	Jeannie R (S9)	1	3	R+rock/sand flats/reef	4 km
Spring & neap tidal range = 2.0 & 0.1 m					

The last several kilometres of the Jeannie River flow behind and in places between a Pleistocene transgressive dune system that occupies the coast between the river mouth and Hummock Creek 10 km to the southeast. The dunes are 5 km wide in the south narrowing to 3 km east of the river mouth. During the Holocene the dunes have been scarped, then partially covered by Holocene transgressive dunes. Beaches Q 586-588 occupy the northern shore of the dunes, while beaches Q 589-594 form the more exposed 8 km long eastern shoreline. There is vehicle access to the southern end of beach Q 594 and up the beaches to beach Q 591 where the track runs in behind the beaches as far as beach Q 587.

Beach **Q 589** commences at the base of a dune-capped 26 m high point, the base of which is composed of Pleistocene coffeerock associated with the older dune system. The coffeerock outcrops along most of the beaches down to Hummock Creek, with fresh water flowing out from the dune aquifers onto the beaches. The beach is a straight 200 m long east-facing high tide beach that terminates in lee of the next small point. It is backed by partially active blowouts that cross the backing 50-200 m wide point. Coffeerock outcrops at both ends and off the beach, together with 100 m wide patchy sand flats, then subtidal seagrass meadows. Beach **Q 590** is located in the next small embayment and is a crenulate 600 m long, east-facing strip of sand, curving between outcrops of coffeerock. It is backed by largely vegetated Holocene dunes overlying the older Pleistocene system. Rocks and sand flats widen to 200 m at the southern point. A continuous fringing reef extends the length of both beaches.

Beach **Q 591** is a 700 m long beach bordered by low sand-capped rocky points, with more open 200 m wide sand flats paralleling the beach and fringing reef at either end. It is backed by some minor climbing blowouts and largely vegetated 20-30 m high transgressive dunes. Beach **Q 592** is a similar 800 m long beach, with reef fringing the base of the beach. It is backed by well vegetated, 1 km wide, Holocene longwalled parabolics, including two areas of Holocene dune activity, then another 1-2 km of Pleistocene dunes (Fig. 6.81).

Figure 6.81 Beach Q 592 and its backing longwalled parabolic dunes.

Beach **Q 593** is a slightly curving, south-trending, 1.8 km long beach also paralleled by fringing reef and terminating in the south at a low coffeerock-over-bedrock point. The entire beach is backed by Holocene dunes up to 60 m high and inner Pleistocene transgressive dunes up to 3 km wide, then the Jeannie River. Beach **Q 594** commences at the rocky point and trends to the south for 4 km to the mangroves of Hummock Creek, with older spits extending for another 2 km into the mangroves. The northern 1 km of the beach is free of reef and has a 50 m wide low tide bar, with reefs and tidal flats widening to 1 km and dominating the southern 3 km. The beach is backed by a continuation of the Holocene and Pleistocene dunes and the river, with the southernmost dunes extending 5 km to the river, and then for another 2.5 km to the west of the river.

The southern boundary of Cape Melville National Park crosses beach Q 594, 1 km north of Hummock Creek, with a vehicle access track from Starcke Homestead reaching the southern end of the beach. Vehicles have to drive up the beach to reach the northern beaches.

Q 595 STARCKE RIVER

No.	Beach	Rating HT LT	Type	Length
Q595	Starcke R	1 2	R+mud flats	800 m
Spring & neap tidal range = 2.0 & 0.1 m				

Beach **Q 595** is located at the mouth of the otherwise mangrove-lined **Starcke River**. The beach extends from the southern side of the 100 m wide river mouth for 800 m as a curving northeast-facing low energy beach ridge, fringed by mangroves and fronted by 2 km wide mud flats. The beach is the outermost ridge in a 400 m wide sequence. A larger beach ridge complex also extends for 5.5 km west of the river mouth, widening in the west to a 2.5 km wide splay containing four ridges separated by salt flats. There is a fishing camp, consisting of a couple of sheds, on the eastern side of the river mouth, connected by a vehicle track to the Mount Webb-Wakooka road, 3 km to the south. The prominent 176 m high Round Hill is located 1 km southwest of the beach.

Lizard Island

Lizard Island is a 990 ha high island located 15 km northeast of Lookout Point. The entire island is a national park island and rises to a central peak of 359 m at Cooks Lookout. There is a landing ground on the island, which services a research station run by the Australian Museum and a tourist resort. The island has extensive fringing reefs particularly around its southern half, as well as 400 ha Blue Lagoon rimmed by the island, and Palfrey and South islands and the reef. There are 18 pocket beaches located around the island, which occupy about half of the 14 km of shoreline.

Hope Vale Aboriginal Land

Coast length: 123 km (2377-2500 km)
Beaches: 45 beaches (Q 596-640)

Hope Vale Aboriginal Land extends from Lookout Point south for 65 km to the Endeavour River, and up to 30 km inland. It encompasses Lookout Point, Cape Flattery and Cape Bedford and their intervening beaches and massive dune fields, as well as the low energy systems to the lee of each of the points. The main community is at Hope Vale, located 20 km inland, with smaller coastal communities at Elim and Nob Point. The Cape Flattery sand mining operation is located in lee of the cape.

Q 596-597 LOOKOUT POINT

No.	Beach	Rating HT LT	Type	Length
Q596	Lookout Pt (W)	1 2	R+ridged sand flats	2.5 km
Q597	Lookout Pt	1 3	R+LT rips	5.4 km
Spring & neap tidal range = 2.0 & 0.1 m				

Lookout Point is an isolated outcrop of Palaeozoic metasediments that rises to 84 m and is connected to the mainland by the extensive Pleistocene and Holocene beach and dune systems. Low energy beach Q 596 extends west of the point, while higher energy beach Q 597 trends to the south of the point. There is vehicle access through the largely vegetated dunes to both beaches.

Beach **Q 596** is a 2.5 km long north-facing beach that is part of the sand transport that moves around and to the west of the point as a series of intricate inter to subtidal sand ridges and subaerial recurved spits, the beach part of the present spit system. Older spits extend up to 1 km inland at the western end of the beach, while an older inactive series of spits continues to the west for 9 km, all now lying in lee of a 0.5-1 km wide band of mangroves. The present beach is very low energy and narrow with mangroves scattered along the shore and a solitary shack located toward the eastern end. It is backed by vegetated truncated transgressive dunes in the east, grading into the low vegetated recurved spits to the west. A pattern of intertidal shore parallel and transverse 'cross ridges' extends up to 1 km north of the beach.

Beach **Q 597** commences at the point and curves slightly to the south for 5.4 km to a mangrove-fronted creek that drains the backing dune systems, with a vehicle track reaching the southern side of the creek mouth. The beach is well exposed to the easterly winds and waves and is backed by moderately active blowouts and larger longwalled parabolic dunes, extending in the south up to 4.5 km inland and to elevations of 35 m (Fig. 6.82). Two vehicle tracks wind down through the dunes to the beach. This barrier-dune system represents the 'tip' of the massive Cape Flattery barrier-dune system that is located immediately to the south. The beach is paralleled by a 100 m wide low tide bar and receives sufficiently high waves to maintain low tide rip channels spaced about every 100 m. In the south the bar widens to 500 m wide tidal shoals associated with the boundary creek.

Figure 6.82 Part of beach Q 597 and its massive dune field.

Q 598-600 **FLATTERY HARBOUR (N)**

No.	Beach	Rating HT LT	Type	Length
Q598	Lookout Pt (S1)	1 1	R+ridged sand flats	500 m
Q599	Lookout Pt (S2)	1 1	R+ridged sand flats	800 m
Q600	Lookout Pt (S3)	1 1	R+ridged sand flats	800 m
Spring & neap tidal range = 2.0 & 0.1 m				

Flattery Harbour is an open 8.5 km wide northeast-facing embayment bordered by Lookout Point in the north and Cape Flattery to the south. While the Lookout Point beach (Q 597) is well exposed, the remainder (Q 598-609) are partly sheltered from waves by the 6 km long protrusion of Cape Flattery, resulting in low beach energy systems fronted by 1 km wide sand flats.
Beaches Q 598 and Q 599 are adjacent migratory recurved spits and part of the sand transport from east to west along the harbour shore, which terminates at beach Q 597 where some sand is transferred into the backing dune systems and some migrates around the point to the sand shoals and spits of beach Q 596. Beach **Q 598** is a 500 m long spit which is slowly migrating along the backing mangrove-lined shore. The spit faces east and is tangential to the more northeast-facing shoreline. It has a dense band of mangroves in the protected 'lagoon' behind the beach, as well as mangroves dominating the eastern end of the beach, which is being overlapped by beach Q 599. It is fronted by 500 m wide sand flats containing transverse ridges.

Beach **Q 599** is a similar 800 m long migrating recurved spit that originates at the mouth of a tidal creek, which drains part of the backing Cape Flattery dune field. The beach emerges from the mangrove-lined creek mouth and trends to the north, recurving round to the west at its northern end. It has enclosed a mangrove-filled lagoon between the spit and the mainland, which also contains older inner spits. It is fronted by a 1 km wide combination of sand ridges and tidal sand shoals associated with the creek mouth.

Beach **Q 600** is a curving, northeast-facing beach that extends to the 800 m east of the creek mouth to the first of the outcrops of coffeerock, which dominate much of the shoreline to the east. It is a narrow low energy high tide beach, with scattered mangroves the length of the beach, and ridged sand flats extending 1 km offshore.

Q 601-608 **FLATTERY HARBOUR (S**

No.	Beach	Rating HT LT	Type	Length
Q601	Flattery Hbr (1)	1 1	R+sand flats	500 m
Q602	Flattery Hbr (2)	1 1	R+sand flats	500 m
Q603	Flattery Hbr (3)	1 1	R+sand flats	400 m
Q604	Flattery Hbr (4)	1 1	R+sand flats	250 m
Q605	Flattery Hbr (5)	1 1	R+sand flats	300 m
Q606	Flattery Hbr (6)	1 1	R+sand flats	600 m
Q607	Flattery Hbr (7)	1 1	R+sand flats	700 m
Q608	Flattery Hbr (8)	1 1	R+sand flats	800 m
Spring & neap tidal range = 2.0 & 0.1 m				

Between the boundary coffeerock of beach Q 600 and the southern corner of **Flattery Harbour** is 8 km of low energy northeast-facing shoreline, all backed by the truncated remnants of Pleistocene transgressive dunes, and fronted by 1 km wide sand flats, with some of the higher dunes and a 2-3 m thick band of dune coffeerock producing crenulations along the shore and separating the eight beaches (Q 601-608) that occupy most of the shoreline. A vehicle track parallels the shoreline about 1 km inland with no access to the shore.

Beach **Q 601** is a curving 500 m long narrow beach bordered by coffeerock to either end, with the rock outcropping along much of the beach. Beach **Q 602** is a relatively straight 500 m long section of coffeerock and high tide sand. Beach **Q 603**, a similar 400 m long straighter section of beach and coffeerock, with a clear patch of sand at its eastern end. Beach **Q 604** lies on the eastern side of a 15 m high scarped dunes and is a curving 250 m long beach, bordered by the coffeerock and backed by the bare scarped dune. Beach **Q 605** extends along the base of the next vegetated dune ridge for 300 m, with coffeerock to either end.

The next 2 km of shore in dominated by coffeerock, with beach **Q 606** a narrow 600 m long high tide beach, also bordered by a longer section of rocks. Beach **Q 607** commences 500 m to the south and is a relatively straight 700 m long beach that terminates at a sand-capped coffeerock point. Beach **Q 608** lies at the very southern extremity of the bay and is a slightly curving 800 m long strip of high tide sand bordered by both rocks and mangroves, together with some scattered mangroves along the beach.

Q 609-615 **CAPE FLATTERY**

No.	Beach	Rating HT	Rating LT	Type	Length
Q609	Cape Flattery (W6)	1	1	R+sand flats	300 m
Q610	Cape Flattery (W5)	1	1	R+sand flats	300 m
Q611	Cape Flattery (W4)	1	1	R+sand/rock flats	100 m
Q612	Cape Flattery (W3)	1	1	R+sand/rock flats	100 m
Q613	Cape Flattery (W2)	1	1	R+LTT-sand flats	1.8 km
Q614	Cape Flattery (W1)	1	1	R+sand/rock flats	100 m
Q615	Cape Flattery	1	1	R+LTT+rock	100 m
Spring & neap tidal range = 2.0 & 0.1 m					

Cape Flattery is a 281 m high, deeply weathered outcrop of Jurassic sandstone, overlying Palaeozoic metasediments. It is one of the larger and more prominent headlands on the entire Cape coast, occupying about 1000 ha, and is connected to the mainland by the extensive transgressive dune field that extends for more than 20 km to the west. Parts of the dunes are mined for silica and the mining operations and jetty are located in lee of the cape. Between the southernmost section of Flattery Harbour and the cape are 7 km of moderately to well protected north-facing shoreline, containing seven beaches (Q 609-615) and the port facilities for the mine, as well as the Cape Flattery anchorage, which is used by both the fishing fleet and passing vessels and yachts.

Beaches Q 609 and Q 610 are located at the southeast section of Flattery Harbour and are both backed by the small Cape Flattery mining community and the mine offices. Beach **Q 609** is a 300 m long strip of high tide sand with mangroves growing to either end. It is backed by a few buildings including the community school and roads. Transverse and outer ridged sand flats extend 600 m off the beach. Four hundred metres of mangrove-lined shore separate it from beach **Q 610**, a similar 300 m long beach that terminates at the westernmost rocky shore of the cape, which is also the site of the 50 m long mine jetty, with boats often anchored off the jetty (Fig. 6.83). It is a low crenulate beach, backed by a grassy beach ridge and fronted by sand flats, which narrow to 200 m in the north. The jetty road parallels the back of the beach leading to the main mine facilities just south of the beach including a landing ground. The steep hillside behind the road is quarried for road base.

Large tabular sandstone outcrops along the rocky shore adjacent to the 50 m long jetty. Immediately east of the jetty is beach **Q 611**, a 100 m long pocket of sand, backed by an informal port storage area, with scattered rocks then a 100 m wide mixture of sand and rocks off the beach. It terminates at a 100 m long old barge landing, with beach **Q 612** located between the barge landing and the bedrock shore to the east. This is also a curving 100 m long pocket beach backed by steep tree-covered slopes which rise to 213 m.

Figure 6.83 The Cape Flattery jetty with beaches Q 610 & 611 to either side.

Beach **Q 613** occupies a 1.5 km wide north-facing embayment located between the jetty hill and the main cape. The beach curves for 1.8 km, with a low tide bar in the west grading to ridged sand flats in the east as wave energy decreases. A rough vehicle track links the western end with the port area, while steep wooded slopes back the beach. Well vegetated dunes that originated from beach Q 621 and migrated 4 km across the cape back the beach. Fishing boats usually anchor off the moderately protected beach.

Beaches Q 614 and Q 615 are two small pockets of sand located on the northwestern side of the cape. Beach **Q 614** is a 100 m long strip of high tide sand lying partially in lee of large boulders, with 200 m wide sand and rock flats off the beach. It is backed by steeply rising vegetated slopes with a navigation marker on the western slopes. Beach **Q 615** lies 500 m to the east at the northern tip of the cape and is a 100 m long strip of north-facing sand, bordered by rocks, and fronted by a 50 m wide low tide bar and some scattered rocks and a couple of mangroves on the boundary rocks. A rough vehicle track winds around the backing slopes to above beach Q 615, with access down to beach Q 614.

Q 616-620 **CAPE FLATTERY (S)**

No.	Beach	Rating HT	LT	Type	Length
Q616	Cape Flattery (S1)	1	2	R+rock flats	100 m
Q617	Cape Flattery (S2)	1	2	R+rock/sand flats	200 m
Q618	Cape Flattery (S3)	1	2	R (cobble/boulder)	150 m
Q619	Cape Flattery (S4)	1	2	R (boulder)	100 m
Q620	Cape Flattery (S5)	1	2	R+LTT	200 m
Spring & neap tidal range = 2.0 & 0.1 m					

The eastern side of **Cape Flattery** consists of a 4 km long east-facing section of vegetated slopes descending steeply from 100-200 m, with a narrow valley dissecting the cape and occupied by beach Q 620. Beaches Q 616-620 occupy a total of 750 m of the shore, all located in small indentations and valleys and dominated by the surrounding bedrock, and with most beaches dominated by cobbles and boulders. There is a vehicle access via the valley to the rear of beach Q 620.

Beach **Q 616** is a 100 m long veneer of high tide sand and cobble wedged between backing slopes rising steeply to 60 m, while it is fronted by 50-100 m wide irregular rock flats. Beach **Q 617** lies 500 m to the south and is a 200 m long slightly more substantial sand and cobble high tide beach fronted by a mixture of rocks and 50 m wide low tide bar, with rocks and rock reefs also bordering and scattered off the beach.

Beach **Q 618** occupies the mouth of a steep, V-shaped valley. It is a curving 150 m long cobble-boulder beach with the boulders extending into the subtidal, where they become covered by sand (Fig. 6.84). Beach **Q 619** is a narrow 100 m long strip of boulders interrupted by protruding rocks and backed by cliffs rising to 80 m.

Beach **Q 620** curves for 200 m across the mouth of the steep V-shaped valley, with a usually dry creek located against the northern valley side. The low gradient sandy beach is backed by a low grassy foredune and fronted by a 100 m wide low tide bar, with rocky points to either side, including a well developed rock platform around the northern point.

Figure 6.84 Beaches Q 618-620 are located at the mouth of steep valleys extending south of Cape Flattery.

Q 621-622 **CAPE FLATTERY-MCIVOR RIVER**

No.	Beach	Rating HT	LT	Type	Length
Q621	Cape Flattery-Red Hill	1	3	R+LT rips	15.8 km
Q622	Red Hill-McIvor R	1	2	R+LTT	5.7 km
Spring & neap tidal range = 2.0 & 0.1 m					

The southern rocks of Cape Flattery form the northern boundary of a 14 km wide embayment, bordered to the south by the equally prominent Cape Bedford. In between are 45 km of sandy shoreline containing nine beaches (Q 621-629). The northern two beaches (Q 621 & Q 622) are backed by one of the largest transgressive dune systems in Australia, extending in places up to 28 km inland and to heights of 140 m, and with a total area of approximately 43 500 ha. The system is composed of northwest-trending Pleistocene and Holocene longwalled parabolics, with some of the dunes several kilometres long and up to 1 km in width. The dune field is a mixture of stable vegetated dunes, brilliant white active blowouts and parabolics, and in places exposed red Pleistocene surfaces, including the 100 m high Red Hill to the lee of Murray Point, a slight sand foreland formed to the rear of Murray Reefs. In the north the pure white silica Pleistocene dunes are mined for export for use in glass manufacturing. There is access at the very northern end of beach Q 621 to the main Cape Flattery jetty located on the southern rocks of the cape, and in the south via the dunes to beach Q 622, 2 km north of the McIvor River.

Beach **Q 621** commences against the southern rocks of the cape, with the road and sand conveyor belt to the 300 m long jetty clipping the northern end of the beach. The beach curves to the southeast for 15.8 km to the cuspate foreland at Murray Point formed in lee of the Murray Reefs located 1 km offshore. The beach is well exposed to the prevailing southeast winds and waves. The waves have delivered the sand to the beach, which has been blown by the winds inland as multiple longwalled parabolic dunes during multiple episodes of rising and high sea level stillstands. There is an average of one parabolic dune for every 500 m of beach. Many are still active some kilometres inland and leave behind their vegetated trailing arms (Fig. 6.85). Smaller blowouts and some parabolics are presently active along most of the length of the beach. The beach itself has a low gradient intertidal beach, with a continuous low tide bar cut by up to 150 rips spaced approximately every 100 m (Fig. 6.86).

Figure 6.85 The massive longwalled parabolics extending in lee of beach Q 621.

Beach **Q 622** commences at Murray Point and curves to the south-southeast for 5.7 km to the 300 m wide mouth of the McIvor River, a large mangrove-lined tidal creek that drains the backing dunes as well as some upland regions. The beach is backed by a continuation of the dunes, which extend up to 15 km inland. The protection afforded by the Murray Reefs and the 1 km wide river mouth shoals and fringing reefs lowers waves at the beach, resulting in a continuous low tide bar and rips in the north, grading into a continuous bar in the centre and river mouth shoals in the south.

Figure 6.86 View south along beach Q 621, with the rip channels visible in the bar morphology.

Q 623-629 **MCIVOR RIVER-ELIM**

No.	Beach	Rating HT	Rating LT	Type	Length
Q623	McIvor R (S1)	1	2	R+ridged sand flats/ reef	2.7 km
Q624	McIvor R (S2)	1	2	R+ridged sand flats/ reef	4.2 km
Q625	McIvor R (S 3)	1	2	R+sand flats/reef	2.2 km
Q626	Elim (W)	1	2	R+sand flats	4 km
Q627	Elim	1	2	R+sand flats	4.5 km
Q628	Elim (E 1)	1	2	R+sand flats	400 m
Q629	Elim (E 2)	1	2	R+sand flats	200 m

Spring & neap tidal range = 1.8 & 0.1 m

The **McIvor River** marks the beginning of a 14 km wide northeast-facing embayment, bordered in the south by the prominent Cape Bedford. The cape, together with the shoreline orientation and offshore and fringing reefs, lowers waves within the embayment, with wave energy decreasing south of the river and mangroves growing in lee of the cape. The Elim community is located along the southern shores of the embayment, with vehicle access through the dunes from the Hope Vale community. The entire shoreline is part of the leeward shore of the massive Cape Bedford transgressive dune system, which primarily originates from beaches Q 630 and Q 631.

Beach **Q 623** commences on the southern side of the river mouth mangroves, in lee of both 500 m wide ridged sand flats and fringing reef extending another 600 m offshore. These induce a protrusion in the shore, with the 2.7 km long beach curving to the south to protrude in lee of a second fringing reef, with more open water in between. A small creek and scattered mangroves reach the shore to the lee of the southern reef and form the southern boundary of the beach. The beach is backed by part of the Cape Bedford dune field, and while there is a series of active small blowouts along the shore, the backing 7 km long longwalled parabolics are predominantly well vegetated and stable. Ridged sand flats extend off each foreland to the reefs, with a continuous low tide bar in between.

Beach **Q 624** commences at the foreland and also curves to the south for 4.2 km to the next reef-anchored, cuspate foreland, with fringing reef also extending from the south part way along the centre of the beach. To the lee of the beach is a 500 m wide series of Holocene foredune ridges, then older well vegetated transgressive dunes that have migrated up to 6 km to the northeast to the base of the 270 m high Gubbins Range. The foredunes give way to a 400 m wide wetland in lee of the southern foreland. Ridged sand flats back the northern reef, with a section of low tide bar, then fringing reef to the south, while, on the southern foreland, migratory recurved spits and attached sand waves are moving between the reef and the shore.

Beach **Q 625** is a third 2.2 km long reef-controlled beach that curves to the south between two forelands. The reef impinges on the northern foreland, while inner transverse and outer ridged sand flats widen to 500 m off the southern foreland. Sand is actively moving around the southern foreland as a couple of low recurved spits. In between the forelands is a vegetated foredune ridge, backed by a 300-500 m wide wetland drained by a creek that exits onto the southern end of beach Q 624. Beyond the wetland are massive, northwest-trending, vegetated longwalled parabolics up to 110 m high, including one large active parabolic 1 km south of the beach.

Beach **Q 626** commences on the southern side of the foreland and curves to the southeast for 4 km to an easterly turn in the shore that marks the beginning of Elim beach (Q 627). The beach is backed by scarped coloured Pleistocene dunes rising to 60-80 m, with the beach located along the base of the partially vegetated dune scarp. The shape and width of the beach is controlled by northward migrating sand waves which become increasingly prominent to the south as they merge with 1 km wide transverse sand ridges.

The beachfront **Elim** community is located along the shores of 4.5 km long beach **Q 627**. This beach occupies the relatively straight, north-facing southern shore of the embayment. It consists of a narrow, crenulate, low energy beach with mangroves dominating the western end, as well as scattered along the beach. It is backed by a low irregular 100-300 m wide beach ridge-wetland plain, that abuts the rear of Green and Red hills, a 121 m high ridge of Carboniferous metasediments, which are partly covered by older transgressive dunes. There is vehicle access to and along the beach to the houses, which are scattered along about 2 km of the shore. The beach is fronted by 1 km wide transverse sand flats in the west narrowing to 500 m wide in the east (Fig. 6.87).

Figure 6.87 Section of Elim beach (Q 627) with the intertidal transverse sand ridges.

Beach **Q 628** is a 400 m long, slightly protruding section of sand, bordered by mangroves, with some also scattered along the beach. There are houses located at either end of the beach, and a road running behind. It is fronted by a continuation of the 500 m wide transverse sand flats.

Finally, beach **Q 629** is a mangrove-dominated 200 m long strip of sand located immediately east, that represents the transition between the beach and the 2.5 km of mangrove-lined shore that links with the western side of Cape Bedford. A house is located in the centre of the beach in a small section clear of mangroves.

Q 630-634 **CAPE BEDFORD-NOB POINT**

No.	Beach	Rating HT	Rating LT	Type	Length
Q630	Cape Bedford	3	4	R+LT rips	3.4 km
Q631	Cape Bedford (S)	3	4	R+LT rips	5.9 km
Q632	Nob Pt (N3)	1	2	R+ridged sand flats	400 m
Q633	Nob Pt (N2)	1	2	R+ridged sand flats	100 m
Q634	Nob Pt (N1)	1	2	R+rock flats	50 m

Spring & neap tidal range = 1.8 & 0.1 m

Cape Bedford is a 248 km high 40 ha mesa (Fig. 6.88), linked in the south to 234 m high Mound Stone, a narrow ridge that terminates at 179 m high South Cape Bedford, the three occupying 6 km of coast. There are three boulder beaches located at the base of the saddle between the cape and the mound. To the west and north of the cape are the 27 km wide embayment and beaches that connect to Cape Flattery, while to the south is a 20 km long southeast–facing embayment that connects with Nob Point, and is occupied by beaches Q 630-634. The main beaches (Q 630 and Q 631) are backed by the massive Cape Bedford transgressive dune system, which extends up to 22 km inland and covers an area of approximately 17,000 ha. There are several vehicle tracks through the dunes to the beaches and along the base of Nob Point to the three point beaches. Several houses are located at the very southern end of beach Q 631 and along beach Q 632.

Figure 6.88 The 248 m high sandstone capped Cape Bedford.

Beach **Q 630** commences in lee of the western tip of South Cape Bedford. It initially faces south and is afforded some protection by the cape and a submerged cobble-boulder spit, with a clump of mangroves growing in the northern corner. It then swings to face southeast into the trade wind and waves and remains a high energy beach down to the rocks of Red Hill, which separate it from beach Q 631. The beach is backed by generally low (<20 m high), deflated, partially active transgressive dunes, and deflated wetlands that extend 2.5 km to the Elim shore (beaches Q 627-629) and also partly climb the sides of 101 m high Red Hill. The 3.4 km long beach consists of a 100 m wide low gradient intertidal beach, fronted by a 100 m wide low tide bar cut by rips approximately every 100 m. It can be reached from Elim via a few vehicle tracks across the dunes and lower slopes of the cape.

Beach **Q 631** commences on the southern side of the 150 m long Red Hill rocks and continues southeast, then south for 5.9 km to the base of Nob Point. For most of its length the beach is well exposed and has a wide beach and low tide, rip-dominated bar. In the southern corner protection from Nob Point lowers the waves, and sand flats extend off the beach. Also cobbles from Nob Point are scattered along the southern corner of the beach. There are vehicle tracks from Elim via the dunes to the beach, as well as the main track from Hope Vale which reaches the score of houses at the southern end of the beach. The beach is backed by a 2 km wide zone of generally deflated partially active dunes, then by the larger partly active longwalled parabolic dunes that extend up to 22 km inland.

Beach **Q 632** is a 400 m long east-facing lower energy beach located between the western slopes of Nob Point and a cobble foreland containing a few stunted mangroves. Several houses back the beach, which is fronted by ridged sand flats, which widen to 300 m in the south.

Beach **Q 633** lies 200 m to the east and is a 100 m long pocket of sand and cobbles, bordered by mangrove-covered rocks and backed by the densely vegetated slopes of the point. It is fronted by rock flats and the eastern end of the ridged sand flats which narrow to 150 m. Beach **Q 634** is located 1 km to the east on the northern side of the point, and is a curving 50 m pocket of sand, bordered by rocks to the north together with a cobble recurved spit which has partly grown across the southern end of the beach and is covered with low mangroves. Densely vegetated slopes back the beach, which is fronted by a patch of intertidal sand, and patchy intertidal and subtidal rocks.

Q 635-638 **NOB POINT-INDIAN HEAD**

No.	Beach	Rating HT	LT	Type	Length
Q635	Nob Pt (S1)	3	4	R (boulder)	200 m
Q636	Nob Pt (S2)	3	4	R+LTT/reef	1.3 km
Q637	Indian Hd (N)	2	3	R+rock flats	150 m
Q638	Indian Hd	2	3	R+rock flats	200 m

Spring & neap tidal range = 1.8 & 0.1 m

Nob Point and Indian Head are part of a ridge of lower Palaeozoic metasediments and upper Jurassic sandstone, which rises to 300 m high sandstone plateaus. The two headlands are 4 km apart and combined have 9 km of steep rocky shoreline. Beaches Q 635-638 lie between the tip of Nob Point and the southern side of Indian Head.

Beach **Q 635** is a curving 200 m long pocket cobble and boulder beach located at the base of the vertically bedded, steep, 40 m high slopes of Nob Point. The boulders extend 50 m off the shoreline where they grade into subtidal sands. The beach has partially filled in the small embayment with a grassy cobble overwash plain.

Beach **Q 636** commences 1.5 km south of the point and trends south for 1.3 km to the southern side of Indian Head. This relatively straight beach is backed by vegetated slopes rising steeply to 300 m. The beach consists of a cobble and boulder high tide beach fronted by a 50 m wide low tide sand bar, with boulders increasing towards the southern end. The bar is in turn paralleled by a 100 m wide band of subtidal boulders, then deeper sand.

Beach **Q 637** lies 1 km to the south on the eastern tip of Indian Head. It is a slightly curving, 150 m long, sand and boulder beach, fringed by rocks and boulders, with a narrow section of intertidal sand in the centre. It is backed by slopes rising steeply to 300 m. Beach **Q 638** is located at the southern extremity of Indian Head and is a 200 m long pocket of sand bordered by the steep slopes of the head. It has low rocky points and inter to subtidal rocks extending off each end, with a central low tide bar against the base of the beach.

Q 639-640 **INDIAN HEAD-POINT SAUNDERS**

No.	Beach	Rating HT	LT	Type	Length
Q639	Indian Head	1	3	R+LT rips	4.8 km
Q640	Pt Saunders	1	2	R+LTT-sand flats	4.5 km

Spring & neap tidal range = 1.6 & 0.4 m

To the south of Indian Head two near continuous beaches sweep for 10 km to Point Saunders and the mouth of the Endeavour River, opposite Cooktown (Fig. 6.89). The base of Mount Saunders separates the two beaches with a 700 m long section of rocky shore.

Beach **Q 639** commences just to the south of beach Q 638 and trends relatively straight to the south-southwest for 4.8 km to the northern base of Mount Saunders. The beach is backed by 100-300 m of generally vegetated dunes, that have climbed 20-30 m up the backing slopes, which continue to rise to 288 m high Barnett Hill. Today most dunes are stable and a low grassy foredune and occasional coconut tree back the beach. The beach is well exposed to the southeast trades, but afforded some protection from the higher waves by the large Boulder and Egret reefs located 17 km offshore.

The result is a wide, low gradient, intertidal beach, and low tide bar cut by up to 40 rips spaced every 100 m.

Mount Saunders rises to 286 m and is surrounded by steep grassy slopes, with termite mounds scattered over the northern slopes. Beach **Q 640** commences on the southern side of the rocks and curves to the south for 4.5 km, the final kilometre a vegetated sand spit that extends part way across the 500 m wide mouth of the Endeavour River, with **Cooktown** located on the southern banks of the river and 1 km east of the southern end of the beach. The beach is backed by some minor 10-15 m high blowout activity, then a low (< 20 m), 1-2 km wide, 700 ha transgressive dune plain covered by dense rain forest. The plain is likely to be mid-Holocene in origin. The beach becomes increasingly protected to the south by the slopes of Mount Cook, and consists of a low gradient beach and low tide bar cut by about 10 rips in the north and grading in the south to a continuous bar and then the 1 km wide tidal shoals of the river mouth. The Endeavour River National Park includes the river mouth and southern 2 km of the beach and backing rain forests, and forms the southern boundary of the Hope Vale Aboriginal Land.

Figure 6.89 The mouth of the Endeavour River, with beaches Q 640 & 639 curving to the north, and Cooktown in the foreground.

GLOSSARY

bar (sand bar) - an area of relatively shallow sand upon which waves break. It may be attached to or detached from the beach, and may be parallel (longshore bar) or perpendicular (transverse bar) to the beach.

barrier - a long term (1000s of years) shore-parallel accumulation of wave, tide and wind deposited sand, that includes the beach and backing sand dunes. It may be 100s to 1000s of metres wide and backed by a lagoon or estuary. The beach is the seaward boundary of all barriers.

beach - a wave deposited accumulation of sediment (sand, cobbles or boulders) lying between modal wave base and the upper limit of wave swash.

beach face - the seaward dipping portion of the beach over which the wave swash and backwash operate.

beach type - refers to the type of beach that occurs under wave dominated (6 types), tide-modified (3 types) and tide-dominated (4 types) beach systems. Each possesses a characteristic combination of hydrodynamic processes and morphological character, as discussed in chapter 2.

beach types

wave dominated (abbreviations, see Figure 2.1)

- R - reflective
- LTT low tide terrace
- TBR transverse bar and rip
- RBB rhythmic bar and beach
- LBT longshore bar and trough
- D dissipative

tide-modified (abbreviations, see Figure 2.3)

- R+LT reflective + low tide terrace
- R+BR R + low tide bar and rip
- UD ultradissipative

tide-dominated (abbreviations, see Figure 2.6)

- R+RSF beach + ridged sand flats
- R+SF beach + sand flats
- R+TSF beach + tidal sand flats
- R+MF beach + mud flats

beach plus (intertidal) rock flats/fringing coral reef/sand waves

- R+RF beach + rock flats
- R+FR beach+fringing reef
- R+SW beach+sand waves

beach hazards - elements of the beach environment that expose the public to danger or harm. Specifically: water depth, breaking waves, variable surf zone topography, and surf zone currents, particularly rip currents. Also include local hazards such as rocks, reefs and inlets.

beach hazard rating - the scaling of a beach according to the hazards associated with its beach type as well as any local hazards.

berm - the nearly horizontal portion of the beach, deposited by wave action, lying immediately landward of the beach face. The rear of the berm marks the limit of spring high tide wave action.

blowout - a section of dune that has been destabilised and is now moving inland. Caused by strong onshore winds breaching the dune.

crocodile – two species of crocodiles inhabit the entire northern Australian coast. The larger saltwater crocodile *Crocodlyus porsus* is extremely dangerous, while the smaller freshwater crocodile *Crocodylus johnstoni* is less of a threat unless provoked.

cusp - a regular undulation in the high tide swash zone (upper beach face), usually occurring in series with spacing of 10-40 m. Produced during beach accretion by the interaction of swash and sub-harmonic edge waves.

foredune - the first sand dune behind the beach. In Queensland it is usually vegetated by spinifex grass and ipomoea, then casuarina thickets.

fringing reef – intertidal coral reef extending from the shoreline seaward.

hole - a localised, deeper part of the surf zone, usually close to shore. It may also be part of a rip channel.

Holocene - the geological time period (or epoch) beginning 10 000 years ago (at the end of the last Glacial or Ice Age period) and extending to the present.

lifeguard - in Australia this refers to a professional person charged with maintaining public safety on the beaches and surf area that they patrol. Sometimes called *beach inspectors*.

lifesaver - an Australian term referring to an active volunteer member of Surf Life Saving Australia, who patrol the beach to maintain public safety in the surf.

laterite - a deeply weather tropical soil what has an thin upper red ferricrete layer, underlain by deeper (to 30 m) mottled white clay hoziron. The ferricrete layer is mined for bauxite.

mangrove – a species of saltwater tolerant trees and some palms that grows in the intertidal zone, usually between neap high tide and mean sea level. There are 39 species of mangroves across northern Australia.

megacusp - a longshore undulation in the shoreline and swash zone, with regular spacings between 100 and 500 m, which match the adjacent rips and bars. Produced by wave scouring in lee of rips (megacusp or rip embayment) and shoreline accretion in lee of bars (megacusp horn).

mud flats – intertidal area of mud extending from the base of the high tide beach to the low tide limit. Extends from tens to hundreds of metres seaward.

parabolic dune - a blowout that has extended beyond the foredune and has a U shape when viewed from above.

Pleistocene - the earlier of the two geological epochs comprising the Quaternary Period. It began 2 million years ago and extends to the beginning of the Holocene Epoch, 10 000 years ago.

reef – general term applied to hard intertidal to submerged feature. Coral, laterite and beachrock reefs are common across northern Australia.

rip channel - an elongate area of relatively deep water (1 to 3 m), running seaward, either directly or at an angle, and occupied by a rip current.

rip current - a relatively narrow, concentrated seaward return flow of water. It consists of three parts: the *rip feeder current* flowing inside the breakers, usually close to shore; the *rip neck*, where the feeder currents converge and flow seaward through the breakers in a narrow 'rip'; and the *rip head*, where the current widens and slows as a series of vortices seaward of the breakers.

rip embayment - see megacusp

rip feeder current - a current flowing along and close to shore, which converges with a feeder current arriving from the other direction, to form the basis of a rip current. The two currents converge in the rip (megacusp) embayment, then pulse seaward as a rip current.

rock platform - a relatively horizontal area of rock, lying at the base of sea cliffs, usually lying above mean sea level and often awash at high tide and in storms. The platforms are commonly fronted by deep water (2 to 20 m)

rock pool - a wading or swimming pool constructed on a rock platform and containing sea water.

sand flats – intertidal area of sand extending from the base of the high tide beach to the low tide limit. Extends from tens to hundreds of metres seaward.

sand ridges – low, subdued ridges on intertidal sand flats, that tend to be equally spaced and parallel the shore. Numbers range from a few to 40 and average 80 m in spacing and a few centimeters-decimetre in elevation.

seagrass – marine plants that produce flowers, pollen and seeds, grown on shallow sandy and muddy substrate and obtain nutrients from sediment. Australia has 30 species of tropical and temperate seagrasses, with 14 tropic species occurring across northern Australia..

sea waves - ocean waves actively forming under the influence of wind. Usually relatively short, steep and variable in shape.

set-up - rise in the water level at the beach face resulting from low frequency accumulations of water in the inner surf zone. Seaward return flow results in a *set-down*. Frequency ranges from 30 to 200 seconds.

shore platform - as per rock platform.

stinger – general term applied to a range of poisonous marine animals including bluebottles, Irukandji and the box jellyfish.

swash - the broken part of a wave as it runs up the beach face or swash zone. The return flow is called *backwash*.

swell - ocean waves that have travelled outside the area of wave generation (sea). Compared to sea waves, they are lower, longer and more regular.

trough - an area of deeper water in the surf zone. May be parallel to shore or at an angle.

tide – the periodic rise and fall of sea level owing to the gravitational force of the Moon and Sun. Has a period of 12.5 hours and across northern Australia ranges up to 11 m.

tidal flat – low gradient sand and/or muddy flats, extending from the based of the beach or shoreline to the low tide limit. May be up to several hundred metres wide.

tidal pool - a naturally occurring hole, depression or channel in a shore platform, that may retain its water during low tide.

wave (ocean) - a regular undulation in the ocean surface produced by wind blowing over the surface. While being formed by the wind it is called a *sea* wave; once it leaves the area of formation or the wind stops blowing it is called a *swell* wave.

wave refraction - the process by which waves moving in shallow water at an angle to the seabed are changed. The part of the wave crest moving in shallower water moves more slowly than other parts moving in deeper water, causing the wave crest to bend toward the shallower seabed.

wave shoaling - the process by which waves moving into shallow water interact with the seabed causing the waves to refract, slow, shorten and increase in height.

wave bore - the turbulent part of a wave produced by wave breaking that advances shoreward across the surf zone. This is the part between the wave breaking and the wave swash and also that part caught by bodysurfers. Also called *whitewater*.

Northern Australia references:

Day, R W, Whitaker, W G, Murray, C G, Wilson, I H and Grimes, K G, 1983, *Queensland Geology*. Geological Survey of Queensland Publication 383, 194pp, plus maps.

Hill, D and Denmead, A K (eds), 1960, *The Geology of Queensland*. Melbourne University Press, Melbourne, 474 pp.

Hopley, D, 1982, *The Geomorphology of the Great Barrier Reef.* Wiley-Interscience, New York, 453pp. The definitive scientific text on the Great Barrier Reef.

Kirkman, H, 1997, *Seagrasses of Australia*. Australia: State of the Environment Technical Paper Series, Estuaries and the Sea. Department of Environment, Canberra, 36 pp.

Laughlin, G, 1997, *The Users Guide to the Australian Coast*. Reed New Holland, Sydney, 213 pp. An excellent overview of the Australian coastal climate, winds, waves and weather.

Lee, G P, 2003, *Mangroves in the Northern Territory*. Report 25/2003D, Department of Infrastructure, Planning and Environment, Darwin, 50 pp.

Northern Territory Government, 2000, *Northern Territory Aboriginal Communities*. NT Dept of Lands, Planning & Environment, Darwin, 112 pp.

Nott, J, 1994, The influence of deep weathering on coastal landscape and landform development in the monsoonal tropics of northern Australia. *Journal of Geology*, 102, 509-522.

Ross, J (editor), 1995, *Fish Australia*. Viking, Melbourne, 498 pp. An excellent coverage of all northern Australian coastal fishing spots.

Semeniuk, V, Keaneally, K F and Wilson, P G, 1978, *Mangroves of Western Australia*. Western Australian Naturalists Club, Perth, Handbook 12, 92 pp.

Short, A D, 1993, *Beaches of the New South Wales Coast*. Australian Beach Safety and Management Program, Sydney, 358 pp. The New South Wales version of this book.

Short, A D, 1996, *Beaches of the Victorian Coast and Port Phillip Bay*. Sydney University Press, Sydney, 298 pp. The Victorian version of this book.

Short, A D (editor), 1999, *Beach and Shoreface Morphodynamics*. John Wiley & Sons, Chichester, 379 pp. For those who are interested in the science of the surf.

Short, A D, 2000, *Beaches of the Queensland Coast: Cooktown to Coolangatta*. Sydney University Press, Sydney, 360 pp. The Queensland version of this book.

Short, A D, 2001, *Beaches of the South Australia Coast and Kangaroo Island*. Sydney University Press, Sydney, 346 pp. The South Australian version of this book.

Short, A D, 2005, *Beaches of the Western Australia Coast: Eucla to Roebuck Bay*. Sydney University Press,, Sydney, 433 pp. The Western Australian version of this book.

Short, A D, 2006, *Beaches of the Tasmanian Coast and Islands*. of Sydney University Press, Sydney, 353 pp. The Tasmanian version of this book.

Short, A D, 2006, Australian beach systems – nature and distribution. Journal of Coastal Research, 22, 11-27.

Surf Life Saving Australia, 1991, *Surf Survival*; The Complete Guide to Ocean Safety. Surf Life Saving Australia, Sydney, 88 pp. An excellent guide for anyone using the surf zone for swimming or surfing.

Surf Life Saving Australia, 2003, Surf Lifesaving Training Manual, 32nd edition, Elsevier (Australia), Sydney, 157 pp.

Williamson, J A, Fenner, P J, Burnett, J W and Rifkin, J F, 1996, *Venomous and Poisonous Marine Animals - a Medical and Biological Handbook*, University of New South Wales Press, Sydney, 504 pp. The definitive book on marine stingers and other nasties.

BEACH INDEX

See also
GENERAL INDEX page 454
SURF INDEX page 464

GENERAL INDEX

A

B

D

I

SURFING INDEX

Also see beach listing in BEACH INDEX

AUSTRALIAN BEACHES

Published by the Sydney University Press for the
Australian Beach Safety and Management Program
a joint project of
Coastal Studies Unit, University of Sydney and Surf Life Saving Australia Ltd

by

Andrew D Short
Coastal Studies Unit, University of Sydney

BEACHES OF THE NEW SOUTH WALES COAST (2nd edition)
Publication: 1993 **ISBN:** 0 646 15055 3
358 pages, 167 original figures, including 18 photographs; glossary, general index, beach index, surf index.

BEACHES OF THE VICTORIAN COAST & PORT PHILLIP BAY
Publication: 1996 **ISBN:** 0 9586504 0 3
298 pages, 132 original figures, including 41 photographs; glossary, general index, beach index, surf index.

BEACHES OF THE QUEENSLAND COAST: COOKTOWN TO COOLANGATTA
Publication: 2000 **ISBN** 0 9586504 1 1
369 pages, 174 original figures, including 137 photographs, glossary, general index, beach index, surfing index.

BEACHES OF THE SOUTH AUSTRALIAN COAST & KANGAROO ISLAND
Publication: 2001 **ISBN** 0 9586504 2 X
346 pages, 286 original figures, including 238 photographs, glossary, general index, beach index, surfing index.

BEACHES OF THE WESTERN AUSTRALIAN COAST : EUCLA TO ROEBUCK BAY
Publication: 2005 **ISBN** 0 9586504 3 8
433 pages, 517 original figures, including 408 photographs, glossary, general index, beach index, surf index.

BEACHES OF THE TASMANIAN COAST AND ISLANDS
Publication: March 2006 **ISBN** 1 920898-12-3
353 pages, 367 original figures, including 314 photographs, glossary, general index, beach index, surfing index.

BEACHES OF THE NORTHERN AUSTRALIA: THE KIMBERLEY, NORTHERN TERRITORY & CAPE YORK
Publication: May 2006 **ISBN** 1-920898-16-6
463 pages, 421 original figures, including 365 photographs, glossary, general index, beach index, surfing index.

Order online from Sydney University Press at

http://www.sup.usyd.edu.au/marine

Forthcoming title:

BEACHES OF THE NEW SOUTH WALES COAST (2nd edition) 1-920898-15-8